Addison-Wesley Wants Your Feedback!!

We'd like to learn *your* impressions of this text to ensure that we are providing the best product for your use. Please take a few minutes to fill out this survey and drop it in the mail to Addison-Wesley once you have used the book for a while.

Author/Text: **Knight/*Physics*** _____   Your college or university: _____

1. How far along in the course are you as you fill out this survey?

    [ ] Just starting   [ ] Midway   [ ] Completed

2. What is your major?   [ ] Physics   [ ] Engineering   [ ] Other: _____

3. How much have you used this book?

    [ ] Read all chapters   [ ] Read selected chapters   [ ] Skimmed

4. Did you:

    [ ] Start calculus before physics? [ ] Start calculus with physics? [ ] Haven't had calculus.

5. What were your perceptions of the book on the following criteria:

a. Readability	[ ] Excellent	[ ] Good	[ ] Neutral	[ ] Poor
b. Level of interest	[ ] Excellent	[ ] Good	[ ] Neutral	[ ] Poor
c. Level of difficulty	[ ] Much too hard	[ ] Hard	[ ] Appropriate	[ ] Easy
d. Worked examples	[ ] Excellent	[ ] Good	[ ] Neutral	[ ] Poor
e. Homework problems	[ ] Excellent	[ ] Good	[ ] Neutral	[ ] Poor
f. Learning tips in the text	[ ] Excellent	[ ] Good	[ ] Neutral	[ ] Poor
g. Illustrations	[ ] Excellent	[ ] Good	[ ] Neutral	[ ] Poor
h. Chapter summaries	[ ] Excellent	[ ] Good	[ ] Neutral	[ ] Poor
i. Value as a learning tool	[ ] Excellent	[ ] Good	[ ] Neutral	[ ] Poor
j. Value as a reference text	[ ] Excellent	[ ] Good	[ ] Neutral	[ ] Poor
k. OVERALL IMPRESSION	[ ] Excellent	[ ] Good	[ ] Neutral	[ ] Poor

6. Did you read the Preface and "An Invitation to Faculty and Students?"   [ ] Yes   [ ] No

7. What three key things do you like best about the book?

    1.
    2.
    3.

8. What three key things do you like least about the book?

    1.
    2.
    3.

9. Did you buy the book: [ ] For yourself?   [ ] For yourself and at least one other student?

10. What other materials were you required to buy for the course?

    _____

    _____

MANY THANKS FOR YOUR INPUT!

Fold and tape at top before mailing.

BUSINESS REPLY MAIL
FIRST-CLASS MAIL   PERMIT NO. 11   READING, MA

POSTAGE WILL BE PAID BY ADDRESSEE

Addison Wesley Longman
Attn: Physics Editorial
1 Jacob Way
Reading, MA 01867-9903

NO POSTAGE
NECESSARY
IF MAILED
IN THE
UNITED STATES

PHYSICS
A CONTEMPORARY PERSPECTIVE

**Volume Two
Preliminary Edition**

RANDALL D. KNIGHT
California Polytechnic State University—San Luis Obispo

ADDISON-WESLEY

An imprint of Addison Wesley Longman, Inc.

Reading, Massachusetts • Menlo Park, California • New York • Harlow, England
Don Mills, Ontario • Sydney • Mexico City • Madrid • Amsterdam

Senior Sponsoring Editor   Julia Berrisford
Development Editor   Margy Kuntz
Production Supervisor   Kathleen A. Manley
Executive Marketing Manager   Kate Derrick
Photo Researcher   Sara Peterson
Text Designer   Melinda Grosser
Cover Designer   Diana C. Coe
Prepress Buyer   Caroline Fell
Compositor   Sally Simpson
Art Editor   Susan London Payne
Technical Art Consultant   Joseph Vetere
Technical Illustrator   Scientific Illustrators
Manufacturing Supervisor   Hugh Crawford

Photo Credits
Page 132, Chrysler Corporation. Page 188 and page 336, Richard Megna, Fundamental Photographs, New York. Page 353, Ernest Orlando Lawrence, Berkeley National Laboratory, University of California. Page 369, National Portrait Gallery, London. Page 435, Fig. 32-18a, from Gerhard Herzberg, *Atomic Spectra and Atomic Structures*, NJ: Prentice-Hall, Inc., 1937. Reprinted with permission. Fig. 32-18b, Bausch & Lomb. Page 421, Science Museum/Science & Society Picture Library. Page 455, used with permission of Hamamatsu Corporation, USA. Page 460, from "The Role of Gravity in Quantum Theory" by Daniel M. Greenberger and Albert W. Overhauser, *Scientific American*, May 1980. Illustration by Dan Todd. Reprinted with permission. Page 464, IBM Corporation, Research Division, Almaden Research Center. Page 448, ETH Bibliothek. Page 458, from P. Merli and G. Missiroli, *American Journal of Physics*, 1976. Page 471, courtesy AIP Emilio Segrè Visual Archives. Page 508, Irish Press, courtesy AIP Emilio Segrè Visual Archives. Page 509, IBM Corporation, Research Division, Almaden Research Center. Page 535, Photograph by Louise Barker, courtesy AIP Emilio Segrè Visual Archives. Page 539, Sadtler Research Laboratory. Page 542, from G. Binnign and H. Rohres, *Surface Science*, Vol. 144, p. 321, 1984. Reprinted with permission. Page 545, from L. Change, "Semiconductor Quantum Heterostructures," *Physics Today*, October 1992, p. 38. Reprinted with permission. Page 598, MIT, courtesy AIP Emilio Segrè Visual Archives.

Copyright © 1997 Addison Wesley Longman, Inc. All rights reserved. No part of this publication may be reproduced, stored in a retrieval system, or transmitted, in any form or by any means, electronic, mechanical, photocopying, recording, or otherwise, without the prior written permission of the publisher. Printed in the United States of America.

ISBN 0–201–43165–3

1 2 3 4 5 6 7 8 9 10–CRS–99989796

An Invitation to Faculty and Students

Welcome to the Preliminary Edition of *Physics: A Contemporary Perspective*. This is very much a "work in progress," and we invite you to help us shape it into final form. Although these materials have undergone two full years of classroom testing by the author at California Polytechnic State University, where the assessments have been extremely positive, a change in the foundation course of a discipline cannot take place without consulting the community of users. The purpose of this Preliminary Edition is the first of several steps to obtain extensive feedback from both faculty and students prior to the publication of the First Edition.

We expect that concerns will be raised, that omissions and oversights will be discovered, and that many helpful suggestions will be made. Our intentions, and those of the author, are to address the concerns and to heed your suggestions to the fullest extent possible.

Feedback questionnaires for each chapter are provided at the back of the book. We urge instructors to collect these from students and to return them, with their own comments and perspectives, to Addison-Wesley. A more comprehensive survey is also provided as a tear-out card. Please fill out this card at the end of the first term and mail it, postage paid, to Addison-Wesley.

What Is *Physics: A Contemporary Perspective?*

Physics: A Contemporary Perspective is the first comprehensive textbook that

- is based extensively on the results of physics education research, and
- follows the general guidelines of the Introductory University Physics Project.

This is a calculus-based textbook intended for a one-year introductory physics course. Care has been taken from the earliest stages of development to ensure that it meets the needs of a wide variety of classroom settings—large universities, mid-size state universities, liberal arts colleges, and community colleges.

The principal objectives of *Physics: A Contemporary Perspective* are:

- To meet the needs and interests of contemporary science and engineering students in a one-year course.

- To move the results of physics education research into the classroom.
- To replace the encyclopedic scope of existing textbooks with a manageable and coherent group of topics while still covering the essential physics needed by science and engineering students.
- To balance qualitative reasoning and conceptual understanding with quantitative reasoning and problem solving.
- To provide an explicit discussion of and practice with modeling, assumptions, skill development, and multiple representations of knowledge.
- To include a significant component of 20th century physics and modern ideas while keeping in mind the practical needs of much of the audience.
- To develop in students an awareness and appreciation of the fact that physics is a process, not just a list of facts and formulas, and that it has a historical context.

In the course of meeting these objectives, the author has drawn heavily upon the research findings of Arons, McDermott, Hestenes, Van Heuvelen, Hake, Reif, Larkin, Ganiel, Sherwood, Thornton, Laws, Sokoloff, and others. It is very much a research-based textbook that attempts to move the research findings of "what works" into the classroom.

Active learning in the classroom is an essential ingredient for the success of this approach. The text explanations are full and complete, so there is no need to use class time for derivations or for presenting material that the text omitted. The author has designed these materials to support a classroom format based on the effective use of demonstrations, questions and discussions, group activities, and example problem solving.

These materials do not make explicit use of computers or numerical calculations. They will, however, fully support a computer-oriented "workshop" approach to introductory physics. Neither is any specific laboratory component called for. The author, however, suggests that "guided discovery" labs, such as the Thornton/Laws/Sokoloff microcomputer-based labs or Steinberg's batteries-and-bulbs labs, provide the best pedagogical match to the text materials and thus maximize student learning potential.

Major Features of *Physics: A Contemporary Perspective*

A Story Line: Most texts provide little motivation for topics as they are introduced. Therefore, it is not surprising that most students view physics as a collection of loosely-related facts and formulas rather than a few major principles from which many implications are drawn. To combat this fragmentation of knowledge, *Physics: A Contemporary Perspective* adopts the story line of "Physics and the Atomic Structure of Matter." Discovering the link between the microscopic properties of atoms and the macroscopic properties of bulk matter has been a triumph of physics. Further, in this age of nanostructures and quantum well lasers, it is a topic that is relevant to the interests of engineers as well as scientists. The telling of this story is broad enough to visit all the major areas of introductory physics, yet it provides criteria by which specific topics are selected for inclusion.

Works from the Concrete to the Abstract: Arnold Arons, long-time leader in physics education research, stresses the importance of leading the students to answer "How do we know …?" and "Why do we believe …?" questions. The ability to answer such questions represents real knowledge, not just memorized and quickly-forgotten facts. To meet this goal, considerable attention is given throughout to working "from the concrete to the abstract." General principles are arrived at inductively, from evidence, rather than presented *a priori*.

A Workbook: A unique aspect of these materials is a separate Student Workbook. While the text establishes the content and provides the necessary information, there is a limit to how effective a textbook alone can be, no matter how well intentioned, researched, and written. To develop *skills*, such as interpreting graphs, reasoning with ratios, drawing free-body diagrams, drawing electric field maps, or interpreting wave functions, students need opportunities to *practice*. Quantitative, end-of-chapter problems call upon students to assemble and use various skills, but they rarely give students seeing these ideas for the first time an opportunity to develop the individual skills through focused practice.

The Workbook bridges the gap between textbook and problem solving, providing the needed opportunities to learn and practice skills separately—much as a musician practices technique separately from performance pieces. The Workbook exercises are keyed to each section of the text. The exercises are generally graphical or qualitative, letting students practice using the concepts introduced in that section of the text. The Workbook can be used for in-class active learning, in recitation sections, or for assigned homework.

Multiple Representations of Knowledge: *Physics: A Contemporary Perspective* emphasizes multiple representations of knowledge, such as word descriptions, pictorial descriptions, graphical descriptions, and mathematical descriptions. The text provides specific instruction and examples of translating back and forth between different representations, and both the Workbook and the end-of-chapter problems frequently call upon students to do so. This process, called "modeling" by some, is the heart of problem solving in physics. *Physics: A Contemporary Perspective* is unique in the attention it gives to teaching students these skills.

Explicit Problem-Solving Strategies: In conjunction with multiple representation skills, the text develops explicit problem-solving strategies in which students analyze a problem from several qualitative perspectives before approaching it mathematically. Equally important, the worked examples all follow this strategy in detail, with careful explanations of the underlying, and often unstated, reasoning. Students are led to develop a hierarchical knowledge structure, based on general principles, rather than a fragmented knowledge structure of loosely-connected facts and formulas. Optional worksheets are used in some portions of the text to guide students through the difficult process of converting the word representation of a problem statement into intermediate pictorial and graphical representations and, ultimately, to a solvable mathematical representation. Comments from student evaluations find many students saying things such as "I never before realized there is a *method* for solving problems. I wish I had been taught this in my other classes."

Course Content

The textbook has a six-part structure. It can be covered in two semesters or three quarters. The first five parts are generally consistent with a "standard order" to meet the needs of transfer students. The contents of the two volumes are:

Volume One: Part I   Single Particle Dynamics
                   Part II   Interacting Particles and Conservation Laws
                   Part III   Oscillations and Waves
Volume Two: Part IV: Thermal and Statistical Physics
                   Part V: Electric and Magnetic Fields
                   Part VI: Quantum Physics and the Structure of Atoms

While all the major concepts of introductory physics are here, the author has endeavored to reduce the encyclopedic scope of other textbooks by omitting some topics and scaling back the level of detail for others. Feedback from users of the Preliminary Edition will be especially critical to judge whether the author's choices meet with widespread agreement.

Part VI: Quantum Physics and the Structure of Atoms places introductory quantum physics on an equal footing with thermal and statistical physics and with electricity and magnetism. The subject matter of quantum physics is rapidly becoming important to engineering students and to scientists in other fields. The goals of Part VI are for students:

- To recognize the experimental basis for quantum physics.
- To understand the primary concepts of energy levels and wave functions.
- To see how one-dimensional quantum mechanics applies to scanning tunneling microscopes, quantum well devices, molecular bonds, radioactivity, and other topics.
- To learn how the atomic shell model, the periodic table of the elements, and the emission and absorption of light are important consequences of quantum mechanics.

These are much less ambitious goals than a full modern physics course, but they are a significant step up from the conventional presentation of modern physics topics. Students who go no farther in physics will have been introduced, in a rigorous rather than a handwaving fashion, to the basic ideas of quantum physics. And those students who continue to a modern physics course will be better prepared for the abstract formalism of quantum mechanics.

Initial Classroom Testing

These materials have undergone two full rounds of classroom testing by the author at California Polytechnic State University. Student performance, measured using such assessment tools as the Hestenes and Halloun *Force Concept Inventory*, has been outstanding. Details of the initial classroom testing are available upon request.

For more information on classroom test results, or for a sample copy of the text, please contact Addison-Wesley at the address below.

With your help, *Physics: A Contemporary Perspective* will become the new standard for introductory physics.

Addison-Wesley Publishing Company
Physics Editorial
One Jacob Way
Reading, MA 01867
physics@aw.com

Acknowledgments

I have relied upon conversations with and, especially, the written publications of many members of the physics education research community. Those who may recognize their influence include: Arnold Arons, Uri Ganiel, Ibrahim Halloun, David Hestenes, Leonard Jossem, Priscilla Laws, John Mallinckrodt, Lillian McDermott, Edward "Joe" Reddish, Bruce Sherwood, David Sokoloff, Ronald Thornton, Shelia Tobias, and Alan Van Heuvelen. John Rigden, founder and director of the Introductory University Physics Project, provided the impetus that started me down this path.

Valuable review comments have been contributed by Edward Adelson (The Ohio State University), Ronald Bieniek (University of Missouri–Rolla), S. Leslie Blatt (Clark University), David Jenkins (Virginia Polytechnic Institute), John Mallinckrodt (California State Polytechnic University-Pomona), Robert Marchini (Memphis State University), John Risley (North Carolina State University), Cindy Schwarz (Vassar College), and Judy Tavel (Dutchess Community College). My students at Cal Poly also provided vast amounts of feedback as these materials were developed.

I especially want to thank Julia Berrisford, my sponsoring editor; Margy Kuntz, my development editor; Gordon Wong, who developed much of the artwork; and the staff at Addison-Wesley Longman for their enthusiasm toward this project and their efforts to have it published in a timely manner. I am grateful to my wife, Sally, for her encouragement and patience, and to our cats for their complete indifference to this project.

The development of these materials has been supported by the National Science Foundation as the *Physics for the Year 2000* project. Their support is gratefully acknowledged.

Contents

PART IV THERMAL AND STATISTICAL PHYSICS

OVERVIEW: THE SCIENCE OF ENERGY 3

19 A Macroscopic Description of Matter 7

19.1 Macroscopic Systems 7
19.2 Volume, Mass, and Density 9
19.3 Temperature 14
19.4 Pressure 15
19.5 Ideal Gases 22
19.6 The Equation of State for an Ideal Gas 27
19.7 Ideal Gas Processes 31
Summary 37
Exercises and Problems 38

20 Work, Heat, and the First Law of Thermodynamics 41

20.1 State Variables and Interaction Parameters 41
20.2 Internal Energy 43
20.3 Work 44
20.4 Heat 45
20.5 The First Law of Thermodynamics 48
20.6 Specific Heat and Molar Heat Capacity 49
20.7 Calorimetry 53
Summary 56
Exercises and Problems 57

21 A Statistical View of Gases 59

21.1 The Microstate of a Gas 59
21.2 Configurations and Fluctuations 60
21.3 Probability and the Binomial Distribution 66
21.4 The Distribution of Molecules in a Gas 73
21.5 Macroscopic Fluctuations 76
21.6 Randomness, Order, and Equilibrium 80
Summary 82
Exercises and Problems 83

22 The Kinetic Theory of Gases 86

22.1 The Micro/Macro Connection 86
22.2 A Microscopic View of Pressure 86
22.3 A Microscopic View of Temperature 90
22.4 Molecular Speeds and Molecular Collisions 91
22.5 Thermal Energy 96
22.6 Thermal Interactions and Heat 100
22.7 Irreversible Processes, Entropy, and the Second Law of Thermodynamics 103
Summary 110
Exercises and Problems 111

23 Thermodynamics of Ideal Gases 114

23.1 Thermodynamics: The Transformation of Energy 114
23.2 Thermodynamic Relationships 115
23.3 Four Basic Thermodynamic Processes 119
23.4 Turning Heat into Work 127
23.5 Examples of Heat Engines 132
23.6 Pictorial Model of a Heat Engine 140
23.7 The Limits of Efficiency 142
Summary 148
Exercises and Problems 150

SCENIC VISTA: ORDER, DISORDER, AND THE DIRECTION OF TIME 153

PART V ELECTRICITY AND MAGNETISM

OVERVIEW: ELECTRICITY AND MAGNETISM: THE ATOMIC GLUE 163

24 Electric Forces and Fields 166

24.1 Charges and Forces: A Laboratory 166
24.2 Charges in Matter 172
24.3 Insulators and Conductors 175
24.4 Polarization 178
24.5 Coulomb's Law 181
24.6 The Field Model 186
24.7 The Electric Field of a Point Charge 193
      Summary 196
      Exercises and Problems 198

25 Calculating the Electric Field 202

25.1 Sources of the Electric Field 202
25.2 The Field of Multiple Point Charges 203
25.3 The Field of a Dipole 206
25.4 Calculating the Field of a Continuous Charge 210
25.5 The Field of a Line of Charge 212
25.6 The Field of a Plane of Charge 217
25.7 The Field of a Capacitor 221
25.8 The Field of a Sphere of Charge 223
25.9 Motion of a Charged Particle in a Uniform Field 224
25.10 Motion of a Dipole in an Electric Field 228
      Summary 231
      Exercises and Problems 233

26 Current and Conductivity 237

26.1 Current and Charge Carriers 237
26.2 The Electron Current 239
26.3 The Field Inside the Wire 246
26.4 Current and Current Density 247
26.5 Conductivity and Resistivity 250
      Summary 252
      Exercises and Problems 253

27 The Electric Potential 255

27.1 An Alternative View of Electrical Interactions 255
27.2 Electric Potential Energy 255
27.3 The Potential Energy of Point Charges 258
27.4 The Electric Potential 261
27.5 The Electric Potential of a Point Charge 263
27.6 The Electric Potential Inside a Capacitor 266
27.7 The Electric Potential of a Continuous Charge 270
Summary 273
Exercises and Problems 274

28 Potential and Field 276

28.1 Connecting Potential and Field 276
28.2 Equipotential Surfaces 280
28.3 The Potential and Field of a Conductor in Electrostatic Equilibrium 283
28.4 Sources of Potential 286
28.5 Connecting Potential and Current 288
28.6 Capacitance 291
Summary 293
Exercises and Problems 295

29 Fundamentals of Circuits 297

29.1 Practical Electricity 297
29.2 Resistors and Ohm's Law 298
29.3 The Basic Circuit 301
29.4 Kirchhoff's Laws and Simple Circuits 303
29.5 Energy and Power 307
29.6 Combinations of Resistors 311
29.7 Real Batteries 316
29.8 Resistive Circuits 318
29.9 Getting Grounded 322
Summary 324
Exercises and Problems 326

30 The Magnetic Field 330

30.1 Observing Magnetism 330
30.2 Oersted's Discovery 333
30.3 The Magnetic Field 335
30.4 The Source of Magnetic Fields: Moving Charges 336

30.5 The Magnetic Field of a Long Straight Wire **339**
30.6 The Magnetic Field of a Current Loop **342**
30.7 Uniform Magnetic Fields **346**
30.8 The Magnetic Force on a Moving Charge **348**
30.9 Magnetic Forces on Current-Carrying Wires **354**
30.10 Forces and Torques on Current Loops **356**
30.11 Magnetic Properties of Matter **358**
Summary **363**
Exercises and Problems **364**

31 Electromagnetic Induction **369**

31.1 Faraday and Henry **369**
31.2 Motional emf **372**
31.3 Magnetic Flux **378**
31.4 Lenz's Law **381**
31.5 Faraday's Law **385**
31.6 Induced Currents: Four Applications **388**
31.7 Induced Fields and Electromagnetic Waves **392**
Summary **395**
Exercises and Problems **396**

SCENIC VISTA: THE GLOBAL VILLAGE **399**

PART VI QUANTUM PHYSICS AND THE STRUCTURE OF ATOMS

OVERVIEW: THE ATOMIC STRUCTURE OF MATTER **409**

32 The End of Classical Physics **412**

32.1 Physics in the 1800s **412**
32.2 Faraday **415**
32.3 Cathode Rays **416**
32.4 J. J. Thompson and the Discovery of the Electron **419**
32.5 Millikan and the Fundamental Unit of Charge **423**
32.6 Rutherford and the Discovery of the Nucleus **425**
32.7 Into the Nucleus **432**
32.8 The Emission and Absorption of Light **434**
32.9 Classical Physics at the Limit **437**
Summary **438**
Exercises and Problems **439**

33 Waves, Particles, and Quanta 442

- 33.1 Particles and Waves 442
- 33.2 The Photoelectric Effect 443
- 33.3 Einstein's Explanation 448
- 33.4 Photons 453
- 33.5 De Broglie's Hypothesis: Matter Waves 456
- 33.6 Quantization of Energy 461
- 33.7 Wave–Particle Duality 464
- Summary 465
- Exercises and Problems 466

34 The Bohr Model of the Atom 470

- 34.1 The Atomic Structure Enigma 470
- 34.2 Bohr's Model 471
- 34.3 The Bohr Atom 474
- 34.4 The Hydrogen Spectrum 480
- 34.5 Hydrogen-like Ions 483
- 34.6 The Quantization of Angular Momentum 483
- 34.7 Success and Failure 485
- Summary 486
- Exercises and Problems 487

35 Wave Functions and Probabilities 490

- 35.1 First Steps Toward a Quantum Theory 490
- 35.2 Waves, Particles, and the Double-Slit Experiment 492
- 35.3 The Wave Function 497
- 35.4 Normalization 499
- Summary 504
- Exercises and Problems 505

36 One-Dimensional Quantum Mechanics 508

- 36.1 Schrödinger and Quantum Mechanics 508
- 36.2 Schrödinger's Equation: The Law of Psi 509
- 36.3 Problem Solving in Quantum Mechanics 514
- 36.4 A Particle in a Rigid Box 515
- 36.5 The Correspondence Principle 523
- 36.6 Quantum-Mechanical Models 525
- 36.7 A Particle in a Capacitor 526

- 36.8 Finite Potential Wells **529**
- 36.9 The Quantum Harmonic Oscillator **535**
- 36.10 Quantum-Mechanical Tunneling **540**
  Summary **545**
  Exercises and Problems **548**

37 The Structure of Atoms **551**

- 37.1 The Atomic Structure Problem **551**
- 37.2 A One-Dimensional Hydrogen Atom **552**
- 37.3 The Three-Dimensional Hydrogen Atom **554**
- 37.4 Visualizing Hydrogen **560**
- 37.5 The Electron's Spin **564**
- 37.6 Multielectron Atoms **568**
- 37.7 The Periodic Table of the Elements **572**
  Summary **577**
  Exercises and Problems **578**

38 Atomic Spectra **581**

- 38.1 Light at the End of the Tunnel **581**
- 38.2 Excited States of Atoms **581**
- 38.3 Excitation **584**
- 38.4 Emission **586**
- 38.5 Lifetimes of Excited States **592**
- 38.6 The Stimulated Emission of Radiation **596**
- 38.7 Lasers **597**
- 38.8 The Final Word **602**
  Summary **603**
  Exercises and Problems **604**

SCENIC VISTA: THE JOURNEY NEVER ENDS 606

ANSWERS TO SELECTED EXERCISES AND PROBLEMS   A-1

INDEX   A-3

PART IV

Thermal and Statistical Physics

PART **IV** Overview

The Science of Energy

If there is one area of science that has universal applicability—from astrophysics to zoology and from civil engineering to economics to theology—it is thermodynamics. **Thermodynamics** is the science of energy in its broadest context, especially the transfer and conversion of energy from one form to another. The word *energy* itself is from a Greek root meaning "to do work," and much of thermodynamics is concerned with converting "stored energy" of various forms, such as the energy in fuels, into the energy called "useful work."

Thermodynamics arose hand-in-hand with the industrial revolution as the systematic study of converting heat energy into mechanical motion and work. Hence the name *thermo + dynamics*. Indeed, the analysis of engines and generators of various kinds remains the focus of engineering thermodynamics. But as a science, thermodynamics now extends to all forms of energy conversions, including those involving living organisms. As a few examples:

- Gasoline, diesel, and steam engines convert the energy of a fuel into the mechanical energy of moving gears, wheels, and pistons.
- A generator converts the mechanical energy of a spinning turbine into electrical energy.
- A fuel cell converts chemical energy into electrical energy.
- A photocell converts the electromagnetic energy of light into electrical energy.
- A laser converts electrical energy into the electromagnetic energy of light.
- An organism converts the chemical energy of food into a variety of other forms of energy, including kinetic energy, sound energy, and heat energy.

Our goal in Part IV is to reach an understanding of the basic concepts and laws of thermodynamics. We will be especially interested in how energy is converted from one form to another. Although we can only look at a few applications, this study will prepare you to use the ideas of thermodynamics in your own field of study.

Thermodynamics is about *macroscopic* objects, where *macro* is a prefix (the opposite of *micro*) meaning "large." We will be concerned with systems that are solids, liquids, or gases rather than the "particles" to which we have grown accustomed. The language of thermodynamics will be of temperature, pressure, volume, and moles rather than of position, velocity, and force. One major task will be to learn this new language.

The types of questions that we will be asking are also quite different from those that have occupied us until now. Typical questions of thermodynamics include:

- How do the temperature and pressure change during certain "processes"?
- How does a physical system such as an "engine" do mechanical work? How is the work related to the temperature and pressure of the engine?
- If a system "burns fuel" to do work, how is the amount of work done related to the fuel energy that was used? Are there limits on how "efficient" an engine can be?

As you can see, there are a lot of very new ideas to learn.

A Road Map: The Micro/Macro Connection

Thermodynamics is a science that describes and characterizes a macroscopic system as a whole, regardless of whether the system is a solid crystal, a beaker of liquid, or a container of gas. But as you know, one of the major streams of scientific thought of the last century is that solids, liquids, and gases all consist of atoms. These atoms are ceaselessly moving about and colliding with one another, all acting as microscopic particles. Their motion and interactions are described by Newtonian mechanics and, in some cases, quantum physics.

We will assert, and the major goal of this section of the text will be to demonstrate, that there is a *connection* between the *microphysics* of the atoms and molecules in the system and the *macrophysics* of the system as a whole. That is, the macroscopic quantities of thermodynamics can be understood in terms of the microscopic physics of moving atoms and molecules. This link between the realms of the small and the large is called the **micro/macro connection**.

The discovery of this link between the microscopic and the macroscopic is an exciting story and a major success of physics. You will find that the "familiar" concepts of thermodynamics, such as temperature and pressure, have their roots in the turmoil of motion and collisions at the atomic level. In fact, you will find it possible to learn a great deal about the properties of molecules—such as their speeds—on the basis of purely macroscopic measurements. A careful study of the micro/macro connection will also lead to a new concept called *entropy* and to the second law of thermodynamics—one of the most subtle but also one of the most profound and far-reaching statements in physics.

Our ultimate destination for this section of the text is an understanding of the thermodynamics of heat engines—devices, such as gasoline-powered internal combustion engines, that convert heat energy into useful work. This will be a significant achievement, but we first have many steps to take along the way. The starting point for our journey will be a review of ideas from Chapters 9–11: kinetic energy, potential energy, work, and energy conservation. We will now have to expand these ideas to include the internal energy of a system, and we will have to introduce a new form of energy transfer called *heat*. We will need to understand the concepts of temperature, pressure, and equilibrium. At a deeper level, we will need to see how these concepts are connected to the random motion and collisions of the molecules—that is, how they are connected to the underlying microphysics. Only after all these steps have been taken will we be able to analyze a real heat engine.

This is an ambitious goal, but one we can achieve. You will find it helpful to keep in mind how the various pieces of the puzzle fit together. Figure OIV-1 shows how many of the topics to be studied will help to move us toward our destination. As we go along, you may want to refer back to this figure to review how the topics are interconnected.

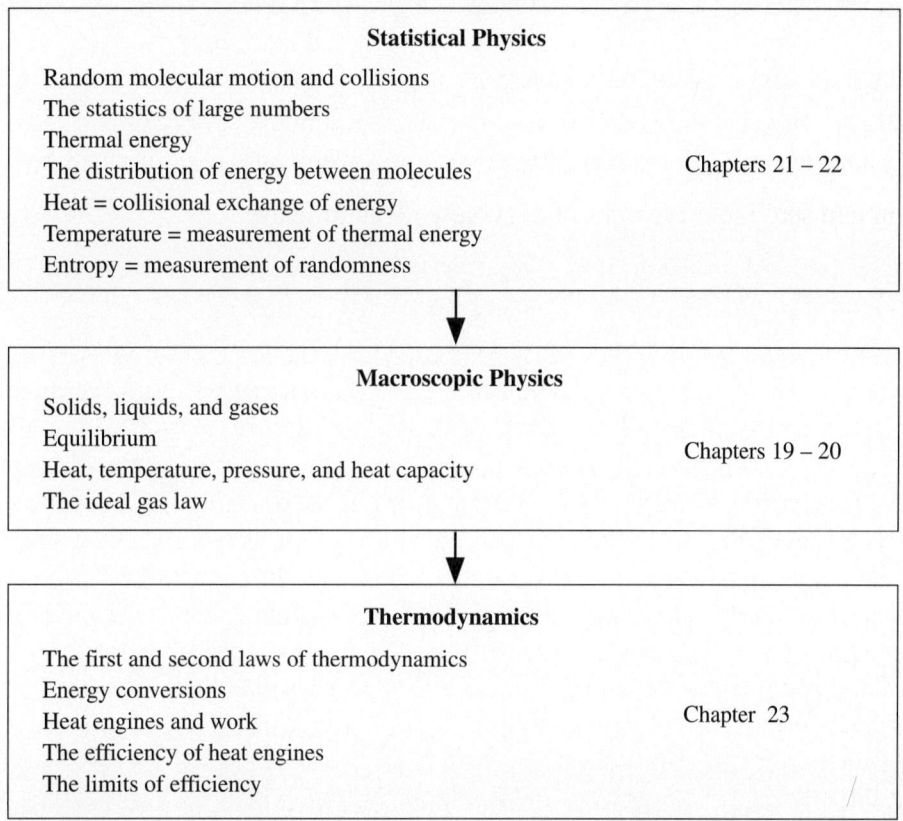

FIGURE OIV-1 The logical structure of statistical and thermal physics.

Although the chart in Fig. OIV-1 shows the "logical structure" of the ideas, we actually will start in the middle. Macroscopic physics deals with familiar ideas, many of which you have learned about in chemistry (e.g., temperature scales, ideal gases, etc.). Much of the first two chapters will be a review, but a few new topics will be introduced. These chapters will develop the language of thermodynamics and will focus on interesting observations that physics will be able to "explain."

We will then go back and develop the statistical physics foundations that underlie these macroscopic concepts. Statistical physics is the branch of physics that applies statistical ideas to the motion and dynamics of particles. From here we will build the micro/macro connection that will "explain" what temperature is, why the ideal gas law exists, and much more. The approach we will follow is quite unlike anything we have done before and, most likely, quite unlike anything you have studied previously.

Finally, we will demonstrate the power of our method with a concluding chapter on thermodynamics and its applications. More advanced science and engineering courses will further extend the applications from this point, but we will have achieved our goal of having placed the ideas of thermodynamics on a firm foundation of physical laws.

Ultimately, the success of the micro/macro connection will provide you with substantial evidence for the atomic structure of matter. Only quite recently have scientists been able to "see" individual atoms, with ultra-sensitive instruments, but the atomic theory of matter had long since been convincingly demonstrated by, among other things, the success of statistical physics in providing an explanation of the phenomena of thermodynamics.

Chapter 19

A Macroscopic Description of Matter

19.1 Macroscopic Systems

A block of steel, a beaker of water, and a room full of air. All are examples of **macroscopic systems**—systems that are large enough to be visible to the unaided eye and that contain an unbelievably large number of atoms. Our interest in this chapter is in finding what kind of physical properties we need to describe and characterize macroscopic systems such as these. The properties of a macroscopic system as a whole are sometimes called its **bulk properties**. A fairly obvious example of such a property is the system's mass. Other properties we will discuss in this chapter include volume, density, temperature, and pressure.

Our ultimate goal is to understand the bulk properties of matter in terms of its atomic structure. As a step toward that goal, we will introduce a few ideas in this chapter that will begin to establish the micro/macro connection. To begin, it will be useful to define the different forms in which macroscopic systems can exist.

Macroscopic systems are generally recognized as existing in one of three **phases**: solid, liquid, or gas. (A gas that becomes ionized at extremely high temperatures, such as in the sun, is sometimes referred to as a fourth phase of matter called a *plasma*.) Changes from liquid to solid (freezing) or liquid to gas (boiling) are called **phase changes**. (Note that this use of the word *phase* has no relationship at all to the phase of a wave!) The phase in which a substance exists depends on the concentration of atoms and molecules within the system. This concentration, in turn, is determined by the interplay between atomic interaction forces, which attract atoms toward one another, and thermal effects, which provide the atoms with kinetic energy and encourage them to break away from their neighbors.

[**Photo suggestion: Three phases of water—ice, liquid water, and steam.**]

Solids

A **solid** is a macroscopic system characterized as being rigid, with a definite shape and volume. At the atomic level, a solid consists of atoms and molecules occupying fixed positions in space. As in all phases of matter, there is a competition between the attractive atomic interaction forces and thermal effects. At sufficiently low temperatures, and when the atoms are densely packed, the interaction forces win out and the atoms "freeze" into the fixed positions of a solid. All substances become solids at some temperature—the *freezing point*—except for helium. Helium can be solidified only by being placed under high pressure, as well as by being cooled to nearly absolute zero.

Liquids

A **liquid** is a macroscopic system in which the attractive interactive forces and the thermal effects are roughly balanced. The atoms have enough thermal kinetic energy to break the fixed bonds of the solid, but not enough to overcome the interaction forces entirely and escape from one another. As a consequence, a liquid is a *deformable* substance—it takes the shape of its container—and it is able to flow. Like solids, liquids are nearly *incompressible* because the atoms in a liquid, as in a solid, are about as close together as they can get.

Gases

A **gas** is a system in which thermal motions dominate. This property allows the atoms to move well apart from one another and travel freely through space until, on occasion, they collide with another atom or the wall of the container. The idealized limit in which there are *no* interaction forces between the atoms is called an **ideal gas**. Although real gases do exhibit small interaction forces, they behave very much like an ideal gas under conditions of low pressure (very dilute gas). A gas, like a liquid, is deformable and flows—like the wind! Note that *both* gases and liquids are called **fluids** because of their ability to flow. As a gas is cooled or as its atoms become more densely packed, the interaction forces increase in importance until, suddenly, the system undergoes a phase change to a liquid.

State Variables

As we mentioned earlier, this chapter will introduce a number of properties that are used to characterize or describe a macroscopic system. They include volume, mass, density, temperature, and pressure. These properties, or parameters, are referred to as **state variables** because, taken all together, they describe the *state* of the macroscopic system. If we change the value of any of these properties, then we are changing the state of the system.

No macroscopic system remains in a single state for long. Its state is constantly changing as it interacts with its environment: A liquid is heated to a higher temperature or a gas is compressed to a higher pressure and smaller volume. Practical devices for doing work—engines and machines—function by having their state changed in a controlled manner. When the spark plug fires in your car engine, it ignites the fuel–air mixture and changes its temperature. This raises its pressure, causing it to exert a force on the piston. As the piston is pushed down by the pressure, the gas volume increases and the temperature falls. This is a complex sequence of state changes, but one that you will be able to analyze by the time you finish this part of the text.

Because we will be interested in changes in the state of a macroscopic system, we will make extensive use of the symbol Δ to represent a *change* in the value of a state variable. For example, ΔT is a change of temperature and Δp is a change of pressure. Make sure you keep in mind that for any quantity X, ΔX is *always* $X_f - X_i$, the final value minus the initial value.

19.2 Volume, Mass, and Density

The first three state variables we will look at are properties with which you are probably already familiar: volume, mass, and density. These properties let us define the amount of substance contained within a macroscopic system.

Volume and Number Density

The first property we can use to characterize a macroscopic system is its volume, which is the amount of space the system occupies. The SI unit of volume is m^3. Solids and liquids have a well-defined volume that is difficult to change—they are *incompressible*. A gas, however, is easy to compress, so the volume of a gas is a variable.

Although volume is a macroscopic property, it is closely related to the microscopic property N, the total number of atoms or molecules in the system. Because N is determined simply by counting, it is a number with no units. A macroscopic system has typically $N \sim 10^{24}$ atoms, an incredibly large number.

The symbol $\sim$, if you are not familiar with it, stands for "has the order of magnitude." It means that the number is only known to within a factor of 10 or so. The statement $N \sim 10^{24}$, which is read "N is of order 10^{24}," implies that N is likely to be somewhere in the range 10^{22} to 10^{26} or so. It is far less precise than the "approximately equal" symbol, $\approx$. As we begin to deal with large numbers, it will often be necessary to distinguish "really large" numbers, such as 10^{24}, from "small" numbers, like merely 10^4. Saying $N \sim 10^{24}$ gives us a rough idea of how large N is and allows us to know that it differs significantly from 10^4 or even 10^{14}.

Another related quantity of particular interest is the number of atoms per cubic meter in a system. We call this quantity, which we will symbolize as d, the **number density**. In a sample of volume V containing N atoms, the number density is

$$d = \frac{N}{V}. \tag{19-1}$$

The number density characterizes how densely the atoms are packed together within the system. The SI units of d are m^{-3}. A "typical" macroscopic system has $d \sim 10^{25}$ m^{-3}.

Because d is a ratio of particle number to volume, its value in a *uniform* system is independent of the size of the volume V. That is, the number density is the same whether you look at the whole system or just a portion of it. For example, suppose a 100 m^3 room has 10,000 tennis balls bouncing around. The number density of tennis balls in the room is $d = 10{,}000/100$ m$^3 = 100$ m^{-3}. If we were to look at only half the room, we would find 5000 balls in 50 m^3, again giving $d = 5000/50$ m$^3 = 100$ m^{-3}. In one-tenth the room, we would find 1000 balls in 10 m^3 and again we would find $d = 1000/10$ m$^3 = 100$ m^{-3}. The value of the number density is the same in all three cases. Also note that although we might say, "There are 100 tennis balls per cubic meter," tennis balls are not units. The proper units of d are simply m^{-3}.

Caution: Although it is true that 1 m = 100 cm, it is *not* true that 1 m^3 = 100 cm^3. Think of a cube 1 m on a side, having $V = 1$ m^3. If you subdivide this cube into little cubes 1 cm on a side, you will have 100 subdivisions along each edge. Thus there are $100 \times 100 \times 100 = 10^6$

of these little 1 cm³ cubes. So 1 m³ = 10⁶ cm³. You can think of this process as cubing the linear conversion factor:

$$1 \text{ m}^3 = 1 \text{ m}^3 \times \left(\frac{100 \text{ cm}}{1 \text{ m}}\right)^3 = 10^6 \text{ cm}^3.$$

Mass

Another important characteristic of a macroscopic system is its mass M. We will use an uppercase M for the mass of an entire system to distinguish it from the mass of the atoms, which we will call m. If the system consists of N atoms, each of mass m, then not surprisingly

$$M = Nm. \tag{19-2}$$

In this text, we will restrict ourselves to considering only systems where N is constant (no atoms added or removed) and where all the atoms are of one type (only one m). Both M and m are measured in kg.

Density

Suppose you have several blocks of copper, each of different size. Each block would have its own unique values of mass M and volume V. Nonetheless, all the blocks are copper, so there should be some parameter having the *same* value for all the blocks, saying, "This is copper, not some other material." The most important such parameter is the *ratio* of mass to volume, which we call the **mass density** ρ (lowercase Greek "rho"):

$$\text{mass density } \rho = \frac{M}{V}. \tag{19-3}$$

Table 19-1 provides a short list of mass densities of various substances. Extensive tables can be found in the *Handbook of Chemistry and Physics* and various other handbooks.

The mass density, like the number density, is independent of the sample size. If a 10 cm³ sample has a mass of 40 g, then cutting it in half gives a 5 cm³ sample with a mass of 20 g. Each sample is unique, but both have the *same* mass density $\rho = 4$ g/cm³. Stated another way, mass and volume are parameters that characterize a *specific piece* of some substance—say copper—whereas the mass density characterizes the substance itself. All pieces of copper have the same mass density, which differs from the mass density of any other substance. This distinction is useful. Mass density allows us to talk about the properties of copper in general, without having to refer to any specific piece of copper.

Several comments are worthwhile. First, mass density is often referred to as

TABLE 19-1 Some typical densities.

Substance	ρ (g/cm³)	ρ (kg/m³)
Air at STP*	0.00121	1.21
Ethyl alcohol	0.79	790
Water (solid)	0.92	920
Water (liquid)	1.00	1000
Aluminum	2.70	2700
Copper	8.92	8920
Lead	11.3	11,300
Mercury	13.6	13,600

*Standard temperature (0°C) and pressure (1 atm).

simply "the density," yet there are several different kinds of densities. You have already seen linear density for strings and number density, and we will later introduce charge density. Thus specifying the type of density is worthwhile. Second, the proper SI units of density are kg/m³. Nonetheless, units of g/cm³ are widely used. You need to convert these to SI units before doing calculations, which means that you have to convert *both* the grams to kilograms *and* the cubic centimeters to cubic meters. The net result is the conversion factor

$$1 \text{ g/cm}^3 = 1000 \text{ kg/m}^3.$$

Third, notice in Table 19-1 that the densities of liquid and solid water are not very different. The reason is that the atoms in liquids and solids are pushed about as close together as possible, which is why they are incompressible. A gas such as air, by contrast, has a much lower density because the atoms are more spread out.

EXAMPLE 19-1 What is the average density of the earth?

SOLUTION Although the earth is hardly uniform, Eq. 19-3 will give us the *average* density ρ_{avg}. Because the earth is nearly spherical,

$$\rho_{avg} = \frac{M}{V} = \frac{M_e}{\frac{4}{3}\pi R_e^3} = 5520 \text{ kg/m}^3.$$

This value is substantially higher than the density of rocks in the earth's crust (≈2800 kg/m³). Such a high average density was one of the early pieces of evidence that the core of the earth must consist of a much higher density substance than the crust. We now know that the core of the earth is mostly iron.

EXAMPLE 19-2 A project on which you are working needs to use a cylindrical lead pipe having outer and inner diameters of 4.0 cm and 3.5 cm, respectively, and a length of 50 cm. What is its mass?

SOLUTION We need to rearrange Eq. 19-3 to give $M = \rho V$, where the density of lead from Table 19-1 is $\rho_{lead} = 11.3$ g/cm³. The volume of a right circular cylinder of height h is $V_{cyl} = \pi r^2 h$. In this case, we need to find the volume of the outer cylinder, of radius r_2, *minus* the volume of air in the inner cylinder, of radius r_1. Thus

$$V = \pi r_2^2 h - \pi r_1^2 h = \pi(r_2^2 - r_1^2)h = 147 \text{ cm}^3,$$

from which we find $M = \rho_{lead} V = 1660$ g = 1.66 kg.

Atoms and Moles

While mass provides one way of determining the amount of substance, another commonly used measure is that of *moles*. To understand what we mean by a mole of substance, we need to start by looking at atoms.

As you may recall, atoms of different elements have different masses. The mass of an atom is determined primarily by its most massive constituents—the protons and neutrons in its nucleus. The *sum* of the number of protons and neutrons is called the *isotope number* and is written as a leading superscript on the atomic symbol. For example, the common isotope of hydrogen, having one proton and no neutrons, is ^{1}H, and the "heavy hydrogen" isotope called *deuterium*, which includes one neutron, is ^{2}H. The primary isotope of carbon has six protons (which is what determines it to be carbon) and six neutrons and is ^{12}C. ^{14}C is a rarer radioactive isotope, containing six protons and eight neutrons, that is used for carbon dating of archeological finds.

Atomic masses are determined by establishing a *secondary mass standard* in which a single atom of ^{12}C has, by definition, a mass of exactly 12.000 *unified atomic mass units*. The symbol for unified atomic mass units is "u," so ^{12}C has a mass of 12 u, exactly. We will use the symbol A to represent the **atomic mass** when expressed in units of unified atomic mass units. Thus $A(^{12}\text{C}) = 12$ u. (Note that A is often incorrectly called the "atomic weight.")

Table 19-2 shows the atomic masses of a few isotopes. Note that the atomic mass A is *nearly* equal to the isotope number. The small differences arise when the electron masses as well as various "relativistic effects" are included. Nonetheless, the isotope number alone is accurate to three significant figures for nearly every isotope, and for the purposes of this text we will accept the isotope number as a sufficiently accurate value of A. (The periodic table of the elements inside the front cover of the book shows the atomic number and the average atomic mass of each element.)

TABLE 19-2 Some atomic masses.

Isotope	A (in u)
Proton	1.0073
^{1}H	1.0078
^{2}H	2.0141
^{4}He	4.0026
^{12}C	12.0000 exactly
^{14}C	14.0032
^{14}N	14.0031
^{16}O	15.9949
^{238}U	238.0508

For molecules, the **molecular mass** (often called the molecular weight) is just the sum of the atomic masses of the atoms forming the molecule. We will continue to use the symbol A for molecules. Thus the molecular mass of ordinary oxygen (^{16}O$_2$) is $A = 31.9898$ u ≈ 32 u.

It is often necessary to convert unified atomic mass units into the SI standard of kilograms. This involves establishing a connection between the microscopic realm of atoms and the macroscopic world of grams and kilograms. To do so, we will define one **mole** to be the amount of any substance containing the same number of basic particles (atoms or molecules) as there are atoms in 12 grams of ^{12}C. The abbreviation for mole is "mol." This unit, along with the meter, second, and kilogram, is one of the basic units in the SI system. Many decades of ingenious experiments have determined that there are 6.022×10^{23} atoms in 12 g of ^{12}C.

One mole of any substance—be it solid, liquid, or gas—is then the amount of matter represented by 6.022×10^{23} atoms or molecules of that substance. The basic particle of helium gas is the helium atom, so 1 mol of helium is 6.022×10^{23} helium atoms. Oxygen has a basic particle of the O$_2$ molecule, so 1 mol of oxygen gas contains 6.022×10^{23} *molecules* of O$_2$ (and thus $2 \times 6.022 \times 10^{23}$ oxygen atoms).

The number of particles per mole of substance is called **Avogadro's number**, which we write as N_A. As you see, the value of Avogadro's number is

$$N_A = 6.022 \times 10^{23} \text{ mol}^{-1}.$$

VOLUME, MASS, AND DENSITY   13

Notice that Avogadro's number, despite its name, is not simply "a number"; it has units. Avogadro's number, like the gravitational constant G or Planck's constant h, is one of the basic constants of nature.

Because there are N_A particles per mole, the number of moles in a substance containing N particles is

$$n = \frac{N}{N_A}, \qquad (19\text{-}4)$$

where n is symbol for moles. Macroscopic systems have typically $N \sim 10^{24}$ atoms or molecules, so they contain typically $n \sim 1$ mol of substance.

Having established the size of one mole, we can now determine the conversion from unified atomic mass units to kilograms. Consider one mole (N_A atoms) of ^{12}C atoms, each having an atomic mass $A = 12$ u exactly. The total mass of this sample of atoms is then $M_{1\,\text{mol}} = N_A A = N_A \cdot 12$ u. But by the definition of the mole, the mass of N_A ^{12}C atoms is exactly 12 g = 0.012 kg. Thus

$$M_{1\,\text{mol}} = N_A \cdot 12\text{ u} = 12\text{ g} = 0.012\text{ kg}.$$

Solving this for u gives

$$1\text{ u} = \frac{0.012\text{ kg}}{12 N_A} = \frac{0.001}{N_A}\text{ kg} = 1.661 \times 10^{-27}\text{ kg}. \qquad (19\text{-}5)$$

This is the conversion between the unified atomic mass unit scale and the kilogram scale. By measuring Avogadro's number we have, in effect, "weighed the atom."

Equation 19-4 allows you to determine the number of moles of a substance if you know how many atoms there are. It is often more convenient to be able to determine the number of moles based on the mass of a substance. The mass of 1 mol of ^{12}C atoms, with $A = 12$, is 12 grams by definition. If 6.02×10^{23} atoms, each with mass 12 u, have a total mass of 12 g, then 6.02×10^{23} atoms of mass 6 u have a total mass of 6 g and 6.02×10^{23} atoms of mass 24 u have a total mass of 24 g. In general, 1 mol of a substance with atomic or molecular mass A has a mass of A grams. Thus M grams of the substance contains

$$n = \frac{M\text{ (in grams)}}{A}. \qquad (19\text{-}6)$$

The symbol A in Eq. 19-6 has the special sense of "grams per mole." Note that this is the one instance in this text that the proper units to use in a calculation are *grams* rather than kilograms. Equations 19-4 and 19-6 are equivalent.

EXAMPLE 19-3   How many moles are contained in 100 g of oxygen gas?

SOLUTION   We can do the calculation two ways, using either Eq. 19-4 or Eq. 19-6. With Eq. 19-4, we need to determine the number of molecules present. The oxygen molecule $^{16}O_2$ has a molecular mass $A = 32$ u. Converting to kg gives

$$m = 32\text{ u} \cdot \frac{1.66 \times 10^{-27}\text{ kg}}{1\text{ u}} = 5.31 \times 10^{-26}\text{ kg}.$$

The number of molecules in 100 g = 0.10 kg is then
$$N = \frac{M}{m} = \frac{0.100 \text{ kg}}{5.31 \times 10^{-26} \text{ kg}} = 1.88 \times 10^{24},$$
from which we find
$$n = \frac{N}{N_A} = 3.13 \text{ mol}.$$
Alternatively, using Eq. 19-6,
$$n = \frac{M \text{ (in grams)}}{A} = \frac{100 \text{ g}}{32 \text{ g/mol}} = 3.13 \text{ mol}.$$

19.3 Temperature

So far there has been nothing fundamentally new in the quantities and definitions we have introduced in this chapter. All of them have been extensions of ideas from single-particle dynamics. The idea of *temperature*, however, is entirely new. It applies *only* to macroscopic systems and has no meaning at all in the realm of single-particle dynamics. This is the first property you will see that truly characterizes the *aggregate* behavior of a macroscopic system.

[**Photo suggestion: Several different kinds of thermometers.**]

We will explore in coming chapters just what temperature means and what it measures. You will find, ultimately, that **temperature** measures the microscopic energy of the atoms and molecules in the system. For now, we simply want to introduce temperature as a means of measuring the extent to which a macroscopic system is "hot" or "cold," properties that we can judge without needing an elaborate theory. This is the beginning of our efforts to describe the *thermal properties* of a system—that is, how the behavior of the system varies with temperature. To start, we need a means to measure the temperature of a system—and to do that, we will use a thermometer.

A *thermometer* can be any small macroscopic system that undergoes a measurable change as it absorbs or gives off heat. It is placed in contact with a larger system whose temperature it will measure. In a common glass-tube thermometer, for example, a small volume of mercury or alcohol expands when it absorbs heat from a hot object or contracts if it gives off heat to a cold object. The amount of expansion or contraction can be measured as the liquid column moves up and down the glass tube. Thus we have a thermometer that measures the temperature of the object with which it is in contact.

There are many other kinds of thermometers. Bimetallic strips (two strips of different metals sandwiched together) that curl and uncurl as the temperature changes are used in thermostats, such as the one in your house. Thermocouples, which generate a small voltage depending on the temperature, are widely used for sensing temperatures in inhospitable environments, such as in your car engine. Ideal gases, whose pressure varies with the temperature, are used in still another thermometer. We will look at an example of a gas thermometer later in the chapter.

To be a useful measuring device, a thermometer needs a *temperature scale*. In 1742 Swedish astronomer Anders Celsius sealed mercury into a small capillary tube and observed how it moved up and down the tube as the temperature changed. He selected two temperatures that anyone could reproduce—the freezing and boiling points of pure water—and he arbitrarily called the freezing point of water 0 and the boiling point 100. He then marked off the glass tube into one hundred equal intervals between these two reference points. By doing so, he invented the temperature scale that we today call the *Celsius scale*. It is the most widely used temperature scale.

Temperature is a quantity with units, just like mass or length, and the units of the Celsius temperature scale are called "degrees Celsius," which we abbreviate "°C." Notice that the degree symbol ° is part of the units, not part of the number. Water freezes at 0°C and boils at 100°C. A comfortable room temperature is 20°C, and normal human body temperature is 37.0°C. The symbol for temperature is T.

In later chapters you will discover that it is impossible for any object to have a temperature lower than −273.15°C. Because this is the lowest temperature physically possible, it is useful to adopt a temperature scale with that as the zero point. Such a temperature scale is called an **absolute temperature scale**. Any physical object whose temperature is measured on an absolute scale will be found to have $T > 0$. Negative absolute temperatures are not physically possible (you will see why later), and even $T = 0$ can only be approached as a limit. The zero on an absolute temperature scale is called **absolute zero**. Its value is $T_{\text{abs zero}} = -273.15°C \approx -460°F$.

We can define an absolute temperature scale having the same unit size as the Celsius scale by moving the zero-point down by 273.15°. This absolute temperature scale, which is called the Kelvin scale, is the SI scale of temperature. The units of the Kelvin scale are *kelvins*, abbreviated as K. The conversion between the Celsius scale and the Kelvin scale is

$$T(\text{K}) = T(°\text{C}) + 273.15.$$

With our standard in this text of three significant figures, you can use simply $T(\text{K}) = T(°\text{C}) + 273$. Notice that the units are *not* "degrees Kelvin," but simply "kelvins."

On the Kelvin scale, absolute zero is 0 K, the freezing point of water is 273 K, and the boiling point of water is 373 K. We will adopt the phrase *room temperature* to mean 20°C or 293 K unless given a more precise temperature. While most practical macroscopic devices utilize temperatures in the range 100 K to 1000 K, it is worth noting that scientists study the properties of matter from temperatures as low as $\approx 1\ \mu\text{K} = 10^{-6}$ K on the one extreme to as high as $\approx 10^7$ K on the other!

19.4 Pressure

[**Photo suggestion: An air tank with a pressure gauge.**]

When we treat a macroscopic object as a point particle, the only property in which we are interested is its mass M. But if we consider the object to be an extended system—having volume and consisting of microscopic particles—then the *density* ρ of the system is often a more useful quantity than the total mass.

Similarly, a particle's behavior is determined by the forces acting upon it. In an extended, continuous system, the number of forces is extremely large. There is also the possibility that forces may vary from point to point or that there are internal motions of

16 CHAPTER 19 A MACROSCOPIC DESCRIPTION OF MATTER

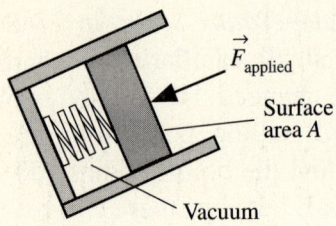

FIGURE 19-1 A pressure measuring device. The pressure can be determined from the compression of the spring.

the system (flows) due to internal forces. We need a property, analogous to density, that is more useful than simply "force" for characterizing a macroscopic system. For fluids—liquids and gases—such a property is the *pressure* of the fluid.

Consider a measuring device such as the one shown in Fig. 19-1. It consists of a tight-fitting piston, whose face has surface area A, that is free to slide back and forth in a cylinder while compressing a spring. All of the air has been removed from the spring end of the cylinder, creating a vacuum. A force applied to the face of the piston compresses the spring to a length L. By knowing the spring constant (we will assume it has been previously measured), we can use the measured length L to determine the magnitude $|\vec{F}_{applied}|$ of this force.

Now place the force measuring device into a beaker of liquid. What happens? The first thing we discover is that a force pushes against the piston, compressing the spring. The force recorded by our device depends on its depth in the liquid—greater depth, more force. Is this surprising? When you swim underwater, you can feel the force of the water against your eardrum. Your eardrum, in fact, is much like the device of Fig. 19-1. As you probably know, the force you experience gets larger as you swim deeper.

In addition, we find that the magnitude of the force is *independent of the orientation* of the device, as shown in Fig. 19-2. The device measures the same force regardless of whether it is pointed up, down, or sideways. This does seem somewhat surprising, although you have probably noticed, as you swim underwater, that your eardrum feels the same force regardless of the orientation of your head. You also know that water squirts out of a hole in the *side* of a container. For this to happen, there must be a sideways force of the liquid against the wall of the container. These examples, and others you can probably think of, show that the forces exerted by a fluid push equally in *all* directions, not just down.

What else can we learn? If we build a second measuring device with twice the piston surface area as the first and place them side by side, the second device records a force twice as large. Another device with half the surface area records a force only half as large. The magnitude of the force is directly proportional to the *area* that the liquid pushes on: $|\vec{F}| = pA$, where p is a proportionality constant.

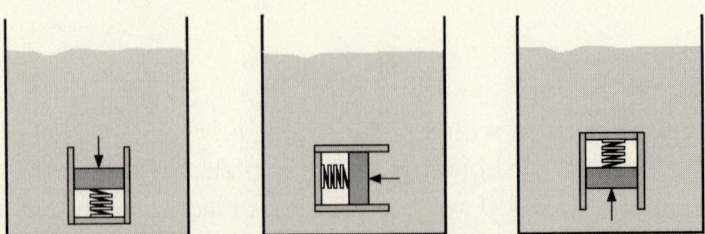

FIGURE 19-2 The pressure of a liquid is the same in all directions if the measuring device is kept at the same depth.

This proportionality constant between force and surface area is what we call **pressure**. The pressure p in a fluid is thus

$$p = \frac{|\vec{F}|}{A}, \qquad (19\text{-}7)$$

where $|\vec{F}|$ is the force exerted by the fluid on a surface of area A. In other words, the pressure of a fluid is the force per unit area exerted by the fluid. Note that pressure itself is *not* a force, even though we often talk about "the force exerted by the pressure." The correct statement is that the *fluid* exerts a force on a surface. If the surface has area A and the fluid has pressure p, then the force has magnitude $|\vec{F}| = pA$ and is directed perpendicular to the surface.

Pressure is a property of a fluid, independent of the device used to measure it. Further, pressure exists at *all* points within a fluid, regardless of whether or not there is a physical surface to push on. You may recall that tension exists at all points in a string, not just where the string is tied to an object, because the different parts of the string pull against each other. Similarly, the different parts of a fluid push against each other, not just against the sides of the container.

From its definition, pressure must have units of N/m². We will define the SI unit of pressure as the **pascal**, where 1 pascal is

$$1 \, \frac{\text{N}}{\text{m}^2} = 1 \text{ pascal} = 1 \text{ Pa}.$$

This unit is named after the seventeenth-century scientist Blaise Pascal, who was one of the first to study the properties of pressure.

So far so good, but why should there be a pressure at all in a fluid? Let's first concentrate just on liquids, where we can answer the question with the experiment shown in Fig. 19-3. In this experiment the piston of our force-measuring device is the bottom of a cylindrical container of liquid. The liquid is pressing directly against the piston face—there is no other "bottom" to the container. The spring is compressed until a static equilibrium situation is reached in which the upward push of the spring balances the downward weight force of the liquid. The force applied to the piston is simply $|\vec{F}| = Mg$, where M is the total mass of the liquid. In other words, the force measured by the device is just the weight of the liquid. The pressure at the bottom of the liquid, where it is being measured, is thus $p = Mg/A$. (We are, for just a moment, ignoring any effect of the atmosphere above the liquid.) The pressure occurs because of gravity—it is a consequence of the weight of the liquid above the surface on which it is pressing.

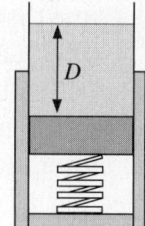

FIGURE 19-3 Measuring pressure at the bottom of a liquid of depth D.

Now the mass of the liquid is $M = \rho V$, where ρ is the liquid density. If the depth of the liquid is D, then the cylinder of liquid has volume $V = AD$ and the mass is $M = \rho AD$. With this result for the mass of the liquid, we find

$$p = \frac{Mg}{A} = \frac{\rho AD \cdot g}{A} = \rho g D.$$

The pressure depends only on the density and the depth of the liquid, not on the surface area of the bottom.

The pressure at depth D should be the same regardless of whether D is the bottom of the container or whether a measuring instrument happens to be there. Pressure is a property of the fluid, not of the measuring instrument. So it must be the case that the pressure at depth D in a liquid, if atmospheric effects are neglected, is $p = \rho g D$. This is true regardless of how narrow or wide the container is.

Now suppose we rise to the surface of the liquid, so that $D \to 0$. Can we conclude that the pressure is zero at the surface? A region of space from which all the matter has been removed and in which the pressure is $p = 0$ is called a **vacuum**. Although there is no pressure at the surface due to the weight of the liquid, it is not a vacuum. There is still a pressure due to the weight of the column of *air* above the surface. If the column of air above a surface of area A has weight W_{air}, it exerts a pressure on the surface of $p_{atmos} = W_{air}/A$. This is what we call the **atmospheric pressure**.

Our measuring device in Fig. 19-3 experiences not only the weight of the liquid above it, but also the weight of the column of air above the liquid. Thus the pressure p_{liq} at depth D below the surface of a liquid needs to include the pressure of the atmosphere:

$$p_{liq} = p_{atmos} + \rho g D \qquad \text{(pressure at depth } D \text{ in a liquid)}, \qquad (19\text{-}8)$$

where p_{atmos} is the pressure of the atmosphere immediately above the fluid's surface. Now we have the proper result at the surface: $p(D = 0) = p_{atmos}$.

The atmospheric pressure varies slightly from day to day as the weather changes. It also decreases with altitude. The average value of the atmospheric pressure at sea level is called **standard atmospheric pressure**. It is measured to be

$$1 \text{ standard atmospheric pressure} = 1 \text{ atm} = 1.013 \times 10^5 \text{ Pa} = 101.3 \text{ kPa}.$$

Note the commonly used units of kPa, where 1 kPa = 1000 Pa. In this text we will always assume that $p_{atmos} = 1$ atm in Eq. 19-8.

The analysis leading to Eq. 19-8 applies *only* to liquids. Gas pressures change only very slightly with depth because their density is so much less than that of liquids. For all practical purposes, we can consider the pressure in any closed container of gas to be *constant* throughout the entire container. The value of the gas pressure, as you will learn in Section 19.6, depends on the temperature of the gas.

EXAMPLE 19-4 A submarine cruises at a depth of 300 m. What is the pressure at the submarine's depth? Give the answer both in pascals and in atmospheres.

SOLUTION From Table 19-1, the density of water is found to be $\rho = 1000$ kg/m^3. The pressure at depth $D = 300$ m is found from Eq. 19-8 to be

$$p = p_{atmos} + \rho g D = 1.013 \times 10^5 \text{ Pa} + (1000 \text{ kg/m}^3)(9.8 \text{ m/s}^2)(300 \text{ m})$$
$$= 3.04 \times 10^6 \text{ Pa}.$$

Converting the answer to atmospheres gives

$$p = 3.04 \times 10^6 \text{ Pa} \times \frac{1 \text{ atm}}{1.013 \times 10^5 \text{ Pa}} = 30.0 \text{ atm}.$$

Measuring Pressure

The pressure in a liquid is usually measured by a *pressure gauge,* a device very similar to that in Fig. 19-1. The liquid pushes against some sort of spring, usually a diaphragm, and the spring's displacement is registered by a pointer on a dial. Gas pressures can also be measured with a spring-loaded pressure gauge.

Pressure gauges usually measure not the actual pressure p but what is called **gauge pressure**. The gauge pressure, denoted p_g, is the pressure *in excess* of 1 atm. Thus

$$p_g = p - 1 \text{ atm.}$$

The gauge on an air tank, such as a tank used for scuba diving, reads zero when the tank is "empty." The tank is not actually empty, nor is there a vacuum inside. Once the pressure inside the tank drops to 1 atm, there is no longer a pressure *difference* between inside and outside, so no more air will flow out. Similarly, a tire gauge reads zero if you have a flat tire. There is not a vacuum inside a flat tire, but simply the same 1 atm pressure inside as there is outside. These gauges are reading gauge pressure, not the actual pressure p. You need to add 1 atm = 101.3 kPa to the reading of a pressure gauge to find the true pressure p that you need for doing science or engineering calculations: $p = p_g + 1$ atm.

EXAMPLE 19-5 An underwater pressure gauge reads 0.60 atm. What is its depth?

SOLUTION The gauge reads gauge pressure. Note, however, that Eq. 19-8 can be rearranged to give

$$\rho g D = p_{\text{liq}} - p_{\text{atmos}} = p_{\text{liq}} - 1 \text{ atm} = p_g.$$

That is, the term $\rho g D$ due to the weight of the liquid is the pressure in excess of atmospheric pressure. So $\rho g D$ *is* the gauge pressure p_g. In this case,

$$D = \frac{p_g}{\rho g} = 6.2 \text{ m.}$$

Note that we had to make the conversion $p_g = 60.8$ kPa before doing the calculation.

[**Photo suggestion: A mercury manometer or a mercury barometer.**]

Accurate gas pressures are often measured with a device called a *manometer*. A manometer, as shown in Fig. 19-4, is a U-shaped tube that is connected to the gas at one end and is open to the air at the other end. The tube is filled with a liquid—usually mercury—of density ρ. The liquid is in static equilibrium, with the pressure of the atmosphere pushing down on the right side and the pressure of the gas pushing down on the left. A scale allows the user to measure the height h of the right side above the left side.

To analyze the manometer, consider the two points in Fig. 19-4 labeled 1 and 2. These two points are both on a horizontal line that is level with the top of the liquid on the left. Because the liquid beneath this line is in static equilibrium,

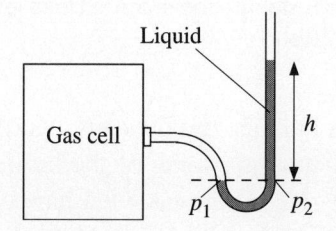

FIGURE 19-4 A manometer is often used to measure gas pressure.

the two pressures must be equal: $p_1 = p_2$. If the pressures differed, there would be a net force causing the liquid to shift one way or the other. Pressure p_1 is simply the gas pressure ($p_1 = p_{gas}$) and pressure p_2 is the pressure at depth $D = h$ in the liquids ($p_2 = \rho g h +$ 1atm). Equating the two pressures gives:

$$p_{gas} = 1 \text{ atm} + \rho g h. \tag{19-9}$$

Figure 19-4 assumed that $p_{gas} > 1$ atm, so the right side of the liquid is pushed higher than the left. However, Eq. 19-9 is also valid for $p_{gas} < 1$ atm if the distance of the right side *below* the left side is considered as a negative value of h.

EXAMPLE 19-6 A mercury manometer is used to measure the pressure of a gas cell. The mercury is 36.2 cm higher in the outside arm than in the arm connected to the gas cell. a) What is the gas pressure? b) What is the reading of a pressure gauge attached to the cell?

SOLUTION a) The density of mercury is found from Table 19-1 to be $\rho = 13{,}600$ kg/m^3. Using $h = 0.362$ m, Eq. 19-9 gives

$$p_{gas} = 1 \text{ atm} + \rho g h = 149.5 \text{ kPa} = 1.476 \text{ atm}.$$

b) The pressure gauge reads gauge pressure $p_g = p - 1$ atm $= 0.476$ atm $= 48.2$ kPa.

Another important pressure measuring instrument is the *barometer*, which is used to measure the atmospheric pressure p_{atmos}. Figure 19-5a shows a glass tube that is sealed at the bottom and that has been completely filled with a liquid. If we temporarily seal the top end—by, for example, placing a finger over it—we can invert the tube and place it in a beaker of the same liquid, as shown in Fig. 19-5b. Upon removing the temporary seal, we find that some, but not all, of the liquid runs out, leaving a liquid column in the tube that is a height h above the surface of the liquid in the beaker. This device is a barometer. What does it measure? And why doesn't *all* the liquid in the tube run out?

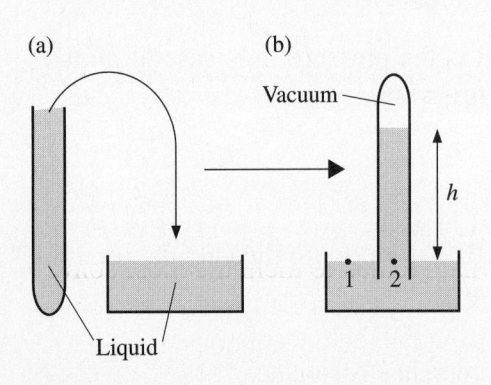

FIGURE 19-5 Construction of a barometer. a) A liquid-filled tube is inverted into a dish of liquid. b) A liquid column of height h stands above the surface level of the dish.

We can analyze the barometer much like we did the manometer. Consider points 1 and 2, in Fig. 19-5b, on a horizontal line drawn even with the surface of the liquid. The liquid is in static equilibrium, so the atmospheric pressure pushing down at point 1 must exactly balance the pressure inside the tube as it pushes down at point 2. If pressure p_2 were higher than p_1, some of the liquid would run out of the tube and force the liquid in the beaker to rise. That was what happened when the inverted tube was first placed in the beaker. But if p_1 were higher than p_2, the pressure of the atmosphere would push liquid from the beaker back *into* the tube. So liquid runs out of the tube only until a balance is reached between the pressure in the tube and the pressure of the air.

The pressure at point 1 is simply the pressure of the atmosphere: $p_1 = p_{atmos}$. The pressure at point 2 is the pressure due to the weight of the liquid plus the pressure of the air above the liquid. But in this case there is no air above the liquid! Because the tube had been completely full of liquid when it was inverted, the space left behind when the liquid partly ran out contains no matter at all—it is a *vacuum*. Thus pressure p_2 is simply $p_2 = \rho gh$.

Equating the two pressures gives:

$$p_{atmos} = \rho gh. \qquad (19\text{-}10)$$

We can measure the atmosphere's pressure by measuring the height of a liquid column in a barometer!

Although the pressure of the air varies somewhat from day to day as weather conditions change, the average pressure at sea level causes a column of mercury in a mercury barometer to stand 760 mm above the surface. Because the density of mercury is 13,600 kg/m^3 (at 0°C), we can use Eq. 19-10 to find the atmospheric pressure to be

$$p_{atmos} = \rho_{Hg} gh = (13{,}600 \text{ kg}/\text{m}^3)(9.80 \text{ m}/\text{s}^2)(0.760 \text{ m})$$

$$= 1.013 \times 10^5 \text{ Pa} = 101.3 \text{ kPa}.$$

This is the value given earlier as being one standard atmospheric pressure. This is the definition of standard atmospheric pressure *because*, on average, a mercury barometer gives a reading of 760 mm.

In practice, pressure is measured in a number of different units and it is necessary to be familiar with all of them. The proper SI unit, and the one you need to use in doing numerical calculations, is the pascal. Pressures of gases, when measured with a barometer or a manometer, are usually given in units of "mm of Hg." Gas pressures below 1 atm, such as when a vacuum pump lowers the pressure, are generally measured in units called "torr," which are exactly equivalent to "mm of Hg" despite the fact that a different name is used. Pressures are also often given in "atmospheres," as in "$p = 3$ atm." And, of course, American industry still widely uses the English units of "lb/in^2," often abbreviated "psi" for "pounds per square inch." This plethora of units and abbreviations has arisen historically as scientists and engineers working on different subjects (liquids, high-pressure gases, low-pressure gases, weather, etc.) developed what seemed to them the most convenient units. These units continue in use through tradition, so it is necessary to become familiar with converting back and forth between them. The basic conversions are

1 atm = 101.3 kPa = 760 mm of Hg = 760 torr = 14.7 psi.

Note that the *barometric pressure* of weather forecasts is given in units of "inches," such as "the barometric pressure is 30.06 inches." This is the pressure recorded by a mercury barometer, but measured in inches rather than millimeters. One standard atmosphere (760 mm) is 29.92 inches. The barometric pressure varies slightly from day to day as the weather changes. Weather systems are called *high-pressure systems* or *low-pressure systems*, depending upon whether the local sea-level pressure (as measured by a barometer) is higher or lower than one standard atmosphere. Higher pressure is usually associated with fair weather, but lower pressure portends rain.

Given the value of the air pressure as ≈15 pounds per square inch, you might wonder why your forearm isn't crushed by the weight of the air pressing down on it when you rest it on a table. Your forearm has a surface area of ≈30 in^2, so there is ≈450 pounds of air pressing against it. How can you even lift it? The reason, as we saw earlier, is that pressure exerts forces in *all* directions. There *is* a downward force of ≈450 pounds on your forearm, but the air underneath your arm exerts an *upward* force of the same magnitude, giving no *net* force.

But, you say, there isn't any air under my arm if I rest it on a table. Actually, there is. There would be a *vacuum* under your arm if there were no air. Imagine placing your arm on the top of a large vacuum cleaner suction tube. What happens? You feel a downward force as the vacuum cleaner "tries to suck your arm in." However, the downward force you feel is not a *pulling* force from the vacuum cleaner; it is the *pushing* force of the air above your arm *when the air beneath your arm is removed and cannot push back!* Air molecules have no ability to "pull" on your arm. They do not have hooks! Vacuum cleaners, suction cups, and other similar devices are powerful examples of how strong atmospheric pressure forces can be *if* the air is removed from the other side so as to produce an unbalanced force. But we can move around in the air, just like we can swim underwater, oblivious to these strong forces. The fact that we are *surrounded* by the fluid means that we experience no *net* force.

EXAMPLE 19-7 A hemispherical suction cup 10 cm in diameter is placed against a smooth ceiling. All the air is then pumped out of the cup to create a vacuum. If an object is suspended from the suction cup, what is the maximum weight the object can have without pulling the suction cup off the ceiling?

SOLUTION Figure 19-6 shows that the air pressure exerts upward forces against the outside surface of the cup, but this time there is no compensating downward force because the air inside has been removed. The horizontal components of the forces will cancel, so the net force will be upward against an effective surface area $A = \pi r^2$. Thus the net upward force (assuming that the suction cup's weight is negligible) is

$$|\vec{F}| = pA = p\pi r^2 = 796 \text{ N} \ (\approx 180 \text{ pounds}).$$

This suction cup will support up to 796 N ≈ 180 pounds of weight! That is, you would need to exert an external force > 796 N to overcome the atmospheric pressure force and pull the suction cup off the ceiling.

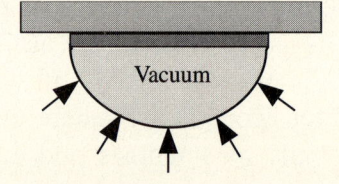

FIGURE 19-6 A suction cup is held to the ceiling by air pressure pushing from the bottom.

19.5 Ideal Gases

Gases are the simplest macroscopic systems. Our emphasis for the rest of this chapter, and in fact for the rest of Part IV, will be on gases. In this chapter our goal is to understand the macroscopic properties of gases and how they change as the state of the gas changes. Even so, it will be useful to start from an atomic description of a gas. Chapters 21 and 22 will then develop this microscopic description of gases in more detail.

While we today take it for granted that matter consists of atoms, this is by no means obvious. How do we know this? What is the evidence for our belief? The concept of atoms is due to two Greek philosophers, Leucippus and Democritus, who flourished about 440–420 B.C. Little is known of Leucippus, and most of our knowledge is traced to the writings of Democritus, who was a contemporary of Socrates. Leucippus and Democritus suggested that all matter consists of small, hard, indivisible, and indestructible particles they called *atoms*. Atoms, according to them, are always in motion and the space between atoms are matter-free regions called the *void*. Their suggestions, needless to say, were pure speculations and had no empirical basis at all. It is thus surprising how close their ideas were to our modern concept of the atom.

Aristotle and most other "mainstream thinkers" were not atomists because they could not accept the idea of the void—or what we, today, would call a vacuum. They did not separate the idea of "space" from the idea of "matter," so to them the idea of space without matter was a logical impossibility. Because the ideas of Aristotle and his followers were absorbed into the theology of the early Catholic Church, the ideas of atoms essentially disappeared for nearly 2000 years. Only with the beginnings of modern chemistry, during the Renaissance, did atomic thinking reappear.

Newton was an atomist (although Descartes was not), but he held the view that atoms in a gas are static, rather than in motion, and that they expand and contract to stay in contact with other atoms. Gases in motion, like the wind, would then be a collective motion of all the atoms moving together with the same velocity.

Daniel Bernoulli advanced a quite different model of gases in a book published in 1738. He suggested that gases consist of small, hard atoms moving randomly at fairly high velocities and, on occasion, colliding with each other or the walls of the container (Fig. 19-7). As a consequence of these collisions, the atoms exert a *pressure* on the walls. Bernoulli, in other words, reintroduced Democritus's idea of atoms—but updated it to include the scientific discoveries of Galileo and Newton!

It may seem surprising today, but Bernoulli's ideas were not accepted for nearly a century. He was too much ahead of his time, and the value of his postulates was not recognized until a complete understanding of energy conservation was achieved in the mid-nineteenth century. Numerous scientists at that time further developed Bernoulli's ideas into the kinetic theory of gases, which we will look at closely in Chapter 22. It forms our present atomic-level understanding of gases.

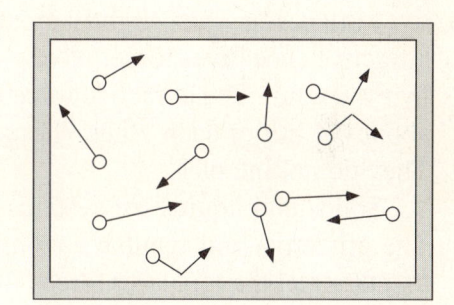

FIGURE 19-7 Bernoulli's atomic model of a gas as atoms in motion.

If, for the moment, we accept the existence and reality of atoms (pending further evidence as we go along), what can macroscopic observations suggest to us about their properties? One such observation, which we noted in the last chapter, is that solids and liquids are nearly incompressible. From this observation we can infer that atoms are fairly "hard" and cannot be pressed together once they come into contact with each other. Today we recognize this behavior as being a consequence of the electron shells coming very close together and exerting very strong repulsive electrical forces on each other. Those "details," however, are not needed for drawing the conclusion, from the incompressibility, that atoms have a hard core.

While atoms resist being pushed too close together, they also resist being pulled apart. Solids would not be solid if the atoms did not have attractive forces holding them together. These attractive forces are responsible for the *tensile strength* of solids—how hard you have to pull to break the solid—as well as for the cohesion of liquid droplets. Nonetheless, it is far easier to break a solid or disperse a liquid than it is to compress it, so these attractive forces must be weak in comparison to the repulsive forces that occur when we push the atoms too close together. The attractive force between two nearby atoms is also an electrical force, but it takes a more advanced study of electricity to see how it arises.

Our "picture" of an atom, then, is of a small particle that is weakly attracted to other fairly nearby atoms but strongly repelled by them if it tries to get too close. This is precisely the view of molecular bonds that we developed back in Chapter 10. Figure 19-8 shows the potential-energy diagram of two atoms separated by distance r. Recall, from Chapter 10, that the force exerted by one atom on the other is the negative of the *slope* of this graph. For values of r less than the equilibrium value r_{eq}, the slope is negative with a very large value. Thus the force, for $r < r_{eq}$, has a large repulsive (positive) value. For r just slightly greater than r_{eq}, the modest positive slope indicates a weak attractive (negative) force. Note—and this is important—that the slope has become zero by $r \approx 0.4$ nm. The attractive force between atoms is only for "nearby" atoms, those within about 0.4 nm of each other. Atoms separated by >0.4 nm exert no forces on each other. They do not interact.

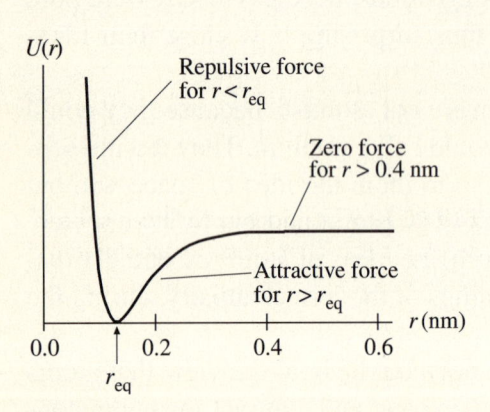

FIGURE 19-8 The potential energy diagram for the interaction of two atoms.

Solids and liquids are systems in which the average separation of atoms is $r_{avg} \approx r_{eq}$. The attractive and repulsive atomic forces are balanced. Try to press the atoms closer together and the repulsive forces for $r < r_{eq}$ resist. Try to pull them apart and the attractive forces for $r > r_{eq}$ resist.

A gas, by contrast, is much less dense, with an average spacing of atoms of $r_{avg} \gg r_{eq}$. This means that the atoms are usually *not interacting* with each other at all! Instead, they spend most of their time moving freely through space in straight lines, only occasionally coming close enough to another atom to interact with it—a collision. When such a collision occurs, it is the steep "wall" of the potential energy curve that is important. That wall is the potential energy of the repulsive electrical force pushing the atoms away as they collide. The small distance over which the atoms experience a weak attractive force is of essentially no importance because the atoms spend so little of their time at those distances.

The Ideal Gas Model

With these ideas in mind, suppose we were to replace the actual potential-energy curve of Fig. 19-8 with the approximate potential-energy curve of Fig. 19-9. This is the potential-energy curve for the interaction of two "hard spheres"—two atoms that have *no* interaction at all until they come into actual contact, at separation $r_{contact}$, and then bounce.

It is not possible, for energy conservation reasons, for the atoms to have $r < r_{contact}$.

Figure 19-9, the *hard sphere model* of the atom, represents what we could call the *ideal atom*. It is, largely, Democritus's idea of a small, hard particle. A gas of such ideal atoms we will then call an *ideal gas*. It is a collection of small, hard, randomly moving atoms that occasionally collide and bounce off each other but otherwise do not interact. The ideal gas is a *model* of a real gas and, as with any other model, a simplified description. Nonetheless, experiments show that the ideal gas model is quite good for gases if two conditions are met:

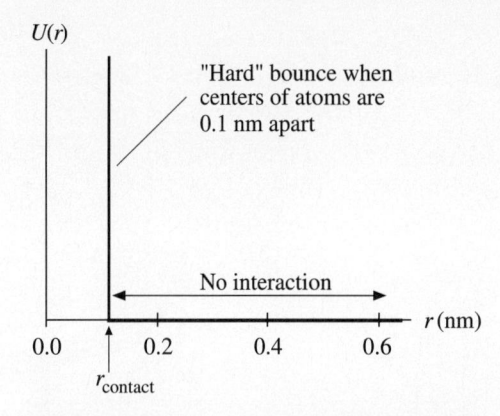

FIGURE 19-9 An idealized "hard sphere" model of the interaction potential energy of two atoms.

1. The density is low.
2. The temperature is well above the condensation temperature.

If the density gets too high, or the temperature too low, then the attractive forces between the atoms begin to play a nonnegligible role and our model, which ignores those attractive forces, fails. These are the forces that are responsible, under the right conditions, for the gas condensing into a liquid. Nonetheless, the ideal gas model works extremely well for many gases under a wide range of conditions.

We've been using the term *atoms*, but many gases, as you know, consist of molecules rather than atoms. Only helium, neon, argon, and the other inert elements in the right column of the periodic table of the elements form atomic gases, which are called **monatomic gases**. Gases such as hydrogen (H_2), nitrogen (N_2), and oxygen (O_2) consist of diatomic molecules and are called **diatomic gases**. There is no distinction in the ideal gas model between a monatomic gas and a diatomic gas; both are considered as simply small, hard spheres. So the terms *atoms* and *molecules* can be used interchangeably to mean the basic constituents of the gas.

The atoms or molecules in a gas exchange energy as they collide with each other—just like two marbles colliding and then going off with altered speeds. The atoms also collide with the walls of the container and thus exert forces on the walls. It is this force that causes the *pressure* of the gas. (Note that the *mechanism* of pressure in a gas—collisions—is different from the mechanism of pressure in a liquid.) Because the atoms are moving randomly in all directions, this atomic-level view of the gas helps us to understand why pressure is exerted on all the walls, not just the bottom.

This extended introduction, it is hoped, will give you a better understanding of what we "mean" when we talk about an ideal gas, some of the assumptions behind the model, and some of the evidence that supports it. The next section will look even more closely at some experimental evidence that gases really are collections of randomly moving atoms.

Molecular Speeds

What kind of experimental evidence could distinguish between Newton's view of a gas of static atoms and Bernoulli's model of a gas as small atoms in random motion? Suppose

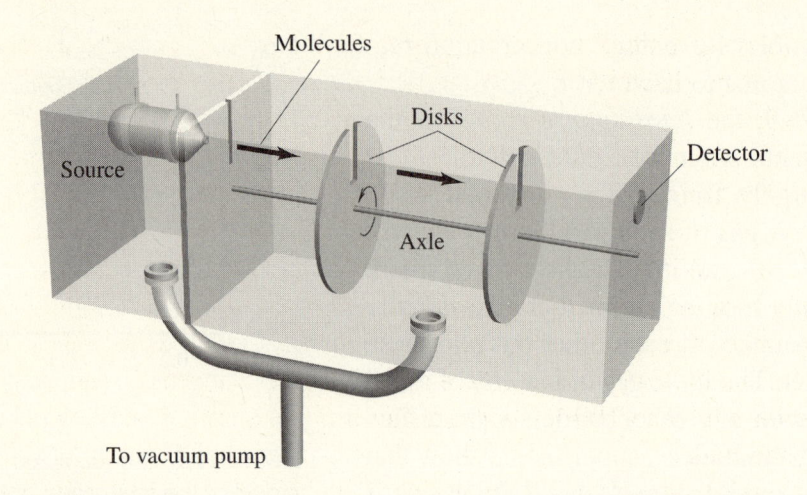

FIGURE 19-10 An apparatus to measure the speeds of molecules in a gas.

we had a container of gas with a very small hole in the side, allowing some of the gas to flow out. If Newton was correct, the atoms would move out uniformly, all touching and moving with the *same speed*—like toothpaste out of a tube as you squeeze it! However, if Bernoulli was right, the individual atoms would shoot through the hole and emerge with many *different* speeds. We can perform an experimental test of these two models if we can find a way to *measure* the speed or speeds of gas atoms, or molecules, emerging from a small hole.

Figure 19-10 shows an experimental apparatus that does exactly what we need—it measures the speeds of molecules emerging from a gas. At the left end is a *source* of molecules—a container with a small hole from which the molecules emerge to form what is called a *molecular beam*. At the right end is a *detector* that records the arrival of a molecule. The entire apparatus (which might be ≈1 m long) is placed in a *vacuum chamber* from which the air is removed by pumps. This allows the test molecules to travel the entire distance from source to detector without undergoing any collisions.

Between the source and the detector is a device called a *velocity selector*. It consists of two (or sometimes more) disks on a rotating axle, with each disk having a small hole or slot through which the molecules can pass. The source is constantly shooting out molecules, but most of them hit the first disk and do not continue any farther. Once every revolution, the slot in the first disk passes by and allows a small pulse of molecules to pass through. By the time the molecules reach the second disk, however, the slots have rotated and the molecules cannot pass through *unless* they happen to have exactly the right speed v such that they travel the distance L between the two disks in exactly the time interval Δt it takes the axle to complete one full revolution: $v = L/\Delta t$.

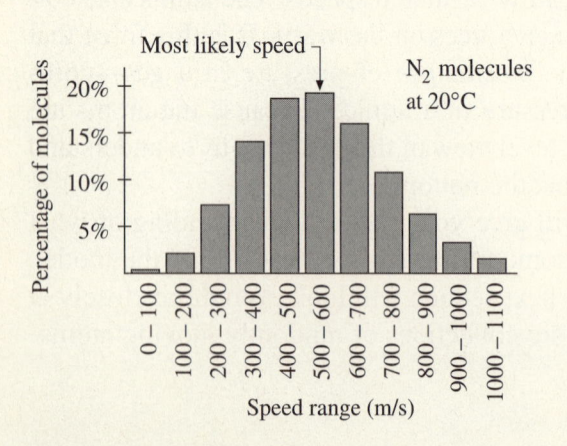

FIGURE 19-11 The distribution of speeds in a beam of N_2 molecules at $T = 20°C$.

Molecules with this particular speed can pass through the slot in the second disk, as it comes past on the next revolution, and can be detected. Molecules having any other speed are blocked by the second disk and can't be detected. By systematically changing the rotation rate of the axle, and thus changing Δt, this apparatus can measure how many molecules have each of many possible speeds.

Figure 19-11 shows the results of such an experiment performed with nitrogen gas (N_2) at room temperature ($T = 20°C$). The results are presented in the form of a histogram. A *histogram* is a bar chart in which the height of each bar tells how many (or, in this case, what percentage) of the molecules have a speed in the *range* of speeds shown below the bar. We see, for example, that 16% of the molecules have speeds in the range 600 m/s to 700 m/s. The bars total to 100%, showing that this histogram describes *all* of the molecules leaving the hole.

What do these results tell us? If the gas consisted of Newtonian atoms, we would have expected to see all the molecules with the same speed. The histogram in that case would show a single bar with 100% of the molecules. But that is *not* what was experimentally observed. Instead, we see what is called a *distribution* of speeds, ranging from as low as ≈0 m/s to as high as ≈1100 m/s. This is, at least qualitatively, exactly what we expected for a Bernoulli gas of individual atoms moving about randomly at many different speeds. Here we have convincing evidence that the basic assumptions behind the ideal gas model are correct.

But we can learn more and thereby improve our understanding of what is going on in a gas at the atomic level. Notice, in Fig. 19-11, that not all speeds are equally likely. There is a most probable speed—the column with the highest percentage—of ≈550 m/s ≈ 1200 mph. This is really *fast*! (Do you have any sensation that the air molecules colliding with you are moving this fast?) Speeds above or below this are not as likely to occur. This is an interesting observation. Also note that most of the speeds do not differ much from the most probable speed. There are, to be sure, a few with very high or very low speeds, but well over 60% (sum of the center four bars) of the molecules have speeds within the range 300 m/s to 700 m/s. Lastly, notice that the most probable speed is not too different from the speed of sound in air (which is 80% N_2) of 343 m/s. Why does sound have that particular speed? Likely because it is somehow connected to how fast the air molecules are moving—as, indeed, is confirmed in a more advanced treatment.

The gas in the source consists of a very large number (~N_A) of molecules, each moving randomly and with a different speed. Yet the *collective* behavior of this sample of atoms is exceedingly stable and predictable. We could repeat this experiment over and over, and we would get the results of Fig. 19-11 every time. This idea—that the *average* of properties of a very large number of molecules is a predictable quantity—will underlie our statistical approach to the micro/macro connection.

19.6 The Equation of State for an Ideal Gas

In Section 19.1 we introduced the idea of *state variables*—those parameters that describe the state of a macroscopic system. The state variables for an ideal gas are the volume V of its container, the number of moles n of the gas present in the container, the temperature T of the gas and its container, and the pressure p that the gas exerts on the walls of the container.

These four state parameters are not independent of one another. If you change the value of one—by, say, raising the temperature—then one or more of the others will change as well. Each change of the parameters represents a *change of state* of the system. Many such changes are possible for a gas. For example, the gas can be heated or cooled, it can be expanded or compressed, or it can have atoms added or removed.

Numerous experiments during the seventeenth and eighteenth centuries led to the discovery that there is a very precise quantitative relationship between the four state variables. For *any* dilute gas, it was found that the product of the pressure and volume (pV) is directly proportional to the product of the quantity of gas and its temperature (nT). That is,

$$pV \propto nT, \qquad (19\text{-}11)$$

where, you will recall, the symbol $\propto$ means "is proportional to." The temperature in this relationship is the *absolute temperature*.

Figure 19-12 shows the results of graphing pV versus nT. Each data point is one state of the gas, with specific measured values for p, V, n, and T. The gas is then changed to a different state—by heating it or compressing it or adding more gas—and the measured values of the state variables in the new state then provide another data point on the graph. As you can see, there is a very clear relationship between the quantity pV and the quantity nT. If we designate the slope of the line in this graph as R, then we can write the relationship as

$$pV = R(nT).$$

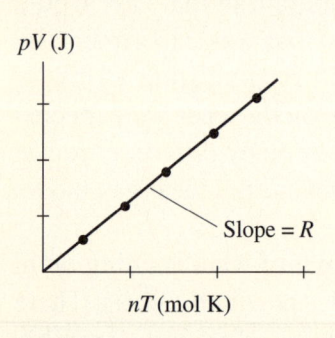

FIGURE 19-12 A graph of pV versus nT as measured for an ideal gas.

The slope R is the proportionality constant for Eq. 19-11. It is customary to write this relationship in a slightly different form, namely

$$pV = nRT \qquad \text{(ideal gas law)}. \qquad (19\text{-}12)$$

Equation 19-12 is formally called an **equation of state** because it is the mathematical relationship between the state variables of a dilute gas. It is more commonly known, however, as the **ideal gas law**.

The constant R, which is determined experimentally as the slope of the graph in Fig. 19-12, is called the *universal gas constant*. Its value, in SI units, is

$$R = 8.31 \text{ J/mol K}.$$

The units of R seem puzzling. The denominator of mol K is clear because R multiplies nT. But what about the joules? The left-hand side of the ideal gas law, pV, has units

$$\text{Pa m}^3 = \frac{\text{N}}{\text{m}^2}\text{m}^3 = \text{N m} = \text{Joules}.$$

You have perhaps learned in chemistry to work gas problems with units of atmospheres and liters—and to do so, you had a different numerical value of R that was expressed in those units. In physics, however, we will always work gas problem in SI units. That means pressures *must* be in Pa, volumes in m^3, and temperatures in K before you compute. Calculations using temperatures in °C will give wildly incorrect answers.

THE EQUATION OF STATE FOR AN IDEAL GAS   29

A remarkable fact, and one worth commenting on, is that *all* gases have the *same* value of R. This is surprising. Even if gases have a direct proportionality between pV and nT, we might expect that the slope of the graph, and hence the value of R, would depend on the particular gas being studied. There is no obvious reason why the slope for a very simple atomic gas such as helium should be the same as the slope for a more complex gas such as methane (CH_4). Nonetheless, it turns out to be true that they both have the same value for R. That is why Eq. 19-12 can be called a "law" of nature. It describes *all* gases (if dilute) with a single value of the constant R.

EXAMPLE 19-8 100 g of oxygen gas are distilled into an evacuated container of 600 cm^3. What is the gas pressure at a temperature of 150°C?

SOLUTION From the ideal gas law, the pressure will be $p = nRT/V$. The problem of converting 100 g of O_2 to moles was the subject of Example 19-3. The result was $n = 3.13$ mol. We also need to do a couple of conversions (gas problems typically involve several conversions):

$$600 \text{ cm}^3 \cdot \left(\frac{1 \text{ m}}{100 \text{ cm}}\right)^3 = 6.00 \times 10^{-4} \text{ m}^3$$

$$150°C + 273°C = 423 \text{ K}.$$

With this information, we find the pressure to be

$$p = \frac{nRT}{V} = \frac{(3.13 \text{ mol})(8.31 \text{ J / mol K})(423 \text{ K})}{6.00 \times 10^{-4} \text{ m}^3}$$

$$= 1.84 \times 10^7 \text{ Pa}$$

$$= 182 \text{ atm}.$$

In this text we will only consider gases in sealed containers, so that the number of moles (and number of molecules) will be a constant. In that case,

$$\frac{pV}{T} = nR = \text{constant}. \tag{19-13}$$

If the gas is initially in state 1, having values of the state variables p_1, V_1, and T_1, and at some later time in state 2, the variables for these two states are related by

$$\frac{p_1 V_1}{T_1} = \frac{p_2 V_2}{T_2}. \tag{19-14}$$

The relationship for these two states will be especially valuable for many problems.

EXAMPLE 19-9 A cylinder of gas is initially at a temperature and pressure of 0°C and 1 atm. A piston compresses the gas to half its original volume and three times its original pressure. What is the final gas temperature?

SOLUTION Equation 19-14, which relates the initial and final values of the state variables, can be written

$$T_2 = T_1 \cdot \frac{p_2}{p_1} \cdot \frac{V_2}{V_1}$$

In this problem, the compression of the gas results in $V_2/V_1 = 1/2$ and $p_2/p_1 = 3$. Notice that we do not need to know actual values of the pressure and volume, just the *ratios* by which they change. The initial temperature is $T_1 = 273$ K. With this information,

$$T_2 = 273 \text{ K} \cdot 3 \cdot \frac{1}{2} = 409 \text{ K} = 136°\text{C}$$

The conditions $T = 0°\text{C} = 273$ K and $p = 1$ atm $= 101.3$ kPa are called *standard temperature and pressure*. These conditions are designated with the abbreviation **STP**.

As we make the micro/macro connection, we often will want to refer to the number of molecules N in a gas rather than the number of moles. This change is easy to make because $N = nN_A$. We can write the ideal gas law as

$$pV = nRT = \frac{N}{N_A}RT = N\frac{R}{N_A}T = NkT, \tag{19-15}$$

where

$$k = \frac{R}{N_A} = 1.38 \times 10^{-23} \text{ J/K} = \text{Boltzmann's constant}.$$

Ludwig Boltzmann was an Austrian physicist who did some of the pioneering work in statistical physics during the mid-nineteenth century. Boltzmann's constant is not really a new constant, because it just combines two previously known constants, but it does express a new point of view about gases. The universal gas constant R can be thought of as the "gas constant per mole," while Boltzmann's constant k is interpreted as the "gas constant per molecule."

Recall our definition of the number density (molecules per m³) as $d = N/V$. If we divide both sides of Eq. 19-15 by V, we arrive at

$$p = dkT. \tag{19-16}$$

Equations 19-12, 19-15, and 19-16 are *all* the ideal gas law, just expressed in terms of different properties of the gas. All will be useful under different circumstances.

EXAMPLE 19-10 a) What is the number density of gas atoms at STP? b) What is the average distance between gas atoms at STP?

SOLUTION a) From Eq. 19-16, the number density is related directly to temperature and pressure as

$$d = \frac{p}{kT} = \frac{1.01 \times 10^5 \text{ Pa}}{(1.38 \times 10^{-23} \text{ J/K})(273 \text{ K})} = 2.69 \times 10^{25} \text{ atoms/m}^3.$$

You may have learned in chemistry that a gas at STP occupies 22.4 liters per mole. Our calculation is making an equivalent statement but, expressed in the quantities of atoms and m³, is more useful for physics. Note that this density is independent of the type of gas we are considering. We can take $d \sim 10^{25}$ m^{-3} to be a "typical" gas density.

b) Imagine freezing all the atoms at some instant of time. After doing so, place a little volume around each atom to separate it from all its neighbors. The total volume V of the gas has thus been divided into N small volumes v_i such that the sum of all these small volumes v_i equals the full volume V. That is, $\Sigma v_i = V$. Although all of these volumes are somewhat different, we can define an *average* little volume $v_{avg} = \frac{1}{N}\Sigma v_i = V/N$. But because $N/V = d$, the average volume surrounding each atom is $v_{avg} = 1/d$. Note that this is not the volume of the atom itself, which is much smaller, but the average surrounding volume of space that each atom can claim as its own, separating it from the other atoms.

Now each of the individual small volumes has a somewhat different shape, but *on average* the shape is likely to be very nearly spherical. Because we know the volume of this average little sphere, we can find its radius:

$$v_{avg} = \frac{1}{d} = \frac{4}{3}\pi r_{avg}^3$$

$$\Rightarrow r_{avg} = \left(\frac{3}{4\pi d}\right)^{1/3} = 2.08 \times 10^{-9} \text{ m} = 2.08 \text{ nm},$$

where we have used the value of d from part a). Because the average atom sits at the center of a sphere of radius 2.08 nm while the average nearby atom is also in the center of a sphere of radius 2.08 nm, the average distance between any two atoms is

$$\text{average distance} = 2r_{avg} = 4.16 \text{ nm}.$$

The results of the preceding example are important. One of our basic assumptions in the ideal gas model is that the atoms are "far apart" in comparison to the distances over which atoms exert attractive forces on each other. That distance, as illustrated in Fig. 19-8, is about 0.4 nm. A gas at STP, which is a fairly typical gas, has an average distance between atoms roughly ten times the interaction distance. This suggests, correctly, that the ideal gas model should be pretty good for gases under "typical" circumstances.

19.7 Ideal Gas Processes

The ideal gas law shows the connection between the state variables pressure, temperature, and volume. When one property is changed—for example, by heating or by compressing the gas—the state of the gas changes. The change is a *process* that the gas undergoes, and we will refer to any such change as an *ideal gas process*.

The *pV* Diagram

It is often convenient to represent ideal gas processes on a graph called the ***pV* diagram**. This is nothing more than a graph of pressure-versus-volume. The important idea behind the *pV* diagram is that *each point* on the graph represents a single, unique state of the gas.

(Although a point on the graph only directly specifies the values of p and V, we can find the temperature by using the ideal gas law.) That is, each point actually represents a triplet of values (p, V, T) specifying the state of the gas. Figure 19-13a is a pV diagram showing three specific states of a system consisting of 1 mol of gas. The values of p and V can be read from the axes for each point, then the temperature at each point can be determined from the ideal gas law.

We will usually be interested in knowing what happens to a gas as it changes from one state to another. A change of state of the gas can be represented by a "trajectory" in the pV diagram. The trajectory shows all the intermediate states through which the gas passes. Figure 19-13b, for example, shows two different ways the gas of Fig. 19-13a can be changed from state 1 to state 3. The left trajectory in Fig. 19-13b shows that only the pressure changes at first, with the volume staying constant (vertical line), and then only the volume changes while the pressure remains constant (horizontal line). The right trajectory shows the state of the gas changing rather erratically, although at some instant of time the gas passes through state 2 on its way to state 1.

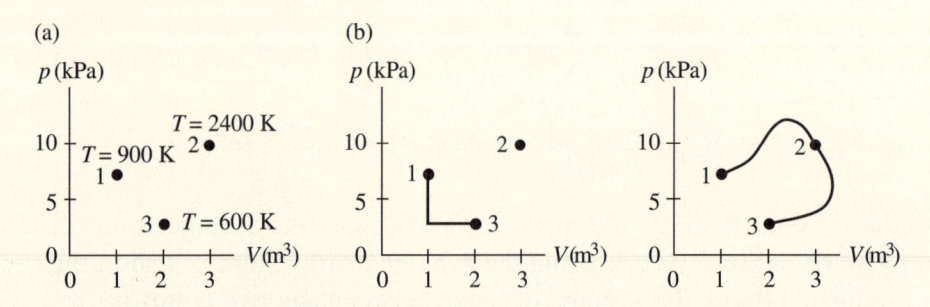

FIGURE 19-13 a) Each state of an ideal gas is represented as a single point in a pV diagram. b) Two different ways to change the gas from state 1 to state 3.

There are, clearly, many different ways to change the gas from state 1 to state 3. Although the initial and final state are the same for all of them, the particular means by which the gas changes—that is, the particular trajectory—will turn out to have very real consequences. For example, you will learn later that the work done by an expanding gas— a quantity of very practical importance in various engines—will depend on the trajectory followed. The pV diagram provides an easy graphical means to show how a gas changes from one state to another. It is a tool used widely in practical thermodynamics.

Constant Pressure Process

Many important gas processes take place at a constant, unchanging pressure. A constant pressure process is called an **isobaric process**, where *iso* is a prefix meaning "constant" or "uniform" and *baric* is from the same root as "barometer" and means "pressure." Figure 19-14a shows one method of changing the state of a gas while keeping the pressure constant. The figure shows a cylinder of gas that has a tight-fitting piston. The piston can slide up and down, but it seals the container so that no atoms enter or escape. A constant downward force is applied to the piston by placing a mass on top. (In practice, other means are used to apply a constant force.) In equilibrium, the upward pressure force on the piston $|\vec{F}_{up}| = pA$, where A is the piston's area, must exactly balance the downward

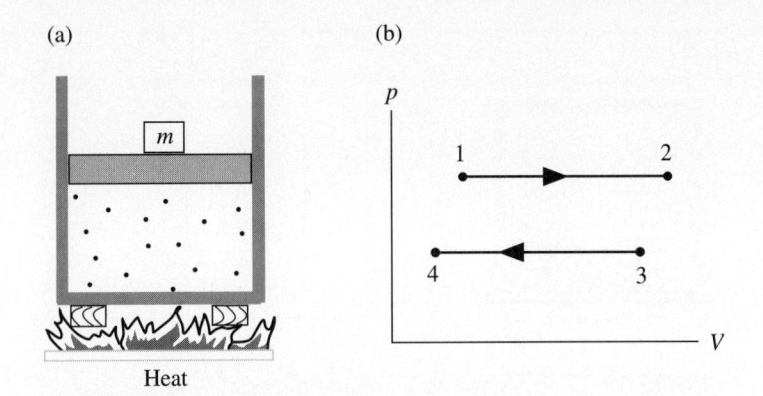

FIGURE 19-14 a) A constant pressure process. The piston moves as heat is added, but the pressure remains constant. b) Isobaric expansion (1 → 2) and isobaric compression (3 → 4) as seen on a *pV* diagram.

weight force $|\vec{F}_{\text{down}}| = mg$, so the pressure is $p = mg/A = $ constant. If the cylinder is heated, the gas will expand and push the piston up—but the pressure will not change! Such a process is shown on the *pV* diagram of Fig. 19-14b as process 1 → 2. We call this an *isobaric expansion*. Process 3 → 4, an *isobaric compression*, would occur if the gas were cooled and contracted, lowering the piston.

EXAMPLE 19-11 A gas occupying 50 cm³ at 50°C is cooled at constant pressure until the temperature is 10°C. What is its final volume?

SOLUTION By definition of an isobaric process, $p_1/p_2 = 1$. Using Eq. 19-14, the ideal gas law for constant *n*, we have

$$V_2 = V_1 \cdot \frac{p_1}{p_2} \cdot \frac{T_2}{T_1} = V_1 \cdot \frac{T_2}{T_1}.$$

Converting both temperatures to kelvins (essential!), we find $V_2 = 43.8$ cm³. Note that as long as we only use *ratios*, we do not need to convert *V* to SI units. Units of cm³ on the right, multiplied by a dimensionless ratio, give units of cm³ on the left.

Constant Volume Process

A constant volume process occurs when the state of a gas is changed while keeping the volume fixed. A constant volume process is called an **isochoric process**. Suppose, for example, that you have a gas in a closed, rigid container, as shown in Fig. 19-15a. Heating the gas will raise its pressure without changing its volume—a process shown as the vertical line 1 → 2 on the *pV* diagram of Fig. 19-15b. A constant volume cooling, shown as process 3 → 4, lowers the pressure. Although it is true that thermal expansion and contraction may change the container's volume slightly, these changes are extremely small in comparison with the temperature and pressure changes and can be safely ignored in all but high-precision measurements.

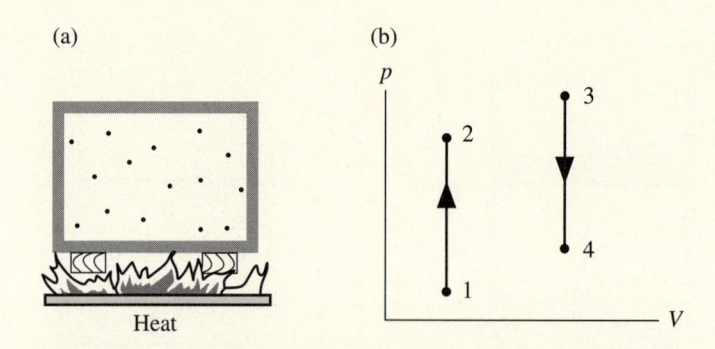

FIGURE 19-15 a) A constant volume process, where a gas is heated in a rigid container. b) Constant volume heating and cooling, as seen on a pV diagram.

EXAMPLE 19-12 Laboratories needing to make very accurate temperature measurements often use a device called a *constant volume gas thermometer*. A sealed bulb of gas, usually helium, is first placed in contact with a "reference temperature" and its pressure is measured. It is then placed in contact with the system at an unknown temperature. The new pressure can be used, with the ideal gas law, to determine the temperature.

A commonly used reference is the *triple point* of water, at which all three phases of water—solid, liquid, and gas—coexist in equilibrium. The triple point of water occurs only at the unique pressure and temperature of 0.0060 atm and 273.16 K (rather unusual conditions, which is why you aren't familiar with the triple point from everyday experience). The triple point is a useful reference because any laboratory bringing water to the triple point knows that the temperature is exactly 273.16 K.

Suppose that a constant volume gas thermometer is placed in contact with a reference cell containing water at the triple point. Its pressure, after coming to equilibrium, is recorded as 55.78 kPa. The thermometer is then placed in contact with a sample of unknown temperature and, after reaching a new equilibrium, has a final pressure of 65.12 kPa. What is the temperature of this sample?

SOLUTION Similar to the previous example, we use the ideal gas law of Eq. 19-14, along with $V_2/V_1 = 1$ for an isochoric process, to find

$$T_2 = T_1 \cdot \frac{V_2}{V_1} \cdot \frac{p_2}{p_1} = T_1 \cdot \frac{p_2}{p_1}$$

$$= 318.90 \text{ K} = 45.75°\text{C}.$$

Constant Temperature Process

The last process we will look at for now is one that takes place at a constant temperature. A constant temperature process is called an **isothermal process**. One way in which an isothermal process might occur is illustrated in Fig. 19-16a. In this figure a piston is being pushed down to compress a gas, but the gas cylinder is floating in a large container of water, or another liquid, that is held at a constant temperature. If the piston is pushed

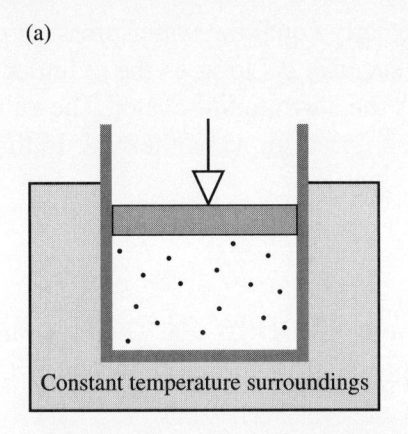

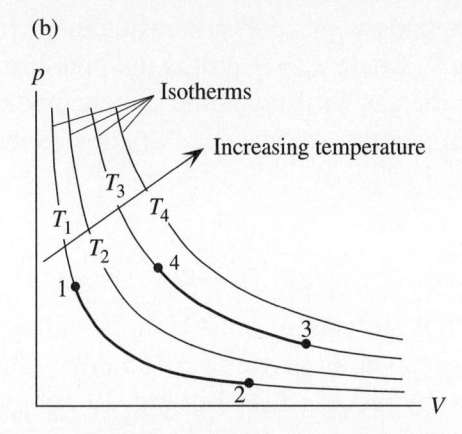

FIGURE 19-16 a) A constant temperature process, with the temperature controlled by the surroundings as the gas is compressed. b) Constant temperature processes are seen as hyperbolic isotherms in the pV diagram.

slowly, then heat transfer through the walls of the cylinder will keep the gas at the same fixed temperature as the surrounding liquid. This would be an *isothermal compression*. The reverse process, with the piston pulled out, would be an *isothermal expansion*.

Representing an isothermal process on the pV diagram is a little harder than the two previous processes because both p and V change. As long as T remains fixed, then we have the relationship

$$pV = nRT = \text{constant}.$$

This is the condition for a p-versus-V graph to be a *hyperbola*. For their product to remain constant, a large value of p requires V to be small whereas a large V requires p to be small. The specific hyperbola depends on the value of T, so isothermal processes can be represented on the pV diagram as a family of hyperbolas, each representing a different temperature.

Several of these hyperbolas are shown in Fig. 19-16b. They are called **isotherms**, and a gas undergoing an isothermal process will move along the isotherm of the appropriate temperature. The process shown as $1 \rightarrow 2$ is an isothermal expansion at temperature T_1, while process $3 \rightarrow 4$ is an isothermal compression (such as the one shown in Fig. 19-16a) at a different temperature, T_3. Isotherms farther away from the origin have a larger value of T, so in this example $T_4 > T_3 > T_2 > T_1$.

EXAMPLE 19-13 A gas cylinder with a tight-fitting piston contains 200 cm³ of air at 1 atm. It floats on the surface of a swimming pool of 15°C water. The cylinder is then slowly pulled to a depth of 3 m. What is the volume of gas at this depth?

SOLUTION At the surface, the pressure inside the cylinder must exactly equal the outside air pressure of 1 atm if the gas is in equilibrium. If they were not equal, the piston would experience a net force that would push it in or out until the pressures balanced and equilibrium was achieved. As the cylinder is pulled underwater, the increasing water pressure will press the piston inward and compress the gas. Equilibrium at depth D

requires that the gas pressure inside the cylinder exactly equal the water pressure $p_{water} = p_0 + \rho g D$, where $p_0 = 1$ atm is the pressure at the surface. As long as the cylinder moves slowly, the gas will stay at the same temperature as the surrounding water. The value of T is not important—all we need to know is that the compression is isothermal. In that case, because $T_2/T_1 = 1$,

$$V_2 = V_1 \cdot \frac{T_2}{T_1} \cdot \frac{p_1}{p_2} = V_1 \cdot \frac{p_1}{p_2} = V_1 \cdot \frac{p_0}{p_0 + \rho g D}.$$

The initial pressure p_0 must be in SI units: $p_0 = 1$ atm $= 1.01 \times 10^5$ Pa. Then a straightforward computation gives $V_2 = 155$ cm^3. This compression would be represented on a pV diagram by process $3 \to 4$ in Fig. 19-16b.

Multistep Processes

Most practical thermodynamics processes involving gases consist of several basic processes performed in series. We will look at one such case as a final example.

EXAMPLE 19-14 A gas at 2 atm pressure and a temperature of 200°C is first expanded isothermally until its volume has doubled. It then undergoes an isobaric compression until it returns to its original volume. a) Show this process on a pV diagram. b) What are the final temperature and pressure?

SOLUTION a) Figure 19-17 shows the process. The gas starts in state 1 at pressure $p_1 = 2$ atm and volume V_1. As the gas expands isothermally, it moves downward along an isotherm until it reaches volume $V_2 = 2V_1$. The pressure has decreased in this process to a lower value, p_2. The gas is then compressed at constant pressure p_2 until its final volume V_3 equals its original volume V_1. State 3 is on a different isotherm having temperature T_3. Because this isotherm is closer to the origin, we expect to find $T_3 < T_1$.

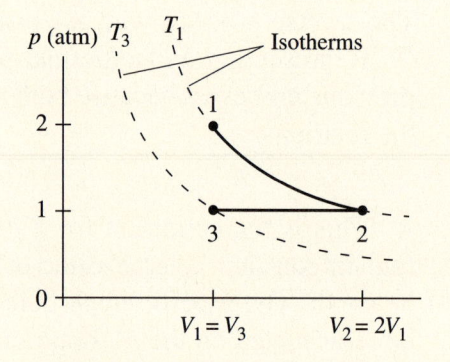

FIGURE 19-17 A pV diagram for the process of Example 19-14.

b) Because $T_1 = T_2$ in the isothermal expansion, $p_1 V_1 = p_2 V_2$. Thus

$$p_2 = p_1 \cdot \frac{V_1}{V_2} = p_1 \cdot \frac{V_1}{2V_1} = \tfrac{1}{2} p_1 = 1 \text{ atm}.$$

During the isobaric compression, $p_2 = p_3$ and so $V_2/T_2 = V_3/T_3$. Thus

$$T_3 = T_2 \cdot \frac{V_3}{V_2} = T_2 \cdot \frac{V_3}{2V_3} = \tfrac{1}{2} T_2 = 236.5 \text{ K} = -36.5°\text{C},$$

where we converted $T_2 = 473$ K before doing calculations. The final state is one in which both the pressure and the absolute temperature are half their original values.

We will continue to use gases as our most important macroscopic system. After developing a better microscopic understanding of gases and building the micro/macro connection, we will return in Chapter 23 to look more thoroughly at the thermodynamics of ideal gas processes.

Summary

Important Concepts and Terms

macroscopic system	absolute zero
bulk properties	pressure
phase	pascal
phase change	vacuum
solid	atmospheric pressure
liquid	standard atmospheric pressure
gas	gauge pressure
ideal gas	monatomic gas
fluid	diatomic gas
state variable	equation of state
number density	ideal gas law
mass density	STP
atomic mass	pV diagram
molecular mass	isobaric process
mole	isochoric process
Avogadro's number	isothermal process
temperature	isotherm
absolute temperature scale	

This chapter, far more than most, has been about presenting facts, definitions, and symbols rather than concepts. While physics as a whole deals with concepts and relationships, we do need facts to analyze, words and phrases with which to speak, and symbols with which to communicate. Sometimes there's no better way than to just lay them out, and that is largely what this chapter has done.

Our goal has been to provide a macroscopic description of matter. A macroscopic object is characterized by a number of physical properties that we call state variables. These include mass, volume, mass density, number density, number of moles, temperature, and pressure.

Pressure is the force per unit area that a fluid exerts on a surface. The pressure pushes equally in all directions, not just down. The pressure in a liquid is due to the weight of the liquid, whereas the pressure in a gas is due to collisions of the atoms or molecules with the walls of the container. The pressure at depth D in a liquid is given by

$$p_{liq} = p_{atomos} + \rho g D.$$

The pressure in a gas is determined by the volume and temperature of the gas. It is found from the ideal gas law:

$$pV = nRT = NkT.$$

The ideal gas model assumes that atoms and molecules are small, noninteracting hard spheres that move about randomly and occasionally collide with each other or with the walls of the container. Real gases are well described by the ideal gas model as long as their density is low and their temperature is well above the condensation temperature. For an ideal gas in a closed container, in which the number of molecules cannot change, the pressure, volume, and temperature at two different instants of time are related by

$$\frac{p_1 V_1}{T_1} = \frac{p_2 V_2}{T_2}.$$

Isobaric, isothermal, and isochoric ideal gas processes were defined as processes in which the pressure, the temperature, and the volume, respectively, are held constant. These processes are represented on a pV diagram by distinctly shaped graphs.

Exercises and Problems

Exercises

1. A circular cover 20 cm in diameter has been placed over a 10 cm diameter opening that leads into a chamber (Fig. 19-18). The chamber has an interior pressure of 20 kPa. How much force would be required to pull the cover off?

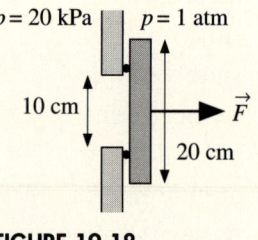

   FIGURE 19-18

2. a. What volume of water has the same mass as 2 m³ of lead?
   b. If this volume of water is cube-shaped, what is the pressure at the bottom?

3. a. Based on data from Fig. 19-11, estimate the average kinetic energy of a molecule of nitrogen at room temperature.
   b. You will learn later that the average kinetic energy of a gas a temperature T is *independent* of the composition of the gas. Assuming this, estimate the average speed of a hydrogen molecule H_2 at room temperature.

4. A cylinder 20 cm in diameter and 40 cm long contains 50 g of oxygen (O_2) gas at room temperature.
   a. How many moles of oxygen does the cylinder contain?
   b. How many oxygen molecules are in the cylinder?
   c. What is the number density of the oxygen?
   d. What is the mass density of the oxygen?
   e. What is the reading of a pressure gauge attached to the tank?

5. A 200 cm³ gas cell containing helium at 50°C is connected to a mercury manometer. The mercury in the outside arm of the manometer stands 25.0 cm below the mercury in the arm attached to the gas cell.
   a. How many helium atoms are in the gas cell?
   b. What is the mass of the helium?

6. 0.1 mol of argon gas is admitted to a 50 cm³ container at 20°C. The gas undergoes an isochoric heating to a temperature of 300°C.
   a. What is the final pressure of the gas?
   b. Show the process on a pV diagram. Include a proper scale on both axes.

7. 0.1 mol of argon gas is admitted to a 50 cm³ container at 20°C. The gas undergoes an isobaric heating to a temperature of 300°C.
   a. What is the final volume of the gas?
   b. Show the process on a pV diagram. Include a proper scale on both axes.

Problems

8. The atomic mass of copper is $A = 64$. If solid copper exists in a cubic crystal lattice, what is the distance between atoms? (**Hint:** Consider a 1 m × 1 m × 1 m cube.)

9. It is often said that 90% of an iceberg is submerged. Is this true? To find out, consider a block of ice of thickness t floating on a lake of depth D, as shown in Fig. 19-19.
   a. What is the pressure at the bottom of the lake, at depth d, at a point that is *not* directly under the ice?
   b. Let h be the height of the ice above the water's surface. Write an expression for the pressure at the bottom of the lake at a point directly under the ice.
   c. By comparing your answers to parts (a) and (b), determine what percentage of the block's volume is submerged.

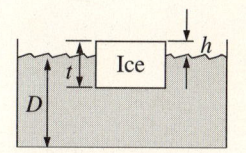

   FIGURE 19-19

10. Water stands at depth D behind a dam of width W.
    a. What is the net force of the water on the dam? (**Hint:** This problem requires an integration.)
    b. Evaluate the net force for a 100 m wide dam with a 60 m water depth.

11. A U-shaped tube, open to the air on both ends, contains mercury. Water is poured into the left arm until the water column is 10.0 cm deep. How far upward from its initial position does the mercury in the right arm move?

12. A container of volume 100 cm³ is pressurized at constant volume until the gas pressure triples. It is then expanded at constant temperature until the pressure is one-half the pressure of the gas before the experiment started.
    a. What is the final volume of the gas?
    b. Show this process on a pV diagram.

13. 0.1 mol of gas undergoes the process 1 → 2 shown in Fig. 19-20.
    a. What are temperatures T_1 and T_2?
    b. What type of process is this?
    c. The gas undergoes an isochoric heating from point 2 until the pressure is restored to the value it had at point 1. What is the final temperature of the gas?

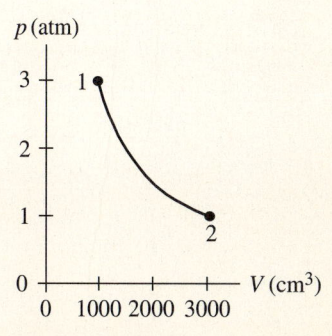

    FIGURE 19-20

14. Five grams of nitrogen gas (N_2) at an initial pressure of 3 atm and at room temperature undergo an isobaric expansion until the volume has tripled.
    a. What is the gas volume after the expansion?
    b. What is the gas temperature after the expansion?

    The gas pressure is then decreased at constant volume until the original temperature is reached.

    c. What is the gas pressure after the decrease?

    Finally, the gas is isothermally compressed until it returns to its initial volume.

    d. What is the final gas pressure?
    e. Show the full three-step process on a pV diagram. Use appropriate scales on both axes.

15. A 3 m long pipe, closed at the top end, is slowly pushed vertically downward into water until the top end of the pipe is level with the water's surface (Fig. 19-21). What is the length L of the trapped volume of air?

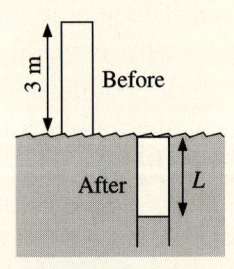

FIGURE 19-21

16. A 7 g piece of dry ice (solid CO_2) is placed in a 10,000 cm³ container, then all the air is quickly pumped out and the container sealed. The container is warmed to 0°C, a temperature at which CO_2 is a gas.
    a. What is the gas pressure? Give your answer in atm.
    b. The gas then undergoes an isothermal compression until the pressure is 3.0 atm, immediately followed by an isobaric compression until the volume is 1000 cm³. What is the final temperature of the gas?

[**Estimated 10 additional problems for the final edition.**]

Chapter 20

Work, Heat, and the First Law of Thermodynamics

> Being engaged lately in superintending the boring of cannon, in the workshops of the military arsenal at Munich, I was struck with the very considerable degree of Heat which a brass gun acquires, in a short time, in being bored; and with the still more intense Heat (much greater than that of boiling water as I found by experiment) of the metallic chips separated from it by the borer. The more I meditated on these phenomena, the more they appeared to me to be curious and interesting. A thorough investigation of them seemed even to bid fair to give a farther insight into the hidden nature of Heat.
>
> Count Rumford before the Royal Society of London, 1798

LOOKING BACK Sections 9.5; 11.1–11.5; 19.6; 19.7

20.1 State Variables and Interaction Parameters

In the last chapter you saw that macroscopic systems are characterized by a number of *state variables*, such as pressure, temperature, and volume. Any state of the system can be described by providing the values of all the state variables. However, our interest in macroscopic systems goes far beyond merely describing their state. Macroscopic systems of importance in science and engineering are almost always changing their state, and our real goal is to understand *how* it is that changes occur. Gases, for example, get compressed, water gets boiled, cars come to a halt and heat up the brakes, electrical circuits heat up when used, and so on.

A system that is in equilibrium does not spontaneously change its state. Instead, the state of a system changes as a consequence of its *interaction* with the surrounding environment. It is some force from the environment that pushes on and compresses the gas, and some source of energy from the environment that heats and boils the water. These are interactions. We need to understand how a macroscopic system interacts with its environment in sufficient detail that we can *predict* what the final state will be.

Our subject matter will be a continuation of the discussion we started back in Chapter 11. (A careful review of that chapter, especially Sections 11.5, is an important prerequisite for mastering the new material to be presented here.) The primary result of that chapter was the law of conservation of energy:

$$W_{ext} + Q = \Delta E_{mech} + \Delta E_{int} . \qquad (20\text{-}1)$$

The quantities E_{mech} and E_{int} are state variables—they characterize the system's center-of-mass mechanical energy ($E_{mech} = K_{cm} + U_{cm}$) and its internal energy. Because of the Δ's, the law of conservation of energy deals with the *changes* in these state variables. A system is in equilibrium, a state in which no changes are occurring, if and only if the right-hand side of Eq. 20-1 is zero.

The left-hand side of the equation describes the interactions of the system with its environment. W_{ext} is the work done on the system by external forces, transferring energy into or out of the system, and Q is the heat energy transferred to or from the system. W_{ext} and Q are what we call **interaction parameters**. Their values characterize the strength of the interactions. Notice that they do *not* have a Δ. Unlike state variables, which are properties of the system, the interaction parameters are not changing. They are simply numbers describing the interaction. We will return to this point for further discussion.

At this point in the text we are not interested in systems that have a macroscopic motion of the system as a whole. Moving macroscopic systems were important to us for many chapters, but now—as we inquire about the thermodynamical and statistical properties of a system—we would like the system as a whole to rest peacefully on the laboratory bench while we study it. So we will assume, throughout Part IV, that $E_{mech} = 0$. In that case, the law of energy conservation is somewhat simplified:

$$W_{ext} + Q = \Delta E_{int} . \qquad (20\text{-}2)$$

Figure 20-1 provides a schematic look at the interplay between interaction parameters and state variables. The system is characterized by the state variable E_{int} as well as other state variables such as temperature and pressure. The values of the state variables change as the system interacts with its environment. These interactions are responsible for the *transfer* of energy into and out of the system. The environment can transfer energy to the system either by doing work on it or by heating it (inward-pointing arrows). Conversely, energy can be transferred back to the environment if the system does work on the environment or if the system is cooled (outward pointing arrows). Regardless of the direction of the energy transfer, conservation of energy *requires* that the change in the internal energy be related, via Eq. 20-2, to the work done and heat transferred. The following three sections will look more closely at internal energy, work, and heat.

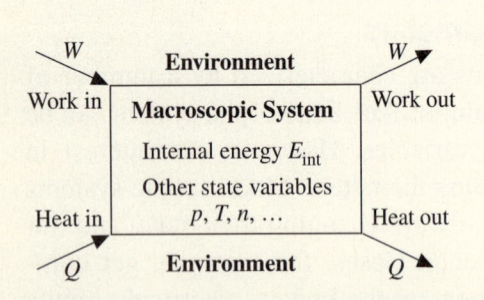

FIGURE 20-1 A schematic look at the interaction between a macroscopic system and the environment.

20.2 Internal Energy

As you learned earlier, the internal energy of a system consists of a number of different contributions:

$$E_{int} = E_{therm} + E_{chem} + E_{nuc} + E_{sound} + E_{light} + \cdots \quad (20\text{-}3)$$

In addition to the thermal energy of the molecules, the system might have chemical or nuclear energy as well as energy from sound or light waves. For our purposes, we will make the simplification that $E_{int} = E_{therm}$ and that all other possible sources of internal energy are negligible. The chemical energy term is quite important in engineering thermodynamics, where it is needed to characterize combustion processes, but we will leave that to more advanced courses. We will also, for ease of notation, shorten the subscript and call the thermal energy E_{th}.

The thermal energy itself has two contributions:

$$E_{th} = K_{int} + U_{int}. \quad (20\text{-}4)$$

Here K_{int} is the internal microscopic kinetic energy of all the moving atoms and molecules and U_{int} is the potential energy stored in molecular bonds and other interactions between the atoms and molecules. In other words, the thermal energy is the energy of the atoms and molecules within the system considered as individual, randomly moving particles, as distinct from the center-of-mass energy of the system as a whole. The thermal energy is "hidden" from our macroscopic view but nonetheless quite real.

You will see later that the thermal energy E_{th} is directly proportional to the system's absolute temperature. In fact, temperature really is just a macroscopic measurement of the thermal energy. For now, we can associate an increase of thermal energy ($\Delta E_{th} > 0$) with a system that is "getting hotter," whereas a decrease of thermal energy ($\Delta E_{th} < 0$) characterizes a system "getting colder." It is important to note, and we will look at this more closely later, that the system may *or may not* have heat added to get hotter or colder. There are ways other than heat addition or removal to change the system's temperature and thermal energy.

EXAMPLE 20-1 Estimate the internal kinetic energy of one mole of N_2 gas at 20°C.

SOLUTION In Chapter 19 we looked at the distribution of speeds of N_2 molecules at 20°C (see Fig. 19-11). Although there were a wide range of speeds, it appeared that the most likely speed was $v_{avg} \approx 550$ m/s. The *average* molecule in the gas will have a kinetic energy $K_{avg} = \frac{1}{2}mv_{avg}^2$. Nitrogen has a molecular weight $A = 28$, so each molecule has mass $m = 28$ u $= 4.65 \times 10^{-26}$ kg. Thus

$$K_{avg} = \tfrac{1}{2}mv_{avg}^2 = \tfrac{1}{2}(4.65 \times 10^{-26} \text{ kg})(550 \text{ m/s})^2 = 7.04 \times 10^{-21} \text{ J}.$$

There are N_A molecules in one mole, each, on the average, with this kinetic energy. So the total microscopic kinetic energy is

$$K_{int} = N_A \cdot K_{avg} = 4230 \text{ J}.$$

This is a substantial energy in the gas—far more than most people would suspect. In fact, 1 mol of N₂ has a total mass of 28 g. If all N_A molecules were to move in the same direction at 550 m/s, we would have a 28 g macroscopic "bullet" moving with a speed of 550 m/s. This bullet would have a macroscopic kinetic energy of 4230 J. Even though the motion of the gas is dispersed microscopically, rather than collected into a single macroscopic motion, it still has the same energy as the bullet.

20.3 Work

[**Photo suggestion: A piston compressing a cylinder of gas.**]

Work is one of the two ways by which a system can exchange energy with its environment, so it is one of the two interaction parameters in Eq. 20-2. Recall that W_{ext} is the work performed *by* external forces acting on particles *at the boundary* of the system—hence the subscript in W_{ext}. (Internal forces also do work, but that work has been incorporated into the internal potential energy term U_{int}.) If the external force $\vec{F}_{environ}$ is constant as the boundary moves through displacement $\Delta \vec{s}$, the work done by the force is

$$W_{ext} = \vec{F}_{environ} \cdot \Delta \vec{s}. \qquad (20\text{-}5)$$

The work done is a positive quantity if the displacement is in the same direction as the external force, a negative quantity if the external force and the displacement are in opposite directions.

The external force must displace the boundary of the system to do work on it. We will call this a **mechanical transfer of energy** across the boundary from the environment to the system. In thermodynamics, work is done when a gas is compressed or expands. For example, a force from the environment must push on the piston of Fig. 20-2a to compress the gas—thus moving the boundary of the system a distance Δs and doing work $W_{ext} = |\vec{F}_{environ}|\Delta s$ on it. By our definition of work, this is a *positive* value for W_{ext} because the force and the piston displacement are in the same direction. Positive work is thus a transfer of energy *from* the environment *into* the system.

It is also possible that a system might, because of internal processes, expand and push the piston outward—as shown in Fig. 20-2b. The environment is still exerting force $\vec{F}_{environ}$ on the piston to balance the pressure in the system and keep the piston from being blown out. Now, however, the piston's displacement Δs is *opposite* the direction of the force exerted on it by the environment. By our definition of work, the environment in this case is doing a *negative* amount of work on the system: $W_{ext} < 0$. Negative work corresponds to energy being transferred, by mechanical means, *from* the system *to* the environment. In this case, we say that the system is doing work *on* the environment.

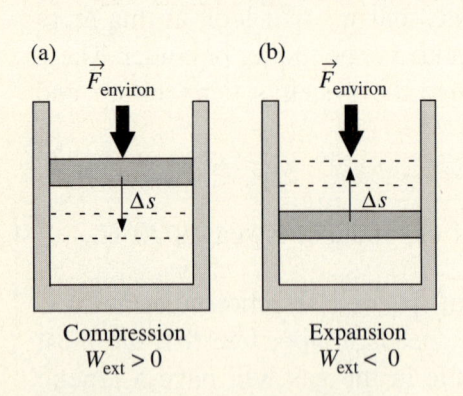

FIGURE 20-2 a) Work is done on the system when an external force compresses it. b) The system does work on the environment ($W_{ext} < 0$) by pushing out against the external force.

An insulated gas—one that cannot absorb or give off heat—has its temperature increased by compression. If you have ever felt the bottom of a bicycle tire pump after pumping for a few seconds then you know it gets quite hot—a compression heating of the air. But this is exactly conservation of energy, as seen from Eq. 20-2. If $Q = 0$ (insulated system), then $\Delta E_{int} = W_{ext}$ and the work done by the external force increases the system's thermal energy, making it hotter. This is an energy transfer into the system. Likewise, an expanding gas that does work *on* the environment ($W_{ext} < 0$) will get cooler as the thermal energy decreases.

It is important to keep in mind that the system does not "possess" a value of W_{ext}. The work is not a state variable—that is, work is not a number characterizing the system. Rather, it is a parameter that describes how the system *interacts* with its environment. Therefore, it would not be meaningful to talk about the "initial value of the work" or the "final value of the work" as we do for state variables such as E_{th}. As a consequence, the work appears in the conservation of energy equation as simply W_{ext}, never as ΔW_{ext}.

20.4 Heat

[**Photo suggestion: A flame heating a beaker of water.**]

Heat is a more elusive concept than work. The reason is that we use the word *heat* very loosely in the English language, often as synonymous with *hot*. We might say, on a very hot day, that "This heat is oppressive." In fact, the very first definition of heat in a popular dictionary is "a condition of being hot." Such a definition makes heat a *property* of the object that is hot. This is *not* how we want to use the term *heat* in physics and engineering.

You have seen that it is possible to transfer energy into a system, and thereby increase its thermal energy, by doing work on the system—a mechanical means of energy transfer. But there are other ways to increase a system's thermal energy, and raise its temperature, without performing any mechanical work. When you place a pan of water on the stove and light the burner, the water gets hotter—its thermal energy increases—without work being done. Figure 20-3 shows a rigid, sealed container of gas over a flame, with the temperature of the gas increasing. No work is being done because the boundary of the system is not moving. We would say—correctly—that the gas in the container, or the pan of water on the stove, is "being heated." But how? What is heat?

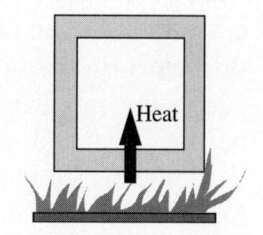

FIGURE 20-3 Heat is a nonmechanical transfer of energy. No work is done.

Heat was long thought to be a *substance* of some kind that had fluidlike properties. If you place a hot object and a cold object together, where they can interact, they evolve toward a common final temperature. Common sense suggests that something is flowing from the hot object to the cold object until equilibrium is achieved. This "heat fluid" was called *caloric*.

One of the first to disagree with this notion, in the late 1700s, was the American-born Benjamin Thompson. Thompson was a British sympathizer who fled to Europe during the American Revolution, eventually settling in Bavaria and later receiving the title Count Rumford. Count Rumford became a supervisor of weapons manufacture in Bavaria. We opened this chapter with a quote of Rumford regarding his observations of heat generation during the boring of cannons.

Rumford conducted experiments with the hot metal chips thrown off by the borer, and he concluded that the heat generation was not consistent with the expectations of the caloric theory. In 1798 he wrote, perceptively, that

> the source of the Heat generated by friction in these Experiments appeared evidently to be *inexhaustible*. It is hardly necessary to add that any thing which...can continue to furnish *without limitation* cannot possibly be a *material substance*: and it appears to me to be extremely difficult, if not quite impossible, to form any distinct idea of anything capable of being excited and communicated in the manner the Heat was excited and communicated in these Experiments, except it be MOTION. [Emphasis in original.]

In other words, if caloric is a substance, the cannon and borer should run out of caloric as the boring continues and the heat generation ought to decrease with time. But it does not. The heat generation appears to be "inexhaustible," which Rumford notes is not consistent with the idea of heat as a substance. He thus concluded that heat is not a substance—it is *motion*!

Rumford was beginning to think along the same lines as had Bernoulli, sixty years earlier. But Rumford's ideas were speculative and qualitative, hardly a scientific theory, and their implications were not immediately grasped by others. Like Bernoulli, it would be some time before his insight was recognized and validated.

The turning point was the work of British physicist James Joule in the 1840s. Unlike Bernoulli and Count Rumford, Joule carried out careful experiments to learn how it is that systems change their temperature. His research led him to conclude that heat is not a substance at all but is simply another form of energy. Joule established that you can raise the temperature of an object *either* by heating it *or* by doing work on it, and the final state of the object is *exactly* the same in both cases. This implies that heat and work are essentially equivalent to each other, and Joule even went so far as to find the conversion factor between heat and work.

Joule placed the idea of conservation of energy on a firm scientific footing and established it as a law of nature. Heat and work, which had been previously regarded as two completely different phenomena, were now seen to be simply two different ways of transferring energy to or from a system. Joule's discoveries fully vindicated the earlier ideas of Bernoulli and Count Rumford, and they opened the door for a rapid advancement of the subjects of thermodynamics and statistical physics during the second half of the nineteenth century. It is only appropriate that the SI unit of energy be named after Joule.

Today we regard **heat** as being the energy transferred between the system and the environment as a consequence of a *temperature difference* between them. Unlike the energy transfer we call work, heat involves no macroscopic motion of the system—it is a **nonmechanical transfer of energy**. When you place a pan of water on the stove, heat is the energy transferred *from* the hotter flame *to* the cooler water. If, however, you place the water in a freezer, heat is the energy transferred from the warmer water to the colder air in the freezer. No *macroscopic* motion occurs, yet energy—heat—is transferred from one system to the other.

Like the work W_{ext}, heat is an interaction parameter, *not* a property of the system. So heat enters the law of conservation of energy always as a value Q, never as a difference ΔQ. We will define Q such that a positive value of Q indicates heat flow *into* the system from the environment, implying that $T_{\text{environ}} > T_{\text{system}}$. A negative Q represents heat flow *from* the system back to the environment and thus $T_{\text{environ}} < T_{\text{system}}$. This is consistent with our definition of work, where in either case a positive value indicates energy being transferred from the environment to the system. It is interesting to note that we still, as a vestige of history, use the phrase *heat flow* to describe the energy transfer process called heat. Make sure, however, that you understand that no substance is actually flowing. Table 20-1 provides a summary of the signs of the work W_{ext} and the heat Q.

TABLE 20-1 Interpreting the signs of work and heat.

Positive Value	**Negative Value**
$W > 0$: Environment does work on the system (compression). Energy is transferred *in*.	$W < 0$: System does work on the environment (expansion). Energy is transferred *out*.
$Q > 0$: Environment is at a higher temperature than the system. Energy is transferred *in*.	$Q < 0$: System is at a higher temperature than the environment. Energy is transferred *out*.

It is particularly important to make a clear distinction between *heat* and *thermal energy*. The temperature of a system describes its thermal energy—the unseen motion of all the atoms and molecules in the system. Temperature does *not* describe heat. Common language can easily mislead you. When an object slides to a halt because of friction, most people say that the object's kinetic energy is "converted into heat." In fact, no heat at all is involved in this process. As you saw in Chapter 11, internal forces of the object-plus-surface system convert the macroscopic motion of the object into microscopic thermal energy of the object's atoms and molecules. The increased thermal energy shows up as an increased temperature of the object and the surface. But nowhere was there a nonmechanical transfer of energy due to a temperature difference. So we should say, correctly, that the object's original energy was "dissipated" into thermal energy.

The unit for measuring heat, the calorie, had been established long before scientists realized that heat is a form of energy. The calorie was defined as:

1 calorie = 1 cal = the quantity of heat needed to change the temperature of 1 gram of water by 1 degree Celsius.

Once Joule had established that heat is a form of energy, it was apparent that the calorie is really a unit of energy. Joule was able, through experiment, to determine the conversion factor between heat units and more conventional energy units. In today's SI units, that conversion is

$$1 \text{ cal} = 4.186 \text{ J}.$$

The calorie you know in relation to food is not the same as the heat calorie. It is called the *food calorie* and is abbreviated Cal, with a capital C:

$$1 \text{ food calorie} = 1 \text{ Cal} = 1000 \text{ cal} = 4186 \text{ J}.$$

The food calorie measures the food's *internal* energy, stored as chemical energy, that is available for doing work or for keeping your body warm. That extra dessert you ate last night containing 300 Cal has a chemical energy

$$E_{\text{chem}} = 300 \text{ Cal} = 3 \times 10^5 \text{ cal} = 1.26 \times 10^6 \text{ J}.$$

The burner on the stove transfers heat energy to a pan of water because of their temperature difference. Although true, this statement is still somewhat unsatisfactory because it doesn't explain *how* the energy transfer occurs. With work, you can see how the energy is transferred as a force pushes the piston through a distance. The "how" of heat transfer is more complicated. We will soon come to see that heat is an energy exchange among atoms as a consequence of their *collisions* with one another at the boundary between the system and the environment.

20.5 The First Law of Thermodynamics

Heat transfer was the missing piece that we needed for a completely general statement of the law of conservation of energy. As we noted earlier, we are now interested only in systems that are macroscopically at rest: $E_{\text{mech}} = 0$. We will now also drop the subscript from W_{ext} and use the simpler W, keeping in mind that it refers to the work done on the system by the environment.

With these assumptions clearly stated, the law of conservation of energy becomes

$$\boxed{\Delta E_{\text{th}} = W + Q,} \qquad (20\text{-}6)$$

which, in this form, is called the **first law of thermodynamics**. We will frequently call Eq. 20-6 simply "the first law." Although we are giving it a new name, it is important to keep in mind that the first law is nothing more than conservation of energy, now generalized to add the new idea of heat energy to the Newtonian concepts of work and thermal energy.

[**Photo suggestion: A steam engine.**]

The first law is one of the most important analysis tools of practical thermodynamics. Consider, for example, the simplified view of an "engine" shown in Fig. 20-4. Fuel is burned, providing a transfer of heat energy into the gas within the cylinder: $Q > 0$. As a result, the gas expands and pushes the piston outward. The *system does work* on the environment as the piston pushes out, so the work done *by* the environment is negative in this case: $W < 0$. Even though the gas cools during the expansion, as the piston moves outward, the exhaust gas is still hotter than when it was first drawn into the cylinder. Thus the internal energy of the gas has been increased: $\Delta E_{\text{th}} > 0$. Notice that this is the change of a state variable characterizing the gas, and is thus a Δ, whereas W and Q are numbers describing the interaction of the gas with the surrounding environment.

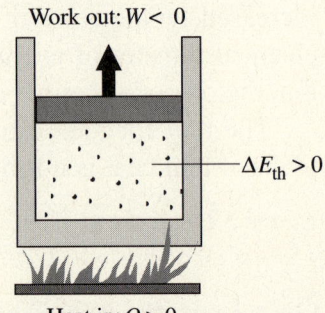

FIGURE 20-4 In an engine, heat energy is used both to do work and to increase the thermal energy of the gas. These quantities are related by the first law: $\Delta E_{\text{th}} = W + Q$.

According to the first law, $\Delta E_{th} = W + Q$. From a practical viewpoint, however, we really want to know not how much work the environment does on the gas but how much work the gas does as it pushes the piston out—or the *output* of the engine. This is simply

$$W_{out} = -W_{environ\ on\ system} = Q - \Delta E_{th}. \qquad (20\text{-}7)$$

The engine is a mechanism for energy transfer. The energy transferred *out* of the system as work is simply the energy *input* to the system, in the form of heat, minus the energy "lost" to heating the gas and thus raising its internal energy. In an ideal engine, no energy would go into heating the gas and the entire energy released from the fuel in the form of heat would ultimately be provided as work done by the engine. That is, an ideal engine has $\Delta E_{th} = 0$ and $W_{out} = Q_{in}$. However, you will see in later chapters that another law of physics—the second law of thermodynamics—prevents us from building an ideal engine.

EXAMPLE 20-2 An engine receives 15,000 J of heat energy by burning fuel. As the gas is heated and then expands to push the piston out, its internal energy increases by 6000 J. What is the efficiency of this engine?

SOLUTION A reasonable definition of *efficiency*—and one that we will confirm later—is the ratio of output energy to input energy, or

$$\text{efficiency} = \frac{\text{energy out}}{\text{energy in}} = \frac{W_{out}}{Q_{in}}.$$

The work done by this engine is found from the first law (Eq. 20-7) to be

$$W_{out} = Q_{in} - \Delta E_{th} = 15{,}000\ \text{J} - 6000\ \text{J} = 9000\ \text{J}.$$

The engine's efficiency is thus

$$\text{efficiency} = \frac{W_{out}}{Q_{in}} = \frac{9000\ \text{J}}{15{,}000\ \text{J}} = 0.60 = 60\%.$$

20.6 Specific Heat and Molar Heat Capacity

What happens to a system when heat energy is transferred in or out? There are three distinct possibilities:

1. The temperature of the system changes,
2. The system undergoes an isothermal expansion or compression, or
3. The system undergoes a phase change, such as melting or freezing.

In this section we will look at situations in which heat causes the temperature of a system to change. The connection between heat and isothermal processes will be investigated in Chapter 23. It is important to know that phase changes can occur, but we will not analyze them in this text.

It is found experimentally that the heat energy Q needed to change the temperature of an object is proportional both to the temperature change ΔT and to the mass M of the object. We can write this as

$$Q = Mc\Delta T, \qquad (20\text{-}8)$$

where c is a constant called the **specific heat** of the object. Note that Q can be either positive (heat energy added to raise the temperature) or negative (heat energy removed to lower the temperature). Equation 20-8 assumes that the temperature change is not so large as to cause the object to undergo a phase change. (Recall that uppercase M is used for the mass of an entire system while lowercase m is reserved for the mass of an atom or molecule.)

The specific heat of an object is found to depend *only* on the substance from which the object is made. Aluminum and copper have different specific heats, but all pieces of copper—no matter what their shape or mass—have the same specific heat. Specific heat is a property of a substance that characterizes its thermal behavior. Table 20-2 provides some typical specific heats for common substances. (Extensive tables can be found in handbooks of chemical and physical data.) The SI units of specific heat are J/kg K, although the "old-fashioned" units of cal/g °C are still widely used for heat problems.

The values of specific heat vary widely. Water, you should notice, has a much larger specific heat than most substances. (Its value of 1.00 cal/g °C reflects the definition of the calorie.) Because $\Delta T = Q/Mc$, it takes more heat energy to change the temperature of a substance with a large specific heat than to change the temperature of a substance with a small specific heat. You can think of specific heat as measuring the "thermal inertia" of a substance. Metals, with small specific heats, heat up and cool down quickly. Water, as you have probably noticed, is slow to heat up and slow to cool down.

TABLE 20-2 Some typical specific heats.

Substance	c (J/kg K)	c (cal/g°C)
Solids		
Aluminum	900	0.215
Copper	387	0.0924
Gold	129	0.0308
Silicon	703	0.168
Ice	2090	0.50
Liquids		
Ethyl alcohol	2400	0.58
Mercury	140	0.033
Water	4190	1.00

EXAMPLE 20-3 a) How much heat is required to raise the temperature of 500 g of copper from 0°C to 40°C? b) How much heat is required to raise the temperature of 500 g of water from 0°C to 40°C?

SOLUTION a) This is a straightforward calculation intended to illustrate the typical amount of heat energy needed to change the temperature of a common substance. For copper,
$$Q = Mc\Delta T = (0.500 \text{ kg})(387 \text{ J / kg K})(40°C - 0°C) = 7740 \text{ J}.$$

b) Water has a specific heat that is larger than copper's by the ratio 4190/387 = 10.8, so the quantity of heat required will be 10.8 times larger than that for copper, or 83,600 J. It is interesting to note that 83,600 J = 20,000 cal = 20 Cal. The energy of much of the food you eat is used to maintain your body temperature. 500 g of water is ≈1 pound, which is roughly the weight of a 16-ounce soda that you might drink at ≈0°C and then need, internally, to warm to ≈40°C (37°C is body temperature, but we are just making an estimate). The energy expenditure to do this requires "burning" ≈20 Cal of your lunch.

Two notes about Example 20-3. First, we used ΔT because the temperature (a state variable) changed, but on the left we used simply Q, *not* ΔQ. Heat is an interaction parameter and is just a number, not the change of anything. Second, we did not need to convert the temperature into kelvins. The *interval* of 1°C is exactly the same as the interval of 1 K, so as long as we are taking *differences* (and *only* then!), it does not matter whether we use temperatures in kelvins or degrees Celsius.

EXAMPLE 20-4 A 50 g aluminum disk at 300°C is inserted into 200 cm³ of ethyl alcohol at 10°C, then quickly removed. The aluminum disk is found to have dropped in temperature to 120°C. What is the new temperature of the ethyl alcohol?

SOLUTION Heat is an energy transfer due to a temperature difference. From the disk's perspective, the ethyl alcohol is the environment and heat energy flows out of the disk to the cooler alcohol. But from the alcohol's perspective, the disk is the environment and heat energy flows into the alcohol from the hotter disk. Because the disk and the alcohol interact with each other, but nothing else, conservation of energy tells us that the heat lost by the disk becomes the heat gained by the alcohol. The aluminum disk loses

$$Q_{Al} = Mc\Delta T = (50 \text{ g})(0.215 \text{ cal} / \text{g} \, °C)(120°C - 300°C) = -1935 \text{ cal}.$$

Q_{Al} is negative because heat is flowing from the aluminum to the environment. The ethyl alcohol, however, *gains* 1935 cal of heat and thus Q_{ethyl} = +1935 cal. We need to find the mass of the ethyl alcohol. Its density was given in Table 19-1 as ρ = 0.79 g/cm³, so the mass is

$$M = \rho V = (0.79 \text{ g} / \text{cm}^3)(200 \text{ cm}^3) = 158 \text{ g}.$$

The heat input from the aluminum thus causes a temperature change

$$\Delta T = \frac{Q_{ethyl}}{Mc} = \frac{1935 \text{ cal}}{(158 \text{ g})(0.58 \text{ cal} / \text{g} \, °C)} = 21.1 \, °C.$$

The ethyl alcohol ends up at temperature

$$T_f = T_i + \Delta T = 10.0°C + 21.1°C = 31.1°C.$$

Molar Heat Capacity of Gases

As you may have noted, the substances in Table 20-2 are all solids or liquids. Gases, it turns out, are harder to characterize because the heat required to cause a specified temperature change depends on the *process* by which the gas changes state. Figure 20-5 shows two isotherms on the pV diagram for a gas. Any process—such as the two shown—that starts on the T_i isotherm and ends on the T_f isotherm will have the same value of $\Delta T = T_f - T_i$. But process 1, which takes place at constant volume, requires a *different* amount of heat energy Q than does process 2, which occurs at

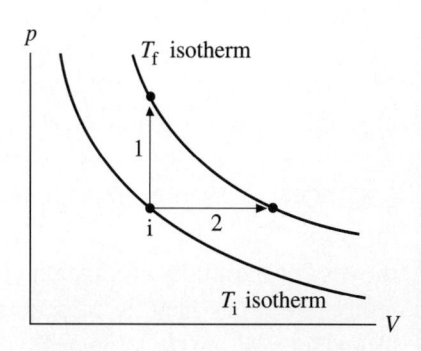

FIGURE 20-5 Processes 1 and 2 have the same ΔT but require different amounts of heat energy.

constant pressure. (You will learn why in Chapter 22.) If there is no unique value for Q then, according to Eq. 20-8, there is no unique value of c. For a gas, the value of the specific heat depends on the process by which the gas temperature is changed.

For gases it turns out to be convenient to define two different versions of the specific heat—one for constant volume (isochoric) processes and one for constant pressure (isobaric) processes. Also, because we usually do gas calculations using moles instead of mass, it will be useful to alter Eq. 20-8 to refer to n moles of gas rather than m kilograms of gas. The quantity of heat needed to change the temperature of n moles of gas by ΔT is

$$Q = nC_V\Delta T \quad \text{(temperature change at constant volume)}$$
$$Q = nC_P\Delta T \quad \text{(temperature change at constant pressure)}. \quad (20\text{-}9)$$

The quantities C_V and C_P are called the **molar heat capacities**, and they play the same role for gases that specific heat does for liquids and solids. In particular, C_V is the *molar heat capacity at constant volume* and C_P is the *molar heat capacity at constant pressure*. Table 20-3 gives the values of C_V and C_P for a few common monatomic and diatomic gases. The units are J/mol K.

TABLE 20–3 Molar heat capacities of typical gases (units are J/mol K).

Gas	C_P	C_V	$C_P - C_V$
Monatomic Gases			
He	20.8	12.5	8.3
Ne	20.8	12.5	8.3
Ar	20.8	12.5	8.3
Diatomic Gases			
H_2	28.7	20.4	8.3
N_2	29.1	20.8	8.3
O_2	29.2	20.9	8.3

EXAMPLE 20-5 Three moles of O_2 gas are initially at a temperature of 20°C. 600 J of heat are added to the gas at constant pressure, then 600 J are removed at constant volume. a) Show the process on a pV diagram. b) What is the final temperature?

SOLUTION a) Figure 20-6 shows the process on a pV diagram. The gas starts at 1 on the 20°C isotherm, moves horizontally (constant pressure) to a higher temperature as heat is added, then moves vertically (constant volume) to a lower final temperature as heat is removed. Because $C_P > C_V$, we anticipate that the final temperature will be lower than the initial temperature.

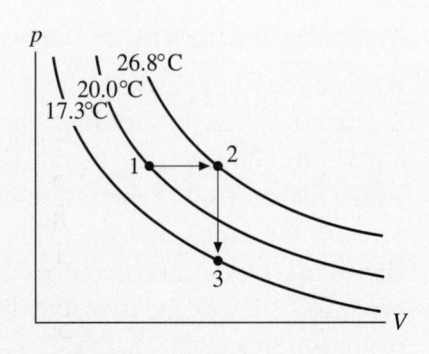

FIGURE 20-6 The process of Example 20-5.

b) The constant pressure heating produces a temperature rise:

$$\Delta T = T_2 - T_1 = \frac{Q}{nC_P} = \frac{600 \text{ J}}{(3 \text{ mol})(29.3 \text{ J / mol K})} = 6.8°\text{C}$$

$$\Rightarrow T_2 = T_1 + \Delta T = 26.8°\text{C}.$$

The temperature falls as heat is removed during the constant volume cooling:

$$\Delta T = T_3 - T_2 = \frac{Q}{nC_V} = \frac{(-600 \text{ J})}{(3 \text{ mol})(21.0 \text{ J / mol K})} = -9.5°\text{C}$$

$$\Rightarrow T_3 = T_2 + \Delta T = 17.3°\text{C}.$$

The fact that $T_3 \neq T_1$ in Example 20-5, despite the fact that just as much heat was removed as added, makes the point that you must, when working with gases, pay careful attention to the type of process. Notice that, as in the last example, we did not need to distinguish between kelvins and Celsius degrees *as long as* we are finding temperature *differences*. Also note that, in the second step, we used a *negative* value for Q because heat energy was being removed.

The net result in Example 20-5 was to lower the temperature, so that $\Delta E_{th} < 0$. The *net* heat input was $Q_{total} = Q(1\rightarrow2) + Q(2\rightarrow3) = 0$, so how did the thermal energy manage to change? According to the first law of thermodynamics, $W + Q = W + 0 = W = \Delta E_{th} < 0$. So somewhere in this process a negative amount of work was done on the system or, equivalently, the system did work on the environment. Where and how this occurred is a topic we will look at closely in Chapter 23.

If you look closely at Table 20-3, you will notice a very curious feature. Unlike the specific heats of solids and liquids, which all appeared quite distinct, the molar heat capacities of monatomic gases are *all alike*. The molar heat capacities of diatomic gases, while different from monatomic gases, are again *very nearly alike*. Furthermore, the *difference* $C_P - C_V = 8.3$ J/mol K is the same in every case. And, most puzzling of all, the value of $C_P - C_V$ appears to be equal to the universal gas constant R! Why should this be? This is a puzzle for which we have no explanation. But a puzzle like this is just what a scientist loves because it suggests there is something important waiting to be discovered that will *explain* the puzzle. And indeed there is, as we will begin to explore in the next chapter.

20.7 Calorimetry

We have introduced specific heats and molar heat capacity because they are going to help us to make the micro/macro connection. But they also have a large practical importance in everyday occurrences where two systems or more at different temperatures are combined. It is worth a brief digression into this practical aspect of heat transfer known as **calorimetry**.

Consider two objects or systems at different temperatures T_1 and T_2. Let the two objects be combined or mixed together while in *thermal isolation* from the rest of their environment. As you know from experience, the two systems will eventually come to a common final temperature T_3, and the temperatures will then change no further.

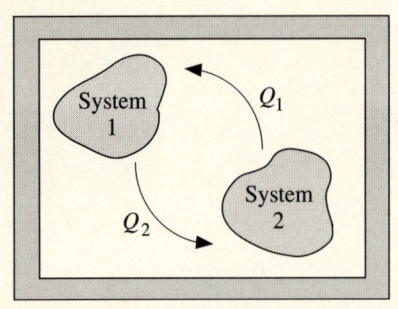

FIGURE 20-7 Two isolated systems that interact have $Q_1 + Q_2 = 0$.

Figure 20-7 shows the two systems interacting with each other but isolated from everything else. Because they have different temperatures, heat energy will flow from the hotter to the colder system. Let Q_1 be the heat input to system 1 and Q_2 be the heat input to system 2. The *total* heat input to the *combined* system is $Q_{tot} = Q_1 + Q_2$. But because the combined system is thermally isolated from the rest of its environment, it must be the case that $Q_{tot} = 0$. There is no *net* heat input or output from the combined system. The two systems can *exchange* heat energy with each other, but the heat gained by one is lost by the other: $Q_1 = -Q_2$.

Each of the heat energies Q_1 and Q_2 can be related, via Eq. 20-8, to a specific heat and a temperature change for the appropriate system. As the two systems interact thermally and come to a final temperature T_3, they are governed by

$$Q_1 + Q_2 = 0$$
$$\Rightarrow M_1 c_1 (T_3 - T_1) + M_2 c_2 (T_3 - T_2) = 0. \tag{20-10}$$

We have written $\Delta T_1 = T_3 - T_1$ and $\Delta T_2 = T_3 - T_2$, where T_1 and T_2 are the initial temperatures of the two systems.

EXAMPLE 20-6 200 g of an unknown metal is heated to a temperature of 200°C, then quickly dropped into 50 g of water at 20.0°C in an insulated container. The water temperature rises within a few seconds to 39.7°C, then changes no further. Identify the metal.

SOLUTION The basic relationship of calorimetry is Eq. 20-10, for which we know all the quantities except the specific heat of the metal—call it c_m. Then

$$M_w c_w (T_3 - T_w) + M_m c_m (T_3 - T_m) = 0$$
$$\Rightarrow c_m = \frac{M_w c_w (T_3 - T_w)}{M_m (T_m - T_3)} = 0.0307 \text{ cal} / \text{g} \,°\text{C}.$$

Referring to Table 20-2, we find we have 200 g of gold!

EXAMPLE 20-7 An industrial process expels 8 g of helium gas at 600°C into a rigid container of volume 10^5 cm^3 containing molecular oxygen at a temperature of 20°C and a pressure of 2 atm. What is the final temperature?

SOLUTION The basic relationship $Q_1 + Q_2 = 0$ is still valid as the two gases mix, but we will need to generalize Eq. 20-10 for the case of gases and molar heat capacities. Because the container is rigid, the gases will interact and change temperatures at constant volume. The relationship we need is

$$Q_{He} + Q_{O_2} = 0$$
$$\Rightarrow n_{He} C_V (\text{He})(T_3 - T_{He}) + n_{O_2} C_V (\text{O}_2)(T_3 - T_{O_2}) = 0.$$

Solving for the final temperature T_3 gives

$$T_3 = \frac{n_{He}C_V(He)T_{He} + n_{O_2}C_V(O_2)T_{O_2}}{n_{He}C_V(He) + n_{O_2}C_V(O_2)}.$$

The initial temperatures are given, and the molar heat capacities we can find from Table 20-3. But we need the number of moles of each gas. Helium has atomic mass $A = 4$, so $n_{He} = M(\text{in grams})/A = 2.00$ mol. For the oxygen, we can use the ideal gas law to find

$$n_{O_2} = \frac{pV}{RT_{O_2}} = 8.32 \text{ mol},$$

where we converted 10^5 cm^3 = 0.10 m^3 and 2 atm = 2.02×10^5 Pa. With this information, we can compute

$$T_3 = \frac{(2 \text{ mol})(12.5 \text{ J / mol K})(873 \text{ K}) + (8.32 \text{ mol})(20.9 \text{ J / mol K})(293 \text{ K})}{(2 \text{ mol})(12.5 \text{ J / mol K}) + (8.32 \text{ mol})(20.9 \text{ J / mol K})}$$

$$= 366 \text{ K}$$
$$= 93°C.$$

It is important to notice that these examples used $Q_1 + Q_2 = 0$. The idea is that the *net* heat flow is zero. A common error is to try to work calorimetry problems by using the *incorrect* relationship $Q_1 = Q_2$. This cannot work because one of the heat terms is positive while the other is negative.

There is nothing that limits the ideas of calorimetry to the combination of only two systems. If three or more systems are combined in isolation from the rest of their environment, each at a different initial temperature, they will all come to a common final temperature that can be found from the relationship

$$Q_{tot} = Q_1 + Q_2 + Q_3 + \cdots = 0. \qquad (20\text{-}11)$$

EXAMPLE 20-8 A 200 g aluminum cylinder of volume 800 cm^3 contains N$_2$ at STP. A 20 cm^3 block of copper at a temperature of 300°C is placed inside the cylinder, and the cylinder is then sealed. What is the final temperature?

SOLUTION This example has three interacting systems—the aluminum cylinder, the nitrogen gas, and the copper block—that will all come to a common final temperature T_4. Call these systems 1, 2, and 3, respectively. The cylinder and gas have the same initial temperature, $T_1 = T_2 = 0°C$, and the container doesn't change size, so this is a constant volume process for the gas. Using Eq. 20-11, we have

$$Q_{tot} = m_{Al}c_{Al}(T_4 - T_1) + n_{N_2}C_V(T_4 - T_2) + m_{Cu}c_{Cu}(T_4 - T_3) = 0$$

$$\Rightarrow T_4 = \frac{m_{Al}c_{Al}T_1 + n_{N_2}C_V T_2 + m_{Cu}c_{Cu}T_3}{m_{Al}c_{Al} + n_{N_2}C_V + m_{Cu}c_{Cu}}.$$

Note that we've used masses and specific heats for the solids and moles and molar heat capacity for the gas. The specific heats and the molar heat capacity at constant volume are found from Tables 20-2 and 20-3. Because C_V is given in J/mol K, the two specific heats will have to be in J/kg K rather than cal/g °C. The mass of the aluminum is given (m_{Al} = 200 g = 0.200 kg) and the mass of the copper is found from its density (in Table 19-1) to be

$$m_{Cu} = \rho_{Cu} V_{Cu} = (8.92 \text{ g/cm}^3)(20 \text{ cm}^3) = 178 \text{ g} = 0.178 \text{ kg}.$$

The number of moles of the gas is found from the ideal gas law, using the initial conditions. Note, however, that inserting the copper block *displaces* 20 cm³ of gas, so the gas volume is only $V = 780 \text{ cm}^3 = 7.80 \times 10^{-4} \text{ m}^3$. Thus

$$n_{N_2} = \frac{pV}{RT} = 0.0348 \text{ mol}.$$

Computing T_4 from the above equation gives a final temperature $T_4 = 83°C$.

Specific heats and molar heat capacities are important parameters that characterize the thermal behavior of a system. Our interest in them is twofold: first, to use them for doing thermal calculations; second, to understand why they have the values they do and, in particular, why the puzzling pattern we see in the molar heat capacities of gases occurs. This will be one of the major goals of our exploration of the micro/macro connection—the task to which we turn next.

Summary
Important Concepts and Terms

interaction parameter	first law of thermodynamics
mechanical transfer of energy	specific heat
heat	molar heat capacity
nonmechanical transfer of energy	calorimetry

Work and heat are the two means by which a system interacts with its environment. Work is a mechanical transfer of energy in which a force displaces the boundary of the system. Heat is a nonmechanical transfer of energy due to a temperature difference. Positive values of W and Q indicate energy transferred from the environment to the system (environment does work on the system or a heating process) while negative values indicate energy transferred from the system to the environment (system does work on the environment or a cooling process).

If the system as a whole is at rest ($E_{mech} = 0$), then the first law of thermodynamics relates the thermal energy of the system to the energy transferred via work and heat by

$$\Delta E_{th} = W + Q.$$

The first law is a general statement about the conservation of energy.

When the transfer of heat energy causes the temperature of a system to change, the temperature change ΔT depends on the specific heat (solid or liquid) or the molar heat capacity (gas). For a solid or liquid, the heat required for a temperature change ΔT is

$$Q = Mc\Delta T$$

while for a gas of n moles the heat required is
$$Q = nC_V\Delta T \quad \text{(temperature change at constant volume)}$$
$$Q = nC_P\Delta T \quad \text{(temperature change at constant pressure)}.$$

When two or more systems interact thermally, the heat loss of one system is the heat gain of another. The final temperature is determined by
$$Q_1 + Q_2 + Q_3 + \cdots = 0.$$

Exercises and Problems

Exercises

1. The average speed of molecules in hydrogen gas (H_2) is 700 m/s.
   a. What is the thermal energy of 1 g of gas?
   b. 500 J of heat are added to the gas, causing it to do 300 J of work. Afterward, what is the average molecular speed?

2. A cylinder contains gas at a pressure of 3 atm. A movable piston in one end of the cylinder has a diameter of 16 cm.
   a. How much force does the gas exert on the piston?
   b. How much force does the environment exert on the piston?
   c. The gas expands at constant pressure and pushes the piston out 10 cm. How much work is done by the gas?
   d. How much work is done by the environment?
   e. The thermal energy of the gas decreases by 250 J in the expansion of part c). Was heat energy added or removed from the gas in this process? How much?

3. How much heat must be removed from a 4 cm × 4 cm × 4 cm block of ice to cool it from 0°C to −30°C?

4. a. 20 cal of heat are added to 20 g of mercury. By how much does the temperature increase?
   b. How many calories of heat would be needed to raise the temperature of 20 g of O_2 in a rigid container by the same amount?

5. A sealed cube 20 cm on each side contains 3 g of helium at room temperature. 1000 J of heat are added to this gas.
   a. If the heating takes place at constant volume, what is the final gas pressure?
   b. If the heating takes place at constant pressure, what is the final gas volume?

Problems

6. Five grams of nitrogen gas (N_2) at an initial pressure of 3 atm and at room temperature undergo an isobaric expansion until the volume has tripled.
   a. What is the gas volume and temperature after the expansion?
   b. How much heat energy was added to the gas to cause this expansion?

   The gas pressure is then decreased at constant volume until the original temperature is reached.

   c. What is the gas pressure after the decrease?
   d. What amount of heat was removed from the gas as its pressure was decreased?

58   CHAPTER 20   WORK, HEAT, AND THE FIRST LAW OF THERMODYNAMICS

7. 10 g of aluminum at 200°C and 20 g of Cu at –50°C are dropped into 50 cm³ of ethyl alcohol at 15°C. What is the final temperature?

8. 512 g of an unknown metal at a temperature of 15°C is dropped into a 100 g aluminum container holding 325 g of water at 98°C. The container of water and metal stabilizes a short time later at a new temperature of 78°C. Use the data in Table 20-2 to identify the metal.

9. A cylinder 10 cm in diameter contains argon gas ($A = 40$) at 10 atm pressure and a temperature of 50°C. A piston can slide in and out of the cylinder. Initially the cylinder's length is 20 cm. 2500 J of heat are added to the gas, causing the piston to move and the gas to expand at constant pressure.
   a. What is the final temperature of the gas?
   b. What is the final length of the cylinder?

10. A cylinder of nitrogen gas 6 cm in diameter is fitted with a tight-sealing but movable copper piston that is 4 cm thick. The cylinder is oriented vertically, as shown in Fig. 20-8, and the air above the piston is evacuated. When the gas temperature is 20°C, the piston "floats" 20 cm above the bottom of the cylinder.
    a. What is the gas pressure?
    b. How many gas molecules are in the cylinder?
    c. If the average molecular speed is 550 m/s, what is the internal energy of the gas?

    Two joules of heat energy are then added to the gas.

    d. What is the subsequent gas temperature?
    e. What is the final height of the piston?
    f. How much work is done *by* the gas as it pushes the piston out?
    g. What is the change in thermal energy of the gas, ΔE_{th}?
    h. What is the average molecular speed after the heat is added?

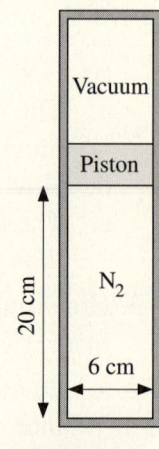

FIGURE 20-8

11. A gas monatomic follows the process $1 \to 2 \to 3$ shown in Fig. 20-9.
    a. How much heat is added or removed during the process $1 \to 2$?
    b. How much heat is added or removed during the process $2 \to 3$?

FIGURE 20-9

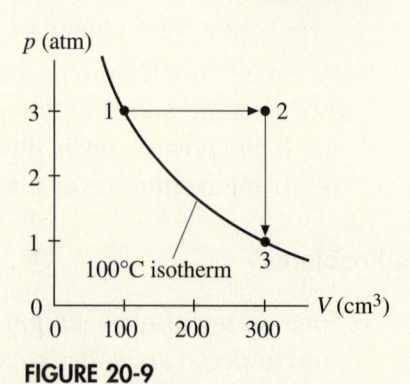

[**Estimated 10 additional problems for the final edition.**]

Chapter 21

A Statistical View of Gases

21.1 The Microstate of a Gas

In the last two chapters, we have been looking mainly at the phenomena and properties of macroscopic systems. For the most part, we were able to draw conclusions about these macroscopic properties *without* having to know any details of the microscopic properties of the system. In this chapter we will begin to look more closely at the microscopic properties of the molecules in a gas. Our goal is to begin to understand how the macroscopic properties of a system are connected to the underlying microscopic behavior of its constituent particles. For example, how can a state variable such as the temperature have a well-defined, unchanging value while the molecules are continuously changing their positions and velocities? We will introduce several statistical methods to help us answer questions such as this. These methods form the basis for a field of physics known as **statistical mechanics**. This chapter will introduce the basic concepts of statistical mechanics, then in the next chapter we will apply these concepts to macroscopic variables such as pressure and temperature.

Let us begin by peering carefully into a container of gas, containing N molecules. The typical number of molecules in a gas is $\sim 10^{24}$, a number so large as to be almost beyond comprehension. Because a real gas is so complex, we will often illustrate ideas with "gases" that have a very small number of molecules—perhaps as few as 4. We do not expect the results of such illustrations to accurately reflect the behavior of real gases. If, however, you can see how a system behaves for 4 molecules, then 64, then 1000—numbers small enough to understand—then you will be able to develop some intuition for how the system will behave in the limit $N \to 10^{24}$.

Let's assume that our container holds an ideal gas of noninteracting molecules. These molecules occasionally collide with each other or the walls, exchanging energy, but otherwise they move in force-free straight lines obeying Newtonian physics. (The molecules, strictly speaking, move in parabolic trajectories under the influence of gravity. But the time and distance between collisions for real gas molecules are so small that the molecules have no time to "fall." Their deviation from straight-line trajectories over these short distances is so small as to be negligible.) We will also assume that the gas is isolated from its environment and that it has been left undisturbed for a long period of time.

Now suppose that we have a very special movie camera that allows us to see all of the molecules inside the container. If we make a movie of the gas and examine sequential frames of film, we will be able to see how the molecules are moving about.

Figure 21-1 shows eight frames from two such movies—one of a gas of 4 molecules and another of a 64-molecule gas. Notice that the positions of the molecules in each frame are completely different from the positions in any other frame. This is not terribly surprising, because all the molecules are constantly moving and colliding. To specify completely the state of the gas in each frame, we would need to specify the position *and* the velocity of each of the N molecules. Such a complete specification is what we will call a **microstate**—the complete description of all the microscopic motion at one instant of time.

It seems possible, for a gas of 4 molecules, to solve the problem of how a given microstate at one instant of time evolves into a different microstate at some later time. The problem is just the Newtonian dynamics of four interacting systems—somewhat harder than anything we have done, but at least imaginable. Even computing the trajectories for a gas of 64 molecules might be a manageable problem using a computer. But a gas of 10^{24} molecules? The amount of information we would need and the number of equations we would have to solve are overwhelming. It is not even remotely possible that we could solve for the dynamics of this system! Every computer on earth speeded up by a factor of a million and all working on the problem in tandem would not make the slightest progress toward determining the dynamics of a realistic gas.

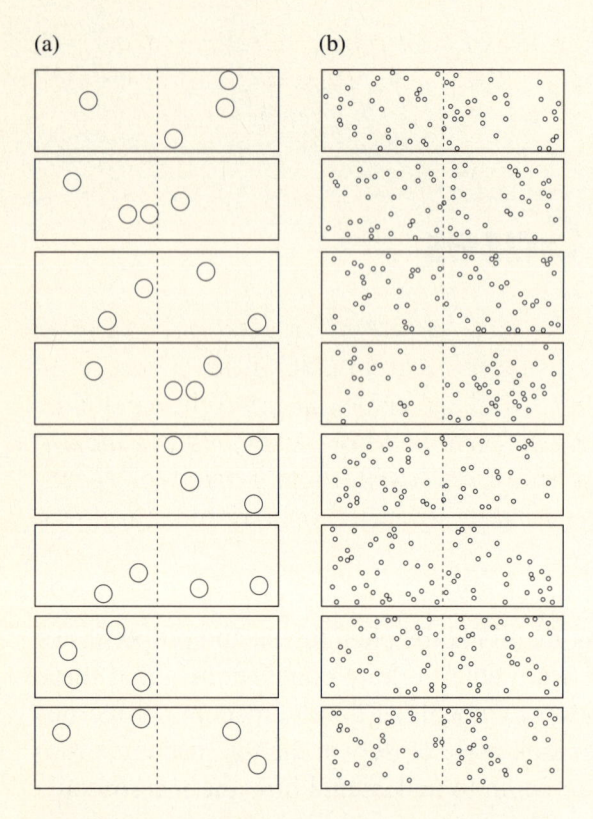

FIGURE 21-1 Eight frames each from "movies" of gases having a) $N = 4$ and b) $N = 64$ molecules. Notice that the number of molecules in each half of the container fluctuates from frame to frame as the molecules move about. The fluctuations occasionally cause one-half of the container to be completely empty for $N = 4$, but for $N = 64$ there is never any significant difference in the number of molecules in each half.

Brute mathematical force will not allow us to understand a macroscopic system on the basis of its microscopic dynamics. Instead, we need a new approach, and some new tools, for trying to deal with very large numbers of particles.

21.2 Configurations and Fluctuations

Our goal is to find a connection between the microscopic behavior of the molecules in a gas and the macroscopic properties of the gas. To do so, we are going to answer a rather artificial question: How many molecules are found in one half of a container? It is the concepts and the analysis tools, rather than the answer, that are of most interest because we will be able to extend these ideas to more significant properties such as temperature and pressure. Even so, we'll find that this simple question has some rather unexpected and profound implications.

Configurations

Imagine dividing the container into two equal halves with an imaginary partition—shown as dotted lines in Fig. 21-1. This is not a real partition, but just a boundary that allows us to determine whether a molecule is in the left side or the right side of the container. In each frame of film, we can *count* the number N_R of molecules in the right-hand side. Because we do not care *which* molecules they are, but only how many, N_R is a *macroscopic* state variable, similar to temperature or pressure. Each different value of N_R represents a different macroscopic state of the gas, just as would a different value of the temperature.

It is quite apparent in Fig. 21-1 that N_R is not constant but, instead, varies from frame to frame. In principle, the value of N_R in a given frame could have any value from 0 (the state with no molecules on the right side) to N (the state with all N molecules on the right side). For example, a gas with $N = 4$ molecules has five possible states: $N_R = 0$, $N_R = 1, \ldots, N_R = 4$. The gas is continually changing from one state to another as the molecules move about.

Let's define a **configuration** of the gas to be a distinct arrangement or distribution of the molecules between the two halves of the container. One configuration, for example, would be where molecule 1 is on the right, molecule 2 is on the left, molecule 3 is also on the left, and so on for all N molecules. Each rearrangement of the molecules provides a different configuration. Each configuration, which is a *microscopic* description of the gas, "belongs" to a particular macroscopic state N_R. In general, there are many different configurations corresponding to each state. For example, the $N_R = 1$ state, which has one molecule on the right, could have molecule 1 on the right and all others on the left, or molecule 2 on the right and all others on the left, or molecule N on the right and all others on the left. Because each of these configurations is different, there are N different microscopic configurations

TABLE 21-1 Listing of all configurations of a $N = 4$ gas that can be distributed in the left or right half of a container.

Molecule 1	2	3	4	State N_R	Number of Configurations $C(N_R)$
L	L	L	L	0	1
R	L	L	L	1	
L	R	L	L	1	
L	L	R	L	1	4
L	L	L	R	1	
R	R	L	L	2	
R	L	R	L	2	
R	L	L	R	2	
L	R	R	L	2	6
L	R	L	R	2	
L	L	R	R	2	
R	R	R	L	3	
R	R	L	R	3	
R	L	R	R	3	4
L	R	R	R	3	
R	R	R	R	4	1

belonging to the macroscopic state $N_R = 1$. This idea—that there are many possible microstates corresponding to a single macroscopic state—is at the heart of our statistical view of gases.

Determining the number of configurations in each macroscopic state is simply a process of counting. Let us illustrate this idea for $N = 4$, where we can do the counting explicitly by listing all possible configurations. We can designate each of the four molecules as being either R if it is found in the right side of the container in that frame of film or L if it is in the left. Table 21-1 shows that there are 16 different configurations of an $N = 4$ gas, or 16 different ways we can assign the values R or L to the four molecules. There are, however, only 5 values for N_R, which can range from 0 to 4, and so only 5 different states. You can see, for example, that there are 4 configurations in the $N_R = 1$ state but 6 having $N_R = 2$. We will use the symbol $C(N_R)$ to represent the number of configurations in the N_R state. Thus $C(1) = 4$ and $C(2) = 6$ for a 4-molecule gas.

The Most Probable State

As you've seen, there are multiple ways in which a particular macroscopic state N_R can occur. To make this idea more precise, we would like to determine the likelihood, or probability, that an observation of the gas will find it in state N_R. We will assume that the probability of finding a given molecule in the right half of the container is not affected by the locations of the other molecules. That is, the location of each molecule is independent of the locations of the other molecules. This is a quite reasonable assumption for a dilute, non-interacting system like a gas. If this assumption is valid (and its validity will be justified if our conclusions agree with reality), then there is no reason to prefer any one configuration over any other. *Each of the configurations is equally likely to occur* or equally likely to be seen in any given frame of film. This simple statement will turn out to have profound consequences, so make sure you understand what it says and how it is related to the assumption that the molecules are all independent.

For a gas of N molecules, there are 2^N configurations for the distribution of molecules between the right and left halves of the container. ($2^3 = 8$ configurations for $N = 3$, $2^4 = 16$ configurations for $N = 4$, etc.) Our assumption is that each of these configurations is equally likely to occur. If we pick a frame of film at random, the probability of seeing any particular configuration is $P = 1/2^N$, the same as the probability of seeing any other configuration. For example, if you observe a 4-molecule gas at a randomly chosen instant, the probability is 1/16 that the gas will be in the configuration LRLL.

Although each of the configurations is equally likely to occur, it is *not* the case that each macroscopic state N_R is equally likely. This is a consequence of the fact that there are differing numbers of configurations in each state. As we've already noted, the number of configurations in state N_R is $C(N_R)$. If you were to observe 2^N frames of film, you would expect, on average, to see each configuration once and thus to see $C(N_R)$ frames in which there are N_R molecules on the right. We can define the *probability* that a randomly chosen frame of film will show the gas to be in macroscopic state N_R as

$$P(N_R) = \text{probability of finding the gas in state } N_R = \frac{C(N_R)}{2^N}. \qquad (21\text{-}1)$$

Note how this probability depends on our assumption that each of the configurations is equally likely to occur.

It is a simple matter to use Eq. 21-1 and Table 21-1 to determine the probability of occurrence for each of the five states of a $N = 4$ gas. The results are shown in Table 21-2. We find that the gas is *most likely* to be found in the state $N_R = 2$, where $P(N_R = 2) = 6/16 = 0.375$. That is, there is a 37.5% chance that a random observation of a 4-molecule gas will find 2 molecules on the right side (and thus 2 molecules on the left side). Note that the *sum* of the probabilities in Table 21-2 equals 1. This is because the gas *has* to be in one of these five possible states.

TABLE 21-2 Probability of occurrence of each of the states of a $N = 4$ gas.

State	$C(N_R)$	$P(N_R)$
$N_R = 0$	1	1/16
$N_R = 1$	4	4/16
$N_R = 2$	6	6/16
$N_R = 3$	4	4/16
$N_R = 4$	1	1/16

The macroscopic state of the gas with the largest probability is called the **most probable state**. The value of N_R for the most probable state is indicated by the symbol $\overline{N_R}$. For a 4-molecules gas, Table 21-2 shows that $\overline{N_R} = 2$. The most likely state of the gas, it comes as no surprise, has the molecules equally divided between the two halves of the container.

Fluctuations

Figure 21-2 shows a graph of N_R-versus-time for several frames of film, starting with the 8 frames from Fig. 21-1 and then continuing. Notice that N_R *fluctuates* around the most likely value $\overline{N_R}$. The $N = 4$ gas spends most of its time in the state with $N_R = \overline{N_R} = 2$, but it is also likely to be found with $N_R = 1$ or $N_R = 3$. There are even a few frames in which $N_R = 0$ or 4 (all the molecules in one half of the container or the other). For $N = 64$, where $\overline{N_R} = 32$, the value of N_R fluctuates between about 24 and 40, but N_R never even comes close to 0 or 64. This suggests—as we will, indeed, confirm—that the macroscopic parameter N_R becomes more and more "well defined" as the number of molecules increases.

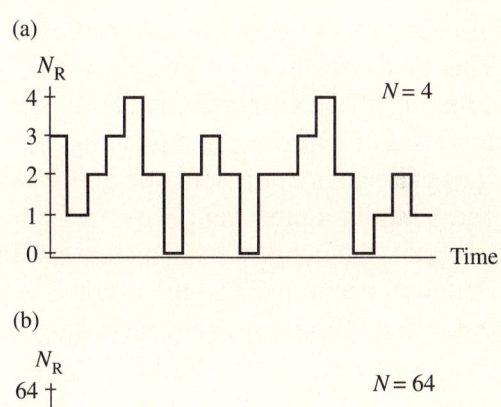

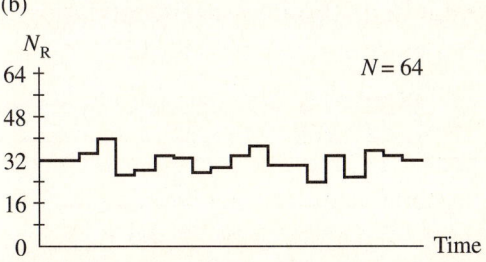

FIGURE 21-2 Graphs showing the variation with time of N_R for gases having a) $N = 4$ and b) $N = 64$ molecules.

Figure 21-2 shows that the *density* in the right half of the container undergoes enormous fluctuations for the 4-molecule gas. The 64-molecule gas has significant density fluctuations, although not as extreme as for $N = 4$. But in a realistic gas, such as the air in the room, there are no noticeable, or even measurable, variations in the density. It simply does not happen that we are breathing comfortably one instant but gasping the next because all of the air molecules have moved to the other half of the room.

Why not? Is there something that "prevents" all of the air molecules from gathering in one half of the room, leaving a vacuum in the other? Figure 21-2 showed that N_R spends most of its time near $\overline{N_R}$ but, on occasion, fluctuates to either extreme. For our $N = 4$ gas there is a 1-in-16 probability that, at any instant of time, all the molecules are in the left side of the container while the right side is a vacuum. Why doesn't the same thing happen in a room of air molecules?

To answer this question, we need to approach it from a statistical perspective. If we ask, "How many molecules are in the right side of a container?" there is not a single answer—N_R is always changing. The best we can do is to give the *probability*—as in Table 21-2—that N_R will be found to have certain value. How likely is it, then, that any one frame of the film will show *all* of the molecules on the left and $N_R = 0$? We call such an event a **large fluctuation**. This is a statistical question, but one we can answer.

You saw in Table 21-1 that the number of configurations in the state $N_R = 0$, where all the molecules are on the left, is $C(0) = 1$. Although that was for $N = 4$, the result is valid for any N. No matter how many molecules there are, there is only *one* way to arrange them to have $N_R = 0$. We also noted that there are a total of 2^N different configurations for a gas of N molecules because there are 2^N different ways to arrange the two labels R and L for N objects. Thus the probability of finding the gas in the $N_R = 0$ state, according to Eq. 21-1, is

$$P(N_R = 0) = P_0 = \frac{C(0)}{2^N} = \frac{1}{2^N}.$$

This gives, for a $N = 4$ gas, $P_0 = 1/2^4 = 1/16$, as we had already deduced from directly counting the configurations. This is not all that unlikely—once, on average, every 16 frames. For a movie camera taking 30 frames of film a second, we would expect to see all four molecules in the left side and a vacuum on the right approximately every half a second—hardly a rare event.

But what happens as N increases? Table 21-3 shows the probability P_0 that the $N_R = 0$ state will occur and also the average length of time you would have to wait to see such an event in a 30 frames-per-second movie.

TABLE 21-3 The probability P_0 of finding no molecules in the right half of a container.

N	P_0	Waiting Time to See a $N_R = 0$ Event
4	$1/2^4 = 1/16$	0.5 s
16	$1/2^{16} = 1/65,000$	35 min
64	$1/2^{64} = 1/2 \times 10^{19}$	2×10^{10} years = age of universe
1000	$1/2^{1000} = 1/10^{301}$	10^{292} years
10^{24}	$1/2^{10^{24}} = 1/10^{3 \times 10^{23}}$	$10^{10^{23}}$ years = $10^{100000000000000000000000}$ years

Wow! As N increases, the chances of finding all the molecules on the left become astronomically small! Even for a gas as small as $N = 64$, which we observed in Figs. 21-1 and 21-2, we would have to wait a time approximately equal to the age of the universe before we would expect to see a movie frame with $N_R = 0$. For a realistic gas, with $N \sim 10^{24}$, the chances of this ever occurring are so small as to be completely incomprehensible.

What is the point? In small systems, such as $N = 4$, large fluctuations away from the average are not uncommon. But the likelihood of a sizable fluctuation decreases unbelievably rapidly as N increases. The underlying reason is our assumption that all configurations are equally likely to occur. The total number of configurations grows exponentially with N, but the number of configurations corresponding to a "large fluctuation" stays relatively small. The overwhelming majority of configurations have $N_R \approx \overline{N_R}$, so the *probability* of seeing a large fluctuation decreases exponentially with N.

For a realistic gas, with $N \approx 10^{24}$, the probability of a large fluctuation away from the average value is so remote that we can safely say it will "never" occur. We are, in this case, providing an operational definition of "never." While we cannot say with absolutely certain knowledge that it has never occurred, the odds are overwhelmingly in our favor to say that "never," in the history of the universe, has a room-sized container of gas anywhere in the universe ever exhibited a "large fluctuation" in which there has been a significant difference between the number of molecules in one half and the number in the other.

For a large system, the value of N_R (or, as we will see, any other macroscopic parameter) is perfectly steady and constant, at its average value, to any imaginable level of measurement precision. Although the values do fluctuate about the average, just as in Fig. 21-2, the sizes of the fluctuations are completely undetectable and unmeasureable in any macroscopic system. This is simply a consequence of the statistics of the system. We will make these ideas more precise in Section 21-5, where we will make an explicit determination of just how large a fluctuation of N_R we would be likely to observe.

Let's summarize what we have found so far. We have established four major ideas:

1. The probability of occurrence of a macroscopic state is determined by the number of microscopic configurations belonging to the state. This is the basic idea of the micro/macro connection.

2. One particular state is the most probable state. It will be observed more often than any other state. The most probable state is found simply by *counting* the configurations of the system.

3. The value of a macroscopic parameter, such as N_R, is not constant. The value fluctuates in time about the most probable value $\overline{N_R}$.

4. The significance of fluctuations decreases extremely rapidly as the size N of the system increases. The likelihood is remotely small that a realistic macroscopic system will undergo a large fluctuation.

Although most of our illustrations have been for $N = 4$, where we can count the configurations very explicitly, you will see shortly that these ideas apply equally well to any value of N. Our tentative conclusion, from these very small systems, is that the macroscopic parameter N_R is related to the average, or statistical, properties of the microscopic configurations. This is our first micro/macro connection. Although N_R may not be a very interesting property of a gas, the *procedures* we are developing here are applicable to other macroscopic parameters, such as temperature and pressure.

21.3 Probability and the Binomial Distribution

We have been using the idea of probability in a rather loose and intuitive sense. Even though we do not require a detailed knowledge of probability theory, we do need to make our ideas more precise. This section will develop some of the mathematics of probability. These analysis tools will then be used in the rest of the chapter to answer our question about how many molecules are in one side of a container.

Probability of Events

If you flip a coin, you cannot predict the outcome with certainty. All you can say is that there is some probability that the result will be heads rather than tails. Similarly with a gas left undisturbed for a while in a container. If you look for one particular molecule, you cannot predict its location with certainty. All you can say is that there is some probability that the molecule will be found in the right half of the box rather than the left half. But how are such probabilities determined?

Consider a "measurement" on some system that has α distinct possible outcomes. Flipping a coin, for example, has two possible outcomes; rolling a die (singular of "dice") has six possible outcomes; while selecting one card from a deck has 52 possible outcomes. We can label all the possible outcomes with an index i where $i = 1, 2, 3, \ldots, \alpha$. Let us focus on one possible outcome—an *event*—that we will call event r. If event r occurs N_r times out of a very large number N of measurements, then we define the **probability** of event r to be

$$P_r = \lim_{N \to \infty} \frac{N_r}{N}. \tag{21-2}$$

The probability P_r is the expected fraction of measurements that result in event r. For example, if you flip a fair coin a very large number of times, you expect that nearly half will be heads. Thus $P_{\text{heads}} = 1/2$. Similarly, the probability of rolling a 3 with one die is $P_3 = 1/6$. What is the probability of dealing an ace from a complete, shuffled deck of cards? There are four aces in a deck of 52 cards, so $P_{\text{ace}} = 4/52 = 1/13$.

We also expect N measurements to yield $N_r = P_r N$ occurrences of event r. This is, keep in mind, only a probabilistic statement and, for finite N, not a precise prediction. If you roll a die 300 times, you would expect to get $N_3 = 300/6 = 50$ threes. But because 300 is a finite number, you would not be surprised it the actual outcome was not exactly 50. The more rolls you make ($N \to \infty$), the closer the fraction of threes will get to 1/6.

Let's look at what happens if we try to find the probability that one of two possible events occurs. This is an either/or question: What is the probability that either event r or event s occurs? For example, what is the probability that, with one die, you roll either a 3 or a 4? Out of a large number of measurements, the number producing either event r or event s is the number producing event r *plus* the number producing event s: $N_{r \text{ or } s} = N_r + N_s$. Thus, by the probability definition of Eq. 21-2,

$$P_{r \text{ or } s} = \frac{N_{r \text{ or } s}}{N} = \frac{N_r + N_s}{N} = \frac{N_r}{N} + \frac{N_s}{N} = P_r + P_s. \tag{21-3}$$

That is, the probability of either event r or event s occurring is simply the sum of the probability of event r with the probability of event s.

EXAMPLE 21-1 What is the probability of rolling either a 3 or a 4?

SOLUTION The probability of rolling a 3 is $P_3 = 1/6$. Similarly, $P_4 = 1/6$. Thus

$$P_{3 \text{ or } 4} = \frac{1}{6} + \frac{1}{6} = \frac{1}{3}.$$

This idea is easily extended to find the probability of more than two events.

There are α possible events, and one of them *must* occur on each measurement. That is, it is certain ($P = 1$) that either event 1 or event 2 or event 3 or ... or event α will occur. Thus

$$P_1 + P_2 + \cdots + P_\alpha = \sum_{r=1}^{\alpha} P_r = 1. \tag{21-4}$$

In other words, *some* event has to happen, so the sum of the probabilities of all possible events must be one. This is an important requirement of probabilities, and you can see, for example, that it is obeyed by the probabilities of the five different states in Table 21-2.

Next, suppose the system we are measuring has more than one object—such as 2 dice or 3 coins. If, for example, you roll a red die and a blue die, what is the probability that the red die is a 5 *and* that the blue die is a 2? This is a both/and question: What is the probability $P_{r \text{ and } s}$ that both event *r and* event *s* occur simultaneously?

We will make the assumption—valid for our needs—that event *r* and event *s* have no influence over each other. That is, the occurrence of event *r* in no way affects the likelihood of event *s*. We say that *r* and *s* are independent of each other. In that case, by the definition of probability (Eq. 21-2):

$$P_{r \text{ and } s} = \frac{N_{r \text{ and } s}}{N}, \tag{21-5}$$

where we will assume that N is a very large number of measurements.

The reasoning behind determining $N_{r \text{ and } s}$ is a bit tricky. Out of the N measurements, the number in which event *r* happens is

$$N_r = P_r N, \tag{21-6}$$

where P_r is the probability of event *r* by itself. Out of these N_r events, some are also *s* events and some are not. Because event *r* and event *s* are independent of each other, the number of the N_r events that also produce an *s* event is

$$N_s(\text{in } N_r) = P_s N_r. \tag{21-7}$$

But $N_s(\text{in } N_r)$, the number of *s*-events in the group of N_r measurements, is the number of events, out of the original N measurements, that produce both an *r and* an *s* outcome. That is,

$$N_{r \text{ and } s} = N_s(\text{in } N_r) = P_s N_r = P_s P_r N. \tag{21-8}$$

Using this result in Eq. 21-5, we find that the probability of both event *r* and event *s* is

$$P_{r \text{ and } s} = \frac{N_{r \text{ and } s}}{N} = P_r P_s. \tag{21-9}$$

So the probability of both event *r* and event *s* happening is the product of the probabilities of the individual events. This result is easily generalized to say that the probability of any number of independent events happening simultaneously is the product of each of the individual probabilities.

EXAMPLE 21-2 If two coins are tossed, what is the probability that they will both be heads?

SOLUTION This is equivalent to asking what the probability is that both coin 1 is a head *and* coin 2 is a head. This probability, according to Eq. 21-9, is simply

$$P_{2 \text{ heads}} = P_{\text{head}} \cdot P_{\text{head}} = \frac{1}{2} \cdot \frac{1}{2} = \frac{1}{4}.$$

Thus we expect that 25% of a large number of tosses of two coins will result in two heads.

EXAMPLE 21-3 If you roll three dice—red, blue, and green—what is the probability that the red die is a 1, the blue die a 2, and the green die a 3?

SOLUTION The probability of each of these individual events is 1/6. Thus the probability of rolling this specific combination is

$$P(\text{red 1 } and \text{ blue 2 } and \text{ green 3}) = \frac{1}{6} \cdot \frac{1}{6} \cdot \frac{1}{6} = \frac{1}{216}.$$

You should expect to see this specific combination only one out of every 216 rolls.

EXAMPLE 21-4 What is the probability that you will roll at least one 6 in eight rolls of a die?

SOLUTION It is tempting to see this as an either/or situation: What is the probability of a 6 on the first roll or a 6 on the second roll or ... or a 6 on the eighth roll? Then you would add 1/6 eight times to get $P = 8/6 = 1.25$. But that can't be! It is impossible to have a probability greater than 1. The difficulty is that the either/or situation applies to two distinct outcomes of a *single* measurement, such as what is the probability of a 5 or a 6 on a single roll. In this situation, however, you are making eight different measurements. The situation is equivalent to rolling eight dice at once and asking for the probability that at least one of them is a 6. This is not an either/or question for two reasons. First, we don't care which die is a 6. Second, it is possible that *more* than one die is a 6.

Despite these difficulties, we can answer the question if we realize that we can make a clear statement of the opposite situation. The outcome "no sixes" is the opposite of the outcome "at least one six." If you roll eight dice, what is the probability that *none* of them are a 6? This is the specific outcome: Die 1 is not a 6 and die 2 is not a 6 and ... and die 8 is not a 6. The probability that an individual die is not a 6 is 5/6, so the probability of no sixes on eight dice is thus

$$P_{\text{no 6}} = \frac{5}{6} \cdot \frac{5}{6} \cdot \frac{5}{6} \cdot \frac{5}{6} \cdot \frac{5}{6} \cdot \frac{5}{6} \cdot \frac{5}{6} \cdot \frac{5}{6} = \left(\frac{5}{6}\right)^8 = 0.23.$$

Because the total probability of all possible events is 1, the probability we seek is

$$P_{\text{at least one 6}} = 1 - P_{\text{no 6}} = 1 - 0.23 = 0.77.$$

There is a 77% chance—but not certainty—that eight rolls will yield at least one 6. The reasoning used in this example is subtle and worth careful study.

The last kind of probability question we need to ask is of the following type: Suppose we toss three coins. What is the probability that we get one head and two tails? We have to be careful with this one, because now we don't care *which* of the coins is the head. So this question is really: What is the probability that

coin 1 is H and coin 2 is T and coin 3 is T

or

coin 1 is T and coin 2 is H and coin 3 is T

or

coin 1 is T and coin 2 is T and coin 3 is H.

Combining what we know about *or* and *and* probabilities, we have

$$P(1H2T) = P_H \cdot P_T \cdot P_T + P_T \cdot P_H \cdot P_T + P_T \cdot P_T \cdot P_H$$
$$= \tfrac{1}{2} \cdot \tfrac{1}{2} \cdot \tfrac{1}{2} + \tfrac{1}{2} \cdot \tfrac{1}{2} \cdot \tfrac{1}{2} + \tfrac{1}{2} \cdot \tfrac{1}{2} \cdot \tfrac{1}{2} = \tfrac{3}{8}.$$

This is *exactly* what we did in Tables 21-1 and 21-2 to find the probabilities of each of the five states of a $N = 4$ gas. From a mathematical perspective, the problem is equivalent to tossing 4 coins and computing the probability of N_H heads. We first, in Table 21-1, listed every possible outcome. The probability of any specific outcome, according to Eq. 21-9, is $(1/2) \cdot (1/2) \cdot (1/2) \cdot (1/2) = 1/16 = 1/2^N$. Then, to find the probability of the event "1 molecule on right side," we counted the number of different ways we can arrange the molecules to have one on the right—just like counting the number of ways to get "1 head." The probabilities for each distinct way *add*, because this is an *or* situation, so the final probability is $C(N_R)$, the number of arrangements, times $1/2^N$, the probability of each arrangement.

The Binomial Distribution

The statistics of locating molecules in a gas are the same as the statistics of tossing coins or rolling dice. Asking whether a given molecule is in the right side or the left side of the container is the same as asking whether a given coin lands heads or lands tails.

Consider a system that has only *two* possible events that we can monitor. These might be simple, such as a coin landing heads or tails, or a little more complex, such as a die being 1 or not-1. (Although the die has more than two *outcomes*, we want to group them into just two *events*.) Let the probabilities of these two events be designated p and q. Thus

p = probability that {coin = H, die = 1, molecule = R, ...}

q = probability that {coin = T, die = not-1, molecule = L, ...}.

Because these two events exhaust all possibilities, $p + q = 1$.

Section 21.2 looked only at distributions of molecules between the right and left *halves* of the container. That was a situation with $p = q = 1/2$. We can, however, imagine dividing the container into right and left *sides* of different volumes V_R and V_L where $V_R + V_L = V_{total}$. Then the probability of locating a molecule in the right side is $p = V_R/V_{total}$ and the probability that a molecule is in the left side is $q = V_L/V_{total}$. For example, the probability of locating a molecule in the right one-quarter of a container is $p = 1/4$. As you can easily see, $p + q = 1$.

Consider a gas of N molecules. If we look at one randomly chosen frame of film, or one instant of time, there are $N + 1$ possible values we might find for the number N_R of molecules on the right side: $N_R = 0, 1, 2, \ldots, N$. Each value of N_R is a different macroscopic state of the gas. The question we would like to answer is, What is the *probability* of occurrence of each of these $N + 1$ possible states?

We will follow a procedure just like we did to find the probability of one head out of three coins or as we did in Tables 21-1 and 21-2. We will first find the probability of one *specific* configuration giving the desired outcome. This will be a series of *and*s, so we will multiply the probabilities of individual molecules. Second, we will determine, by counting, how many different configurations give rise to the same state. Third, and last, we will add the probabilities of each configurations because any one *or* the other of them will give the desired outcome.

Let's number the molecules $1, 2, 3, \ldots, N$. One specific configuration of these N molecules having N_R molecules in the right side is

$$(1 = R) \text{ and } (2 = R) \text{ and } (3 = R) \text{ and } \ldots \text{ and } (N_R = R) \text{ and } (N_R + 1 = L) \text{ and}$$
$$(N_R + 2 = L) \text{ and } \ldots \text{ and } (N = L).$$

Notice that the list does not stop when we get molecules N_R on the right, because we also have to add "*and* $N - N_R$ molecules on the left." The probability of occurrence of this specific configuration is, from Eq. 21-9,

$$\begin{aligned}
P_{config} &= p \cdot p \cdot p \cdot \ldots \cdot p \cdot q \cdot q \cdot q \cdot \ldots \cdot q \\
&= \{N_R \text{ products of } p\} \cdot \{N_L \text{ products of } q\} \\
&= p^{N_R} \cdot q^{N_L} \\
&= p^{N_R} \cdot q^{N-N_R}.
\end{aligned} \quad (21\text{-}10)$$

This is the probability of observing one specific configuration of macroscopic state N_R.

There are many different configurations having the same value of N_R. For example, moving molecule 1 to the left and molecule N to the right would produce a different configuration belonging to the same macroscopic state. Suppose we number all the configurations in state N_R as $1, 2, 3, \ldots, C$. Observing N_R molecules on the right is thus a matter of observing:

(configuration 1 having N_R) or (configuration 2 having N_R) or ...

or (configuration C having N_R).

These configurations differ by which specific molecules are on the right and which are on the left, but the probability of occurrence of each configuration is the same and is given by Eq. 21-10. Thus the probability of observing N_R molecules on the right, which is what we wanted to find, is

$$P(N_R) = P_{\text{config 1}} + P_{\text{config 2}} + \cdots + P_{\text{config C}}$$
$$= C \cdot P_{\text{config}}. \qquad (21\text{-}11)$$

The value of C depends on N_R because some states have more configurations than others. We earlier introduced the notation $C(N_R)$ to represent the number of configurations in the N_R state. Using this more explicit notation for C and using Eq. 21-10 for P_{config}, the probability that N_R molecules will be found in the side of the container is

$$P(N_R) = C(N_R) \cdot p^{N_R} \cdot q^{N-N_R}. \qquad (21\text{-}12)$$

This is a more general conclusion than Eq. 21-1, which assumed that the container was divided into equal halves. If we set $p = q = 1/2$, then

$$\left(\tfrac{1}{2}\right)^{N_R} \cdot \left(\tfrac{1}{2}\right)^{N-N_R} = \left(\tfrac{1}{2}\right)^N = \frac{1}{2^N}. \qquad (21\text{-}13)$$

Equation 21-1 is recovered if this result is used in Eq. 21-12.

EXAMPLE 21-5 For a 4-molecule gas, what is the probability that 3 molecules will be found in the right one-quarter of the container?

SOLUTION We need to find $P(N_R = 3)$ for a situation in which $p = 1/4$ and $q = 3/4$. The number of configurations $C(N_R)$ for a 4-molecule gas was found in Table 21-1. This number does not depend on where the partition is but only on the fact that there is a right side and a left side. So from Table 21-1 we find $C(N_R = 3) = 4$. The probability of observing a state in which 3 molecules are in the right one-quarter of the container, according to Eq. 21-12, is

$$P(N_R) = C(3) \cdot p^3 \cdot q^{4-3} = 4 \cdot \left(\frac{1}{4}\right)^3 \left(\frac{3}{4}\right)^1 = 0.047 = 4.7\%.$$

For an $N = 4$ gas we determined $C(N_R)$ by explicitly counting the configurations in each state. For an arbitrary value of N, finding $C(N_R)$ is still a counting exercise—but not one we would want to undertake with pencil and paper. While the details are not hard, it serves no purpose to get sidetracked with a derivation of $C(N_R)$. We will assert, without proof, that for *any* value of N

$$C(N_R) = \frac{N!}{N_R!(N-N_R)!}, \qquad (21\text{-}14)$$

where ! is the symbol for a factorial:

$$m! = m \cdot (m-1) \cdot (m-2) \cdot (m-3) \cdot \ldots \cdot 3 \cdot 2 \cdot 1.$$

Combining Eqs. 21-12 and 21-14, we find that the probability of observing N_R molecules on the right side in a gas of N molecules is

$$P(N_R) = \frac{N!}{N_R!(N-N_R)!} \cdot p^{N_R} \cdot q^{N-N_R}. \qquad (21\text{-}15)$$

Although a bit cumbersome, Eq. 21-15 is a completely general result that is valid for any value of N from 1 to 10^{24} and beyond.

Equation 21-15 is called the **binomial distribution**, so named because it describes events that have *two* possible outcomes. For N objects, the binomial distribution tells us the probability that N_R of the objects will be found to have the condition measured by probability p. It is not too hard to show that

$$\sum_{N_R=0}^{N} P(N_R) = 1 \qquad (21\text{-}16)$$

because N_R *must* have one of the values between 0 and N.

EXAMPLE 21-6 For a 4-molecule gas, what is the probability that 3 molecules will be found in the right half?

SOLUTION We already know, from Table 21-2, that the answer is $P(N_R = 3) = 1/4$. The purpose of this example is to see that we can arrive at the same answer using Eq. 21-15 rather than by having to explicitly list and count all the configurations. Because the container is divided into halves, $p = q = 1/2$. Using Eq. 21-13 in Eq. 21-15, we find for this special case that

$$P(N_R) = \frac{N!}{N_R!(N-N_R)!} \cdot \frac{1}{2^N} \qquad (\text{if } p = q = \tfrac{1}{2}). \qquad (21\text{-}17)$$

In this case we have

$$P(N_R = 3) = \frac{4!}{3!1!} \cdot \frac{1}{2^4} = \frac{24}{6 \cdot 1} \cdot \frac{1}{16} = \frac{1}{4}.$$

EXAMPLE 21-7 If six dice are rolled, what is the probability of getting four 3s?

SOLUTION The probability of a 3 is $p = 1/6$ and the probability of a not-3 is $q = 5/6$. The situation is equivalent to asking for the probability that 4 molecules in a 6-molecule gas are in the right one-sixth of the container. The probability is given by Eq. 21-15 as

$$P(\text{four 3s}) = \frac{6!}{4!2!} \cdot \left(\frac{1}{6}\right)^4 \cdot \left(\frac{5}{6}\right)^2 = 0.0080 = 0.80\%.$$

EXAMPLE 21-8 For a 32-molecule gas, calculate the probabilities that 7, 8, or 9 molecules will be in the right one-quarter of the container.

SOLUTION We need to compute $P(N_R)$ for $N = 32$, $p = 1/4$, $q = 3/4$, and N_R ranging from 7 to 9. Equation 21-15 is

$$P(N_R) = \frac{32!}{N_R!(32-N_R)!} \cdot \left(\frac{1}{4}\right)^{N_R} \cdot \left(\frac{3}{4}\right)^{32-N_R}$$

$$= \begin{cases} 0.155 & N_R = 7 \\ 0.161 & N_R = 8 \\ 0.143 & N_R = 9. \end{cases}$$

•

The results of Example 21-8 suggest that $N_R = 8$ is the most probable macroscopic state. This is no surprise because you would expect, on average, that one-quarter of the 32 molecules would be in the right one-quarter of the container. We can generalize this idea to say that the *average value* of N_R in a system with N molecules is

$$\overline{N_R} = pN. \tag{21-18}$$

You may recall that we earlier called $\overline{N_R}$ the *most probable* value of N_R. There are some subtle distinctions between *most probable* and *average* values, but for the situations considered in this text we can consider them to be the same thing.

21.4 The Distribution of Molecules in a Gas

What do all these ideas and calculations about probability have to do with gases? Can we really understand anything of significance about gases by playing with probabilities p and q for finding a gas molecule in one side of a container or the other? In the next two sections of this chapter we will demonstrate that you can, indeed, learn a lot.

Calculations are necessary, but it is often more convenient to give a graphical illustration of the results. This is particularly easy to do with a histogram, such as we used in Chapter 19 to show how many atoms had speeds within different ranges. Figure 21-3 shows a histogram of the probabilities of each of the five states of an equally-divided ($p = q = 1/2$) $N = 4$ gas, using the results of Table 21-2 or numbers that are easily calculated from Eq. 21-17. The figure shows very clearly that $\overline{N_R} = pN = 2$ is the most probable state of the gas. The probability of finding a given value of N_R decreases as the value moves away from $\overline{N_R}$. Notice that the histogram is *symmetrical* about $N/2$. It is not hard to show, using Eq. 21-17, that this is true for any value of N when $p = q = 1/2$.

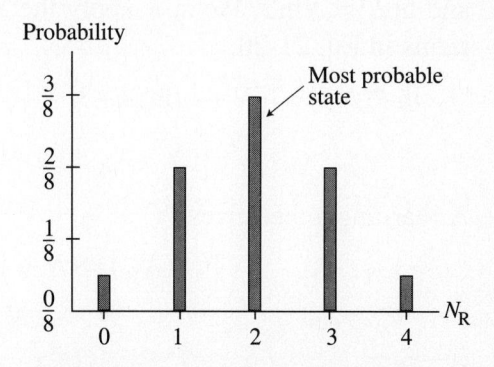

FIGURE 21-3 Histogram showing the probability of occurrence of each of the five possible states of an $N = 4$ gas with $p = q = 1/2$.

What we would like to see is a histogram for every possible value of N. A histogram, such as Fig. 21-3 for an $N = 4$ gas, tells us at a glance not only the most probable state but also the *range* of states that have a reasonable probability of occurrence. Producing such histograms is merely a matter of calculating all $N + 1$ values of Eq. 21-15, for $N_R = 0 - N$, and graphing them. While that is a lot of calculations as N starts to grow, a computer can handle them easily.

The difficulty is not the number of calculations, but the calculation itself. Evaluating Eq. 21-15 for $P(N_R)$ requires the evaluation of three factorials, and the value of factorials grows extremely rapidly with N until fairly soon, even for modest values of N, the result is a number larger than the computer can store. The final result of Eq. 21-15 may be quite modest, because it is a ratio, but it is a ratio of enormous numbers that the computer may not be able to handle. Fortunately, there is an approximate way to calculate factorials that allows us to circumvent this difficulty.

Rather than work directly with the factorial $m!$ of an integer m, suppose we were to consider the natural logarithm $\ln m!$. The value of $\ln x$ grows much less rapidly than the value of x itself. For example $10! = 3{,}628{,}800$ but $\ln 10! = 15.104$. The value of $\ln m!$ can be calculated with good accuracy for $m > 10$ by a formula known as the **Stirling approximation**. This approximation, which is derived in texts on numerical methods, is

$$\ln m! \approx (m + \tfrac{1}{2})\ln m - m + \tfrac{1}{2}\ln 2\pi \quad \text{(Stirling approximation)}. \qquad (21\text{-}19)$$

This approximation gives a value of $\ln 10! \approx 15.095$, remarkably close to the true value, and the accuracy improves as m increases.

Let us apply the Stirling approximation to Eq. 21-17, the binomial distribution for N molecules in a gas for the special case $p = q = 1/2$. First take the natural logarithm of both sides:

$$\ln P(N_R) = \ln N! - \ln N_R! - \ln(N - N_R)! - N\ln 2, \qquad (21\text{-}20)$$

where we have used the logarithmic properties $\ln(a \cdot b) = \ln a + \ln b$, $\ln(a/b) = \ln a - \ln b$, and $\ln 2^N = N\ln 2$. We now apply the Stirling approximation to each of the three factorial terms in Eq. 21-20:

$$\begin{aligned}\ln P(N_R) = &\left[(N + \tfrac{1}{2})\ln N - N + \tfrac{1}{2}\ln 2\pi\right] - \left[(N_R + \tfrac{1}{2})\ln N_R - N_R + \tfrac{1}{2}\ln 2\pi\right] \\ &- \left[(N - N_R + \tfrac{1}{2})\ln(N - N_R) - (N - N_R) + \tfrac{1}{2}\ln 2\pi\right] - N\ln 2.\end{aligned} \qquad (21\text{-}21)$$

A rearrangement gives

$$\begin{aligned}\ln P(N_R) = &(N + \tfrac{1}{2})\ln N - N\ln 2 - (N_R + \tfrac{1}{2})\ln N_R \\ &- (N - N_R + \tfrac{1}{2})\ln(N - N_R) - \tfrac{1}{2}\ln 2\pi.\end{aligned} \qquad (21\text{-}22)$$

Our final result is easily implemented with a computer to calculate many values very quickly, and it can even, with some care, be used with a hand calculator for a few values.

Keep in mind that Eq. 21-22 gives only $\ln P(N_R)$, not $P(N_R)$ itself. After evaluating Eq. 21-22, it is still necessary to complete the final step by computing

$$P(N_R) = e^{\ln P(N_R)}. \qquad (21\text{-}23)$$

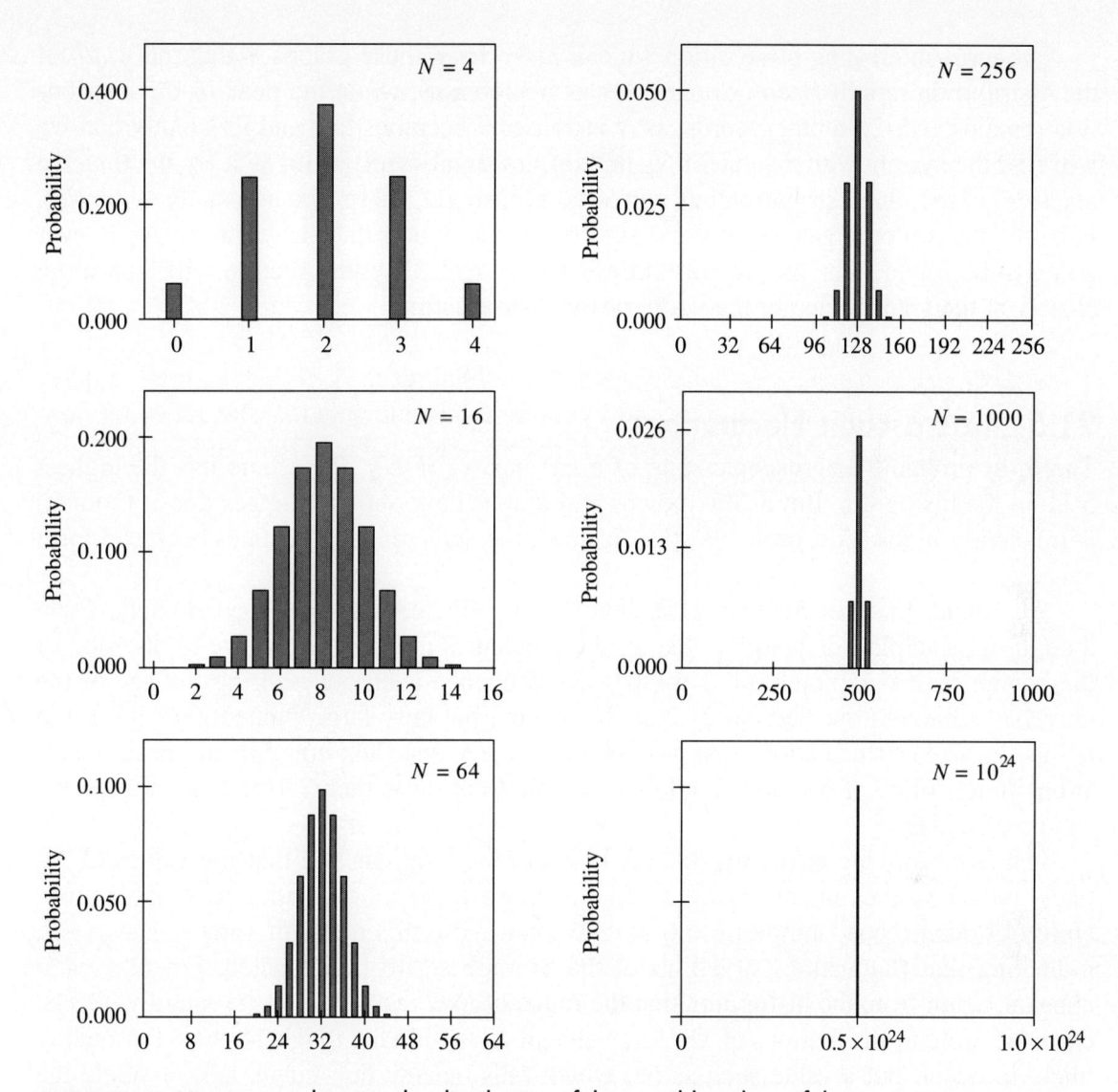

FIGURE 21-4 Histograms showing the distribution of the possible values of the macroscopic quantity N_R for gases having N = 4, 16, 64, 256, 1000, and 10^{24} molecules.

Now that we have a means for computing $P(N_R)$, let us apply it to gases of N = 4, 16, 64, 256, 1000, and 10^{24} molecules. Our goal is to see how the distribution of molecules changes as N increases. Figure 21-4 shows the histograms for the six values of N.

Because our interest is in the *shape* of the distribution, we have drawn the graphs so that all have the same width and the same height. Giving them a common height and width allows us to see the relative changes quite easily just by a visual comparison. Because $\overline{N_R} = pN = N/2$ for all values of N, the "peak" of the distribution is at the midpoint of the axis in every graph. Notice, from the vertical scale, that the probability $P(N/2)$ of the most likely state decreases with N. This is to be expected. As the number of possible states increases, the probability of the gas being in any particular one of them has to decrease if all the probabilities are going to sum to 1. The actual probability of a state is less important than its probability *relative* to other states.

The most interesting observation we can make from these graphs is that the *width* of the distribution rapidly becomes narrower as N increases, while the peak of the distribution remains at $N/2$. In other words, as N increases it becomes less and less likely that we will find the system with a value of N_R that differs significantly from $N/2$. By the time we reach $N = 10^{24}$, the distribution has narrowed to a single, sharp line at exactly $\overline{N_R} = N/2$. For this macroscopic gas, there is no significant likelihood that the value of N_R is ever going to be found to be measurably different from $N/2$. The next section will look more closely at the significance of the width of the distribution.

21.5 Macroscopic Fluctuations

The most probable macroscopic state of a gas has $N_R = \overline{N_R}$. This state has the highest peak in the histogram. But as the movies and graphs have shown, the gas doesn't remain permanently in the most probable state. Instead, the value of N_R fluctuates back and forth near $\overline{N_R}$.

We found, back in Section 21.2, that "large" fluctuations are very unlikely. Even though in principle it is possible that all the molecules in a gas will move to one side of the container, it is remotely unlikely that has happened even once during the age of the universe. Although our Section 21.2 analysis ruled out very large fluctuations, it did not tell us the size of fluctuations that we *should* expect to see. Just how big are these fluctuations in the value of N_R, and should we be able to observe them? This is a question we can now answer.

Just from looking at the histograms in Fig. 21-4, you can see that the values of N_R likely to be observed in an $N = 4$ gas—that is, those values of N_R with a reasonable probability of occurrence—range from 0 to 4. We can write this range of values as $\overline{N_R} \pm 2$, indicating that fluctuations of ± 2 about the average are to be expected. For $N = 64$ it appears, again from the histogram, that the range of *likely* values of N_R is roughly 32 ± 8, corresponding to fluctuations of ± 8. Any value of N_R within the range 24–40 is reasonably likely to occur, but a value such as 52, which falls outside this range, is not. While the absolute size of the fluctuations has increased from ± 2 for $N = 4$ to ± 8 for $N = 64$, *relative* importance has actually decreased. For $N = 4$ the fluctuations are $\pm 100\%$ of $\overline{N_R}$, while for $N = 64$ they are only $\pm 25\%$ of $\overline{N_R}$.

These are simply estimates of the fluctuations, made by looking at the histograms. We made them by estimating the *width* of the distribution—the range of histogram bars that are reasonably tall. Rather than having to guess, we need a more precise criterion for determining the width. Let us define a quantity σ_N such that

$$P(N_R = \overline{N_R} + \sigma_N) = P(N_R = \overline{N_R} - \sigma_N) = \tfrac{1}{2} P(N_R = \overline{N_R}). \qquad (21\text{-}24)$$

In other words, the quantity σ_N is determined by the two histogram bars that are one-half the height of the tallest bar at $N_R = \overline{N_R}$. These two bars are located at distances $\pm \sigma_N$ on either side of $\overline{N_R}$, as is shown in Fig. 21-5. (For small values of N it may not be possible to satisfy this definition exactly, so we will accept the value of σ_N that comes closest to satisfying Eq. 21-24.) For $N = 64$, for example, the histogram shows that the $N_R = 28$ and $N_R = 36$ histogram bars are about half of the tallest, $N_R = \overline{N_R} = 32$, bar (slightly more than half), so $\sigma_{64} = 4$.

Having defined σ_N, what is the probability that the value of N_R for the gas lies *somewhere* between $\overline{N_R} - \sigma_N$ and $\overline{N_R} + \sigma_N$? This is an *or* question: What is the probability that $N_R = \overline{N_R} - \sigma_N$ *or* $\overline{N_R} - \sigma_N + 1$ *or* ... *or* $\overline{N_R} + \sigma_N$? Evaluating this probability, as you have learned, requires adding the probabilities of all the possible outcomes. Thus

$$P(N_R \text{ in range } \overline{N_R} - \sigma_N \text{ to } \overline{N_R} + \sigma_N) = \sum_{N_R = \overline{N_R} - \sigma_N}^{\overline{N_R} + \sigma_N} P(N_R). \qquad (21\text{-}25)$$

It is possible to use the binomial distribution do this sum, although we will mercifully skip the details. The result is

$$P(N_R \text{ in range } \overline{N_R} - \sigma_N \text{ to } \overline{N_R} + \sigma_N) \approx \frac{2}{3}. \qquad (21\text{-}26)$$

In other words, there is about a 2/3 chance that, at any given instant, the value of N_R will lie between $\overline{N_R} - \sigma_N$ and $\overline{N_R} + \sigma_N$. Thus fluctuations of up to $\pm \sigma_N$ away from $\overline{N_R}$ are reasonable and to be expected. On the other hand, the probability is small that N_R will differ from $\overline{N_R}$ by more than σ_N.

Thus it is very likely that the value of N_R will be in the range $\overline{N_R} \pm \sigma_N$. It becomes increasingly unlikely, as you can see from the histograms, that the value of N_R will differ from $\overline{N_R}$ by more than σ_N. So the quantity σ_N does exactly what we want—it measures for us the size of fluctuations we can reasonably expect to see. A fluctuation in N_R away from $\overline{N_R}$ of as much as σ_N is not unusual, but a fluctuation larger than this is unlikely. It is not that a fluctuation larger than this is impossible—these are, keep in mind, only *probability* statements. But fluctuations larger than σ_N are unlikely and, equally important, very rapidly grow increasingly unlikely for ever-larger fluctuations. What we have established, therefore, is that the **macroscopic fluctuations** of the gas—the range of values within which N_R will nearly always be found as it varies naturally—are $\pm \sigma_N$ about the average value of $\overline{N_R}$.

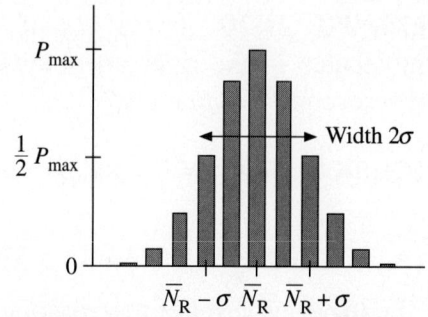

FIGURE 21-5 The two histogram bars that are one-half the height of the tallest are located at distances $\pm \sigma_N$ on either side of $\overline{N_R}$.

The last remaining piece of information we need is a means to determine the value of σ_N. It is possible to calculate the value of σ_N for the binomial distribution of a gas of N molecules. The result[*] is

$$\sigma_N = \sqrt{Npq}. \qquad (21\text{-}27)$$

---

[*]The σ_N of Eq. 21-24 is properly called the *half-width at half-maximum*, whereas the σ_N of Eq. 21-27 is the *standard deviation* of the distribution. These are not exactly the same. However, the difference between them is small and can be ignored for the purposes of this text.

EXAMPLE 21-9 For a gas of 64 molecules, what is the most likely number of molecules in the right half of the container and what is the expected level of macroscopic fluctuations?

SOLUTION $p = q = 1/2$, so the most likely number is given by Eq. 21-18 as

$$\overline{N_R} = pN = \frac{1}{2} \cdot 64 = 32.$$

Equation 21-27 gives the macroscopic fluctuations as

$$\sigma_N = \sqrt{64 \cdot \frac{1}{2} \cdot \frac{1}{2}} = 4.$$

This agrees with the value we estimated for σ_N from the histogram. The value of N_R will be found two-thirds of the time in the range $28 \leq N_R \leq 36$. A fluctuation larger than 4 is unlikely.

EXAMPLE 21-10 For a gas of 1000 molecules, what is the most likely number of molecules in the right one-quarter of the container and what is the expected level of macroscopic fluctuations?

SOLUTION Now $p = 1/4$ and $q = 3/4$, so the most likely number is

$$\overline{N_R} = pN = \frac{1}{4} \cdot 1000 = 250.$$

The expected level of macroscopic fluctuations is

$$\sigma_N = \sqrt{1000 \cdot \frac{1}{4} \cdot \frac{3}{4}} = 14.$$

The value of N_R will nearly always be in the range $236 \leq N_R \leq 264$.

As we noted earlier in this section, the absolute size of the fluctuations grows with N, but the *relative importance* decreases. What we really want to know is the *percentage* by which N_R fluctuates. Does N_R vary by 10% or 1% or 0.00001%? The **fractional variation** of N_R is given by

$$f = \text{fractional variation of } N_R = \frac{\sigma_N}{\overline{N_R}}, \qquad (21\text{-}28)$$

and the percentage variation is just 100 times this. In Example 21-9, the fractional variation of N_R is $f = 4/32 = 12.5\%$ while in Example 21-10 it is $f = 14/250 = 5.6\%$. The fluctuation of Example 21-10 is larger in magnitude, but its significance is less.

If we focus on a gas with $p = q = 1/2$, for an equally-divided container, then we have $\overline{N_R} = N/2$. σ_N is given by Eq. 21-27 as

$$\sigma_N = \sqrt{N \cdot \frac{1}{2} \cdot \frac{1}{2}} = \frac{1}{2} \cdot \sqrt{N} \qquad \left(\text{if } p = q = \frac{1}{2}\right). \qquad (21\text{-}29)$$

Combining these, we find the fractional variation to be

$$f = \frac{\sigma_N}{\overline{N_R}} = \frac{\sqrt{N}/2}{N/2} = \frac{1}{\sqrt{N}}.\qquad(21\text{-}30)$$

This is one of the most important results of statistical physics because it demonstrates the feasibility of establishing a meaningful micro/macro connection. This can best be seen with some examples.

TABLE 21-4 Fluctuations of a gas of N molecules ($p = q = 1/2$).

N	$\overline{N_R}$	σ_N	$f = \sigma_N / \overline{N_R}$	Likely Range of N_R
4	2	1	0.50 = 50%	1–3
16	8	2	0.25 = 25%	6–10
64	32	4	0.12 = 12%	28–36
1000	500	16	0.03 = 3%	484–516
1,000,000	500,000	500	0.001 = 0.1%	499,500–500,500
10^{24}	5×10^{23}	5×10^{11}	1×10^{-12}	$49{,}999{,}999{,}999{,}995 \times 10^{11}$– $50{,}000{,}000{,}000{,}005 \times 10^{11}$

Table 21-4 looks at several gases ranging from $N = 4$ up to a macroscopic gas of $N = 10^{24}$. We have calculated, for each gas, the average value of N_R, the range σ_N of likely fluctuations, and the fractional variation f. We have also shown explicitly the range of "likely" values of N_R—where the value will be found 2/3 of the time. The results are striking. Fluctuations of ±50% are quite common in an $N = 4$ gas, decreasing to ±12.5% by $N = 64$. The relative importance of fluctuations continues to decrease as N increases, dropping to a mere ±0.1%, or one part in a thousand, by $N = 1{,}000{,}000$. While a million molecules may seem like a lot to us, this is still a microscopic system as far as a gas is concerned—many, many factors of 10 away from the size of a macroscopic system.

You should compare Table 21-4 with the histograms of Fig. 21-4 and make sure you understand how the graphical presentation and the calculated results fit together. The figure showed graphically how the relative width of the distribution decreases with N, and now we have the same results, more precisely, from a calculation.

By the time we reach a true macroscopic system, with $N = 10^{24}$ molecules, the relative importance of fluctuations has, for any practical purposes, completely vanished! That is why the Fig. 21-4 histogram for $N = 10^{24}$ is simply a "spike" with no width. It is true that the absolute size of the fluctuations has become quite large, $\pm 5 \times 10^{11}$, but that simply reflects the enormously large number of states available to a macroscopic gas. The important point is that these fluctuations are only one part in a trillion (10^{12}) of the average value $\overline{N_R}$. To observe the fluctuations of a macroscopic gas, you would need measuring instruments sensitive to changes of one part in a trillion—equivalent to trying to distinguish between a mass of 1.000000000000 kg and a mass of 1.000000000001 kg. It cannot be done!

For any practical purpose at all, the value of N_R for a macroscopic system is perfectly constant. This conclusion for N_R will apply equally well to any of the other macroscopic state variables, such as pressure or temperature, that we will soon come to recognize as a quantity averaged over all the microstates of the system. The pressure and temperature of a system are not, strictly speaking, constant. Just like N_R, they fluctuate about some average value. But the relative size of those fluctuations decreases as $1/\sqrt{N}$ and thus, for a macroscopic system, are on the order of one part in a trillion.

For a small system—say $N = 1000$—we could not talk about "the" temperature or "the" pressure. The relative size of the fluctuations would be large enough that temperature, pressure, and other state variables would not have definite values. But the relative importance of fluctuations in a truly macroscopic system is, for all practical purposes, completely negligible and it then *does* become meaningful to refer to "the" temperature or "the" pressure.

21.6 Randomness, Order, and Equilibrium

Let's conclude this chapter by taking a look at an important situation in which probability plays a role. Consider the two containers of gas shown in Fig. 21-6. The container in part a) does not look "natural." This is an extremely improbable state, and all of the molecules in the gas would have to move in very special ways to achieve this on their own. If we were to see such a state of the gas, it is overwhelmingly likely that the gas has been "prepared" this way. Perhaps, for example, there had been a real partition or barrier in the middle of the container and we had used an air pump to move all the molecules into the left side. Then, just a nanosecond before this frame of film was exposed, we quickly pulled the barrier out. In that case we have prepared the gas to be in what is now, with the barrier gone, an extremely improbable state. It will not remain in this state for long, as the gas expands to fill the empty half.

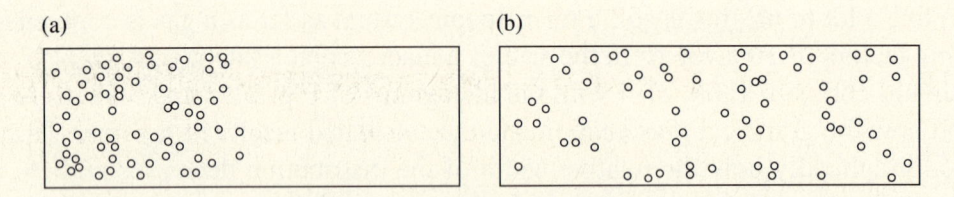

FIGURE 21-6 a) A nonrandom or ordered stated of the gas. b) A random or disordered state of the gas, corresponding to the equilibrium state.

Figure 21-6b, however, looks perfectly normal and natural. It is, more or less, what we would expect the gas to achieve by itself if left undisturbed for a long period of time. There are a great many configurations that look about like this, with $N_R \approx N/2$, and it is a state of high probability.

In comparing these two gases, we say that the molecules in the container in part b) are *randomly* distributed. That is, the location of each molecule is completely random, and so on the average, half the molecules are in each half of the container. The container in part

a) however, is clearly not a random occurrence. It is a very *nonrandom* distribution, or one that we will call *ordered*. To be more precise:

1. A situation or event that can be produced in a great many ways (a very large number of configurations) is called *random* or **disordered**. The gas in Fig. 21-6b is quite random. A system left undisturbed is overwhelmingly likely to be found in a highly random and disordered state.

2. A situation or event that can be produced in only a very small number of ways (a very small number of configurations) is called *nonrandom* or **ordered**. The gas in Fig. 21-6a is nonrandom, or ordered. An event exhibiting order has most likely been specially prepared because it is exceedingly unlikely to have occurred spontaneously.

An ordered state cannot be maintained. Left to itself, the gas will "evolve" until it reaches a state with $\overline{N_R} = N/2$. Once the system reaches the $N_R = N/2$ state it will evolve no further and will not, other than for expected fluctuations about the average, change further with time. At this point, we say that the gas is "in equilibrium." Although we have used the term *equilibrium* before, we can now be more precise in defining **equilibrium** to be the state of maximum randomness or, equivalently, the most probable state, the state having the maximum number of configurations.

Figure 21-7 shows how the state of the system evolves in time from the nonequilibrium state $N_R = 0$, when the barrier separating the halves of the container is removed, to the equilibrium state $N_R = N/2$. After the system reaches equilibrium, all we see are small fluctuations about the average value of N_R.

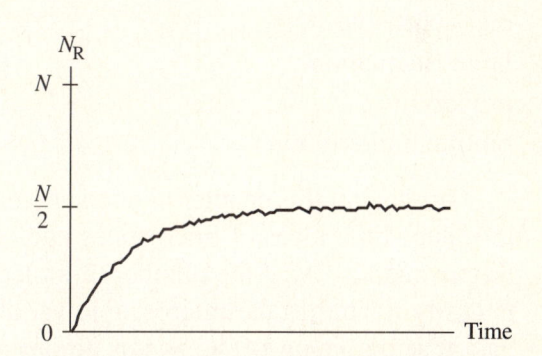

FIGURE 21-7 The evolution from a nonequilibrium $N_R = 0$ state to the equilibrium state $N_R = N/2$. Notice the small fluctuations in the value of N_R.

If an *isolated* system is in a significantly nonrandom or ordered state, it will change in time (evolve) until it reaches the most random or most probable state where it will then be in equilibrium. There is, in other words, a general tendency for isolated systems to evolve such as to turn order into disorder. The word *isolated* is important, however, because an outside intervention could prevent this from happening or even, as in the case of the air pump that prepared the initial $N_R = 0$ state, could generate order from disorder. But left to itself, a system inevitably loses order and becomes more random and disordered.

This is a topic we will return to explore in the next chapter, where it will form the idea behind the second law of thermodynamics. But one more thing is worth noting at this point. If you were shown the pictures of Fig. 21-6 and told that they were both made of an isolated system, could you place them in the proper time order? That is, which occurred earlier in time and which later? This is not difficult—Fig. 21-6a "clearly" was an initial state that evolved toward Fig. 21-6b. You have never seen a system spontaneously move itself into one half of its container—the issue, again, of how incredibly unlikely such an event would be—so the system would never go from Fig. 21-6b to Fig. 21-6a on its own.

This implies that, somehow, the system knows which direction in time is "future" and which is "past." That information is *not* contained in Newton's laws, where particles can move backward as easily as forward and we cannot discern a past or a future. That is, two pictures of a pair of interacting particles cannot be time-ordered as we did Fig. 21-6. Macroscopic systems have a built-in "arrow of time" that is not present in the microscopic physics. How? This is one of the most profound questions in physics, and one that we will address further in Chapter 22.

Summary

Important Concepts and Terms

statistical mechanics
microstate
configuration
most probable state
large fluctuation
probability
binomial distribution

Stirling approximation
macroscopic fluctuation
fractional variation
disordered
ordered
equilibrium

Our goal in this chapter has been to make an explicit micro/macro connection. We first introduced the ideas of microstates and configurations, and we then introduced a *macroscopic* quantity N_R—the number of molecules in the right side of a container of gas—that is easily related to the microscopic variables. We noticed, by analyzing "movies" of the gas, that the value of N_R is not a constant but, instead, fluctuates about some average value. We could also observe from the movies that the relative importance of the fluctuations decreases as the number of molecules N increases.

The mathematical analysis of this chapter has been for the purpose of quantifying these observations. In particular, what is the average value $\overline{N_R}$? How large are the likely fluctuations σ_N? These are statistical questions, because the molecules are moving randomly, and we had to introduce some tools from probability theory to answer these questions. For a gas of N particles, each of which has probability p of being in the right side of the container, the most probable state is

$$\overline{N_R} = pN.$$

Other values of N_R can occur with a probability given by the binomial distribution:

$$P(N_R) = \frac{N!}{N_R!(N-N_R)!} \cdot p^{N_R} \cdot q^{N-N_R}.$$

The probability is ≈2/3 that N_R will be found in the range $\overline{N_R} \pm \sigma_N$, where

$$\sigma_N = \sqrt{Npq}.$$

Fluctuations of size σ_N are to be expected, but fluctuations larger than this rapidly decrease in probability. The *relative importance* of the fluctuations decreases as N

increases. In particular, the fractional variation of N_R in an equally divided container is

$$f = \frac{\sigma_N}{\overline{N_R}} = \frac{1}{\sqrt{N}}$$

For a macroscopic system, the fractional variation is so small as to be insignificant. For any practical purpose, we can consider the value of N_R for a macroscopic system to be *constant* and equal to $\overline{N_R}$. Large fluctuations are so remotely unlikely that we can safely say they never occur.

If we generalize from this analysis, we have established the following:

1. The macroscopic state variables of a system can be defined in terms of the microstates and configurations. This is the micro/macro connection. We still have to make the appropriate definitions for other state variables, such as temperature or pressure, and we will do so in coming chapters.

2. The state variables are not perfectly constant but, like N_R, fluctuate around some average value.

3. The relative importance of the fluctuations decreases as $1/\sqrt{N}$. For macroscopic systems, the fluctuations are completely insignificant and undetectable. For any practical purpose, the values of the state variables are constant and equal to their average values. This is the really important conclusion, because without this the micro/macro connection would not have much significance.

4. *Equilibrium* corresponds to the most probable state of the system or, equivalently, the tallest peak in the histogram at $(N_R)_{eq} = \overline{N_R}$. Whereas nonequilibrium states can be specially prepared by outside influences, a isolated system will always evolve toward the equilibrium state. This is simply a probabilistic statement, because the system is vastly more likely to be found within $\pm \sigma_N$ of $\overline{N_R}$ than it is to remain in a nonequilibrium state. The net result, for an isolated system, is that order evolves toward disorder and randomness. This idea will form the basis for the second law of thermodynamics.

These ways of thinking about and doing physics are very different from anything you have done before, and they will take getting used to. Nonetheless, they lead to results of great significance in physics and in the application of physics to engineering and other sciences. In the next chapter we will begin to apply these ideas to more important state variables.

Exercises and Problems

▲ Exercises

1. If you toss 10 coins, what is the probability that 3 of them will be heads? That 5 of them will be heads?
2. For a toss of 6 coins, calculate the probabilities of 0, 1, 2, 3, 4, 5, and 6 heads. Then make a histogram of your results.
3. Five boxes each contain 16 red balls and 4 blue balls. If you randomly draw one ball from each box,

a. What is the probability of drawing 5 red balls?
b. What is the probability of drawing 5 blue balls?
c. What is the probability of drawing 4 red balls and 1 blue ball?

(**Hint:** Let N_R represent the number of red balls.)

4. For a 3000-molecule gas,
   a. What is the most likely number of molecules in the right one-third of the box?
   b. What is the likely size of macroscopic fluctuations in this number?
   c. What is the fractional variation of the fluctuations?

5. Compute 12! both exactly and by using the Stirling approximation.

6. Compute 1,000,000! Give your answer both as a power of e and as a power of 10.

Problems

7. a. List all possible outcomes for rolling two dice. (That is, list all possible pairs of numbers.)
   b. Determine, from your list, the probabilities for each possible *total* of the two dice, from 2 to 12.
   c. What is the sum of the probabilities you found in part b)?
   d. What is the probability of throwing a 7 *or* an 11?
   e. What is the probability of throwing a 12 three times in a row?
   f. If you roll the dice three times, what is the probability that you will *not* roll a 7?
   g. How many times must you roll the dice for the chances to be ≥50% that at least one roll will be a 7?

8. a. List all the configurations for a $N = 5$ gas that can be distributed in either the left or right half of a container.
   b. Determine, from your list, the number of configurations $C(N_R)$ for each possible value of N_R.
   c. Determine, from your table, the probability for each of the possible states of this gas. Then put your results from a) – c) together in a single table, similar to Table 21-1.
   d. Calculate, using the binomial distribution, the probabilities of occurrence of each possible state of this gas.
   e. Do your answers for parts c) and d) agree?
   f. What is $\overline{N_R}$ for this gas?
   g. Make a histogram, similar to Fig. 21-3, for the probabilities of each state.
   h. Estimate, from your histogram, the value of σ_N. Explain your reasoning!
   i. What is the calculated value for σ_N?

9. Consider a gas of 32 molecules in a container.
   a. If you observe one particular molecule, what is the probability that it will be found in the right three-quarters of the container?
   b. What is the most probable number of molecules in the right three-quarters of the container?
   c. What is the probability that, at any instant of time, $N_R = \overline{N_R}$?
   d. Calculate $P(N_R)$ for *even* values of N_R from $N_R = 16$ to $N_R = 32$.

e. Make a histogram of $P(N_R)$ versus N_R, from 0 to 32.
f. Estimate the value of σ_N from your histogram. Keep in mind that you have only graphed every other bar.
g. What is the calculated value for σ_N?
h. What is the range of values for N_R that you would be likely to observe?
i. What is the probability that a random observation of the container would find *no* molecules in the right three-quarters?
j. If you were to make 100 observations each second, how many years would you expect to wait before seeing the fluctuation of part i)?

10. Consider a gas of 10,000 molecules in a container.
    a. Calculate the probabilities that you will find N_R = 4900, 4925, 4950, 4975, 5000, 5025, 5050, 5075, and 5100 molecules in the right half of the container. (**Hint:** Use the symmetry of the histogram about $N/2$ to avoid duplicate calculations.)
    b. Make a histogram of $P(N_R)$ versus N_R.
    c. Estimate σ_N from your histogram.
    d. Calculate σ_N.
    e. What is the fractional variation of N_R in this gas?

[**Estimated 10 additional problems for the final edition.**]

Chapter 22

The Kinetic Theory of Gases

LOOKING BACK Sections 8.2; 19.5; 19.6; 20.6; 21.6

22.1 The Micro/Macro Connection

A gas consists of a vast number of molecules ceaselessly colliding with one another and the walls of their container as they move about. These collisions exert forces on the walls, and they continuously redistribute and rearrange the microscopic thermal energy among all the molecules in the gas. Our goal in this chapter is to show how this turmoil and change at the molecular level give rise to predictable and unchanging values of macroscopic variables such as pressure, temperature, and heat capacity.

This micro/macro connection is called the **kinetic theory of gases**. Chapter 21 was an introduction to the statistical ideas that underlie the kinetic theory of gases. Three major ideas were developed there. First, macroscopic parameters, such as N_R, can be computed by averaging over the many possible microstates of a system. Second, a system evolves, by random processes, toward its most probable configuration, which is called *equilibrium*. And third, the equilibrium values of the macroscopic parameters fluctuate about their average values but, for large N, the fluctuations are so small as to be meaningless.

Now we want to use these basic ideas to describe more interesting macroscopic parameters, such as pressure and temperature. We also want to explore the concept of *heat* in more detail and to show what happens when two systems, at different temperatures, interact thermally.

22.2 A Microscopic View of Pressure

Why does a gas have pressure? That is, why does it exert forces on the walls of its container? All the many molecules are colliding with the walls, and each time a collision occurs a molecule exerts an *impulsive force* on the wall. Though the force due to one such

A MICROSCOPIC VIEW OF PRESSURE

collision may be unmeasurably tiny, the steady "rain" of vast numbers of molecules on the walls exerts a measurable macroscopic force. The gas pressure is the force per unit area ($p = F/A$) resulting from these molecular collisions.

Our main task in this section is to calculate the pressure by doing the appropriate averaging over molecular motions and collisions. We will begin by finding how much force a single molecule exerts on the wall during a collision, summing this over all the collisions that occur, and then dividing the force by the surface area to find the pressure.

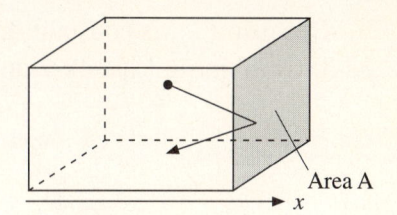

FIGURE 22-1 A molecule colliding with the wall exerts an impulse J on it.

Consider a molecule approaching a wall along the x-axis with an x-component of velocity v_x, as shown in Fig. 22-1. Because the collision with the wall is elastic, the molecule will rebound from the wall with its x-component of velocity changed from $+v_x$ to $-v_x$. This molecule experiences an impulse given by the impulse-momentum theorem as

$$J_{\text{molecule}} = \Delta p = m(-v_x) - mv_x = -2mv_x. \tag{22-1}$$

According to Newton's third law the wall experiences an equal but opposite impulse. Thus

$$J_{\text{wall}} = +2mv_x \tag{22-2}$$

as a result of this single collision.

Now impulse was defined as

$$J = \int_0^{\Delta t_{\text{coll}}} F(t)\,dt, \tag{22-3}$$

where $F(t)$ is the time-varying force the molecule exerts on the wall during the collision and Δt_{coll} is the duration of the collisions. A typical collision force is shown in Fig. 22-2a. Suppose we replace the actual collision force with an *average* force F_{avg} defined, in Fig. 22-2b, such that the area under the curve, or the impulse, is the same as for the true force. This average force imparts the correct momentum change to the molecule during the collision.

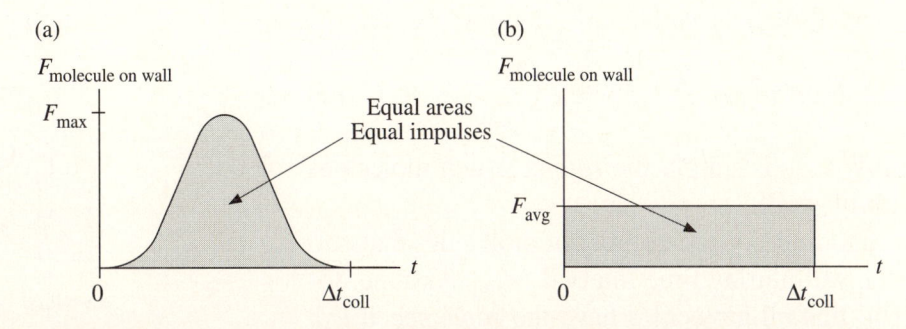

FIGURE 22-2 a) The time-varying force of the real collision. b) A constant *average* force with the same duration and exerting the same impulse as the real force.

88 CHAPTER 22 THE KINETIC THEORY OF GASES

Because F_{avg} is constant during the collision, it can be taken outside the integral in Eq. 22-3. Combining Eqs. 22-2 and 22-3 gives

$$J_{wall} = \int_0^{\Delta t_{coll}} F_{avg}\, dt = F_{avg} \cdot \Delta t_{coll} = +2mv_x$$

$$\Rightarrow F_{avg} = \frac{2mv_x}{\Delta t_{coll}}.$$

(22-4)

In other words, the *average* force exerted on the wall by one molecule during the time of the collision is just the magnitude of the molecule's momentum change divided by Δt_{coll}.

EXAMPLE 22-1 Estimate the average force exerted on a wall during the collision of a room-temperature nitrogen molecule.

SOLUTION The mass of one molecule is $m = 28$ u $= 4.68 \times 10^{-26}$ kg. We found in Chapter 19 that room-temperature nitrogen molecules have a typical speed $v_x \approx 500$ m/s. Molecular diameters are $\approx 10^{-10}$ m. For estimate purposes, let's assume that a collision causes a molecule to be "flattened" against the wall by 25%. Thus the molecule moves through a total distance—in and out—of $\approx 0.5 \times 10^{-10}$ m at a speed of ≈ 500 m/s. The duration of the collision is then

$$\Delta t_{coll} = \frac{\Delta x}{v} \approx \frac{0.5 \times 10^{-10} \text{ m}}{500 \text{ m/s}} = 10^{-13} \text{ s}.$$

Using these values in Eq. 22-4, we get:

$$F_{avg} = \frac{2mv_x}{\Delta t_{coll}} \approx 5 \times 10^{-10} \text{ N}.$$

The duration of a collision is extremely short, and the force exerted by one molecule is very tiny. But there are a very large number of molecules all colliding simultaneously. The net force exerted on the wall by *all* the molecules is the average force per molecule, given by Eq. 22-4, multiplied by the total number ΔN of such collisions that occur during the time interval Δt_{coll}. That is,

$$F_{net} = F_{avg} \cdot \Delta N = 2mv_x \frac{\Delta N}{\Delta t_{coll}}.$$

(22-5)

The ratio $\Delta N/\Delta t_{coll}$ is simply the *rate* at which molecules collide with the wall.

We need to find ΔN, the number of molecules that collide with the wall during time interval Δt_{coll}. Assume, for the moment, that all molecules have the *same* speed $|v_x|$. During the interval Δt_{coll} all will move a distance $\Delta x = |v_x|\Delta t_{coll}$ along the x-axis. Figure 22-3 shows the end of the box near the wall with a section of length Δx shaded. *Every* one of the molecules in this shaded region that is moving to the right will reach and collide with the wall

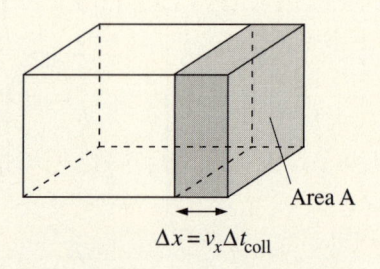

FIGURE 22-3 Every molecule in the shaded volume that is moving to the right will collide with the wall during interval Δt_{coll}.

during time Δt_{coll}. Molecules outside this region will not reach the wall during Δt_{coll}, because they are not moving fast enough. This shaded region has a volume $V = A\Delta x$, where A is the surface area of the wall. The number of molecules in this volume is

$$N = dV = dA\Delta x = dAv_x\Delta t_{\text{coll}}, \tag{22-6}$$

where d is the number density. However, we must keep in mind that only half of the molecules are moving to the right while the other half are moving to the left. The number of collisions ΔN during Δt_{coll} will be just half of N, or

$$\Delta N = \tfrac{1}{2}dAv_x\Delta t_{\text{coll}}. \tag{22-7}$$

The *rate* of molecular collisions with a wall of area A is thus

$$\text{rate of collisions} = \frac{\Delta N}{\Delta t_{\text{coll}}} = \tfrac{1}{2}dAv_x. \tag{22-8}$$

Now that we know the rate of collisions with the wall, we can incorporate this result in Eq. 22-5 to find the net force on the wall:

$$F_{\text{net}} = 2mv_x \cdot \tfrac{1}{2}dAv_x = dmv_x^2 A. \tag{22-9}$$

We made the assumption in reaching Eq. 22-9 that collisions with other molecules are negligible during interval Δt_{coll}. This is quite reasonable because Δx is much less than the mean spacing between molecules in a dilute gas. We also assumed that all molecules had the same speed. This is a condition we need to relax if our result for the force is to apply to real gases. To correct for this, we need to replace the squared velocity v_x^2 in Eq. 22-9 with the average value $\overline{v_x^2}$. That is,

$$F_{\text{net}} = dm\overline{v_x^2}A, \tag{22-10}$$

where $\overline{v_x^2}$ is the quantity v_x^2 averaged over all molecules in the container.

Pressure, you will recall, is defined as $p = F/A$. Thus the pressure exerted on the wall of the container by the gas molecules is

$$p = \frac{F_{\text{net}}}{A} = dm\overline{v_x^2}. \tag{22-11}$$

Equation 22-11 gives an expression for the pressure of a gas in terms of v_x, the x-component of the velocity. Now the coordinate system is something that *we* impose on the problem. Nothing about the situation distinguishes one axis or one direction in space from any other, so it must be the case that

$$\overline{v_x^2} = \overline{v_y^2} = \overline{v_z^2}. \tag{22-12}$$

Because the total velocity squared is $v^2 = v_x^2 + v_y^2 + v_z^2$, we can use Eq. 22-12 to write

$$\overline{v^2} = \overline{v_x^2} + \overline{v_y^2} + \overline{v_z^2} = 3\overline{v_x^2}$$
$$\Rightarrow \overline{v_x^2} = \tfrac{1}{3}\overline{v^2}. \tag{22-13}$$

This result for $\overline{v_x^2}$ in terms of $\overline{v^2}$ allows us to write Eq. 22-11 for the pressure as

$$p = \tfrac{1}{3} dm\overline{v^2} = \tfrac{2}{3}\frac{N}{V}\left(\tfrac{1}{2} m\overline{v^2}\right), \tag{22-14}$$

where we have used the definition $d = N/V$.

We have succeeded in our goal of finding an expression for the pressure, a macroscopic parameter, in terms of the microscopic physics. Note that the pressure depends on the density of molecules in the container, on how fast the molecules are moving, and on the molecular mass. Strictly speaking, Eq. 22-14 is the *average* pressure because it depends on averaging v^2 for all the molecules in the gas. As we found in Chapter 21 for N_R, the actual pressure will fluctuate about this average value. But for any realistic gas the relative size of the fluctuations will be so small as to be unmeasurable. We can, for any practical purpose, consider the value of p to be perfectly steady.

22.3 A Microscopic View of Temperature

Notice that the expression $\tfrac{1}{2} m\overline{v^2}$ in Eq. 22-14 is simply the average kinetic energy of a molecule in the gas. It will be convenient to define the average energy per molecule as

$$\overline{\varepsilon} = \text{average kinetic energy per molecule} = \frac{K_{\text{int}}}{N} = \tfrac{1}{2} m\overline{v^2}, \tag{22-15}$$

where $K_{\text{int}} = N\overline{\varepsilon}$ is the total kinetic energy of all the molecules in the gas. Thus Eq. 22-14 can be written

$$p = \tfrac{2}{3}\frac{K_{\text{int}}}{V}$$
$$\Rightarrow K_{\text{int}} = \tfrac{3}{2} pV. \tag{22-16}$$

But we also know, from the ideal gas law, that

$$pV = nRT = NkT,$$

from which we reach the significant conclusion that the microscopic kinetic energy of a gas is

$$K_{\text{int}} = \tfrac{3}{2} nRT = \tfrac{3}{2} NkT. \tag{22-17}$$

Finally, using Eq. 22-17 in Eq. 22-15, we can write the average kinetic energy per molecule as

$$\overline{\varepsilon} = \frac{K_{\text{int}}}{N} = \tfrac{3}{2} kT. \tag{22-18}$$

This is an especially satisfying discovery because it gives new meaning to the idea of temperature: Temperature is a state variable that measures the average energy per molecule in a gas. Thus the thing we call *temperature* is a direct measurement of the energy stored in the microscopic motions of the molecules in a system. The reading of a thermometer is determined by collisions of the system's molecules against it. A higher temperature corresponds to a larger value of $\overline{\varepsilon}$ and thus to higher molecular speeds and more energetic collisions, causing the thermometer's reading to increase. This concept of temperature also gives meaning to *absolute zero* as the temperature at which $\overline{\varepsilon} = 0$ and all molecular motion ceases. (Quantum effects at very low temperatures prevent the motions from actually stopping, which would violate the Heisenberg uncertainty principle.)

Note that the average energy per molecule depends *only* on the temperature, not on the molecule's mass. If two gases have the same temperature, their molecules will have the same average energy even if the two gases are different. This will be an important idea when we look at the thermal interaction between two systems.

EXAMPLE 22-2 A 100 cm³ container holds a gas at 1 atm pressure and 20°C. a) What is the average kinetic energy of a molecule in the gas? b) What is the total microscopic kinetic energy of the gas?

SOLUTION a) First we need to convert the temperature of the gas to the Kelvin scale: 20°C = 293 K. Now we can use Eq. 22-18 to find the average kinetic energy per molecule:

$$\bar{\varepsilon} = \tfrac{3}{2}kT = \tfrac{3}{2}(1.38 \times 10^{-23} \text{ J/K})(293 \text{ K}) = 6.06 \times 10^{-21} \text{ J}.$$

b) The number of molecules in the container is found from the ideal gas law to be

$$N = \frac{pV}{kT} = \frac{(1.013 \times 10^5 \text{ Pa})(10^{-4} \text{ m}^3)}{(1.38 \times 10^{-23} \text{ J})(293 \text{ K})} = 2.50 \times 10^{21}.$$

The total microscopic kinetic energy is thus

$$K_{\text{int}} = N\bar{\varepsilon} = 15.2 \text{ J}.$$

The total energy is significant even though the energy per molecule is extremely small.

22.4 Molecular Speeds and Molecular Collisions

In the previous two sections we showed that the macroscopic properties of pressure and temperature depend on the microscopic motion and kinetic energy of the gas molecules. In this section, we will take a closer look at the motion and collisions of the molecules within the gas.

The Root-Mean-Square Velocity

How fast do molecules move about in a gas? You saw experimental evidence, in Chapter 19, that the typical molecular speed in room-temperature nitrogen is $v \approx 500$ m/s. How does this speed depend on the temperature or on the specific type of molecule? To answer this question, you might think of computing the average velocity of the molecules in the gas. But velocity is a *signed* quantity. Half the molecules in a container have a positive velocity and the other half are negative. Thus the *average* velocity is $v_{\text{avg}} = 0$. (If this weren't true, the entire container of gas would move away!)

It is possible to average the speeds $|\vec{v}|$ of all the molecules, but that calculation turns out to be difficult. Instead, scientists prefer to use the **root-mean-square velocity** v_{rms}, which is defined as

$$v_{\text{rms}} = \sqrt{\overline{v^2}} = \text{ root mean square velocity.} \tag{22-19}$$

This quantity is usually called the "rms velocity." You can remember its definition by noting that its name is the *opposite* of the sequence of operations: First you square all the velocities, then you average the squares (find the mean), and then you take the square root.

CHAPTER 22 THE KINETIC THEORY OF GASES

Because the square root "undoes" the square, v_{rms} must, in some sense, give an "average" velocity. This time, however, the average is nonzero because you are averaging squared, and thus positive, quantities. It turns out that v_{rms} differs from the average speed $|\vec{v}|_{avg}$ by less than 10%. So for practical purposes we can interpret v_{rms} as being essentially the average speed of a molecule in a gas.

The root-mean-square velocity is preferred over the average speed because it tends to arise naturally in calculations. By combining Eqs. 22-15 and 22-18 we can find an explicit expression for v_{rms}:

$$\tfrac{1}{2} m \overline{v^2} = \tfrac{3}{2} kT$$

$$\Rightarrow v_{rms} = \sqrt{\overline{v^2}} = \sqrt{\frac{3kT}{m}}. \tag{22-20}$$

Now you can see that the average speed of a molecule in a gas depends on the square root of the temperature and inversely on the square root of the mass.

EXAMPLE 22-3 Figure 22-4 shows the velocities of all the molecules in a 6-molecule two-dimensional gas. Calculate and compare both v_{rms} and the average speed $|\vec{v}|_{avg}$.

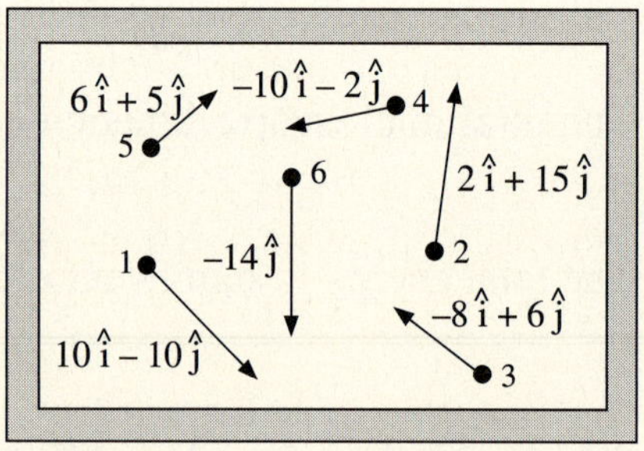

SOLUTION Table 22-1 shows the velocity components v_x and v_y for each molecule, the squares v_x^2 and v_y^2, their sum $v^2 = v_x^2 + v_y^2$, and the speed $|\vec{v}| = (v_x^2 + v_y^2)^{1/2}$. Averages of all the values in each column are shown at the bottom. You can see that the average speed is $|\vec{v}|_{avg} = 11.9$ m/s. The rms velocity is

FIGURE 22-4 The molecular velocities of Example 22-3. Units are m/s.

$$v_{rms} = \sqrt{\overline{v^2}} = \sqrt{148.3 \text{ m}^2/\text{s}^2} = 12.2 \text{ m}/\text{s}.$$

The rms velocity is larger than the average speed by only 2.5%. Notice that the average values of v_x and v_y are zero, as expected.

TABLE 22-1 Calculation of rms velocity and average speed for the molecules of Example 22-3.

Molecule	v_x	v_y	v_x^2	v_y^2	v^2	$|\vec{v}|$
1	10	−10	100	100	200	14.1
2	2	15	4	225	229	15.1
3	−8	6	64	36	100	10.0
4	−10	−2	100	4	104	10.2
5	6	5	36	25	61	7.8
6	0	−14	0	196	196	14.0
Average:	0	0			148.3	11.9

EXAMPLE 22-4 What is v_{rms} for molecular nitrogen at room temperature?

SOLUTION The molecular mass is $m = 28$ u $= 4.68 \times 10^{-26}$ kg, and we will assume that $T = 20°C = 293$ K. It is then a simple calculation to find

$$v_{rms} = \sqrt{\frac{3(1.38 \times 10^{-23} \text{ J/K})(293 \text{ K})}{4.68 \times 10^{-26} \text{ kg}}} = 509 \text{ m/s}.$$

Some speeds will be greater than this and others less, but 509 m/s will be a typical or fairly average speed. This is in excellent agreement with the experimental results of Fig. 19-11.

Notice how fast molecular speeds are. We have no sensation that the air molecules in the room are moving about this quickly.

EXAMPLE 22-5 It is possible to "cool" atoms by letting them interact with a laser beam under the proper, carefully controlled conditions. The atoms' kinetic energy is transferred to the energy of photons they radiate away. Laser cooling is currently a subject of intense research activity, and it is now possible to cool a dilute gas of atoms to a temperature less than one *micro*kelvin! (The atoms are kept from solidifying by their extremely low density.) Various novel quantum effects appear at these incredibly low temperatures. What is v_{rms} for a cesium atom at a temperature of 1 μK?

SOLUTION Referring to the periodic table of the elements shows that the mass of cesium is $m = 133$ u $= 2.22 \times 10^{-25}$ kg. At $T = 1$ μK $= 1 \times 10^{-6}$ K the rms velocity is found to be $v_{rms} = 1.37 \times 10^{-2}$ m/s $= 1.37$ cm/s. This is slow enough to "watch" the atoms moving about!

Equation 20-20 is a theoretical prediction, a consequence of the statistical theory of gases. The predictions of molecular speeds, which range over a factor of a million, are all amply confirmed by experimental measurement.

The Mean Free Path

Atoms and molecules have a finite size and thus collide with each other as well as with the walls. If you were to follow the motion of one molecule, you would see it moving along a very zig-zag course—straight line motion for a short distance followed by a "bend" where it undergoes a collision with another molecule. Figure 22-5 illustrates this idea. You can easily see that the straight-line segments between collisions are of unequal lengths, a consequence of the random distribution of the molecules. A question we could ask is, "What is the *average* distance between collisions?" This average distance between collisions is

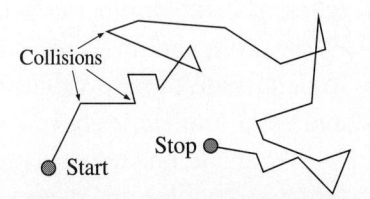

FIGURE 22-5 A molecule follows a zig-zag path through the gas as it collides with other molecules.

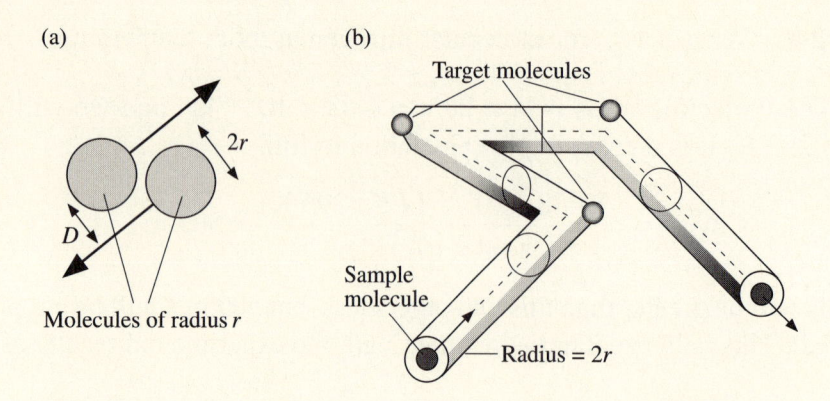

FIGURE 22-6 a) Two molecules of radius r will collide if the distance D between their centers is less than $2r$. b) A sample molecule will collide with all target molecules whose centers are within a cylinder of radius $2r$ centered on the path of the sample molecule. The "bent cylinder" has a total length L.

called the **mean free path** of the molecule. The concept of mean free path is widely used, not only in gases but also to talk about electrons moving through semiconductors or light passing through a medium that scatters the photons.

Figure 22-6a shows two molecules approaching each other. We will make the simplifying assumption that the molecules are spherical and of radius r. We will also continue our ideal gas assumption that the molecules undergo "hard sphere" collisions as do billiard balls, with no interaction before they touch. In that case, we see that the molecules will collide if the distance D between their *centers* is $D < 2r$ and they will miss if $D > 2r$.

Consider a cylindrical tube of radius $2r$ around the trajectory of a "sample" molecule. Figure 22-6b shows a length L of this cylinder. As you just saw, any "target" molecules whose centers are located within the cylinder are going to have a collision with the sample molecule. Thus the number N of collisions is just the number of molecules in a length L of this cylinder. If the gas's number density is d, then this number is

$$N = d \cdot V = d \cdot AL = d \cdot \pi(2r)^2 L = 4\pi dr^2 L. \tag{22-21}$$

The mean free path λ, which is the average distance between collisions, is then

$$\lambda = \frac{L}{N} = \frac{1}{4\pi dr^2}, \tag{22-22}$$

where d is the number density of the gas (molecules per m^3) and r is the molecular radius. Every atom and molecule has a different radius, and chemical measurements are necessary to determine precise values. As a reasonable rule of thumb, atoms have $r \approx 0.5 \times 10^{-10}$ m and small molecules have $r \approx 1.0 \times 10^{-10}$ m.

We made one tacit assumption in this derivation that is not really valid—that all of the target molecules are at rest. Although our general idea is correct, a more careful calculation in which all molecules are moving introduces an extra factor of $\sqrt{2}$, giving

$$\lambda = \frac{1}{4\sqrt{2}\pi dr^2} \quad \text{(corrected version)}. \tag{22-23}$$

This result is generally in good agreement with experimental measurements.

Because we know that the rms velocity is essentially the average molecular speed, we can find the **mean time between collisions** as the time to travel distance λ at speed v_{rms}. This is

$$\tau = \text{mean time between collisions} = \frac{\lambda}{v_{rms}}. \quad (22\text{-}24)$$

Lastly, the inverse of τ is going to be the average number of collisions per second or, equivalently, the **collision frequency**:

$$f_{coll} = \text{collision frequency} = \frac{1}{\tau} = \frac{v_{rms}}{\lambda}. \quad (22\text{-}25)$$

EXAMPLE 22-6 What are the a) mean free path, b) mean time between collisions, and c) collision frequency for a nitrogen molecule at 1 atm pressure and room temperature?

SOLUTION a) Nitrogen is a small molecule, so we can assume that $r \approx 1.0 \times 10^{-10}$ m. We can use the ideal gas law to determine the density:

$$d = \frac{N}{V} = \frac{p}{kT} = 2.50 \times 10^{25} \text{ m}^{-3},$$

where we have used $T = 293$ K and $p = 1$ atm $= 101{,}300$ Pa. From Eq. 22-23 we find the mean free path to be $\lambda = 2.25 \times 10^{-7}$ m $= 225$ nm. You may recall from Chapter 19 that the average separation between molecules is ≈ 2 nm. It seems that any given molecule can slip between its neighbors, which are spread out in three dimensions, and travel—on average—about 100 times the average spacing before it collides with another molecule.

b and c) We calculated $v_{rms} = 509$ m/s for a nitrogen molecule in Example 22-4. That is all we need to find $\tau = \lambda/v_{rms} = 4.42 \times 10^{-10}$ s. The inverse of this gives $f_{coll} = 2.26 \times 10^9$ collisions per second. The results of this calculation are astonishing! The molecules in the air around us are colliding with their neighbors about two *billion* times every second, managing to move, on average, only about 225 nm between collisions—50% of one wavelength of light. The frequency of collisions helps us better to understand how easily systems of particles exchange energy via molecular collisions.

EXAMPLE 22-7 At what pressure will a room temperature helium atom have a mean free path of 1 m?

SOLUTION A helium atom has a radius $r \approx 0.5 \times 10^{-10}$ m. From Eq. 22-23 we can calculate that the mean free path will be 1 m at a gas density $d = 2.25 \times 10^{19}$ m^{-3}. While the required density is temperature independent, the corresponding pressure is not. From the ideal gas law,

$$p = \frac{N}{V}kT = dkT = 9.10 \times 10^{-2} \text{ Pa} = 9.01 \times 10^{-7} \text{ atm} = 6.84 \times 10^{-4} \text{ mm of Hg}.$$

This pressure is not especially low by the standards of today's vacuum technology. Pressures of 10^{-8} mm are routinely generated in science and engineering laboratories, and semiconductor manufacturing and processing can require pressures of 10^{-10} mm. Mean free paths at such pressures are $\approx 2 \times 10^6$ m $\approx$ 1000 miles! This implies that the gas molecules remaining in a vacuum chamber at this pressure bounce back and forth off the walls but only collide with other molecules on extremely rare occasions.

22.5 Thermal Energy

We previously defined the thermal energy of a system to be

$$E_{th} = K_{int} + U_{int},$$

where K_{int} and U_{int} are the microscopic kinetic energy and the potential energy associated with stretching and compressing molecular bonds. The molecules in an ideal gas have no molecular bonds with their neighbors, so $U_{int} = 0$ and we have simply

$$E_{th} = K_{int} = \varepsilon_1 + \varepsilon_2 + \varepsilon_3 + \varepsilon_4 + \cdots + \varepsilon_N = N\bar{\varepsilon}, \quad (22\text{-}26)$$

where ε_i is the kinetic energy of molecule i.

The total thermal energy E_{th} of an isolated system remains constant, to conserve energy, but the individual molecular energies ε_i are constantly changing due to collisions between the molecules. It is convenient to say that the total available energy E_{th} is *distributed* among the N molecules. The situation is similar to the problem we studied in Chapter 21 of distributing N molecules into the two sides of a box. Although it is possible that nearly all the particles will be in one side of the box, with very few in the other, we found that situation to be incredibly unlikely when N is large. It is overwhelmingly probable that each side will have the same density. Likewise with distributing energy among molecules. You could imagine that just a handful of molecules take nearly all the energy, zooming around at high speed, while most of the molecules are nearly at rest. Possible, but incredibly unlikely. Instead, it is overwhelmingly probable that the energy will be shared nearly equally among all molecules. That is, each molecule's energy will be approximately the average energy per molecule ($\varepsilon_i \approx \bar{\varepsilon}$), and none will have either exceptionally small or exceptionally large energies.

You saw evidence for this conclusion in Chapter 19 in the measured distribution of the speeds of nitrogen molecules. Figure 22-7 shows the data again, this time with the calculated value of v_{rms} indicated. Previously we only noted the average speed, but now you should look at the *distribution* of speeds. The large majority of molecules have speeds within a factor of 2 of average, and essentially no molecules have exceptionally small or exceptionally large speeds. Figure 22-7 confirms that all the molecules have roughly equal energies.

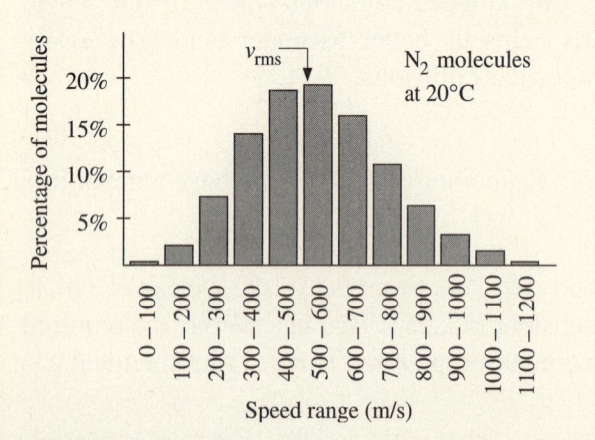

FIGURE 22-7 The distribution of molecular speeds is a consequence of the distribution of molecular energies. Nearly all molecules have a speed within a factor of 2 of v_{rms}.

This distribution of energies, and speeds, comes about by the continual collisions of molecules. On average, a collision between two molecules speeds up the slower one and slows down the faster one. Thus collisions have an equalizing effect on the energies, and after many collisions all the molecules have nearly equal energies. A statistical analysis of the collisions, similar to the analysis we did in Chapter 21, can predict what fraction of the molecules have each speed. The prediction is in perfect agreement with the experimental data. Figure 22-7 is the *most probable distribution* of the thermal energy among the molecules, and thus it represents the equilibrium state of the gas at temperature $T = 20°C$.

The Equipartition Theorem

We have made a tacit assumption that the energy of a molecule consists exclusively of its kinetic energy of motion back and forth in the container. This form of molecular energy is called **translational kinetic energy**. However, it is also possible for a molecule to have kinetic energy if it rotates end over end, like a dumbbell, while staying in one place. This energy is the molecule's **rotational kinetic energy**. A *monatomic gas*, such as He, Ne, or Ar, has no rotational kinetic energy, so our analysis to this point is valid. But we need to consider the contribution of rotational kinetic energy to the thermal energy if we wish to understand a *diatomic gas* such as H_2, N_2, or O_2.

For a monatomic gas, the kinetic energy of a molecule is

$$\varepsilon = \tfrac{1}{2}mv^2 = \tfrac{1}{2}mv_x^2 + \tfrac{1}{2}mv_y^2 + \tfrac{1}{2}mv_z^2 = \varepsilon_x + \varepsilon_y + \varepsilon_z, \qquad (22\text{-}27)$$

where we have written separately the energy associated with translational motion along the three axes. Each of these is a separate and independent *mode* of energy. On average, because of collisions, each of the modes has the same energy. If you fill a container of gas by shooting all the molecules in along the *x*-axis, collisions will quickly change the velocities until the average values of ε_x, ε_y, and ε_z are equal. Because the total energy of a monatomic gas is $E_{th} = K_{int} = \tfrac{3}{2}NkT$, as given by Eq. 22-17, the energy associated with each of the three modes of translational kinetic energy is $\tfrac{1}{2}NkT$.

Figure 22-8 shows a diatomic molecule, such as molecular nitrogen N_2, oriented along the *z*-axis. The molecule can rotate end over end about either the *x*-axis or the *y*-axis, as shown. These rotations have kinetic energy because the atoms are in motion. The two rotations shown, with kinetic energies $\varepsilon_{rot\,x}$ and $\varepsilon_{rot\,y}$, are distinct modes of molecular energy. For a diatomic molecule, however, rotation about the *z*-axis (the molecular axis) does *not* contain kinetic energy. The reason is that, unlike rotation about the other two axes, the atoms are not moving end over end. So a diatomic molecule has two modes of rotational kinetic energy in addition to three modes of translational kinetic energy.

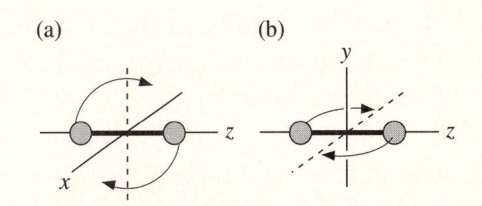

FIGURE 22-8 A diatomic molecule has rotational kinetic energy for rotation about a) the *x*-axis and b) the *y*-axis.

In a gas of diatomic molecules, collisions can change either a molecule's speed *or* its rotation. An off-center collision can cause the molecule to spin or, if it's already spinning, can perhaps slow down the spin. If all the molecules are initially spinning slowly, the collisions will transfer translational kinetic energy into rotational kinetic energy. Conversely, if all the molecules are rotating very rapidly their collisions will tend to slow the rotations while batting other molecules away with increased speeds and increased translational kinetic energy. On average, collisions will equalize the energies of the translational energy modes and the rotational energy modes. It is remotely unlikely, because of the constant rapid collisions, that any one energy mode will have energy significantly different from those of the other modes.

This qualitative discussion can be made more precise with a statistical analysis. The result is known as the **equipartition theorem**, meaning that the energy is "equally divided." The theorem is stated as follows:

> *Equipartition theorem*: The thermal energy of a system of particles is equally divided among all the possible energy modes. For a system of N particles at temperature T, the total energy of each mode is $\frac{1}{2}NkT$.

A monatomic gas has three modes and thus $E_{th} = \frac{3}{2}NkT$. A diatomic gas has *translational* kinetic energy of $\frac{3}{2}NkT$, which is the result we used in Section 22.3 to understand pressure, but its thermal energy also includes the *rotational* kinetic energy of the molecules. There are, altogether, *five* modes of energy for a diatomic gas, so $E_{th} = \frac{5}{2}NkT$. Summarizing these ideas, the thermal energy of a gas of N particles at temperature T is

$$E_{th} = \begin{cases} \frac{3}{2}NkT = \frac{3}{2}nRT & \text{monatomic gas} \\ \frac{5}{2}NkT = \frac{5}{2}nRT & \text{diatomic gas.} \end{cases} \quad (22\text{-}28)$$

We've used the relationship $nR = Nk$, from Chapter 19, to write the result in two alternative forms. A diatomic gas has more thermal energy at the same temperature because the molecules are rotating as well as translating.

Molar Heat Capacity

We can use this information about thermal energy to solve a puzzle that was discovered in Chapter 20. Recall that the molar heat capacity relates the temperature change of a gas to the amount of heat energy used to change it: $Q = nC_V\Delta T$ for a constant volume process involving n moles of gas. The puzzle is that *all* monatomic gases have the same value for C_V, namely 12.5 J/mol K, and that *all* diatomic molecules have $C_V \approx 20.8$ J/mol K. That was an empirical finding. Now we are in a position to understand *why* this is the case.

Consider adding heat to a gas while keeping its volume constant. You could do this by putting a burner under a sealed, rigid container of gas. No external work ($W = 0$) is done on the system, because its boundaries do not change, so the first law of thermodynamics tells us that $\Delta E_{th} = Q$ in a constant volume process. But $Q = nC_V\Delta T$, so we have the constant volume relationship

$$C_V = \frac{Q}{n\Delta T} = \frac{\Delta E_{th}}{n\Delta T}. \quad (22\text{-}29)$$

We can now compute ΔE_{th} directly from Eq. 22-28. For a monatomic gas that undergoes a temperature change ΔT, due to the addition of heat, the thermal energy changes by

$$E_{th} = \tfrac{3}{2}nRT \quad \Rightarrow \quad \Delta E_{th} = \tfrac{3}{2}nR\Delta T \quad \text{(monatomic gas)}. \quad (22\text{-}30)$$

The result for a diatomic gas is

$$E_{th} = \tfrac{5}{2}nRT \quad \Rightarrow \quad \Delta E_{th} = \tfrac{5}{2}nR\Delta T \quad \text{(diatomic gas)}. \quad (22\text{-}31)$$

Using these expressions for ΔE_{th} in Eq. 22-29 gives

$$C_V = \begin{cases} \tfrac{3}{2}R = 12.5 \text{ J/mol K} & \text{monatomic gas} \\ \tfrac{5}{2}R = 20.8 \text{ J/mol K} & \text{diatomic gas,} \end{cases} \quad (22\text{-}32)$$

where $R = 8.31$ J/mol K. The kinetic theory of gases has given us a precise prediction for the molar heat capacity at constant volume. The prediction agrees *perfectly* with the experimental values for monatomic gases! It is extremely close (within 1%) for diatomic gases, with the slight difference being due to molecular vibrations that we haven't considered. The micro/macro connection has thus given us a deeper understanding of gases.

The molar heat capacities of more complex gases as well as solids can be nicely understood from straightforward extensions of our statistical theory. We will leave such calculations to more advanced courses in thermodynamics. The important point is that we have built the framework of a highly successful theory of the micro/macro connection. It is a view of nature very different from the mechanical Newtonian physics that kept us busy for the first half of this text. The statistical approach to physics is a contemporary perspective, and the application of statistical theories to complex systems is very much a current topic of research in science and engineering.

EXAMPLE 22-8 Molecular nitrogen N_2 has a bond length of 0.12 nm. Estimate the rotational frequency of N_2 at room temperature.

SOLUTION The molecule rotates end over end, as shown in Fig. 22-8. Because the molecule rotates about its midpoint, each of the nitrogen atoms undergoes uniform circular motion in a circle of radius $r = 0.06$ nm. If the rotation frequency is f, the speed of each atom is $v = 2\pi r f$. Each atom has kinetic energy, so the rotational kinetic energy of the molecule is

$$\varepsilon_{rot} = \tfrac{1}{2}mv^2 + \tfrac{1}{2}mv^2 = m(2\pi r f)^2 = 4\pi^2 m r^2 f^2,$$

where m is the mass of each *atom*, not the whole molecule. The energy associated with this mode is $\tfrac{1}{2}NkT$, so the *average* rotational kinetic energy per molecule is

$$\overline{\varepsilon_{rot}} = \tfrac{1}{2}kT.$$

Equating these two expressions for ε_{rot} will give the average rotational frequency:

$$4\pi^2 m r^2 f_{avg}^2 = \tfrac{1}{2}kT$$

$$\Rightarrow f_{avg} = \sqrt{\frac{kT}{8\pi^2 m r^2}}.$$

Room temperature is $T = 20°C = 293$ K, and the mass of a nitrogen atom is $m = 14$ u $= 2.34 \times 10^{-26}$ kg. Evaluating f_{avg} gives

$$f_{avg} = 7.80 \times 10^{11} \text{ s}^{-1}.$$

This is an extremely high frequency, with a rotational period of only 1.3×10^{-12} s $= 1.3$ ps, but it is typical of molecular rotations.

22.6 Thermal Interactions and Heat

Now we're in a position to look at what happens when two systems at different temperatures interact with each other. Figure 22-9 shows a rigid, insulated container that is divided into two parts (not necessarily halves) by a very thin membrane. The insulation prevents any heat from entering or leaving the container. The left side, which we'll call system 1, has N_1 atoms at an initial temperature T_{1i} and system 2 on the right has N_2 atoms at an initial temperature T_{2i}. We'll consider the gases to be monatomic (hence atoms rather than molecules) to keep things straightforward. (A homework problem will let you consider what happens if one of the gases is diatomic.) The membrane is so thin that atoms can collide at the boundary as if the membrane were not there, yet it is a barrier that prevents atoms from moving from one side to the other. The situation is analogous, on an atomic scale, to basketballs colliding through a shower curtain.

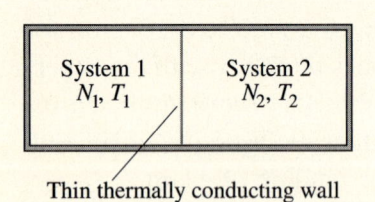

FIGURE 22-9 Two gases can interact thermally through a very thin barrier.

Suppose that system 1 is initially at a higher temperature: $T_{1i} > T_{2i}$. This is not an equilibrium situation. You probably know, from much experience, that the temperatures will change with time until they eventually reach a common final temperature: $T_{1f} = T_{2f} = T_f$. If you *watch* the gases as one warms and the other cools, you see nothing happening. Yet the gases must somehow be interacting with each other for change to occur. This interaction is quite different from a mechanical interaction in which, for example, a piston moves from one side toward the other. The only way in which the gases can interact is via molecular collisions at the boundary. This type of interaction is called a **thermal interaction**, and our goal is to understand how thermal interactions bring the systems to thermal equilibrium.

System 1 and system 2 begin with thermal energies

$$E_{1i} = \tfrac{3}{2} N_1 k T_{1i} = \tfrac{3}{2} n_1 R T_{1i}$$
$$E_{2i} = \tfrac{3}{2} N_2 k T_{2i} = \tfrac{3}{2} n_2 R T_{2i}, \tag{22-33}$$

where we have used $nR = Nk$ to write the thermal energy in terms of both atomic numbers N and moles n. The coefficient $\tfrac{3}{2}$ reflects our assumption that both gases are monatomic.

The total energy of the combined systems is

$$E_{tot} = E_{1i} + E_{2i}. \tag{22-34}$$

As systems 1 and 2 interact, their individual thermal energies E_1 and E_2 can change but the total energy E_{tot} is conserved. There are many ways in which the total energy can be divided between the two systems, just as there were many ways in which the total number of particles in a box could be divided between the right side and the left side. Thermal equilibrium corresponds to the *most probable* distribution of the energy.

Let's look more closely at how the thermal interaction happens. Figure 22-10 shows a fast atom and a slow atom approaching the barrier from opposite sides. They undergo a perfectly elastic collision at the barrier. Although no net energy is lost in a perfectly elastic collision, the faster atom loses energy while the slower one gains energy. In other words, there is an energy *transfer* from the faster atom's side to the slower atom's side.

Because the average kinetic energy per molecule is directly proportional to the temperature, as given by $\bar{\varepsilon} = \frac{3}{2}kT$, the atoms in system 1 are, on average, more energetic than the atoms in system 2. As collisions occur at the boundary, *on average* there is a transfer of energy from system 1 to system 2. This collisional transfer of energy at the boundary, due to a temperature difference between the systems, is what we call *heat*. We previously defined heat simply as an energy transfer due to a temperature difference, but we didn't say how it happened. Now that we're examining the systems on a microscopic scale we can understand heat as the energy transferred *during collisions* between the more energetic (hotter) atoms on one side and the less energetic (cooler) atoms on the other.

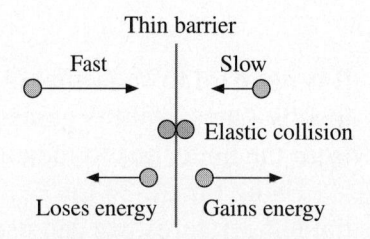

FIGURE 22-10 Collisions at the barrier transfer energy from faster molecules to slower molecules.

Energy transfer will continue until the atoms on each side of the barrier have the same average energy. Once the average energies are the same there will be no tendency for heat energy to flow in either direction. This is the state of thermal equilibrium, so the condition for thermal equilibrium is

$$\bar{\varepsilon}_1 = \bar{\varepsilon}_2 \quad \text{(thermal equilibrium)}. \tag{22-35}$$

But the average energies are directly proportional to the final temperatures: $\bar{\varepsilon} = \frac{3}{2}kT_f$. Thus thermal equilibrium is characterized by the condition

$$T_{1f} = T_{2f} = T_f \quad \text{(thermal equilibrium)}. \tag{22-36}$$

In other words, the common experience that two thermally interacting systems reach a common final temperature occurs *because* they exchange heat energy via molecular collisions until the atoms on each side have, on average, equal energies.

Equation 22-35 can be used to determine the final, equilibrium energies and temperatures. If the final average molecular energies of systems 1 and 2 are the same, then

$$\left[\bar{\varepsilon}_1 = \frac{E_{1f}}{N_1}\right] = \left[\bar{\varepsilon}_2 = \frac{E_{2f}}{N_2}\right] = \left[\bar{\varepsilon}_{tot} = \frac{E_{tot}}{N_1 + N_2}\right], \tag{22-37}$$

from which we can conclude:

$$E_{1f} = \frac{N_1}{N_1 + N_2} E_{tot} = \frac{n_1}{n_1 + n_2} E_{tot}$$

$$E_{2f} = \frac{N_2}{N_1 + N_2} E_{tot} = \frac{n_2}{n_1 + n_2} E_{tot}. \tag{22-38}$$

The result can be written in terms of either N's or n's because $N = N_A n$ and the N_A's cancel. Notice that $E_{1f} + E_{2f} = E_{tot}$, verifying that energy has been conserved even while being redistributed between the systems.

No work is done on either system, because the barrier does not move, so the first law of thermodynamics relates the heat flow to the thermal energy change as

$$Q_1 = \Delta E_1 = E_{1f} - E_{1i}$$
$$Q_2 = \Delta E_2 = E_{2f} - E_{2i}. \qquad (22\text{-}39)$$

It is not hard to show that $Q_1 = -Q_2$, as required by energy conservation. That is, the heat lost by one system is gained by the other. We can say that a quantity of heat $|Q_1|$ flows from the hotter gas to the cooler gas during the thermal interaction.

Using Eq. 22-34 for E_{tot} and Eqs. 22-33 to relate the initial thermal energies to the initial temperatures, we can use Eq. 22-38 to find the final, equilibrium temperature:

$$E_{1f} = \tfrac{3}{2} N_1 k T_{1f} = \frac{N_1}{N_1 + N_2}\left(\tfrac{3}{2} N_1 k T_{1i} + \tfrac{3}{2} N_2 k T_{2i}\right)$$
$$\Rightarrow T_f = T_{1f} = \frac{N_1 T_{1i} + N_2 T_{2i}}{N_1 + N_2} = \frac{n_1 T_{1i} + n_2 T_{2i}}{n_1 + n_2}. \qquad (22\text{-}40)$$

The same result is obtained for T_{2f}, verifying that this is a *common* final temperature T_f.

Notice that, in general, the equilibrium thermal energies of the system are *not* equal. That is, $E_{1f} \neq E_{2f}$. They will be equal only if $N_1 = N_2$. A common misconception is that thermal equilibrium is reached when the thermal energies of the two systems are equal. The correct statement, Eq. 22-35, is that equilibrium is reached when the average molecular energies of the two systems are equal. Make sure you understand the distinction.

EXAMPLE 22-9 A sealed, insulated container has 2 g of helium at an initial temperature of 300 K on one side of a barrier and 10 g of argon at an initial temperature of 600 K on the other side. a) What is the final temperature? b) How much heat flows and in which direction?

SOLUTION a) This problem will be easier to work using moles rather than numbers of atoms. Helium has atomic mass $A = 4$, so $n_1 = m/A = 0.5$ mol. Similarly, argon has $A = 40$, so $n_2 = 0.25$ mol. The final temperature of the gases, according to Eq. 22-40, is

$$T_f = \frac{n_1 T_{1i} + n_2 T_{2i}}{n_1 + n_2} = \frac{(0.50 \text{ mol})(300 \text{ K}) + (0.25 \text{ mol})(600 \text{ K})}{0.50 \text{ mol} + 0.25 \text{ mol}}$$
$$= 400 \text{ K}.$$

b) The initial thermal energies of the two gases are

$$E_{1i} = \tfrac{3}{2} n_1 R T_{1i} = 225 R = 1871 \text{ J},$$
$$E_{2i} = \tfrac{3}{2} n_2 R T_{2i} = 225 R = 1871 \text{ J}.$$

Notice that the systems start with *equal* thermal energies, but they are not in equilibrium. The total energy is $E_{tot} = 3742$ J. Equation 22-38 can be used to find the final, equilibrium thermal energies:

$$E_{1f} = \frac{n_1}{n_1 + n_2} E_{tot} = \frac{0.50}{0.75} \cdot 3742 \text{ J} = 2495 \text{ J},$$

$$E_{2f} = \frac{n_2}{n_1 + n_2} E_{tot} = \frac{0.25}{0.75} \cdot 3742 \text{ J} = 1247 \text{ J}.$$

Thus the heat energy entering or leaving each system is
$$Q_1 = E_{1f} - E_{1i} = 624 \text{ J},$$
$$Q_2 = E_{2f} - E_{2i} = -624 \text{ J}.$$

We can summarize the situation by saying that the helium and the argon interacted thermally via collisions at the boundary, causing 624 J of heat to flow from the hotter argon to the cooler helium until they reached a common final temperature of 400 K.

The main idea of this section is that two interacting systems reach a common final temperature not by magic or by a prearranged agreement but simply through the energy exchange of vast numbers of molecular collisions. Real interacting systems, of course, are separated by real walls rather than our unrealistic thin membrane. As the systems interact, the energy is first transferred from system 1, via collisions, into the wall between the gases and subsequently, as the cooler molecules collide with a warm wall, into system 2. The energy transfer is $E_1 \to E_{wall} \to E_2$. This is still heat flow because the energy transfer is occurring via molecular collisions rather than mechanical motion.

The statistics of the interaction dictate that the most probable way for two systems to share energy E_{tot} is for the average molecular energies of the two systems to be equal: $\bar{\varepsilon}_1 = \bar{\varepsilon}_2$. Any other distribution of the energy is incredibly less probable. But even though other distributions are improbable, they are not impossible. In Chapter 21 we found that the number of molecules N_R in the right side of the box is not a fixed number but that it fluctuates about the average value. Similarly, the average thermal energy in system 1 is not a fixed number but, instead, fluctuates about the average value E_{1f} given in Eq. 22-38. This means that the equilibrium temperature of system 1 fluctuates about the average value T_f. A more advanced analysis shows that the σ_T, the range of expected fluctuations in T, is

$$\sigma_T \approx \frac{T}{\sqrt{N}}. \tag{22-41}$$

Although the temperature does fluctuate, the size of the fluctuations for any realistic system ($N > 10^{20}$) is completely insignificant and unmeasurable. As such, we are perfectly safe in talking about "the" temperature of a system.

22.7 Irreversible Processes, Entropy, and the Second Law of Thermodynamics

In the last section we considered the thermal interaction between a hot gas and a cold gas. Heat energy flows from the hot gas to the cold gas until they reach a common final temperature. But why doesn't heat energy flow from the cold gas to the hot gas, making the cold side colder and the hot side hotter? Such a process would still conserve energy, but it never happens. The flow of heat energy from hot to cold is an example of an **irreversible process**—a process that can only happen in one direction.

Examples of irreversible processes abound. Stirring the cream in your coffee mixes the cream and coffee together. No amount of stirring ever unmixes them, although none of the laws of physics we have studied thus far would be violated if that did happen. If you shook a jar that had red marbles on the top and blue marbles on the bottom, the two colors would quickly mix together. No amount of shaking would ever separate them again. If you watched a movie of someone shaking the jar and saw the red and blue marbles separating, you would be certain that the movie was running backward. In fact, a reasonable definition of an irreversible process is one for which a backward-running movie shows a physically impossible process.

There's an old saying that you "can't unscramble an egg." Well, why not? We want to go beyond commonsense observations and try to understand *why* macroscopic systems, like eggs, exhibit irreversible behavior. The reason this needs an explanation is that the microscopic behaviors of the molecules in the egg *are* **reversible processes**; that is, any change can be reversed so that the system returns to its original state.

Figure 22-11a shows a movie of two particles—perhaps two molecules in a gas—undergoing a collision. Suppose that sometime after the collision is over we could reach in and reverse the velocities of both particles—that is, to replace vector $\vec{v}$ with vector $-\vec{v}$. Then the collision would happen in reverse, as seen in Fig. 22-11b. This is equivalent to playing a movie backward.

Is there any way you can tell, just by looking at the two movies, which is going forward and which is being played backward? You cannot! Maybe Fig. 22-11b was the original collision and Fig. 22-11a is the backward version. Nothing that you see in either the forward or the reversed collisions looks "wrong," and no measurements you might make on either would reveal any violations of Newton's laws. In fact, it is possible to prove quite generally that any two-particle interaction obeying Newton's laws is an equally possible and legitimate process if all velocity vectors are reversed and the motion "runs backward." Interactions at the molecular level are said to be reversible processes.

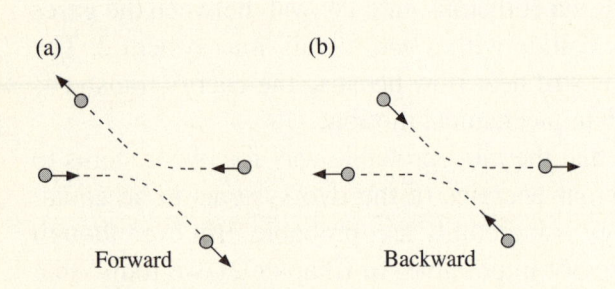

FIGURE 22-11 Molecular collisions are reversible. a) A movie run forward. b) The same movie run backward. Both collisions are valid possibilities.

The reversible processes at the molecular level are in stark contrast with movies of macroscopic phenomena. A car runs into a wall and the front bumper crumples. Would a movie of this event look legitimate run backward? "Past" and "future" are clearly distinct in an irreversible process, and a backward movie is obviously "wrong." But what has been violated in the backward movie? To have the crumpled car spring away from the wall would not violate any laws of physics we have so far discovered—it would simply require converting the thermal energy of the car and the wall back into the macroscopic center-of-mass energy of the car as a whole. It would not violate the law of energy conservation, as stated in the first law of thermodynamics, for heat to flow *from* a cold ice cube *to* hot water.

Furthermore, we have asserted that these macroscopic phenomena can all be understood on the basis of the microscopic molecular motions. If the microscopic motions are all reversible, how can the macroscopic phenomena end up being irreversible? As you can begin to see, asking *why* you can't unscramble an egg is not a simple or silly question at all.

If unscrambling an egg, or having heat flow from cold to hot, is impossible, there must be another law of physics—one we haven't yet found—preventing it. The law of physics we seek must, in some sense, be able to distinguish the past from the future. There is such a law—the *second* law of thermodynamics—and our goal in this section is to understand this law and some of its profound implications.

Which Way to Equilibrium?

Thermally interacting systems, such as a hot and a cold gas, evolve toward equilibrium. The interaction is an irreversible process. But if all the microscopic collisions are random and reversible, then how does the system "know" which way to go to reach equilibrium? Perhaps a mechanical analogy will help to answer this question.

Figure 22-12 shows a ball inside a box that has several partitions, each with one or more holes through which the ball can pass. Within each compartment, the partition toward the center of the box has more holes than the partition toward the outer edge of the box. Place the ball at random in one of the compartments, then gently shake the box so that the ball rolls around. What happens?

At any instant of time the ball is as likely to be moving toward the right as toward the left. Each collision with a wall is a reversible processes, analyzable with Newtonian dynamics. We could, after one such collision, give a precise prediction as to where the ball's next collision will occur. By averaging over many dozens or hundreds of collisions with the partitions, we can safely predict that the ball is *more likely* to pass through a partition with more holes than it is through one with fewer holes. After many collisions and passing through many holes, the ball is far more likely to be found in one of the compartments near the center than in a compartment near the edge. The ball did not in any way "know" that it should go toward the center. On any particular shake of the box there is a chance that it will, at least briefly, move away from the center. But it will spend the majority of its time near the center simply because that is the *most probable* place to be.

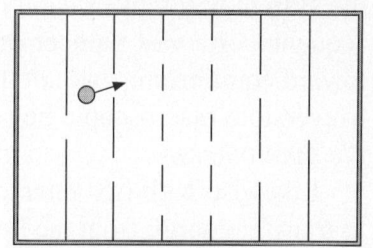

FIGURE 22-12 The ball is more likely to pass through the walls with more holes and to end up near the center of the box.

Now consider another box containing $N = 64$ molecules. In equilibrium, the *average* number of molecules in the right half of the box is $\overline{N_R} = 32$. Suppose at one instant of time you observe $N_R = 19$. This is a situation far from equilibrium. A brief instant later the right side might have either $N_R = 18$ or 20, if one molecule changes sides, or it might still have $N_R = 19$. How does the gas "know" that N_R needs to increase, rather than decrease, to reach equilibrium?

Equation 21-15 provided an explicit expression for computing the probability $P(N_R)$ that there are N_R molecules in the right half. Calculations for $N = 64$ give

$$P(N_R = 18) = 0.00019$$
$$P(N_R = 19) = 0.00047$$
$$P(N_R = 20) = 0.00106.$$

The three probabilities are different because there are more ways (more configurations) to arrange the box with 20 molecules on the right than with 18 molecules on the right. You can see that a macroscopic state with $N_R = 19$ is 5.4 times more likely to evolve to a state with $N_R = 20$ than it is to a state with $N_R = 18$. It is really no different from the ball in the box of Fig. 22-12—there are simply many more options with $N_R = 20$ than there are with $N_R = 18$.

Both the ball in the box and the 64 molecules in the container zig-zag their way toward equilibrium. There are some zigs away from equilibrium, but there are vastly more zags toward it. The system reaches equilibrium not by any "plan" or by outside intervention, but simply because equilibrium is the *most probable* state in which to be. It is *possible* that the system will move steadily away from equilibrium, but remotely improbable. The consequence of a vast number of random events is that the system evolves in one direction, toward equilibrium, and not the other. Reversible microscopic events have given rise to an irreversible macroscopic behavior *because some macroscopic states are far more probable than others*.

Likewise with two interacting systems. The molecules do not "know" how to collide to transfer energy from the hot side to the cold side. There will be some collisions that, at least briefly, move energy from cold to hot. But over the course of billions and quadrillions of such collisions, the net effect is that the system will move toward the most probable distribution of energy, which corresponds to equal temperatures. The system—by random and *reversible* collisions—moves in the single *irreversible* direction in which heat energy flows from the hot side to the cold side until the system is in equilibrium.

Entropy and the Second Law of Thermodynamics

Scientists and engineers use a state variable called **entropy** to measure the probability of occurrence of a macroscopic state. A state with a large probability of occurrence has a large entropy and a state with a small probability of occurrence has a small entropy. For example, the state in which only 1 of 64 molecules is found in the right half of a container is extremely improbable and has a very small entropy. The state in which is 32 of 64 molecules are in the right half is quite probable and has a large entropy. Similarly, it is very improbable to observe a macroscopic state of two thermally interacting gases in which the molecules of one gas are moving very rapidly while those of the other gas are moving very slowly. This state has low entropy. The state in which the gases have equal average energy per molecule is far more likely and has a higher entropy.

We will not need to be concerned in this text with computing actual values for the entropy. All you need to know is

1. Entropy S measures the probability of occurrence of a macroscopic state. More probable states have larger values of entropy.
2. When two systems interact, their entropy is additive. The total entropy S_{tot} is

$$S_{tot} = S_1 + S_2. \tag{22-42}$$

As the interaction proceeds, the change in the total entropy is

$$\Delta S_{tot} = \Delta S_1 + \Delta S_2. \tag{22-43}$$

where ΔS_1 and ΔS_2 are the changes in the entropies of systems 1 and 2.

Our analysis in the preceding subsection indicates that purely random processes cause a system to evolve toward more and more probable macroscopic states until reaching equilibrium—the most probable of all possible states. This idea can be stated in terms of entropy by saying that the entropy of the system continually increases until it reaches a maximum value in the state of equilibrium.

This idea, as simple as it may sound, turns out to be one of the most important scientific discoveries of the last two hundred years. The evolution of macroscopic systems toward more probable states is now recognized as a law of nature that is called the **second law of thermodynamics**. Its formal statement is:

> *The Second Law of Thermodynamics*: For an isolated system that consists of two or more interacting parts or subsystems,
>
> $$\Delta S_{tot} = \Delta S_1 + \Delta S_2 + \cdots \geq 0.$$
>
> Furthermore, $\Delta S_{tot} = 0$ if and only if the system is in thermal equilibrium, in which case S_{tot} has reached its maximum possible value.

In other words, the total entropy of an isolated system can only increase or stay the same, giving $\Delta S \geq 0$. The entropy can never decrease. And it remains unchanged only after the system reaches equilibrium.

Note the important qualification that the total combined system must be *isolated* from its environment, exchanging no energy through either work or heat. A common error in applying the second law is to overlook this insistence on isolation. It is certainly possible that you could reach in and move 63 of 64 molecules in a container to the left side, thus lowering its entropy, or that you could use an external source of energy to make a hot gas hotter and a cold gas colder. Those actions are not forbidden by the second law because the system is not isolated. What the second law says is that such processes will not happen *spontaneously* in an isolated system. Also note that S_j for some *particular* subsystem j might decrease ($\Delta S_j < 0$), which is not forbidden as long as the total entropy S_{tot} increases.

The second law of thermodynamics is an independent statement about nature, separate from the first law. The first law is a precise statement about energy conservation. The second law, by contrast, is a *probabilistic* statement, based on the statistical laws of very large numbers. While it is conceivable that the total entropy of an isolated system could decrease rather than increase, our Chapter 21 calculations about the probabilities of large fluctuations tell us that an entropy decrease will "never" occur in any realistic macroscopic system.

For example, a swinging pendulum eventually slows down and stops. The energy of its swing is not lost. The energy is gradually transferred—via collisions—to the molecules in the air, slightly raising the air temperature. We could imagine a situation in which the air molecules transfer the energy back to the pendulum, causing it to spontaneously start swinging. But such events don't happen. The pendulum *and* the air together are interacting systems that are governed by the second law of thermodynamics.

The pendulum swinging in a room of air is a state of low entropy because it is a very improbable distribution of the energy. It happens only because an *external* force reached in and started the motion. But once the pendulum and air are isolated it is vastly more probable that the energy will be distributed equally among all the molecules rather than concentrated in the pendulum. The pendulum runs down because that is the evolution which increases the *total* entropy $S_{tot} = S_{pend} + S_{air}$. The pendulum will not spontaneously start up from rest, even though energy could be conserved by using the thermal energy of the air, because to do so would require the total entropy to decrease. Such processes are prohibited by the second law of thermodynamics.

It is the second law that requires most macroscopic phenomena to be irreversible processes. This aspect of the second law will be needed in the next chapter to understand some of the practical aspects of the thermodynamics of gases. But the second law also has interesting and profound implications for such issues as biological evolution, the direction of time, and the future of the universe. We'll return to look at those issues in the Scenic Vista of Part IV.

This has been a somewhat lengthy digression, but the answer to our question of *why* you can't unscramble an egg is that to do so would violate the second law of thermodynamics. Perhaps Lewis Carroll gave the best statement of the second law in *Alice in Wonderland* when he observed, "All the King's horses and all the King's men couldn't put Humpty Dumpty together again."

Which Way Does Heat Flow?

We will conclude this chapter by using the second law of thermodynamics to examine once again the situation in which two systems interact thermally. Suppose that heat energy is added to a system. Doing so raises the temperature and speeds up the molecules, thereby increasing the number of microstates the system can access and, as a consequence, increasing the system's entropy. We will assert, without proof, that a small amount of heat dQ added to a system at temperature T changes the system's entropy by

$$dS = \frac{dQ}{T}. \qquad (22\text{-}44)$$

Now consider two thermally interacting, isolated gases at temperatures T_1 and T_2. They have a total entropy $S_{tot} = S_1 + S_2$. The systems can interact only by exchanging heat energy, so $dQ_2 = -dQ_1$ because any heat lost by one side is gained by the other. A small energy exchange causes the entropy of the systems to change by the small amounts dS_1 and dS_2. As a consequence, the total entropy of the combined system changes by

$$dS_{tot} = dS_1 + dS_2 = \frac{dQ_1}{T_1} + \frac{dQ_2}{T_2}$$
$$= dQ_1 \left(\frac{1}{T_1} - \frac{1}{T_2} \right), \qquad (22\text{-}45)$$

where we used $dQ_2 = -dQ_1$. Now according to the second law of thermodynamics,

$$dS_{tot} \geq 0 \quad \Rightarrow \quad dQ_1 \left(\frac{1}{T_1} - \frac{1}{T_2} \right) \geq 0. \qquad (22\text{-}46)$$

There are three ways to satisfy Eq. 22-46:

1. Suppose system 1 is hotter than system 2, so that $T_1 > T_2$. Then $(1/T_1 - 1/T_2) < 0$. Equation 22-46 thus requires $dQ_1 < 0$. That is, system 1 *loses* heat energy while system 2 gains heat energy. Heat flows from the hotter system to the colder system!

2. If $T_1 < T_2$, then the reverse is true. $(1/T_1 - 1/T_2) > 0$ so we must have $dQ_1 > 0$, indicating that system 1 gains heat energy. Again, heat flows from the hotter system 2 to the colder system 1.

3. If $T_1 = T_2$ then $dS_{tot} = 0$. The total entropy is a maximum and the system is in thermal equilibrium.

The conclusion that heat flows from hot to cold is a *consequence* of the second law of thermodynamics. According to the first law, it would be perfectly acceptable for energy to flow "backward" from cold to hot as long as the total amount is conserved. But it does not happen, and the reason is that such a backward flow of energy would violate the second law.

The formal statement of the second law in terms of entropy is a bit abstract. It is customary to state the second law in a variety of alternative, informal versions. The first of these is:

> *Second Law—Informal Statement #1*: When two systems at different temperatures interact, heat always flows from the hotter to the colder and never from the colder to the hotter.

You will encounter other statements of the second law in Chapter 23 and the Scenic Vista.

Summary

Important Concepts and Terms

kinetic theory of gases
root-mean-square velocity
mean free path
mean time between collisions
collision frequency
translational kinetic energy
rotational kinetic energy

equipartition theorem
thermal interaction
irreversible process
reversible process
entropy
second law of thermodynamics

The goal of this chapter has been to connect the invisible microscopic behavior of the atoms and molecules in a gas to the gas's observable macroscopic behavior. We have looked at several different characteristics of gases, including the origins of pressure, temperature, and thermal energy; the nature of heat and thermal interactions; and the concept of entropy and its role in the irreversible evolution of systems toward equilibrium. We have now established that useful and practical results come out of statistical mechanics and the kinetic theory of gases.

The pressure in a gas is due to the vast number of collisions of molecules with the walls of the container, each exerting a small impulsive force. A statistical theory of the collisions finds that the pressure is

$$p = \tfrac{1}{3} d m v_{\text{rms}}^2,$$

where the root-mean-square velocity is given by

$$v_{\text{rms}} = \sqrt{\frac{3kT}{m}}.$$

An unexpected finding of this analysis is that the average translational kinetic energy per molecule is

$$\bar{\varepsilon} = \tfrac{3}{2} kT.$$

This allows us to interpret the temperature of a gas as a parameter that measures the average energy per molecule.

A closer look at the different modes by which the molecules in a gas acquire microscopic energy led to important conclusions about the thermal energy and the molar heat capacity of gases:

Monatomic gas: $E_{\text{th}} = \tfrac{3}{2} NkT = \tfrac{3}{2} nRT = \tfrac{3}{2} pV \quad C_V = \tfrac{3}{2} R = 12.5 \text{ J / mol K}$

Diatomic gas: $E_{\text{th}} = \tfrac{5}{2} NkT = \tfrac{5}{2} nRT = \tfrac{5}{2} pV \quad C_V = \tfrac{5}{2} R = 20.8 \text{ J / mol K}.$

The difference arises because a diatomic gas can have kinetic energy due to rotation of the molecules. The molar heat capacity predictions are in perfect agreement with experiments.

Molecules undergo frequent collisions with other molecules as they move through a gas. The mean free path between collisions and the mean time between collisions are given by

$$\lambda = \frac{1}{4\sqrt{2}\pi dr^2}$$

$$\tau = \frac{\lambda}{v_{rms}},$$

where r is the radius of the molecule.

Two systems interact thermally by molecular collisions at the boundary, causing an exchange of energy. These collisions cause a net flow of energy from the higher temperature side to the lower temperature side. This energy flow is called heat. Heat energy continues to flow until equilibrium is reached when $\bar{\varepsilon}_1 = \bar{\varepsilon}_2$ and $T_1 = T_2$.

The flow of heat energy from hot to cold is an irreversible process. An analysis of how irreversible processes come about, even though the microscopic physics is reversible, led to the concept of entropy S as a measure of the probability of occurrence of a macroscopic state. The second law of thermodynamics says that $\Delta S_{total} \geq 0$ for an isolated system. This law of physics tells us that purely random processes cause a macroscopic system to evolve to more and more probable macroscopic states until reaching equilibrium—the most probable of all possible states. Reversible microscopic processes gives rise to irreversible macroscopic behavior because some macroscopic states are vastly more probable than others.

Exercises and Problems

Exercises

1. During a physics experiment, helium gas is cooled to a temperature of 10 K and a pressure of 0.1 atm.
   a. What is the average energy per atom?
   b. What is the rms velocity of the gas?
   c. What is the mean free path in the gas?
   d. What is the collision frequency in the gas?

2. A cylinder of compressed argon gas ($A = 40$) has a volume of 15,000 cm³ and a pressure of 100 atm.
   a. What is the thermal energy of this gas at room temperature?
   b. What is the collision frequency between the gas atoms?
   c. The valve is opened and the gas is allowed to expand isothermally until it reaches a pressure of 1 atm. What is the change in thermal energy of the gas?

3. A gas at $p = 50$ kPa and $T = 300$ K has a mass density of 8.02×10^{-5} g/cm³.
   a. Identify the gas.
   b. What is v_{rms} for atoms in this gas?
   c. What is the mean free path of atoms in the gas?
   d. What is the collision frequency between atoms in the gas?

4. Two grams of helium ($A = 4$) at an initial temperature of 100°C interact with neon ($A = 20$) at an initial temperature of 300°C. The final temperature is 250 °C. What is the mass of the neon?

Problems

5. A cylinder that is 10 cm in diameter and 20 cm long contains 2×10^{22} atoms of argon ($A = 40$) at a temperature of 50°C.
   a. What is the number density of the gas?
   b. What is the root-mean-square velocity?
   c. What is $(v_x)_{rms}$, the rms value of the x-component of velocity?
   d. What is the rate at which atoms collide with one end of the cylinder?
   e. Determine the pressure (in atm) in the cylinder using the results of the kinetic theory of gases.
   f. Determine the pressure (in atm) in the cylinder using the ideal gas law.

6. A 100 cm³ box contains helium ($A = 4$) at 2 atm pressure and a temperature of 100°C. It is placed in thermal contact with a 200 cm³ box containing argon ($A = 40$) at 4 atm pressure and a temperature of 400°C.
   a. What is the final temperature?
   b. What is the final pressure in each box?
   c. How much heat flows, and in which direction?

7. Specific heats can be measured in the laboratory with calorimetry experiments. The specific heat can then be used to determine the atomic or molecular mass. A monatomic gas is found to have a specific heat at constant volume of 0.0747 cal/g °C. Identify the gas. (**Note:** This is the specific heat, *not* the molar heat capacity. You need to relate the two.)

8. Consider a container like that shown in Fig. 22-9 with N_1 atoms of a monatomic gas on one side and N_2 molecules of a diatomic gas on the other. The monatomic gas has initial temperature T_{1i} and the diatomic gas has initial temperature T_{2i}.
   a. Perform an analysis similar to that of Section 22.6 to find an expression, analogous to Eq. 22-40, for the final temperature.
   b. Analyze the situation as a calorimetry problem, using molar specific heats, to find an expression for the final temperature.
   c. Two grams of helium gas at an initial temperature of 300 K interact thermally with 8 g of oxygen (O_2) at an initial temperature of 600 K. What is the final temperature?
   d. How much heat flows, and in which direction, in the situation described in part c)?

9. A 1 kg ball is at rest on the floor in a 2 m × 2 m × 2 m room of air at STP. Air is 80% nitrogen (N_2) and 20% oxygen (O_2) by volume.
   a. What is the thermal energy of the air in the room?
   b. What fraction of the thermal energy would have to be conveyed to the ball for it to be spontaneously launched to a height of 1 m?
   c. By how much would the air temperature have to decrease to launch the ball to a height of 1 m?
   d. Your answer to part c) is so small as to be unnoticeable. Yet this event never happens. Why not?

10. A 10 cm × 10 cm × 10 cm box contains nitrogen at 20°C.
    a. At what pressure would temperature fluctuations of ±0.1°C be expected? Give your answer in atm.
    b. Estimate the rate of collisions on one wall at this temperature and pressure.

11. An inventor wants you to invest money with his company, offering you 10% of all future profits. He reminds you that the brakes on cars get extremely hot when they stop and that there is a large quantity of thermal energy in the brakes. He has invented a device, he tells you, that converts that thermal energy into the forward motion of the car. This device will take over from the engine after a stop and accelerate the car back to its original speed, thereby saving a tremendous amount of gasoline. Now you're a smart person, so he admits up front that this device is not 100% efficient, that there is some unavoidable heat loss to the air and to friction within the device, but the upcoming research for which he needs your investment will make those losses extremely small. You do also have to start the car with cold brakes after it has been parked a while, so you'll still need a gasoline engine for that. Nonetheless, he tells you, his prototype car gets 500 miles to the gallon and he expects to be at well over 1000 miles to the gallon after the next phase of research.

    Should you invest? Base your answer on an analysis of the *physics* of the situation.

[**Estimated 10 additional problems for the final edition.**]

Chapter 23

Thermodynamics of Ideal Gases

LOOKING BACK Sections 19.6; 19.7; 20.4–20.6; 22.7

23.1 Thermodynamics: The Transformation of Energy

It is important, as we draw toward the close of our study of thermal and statistical physics, to connect the ideas we have developed to some important applications. The science of thermodynamics has historically been concerned with the *transformation of energy* from one form to another, particularly the transformation of heat energy into work. For example, *engines* burn fuel of some sort (heat source) in order to perform useful work, and *refrigerators* have work done on them (electrical work from a compressor) in order to move heat energy "backward" from cold to hot. We are now in a position to analyze and understand the physics of such practical devices.

Heat energy can be transformed to work energy in many different ways, and we will look at several examples. Our primary tool for analysis will the *pV* diagram that we introduced in Chapter 19. Consider a gas that starts in an initial state i and changes to a final state f, as shown in Fig. 23-1. Each point on the "trajectory" in the *pV* diagram gives the pressure and volume of an intermediate state of the gas. The temperature of the gas at that point can then be found using the ideal gas law. A continuous change of the thermodynamic state of the gas, during which heat energy is transferred and work is done, is called a **thermodynamic process**.

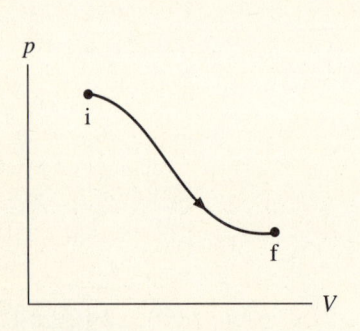

FIGURE 23-1 A thermodynamic process takes the gas from initial state i to final state f.

Thermodynamic processes are changes of the macroscopic state of the gas. The state variables p, V, T, E_{th}, and S change in response to work W being done on the system or by heat Q being transferred between the system and the environment. In Section 23.3 we will analyze four basic thermodynamic processes to learn *how much* work is done and *how much* heat is transferred. We will then combine these processes into a thermodynamic "cycle" that can be used to understand real engines. Our analysis will rely heavily on what you have already learned about ideal gases, including their thermal energy, temperature, and heat capacity.

A major difficulty with thermodynamics is the large number of variables. Thermodynamics can look like an overgrown forest of equations, and for this reason it has a reputation as being a difficult subject. Our goal is to discern the *logic* of thermodynamics, and we will look at numerous examples to see how this logic is applied. Once you grasp the logic, thermodynamics becomes quite elegant and straightforward. So there will, indeed, be lots of equations, but as you study this chapter you should focus on the line of reasoning rather than on "which equation should I use."

23.2 Thermodynamic Relationships

We have introduced thermodynamic ideas throughout the last four chapters. It will be helpful in this section to bring together the major relationships we need to study the thermodynamics of gases. The major scientific principle underlying our analysis will be the first law of thermodynamics—a generalized statement of energy conservation that relates the thermal energy of a gas to the work done and the heat transferred:

$$\Delta E_{th} = Q + W \quad \text{(first law of thermodynamics)}.$$

This law is a statement about energy conservation as one kind of energy is transformed or converted into another.

Work

Work is a mechanical transfer of energy. Real work is done on the system when a force causes the boundary of the system to undergo a displacement. In thermodynamics, work is done when a piston pushes in to compress a gas. If the pushing force is constant, then work = force × piston displacement.

The work W in the first law is work done *on* the system *by* the environment. This is work as we have defined it in mechanics. From a thermodynamics perspective, however, it is more useful to consider the work done *by* the system. After all, you would rather know how much work your engine can do than how much work is done on your engine. To find the work done *by* the system, we need to consider the force $\vec{F}_{\text{system on environment}}$ rather than $\vec{F}_{\text{environment on system}}$. But these are action/reaction pairs, so Newton's third law tells us that $W_{\text{by system}} = -W_{\text{on system}}$. Let's define a new symbol W_s to represent the work done *by* the system. Because this is just the negative of our previous work W, we can rewrite the first law of thermodynamics as

$$\Delta E_{th} = Q - W_s \quad \text{(revised first law of thermodynamics)}. \tag{23-1}$$

This is the form of the first law we will use in this chapter. The subscript s on W will be essential to indicate which work is being used. Equation 23-1 says that the thermal energy change of the system is equal to any heat energy input minus any work that is done *by* the system. This is a significant change in viewpoint.

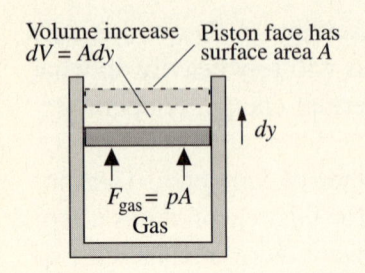

FIGURE 23-2 Work $dW = pdV$ is done by the gas as it expands.

Figure 23-2 shows a cylinder of gas that is closed at one end by a tight-fitting piston—a movable boundary. The piston can slide in or out, but it maintains a tight seal so that the number of molecules N does not change as the volume changes.

The pressure in the gas exerts an upward force $F_{gas} = pA$ on the piston, where p is the gas pressure and A is the surface area of the face of the piston. Suppose the piston undergoes a small, infinitesimal displacement dy. This displacement is a *signed* quantity, with a positive value for an outward displacement (expansion of the gas) and a negative value for an inward displacement (compression of the gas). The volume of the gas changes, as a consequence of the piston's displacement, by an amount

$$dV = Ady.$$

Because the gas exerts a force on the piston as it is displaced, the gas does an infinitesimal amount of work dW_s given by

$$dW_s = F_{gas}dy = pAdy \qquad (23\text{-}2)$$
$$= pdV.$$

A small change of volume dV results in the system doing a small amount of work $dW_s = pdV$. If the gas pushes the piston out (an expansion), then both dV and dW_s are positive. If some force in the environment pushes the piston in (a compression), then both dV and dW_s are negative. We say that the gas does work on the environment as it expands ($W_s > 0$) but that the environment does work on the gas to compress it ($W_s < 0$).

Figure 23-3 shows a thermodynamic process in which the volume of the gas changes from an initial V_i to a final V_f. Because the pressure is *not* constant, we must integrate Eq. 23-2 to find the total work W_s done by the system:

$$W_s = \int_{V_i}^{V_f} pdV = \text{ area under } pV \text{ curve.} \qquad (23\text{-}3)$$

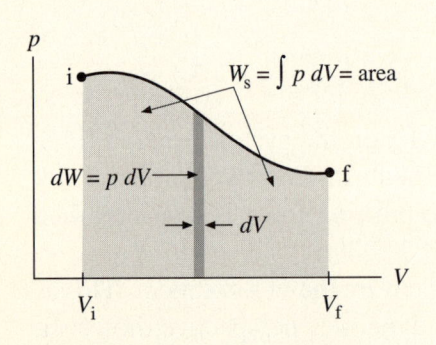

FIGURE 23-3 The work W_s is given by the area under the pV curve.

The geometric interpretation of W_s as the area under the curve in the pV diagram is especially important. We can see that the system does positive work by expanding and pushing on the piston, which, in turn, can perform some useful task in the environment. A gas compression, with the volume decreasing and the pV trajectory moving *to the left*, means that the environment is doing work *on* the system to push the piston in. An integration to the left, as you know from calculus, has $W_s < 0$. No work at all is done if the piston, and thus the system boundary, does not move.

EXAMPLE 23-1 How much work is done by the system in the thermodynamic process shown in Fig. 23-4?

SOLUTION We can determine the work W_s done by the system as the area under the pV curve. Because the volume is expanding, the gas is pushing the piston out and the work is positive. To find the area under the curve, divide the area from 500 to 1000 cm³ into a rectangle (from 0 to 1 atm) and triangle (from 1 to 3 atm). The area is

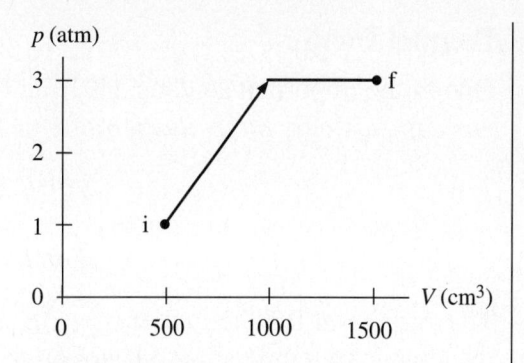

FIGURE 23-4 The pV diagram for Example 23-1.

$$W_s(500 \to 1000 \text{ cm}^3) = (1000 - 500) \times 10^{-6} \text{ m}^3 \times (1-0) \times 1.013 \times 10^5 \text{ Pa}$$
$$+ \frac{1}{2} \times (1000 - 500) \times 10^{-6} \text{ m}^3 \times (3-1) \times 1.013 \times 10^5 \text{ Pa}$$
$$= 101 \text{ J.}$$

From 1000 to 1500 cm³ there is a single rectangle with area

$$W_s(1000 \to 1500 \text{ cm}^3) = (1500 - 1000) \times 10^{-6} \text{ m}^3 \times (3-0) \times 1.013 \times 10^5 \text{ Pa}$$
$$= 152 \text{ J.}$$

Altogether, the work done by the system during the expansion from 500 cm³ to 1500 cm³ is $W_s = 253$ J.

It is important to remember that pressure and volume *must* be in SI units of pascals and m³ before doing calculations. (We noted previously that the product Pa m³ is equivalent to joules.) In practical situations, however, it is common to use units of atm and cm³. Thus unit conversions are an ever-present fact of life in thermodynamics.

Ideal Gases

The equation of state for gases, more commonly called the ideal gas law, is

$$pV = nRT, \tag{23-4}$$

where we have used nR rather than Nk as being more appropriate for our needs in this chapter. The ideal gas law is a relationship between the three major state variables p, V, and T. As a system undergoes a *change* of state, the changes in the state variables are related by

$$\Delta(pV) = nR \cdot \Delta T. \tag{23-5}$$

Note that $\Delta(pV)$ is *not* equal to $p\Delta V + V\Delta p$. That would be true for infinitesimal changes dV and dp, but not for finite changes.

Thermal Energy

One of the important conclusions from Chapter 22 was that the *thermal energy of an ideal gas depends only on its temperature* and not on the pressure:

$$E_{th} = \begin{cases} \frac{3}{2}nRT = nC_V T & \text{monatomic gas} \\ \frac{5}{2}nRT = nC_V T & \text{diatomic gas.} \end{cases} \quad (23\text{-}6)$$

The *change* in the thermal energy ΔE_{th}, which appears in the first law, depends only on the change in temperature ΔT, not on the change in pressure or volume. This is an important conclusion. *Any* thermodynamic process that starts with temperature T_i and ends with temperature T_f will have the *same* value of ΔE_{th}, regardless of how the pressure and volume change. We can see from Eq. 23-6 that an infinitesimal temperature change dT causes an infinitesimal change in the thermal energy given by

$$dE_{th} = nC_V \, dT. \quad (23\text{-}7)$$

This relationship is true for *any* thermodynamic process.

Heat

Finally, we have the heat/temperature relationships. We found, in Section 20.6, that the heat Q required to bring about a temperature change ΔT depends on the thermodynamic process followed. This led us to define both the molar heat capacity at constant volume C_V as well as the molar heat capacity at constant pressure C_P. Thus

$$Q = nC_V \Delta T \quad \text{constant volume process}$$
$$Q = nC_P \Delta T \quad \text{constant pressure process,} \quad (23\text{-}8)$$

where

$$C_V = \begin{cases} \frac{3}{2}R = 12.5 \text{ J/mol K} & \text{monatomic gas} \\ \frac{5}{2}R = 20.8 \text{ J/mol K} & \text{diatomic gas} \end{cases}$$

$$C_P = C_V + R = \begin{cases} \frac{5}{2}R = 20.8 \text{ J/mol K} & \text{monatomic gas} \\ \frac{7}{2}R = 29.1 \text{ J/mol K} & \text{diatomic gas.} \end{cases}$$

This result for C_V was found in Chapter 22, and we will verify the result for C_P in the next section.

Equations 23-2 and 23-7 relate changes in the two *energy* quantities E_{th} and W_s to changes in the state variables T and V. Is there a similar relationship for the third energy quantity, namely Q? Indeed there is—and we have already found it! Recall that in Section 22.6, where we found an explanation as to why heat flows from hotter to colder objects, we discovered the relationship $dS = dQ/T$. This is easily rewritten as

$$dQ = T\,dS, \quad (23\text{-}9)$$

which is analogous to Eqs. 23-2 and 23-7 for work and thermal energy.

Summary of Thermodynamic Relationships

In summary, the three fundamental relationships connecting energy variables to state variables are

$$dQ = T\,dS$$
$$dW_s = p\,dV \quad \Rightarrow \quad W_s = \int_{V_i}^{V_f} p\,dV \tag{23-10}$$
$$dE_{th} = \mu C_V\,dT \quad \Rightarrow \quad \Delta E_{th} = \mu C_V \Delta T.$$

(We will have only minimal need in this text for the heat/entropy relationship, so we have not shown an integration of this equation.) The three energy quantities themselves are connected by the first law of thermodynamics, $\Delta E_{th} = Q - W_s$.

Now let us put these relationships to work and see how powerful they are for analyzing thermodynamic processes and real devices.

23.3 Four Basic Thermodynamic Process

In Chapter 19 we introduced three basic thermodynamic processes, namely the isobaric (constant pressure), isothermal (constant temperature), and isochoric (constant volume) processes. In this section we will analyze these processes in more detail, as well as introduce a fourth process called an *adiabatic process*. Figure 23-5 shows the four basic processes, all starting from the same initial state i. (We will justify the curve for the adiabatic process later in this section.) In the analyses that follow we will consider processes that move along these curves to one of the four final states f shown.

Our goal for each process is to:

1. Find a specialized equation of state for just those variables that are changing.
2. Find expressions for the heat Q and work W_s in terms of the initial state parameters p_i and V_i and the final state parameters p_f and V_f.
3. Find a specialized expression for the first law of thermodynamics.

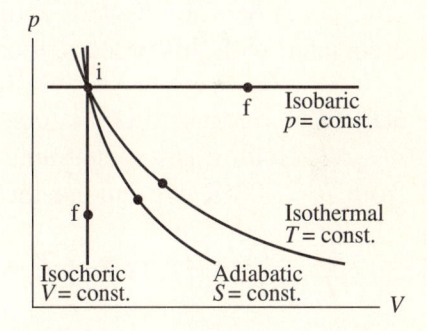

FIGURE 23-5 The four basic thermodynamic processes of an ideal gas. Each starts from the same initial state i and moves along one of the curves to one of the four final states.

Constant Volume Process (Isochoric)

A constant volume process takes place in a rigid container that can be heated, causing the temperature and pressure to change, but that cannot expand or be compressed. By definition, $V = $ constant and $\Delta V = 0$. Figure 23-5 shows that a constant volume process is represented on a pV diagram as a vertical line.

120 CHAPTER 23 THERMODYNAMICS OF IDEAL GASES

1. *Equation of state*: From the ideal gas law with unchanging volume, we have

$$\frac{p_i}{T_i} = \frac{p_f}{T_f} = \text{constant}. \tag{23-11}$$

Because $\Delta(pV) = V\Delta p$ in this situation, we also have $V\Delta p = nR\Delta T$.

2. *Work*: $W_s = \int_{V_i}^{V_f} p\, dV = 0$ because $\Delta V = 0$.

3. *Heat*: $Q = nC_V\Delta T$ for a constant volume process. But $V\Delta p = nR\Delta T$, so we can also write

$$Q = nC_V\, \Delta T = \frac{VC_V}{R}\Delta p. \tag{23-12}$$

Heat input ($Q > 0$) causes both T and p to increase. If heat flows *out* ($Q < 0$), then $\Delta p < 0$ and $\Delta T < 0$.

4. *First law*: Because $W_s = 0$, we have

$$\Delta E_{th} = Q. \tag{23-13}$$

A constant volume process is one in which heat is transformed directly to or from thermal energy ($Q \leftrightarrow E_{th}$) with no work being done.

EXAMPLE 23-2 Two moles of helium at an initial temperature of 27°C in a rigid 1000 cm³ container are heated until the pressure doubles. Calculate the heat and work associated with this process, then show the process on a *pV* diagram.

SOLUTION Because this is a constant volume process (rigid container), $W_s = 0$ and $Q = nC_V\Delta T$. Helium is a monatomic gas, so $C_V = 3R/2 = 12.5$ J/mol K. To find ΔT, starting from $T_i = 300$ K, we can use the relationship

$$\frac{p_f}{T_f} = \frac{p_i}{T_i} \Rightarrow T_f = \frac{p_f}{p_i}T_i = 2T_i = 600 \text{ K}.$$

Thus $\Delta T = 300$ K, from which we find $Q = 7500$ J. We were able to find Q without knowing the pressures, but we will need them to draw a *pV* diagram. From the ideal gas law we find

$$p_i = \frac{nRT_i}{V} = 4.99 \times 10^6 \text{ Pa} = 49.4 \text{ atm,}$$

and then $p_f = 2p_i = 98.8$ atm. We can use these values to draw the *pV* diagram shown in Fig. 23-6. In this process, 7500 J of heat are used to increase the temperature (and thus E_{th}) and pressure without doing any work.

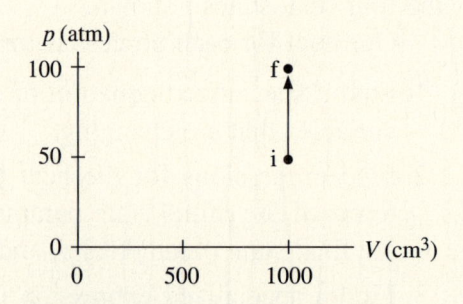

FIGURE 23-6 The *pV* diagram for Example 23-2.

Constant Pressure Process (Isobaric)

A constant pressure process requires some type of *pressure regulator* that moves the piston in or out, as needed, to keep the pressure from changing. In this process, p = constant and $\Delta p = 0$. Figure 23-5 shows that a constant pressure process is represented on a pV diagram as a horizontal line.

1. *Equation of state*: From the ideal gas law, we have

$$\frac{V_i}{T_i} = \frac{V_f}{T_f} = \text{constant}. \tag{23-14}$$

Because $\Delta(pV) = p\Delta V$, we also have $p\Delta V = nR\Delta T$.

2. *Work*: $W_s = \int_{V_i}^{V_f} p\, dV = p \int_{V_i}^{V_f} dV = p\Delta V$ because p = constant. (23-15)

Note that $p\Delta V$ is just the area of the rectangle under the pV curve of an isobaric process.

3. *Heat*: $Q = nC_P \Delta T$ for a constant pressure process. Note that we are using C_P, the molar heat capacity at constant pressure. Because $p\Delta V = nR\Delta T$, we can also write

$$Q = nC_P \Delta T = \left(\frac{C_P}{R}\right) p\Delta V = \left(\frac{C_P}{R}\right) W_s. \tag{23-16}$$

An expansion of the gas ($\Delta V > 0$) does work *on* the environment ($W_s > 0$). But to make the gas expand at constant pressure requires the addition of heat ($Q > 0$) and an increase in the gas temperature ($\Delta T > 0$). To keep the pressure constant as the gas is compressed requires *removing* heat ($Q < 0$) and decreasing the temperature ($\Delta T < 0$).

4. *First law*: Neither Q nor W_s is zero, so $\Delta E_{th} = Q - W_s$.

We can now justify the important relationship $C_P = C_V + R$ that was asserted in the previous section. The first law for an isobaric process can be written as

$$\Delta E = Q - W_s = nC_P \Delta T - p\Delta V = nC_P \Delta T - nR\Delta T = n(C_P - R)\Delta T. \tag{23-17}$$

But for any process, Eq. 23-7 gives $\Delta E_{th} = nC_V \Delta T$. Using this expression for ΔE_{th} on the left-hand side of Eq. 23-17 gives

$$nC_V \Delta T = n(C_P - R)\Delta T$$
$$\Rightarrow C_V = C_P - R,$$

which is easily rearranged to yield

$$C_P = C_V + R = \begin{cases} \frac{5}{2}R & \text{monatomic gas} \\ \frac{7}{2}R & \text{diatomic gas.} \end{cases} \tag{23-18}$$

This is a very general relationship, for any gas, between C_P and C_V. The molar heat capacities in Table 20.3 show that this prediction agrees perfectly with experimental values. Notice in Table 20.3 that $C_P - C_V = 8.3$ J/mol K $= R$ for *all* gases.

EXAMPLE 23-3 707 J of work are done by 0.20 mol of molecular nitrogen as it expands, starting from 500 cm³, at a constant pressure of 10 atm. a) What are the final volume and temperature? b) How much heat is input or output during this process? c) Show the process on a pV diagram.

SOLUTION a) The work done is $W_s = p\Delta V$, from which

$$\Delta V = \frac{W_s}{p} = 7.00 \times 10^{-4} \text{ m}^3 = 700 \text{ cm}^3.$$

Thus $V_f = V_i + \Delta V = 1200 \text{ cm}^3$. The initial temperature can be found from the ideal gas law:

$$T_i = \frac{pV_i}{nR} = 304 \text{ K}.$$

Then $V_i/T_i = V_f/T_f$ gives

$$T_f = \frac{V_f}{V_i} T_i = \frac{1200 \text{ cm}^3}{500 \text{ cm}^3} \cdot 304 \text{ K} = 730 \text{ K} = 457°\text{C}.$$

b) Heat transfer for a constant pressure process is given by Eq. 23-16:

$$Q = nC_P\Delta T = 2480 \text{ J},$$

where $C_P = 29.1$ J/K for N_2 was taken from Table 20.3. $Q > 0$, so heat is input to the system.

c) The pV diagram is shown in Fig. 23-7.

In this process, 2480 J of heat energy are added to the gas, causing it to expand. As a result, 707 J of work are done and the remaining 1773 J of heat energy go into thermal energy as the gas heats from 31°C to 457°C.

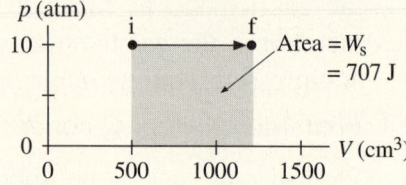

FIGURE 23-7 The pV diagram for Example 23-3.

Constant Temperature Process (Isothermal)

An isothermal process requires that the gas be in thermal contact with a much larger system whose temperature doesn't vary. Heat flows to or from the larger system as necessary to keep the gas at a constant temperature. In other words, T = constant and $\Delta T = 0$. Compressions or expansions that take place very slowly are often isothermal because the gas stays in thermal equilibrium with the air surrounding the cylinder. Because the thermal energy depends *only* on the temperature, an important conclusion about isothermal processes is that the thermal energy does not change: $\Delta E_{th} = 0$.

1. *Equation of state*: From the ideal gas law we have

$$p_i V_i = p_f V_f = \text{constant}. \tag{23-19}$$

2. *Work*: From the ideal gas law, $p = nRT/V$. This is a function of V, allowing an explicit integration to find the work done in a volume change from V_i to V_f:

$$W_s = \int_{V_i}^{V_f} p\,dV = nRT\int_{V_i}^{V_f} \frac{dV}{V} = nRT(\ln V_f - \ln V_i) = nRT\ln\left(\frac{V_f}{V_i}\right). \tag{23-20}$$

As Fig. 23-8 shows, this integral is the area under a hyperbolic isotherm. Because $nRT = p_i V_i = p_f V_f$, we can also write the result as

$$W_s = p_i V_i \left(\ln \frac{V_f}{V_i} \right) = p_f V_f \left(\ln \frac{V_f}{V_i} \right). \quad (23\text{-}21)$$

3. *Heat*: Because $\Delta E_{th} = 0$, the first law gives

$$Q = W_s. \quad (23\text{-}22)$$

If heat flows into the system ($Q > 0$), then the gas expands and does work on the environment ($W_s > 0$). If the environment does work on the system by compressing the gas ($W_s < 0$), then heat must be removed ($Q < 0$) to keep the temperature constant.

4. *First law*: $\Delta E_{th} = 0 = Q - W_s$ and thus $Q = W_s$. An isothermal process is one in which heat is converted directly to or from work ($Q \leftrightarrow W_s$).

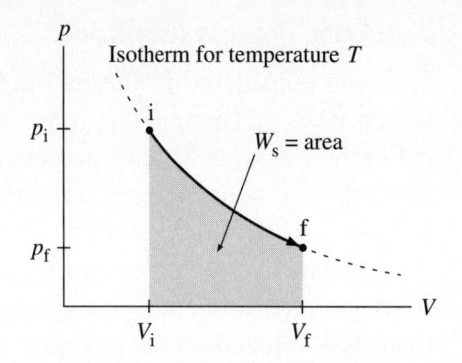

FIGURE 23-8 The work done during an isothermal process is the area beneath a hyperbolic isotherm.

EXAMPLE 23-4 A piston isothermally compresses 0.50 mol of gas at 25°C from an initial volume of 1000 cm³ until the pressure has tripled. How much heat must be removed from the system during this process? Show the process on a *pV* diagram.

SOLUTION The first law for an isothermal process is $Q = W_s$. Work $W_s < 0$ for a compression, so $Q < 0$ and heat energy must be *removed* from the system to keep the temperature constant as the gas is compressed. We can use Eq. 23-19 to find

$$\frac{V_f}{V_i} = \frac{p_i}{p_f} = \frac{1}{3}.$$

Equation 23-21 then gives for the heat and work:

$$Q = W_s = nRT \ln\left(\frac{V_f}{V_i}\right) = -1360 \text{ J}.$$

The minus sign indicates that heat flows from the system to the environment, as expected. The initial pressure is found from the ideal gas law to be

$$p_i = \frac{nRT}{V_i} = 1.24 \times 10^6 \text{ Pa} = 12.3 \text{ atm},$$

from which $p_f = 3p_i = 36.9$ atm. Figure 23-9 shows the *pV* diagram. This is a process in which the environment does 1360 J of work on the gas to compress it while 1360 J of heat are removed from the gas to keep the temperature at a constant 25°C.

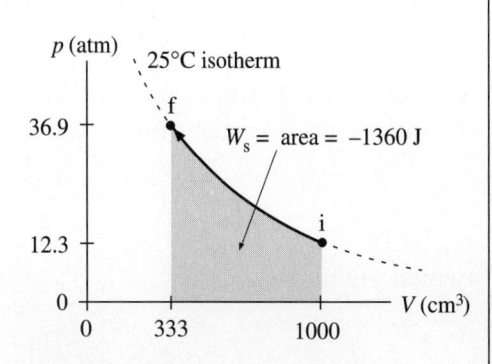

FIGURE 23-9 The *pV* diagram for Example 23-4.

Adiabatic Process (Isentropic)

We have considered isothermal processes, in which $\Delta E_{th} = 0$, and isochoric processes, in which $W_s = 0$. There is one more energy term in the first law, namely Q, so it seems logical to inquire what kind of process has $Q = 0$. Stated differently, what kind of process is it, and how is it characterized, in which no heat flows to or from the system? Such a process is called an **adiabatic process**, from a Greek root meaning "heat does not pass through."

Consider the process shown in Fig. 23-10, in which the system is thermally insulated from the environment—perhaps with very thick pieces of Styrofoam. Even the piston is insulated. "Heat does not pass through" these walls. The system can still do work by motion of the piston, or have work done on it, but no heat will be able to flow in or out of the system. Thus $Q = 0$. This is one way to create an adiabatic process. An alternative method is to compress or expand the gas *very rapidly* so that there is no time for the system to exchange heat energy with the environment. Such processes, called *adiabatic compressions* or *adiabatic expansions*, are important in many types of engines where pistons do, indeed, change the volume quite rapidly.

Because $Q = 0$ for an adiabatic process, the relationship $dQ = TdS$ implies that $dS = 0$ and thus S = constant. So an adiabatic process takes place at *constant entropy*. Adiabatic processes are sometimes called **isentropic**. This completes our "quartet" of basic thermodynamic processes, one each for keeping p, V, T, and S constant.

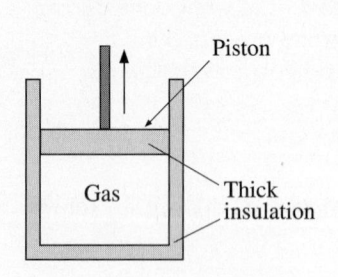

FIGURE 23-10 An adiabatic process is one in which no heat flows to or from the system.

Unlike the previous three processes, p, V, and T all change during an adiabatic process. The ideal gas law is still valid, of course, but we can also deduce a specialized equation of state. The first law for an infinitesimal change of state is

$$dE_{th} = dQ - dW_s. \tag{23-23}$$

An adiabatic process has $dQ = 0$, and we also have the Eq. 23-10 relationships for dE_{th} and dW_s. These allows us to write the first law as

$$\left[dE_{th} = nC_V\, dT \right] = \left[-dW_s = -p\, dV = -\frac{nRT}{V} dV \right]$$
$$\Rightarrow \frac{C_V}{R} \cdot \frac{dT}{T} = -\frac{dV}{V} \tag{23-24}$$

where, in the first line, we used the ideal gas law to eliminate p. Equation 23-24 can now be integrated from the initial state i to the final state f:

$$\frac{C_V}{R} \int_{T_i}^{T_f} \frac{dT}{T} = -\int_{V_i}^{V_f} \frac{dV}{V},$$

which yields

$$\frac{C_V}{R}\left(\ln T_f - \ln T_i \right) = -(\ln V_f - \ln V_i)$$
$$\Rightarrow \ln\left(\frac{T_f}{T_i}\right)^{C_V/R} = -\ln\frac{V_f}{V_i} = \ln\frac{V_i}{V_f}. \tag{23-25}$$

Taking the exponential of the first and last terms gives

$$\left(\frac{T_f}{T_i}\right)^{C_V/R} = \left(\frac{V_i}{V_f}\right) \tag{23-26}$$

$$\Rightarrow V_i T_i^{C_V/R} = V_f T_f^{C_V/R} = \text{constant}.$$

In other words, the product of the volume with the temperature raised to the C_V/R power is constant throughout an adiabatic process.

This is an interesting conclusion, but we need to cast it in terms of p and V in order to use the pV diagram. The T in Eq. 23-26 can be replaced with p and V by using the ideal gas law:

$$V_i\left(\frac{p_i V_i}{nR}\right)^{C_V/R} = V_f\left(\frac{p_f V_f}{nR}\right)^{C_V/R} = \text{constant} \tag{23-27}$$

$$\Rightarrow p_i^{C_V/R} V_i^{C_V/R+1} = p_f^{C_V/R} V_f^{C_V/R+1} = \text{constant}.$$

This expression is awkward, but it can be simplified by raising each side to the power R/C_V. The left side of Eq. 23-27 becomes

$$\left[p_i^{C_V/R} V_i^{(C_V/R)+1}\right]^{R/C_V} = p_i V_i^{1+(R/C_V)} = p_i V_i^{(C_V+R)/C_V} = p_i V_i^{C_P/C_V} \tag{23-28}$$

$$= p_i V_i^{\gamma},$$

where we used $C_P = C_V + R$ and where we have defined the **heat capacity ratio** γ to be

$$\gamma = \frac{C_P}{C_V} = \begin{cases} \frac{5}{3} = 1.67 & \text{monatomic gas} \\ \frac{7}{5} = 1.40 & \text{diatomic gas}. \end{cases} \tag{23-29}$$

Raising both sides of Eq. 23-27 to the power R/C_V then gives our final result:

$$p_i V_i^{\gamma} = p_f V_f^{\gamma} = \text{constant} \quad \text{(adiabatic equation of state)}. \tag{23-30}$$

This has been a lengthy derivation, but the adiabatic equation of state given in Eq. 23-30 is one of great importance in thermodynamics. The heat capacity ratio γ is widely used to characterize gases. The ideal values for monatomic and diatomic gases are 1.67 and 1.40, as given in Eq. 23-29, and real gases come extremely close to these values. The γ-values for more complex gases, such as might be found in an air–fuel mixture, are tabulated in engineering handbooks.

Equation 23-30 is the equation of a curve in the pV-plane. The curves found by graphing p versus V are called **adiabats**, and each adiabat is a trajectory having a specific value of the entropy S. Two such adiabats are shown in Fig. 23-11, where we see that they are *steeper* than the hyperbolic isotherms. An adiabatic process moves along one of these curves in just the same way that an isothermal process moves along an isotherm. Because the adiabats cross the

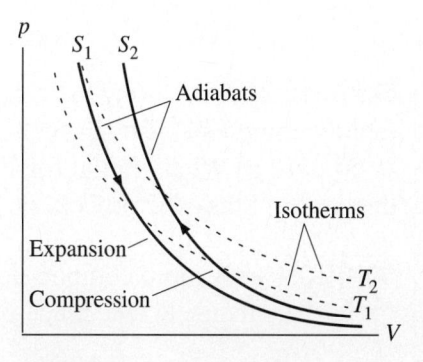

FIGURE 23-11 Two *adiabats*, showing an adiabatic expansion and an adiabatic compression at constant entropies S_1 and S_2.

isotherms, you can see from the figure that an adiabatic compression *increases* the gas temperature while an adiabatic expansion *lowers* its temperature.

From the definition of the heat capacity ratio (Eq. 23-29), we can work backward to find

$$C_V = \frac{R}{\gamma - 1}. \tag{23-31}$$

This result is useful for finding C_V when γ is given for a gas, as is often the case. We can also, with a few algebraic steps that we will omit, now rewrite Eq. 23-26 relating temperature and volume as

$$TV^{\gamma-1} = \text{constant}. \tag{23-32}$$

This result is sometimes more useful than the Eq. 23-30 equation of state.

As we did with the other processes, we can find expressions for the heat and the work of an adiabatic process. By definition, $Q = 0$. Therefore, according to the first law, $W_s = -\Delta E_{th}$. In other words, an adiabatic process is one in which the system's thermal energy is converted directly to or from work ($E_{th} \leftrightarrow W_s$).

But we also know, for *any* process, that $\Delta E_{th} = nC_V \Delta T$. Thus

$$W_s = -\Delta E_{th} = -nC_V \Delta T. \tag{23-33}$$

The ideal gas law allows us to write $n\Delta T = \Delta(pV)/R = (p_f V_f - p_i V_i)/R$. Also, as we found in Eq. 23-31, $C_V = R/(\gamma - 1)$. Combining these pieces gives

$$W_s = \frac{p_f V_f - p_i V_i}{1 - \gamma}. \tag{23-34}$$

We will leave it as a homework problem to show that the same result can be reached by integrating $p dV$.

You can see from Eq. 23-33 that work is done *by* the system ($W_s > 0$) if $\Delta T < 0$. Figure 23-11 shows that the temperature decreases for an adiabatic process in which the volume V increases—an *expansion* of the gas. So in an adiabatic expansion the temperature falls ($\Delta T < 0$), the thermal energy decreases ($\Delta E_{th} < 0$), and that thermal energy is converted to work done *by* the system ($W_s > 0$). Similarly, an adiabatic compression requires work to be done *on* the system ($W_s < 0$), increasing the temperature and the thermal energy ($\Delta T > 0$).

• **EXAMPLE 23-5** During the compression stroke of an internal combustion engine, the piston rapidly compresses the air–fuel mixture, initially at 1 atm pressure and 30°C, from 500 cm³ to 50 cm³. a) What are the final temperature and pressure? b) How much work is done on the gas? c) Show the process on a pV diagram. Assume that γ has its "air value" of 1.40.

SOLUTION a) A rapid compression is an adiabatic process because there is insufficient time for heat exchange to take place with the environment. Thus we can use Eq. 23-30 to find p_f:

$$p_f V_f^{\gamma} = p_i V_i^{\gamma}$$

$$\Rightarrow p_f = 1 \text{ atm} \cdot \left(\frac{V_i}{V_f}\right)^{1.40} = 1 \text{ atm} \cdot (10)^{1.40} = 25.1 \text{ atm}.$$

From the ideal gas law, because only n stays constant,

$$\frac{p_i V_i}{T_i} = \frac{p_f V_f}{T_f} \Rightarrow T_f = T_i \cdot \frac{p_f V_f}{p_i V_i} = 761 \text{ K} = 488°\text{C}.$$

b) The work done *on* the gas to compress it is $W = -W_s$. This is

$$W = -\left(\frac{p_f V_f - p_i V_i}{1 - \gamma}\right) = 191 \text{ J}.$$

In this process, 191 J of work are done on the gas to compress it, thereby raising its temperature to 488°C and increasing its thermal energy by 191 J. Note that *no* heat is transferred, despite the fact that the temperature increases.

c) Figure 23-12 shows the pV diagram.

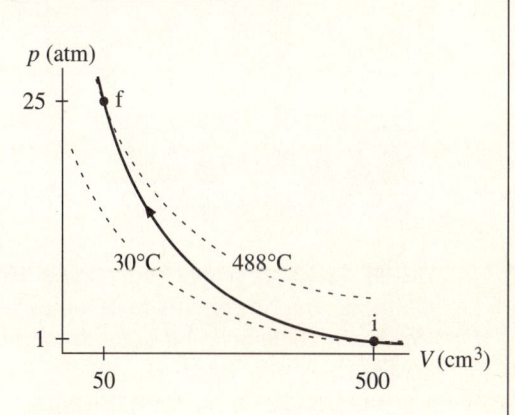

FIGURE 23-12 The pV diagram for Example 23-5.

23.4 Turning Heat into Work

When fuel is burned—be it wood, coal, gasoline, or glucose—its chemical energy is transformed into the thermal energy of the high-temperature reaction products. Much of that thermal energy is then transferred, via molecular collisions, to the cooler surroundings. This is *heat*—the transfer of energy due to a difference in temperature.

The earliest humans learned to use the heat energy from fires to warm themselves and cook their food. This is a transformation of heat energy into thermal energy. But is there a way to transform heat energy into *work*? Can we use the energy released by the fuel to grind corn, pump water, accelerate cars, launch rockets, or do any other task in which a force is exerted through a distance?

The first practical device for turning heat into work was the steam engine, invented in the eighteenth century. A steam engine uses water that is boiled, expanded to push a piston, then condensed back to liquid. The water is called the **working substance** of the engine. Similarly, a gasoline engine moves pistons up and down, alternately compressing and expanding a fuel-air mixture. The periodic nature of these devices causes the working substance to move in a closed, repetitive trajectory in the pV diagram, returning to its starting conditions once per cycle.

A cyclical device that follows a closed pV trajectory as it converts heat energy into useful work is called a **heat engine**. This is a generic name. Specific devices such as steam engines, gasoline engines, diesel engines, gas turbine generators, and many others are all examples of heat engines. Note that the working substance—the substance that pushes on the piston and does the work—is *not* the same as the fuel, which is the material burned to supply the heat energy. The working substance in a steam engine is water, which is heated by burning a fuel that could be wood, coal, oil, old physics textbooks, or anything else that burns. We will restrict our analysis in this text to heat engines using a gas as the working substance.

128 CHAPTER 23 THERMODYNAMICS OF IDEAL GASES

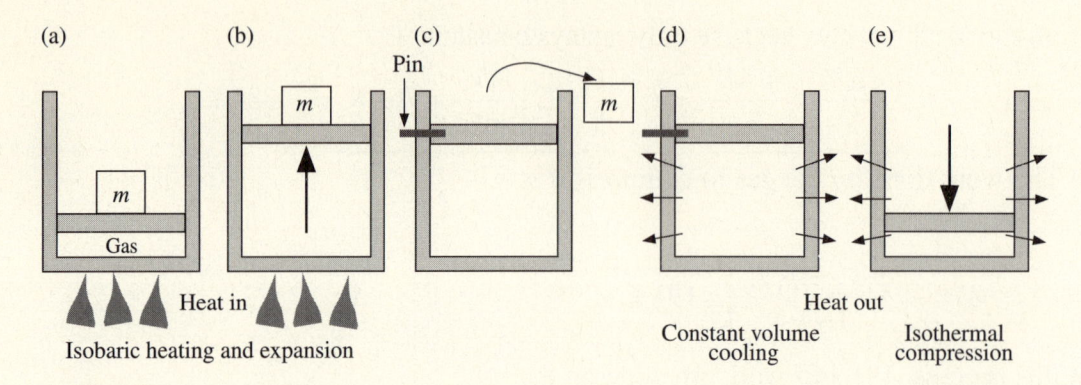

FIGURE 23-13 A simple heat engine. Heat energy, from the flame, does the work of lifting the mass. The gas returns to its initial state at the end and is ready to repeat the process. Work can continue to be done as long as there is a supply of fuel *and* there is source of cooling.

Figure 23-13 shows a simple engine that converts heat into useful work. The device consists of a cylinder of gas supporting a mass m that we wish to lift—that will be the "useful work" done by this engine. In part a), burning fuel supplies heat input to the gas, causing the gas to expand and lift the mass to the position shown in b). If the expansion is always very close to equilibrium, then the upward pressure force on the piston exactly balances the downward *constant* weight force of the mass. This implies that the expansion occurs at constant pressure. In c) we turn off the heat, *lock* the piston in place (with a pin), and slide the mass off. (Locking is necessary, otherwise the gas pressure would blow the piston out once the mass was removed.) None of the state variables change value as the mass is removed. Then, in d), the gas is slowly cooled back to room temperature. This is a constant volume process because the piston is still locked. The pressure after cooling is *less* than the initial pressure. Finally, in e), the pin is removed and the piston is allowed to slowly move back down while the gas is maintained at a constant room temperature. This isothermal compression restores the volume and pressure to their initial values.

The net effect of this multistep process is to convert some of the fuel's energy into work that lifts the mass. There is *no* net change in the gaseous working substance. The gas has followed a closed pV trajectory and returned to its initial state. Now we could imagine starting the whole process over again to lift another mass. We can continue lifting masses (doing work) as long as we have fuel.

Figure 23-14 shows the cycle of the Fig. 23-13 engine on a pV diagram. From 1 to 2 (isobaric expansion), heat is input and work is done. From 2 to 3, heat is removed in a constant volume cooling and no work done. Lastly, from 3 to 4, heat continues to be removed to allow the gas to be compressed at constant temperature. The environment (air pressure) is doing work *on* the gas, to compress it, during this process. State 4 is identical to state 1, so the gas has been returned to its initial state and can begin the process again.

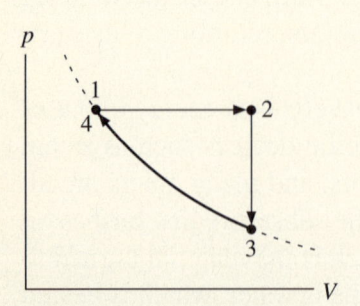

FIGURE 23-14 The closed cycle pV diagram for the heat engine of Fig. 23-13.

Note that the cyclical process of Fig. 23-13 involves a *cooling stage* in which heat is *removed* from the gas. But if heat energy is to flow out of the gas, there must be a surrounding environment that is *colder* than the gas. The key to understanding engines is that they require both a *heat source* (burning fuel) *and* a *heat sink* (cooling water, the air, or something at a lower temperature than the gas). The heat output increases the thermal energy in the environment and is called **waste heat**.

Using this information, we can characterize a **closed cycle** as a thermodynamic process in which

1. Heat is input from a high-temperature heat source in the environment.
2. Work is done.
3. Waste heat is returned to a low-temperature heat sink in the environment.
4. The working substance returns to its initial state.

Because the working substance returns to its initial state, it must have $\Delta T_{cycle} = 0$. The temperature goes up and down during different portions of the cycle, but it has no *net* change. From the relationship between E_{th} and T, this implies that $\Delta E_{cycle} = 0$. When the cycle is complete there has been no *net* change in the internal energy of the gas. If we now consider the first law, we see that for the *entire cycle*

$$\Delta E_{cycle} = 0 = Q_{cycle} - (W_s)_{cycle}$$
$$\Rightarrow (W_s)_{cycle} = Q_{cycle},$$
(23-35)

where $Q_{cycle} = \Sigma Q$ and $(W_s)_{cycle} = \Sigma W_s$ are the *net* heat and *net* work summed over all the individual thermodynamic process that make up the full cycle. The *net* effect of a heat engine is to convert heat into work *without* changing the working substance, and that is what Eq. 23-35 is telling us. Note that this is most emphatically *not* true for an individual process within the cycle, but only for the net heat input and net work done during the *entire* cycle.

We can make an interesting and important geometric interpretation of the work done by the system during a closed cycle. Figure 23-15a shows a closed cycle process. How much work W_s is done during this cycle? As we have seen, work is done only when the volume changes. A *positive* amount of work W_{expand} is done *by* the system as it expands from V_{min} to V_{max}, and this work is the area under the upper part of the curve in Fig. 23-15b. To complete the cycle, the environment must do work *on* the system to compress the

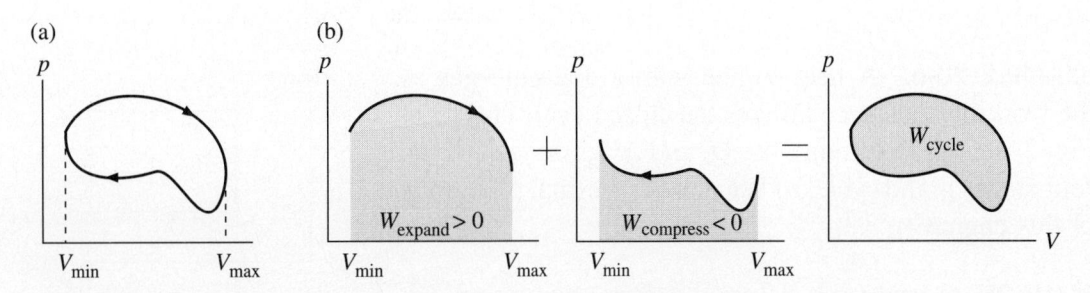

FIGURE 23-15 a) A closed cycle process. b) The work $(W_s)_{cycle}$ done *by* the system in this process is seen to be just the area enclosed *inside* the curve.

gas back to V_{min}. From the system's perspective, a *negative* amount of work $W_{compress}$ is done, whose magnitude is equal to the area under the lower part of the curve. Thus

$$(W_s)_{cycle} = W_{expand} + W_{compress} = \text{area } inside \text{ the closed curve.} \qquad (23\text{-}36)$$

The net work done by a heat engine during one complete cycle is just the area enclosed by the *pV* curve for the cycle. A thermodynamic cycle with a larger enclosed area will do more net work than one with a smaller enclosed area. Notice that the working substance must go around the *pV* trajectory in a *clockwise* direction for $(W_s)_{cycle}$ to be positive—that is, for the heat engine to do net work on the environment.

Heat engines are characterized by their *efficiency* at converting heat energy into work. As the example of Figs. 23-13 and 23-14 illustrated, some processes in the cycle have heat flowing *into* the system ($Q > 0$) and others have heat flowing *out* of the system ($Q < 0$). Define Q_{in} as the sum of all *positive* heat terms and Q_{out} as the sum of all *negative* heat terms. Then

$$Q_{cycle} = Q_{in} + Q_{out} = Q_{in} - |Q_{out}|. \qquad (23\text{-}37)$$

Q_{in} represents heat input to the system by burning fuel, while Q_{out} is heat returned from the system to the environment during any cooling processes. For example, $Q_{in} = Q_{12}$ and $Q_{out} = Q_{23} + Q_{34}$ in the cycle of Fig. 23-14. Note that Q_{out} is a negative number, according to our sign convention for the first law. But we often want to refer to just the magnitude $|Q_{out}|$ of the heat flowing from the system to the environment, and that is why we have used absolute value signs in Eq. 23-37.

Efficiency measures how much work is done per unit of fuel burned, so let us define the **thermal efficiency** η as

$$\eta = \frac{(W_s)_{cycle}}{Q_{in}}. \qquad (23\text{-}38)$$

(The symbol η is the lowercase Greek letter "eta.") But, from the first law, $(W_s)_{cycle} = Q_{cycle}$. Thus

$$\eta = \frac{Q_{cycle}}{Q_{in}} = \frac{Q_{in} - |Q_{out}|}{Q_{in}} = 1 - \frac{|Q_{out}|}{Q_{in}} \leq 1. \qquad (23\text{-}39)$$

A "perfect" heat engine would have $\eta = 1.00$ and would be 100% efficient at converting the fuel's heat energy into work. This would require $|Q_{out}| = 0$ and thus no waste heat. The goal of maximizing the efficiency of engines is achieved by minimizing waste heat.

EXAMPLE 23-6 A heat engine with a diatomic gas as the working substance utilizes the closed cycle shown in Fig. 23-16. a) Determine W_s, Q, and ΔE_{th} for each of the four stages in the cycle. b) What is the thermal efficiency of this engine?

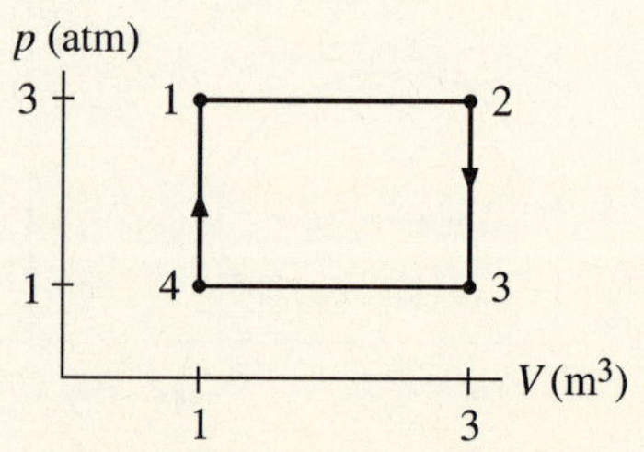

SOLUTION a) Process 1→ 2 is an isobaric expansion, so

$$(W_s)_{12} = p\Delta V = 6.06 \times 10^5 \text{ J},$$

FIGURE 23-16 The closed cycle *pV* diagram of Example 23-6.

where we converted pressure to pascals before computing. The heat for an isobaric expansion is given by Eq. 23-16:

$$Q_{12} = nC_P\Delta T = (C_P/R)p\Delta V = (C_P/R)(W_s)_{12} = 21.21 \times 10^5 \text{ J}.$$

Notice how we related ΔT first to ΔV and then to W_s. This is necessary because we do not know T. Then, using the first law,

$$\Delta E_{12} = Q_{12} - (W_s)_{12} = 15.15 \times 10^5 \text{ J}.$$

Process 2 → 3 is a constant volume cooling, so $(W_s)_{23} = 0$ and $\Delta E_{23} = Q_{23} = nC_V\Delta T$. This time $n\Delta T = V\Delta p/R$, and Δp is *negative* ($\Delta p = -2$ atm), so

$$\Delta E_{23} = Q_{23} = (C_V/R)V\Delta p = -15.15 \times 10^5 \text{ J}.$$

Process 3 → 4 is an isobaric compression. Note, however, that ΔV is *negative*. $(W_s)_{34} = p\Delta V = -2.02 \times 10^5$ J and

$$Q_{34} = nC_P\Delta T = (C_P/R)(W_s)_{34} = -7.07 \times 10^5 \text{ J}.$$

Then $\Delta E_{th} = -5.05 \times 10^5$ J.

Process 4 → 1 is another constant volume process, so again $(W_s)_{41} = 0$ and

$$\Delta E_{23} = Q_{23} = nC_V\Delta T = (C_V/R)V\Delta p = 5.05 \times 10^5 \text{ J}.$$

The results of these four processes are shown in Table 23-1. We can find the net result for $(W_s)_{cycle}$, Q_{cycle}, and ΔE_{cycle} by summing the columns:

$$(W_s)_{cycle} = 4.04 \times 10^5 \text{ J} \qquad Q_{cycle} = 4.04 \times 10^5 \text{ J} \qquad \Delta E_{cycle} = 0.$$

As expected, $(W_s)_{cycle} = Q_{cycle}$ and $\Delta E_{cycle} = 0$.

b) Heat is *input* during processes 1 → 2 and 4 → 1, so summing these gives $Q_{in} = 26.26 \times 10^5$ J. The thermal efficiency is then

$$\eta = \frac{(W_s)_{cycle}}{Q_{in}} = \frac{4.04 \times 10^5 \text{ J}}{26.26 \times 10^5 \text{ J}} = 0.154 = 15.4\%.$$

This may not seem very efficient, but it is quite typical of many real engines.

TABLE 23-1 Summary of energy transfers in Example 23-6. All energies $\times 10^5$ J.

Process	W_s	Q	ΔE
1 → 2	6.06	21.21	15.15
2 → 3	0	−15.15	−15.15
3 → 4	−2.02	−7.07	−5.05
4 → 1	0	5.05	5.05
Net	4.04	4.04	0
$Q_{in} =$		26.26	
$\eta =$		0.154	

The engine in Example 23-6 was not a very realistic heat engine, but it did illustrate the kinds of reasoning and computations involved in the analysis of a heat engine. In the next section we will look at some heat engines of practical and theoretical significance.

23.5 Examples of Heat Engines

In this section we will look at several examples of heat engines. These thermodynamic cycles are *models* of real engines. Although they leave out some details, they do a remarkably good job of characterizing how real devices turn heat into work.

The two most important parameters characterizing an engine are its work per cycle $(W_s)_{\text{cycle}}$ and its thermal efficiency η. Our analysis will be aimed at determining these two parameters in terms of "operating parameters" such as temperatures, compression ratios, and heat of combustion. Along the way we will determine pressure, temperature, and volume at different points in the cycle. Each stage of the analysis will rely heavily on our already completed analyses of the four basic thermodynamic processes.

The Otto Cycle

The heat engine with which you are most familiar is the gasoline-powered engine in your car. Although the details of the engine's thermodynamics are complex, we can achieve a good understanding of the basic operation with an *idealized* engine that follows a thermodynamic cycle called the *Otto cycle*. Figure 23-17 shows a cutaway picture

FIGURE 23-17 A cutaway view of a typical gasoline engine.

of a gasoline engine, where you see that it consists of pistons moving up and down in cylinders. The up-down piston motion is converted, by clever mechanical devices, to the rotational motion of the crankshaft, but we do not need to understand the mechanical couplings to analyze the thermodynamics. Notice that the top of each cylinder has an intake valve and an exhaust valve. These valves open and close to admit the fuel–air mixture and, later, to exhaust the combustion products. Not shown, but also at the top of each cylinder, is the spark plug that ignites the fuel–air mixture.

How does this engine work? We can identify five key steps in the process.

1. When the piston is at the very bottom (V_{max}), the intake value opens and the fuel–air mixture flows in. This occurs at $p \approx 1$ atm and at ambient air temperature.

2. The piston then rapidly compresses the fuel–air mixture during the *compression stroke* until the piston reaches the top of its travel (V_{min}). The ratio $r = V_{max}/V_{min}$ is called the **compression ratio**, and it is one of the most important numbers for characterizing an engine. Because the compression is very rapid, providing little time for heat to be exchanged with the surroundings, the compression stroke is adiabatic.

3. The spark plug fires when the piston reaches its highest point, igniting the gasoline and causing it to release its **heat of combustion**. This is a little different from the heat sources we have considered until now, which were in the outer environment, but this new type of heat is equivalent. The gasoline burns so quickly that there is no time for the piston to move, so the heating occurs at constant volume. The heating increases the temperature and pressure.

4. The increased pressure now pushes the piston back out in the *power stroke* until it reaches the bottom. This is another adiabatic process.

5. Last, the exhaust valve opens and the hot combustion products flow out. This is a cooling process that takes place at constant volume, dropping the pressure and temperature back to their starting values. Now the cycle can start over again.

We can transfer these statements about the processes to the pV diagram of Fig. 23-18. This combination of two constant volume processes and two adiabatic processes is the Otto cycle. Notice that $V_1 = V_4 = V_{max}$ and $V_2 = V_3 = V_{min}$, so the compression ratio r is

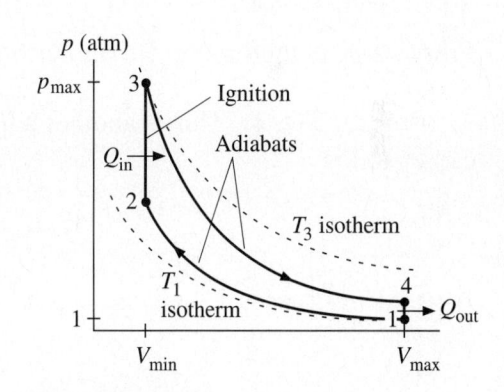

$$r = \frac{V_{max}}{V_{min}} = \frac{V_1}{V_2} = \frac{V_4}{V_3}. \tag{23-40}$$

FIGURE 23-18 The Otto cycle, which is a good model of a gasoline-powered internal combustion engine.

Our analysis of the Otto cycle has three goals:

1. To find an expression for the net work $(W_s)_{cycle}$,
2. To find an expression for the thermal efficiency η, and
3. To find the pressure, temperature, and volume at the four "corners" of the cycle.

The initial conditions (p_1, V_1, T_1), the compression ratio r, and the heat of combustion will be considered known quantities. We will proceed by considering, in turn, each of the four "sides" of the cycle.

Process 1 → 2: $Q_{12} = 0$ for an adiabatic process, and the work done in an adiabatic compression is

$$(W_s)_{12} = \frac{p_2 V_2 - p_1 V_1}{1 - \gamma} < 0.$$

We also know that $p_2 V_2^\gamma = p_1 V_1^\gamma$, from which we can find the pressure p_2:

$$p_2 = p_1 \left(\frac{V_1}{V_2}\right)^\gamma = p_1 r^\gamma,$$

where r is the compression ratio. Similarly $T_2 V_2^{\gamma-1} = T_1 V_1^{\gamma-1}$ from which we find

$$T_2 = T_1 r^{\gamma-1}.$$

Process 2 → 3: A constant volume process ($V_3 = V_2$) has $(W_s)_{23} = 0$ and $Q_{23} = nC_V \Delta T_{23}$. The heat of combustion Q_{23} will be determined by the quantity of gasoline admitted and the chemical properties of gasoline. The temperature rises to

$$T_3 = T_2 + \Delta T_{23} = T_2 + \frac{Q_{23}}{nC_V}.$$

Pressure p_3 is then found for an isochoric process from $p_3 = p_2(T_3/T_2)$.

Process 3 → 4: This is another adiabatic process, so $Q_{34} = 0$ and the work done in the expansion is

$$(W_s)_{34} = \frac{p_4 V_4 - p_3 V_3}{1 - \gamma} > 0.$$

Also, as in Process 1 → 2, $p_3 = p_4 r^\gamma$ and $T_3 = T_4 r^{\gamma-1}$, which can be inverted to find p_4 and T_4.

Process 4 → 1: This is another constant volume process, so $(W_s)_{41} = 0$ and $Q_{41} = nC_V \Delta T_{41} < 0$.

We can now put these results together to find the net work done during the complete cycle:

$$(W_s)_{\text{cycle}} = (W_s)_{12} + (W_s)_{34} = \frac{p_2 V_2 - p_1 V_1 + p_4 V_4 - p_3 V_3}{1 - \gamma}. \tag{23-41}$$

We can put this in a different form by noting that $pV = nRT$, from which we get

$$(W_s)_{\text{cycle}} = \frac{nR}{1 - \gamma} \cdot (T_2 - T_1 + T_4 - T_3). \tag{23-42}$$

Now let's determine the efficiency of the Otto cycle. The thermal efficiency is defined as $\eta = (W_s)_{cycle}/Q_{in}$. The work is given by Eq. 23-42, and the heat input is the positive Q_{23} of the ignition. (Q_{41} is Q_{out}.) Combining these gives

$$\eta = \frac{nR/(1-\gamma) \cdot (T_2 - T_1 + T_4 - T_3)}{nC_V \cdot (T_3 - T_2)} = \frac{R}{C_V(1-\gamma)} \cdot \frac{T_2 - T_1 + T_4 - T_3}{T_3 - T_2}. \quad (23\text{-}43)$$

Now $C_V = R/(\gamma - 1)$, from Eq. 23-31, from which we find

$$\frac{R}{C_V(1-\gamma)} = \frac{R}{(R/(\gamma-1))(1-\gamma)} = -1.$$

Thus the efficiency is

$$\eta_{Otto} = -\frac{T_2 - T_1 + T_4 - T_3}{T_3 - T_2} = \frac{(T_3 - T_2) - (T_4 - T_1)}{T_3 - T_2} = 1 - \frac{T_4 - T_1}{T_3 - T_2}. \quad (23\text{-}44)$$

Thermal efficiency of the Otto cycle, it turns out, can be expressed entirely in terms of the temperatures of the four "corners." We can, however, simplify this further by use of the adiabatic relationships $T_2 = T_1 r^{\gamma-1}$ and $T_3 = T_4 r^{\gamma-1}$ that we found previously. They give

$$\eta_{Otto} = 1 - \frac{T_4 - T_1}{T_4 r^{\gamma-1} - T_3 r^{\gamma-1}} = 1 - \frac{T_4 - T_1}{(T_4 - T_1)r^{\gamma-1}} = 1 - \frac{1}{r^{\gamma-1}}. \quad (23\text{-}45)$$

We end up with the remarkably simple result that the thermal efficiency of the Otto cycle depends only on the engine's compression ratio r and the γ of the fuel–air mixture! That is why the compression ratio is such an important number to car enthusiasts. Figure 23-19 shows a graph of the efficiency as a function of the compression ratio for the air-value $\gamma = 1.40$. You can see that η initially increases quite rapidly, but after a while further increases in the compression ratio give little further improvement in efficiency. Increasing the compression ratio also increases the maximum pressure and temperature in the cylinder, and the designer must make sure that she does not exceed the limits of the materials in use.

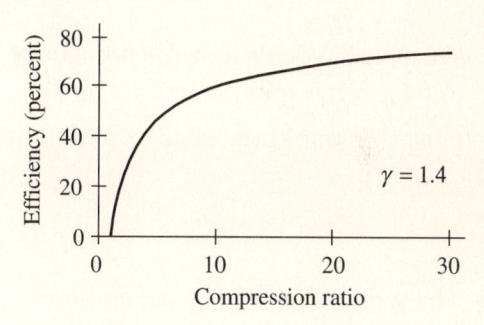

FIGURE 23-19 The thermal efficiency η as a function of the compression ratio for an Otto-cycle engine.

EXAMPLE 23-7 A 4-cylinder automobile engine has a *displacement* of 1800 cm³ and a compression ratio of 9.0. It operates with an air temperature and pressure of 20°C and 1 atm = 101 kPa. The gasoline injected into each cylinder has a heat of combustion of 900 J. The fuel–air mixture in the cylinder has $\gamma = 1.30$. a) Find the pressure, volume, and temperature at each of the four "corners" of the thermodynamic cycle for *one* cylinder of this engine. b) Find the net work done by this cylinder during one cycle and the thermal efficiency. c) Find the engine's power output at 3000 rpm if this is a *four-stroke engine*.

SOLUTION This is realistic data for a mid-size passenger car, but it takes a bit of explanation to see what the "car talk" means. A cylinder's *displacement* is the volume displaced by the piston as it moves from the bottom to the top, or $\Delta V = V_{max} - V_{min}$. The measured value is the total for all four cylinders, so each cylinder has $\Delta V = 450$ cm^3. The heat of combustion is the heat energy Q_{23} supplied by the burning gasoline during the constant volume ignition. Finally, the Otto cycle describes a *two-stroke engine* with a compression stroke and a power stroke. Automobile engines have a more complicated *four-stoke engine* in which two strokes (one up-down piston oscillation) are identical to the Otto cycle while the other two (another up-down oscillation) are used to empty and refill the gas in the cylinder. The spark plug does not fire during the second two strokes. The net effect is that *half* the cylinders fire on one revolution of the crankshaft and the other half fire on the next revolution. Each cylinder fires once for every two revolutions, thus firing 1500 times a minute if the crankshaft is rotating at 3000 rpm. So with this background, we can begin.

a) We will use the idealized Otto cycle, with $P_1 = 1$ atm $= 101$ kPa and $T_1 = 20°C = 293$ K. We know $\Delta V = V_{max} - V_{min} = 450$ cm^3 and the compression ratio $r = V_{max}/V_{min} = 9$. So

$$V_{max} = 9V_{min} \quad \Rightarrow \quad \Delta V = 9V_{min} - V_{min} = 8V_{min} = 450 \text{ cm}^3$$

$$\Rightarrow \begin{cases} V_{min} = V_1 = V_4 = 56.2 \text{ cm}^3 \\ V_{max} = V_2 = V_3 = 506.2 \text{ cm}^3. \end{cases}$$

We can also determine the number of moles of working substance, using the ideal gas law, to be $n = p_1 V_1 / RT_1 = 0.0210$ mol.

Now that our knowledge at point 1 is complete, we can advance to point 2 with

$$p_2 = p_1 r^\gamma = 17.40 \text{ atm} = 1757 \text{ kPa}$$
$$T_2 = T_1 r^{\gamma-1} = 566.4 \text{ K} = 293.4°C.$$

The work done during the compression stroke is

$$(W_s)_{12} = \frac{p_2 V_2 - p_1 V_1}{1 - \gamma} = -158.7 \text{ J}.$$

It is negative because the environment is doing work on the gas during the compression.

To find the temperature at point 3, we need C_V, which we can find from Eq. 23-31 to be $C_V = R/(\gamma - 1) = 27.7$ J/mol K. Now $Q_{23} = Q_{in} = 900$ J, and this increases the temperature to

$$T_3 = T_2 + \frac{Q_{23}}{\mu C_V} = 2113.6 \text{ K} = 1840.6°C.$$

For a constant volume process, $P_3 = P_2(T_3/T_2) = 64.93$ atm $= 6558$ kPa.

Finally, the adiabatic expansion to point 4 gives

$$p_4 = \frac{p_3}{r^\gamma} = 3.73 \text{ atm} = 377 \text{ kPa}$$

$$T_4 = \frac{T_3}{r^{\gamma-1}} = 1093.3 \text{ K} = 820.3°C.$$

The work done during the power stroke expansion is

$$(W_s)_{34} = \frac{p_4 V_4 - p_3 V_3}{1 - \gamma} = 592.4 \text{ J}.$$

b) The net work is $(W_s)_{cycle} = (W_s)_{12} + (W_s)_{34} = 433.7$ J. Because $Q_{in} = 900$ J, the thermal efficiency is

$$\eta = \frac{(W_s)_{cycle}}{Q_{in}} = \frac{433.7 \text{ J}}{900 \text{ J}} = 0.482 = 48.2\%.$$

We could, of course, have calculated this directly as $\eta = 1 - 1/r^{\gamma-1} = 0.482$. However, we would have missed finding the pressure and temperature at each point in the process.

c) Power is the *rate* at which work is done. A four-stroke engine turning at 3000 rpm will fire each cylinder 1500 times per minute or 25 times per second. Each time the cylinder fires, it does 433.7 J of work. Thus the work done per second by *one* cylinder is $P_{cyl} = 25 \cdot (W_s)_{cycle} = 10.85$ kW. Because there are four cylinders doing this, the engine's power output is

$$P_{engine} = 4 \cdot P_{cyl} = 43.4 \text{ kW} \approx 60 \text{ horsepower}.$$

All the numbers in this example, including the power output and the efficiency, are quite realistic. We have come a long way in our study of thermal physics to be making accurate predictions of an automobile engine's power output!

The Diesel Cycle

Another cycle of practical importance is the *Diesel cycle,* which describes a diesel engine. A diesel engine is similar to a gasoline engine in most respects, but there is one significant difference. The piston, during the compression stroke, compresses only air ($\gamma = 1.4$) rather than a fuel–air mixture, and the compression ratio is higher than in a gasoline engine ($r \approx 20$). This increases T_2, at the end of the compression stroke, to ≈ 1000 K. At that point, with the piston at the top, fuel is *injected* into the cylinder where it begins to burn spontaneously because of the high temperature. (There are no spark plugs in a diesel engine!) Unlike the very quick "explosion" in a gasoline engine, the diesel fuel burns more slowly. To keep the combustion going, fuel continues to be injected as the piston begins moving down. These factors make the combustion stage a constant *pressure* process rather than the constant volume process found in a gasoline engine. Once burning terminates, the expansion continues adiabatically.

Figure 23-20 shows the Diesel cycle. You should compare this to the Otto cycle of Fig. 23-18 to see the similarities and differences. In general, a diesel engine has a higher thermal efficiency than a gasoline engine due to its higher compression ratio. That is why they are used for heavy-duty applications in trucks, trains, and farm equipment. However, they have environmental and other practical difficulties that make them less suited for passenger cars or light duty applications.

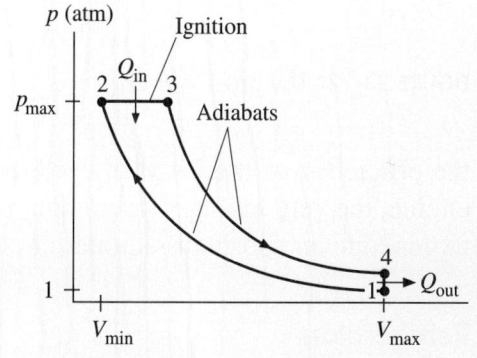

FIGURE 23-20 The Diesel cycle.

The Brayton Cycle

All the heat engines we have looked at thus far have had a *fixed gas* that is alternately compressed and expanded by a moving piston. Many other important applications of heat engines utilize a *flowing gas* operation. A good example is a *gas turbine engine*. These engines are used to generate high-power outputs for industrial processes, for electrical power generation, and as the basis for jet engines in aircraft and rockets. The thermodynamic cycle of a gas turbine is called the *Brayton cycle*.

[**Photo suggestion: A gas turbine engine.**]

Figure 23-21 is a schematic look at a gas turbine engine. Air, starting at room pressure and temperature, is first compressed adiabatically, to raise its temperature. The air then flows into a combustion chamber. Fuel is continuously admitted to the combustion chamber where it mixes with the hot air and is ignited, causing a constant pressure heating of the gas. These first two stages are very much like a diesel engine, although here the gas physically flows from one process to the next. The high pressure gas is then used to spin a turbine that does some form of useful work. This is an adiabatic expansion, which drops the temperature and pressure of the gas. To get the most work, we need the maximum pressure drop, so the pressure at the end of the expansion is back to ≈1 atm but the gas is still quite hot. The gas completes the cycle by flowing through a device called a **heat exchanger**, which transfers heat energy to some kind of cooling fluid. Large power plants are often sited on rivers or coasts in order to use the water for the cooling fluid in the heat exchanger.

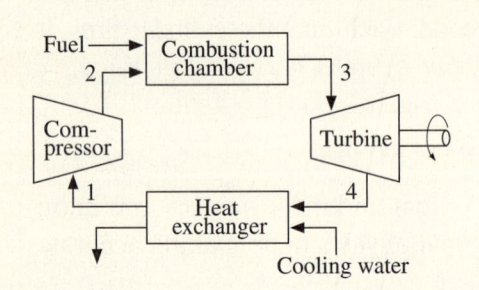

FIGURE 23-21 A gas turbine engine.

Figure 23-22, shows the Brayton cycle. This cycle completes two adiabatic stages—the compression and the expansion through the turbine—using constant pressure combustion and constant pressure cooling. We will leave it as a homework problem for you to show that the thermal efficiency of the Brayton cycle is

$$\eta_{\text{Brayton}} = 1 - \frac{1}{r_p^{(\gamma-1)/\gamma}}, \qquad (23\text{-}46)$$

FIGURE 23-22 The Brayton cycle.

where $r_p = p_{\text{max}}/p_{\text{min}}$ is the *pressure ratio*. Although the efficiency of the Brayton cycle is not significantly different from that of a gasoline engine, the very high gas throughput possible in a flow system allows a gas turbine engine to generate an enormous amount of power.

Refrigerators

So far, all the examples you have seen have had a thermodynamic cycle whose *pV* diagram operates in the clockwise direction. These engines input heat ($Q_{\text{cycle}} > 0$) from a fuel

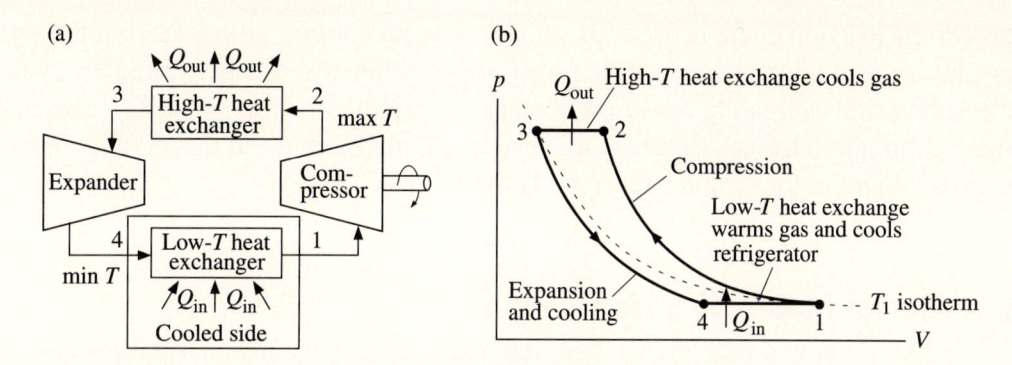

FIGURE 23-23 a) A *refrigerator* that removes heat from the cold side and exhausts heat to the hot side. b) The reversed Brayton cycle for this device. Note that $T_3 \approx T_1$.

source and generate a net positive work $[(W_s)_{\text{cycle}} > 0]$. Suppose that we were to operate a "reversed" Brayton cycle, going counterclockwise in the pV diagram. Figure 23-23a shows a device for doing this, and Fig. 23-23b is its pV diagram. Starting from point 1, the gas is adiabatically compressed to increase its temperature and pressure. It then flows through a high-temperature heat exchanger where the gas cools, at constant pressure, to $T_3 \approx T_1$. After cooling, the gas is adiabatically expanded, which further cools it. Now the gas is significantly colder than when it started. It completes its cycle by flowing through a low-temperature heat exchanger, where it is warmed back to its starting temperature.

Process $4 \rightarrow 1$, while warming the gas of the engine, *cools* the fluid with which it interacts in the heat exchanger. That is, heat is flowing *from* the heat exchanger into the engine's gas, thereby cooling the fluid while warming the gas. Suppose that the low-temperature heat exchanger is a closed container of air surrounding a pipe through which the engine's cold gas is flowing. The heat exchange process $4 \rightarrow 1$ is then going to cool the air in the container as it warms the gas flowing through the pipe. If you were to place eggs and milk inside this closed container, you would call it a refrigerator!

Any engine that runs a counterclockwise cycle is known generically as a **refrigerator**. You can see that heat energy is taken *from* the low-temperature heat exchanger, thus cooling it, and exhausted *to* the high-temperature heat exchanger. The net effect is that heat is flowing from cold to hot! But doesn't this violate the second law of thermodynamics? Not here, because this is not an isolated system. To have heat flow "backward," the environment is doing work *on* the system and, in the process, increasing the entropy of the environment by more than the refrigerator's entropy is decreasing.

The net work $(W_s)_{\text{cycle}}$ is negative for a counterclockwise cycle, meaning that the environment does net work *on* the system. Define $W_{\text{ref}} = -(W_s)_{\text{cycle}} > 0$ as the work done *on* the refrigerator to operate it. From the first law,

$$(W_s)_{\text{cycle}} = -W_{\text{ref}} = Q_{\text{cycle}} = Q_{\text{in}} - |Q_{\text{out}}|$$
$$\Rightarrow |Q_{\text{out}}| = Q_{\text{in}} + W_{\text{ref}} > Q_{\text{in}}.$$

(23-47)

The heat output to the high-temperature heat exchanger exceeds the heat taken in from the low-temperature heat exchanger. The excess is simply the work energy that was needed to "pump" the heat energy from the cold side to the hot side.

A reversed Brayton cycle is used for some industrial cooling processes, but the refrigerators and air conditioners with which you are familiar use a more complex cycle in which compression causes the working substance to liquefy and the expansion causes it to vaporize. Although this takes the analysis beyond what we can do in this course, the *principles* are identical to the simpler reversed Brayton cycle.

23.6 Pictorial Model of a Heat Engine

In Section 23.4 we found that *any* heat engine, regardless of its design, requires both heat input and heat output. Because heat flows only if there is a temperature difference, a heat engine must take in heat from a source that is hotter than the working substance and exhaust heat to a sink that is colder than the working substance. Figure 23-24 shows a pictorial model of a heat engine operating "between" two temperatures, T_H and T_C. The *hot side*, at temperature T_H, is due to burning fuel or to using an existing source of thermal energy (such as in tapping geothermal energy to generate electrical energy). The *cold side*, at temperature T_C, is some kind of cooling substance—such as running water or blowing air. In the Otto cycle, for example, T_H is the maximum temperature in the cycle, after the fuel is ignited, and T_C is the minimum temperature, at the point where the gas is returned to the temperature of the ambient air.

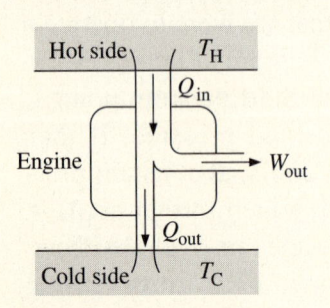

FIGURE 23-24 A heat engine converts input heat energy from the hot side into work and exhausts waste heat energy to the cold side.

As Fig. 23-24 shows, heat energy Q_{in} flows *into* the system (the engine) from the hot side. Some of the heat input is converted to useful work W_{out} and some is rejected or exhausted as "waste heat" Q_{out} to the cold side. The heat engine is simply a device for dividing the input heat energy Q_{in} into work W_{out} and leftover heat energy Q_{out} that must be exhausted from the engine. This is the basic operation of *any* heat engine, regardless of the specific cycle it happens to use, and Fig. 23-24 captures the idea nicely in pictorial form.

Over the course of a complete cycle, the gas returns to its initial temperature and so it has *no* net change in its thermal energy: $(\Delta E_{th})_{cycle} = 0$. The net heat flow during one cycle is $Q_{cycle} = Q_{in} + Q_{out} = Q_{in} - |Q_{out}|$. (Keep in mind that the quantity Q_{out} is negative.) We can use the first law to relate these heat flows to the work done by the engine during one cycle:

$$\Delta E_{th} = 0 = Q - W_{out} = (Q_{in} - |Q_{out}|) - W_{out}$$
$$\Rightarrow Q_{in} = W_{out} + |Q_{out}|, \quad (23\text{-}48)$$

where $W_{out} = (W_s)_{cycle}$. For the rest of this chapter we will replace $(W_s)_{cycle}$ with the more descriptive W_{out}, indicating the useful "output" of the engine in the form of work done.

Equation 23-48 is the essence of energy conservation: The input heat energy is divided into two forms of output energy, but no energy is created or destroyed. Thus the "pipes" in Fig. 23-24 show Q_{in} flowing in and W_{out} and Q_{out} flowing out.

Suppose that the engine is removed and that a direct thermal connection is made between the hot side and the cold side. Not surprisingly, heat energy flows from the higher temperature T_H to the lower temperature T_C, as shown in Fig. 23-25a, with $|Q_{out}| = Q_{in}$ and $W_{out} = 0$. But the second law of thermodynamics, as we have seen, prohibits the spontaneous flow of energy from cold to hot in an isolated system. The "reverse" process shown in Fig. 23-25b does *not* occur, even though energy would be conserved.

On a hot day, you want the air conditioner to move heat energy *from* your cool house *to* the warmer outdoors—thus from cold to hot. We can arrange this, but only by performing external work on the system. That is what a refrigerator does. Figure 23-26 shows a pictorial model of a refrigerator, in which input work W_{in} is used to "lift" heat from the cold side to the hot side. Notice that Q_{in} is coming from the cold side and that the waste heat Q_{out} is being exhausted to the hot side. From the first law, or energy conservation,

$$|Q_{out}| = Q_{in} + W_{in} > Q_{in}, \qquad (23\text{-}49)$$

so *more* heat is exhausted on the hot side than was removed from the cold side—an important conclusion. The refrigerator of Fig. 23-26 does not violate the second law because it is not an isolated system. The environment is doing work on the system to force heat energy to flow from cold to hot.

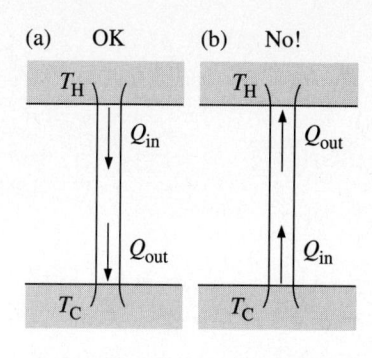

FIGURE 23-25 a) Heat flows spontaneously from T_H to T_C. b) The second law prohibits the reverse flow from T_C to T_H.

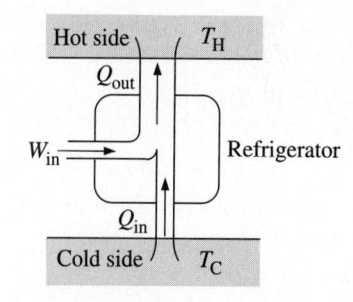

FIGURE 23-26 In a refrigerator, external work is done *on* the system to move heat from T_C to T_H.

A Perfect Engine?

Is it possible to build a perfect engine, one in which Q_{in} is converted entirely to work ($W_{out} = Q_{in}$) with $Q_{out} = 0$. By our definition of thermal efficiency, Eq. 23-38, an engine that converts the input heat energy entirely to useful work would have an efficiency $\eta = 1$. Such an engine would not violate the first law, because energy would be conserved, so perhaps we can invent such an engine if we are clever. Figure 23-27a shows a perfect engine.

Assuming that we have such an engine, let the work that it does power a refrigerator. Figure 23-37b shows the perfect engine doing work W_{out}. Its output becomes the input work W_{in} needed to operate the refrigerator. The refrigerator then removes heat energy $(Q_{ref})_{in}$ from the cold side and exhausts heat energy $(Q_{ref})_{out}$ to the hot side.

Think carefully about what we have done. The heat engine/refrigerator *combination* forms an isolated system with no work on or by the environment. Thus we have built an *isolated system* in which heat flows from the cold side to hot. (Compare this with the refrigerator of Fig. 23-26, where W_{in} comes from the outside environment so that the system is *not* isolated.) But this violates the second law of thermodynamics, so it can't be done! It must be that our *assumption* of a perfect engine, with $\eta = 1$, is invalid. We are

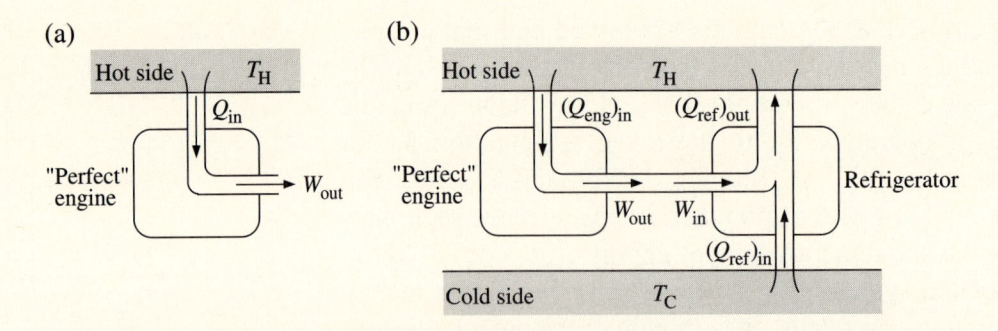

FIGURE 23-27 a) A "perfect engine" converts 100% of Q_{in} to work and has no waste heat energy. b) A perfect engine driving a refrigerator violates the second law by moving heat energy from T_C to T_H without needing any external work.

forced to conclude that there are no perfect heat engines, because if there were we could use one to violate the second law and make heat flow "backward" from cold to hot.

We noted in Chapter 22 that the second law of thermodynamics is often given a variety of more informal statements. Another common version of the second law that we can now give is:

> *Second Law–Informal Statement #2*: There are no perfect heat engines that can convert heat energy to work with 100% efficiency.

23.7 The Limits of Efficiency

Thermodynamics, and especially engineering thermodynamics, has its historical roots in the development of the steam engine and other machines of the early industrial revolution. The engines, for the most part, were designed and built by "tinkerers" on the basis of experience and insight rather than scientific analysis. The early steam engines were not very efficient at converting fuel energy into work.

In 1824, a brilliant young French engineer, Sadi Carnot, published the first major theoretical analysis of heat engines. His analysis, interestingly, *precedes* Joule's discovery of the full statement of conservation of energy. Carnot wrote:

> Everyone knows that heat can produce motion. That it possesses vast motive power no one can doubt, in these days when the steam engine is everywhere so well known ... Notwithstanding the satisfactory condition to which they have been brought today, their theory is very little understood. The question has often been raised whether the motive power of heat is unbounded, or whether the possible improvements in steam engines have an assignable limit.

Stated in more modern terms, Carnot's interest was whether engine efficiency can be improved without limit, if the designer is sufficiently clever, or whether there is some maximum efficiency that cannot be exceeded? If so, what is that maximum? In an effort to answer these questions, Carnot invented a thermodynamic cycle that now bears his name. No real engine uses this thermodynamic cycle, but its theoretical implications are especially important.

The Carnot Cycle

The **Carnot cycle** consists of two adiabatic stages and two isothermal stages, as shown in Fig. 23-28. The gas is first compressed from V_1 to V_2 while being maintained at the cold temperature T_C. Heat has to be removed from the gas as it is compressed ($Q_{12} < 0$) in order to keep the temperature constant. The gas continues to be compressed from V_2 to V_3, but now adiabatically as the gas is "decoupled" from its environment. This compression heats the gas to the hot temperature T_H. After reaching maximum compression at point 3, the gas expands—first isothermally at T_H from V_3 to V_4, then adiabatically from V_4 to V_1. The adiabatic expansion cools the gas back to its initial temperature and pressure, completing the cycle.

We have already analyzed the adiabatic stages as part of the Otto cycle, so we only need to add the four isothermal stages.

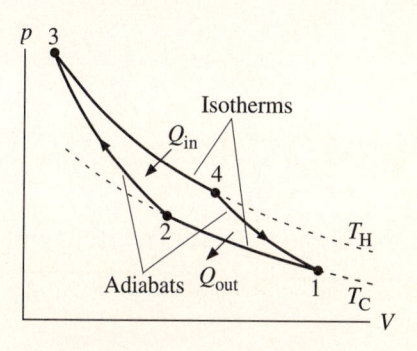

FIGURE 23-28 The Carnot cycle consists of two adiabatic processes and two isothermal processes.

Process 1 → 2: The work done in an isothermal compression at the cold temperature T_C is

$$(W_s)_{12} = nRT_C \ln \frac{V_2}{V_1} < 0.$$

An isothermal process also has $Q_{12} = (W_s)_{12}$ as well as $p_1 V_1 = p_2 V_2$.

Process 2 → 3: As you learned in the Otto cycle, $Q_{23} = 0$ and $(W_s)_{23} = (p_3 V_3 - p_2 V_2)/(1 - \gamma)$.

Process 3 → 4: The work done in the isothermal expansion at the hot temperature T_H is

$$(W_s)_{34} = nRT_H \ln \frac{V_4}{V_3} = Q_{34} > 0.$$

As the only positive heat term in the cycle, $Q_{34} = Q_{in}$. Also $p_3 V_3 = p_4 V_4$.

Process 4 → 1: $Q_{41} = 0$ and $(W_s)_{41} = (p_1 V_1 - p_4 V_4)/(1 - \gamma)$.

Work is done in all four stages. Summing them gives

$$(W_s)_{cycle} = nRT_C \ln \frac{V_2}{V_1} + nRT_H \ln \frac{V_4}{V_3} + \frac{p_3 V_3 - p_2 V_2 + p_1 V_1 - p_4 V_4}{1 - \gamma}. \quad (23\text{-}50)$$

This result seems cumbersome, but notice what we learned from the two isothermal stages—namely, $p_1 V_1 = p_2 V_2$ and $p_3 V_3 = p_4 V_4$. When these are used, the last term—the total work of the two adiabatic stages—becomes zero! Then we have a simpler result,

$$(W_s)_{cycle} = nR \left(T_H \ln \frac{V_4}{V_3} - T_C \ln \frac{V_1}{V_2} \right), \quad (23\text{-}51)$$

where we inverted V_2/V_1 to make both logarithms positive.

This expression for the work done in a complete cycle can be simplified yet further. From the two adiabatic stages we have

$$T_C V_2^{\gamma-1} = T_H V_3^{\gamma-1} \text{ and } T_C V_1^{\gamma-1} = T_H V_4^{\gamma-1}$$

$$\Rightarrow \frac{T_H}{T_C} = \left(\frac{V_2}{V_3}\right)^{\gamma-1} \quad \text{and} \quad \frac{T_H}{T_C} = \left(\frac{V_1}{V_4}\right)^{\gamma-1} \tag{23-52}$$

$$\Rightarrow \frac{V_2}{V_3} = \frac{V_1}{V_4}.$$

A simple algebraic rearrangement gives

$$\frac{V_1}{V_2} = \frac{V_4}{V_3}. \tag{23-53}$$

With this expression for V_1/V_2, Eq. 23-51 becomes a quite simple and useful result:

$$(W_s)_{\text{cycle}} = nR(T_H - T_C)\ln\frac{V_4}{V_3}. \tag{23-54}$$

We can now find the thermal efficiency, using $Q_{\text{in}} = Q_{34}$, to be

$$\eta_{\text{Carnot}} = \frac{(W_s)_{\text{cycle}}}{Q_{\text{in}}} = \frac{nR(T_H - T_C)\ln(V_4/V_3)}{nRT_H \ln(V_4/V_3)} = \frac{T_H - T_C}{T_H}$$
$$= 1 - \frac{T_C}{T_H}. \tag{23-55}$$

This remarkably simply result—that the efficiency depends only on the ratio of the temperatures of the hot side and the cold side of the heat engine—is Carnot's legacy to thermodynamics. You will see shortly that *no* engine, if operating between temperatures T_H and T_C, can exceed the efficiency of the Carnot cycle.

• **EXAMPLE 23-8** A Carnot-cycle engine is cooled by cooling water at $T_C = 10°C$. What temperature must the hot side of the engine maintain to have a thermal efficiency of 70%?

SOLUTION From Eq. 23-55,

$$T_H = \frac{T_C}{1 - \eta} = 943 \text{ K} = 670°C.$$

Note that, as always, Eq. 23-55 refers to *absolute* temperatures. A "real" engine would have to have a higher temperature than this to provide 70% efficiency because no real engine will match the Carnot efficiency.
•

• **EXAMPLE 23-9** The Otto-cycle engine of Example 23-7 operated between a highest temperature of 2144 K and a lowest temperature of 293 K, giving a thermal efficiency of 48.2%. What would be the efficiency of a Carnot cycle operating between these two temperatures?

SOLUTION This is a straightforward application of Eq. 23-55, giving

$$\eta = 1 - \frac{T_C}{T_H} = 0.863 = 86.3\%.$$

This is substantially higher than the Otto-cycle efficiency.

The Maximum Efficiency

The first law places no restrictions on the direction in which energy flows, but the second law does. Energy cannot flow in a direction that would reduce the entropy of an isolated system, and that is the reason that we can convert mechanical energy to thermal energy with 100% efficiency but can *not* make the reverse conversion of thermal to mechanical energy with 100% efficiency. This is a conclusion of major importance in engineering because it places a *limit* on our ability to convert heat energy into work.

Consider a heat engine such as the one depicted in Fig. 23-24. The *total* system consists of the engine, the hot and cold sides, and the object on which the engine is doing work. This total system is isolated from the rest of its environment, so the second law applies to it. The change in entropy of this total system is

$$\Delta S_{system} = \Delta S_{hot} + \Delta S_{cold} + \Delta S_{engine} + \Delta S_{object}, \qquad (23\text{-}56)$$

where S_{hot} and S_{cold} are the entropies of the high-temperature heat source and the low-temperature heat sink.

The engine operates in a *cycle* and always returns back to the same state, so after one cycle $\Delta S_{engine} = 0$. Furthermore, the work done to the object gives it a center-of-mass mechanical motion—such as lifting the entire object—but does not change its internal energy or the number of microstates available to it. Thus $\Delta S_{object} = 0$ and Eq. 23-56 becomes

$$\Delta S_{system} = \Delta S_{hot} + \Delta S_{cold}. \qquad (23\text{-}57)$$

The hot side and the cold side, by definition, remain at constant temperature T_H and T_C as heat is added or removed. We can relate their entropy change to the amount of heat that flows in or out by using the fundamental thermodynamic relationship Eq. 23-9: $dQ = TdS$. For a system in which T remains constant, a direct integration of $dQ = TdS$ gives

$$Q = T\Delta S \quad \Rightarrow \quad \Delta S = \frac{Q}{T}. \qquad (23\text{-}58)$$

Equation 23-58 allows us to evaluate ΔS_{system} in terms of the heat flows and temperatures:

$$\Delta S_{system} = \frac{Q_H}{T_H} + \frac{Q_C}{T_C}. \qquad (23\text{-}59)$$

Q_H is the heat transfer *out* of the hot side and *into* the engine, so by energy conservation $Q_H = -Q_{in}$. (Q_H is a *negative* quantity because it is heat flow *out* of the hot side.) Similarly, the heat flow into the cold side is the positive quantity $Q_C = -Q_{out} = |Q_{out}|$. Thus we have

$$\Delta S_{system} = \frac{-Q_{in}}{T_H} + \frac{|Q_{out}|}{T_C} = \frac{|Q_{out}|}{T_C} - \frac{Q_{in}}{T_H}. \qquad (23\text{-}60)$$

146 CHAPTER 23 THERMODYNAMICS OF IDEAL GASES

In Section 23.6 we tried to invent a perfect engine, having $Q_{out} = 0$. If there were such an engine, the total entropy change of the system would be $\Delta S_{system} = -Q_{in}/T_H < 0$. But this is not possible because the second law demands that $\Delta S_{system} \geq 0$. Here we see formally, in agreement with the less formal analysis of Fig. 23-27, that a perfect engine would violate the second law of thermodynamics. So no engine can have efficiency $\eta = 1$.

But could we design an engine with $\eta = 0.999$ if we are clever? Are there any other limits to the efficiency of a heat engine? The thermal efficiency of a heat engine is, by definition,

$$\eta = \frac{W_{out}}{Q_{in}} = \frac{Q_{in} - |Q_{out}|}{Q_{in}} = 1 - \frac{|Q_{out}|}{Q_{in}}. \qquad (23\text{-}61)$$

Because the second law requires $\Delta S_{system} \geq 0$, Eq. 23-60 can be rearranged to give

$$\frac{|Q_{out}|}{Q_{in}} \geq \frac{T_C}{T_H}. \qquad (23\text{-}62)$$

If we use the inequality of Eq. 23-62 in Eq. 23-61, we are led to conclude that

$$\eta \leq 1 - \frac{T_C}{T_H} \quad \text{(second law limit of efficiency)}. \qquad (23\text{-}63)$$

Equation 23-63 is a major result of this chapter and one with profound implications: The efficiency limit of a heat engine is set by the temperatures of the heating and cooling processes. High efficiency requires $T_C/T_H \ll 1$ and thus $T_H \gg T_C$. Practical realities, however, often necessitate $T_H \approx T_C$, in which case the engine cannot possibly have a large efficiency. This limit on the efficiency of heat engines is a consequence of the second law of thermodynamics.

• **EXAMPLE 23-10** A steam generator for producing electricity boils water under high pressure to produce steam at 300°C. The high-pressure steam spins a turbine as it expands, with the turbine generating electricity, and then the steam is condensed back to water in a heat exchanger, cooled by a river or ocean, at 35°C. What is the *maximum* possible efficiency with which heat energy can be converted to electrical energy?

SOLUTION The maximum possible efficiency is $\eta_{max} = 1 - T_C/T_H$. These are absolute temperatures, so we must find $T_H = 300°C = 553$ K and $T_C = 35°C = 308$ K. Then

$$\eta_{max} = 1 - \frac{308}{553} = 0.44 = 44\%.$$

•

Example 23-10 gives an upper limit, and real electrical generating plants will not reach the limit. In fact, coal, oil, gas, and nuclear-heated steam generators operate at ≈35% efficiency. (The heat *source* has nothing to do with the efficiency. All it does is boil water.) Thus electrical power plants convert about 35% of the heat energy to electrical energy but exhaust about 65% of the heat energy to the environment. Not much can be done to alter the low-temperature limit, while the high-temperature limit is determined by the maximum temperature and pressure the boiler and turbine can withstand. The efficiency of electrical generation is far less than most people imagine, but it is an unavoidable consequence of the second law of thermodynamics.

EXAMPLE 23-11 Popular science writers have long talked about using ocean temperature differences to run heat engines of various kinds. The ocean contains vast quantities of thermal energy. Can that energy be used to do useful work? A heat engine needs a temperature *difference*—a hot side and a cold side—and conveniently, the ocean surface waters are warmer than the deep ocean waters. Consider a power plant built in the tropics, where the surface water temperature is ≈30°C. This would be the hot side of the engine. For the cold side, water would be pumped up from the ocean bottom where it is always ≈5°C. What is the maximum efficiency of such a power plant?

SOLUTION After converting the temperatures to kelvins, a straightforward calculation gives

$$\eta_{max} = 1 - \frac{T_C}{T_H} = 0.082 = 8.2\%.$$

The difference between T_H and T_C, in kelvins, is very small, so no engine operating between these two temperatures can possibly be very efficient. A practical engine might be more like 5% efficient. This is doable, and prototype machines have been built, but they are not financially competitive with other sources of power because their efficiency is so low.

Earlier in this section we found that a Carnot engine has an efficiency

$$\eta_{Carnot} = 1 - \frac{T_C}{T_H}.$$

We see, by comparison with Eq. 23-63, that this is the *maximum possible* efficiency. That is,

$$\eta_{max} = \eta_{Carnot}. \tag{23-64}$$

This is why we analyzed the Carnot cycle in detail—it is *the* maximally efficient cycle for a heat engine operating between temperatures T_H and T_C. It is not difficult to show that a Carnot engine has $\Delta S_{tot} = 0$. In other words, a Carnot engine has the maximum possible efficiency because it causes no change of entropy for the entire system. No other kind of engine can improve on this because it cannot have $\Delta S < 0$. Thus we can state:

> *Second Law—Informal Statement #3*: No heat engine operating between two temperatures T_H and T_C can exceed the Carnot-engine efficiency $\eta_{Carnot} = 1 - T_C/T_H$.

Real engines usually fall well short of the Carnot limit. In Example 23-9, for example, we found that a realistic Otto-cycle efficiency of 48.2% was significantly less than the limiting Carnot-cycle efficiency of 86.3%.

EXAMPLE 23-12 What would be the maximum efficiency of a heat engine using room temperature air, at 20°C, as its heat input and liquid helium, at –269°C, for its cooling?

SOLUTION The cold side temperature $T_C = -269°C = 4$ K is *really* cold. This gives an efficiency of

$$\eta_{max} = 1 - \frac{T_C}{T_H} = 0.986 = 98.6\%.$$

Specialized engines can, indeed, operate with exceptionally high efficiency with the kind of temperature difference found in Example 23-12. But liquid helium does not occur naturally. As you might expect in a world of limits and less than perfect engines, it takes more energy to liquefy the helium than can be recovered as useful work from this super-efficient engine. You can achieve very high efficiency in one engine only by paying the price elsewhere in the system. While that may be desirable in some circumstances (if, for example, you need very high efficiency in some critical manufacturing step and are willing to pay for it by using excess energy elsewhere to lower the temperature), the important thing to remember is that there is "no free lunch." The energy budget for the entire system will always show that energy losses are occurring and that you will never recover in useful work all of the heat energy that you supplied.

This limit on the efficiency of heat engines was not expected. We are used to thinking in terms of energy conservation, so it comes as no surprise that we cannot make an engine with $\eta > 1$. But the limits arising from the second law were not anticipated, nor are they obvious. Nonetheless, they are a very real fact of life and a very real constraint on any practical device. No one has ever invented a machine that exceeds the second law limits, and we have seen that the maximum efficiency for realistic temperatures is surprisingly low.

You likely wondered, when we first introduced the idea of entropy, what possible use there could be for a quantity that measures something about the unseen microstates of a system. Now we find that entropy and its governing law—the second law of thermodynamics—have exceptionally important implications for engineering and applied science. It is through the concept of entropy that we have discovered the existence of a maximum, limiting efficiency for any engine that converts heat energy to work. This limit applies to power generation, propulsion systems, and machines that do mechanical work as well as to the "biological machines" of living organisms.

Summary
Important Concepts and Terms

thermodynamic process
adiabatic process
isentropic
heat capacity ratio
adiabat

working substance
heat engine
waste heat
closed cycle
thermal efficiency

compression ratio
heat of combustion
heat exchanger
refrigerator
Carnot cycle

We have done some rather elaborate and sophisticated analyses in this chapter. In the process, you have seen several examples of how practical devices convert heat energy into work, you have learned how to compute the thermal efficiency of a heat engine, and you have discovered the maximum possible efficiency a heat engine can have. The key to the analysis of a heat engine is to break it down into a series of basic thermodynamic processes. Then you can systematically find p, V, and T at each "corner" of the cycle as well as the heat transferred and work done during each process. During problem solving, it is often helpful to create a table of information as we did in Example 23-6.

An important new idea introduced in this chapter has been the work W_s done *by* the system. We found, for a single thermodynamic process, that

$$W_s = \int_{V_i}^{V_f} p\, dV = \text{ area under the } pV \text{ curve.}$$

The net work done during the entire cycle of a heat engine is then

$$(W_s)_{\text{cycle}} = \text{area } \textit{inside} \text{ the } pV\text{-diagram closed curve.}$$

We also introduced one new process, the adiabatic process, in which $Q = 0$. We found the equation of state for an adiabatic process to be

$$pV^\gamma = \text{ constant and } TV^{\gamma-1} = \text{ constant,}$$

where the heat capacity ratio $\gamma = C_P/C_V$ is an important parameter for characterizing the thermodynamic properties of a gas. The molar heat capacities at constant pressure and at constant volume were found to be related by

$$C_P = C_V + R.$$

Specific results for the four basic thermodynamic processes were given in Section 23.3.

A heat engine is a device that uses a closed-cycle thermodynamic process to convert heat energy into work. The thermal efficiency of a heat engine is

$$\eta = \frac{(W_s)_{\text{cycle}}}{Q_{\text{in}}} = \frac{W_{\text{out}}}{Q_{\text{in}}}.$$

Several specific heat engines were examined in detail, including the Carnot engine, whose efficiency was found to be

$$\eta_{\text{Carnot}} = 1 - \frac{T_C}{T_H}.$$

Whereas the first law of thermodynamics demands that energy be conserved, the second law sets a limit on the maximum possible efficiency of converting heat into work. For any heat engine operating between hot and cold temperatures T_H and T_C, the efficiency is

$$\eta \leq 1 - \frac{T_C}{T_H}.$$

The Carnot engine has the maximum possible efficiency.

Exercises and Problems

Exercises

1. Two grams of helium at an initial temperature of 100°C and an initial pressure of 1 atm undergo an isobaric expansion until the volume is doubled.
   a. What is the final temperature?
   b. What is the work W_s done by the gas?
   c. What is the heat input Q to the gas?
   d. What is the change in thermal energy ΔE_{th} of the gas?
   e. Show the process on a pV diagram, using proper scales on both axes.

2. Two grams of helium at an initial temperature of 100°C and an initial pressure of 1 atm undergo an isothermal expansion until the volume is doubled.
   a. What is the final pressure?
   b. What is the work W_s done by the gas?
   c. What is the heat input Q to the gas?
   d. What is the change in thermal energy ΔE_{th} of the gas?
   e. Show the process on a pV diagram, using proper scales on both axes.

3. Fourteen grams of N_2 gas at STP are adiabatically compressed to 20 atm pressure.
   a. What is the final temperature?
   b. How much work is done by the gas during this compression?
   c. How much heat is added to the gas during this compression?
   d. What is the compression ratio $r = V_{max}/V_{min}$?
   e. Show the process on a pV diagram, using proper scales on both axes.

4. Fourteen grams of N_2 gas at STP are heated to 20 atm pressure in an isochoric process.
   a. What is the final temperature?
   b. How much work is done by the gas during this compression?
   c. How much heat is added to the gas during this compression?
   d. What is the pressure ratio $r_p = p_{max}/p_{min}$?
   e. Show the process on a pV diagram, using proper scales on both axes.

5. a. At what compression ratio would an engine operating with an Otto cycle and a heat capacity ratio 1.4 have an efficiency of 60%?
   b. At what maximum temperature would an engine operating with a Carnot cycle and a cold temperature of 20°C have an efficiency of 60%?

6. A Carnot engine operating between heat reservoirs at temperatures 300 K and 500 K produces a power output of 1000 W.
   a. What is the thermal efficiency of this engine?
   b. What is the *rate* of heat input, in J/s or W?
   c. What is the *rate* of heat output, in J/s or W?

7. Figure 23-29 shows a thermodynamic process followed by 0.015 mol of a diatomic gas.
   a. How much work is done by the gas?
   b. How much heat is added to the gas?
   c. By how much does the thermal energy of the gas change?

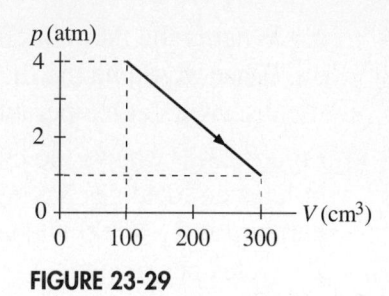

FIGURE 23-29

Problems

8. For an adiabatic process $V_i \to V_f$
   a. Prove that $Tp^{(1-\gamma)/\gamma} = $ constant.
   b. Prove, by direct integration of pdV, that $W_s = (p_f V_f - p_i V_i)/(1-\gamma)$.

9. A diatomic gas undergoes the thermodynamic process shown in Fig. 23-30. Its temperature at point 1 is 20°C.
   a. Determine W_s, Q, and ΔE for each of the three stages in this cycle. Put your results in a table.
   b. Determine the net values $(W_s)_{cycle}$, Q_{cycle}, and ΔE_{cycle} for one complete cycle.
   c. What is the thermal efficiency of this heat engine?
   d. What is the power output of the engine if it runs at 500 rpm?

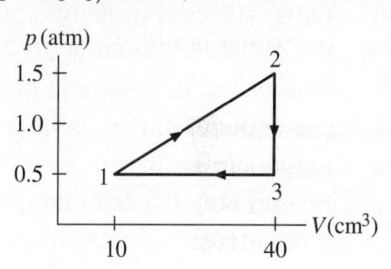

FIGURE 23-30

10. Design a heat engine using a three-step closed cycle that meets the following conditions:

    $p_{min} = 1$ atm $p_{max} = 3$ atm $V_{min} = 100$ cm³ $(W_s)_{cycle} = 40.5$ J.

    Give your answer as a pV diagram showing the cycle.

11. Three engineering students submit their solutions to a design problem. They had been asked to design an engine that operates between temperatures 300 K and 500 K. The heat input/output and work done by their designs are shown in the following table:

Student	Q_{in}	Q_{out}	W_{out}
1	250 J	−140 J	110 J
2	250 J	−170 J	90 J
3	250 J	−160 J	90 J

    What grade will you give to each of these students *and why*? You must fully justify each grade because any dissatisfied student is likely to appeal to the Fairness Board.

12. A diesel engine has a displacement of 1000 cm³ per cylinder and a compression ratio of 21. It operates with intake air ($\gamma = 1.40$) at 25°C and 1 atm pressure. The quantity of fuel injected into each cylinder has a heat of combustion of 1000 J.
    a. Find p, V, and T at each of the four corners of the cycle. Put your results in a table.
    b. What is the net work done by each cylinder during one full cycle?

c. What is the thermal efficiency of this engine?
d. What would be the thermal efficiency of a Carnot engine operating between heat reservoirs at temperatures T_{max} and T_{min}?

13. Figure 23-31 shows the cycle for a heat engine that utilizes an ideal gas having $\gamma = 1.25$. The initial temperature is $T_1 = 300$ K, and this engine operates at 20 cycles per second.
    a. What is the power output of the engine?
    b. What is the engine's thermal efficiency?

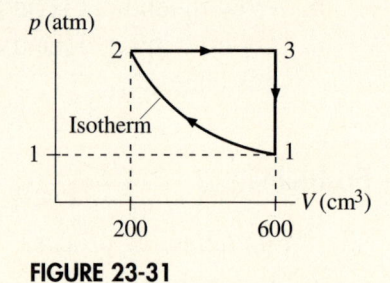

FIGURE 23-31

14. Figure 23-32 shows two compartments separated by a thin wall. The left side contains 0.06 mol of helium at an initial temperature of 600 K and the right side contains 0.03 mol of helium at an initial temperature of 300 K. The compartment on the right is attached to a vertical cylinder. A 2 kg piston can slide without friction up and down the cylinder. Neither the cylinder diameter nor the volumes of the compartments are known.
    a. What is the final temperature? (**Hint:** One gas undergoes a constant volume process while the other undergoes a constant pressure process.)
    b. How much heat is transferred from the left side to the right side?
    c. How high is the piston lifted due to this heat transfer?
    d. What fraction of the heat is converted into work?

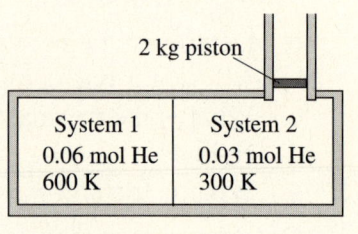

FIGURE 23-32

15. The Brayton cycle, which characterizes a gas turbine engine, was described in Section 23.5 and shown in Fig. 23-22. Show that the thermal efficiency is given by

$$\eta = 1 - \frac{1}{r_p^{(\gamma-1)/\gamma}},$$

where $r_p = p_{max}/p_{min}$ is the pressure ratio. (**Hint:** Follow an analysis similar to that use for the Otto cycle.)

[**Estimated 10 additional problems for the final edition.**]

PART IV Scenic Vista

Order, Disorder, and the Direction of Time

In the past few decades, something very dramatic has been happening in science, something as unexpected as the birth of geometry or the grand vision of the cosmos as expressed in Newton's work. We are becoming more and more conscious of the fact that on all levels, from elementary particles to cosmology, randomness and irreversibility play an ever increasing role. Science is rediscovering time.

Ilya Prigogine
Winner of the Nobel Prize in 1977 for his work in thermodynamics

The primary goal of Part IV was to establish a micro/macro connection, a way in which we can understand the macroscopic behavior of solids, liquids, and especially gases in terms of the microscopic motions of their constituent molecules. We have been quite successful. The following is a list of just some of the important findings we covered:

- You learned that we can understand the macroscopic state of a gas by averaging the motions and collisions of vast numbers of molecules. This approach to describing macroscopic systems is called statistical physics.
- You learned that some macroscopic states are vastly more probable than others and that thermal equilibrium corresponds to the system being in its most probable state.
- You learned that temperature is a measurement of the thermal energy of the molecules in a system and that the average kinetic energy per molecule is simply $\frac{3}{2}kT$. Further, two interacting systems are in equilibrium with each other when they have the *same* average energy per molecule. This implies that they also have the same temperature.
- You learned that heat is a transfer of energy between two systems that have different temperatures. The *mechanism* of heat transfer is molecular collisions at the boundary between the two systems.
- You learned that the pressure of a gas is due to collisions of the molecules with the walls of the container. The ideal gas law, connecting pressure and temperature, is a direct consequence of the microscopic motion and collisions of the molecules.

- You learned that the entropy of a macroscopic state is a measurement of the probability of that state. Macroscopic states with many possible microstates have a larger entropy than do macroscopic states having only a few possible microstates.
- You learned that isolated systems evolve, by random processes, to the most probable macroscopic state—the state of equilibrium. This idea is captured in the second law of thermodynamics: $\Delta S_{total} \geq 0$. Although seeming rather esoteric, the second law turned out to have important practical implications for the efficiencies of engines.

All of this came from the very humble starting point of simply counting the number of molecules in each side of a container. This led us, bit by bit, to an understanding of the thermodynamic concepts of equilibrium, temperature, pressure, thermal energy, and finally entropy.

A knowledge structure of the major ideas and concepts of thermodynamics may be useful to focus your attention on how all the various pieces fit together. Table SVIV-1 is such a knowledge structure.

First Law
$\Delta E_{th} = Q - W_s$

Second Law
$\Delta S_{total} \geq 0$

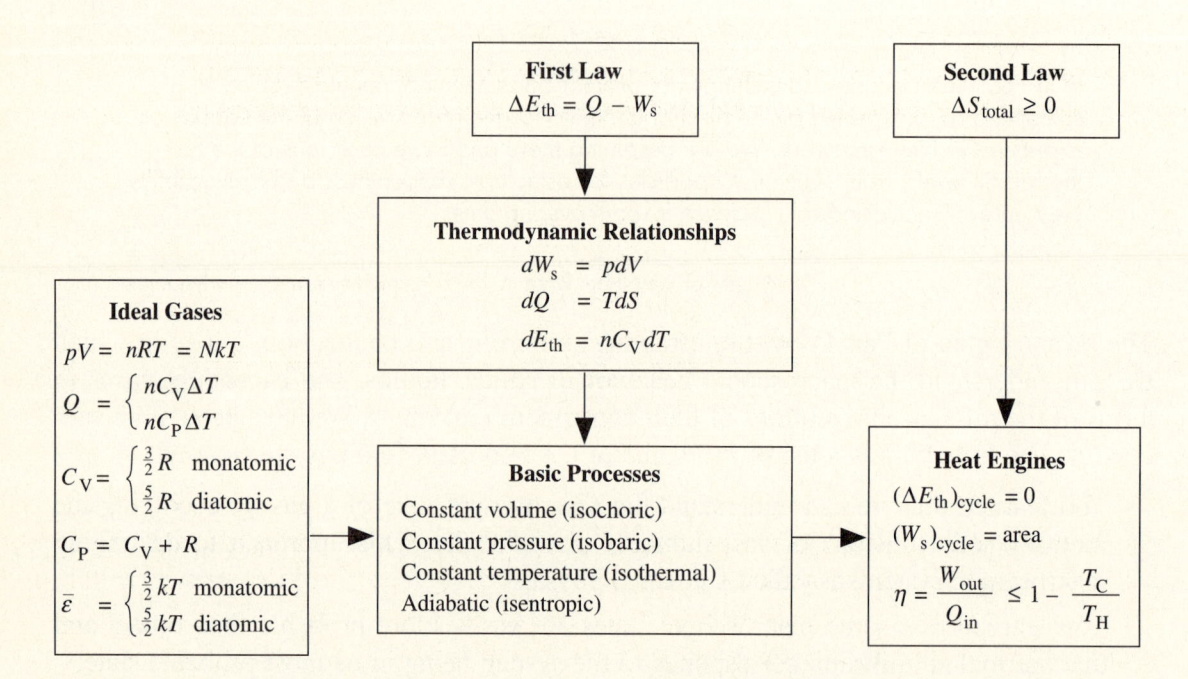

Thermodynamic Relationships
$dW_s = pdV$
$dQ = TdS$
$dE_{th} = nC_V dT$

Ideal Gases
$pV = nRT = NkT$
$Q = \begin{cases} nC_V \Delta T \\ nC_P \Delta T \end{cases}$
$C_V = \begin{cases} \frac{3}{2}R & \text{monatomic} \\ \frac{5}{2}R & \text{diatomic} \end{cases}$
$C_P = C_V + R$
$\bar{\varepsilon} = \begin{cases} \frac{3}{2}kT & \text{monatomic} \\ \frac{5}{2}kT & \text{diatomic} \end{cases}$

Basic Processes
Constant volume (isochoric)
Constant pressure (isobaric)
Constant temperature (isothermal)
Adiabatic (isentropic)

Heat Engines
$(\Delta E_{th})_{cycle} = 0$
$(W_s)_{cycle} = \text{area}$
$\eta = \dfrac{W_{out}}{Q_{in}} \leq 1 - \dfrac{T_C}{T_H}$

TABLE SVIV-1 The knowledge structure of thermal and statistical physics.

We noted at the beginning of Chapter 23 that thermodynamics can seem very "equation oriented." It's undeniable that there are more equations than we used in earlier parts of this text. But focusing on trying to remember all the equations is seeing only the trees, not a forest. A much better strategy is to focus on the relationships of Table SVIV-1. You can find the equations you need if you know how the ideas are connected, but memorizing all the equations won't help if you don't know which ones are relevant to different situations.

Having summarized the major ideas of Part IV, we will spend the rest of this Scenic Vista looking at some of the more interesting and profound implications of the second law of thermodynamics.

The Direction of Time

Why do we remember the past but not the future? This may seem like a silly question, but it is not. Objects can move either direction through space—forward or backward—so why not also through time? Why does time have a well-defined "direction" from past to future? The situation is even more puzzling once you realize, as we discussed in Section 22.7, that the microscopic physics of atoms and molecules does *not* have a direction of time. A movie of molecular motions and collisions looks equally valid played either forward or backward. But past and future are clearly distinct in the macroscopic world in which we humans live. The flow of time, it turns out, is intimately connected with the concepts of order and disorder.

In Chapter 21 we touched on the distinction between randomness and order. Still, this is a rather strange idea and perhaps some examples would help. Figure SVIV-1 shows three different systems. On the left we see a group of atoms arranged into a solid-like lattice. This is a highly ordered and nonrandom system, with each position precisely specified. Contrast this with the system on the right, where there is no order at all. The position of every single molecule is entirely random—like flipping coins or throwing dice. The sense in which we are using the word *order* is that a precisely specified configuration has a high degree of order while a completely unspecified, random configuration has no order.

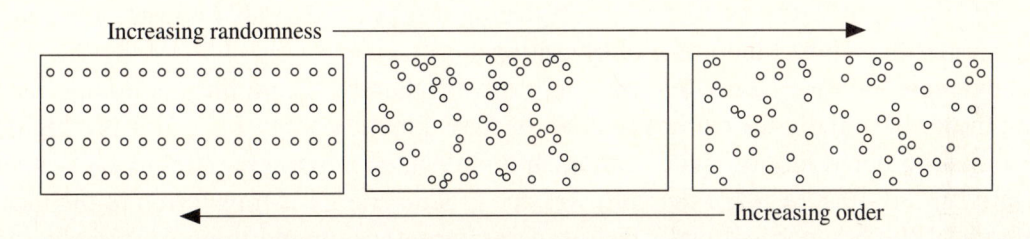

FIGURE SVIV-1 Examples of ordered and random systems.

Out of all the conceivable microstates in which a system might be found, only a very few would look like the lattice on the left of Fig. SVIV-1. It is extremely improbable that the atoms would *spontaneously* arrange themselves in this pattern. By contrast, there are a vast number of microstates that randomly fill the container like the right picture in Fig. SVIV-1. This is a highly probable situation.

You learned in Chapter 21 that entropy measures the probability of a macroscopic state occurring. The ordered lattice has a very low entropy, because it has such a small probability, while the entropy of the randomly filled container is high. We can also say that entropy measures the amount of *disorder* in a system. The more disordered state has a high entropy while the highly ordered lattice has a low entropy.

There is also a sense in which entropy is associated with knowledge or information. It takes a large amount of information to know how to construct a perfect lattice of atoms, but it requires no information at all to place them randomly in a container. Thus we can say that increasing entropy corresponds to a *decreasing* amount of information encoded in the system. This idea forms the basis of an entire branch of mathematics called *information theory*, which is important in computer science and communications.

The middle picture of Fig. SVIV-1 shows an in-between situation, such as might be seen if a gas expands from a smaller volume to a larger one. This is not a completely random situation. It exhibits some degree of order, in the sense that there are only a limited number of microstates with no molecules on the right side. It took some knowledge to arrange the atoms such that the right side is empty. This situation is more probable than the entirely ordered lattice and thus has more entropy. It is less probable, and has less entropy, than the completely random system on the right.

The second law of thermodynamics says that the entropy of an isolated system always increases, never decreases, until the system reaches equilibrium. In other words, the second law tells us that an *isolated* system evolves such that:

>Order turns into disorder and randomness.
>Information is lost rather than gained.
>The system "runs down."

The reverse of these, such as an isolated system that spontaneously generates order out of randomness, simply does not happen. It is not that the system "knows" about order or randomness, but rather that there are vastly more states corresponding to randomness than there are corresponding to order. As collisions occur at the microscopic level, the laws of probability dictate that the system will, on average, move inexorably toward the most probable—and thus most random—configuration.

For example, consider a car that runs into a wall, thereby crumpling its bumper and bending the wall. Initially the car has a center of mass kinetic energy K_{cm}. This is a highly organized and ordered motion, with all the car's molecules moving parallel to each other at the same speed. As a result of the collision, the kinetic energy is transferred to the thermal energy of the bumper and wall. Their temperature goes up and there is an increased microscopic motion of their molecules. This is highly random situation, with all the molecules moving independently of one another. No energy has been lost in the collision, but it has been transferred from an ordered energy to a disordered, random energy.

Why can't the thermal energy be transferred back to the car's center of mass energy, causing the bumper and wall to straighten and the car to race away? The first law would not be violated in this process, but the second law would! Such a process would require the system to spontaneously turn randomness into order, thereby decreasing the system's entropy. And this is precisely what the second law prohibits. Thus:

> *Second Law—Informal Statement #4*: The random thermal energy (micromotion) of a isolated system will not spontaneously be converted into macroscopic kinetic energy of the object as a whole.

This statement of the second law is especially important—it is this prohibition against turning disordered motion into ordered motion that limits the maximum efficiency of engines.

The ideas contained within the second law are applicable to countless other situations. Think about a perfume bottle when the stopper is removed. Having all the molecules in the bottle and none in the room is a much more ordered and less random situation—a low entropy situation—than having the molecules spread uniformly through the room. As the molecules diffuse throughout the room, the disorder and the total entropy are increasing. Could the molecules all manage to diffuse back *into* the bottle? No, because such a process would require a *reduction* of entropy, which would be a violation of the second law. You can also think about this situation in another way. The molecules can leave the bottle entirely independently of one another as they undergo collisions with the air molecules. To get them to go back into the bottle, however, requires that all of the molecules *cooperate* with each other, producing billions upon billions of highly correlated collisions that gradually nudge the perfume molecules back into the bottle. Because we have seen many examples of how many microstates a gas has, the odds of all the molecules in the room reaching this *one* special microstate are astronomically small. You will "never" see the perfume spontaneously go back into the bottle. This allows us to state the second law in yet another way:

> *Second Law—Informal Statement #5*: The molecules of an isolated gas will not spontaneously segregate themselves into only a portion of their container.

A major conclusion of Part IV is that irreversible processes and the directionality of time are not inconsistent with the reversibility of the microscopic motions. Systems of large numbers of particles are simply *vastly more likely* to generate some macroscopic states than others, and the irreversible evolution from less likely macrostates to more likely macrostates is what gives us a macroscopic direction of time.

So molecules expand to fill their container but won't spontaneously return to a smaller volume. Stirring blends your coffee and cream, it never unmixes them. A plant in a sealed jar dies and decomposes to carbon and various gases; the gases and carbon never spontaneously assemble themselves into a flower. These are all examples of irreversible processes. They each show a clear "direction of time," a distinct difference between past and future. Thus we can conclude:

> *Second Law–Informal Statement #6*: The time direction in which the entropy of an isolated macroscopic system increases is "the future."

Establishing the "arrow of time" is perhaps the most profound implication of the second law of thermodynamics.

Order Out of Chaos

The second law predicts that systems will run down, that order will evolve toward randomness, and that complexity will give way to simplicity. But just look around you! Plants grow from simple seeds to complex entities; single-cell fertilized eggs grow into complex adult organisms; electrical current passing through a "soup" of simple random molecules produces such complex chemicals as amino acids; and over the last billion or so years life has evolved from simple unicellular organisms to very complex forms. Knowledge and information seem to grow every year, not to fade away. Everywhere we look, it seems, the second law is being violated. How can this be?

There is an important qualification in the second law—it applies *only* to *isolated* systems that exchange no energy with their environment. The situation is entirely different if energy flows into the system, and we cannot predict what will happen to the entropy of a nonisolated system. The popular science literature is full of arguments and predictions that make incorrect use of the second law by trying to apply it to systems that are not isolated.

The phenomena we just listed of systems that become *more* ordered as time passes, and in which the entropy decreases, are all examples of what are called *self-organizing systems*. One of their major characteristics is a substantial flow of energy *through* the system. Plants and animals, for example, take in energy from the sun or chemical energy from food and give waste heat back to the environment via evaporation, decay, and other means. It is this energy flow that allows the systems to maintain, or even increase, a high degree of order and a very low entropy. But—and this is the important point—the entropy of the *entire* system, including the earth and the sun, undergoes a significant *increase* in order to let selected subsystems decrease their entropy. The second law is not violated at all, but you have to apply it to the combined systems that are interacting and not just to a single subsystem.

Self-organization is closely related to nonlinear mechanics, chaos, and the geometry of fractals. It has important applications in fields ranging from ecology to computer science to aerodynamical engineering. For example, gas flow across a wing gives rise to large-scale turbulence—eddies and whirlpools—in the wake of an airplane. Their formation affects the aerodynamics of the plane and can also create hazards for following aircraft. Whirlpools are ordered, large-scale macroscopic structures and have a low entropy. But they are produced from disordered, random collisions of the air molecules.

Even though the basic ideas of self-organization are in accord with the laws of thermodynamics, it is not easy to understand the details of how the process occurs. This is currently a very active field of research in both science and engineering, and the 1977 Nobel Prize in Chemistry went to Ilya Prigogine for his studies on "nonequilibrium thermodynamics," the basic science underlying self-organizing systems. Prigogine and others have shown how energy flow through a system can, when the conditions are right, bring "order out of chaos!"

The Entropy of the Universe

These ideas lead us to an interesting question. Suppose we consider the universe as a whole to be the system. Unless there are other universes of which we have no knowledge, this would seem to be the ultimate isolated system. If so, then the laws of thermodynamics tell us that the total energy of the universe is conserved but that the total entropy of the universe is increasing. All processes that we can observe in the universe are consistent with these conclusions. What are the implications of applying the second law to the entire universe?

First, the universe must have had a beginning. If not—that is, if the universe has always existed and infinite time has been available—then the entropy $S_{universe}$ would have reached its maximum value and the universe would be in thermal equilibrium. Because it is not in thermal equilibrium, the universe must have had a beginning in a more-ordered state and it must have evolved to the less-ordered and more random state in which we find it at the present.

Second, the universe must be approaching an equilibrium, when $S_{universe}$ reaches its maximum value, in which all objects in the universe are at the same constant temperature. All the stars will burn out and no new stars will form. This fate is often called the "heat death of the universe," although the energy available in the universe is such that the final temperature will actually be extremely cold. This is not an immediate concern for which governments should have a contingency plan, because it is hundreds of billions of years in the future. Nonetheless, the mere existence of such a fate for the universe raises interesting but troublesome philosophical and theological issues.

Concluding Thought

Thermodynamics is one of the major achievements of modern physics. The overwhelming success of thermodynamics provides substantial evidence for the validity of the atomic view of matter that provides its conceptual foundation. Indeed, the success of statistical physics and thermodynamics at the end of the nineteenth century provided some of the most persuasive evidence for the reality of atoms—long before technology had advanced to the point where we could begin to accumulate direct experimental evidence for atoms. And as you have seen in this Scenic Vista, thermodynamics provides deep insights into profound issues about the universe and about human existence.

With that in mind, perhaps we should let an expert on physics have the last word. In 1949, Albert Einstein wrote the following:

> A theory is the more impressive the greater the simplicity of its premises are, the more different kinds of things it relates, and the more extended is its area of applicability. Therefore the deep impression that classical thermodynamics made upon me. It is the only physical theory of universal content concerning which I am convinced that, within the framework of applicability of its basic concepts, it will never be overthrown.

PART V
Electricity and Magnetism

PART V Overview

Electricity and Magnetism: The Atomic Glue

[**Photo suggestion: Pieces of amber.**]

Amber, or fossilized tree resin, has long been prized for the beauty of its lustrous yellow color. Amber is of scientific interest today because biologists have learned how to recover strands of DNA from million-year-old insects that were trapped in the resin. But amber has an ancient scientific connection as well. The Greek word for amber is *elektron*.

It has been known since at least the fifth century B.C. that a piece of amber that has been rubbed briskly can pick up feathers or small pieces of straw—seemingly magical powers to a prescientific society. It was also known to Greeks of the same time period that certain stones from the region they called *Magnesia* (in present-day Turkey) could pick up pieces of iron. It is from these humble beginnings that we today have high-speed computers, lasers, fiber optic communications, and magnetic resonance imaging—as well as such mundane modern-day miracles as the light bulb.

The story of electricity and magnetism is vast. The development of a successful electromagnetic theory, which occupied the leading physicists of Europe for most of the nineteenth century, led to sweeping revolutions in both science and technology. The complete formulation of the theory of the electromagnetic field has been called by no less than Einstein "the most important event in physics since Newton's time." Not surprisingly, all that we can do in this text is to develop some of the basic ideas and concepts, leaving mathematical sophistication and advanced applications to later courses. Even so, our study of electricity and magnetism will require learning many new, important, and sometimes difficult ideas. Foremost among these will be the idea of a *field*.

Phenomena and Theories

The basic phenomena of electricity and magnetism are not as familiar as those of mechanics. You have spent your entire life exerting forces on objects and watching them move, but your experience with electricity and magnetism may be limited to noting that the light comes on when you turn on the light switch or that magnets stick to the door of your refrigerator. This limited experience has both good and bad aspects. On the positive side, you are less likely to have developed serious misconceptions about phenomena that are not a regular part of your life. (This is in contrast to mechanics with which, you will recall, we had to deal with many misconceptions that most people hold about forces and their properties.) But it is hard to motivate the need for a major new theory if you are not aware of the phenomena and behaviors that the theory is going to attempt to explain.

So we will place a large emphasis on the basic *phenomena* of electricity and magnetism. We will begin by looking in detail at the most basic properties of *electric charge* and the process of *charging* an object. It is easy to make systematic observations of how charges behave, and we will be led to consider the forces between charges and how charges behave in different materials. The development of electrical technology, and the dawn of the "electronic age," came about as scientists and engineers learned to *control* the movement of charges. Electric current, whether it be for lighting a light bulb or changing the state of a computer memory element, is simply a controlled motion of charges through conducting materials. One of our goals will be to understand how currents flow through electrical circuits.

When we turn to magnetic behavior, we will start by observing how magnets stick to some metals but not others and how magnets affect compass needles. But our most important observation, which you may have seen, is that an electric current can affect a compass needle in exactly the same way as does a magnet. This observation will suggest to us that there is a close connection between electricity and magnetism, and we will explore this relationship in detail. Our path will eventually lead to the discovery of electromagnetic waves.

Our goal is to develop a *theory* that will explain the phenomena of electricity and magnetism. We will introduce the entirely new concept of a *field* to explain the interactions between charges. Much of our attention will be focused on the interplay between charges and fields: how fields are created by charges and how charges, in return, respond to the fields. Bit by bit, we will assemble a theory—based on the new concepts of electric and magnetic fields—that will allow us to understand, explain, and predict a wide range of electromagnetic behavior.

The Microscopic Model

There are two different aspects to the theory of electromagnetism. The field theory provides a macroscopic perspective on the phenomena, but we can also take a microscopic view. At the microscopic level, we want to know what charges are, how they are related to atoms and molecules, and how they move about through various kinds of materials. We will develop a microscopic model of electrons and ions moving in response to electric and magnetic forces. This microscopic perspective of electricity and magnetism is analogous to the kinetic theory of gases in our study of thermodynamics.

Likely *the* most important scientific discovery of the modern era is that matter consists of atoms. It was found near the end of the nineteenth century that the atoms themselves are not indestructible objects but, instead, have constituents that are *charged particles*. We know them today as electrons and protons. Much of the time these charged constituents all balance and the atoms, as well as macroscopic objects built of these atoms, are electrically neutral. That has been the implicit assumption for all the physics we have done until now. Under some circumstances, however, the charges can become separated and move about. An important goal of our microscopic model will be to understand how charges become separated and how charged particles move through conductors as what we call a *current*.

[**Photo suggestion: MRI scan.**]

Our interest at the microscopic level is more than simply how currents flow. Electricity and magnetism are of particular significance in our quest to understand the atomic structure of matter itself. The electric force is the "atomic glue" that holds the atom together and that binds atoms into molecules and solids. It is really nothing other than electric forces acting at the atomic level that we see in the macroscopic world as friction, tension, adhesion, and other contact forces. Magnetism plays a lesser, but not insignificant, role. Although magnetism is not involved in the forces that bind atoms and molecules together, it does figure prominently in the macroscopic behavior of materials. In addition, the magnetic properties of atoms have become important "windows" by which the interior structure of solids can be probed. Magnetic resonance imaging in medicine is the most well-known example of this technology, but many of the same techniques are used in science and engineering to characterize materials. For all these many reasons, acquiring a knowledge of electricity and magnetism is now an essential part of science and engineering education.

Chapter **24**

Electric Forces and Fields

LOOKING BACK Sections 12.3–12.4

24.1 Charges and Forces: A Laboratory

The theory of electricity and of electric forces was developed during the eighteenth and nineteenth centuries—long before the discovery of the electron or before the atomic structure of matter was understood. Many of the initial ideas about electricity came from casual observations—that rubbed amber could pick up small objects, that walking across a carpet could lead to a spark and shock upon touching a metallic object, that vigorously brushing clean hair could cause "prickly" sensations and make all the hairs stand apart. The common factor to these observations is that objects are *rubbed* together. What does rubbing do? What kind of forces are involved? What causes them? What are their properties? Eighteenth-century scientists were well versed in the successes of Newtonian physics, with its basic concepts of particles and forces, and it was only natural that they would begin to study the forces of electricity. The road to understanding would turn out to be long and difficult, with many detours and dead ends. What seems to us today, after the fact, a fairly straightforward and "obvious" theory was anything but that to scientists facing a bewildering array of conflicting and difficult-to-reproduce phenomena.

In this section we will take a close look at the electric phenomena observed by nineteenth-century scientists and the ideas they developed to explain these phenomena. It is important to note that the basic concepts and theory of electricity make *no* reference to atoms or electrons. The theory of electricity was well established long before the electron was discovered. In later sections, we will use our contemporary knowledge of atoms to understand electricity on a microscopic level. But it is important that you be able to understand and explain basic electric phenomena directly in terms of *charges* and *forces*.

Experimenting with Charges

Let us enter a laboratory where we can make direct observations of electric phenomena. This is a modest laboratory, much like you would have found in 1800, with no modern equipment. The major tools in the lab are:

- A variety of lightweight plastic, glass, and wooden rods, each a few inches long. These pieces can be hand-held or suspended by threads from a support.
- A few metal rods with wooden handles.
- Pieces of wool and silk.
- Small metal spheres, an inch or two in diameter, on wooden stands.

As we manipulate and use these tools, we will try to develop a theory to explain the phenomena we see. Our theory will have nothing to do with those "new-fangled atoms" that a few scientists have been talking about. We are not sure whether they really exist, and, even if they do, we have no idea how they are constructed or what their properties are.

So with this in mind, let us begin.

Experiment 1: We pick up a plastic or glass rod that has been undisturbed for a long period of time and slowly bring it up close to a hanging rod of the same substance. Nothing happens to either rod!

Experiment 2: We vigorously rub both a hanging plastic rod and a hand-held plastic rod with wool. When we bring the two close together, the hanging rod tries to move away from the hand-held rod, as shown in Fig. 24-1a. There appears to be a *repulsive* force between the two rods. This is somewhat surprising because all the other forces we know, with the exception of magnets, are always attractive.

We repeat the procedure using two glass rods rubbed with silk and observe the same result: the two rods repel each other.

Experiment 3: We bring a glass rod that has been rubbed with silk close to a hanging plastic rod that has been rubbed with wool. These two rods *attract* each other, as seen in Fig. 24-1b.

Experiment 4: We rub a hanging plastic rod with wool and then hold the *wool* close to the rod. This causes a weak but definite attraction, as shown in Fig. 24-1c. If we rub a glass rod with silk and then bring the silk near the same hanging plastic rod that was rubbed with wool, we see a weak but definite repulsion (Fig. 24-1d).

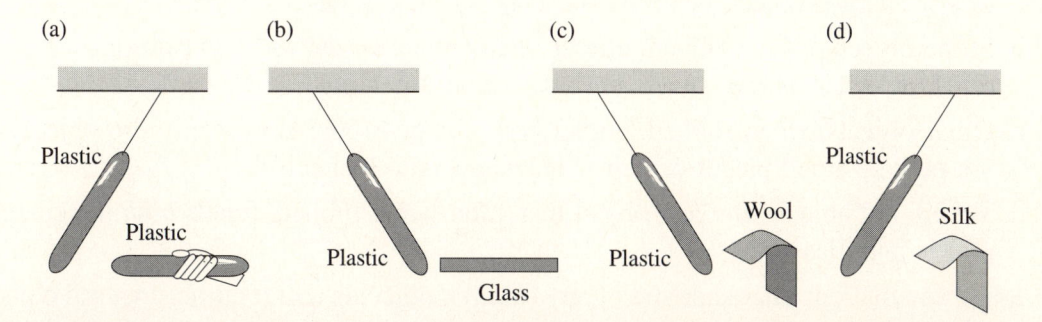

FIGURE 24-1 a) Two plastic rods rubbed with wool repel each other. b) A plastic rod rubbed with wool attracts a glass rod rubbed with silk. c) Wool used to rub a plastic rod is then attracted to the plastic. d) Silk used to rub a glass rod is repelled by the plastic rod of c).

Experiment 5: Further observations show that the strength of these attractive and repulsive forces is less for plastic that is rubbed less vigorously and more for plastic that is rubbed more vigorously; that the strength of attractive and repulsive forces between charged rods decreases as the separation between them increases; and that the size of these effects slowly fades away over a time interval of a few minutes.

What do these experiments tell us? Rubbing the rods changes their properties and causes forces to be exerted between them. These forces did not exist before the rubbing. We will say that the original objects are **neutral**. We will call the rubbing process **charging** and say that the rubbed rod is *charged*. At this time these are simply descriptive names—we have no idea what they "mean." Our initial observations show that there is a repulsive force between two identical objects that have been charged the *same* way, such as between two plastic rods both rubbed with wool. Furthermore, the two objects that were rubbed together during the charging process—the plastic and the wool—seem to want to get back together again. But Experiment 3 is a puzzle: Two rods that seem to have been charged in the same way—by rubbing—*attract* each other rather than repel each other. Why should the outcome of Experiment 3 differ from Experiment 2? Back to the lab.

Experiment 6: We hold a *neutral* (unrubbed) plastic rod close to a hanging plastic rod that was charged by rubbing with wool. The two rods are attracted to each other. Interestingly, the neutral plastic rod is also attracted to a hanging glass rod that was rubbed with silk.

Similarly, a neutral glass rod is attracted to both a charged plastic rod and a charged glass rod. Continued experiments find that the charged rods are attracted to *any* neutral object, such as a finger, a piece of paper, or the metal rod.

Experiment 7: If a charged plastic rod is waved over small pieces of paper on the table, the pieces of paper leap up and stick to the rod. A charged glass rod gives the same result. However, a neutral rod has no effect on the pieces of paper.

Experiment 8: If we rub a *wooden* rod with either wool or silk, it will not pick up pieces of paper. It is weakly attracted to both the charged plastic *and* the charged glass rods.

Experiment 9: Continued experiments show that

a. Neutral objects exert no force on other neutral objects, but they are *always attracted* to *any* charged object. Neutral objects never pick up pieces of paper.

b. Some objects, when rubbed, attract one of the charged rods (plastic or glass) and repel the other. These objects always pick up small pieces of paper.

c. Other objects, when rubbed, attract both charged rods. However, these objects do not pick up small pieces of paper. They appear to be neutral.

d. We are not able to find any object that, after being rubbed, repels *both* the charged plastic and glass rods.

It appears that one characteristic of any *charged* object is that it attracts neutral objects and will pick up small pieces of paper. This seems to be a good test for "is this object charged?" But not all objects can be charged by rubbing them. Objects like wood, for example, seem to be difficult, perhaps impossible, to charge. Experiment 9d is especially important. When combined with all the other observations, it is 9d that really provides the impetus for us to suggest that there are two *and only two* "kinds" of charge.

Based on these observations, let us tentatively advance the first stages of a "theory" of electricity. The postulates of our theory are:

1. Frictional forces, such as rubbing, can add or remove something called **charge** to or from an object. The process itself we called *charging*. More vigorous rubbing produces a larger quantity of charge.
2. There are two and only two kinds of charge. For now we will call them *plastic charge* and *glass charge*.
3. Two charges of the same kind, which we will call **like charges** (plastic and plastic or glass and glass) exert repulsive forces on each other. Two charges of different kinds, which we will call **opposite charges** (plastic and glass), exert attractive forces on each other.
4. The size of the force increases as the quantity of charge increases. The size of the force decreases as the distance between the charges increases.
5. Nonglass and nonplastic objects can sometimes (but not always!) be charged by rubbing, but the charge they receive is either plastic charge or glass charge. We can identify the type of charge by observing the forces it exerts on charged glass and plastic. For example, an object that repels charged glass and attracts charged plastic has glass charge.
6. Because a charged plastic rod is attracted to the wool used to rub it, while silk that has rubbed glass repels the charged plastic, it appears that the wool ends up with glass charge while the silk has plastic charge. This leads us to postulate that *neutral* objects have an *equal mixture* of both plastic charge and glass charge and that, somehow, the rubbing process manages to separate the two.

Although this hypothesis or model is *consistent* with the observations, it is by no means proved. One could easily imagine other hypotheses that are just as consistent with the limited amount of observations we have so far made. We still have some very large unexplained "puzzles," such as why charged objects exert attractive forces on neutral objects and why the charge does not seem to last. The ultimate success of our theory will depend on how well it does at providing an explanation for these puzzles as well as on other observations we have not yet made. Further progress requires more experimentation.

More Experimenting with Charges

Let's try some more experiments to see whether we can clarify the properties of charge.

Experiment 10: We rub just one *end* of a plastic rod with wool. That end is then found to repel a charged hanging plastic rod and also to pick up small pieces of paper. The other end, which was not rubbed, does not pick up paper, *attracts* the hanging rod, and appears to be neutral. We find the same behavior if we substitute glass rods for both plastic rods.

Experiment 11: We rub two plastic rods with wool. One of them we hang, and the other we touch to a metal sphere. The metal sphere is then found to repel the hanging charged rod.

Experiment 12: We again charge the two plastic rods and hang one of them. With the other rod we touch the *end* of a metal rod that is held by a wooden handle. When the metal rod is brought close to the hanging plastic rod, we find that *all* pieces of the metal repel the plastic.

Experiment 13: Two metal spheres are placed close together with a plastic rod connecting them, as shown in Fig. 24-2. We charge two plastic rods, one hanging and one hand-held, and touch the hand-held rod to one of the metal spheres. Afterward, the metal sphere that was touched repels the hanging plastic rod but the other metal sphere does not.

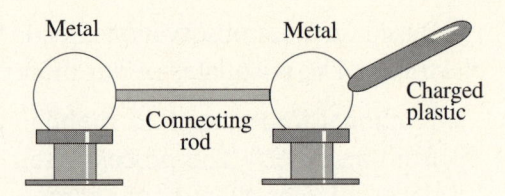

FIGURE 24-2 Charge moves through a metal connecting rod, but not through a plastic one.

We repeat the experiment, but this time a metal rod connects the two metal spheres. Again we touch one metal sphere with a charged plastic rod. Afterward, *both* metal spheres are found to repel the hanging charged plastic rod.

Experiment 14: We charge two plastic rods, then run a finger along both of them. Afterwards, they no longer repel each other or pick up pieces of paper. Similarly, the metal sphere of Experiment 11 no longer repels the plastic rod after being touched.

Our final set of experiments has found new and important behavior. We have found that it is possible to *move* charge from one object to another by *contact*. We could move it from the plastic to the metal. We could also *remove* charge from both the plastic and the metal by contact with our fingers. We call this **discharging**.

Very importantly, we seem to have found two *classes* of materials with regard to their charge properties. Experiment 12 for a metal rod is in sharp contrast to Experiment 10 for glass and plastic. When these are combined with Experiment 13, we see that charge somehow moves through or along the metal rod but remains fixed on the plastic and glass rods. Let us define **conductors** as those materials through or along which charge easily moves and **insulators** as those materials on or in which charge remains fixed and immobile. Glass and plastic are insulators while metal is a conductor.

Based on this new information, we can add two more postulates to our theory about charges:

7. There are two types of electrical materials. Conductors are materials through or along which charge easily moves. Insulators are materials on or in which charge remains fixed in place.
8. Charge can be transferred from one object to another by contact.

We have by no means exhausted the number of experiments and observations we might try. Early scientific investigations were faced with all of these results, plus many others, and, in fact, many of these experiments are hard to reproduce with much accuracy. How should we make sense of it all? The "two-charge model," with a force that grows with charge and decreases with distance and with a distinction between conductors and insulators, seems promising but certainly not proved. We have, as yet, no explanation for how charged objects exert attractive forces on *neutral* objects—not to mention no explanation for what charge "is," how it is transferred, or *why* it should move through some objects but not others. We will nonetheless take advantage of our historical hindsight and continue to pursue this model without providing detailed evidence as to how other, competing models were eliminated. Homework problems will let you practice "explaining" other observations that can be made.

As you no doubt know, the modern names for the two types of charge are *positive charge* and *negative charge*. The names, you may be surprised to learn, were coined by Benjamin Franklin in conjunction with an early, and erroneous, theory of electricity. The names, unfortunately, suggest that one type of charge represents an "excess" and the other a "deficit," and, in fact, that was the idea of Franklin's erroneous theory. What we have actually found is that there are two *types* of charge, and we could call them "plastic and glass" or "red and blue" just as readily as "positive and negative." The words are simply *names*, with no other connotation.

So what is positive and what is negative? Whatever we *define* them to be! By definition, a plastic or rubber rod rubbed with wool or fur is *negatively* charged. That's it. Then any other object that repels a charged plastic rod is also negative, and any charged object that attracts a charged plastic rod is positive. It was only long afterward, with the discovery of electrons and protons, that electrons were found to be repelled by a charged plastic rod while protons were attracted. Thus, simply *by definition*, electrons have a negative charge and protons a positive charge. Most students are surprised by this, thinking that there is some "significance" to electrons being negative. So you may want to think carefully about this paragraph.

EXAMPLE 24-1 Many electricity experiments are carried out with the help of an *electroscope*, which is shown in Fig. 24-3a. An electroscope consists of two very thin and lightweight gold foils, called *leaves*, that are attached to the bottom of a metal post. The leaves and lower half of the post are contained in a glass enclosure, isolated from air currents and fingers, while the top of the post consists of an exposed metal sphere. If the electroscope sphere is touched with a charged plastic rod, which is then withdrawn, the leaves are seen to "jump apart" and remain hanging at an angle, as seen in Fig. 24-3b. Explain why this happens.

SOLUTION To answer this question, we need the following ideas from our theory: a) Charge is transferred from the rod to the metal sphere upon contact, b) metal is a conductor, and c) like charges repel. Once the rod has transferred charge (negative in this case) to the metal, these charges exert repulsive forces on each other and try to move away from each other. Because the electroscope pieces are metal, and because charges can move through or along conductors, the charges *do* move away from each other as far as possible. Some of the charges will be repelled down into the leaves, causing each leaf to become negatively charged. Once charged, the two leaves will exert repulsive forces on each other, which pushes them apart. They will reach some equilibrium position in which the repulsive electric force between them balances the gravitational force pulling downward.

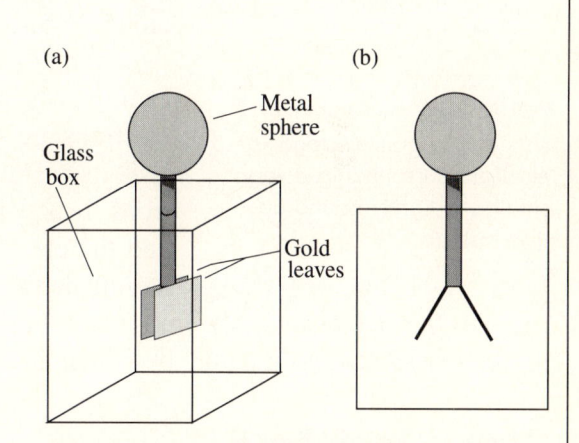

FIGURE 24-3 a) An *electroscope*. b) After contact with a charged plastic rod, the leaves stand apart from each other.

24.2 Charges in Matter

Now let's fast-forward to the twentieth century. Although the theory of electricity was developed without knowledge of atoms, there is no reason for us to continue to overlook this important part of our contemporary perspective. We *do* know that matter consists of atoms and that atoms are constructed from charged particles. We can begin to make use of that knowledge and, as a consequence, gain significant insight into many phenomena and observations of Section 24.1. Some of the relevant characteristics of atoms and matter we will, at this time, simply assert without proof. As we proceed with our study of electricity and magnetism we will have opportunities to learn about the experimental evidence supporting these assertions.

As you learned in Chapter 8, the basic structure of atoms was discovered by Rutherford through the collisional scattering of alpha particles from thin gold foils. An atom consists of a very small and dense *nucleus* (diameter $\sim 10^{-15}$ m) surrounded by much less massive orbiting *electrons*. The electron orbital frequencies are so enormous ($\sim 10^{15}$ revolutions per second) that the electrons seem to form a spherical **electron cloud** of diameter $\sim 10^{-10}$ m— a factor of 10^5 larger than the nucleus. In fact, the wave–particle duality of quantum physics destroys any notion of a well-defined electron trajectory and *all* we know of the electrons is the size and shape of the electron cloud. Figure 24-4 shows a typical atom.

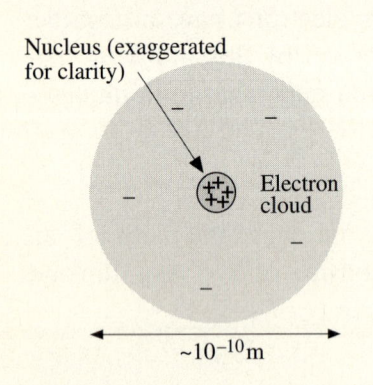

FIGURE 24-4 An atom consists of a negative "cloud" of electrons surrounding a very small and tightly bound positive nucleus.

Atomic experiments at the beginning of the twentieth century, which we will study in detail in Part VI, showed that electrons are negatively charged particles. The nucleus consists of positively charged *protons*, along with neutral neutrons. It is the attractive electric force between the nucleus and the electrons that holds the atom together. These subatomic particles have not "received" a charge, as the plastic rod did. Instead, charge, like mass, is an *inherent property* of electrons and protons. These *are* the basic charges in nature by which macroscopic objects become charged.

Quantization of Charge

Careful experiments have shown that an electron and a proton have charges of *exactly* equal magnitude. The quantity of charge of an object is represented by the symbol q. Because charge can be either positive or negative, q is a signed quantity. The fundamental atomic-level unit of charge is designated by e. Because electrons and protons are oppositely charged, we have

$$q_{\text{proton}} = +e$$

$$q_{\text{electron}} = -e.$$

Every charged object in nature has *always* been found to have a charge that is an integer multiple of e. That is, $q = \pm Ne$ for any object with charge q, where N is an integer. Thus charge, like energy, is *quantized*.

Most atoms have an equal number of protons and electrons and therefore are electrically neutral. Notice that *neutral* does *not* mean "no charges" but rather that there is no *net* charge because the positive and negative charges are equally balanced. Protons are *extremely* tightly bound within the nucleus and cannot be added to or removed from atoms. Electrons, on the other hand, are bound rather loosely and can be removed without great difficulty. The process of removing an electron from the electron cloud of an atom is called **ionization**, and the resulting atom is called an **ion** or, to be specific, a *positive ion*. A positive ion has a net charge, most typically of $+1e$, due to the charge imbalance after an electron is removed. Ions are important in charging processes, in chemical solutions, and in atmospheric processes such as lightening.

Some atoms, it turns out, can accommodate an *extra* electron and thus become a *negative ion* with a net charge of $-1e$. A good example is a salt–water solution. When salt, the chemical sodium chloride (NaCl), dissolves, it separates into positive sodium ions Na^+ and negative chlorine ions Cl^-. Salt water is a good conductor of electricity, and the conduction occurs—as you will see—due to the motion of *both* positive and negative ions.

Nearly all atoms, with the exception of the rare gases, are bound into molecules that can be as simple as H_2 or as complex as large organic and biochemical molecules. In general, molecules are electrically neutral. However, they can also gain or lose an electron and become *molecular ions*. Molecular ions are often created when a large molecule has one of its molecular bonds broken. When this happens, one "half" of the broken molecule ends up as a positive molecular ion and the other "half" as a negative molecular ion.

All the charging processes that we observed in Section 24.1 involved friction. As the two materials slide past each other, the bonds of molecules at the surface often break, creating molecular ions. Chemical processes at the surface then cause the positive molecular ions to remain on one material and the negative ions on the other. The net result is that one material ends up with a net positive charge and the other with a net negative charge! This is how a plastic rod is charged when rubbed with wool or a comb is charged when passed through your hair.

At the atomic level, this frictional charging process moves electrons from one side of the interface to the other because of where they happen to be on the molecule when it breaks. This process works mostly with organic materials, whose large molecules break easily into ions. It is possible to charge inorganic materials by pulling electrons directly from the outer edges of their electron clouds, but this is *much* more difficult than breaking molecular bonds. Consequently, rubbing glass with steel wool produces little charge. It is especially important to note that charging processes—at least the common processes with which we will be concerned—*never* occur due to the transfer of protons. Protons are tightly bound within the nucleus and are not subjected to the external forces that move electrons about.

Most of the materials with which we work in electricity are solids. Solids consist of a regular *array* of atoms, such as shown in Fig. 24-5, with a typical spacing between atoms of $\sim 2 \times 10^{-10}$ m or 2×10^{-8} cm. Because there are $\sim 5 \times 10^7$ atoms along a 1 cm line, there are $\sim (5 \times 10^7)^3$ or $\sim 10^{23}$ atoms in a 1 cm^3 volume. Each atom has typically 10 or more electrons, so a 1 cm^3 solid contains $\geq 10^{24}$ electrons.

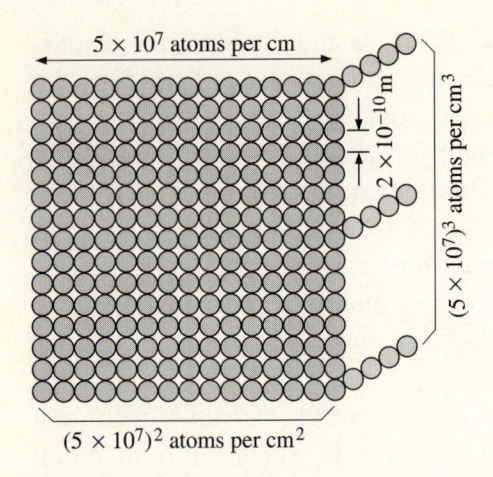

FIGURE 24-5 A solid is a regular array of closely spaced atoms.

Despite the tremendous number of charges present, most solids are electrically neutral or very close to it. This emphasizes the fact that when we talk about an object being charged, we are referring only to its *net charge*:

$$q_{net} = (N_{protons} - N_{electrons}) \cdot e. \quad (24\text{-}1)$$

Most objects have $N_{protons} \approx N_{electrons}$ and thus $q_{net} \approx 0$. A glass rod loses only ~10^{10} electrons as it is positively charged by rubbing. This is a tiny fraction of the total number of electrons, corresponding to only 1 atom out of every 10^{13} losing an electron.

Conservation of Charge

One of the important discoveries about charge is that it is neither created nor destroyed. This is a statement of the **law of conservation of charge**. Charge can be transferred or moved from one object to another, as electrons and ions move about, but the *total* amount of charge remains constant. This is analogous to energy, which can also be transferred but cannot be created or destroyed. For example, charging a plastic rod by rubbing it with wool transfers electrons from the wool to the plastic. The wool is left with a positive charge equal in magnitude but opposite in sign to the negative charge of the rod: $q_{wool} = -q_{plastic}$. The *net* charge, however, remains zero.

Diagrams will be an important tool for understanding and explaining charges and the forces on charged objects. As you begin to use such diagrams, it will be important to make explicit use of charge conservation. The net number of plusses and minuses drawn on your diagrams should *not* change as you show them moving around.

The basic rules are for drawing charge diagrams are as follows:

1. Draw a simplified cross-sectional (two-dimensional) view of the object.

2. Draw *surface* charges *inside* but *very close* to the boundary.

3. Charges that are spread throughout the volume of an object should be spaced uniformly within the interior of the drawing.

4. Show only the *net* charge. For example, a positively charged object is missing electrons. Perhaps they were removed directly from the electron clouds of the atoms, or perhaps positive molecular ions became attached because of friction. Either way, the object now has a net positive charge, so all we want to show is the plusses. A neutral object should show *no* charges, not a lot of plusses and minuses.

Figure 24-6 shows two examples of charge diagrams. Note that although we have not yet distinguished surface charge from inside charge, we will discuss them in the next section.

Net positive charge on surface   Net negative charge in interior

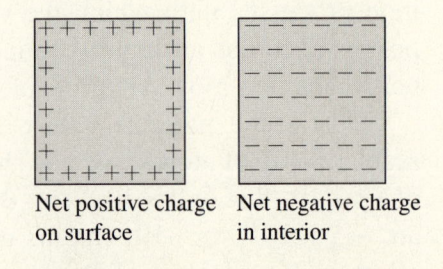

FIGURE 24-6 How to show a *net* charge on the surface or in the interior.

24.3 Insulators and Conductors

In Section 24.1, you saw that there appear to be two classes of materials as far as their electrical properties are concerned: insulators and conductors. It's time for a closer look at these materials.

An insulator is a material whose atomic electrons are all tightly bound to the positive nuclei. As a consequence, no "free" electrons are available to move about and carry charge from one place to another. Many insulators, such as rubber, plastic, and glass, are easily charged by rubbing. In this case, the charge comes from the breaking of molecular bonds in large organic molecules, leaving molecular ions—either positive or negative—on the surface. These charges, which often form "patches" on the surface, are immobile. That is why, as you saw, rubbing one *end* of a plastic rod charges that end while the opposite end remains neutral.

Figure 24-7 shows how an insulator is charged by rubbing. Notice that we have shown a) the charges right at the surfaces and b) that charge is conserved by having the number of positive charges left behind on the wool equal the number of negative charges transferred to the rod.

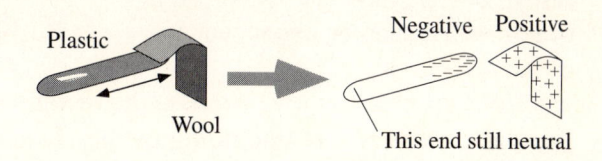

FIGURE 24-7 An insulating plastic rod is charged by rubbing. Note that charge is conserved.

Conductors are a completely different class of materials. The first example that comes to mind is usually something like copper. However, solutions such as salt water are equally good conductors that we need to consider. Let us start, however, with metals.

A metal, like an insulator, is an array of atoms. In metals, however, the outer electrons (the *valence electrons* in chemistry terms) are only weakly bound to the nuclei. As the atoms come together to form a solid, these outer electrons become detached from their parent nucleus and are free to wander about through the entire solid. The solid as a whole remains electrically neutral, because we have not added or removed any electrons, but the electrons are now rather like a liquid. Physicists like to refer to a **sea of electrons** surrounding a lattice of "positive ion cores." The repulsive forces between the negative electrons cause the electron sea to spread uniformly throughout the solid. They cannot, however, leave the solid and enter the air. Figure 24-8 depicts a metal as a regular lattice of positive ions surrounded by a uniform gray sea of electrons.

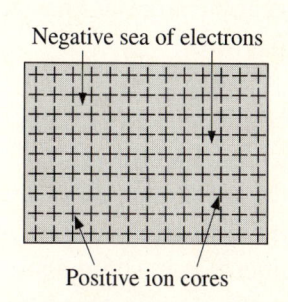

FIGURE 24-8 A metal consists of an array of positive ion cores surrounded by a "sea" of mobile electrons.

The primary consequence of this structure for metals is that electrons are highly mobile. They can quickly and easily move through the metal and along the surface in response to forces exerted on them by other charges. The motion of charges through a material is what we will later call a **current**, and the charges that physically move are called the **charge carriers**. The charge carriers in metals are electrons, but we will see that other kinds of conductors can have different charge carriers. Keep in mind that simply looking at a current-carrying piece of metal reveals absolutely *nothing* about what, if anything, is moving. Only through a wide variety of experiments—some of which we will describe later—have we been able to deduce that metals are structured in this way and have electrons as charge carriers.

This knowledge of metals allows us to understand why it was that charging one *end* of the metal rod resulted in the entire rod becoming charged. Figure 24-9 gives a pictorial explanation. Initially the plastic rod is negatively charged (an excess of electrons) while the metal rod is neutral (electron sea and positive ions in balance). Upon contact, some of the excess electrons from the plastic are transferred to the metal. (Note that the charge leaves the plastic insulator *only* at points in contact with the metal. Other points on the plastic retain their charge.) There is then an excess of electrons on the metal at the point of contact, and these like charges exert repulsive forces on each other. The electrons are free to move because the metal is a conductor, so these repulsive forces cause the entire electron sea (including the newly added electrons) to spread as far apart as possible. The result is that the excess charge is spread over the surface of the *entire* metal.

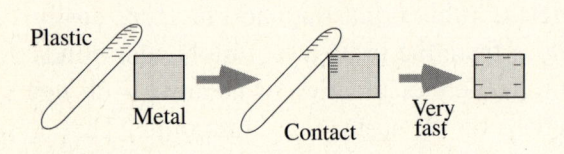

FIGURE 24-9 A conductor is charged by contact with a charged plastic rod.

Note that it is *not* necessary for the newly added electrons themselves to move to the far corners of the metal. Because of the repulsive forces, the newcomers simply "shove" the entire electron sea a little to the side. The time it takes the electron sea to readjust itself to the presence of the added charge is *extremely* short—typically about 1 ns = 10^{-9} s. Thus for all practical purposes we can consider a conductor to respond instantaneously to the addition or removal of charge. Only very sophisticated instruments can detect the currents that flow briefly as this readjustment takes place.

Other than this very brief interval during which the electron sea is undergoing a readjustment, the charges in an *isolated* conductor are in static equilibrium—at rest and with no *net* force being applied to any charge. We call this condition **electrostatic equilibrium**. If any charge *did* have a net force, it would quickly move to an equilibrium point at which the force was zero.

This fact of electrostatic equilibrium has an important consequence: In an isolated conductor, *any* excess charge is located on the surface. To see this, suppose there were an excess electron in the interior of an isolated conductor. The added electron would upset the electrical neutrality of the interior and would exert forces on nearby electrons, causing them to move and currents to flow. But this motion violates the fact that the isolated conductor is in static equilibrium. Thus we conclude that there cannot be any excess electrons in the interior. Because the excess electrons repel each other, they move as far apart as possible and spread out along the surface.

EXAMPLE 24-2 Return to Example 24-1, in which an electroscope was charged by touching it with a charged plastic rod. Provide an explanation of this phenomenon, using a series of diagrams.

SOLUTION The explanation is shown in Fig. 24-10. First, negative charge (excess electrons) is transferred from the rod to the metal sphere upon contact. Because the post and the leaves are conductors, this charge *very quickly* spreads throughout the entire

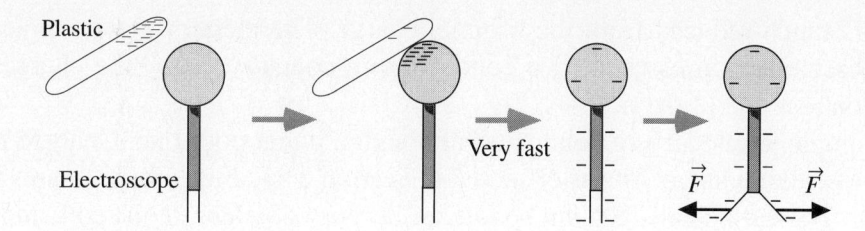

FIGURE 24-10 A detailed look at how an electroscope is charged.

electroscope. As a consequence, the leaves become negatively charged and exert repulsive forces on each other, causing them to spread apart until reaching an equilibrium position in which the electric and gravitational forces are balanced.

Metals are not the only conductors. Solutions are another very important class of materials that conduct electricity. Absolutely pure water is not a terribly good conductor, but nearly all water contains a variety of dissolved minerals that float around as ions. Dissolved salt, as we noted previously, separates into Na$^+$ and Cl$^-$ ions. These ions—*both* of them—are the primary charge carriers in solutions. Free electrons play little or no role.

The human body consists largely of salt water, both as cell and blood fluids and as moisture on the skin. Consequently, and occasionally tragically, humans are generally good conductors. This fact allows us to understand how it is that *touching* charged objects discharges them, as we observed in Section 24.1. Figure 24-11 shows a person touching a positively charged metal—one that is missing electrons. Upon contact, some of the negative Cl$^-$ ions on the skin surface transfer their extra electrons to the metal, neutralizing both the metal and the chlorine atoms. This leaves the body with an excess of positive Na$^+$ ions and, thus, a net positive charge. As in any conductor, these excess positive charges quickly spread as far apart as possible over the surface of the conductor.

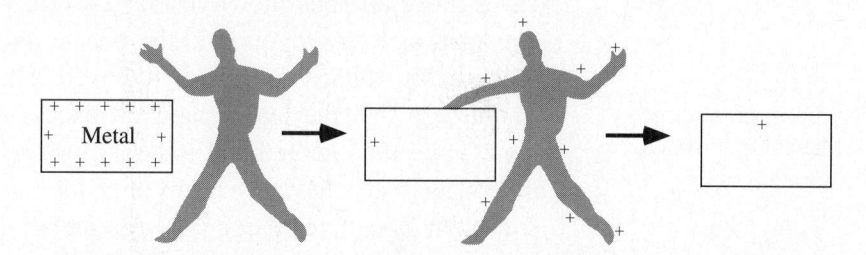

FIGURE 24-11 Touching a metal discharges it because the human body is a good conductor and the charge spreads out through the much larger conductor.

The net effect of touching a charged metal is that the conducting metal and the conducting human together become a much larger conductor than the metal alone. Thus any excess charge that was initially confined to the metal can now spread over the larger metal+human conductor. This may not entirely discharge the metal, but in typical circumstances, where the human is much larger than the metal, the residual charge remaining on the metal is much reduced from the original charge. The metal, for most practical purposes, is discharged. In essence, two conductors in contact "share" the charge that was originally on just one of them.

For example, moist air is a conductor, although a rather poor one. Charged objects in air slowly lose their charge, as was observed in Section 24.1, as the object shares its charge with the air. The earth is also a giant conductor because of water, moist soil, and a variety of ions—not, admittedly, as good a conductor as a piece of copper, but a conductor nonetheless. Any object that is physically connected to the earth through a conductor is said to be **grounded**. The effect of being grounded is that the object shares any excess charge it has with the entire earth! But the earth is so enormous that any conductor attached to the earth will be completely discharged. The purpose of "grounding" objects, which is common in many circuits and for many appliances, is to prevent the buildup of any charge on the objects. As you will see later, this has the effect of preventing a *voltage difference* between the object and the ground. The third prong on appliances and electronics that have a three-prong plug is the ground connection. The building wiring physically connects that third wire deep into the ground somewhere just outside the building, often by attaching it to a metal water pipe that goes underground.

24.4 Polarization

We have made great strides in the last two sections in showing how the atomic structure of matter can explain charging processes and the properties of insulators and conductors. However, there is still one "mystery" that we observed in Section 24.1 that still needs an explanation: How do charged objects of either sign exert an attractive force on a *neutral* object?

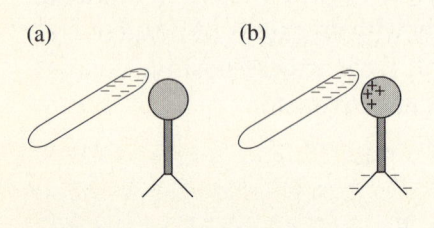

FIGURE 24-12 a) Holding a charged rod close to an electroscope, but not touching it, repels the leaves. b) Location of the charges.

To answer this question, let's again consider an electroscope. If you bring a charged plastic rod close to an electroscope *without* touching the metal sphere, the leaves move apart as shown in Fig. 24-12a. The leaves stay apart as long as you hold the plastic rod near, but they quickly collapse once it is removed. Thus it must be the case that the leaves have an excess charge on them while the charged rod is held close by. Can we understand this behavior in terms of charges and forces?

When a positive charge is near a metal, the charge exerts an attractive force on every electron in the metal and a repulsive force on every positive ion core in the metal. Because the electrons are free to move about, the whole electron sea shifts slightly toward this external charge! As Fig. 24-13 shows, this shift of the electron sea creates an excess of negative charge on the surface closest to the charge and a deficit of negative charge, and thus a net positive charge, on the surface farthest from the charge. Although the metal as a whole is still electrically neutral, we say that we have *polarized* the object. Thus **polarization** is a slight separation of the positive and negative charges in a neutral object.

Polarization explains the behavior of the electroscope in Fig. 24-12a. Because the external charge on the plastic rod is negative, the electrostatic forces pushes the electron sea slightly away from the rod. This causes an excess of negative charge on the leaves, so that they repel each other, and an excess of positive charge at the top. But because the electroscope has no *net* charge, the electron sea quickly readjusts to balance the positive ions once the plastic rod is removed. This reaction is illustrated in Fig. 24-12b.

Why don't *all* the electrons in Fig. 24-13 rush to the near side because of the attractive force of the external charge? Once the electron sea begins to shift slightly to the left, the stationary positive ions begin to exert a force, a restoring force, trying to pull the electrons back to the right. The equilibrium position for the sea of electrons is just far enough to the left that the forces due to the external charge and to the positive ions are in balance. In practice, the displacement of the electron sea is usually *less than* 10^{-15} m! That minuscule shift is all it takes to charge the surfaces by polarization.

Suppose a positively charged object, such as a glass rod that has been rubbed with silk, is held near a neutral piece of metal. It will polarize the metal, so the negative charge in the metal is slightly closer to the external charge than is the metal's positive charge. *Because the electric force decreases with distance*, the attractive

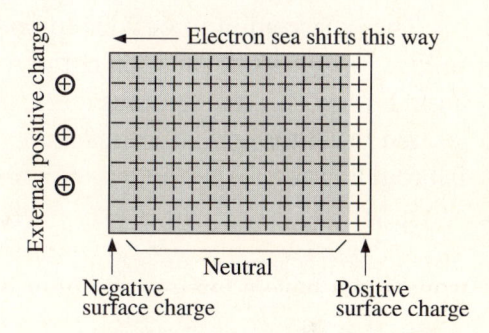

FIGURE 24-13 An external charge shifts the sea of electrons, polarizing the metal and creating surface charges. The amount of shift is greatly exaggerated in the figure.

force that the rod exerts on the electrons is *slightly greater* than the repulsive force it exerts on the positive ions. Thus there is a *net force* on the neutral metal that *attracts* it to the external positive charge! This is called a **polarization force**.

What if the external object is negative? A negatively charged object would push the electron sea slightly away, polarizing the metal to have a positive surface charge near the external charge and a negative surface charge on the far side. Once again, these are the conditions for the charge exerting a *net attractive force* on the metal. Thus our theory of charge has allowed us to construct an explanation for how a charged object of *either* sign attracts neutral pieces of metal.

Now let's consider a slightly trickier situation: Why does a charged rod pick up paper, which is an insulator? First consider what happens if we bring a positive charge near a single neutral atom. As you can see in Fig. 24-14a, the charge will polarize the atom. The electron cloud is attracted to this external charge, so it distorts and shifts slightly to the left. It cannot go far because the force from the positive nucleus is pulling it back. Nonetheless, there is now a slight separation between the center of

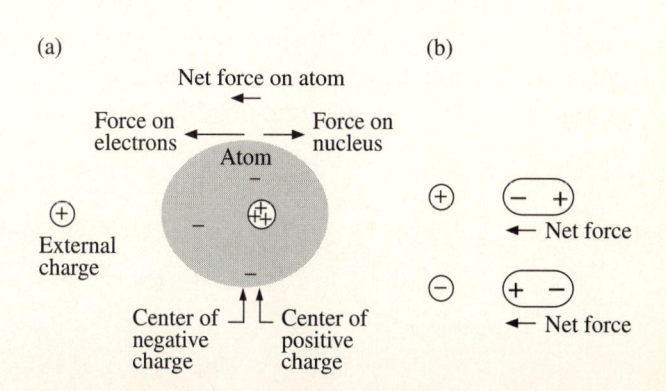

FIGURE 24-14 a) A neutral atom is polarized by an external charge, forming an *electric dipole*. b) The atom is attracted to either a positive or a negative charge by the polarization force.

positive charge and the center of negative charge. This arrangement of equal positive and negative charges with a slight separation between them is called an **electric dipole**. You can see where the name originates because the atom has two "poles" or charges.

The negative side of the dipole is slightly closer to the positive charge than the positive side. Thus the attractive force on the atom's negative charge slightly exceeds the repulsive force on the atom's positive charge. As a consequence, the external charge exerts a *net attractive force* on the neutral atom. This is another case of a polarization force.

The polarization forces due to both a positive charge and a negative charge are shown in Fig. 24-14b. Both cause an attractive force. Notice how dipoles are drawn, as a distorted atom with a plus and a minus at the ends.

An insulator, unlike a conductor, has no sea of electrons to shift if an external charge is brought close. However, all the individual atoms inside the insulator can become polarized. As Fig. 24-15 shows, this still has the effect of shifting a small amount of negative charge toward or away from the external charge, creating a surface charge. So insulators can be polarized just like metals, although the effect is usually not as strong. Once polarized, there is again a net polarization force attracting the insulator toward the external charge.

So we have solved the "puzzle" of how charged objects attract neutral objects. A charged rod picks up pieces of paper by first polarizing them, then exerting an attractive polarization force. This conclusion is significant, so be sure you understand all the steps in the reasoning.

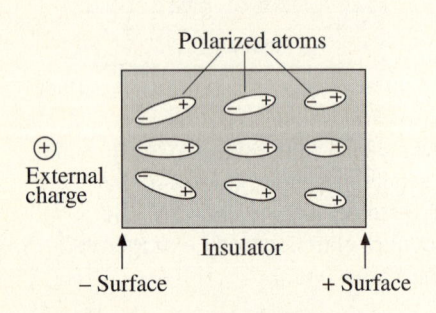

FIGURE 24-15 The atoms in an insulator are polarized by an external charge, creating a surface charge and a polarization force.

Let us look at one more puzzle. Charge a plastic rod negatively and bring it close to, without actually touching, the top of an electroscope, as shown in Fig. 24-16. At the same time, place your finger on the electroscope. The leaves do not move. Next, while the rod is still held close, remove your finger. The leaves remain down. However, when you take the plastic rod away, the leaves spring outward! The behavior of the leaves indicates that the electroscope is charged, but how did the charge get there when the rod never touched the electroscope? And how is it charged, positively or negatively? If you bring the negatively charged rod close to but not touching the electroscope, the leaves drop. If you bring a *positively* charged rod close but not touching, the leaves spread apart even farther. These observations

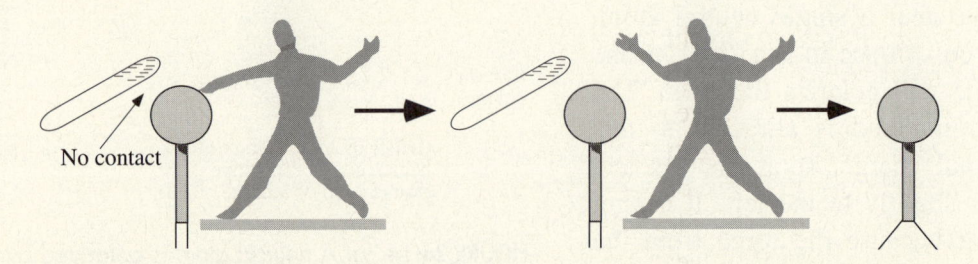

FIGURE 24-16 Charging by induction. The rod never touches the electroscope, but it nonetheless ends up charged. How?

tell us that the electroscope has a *positive* charge! This process is called **charging by induction**. We will leave it as a homework problem for you to explain how it happens, using the properties of charges and the properties of insulators and conductors that have been discussed in the last three sections.

24.5 Coulomb's Law

The last few sections have established a *model* of charges and forces. This model is very good at explaining various electric phenomena and with providing a general understanding of electricity. Now we need to become more quantitative and learn how the *size* of the force is related to the *quantity* of charge.

Charles Coulomb was one of many scientists investigating charges and their properties in the late eighteenth century. Coulomb had the idea of studying electric forces using the torsion balance scheme by which Cavendish had measured the value of the gravitational constant G (see Section 12.4). This was an exceptionally difficult experiment. Cavendish's masses could be placed in position and did not change, but Coulomb was constantly having to recharge the ends of his balance. How could he do this reproducibly? How could he know whether two objects were "equally charged"? How could he know for sure where the charge was located?

Despite these obstacles, in 1785 Coulomb announced that the electric force obeys an inverse-square law analogous to Newton's law of gravity. Historians of science debate whether Coulomb really discovered this law from his data or, perhaps, whether he leaped to unwarranted conclusions because he so wanted his discovery to match that of the great Newton. Nonetheless, Coulomb's discovery or lucky guess, whichever it was, was subsequently confirmed, and the basic law of electric force bears his name.

Coulomb's law can be stated as follows:

1. If two charged objects having charges q_1 and q_2 are a distance r apart, the objects exert forces on each other of magnitude

$$|\vec{F}_{1 \text{ on } 2}| = |\vec{F}_{2 \text{ on } 1}| = \frac{K|q_1||q_2|}{r^2}, \qquad (24\text{-}2)$$

where K is called the **electrostatic constant**. These forces are an action/reaction pair, equal in magnitude and opposite in direction.

2. The forces are directed along the line joining the two charges. The forces are *repulsive* for two like charges and *attractive* for two opposite charges.

Because charge can be either positive or negative, unlike mass, q is a *signed* quantity. Thus the absolute value signs in Eq. 24-2 are especially important. The first part of Coulomb's law gives only the *magnitude* of the force, which is always positive. The direction must be determined from the second part of the law. Figure 24-17 shows the forces between different combinations of positive and negative charges.

Although it is customary to speak of the "force between charge q_1 and charge q_2," keep in mind that we are really dealing with charged *objects*, which also have a mass, a size, and other parameters. Charge is not some disembodied entity that exists apart from matter. We will frequently be interested in charged *particles*, which are an extension of the

182 CHAPTER 24 ELECTRIC FORCES AND FIELDS

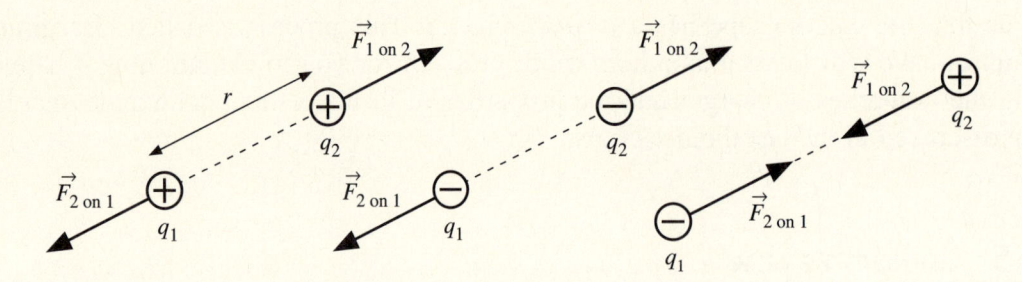

FIGURE 24-17 Attractive and repulsive forces between charges.

particle model from Part I. A charged particle—also called a **point charge**—also has a mass but has no size.

Coulomb had no *unit* of charge, so he was unable to determine a value for K, whose numerical value depends upon the units of both charge and distance. The SI unit of charge —the **coulomb**—is derived from the SI unit of *current*, which is the *ampere*. For now, all you need to know is that a current of 1 ampere delivers 1 coulomb (1 C) of charge during a time interval of 1 s. One coulomb is a rather large amount of charge. A "typical" electrostatic charge produced by rubbing is in the range 1 nC (10^{-9} C) to 1 μC (10^{-6} C).

Once the unit of charge is established, torsion balance experiments such as Coulomb's can be used to measure the electrostatic constant K. In SI units we find

$$K = 8.99 \times 10^9 \text{ N m}^2/\text{C}^2.$$

It is customary to round this to $K = 9.0 \times 10^9$ N m^2/C^2 for all but precision calculations, and we will do so.

Although Coulomb's law is most easily written in terms of the electrostatic constant K, we will find that Coulomb's law is not explicitly used in much of the theory of electricity. Although it *is* the basic force law, most of our future discussion and calculations will be of things called *fields* and *potentials*. It turns out that we can make many future equations easier to use if we rewrite Coulomb's law in a somewhat more complicated way. Let's define a new constant, called the **permittivity constant** ε_0 (pronounced "epsilon zero" or "epsilon naught"), as

$$\varepsilon_0 = \frac{1}{4\pi K} = 8.85 \times 10^{-12} \text{ C}^2 / \text{N m}^2.$$

In terms of ε_0, $K = 1/4\pi\varepsilon_0$ and Coulomb's law becomes

$$|\vec{F}| = \frac{1}{4\pi\varepsilon_0} \cdot \frac{|q_1||q_2|}{r^2}. \tag{24-3}$$

It will be easiest, as long as we are using Coulomb's law directly, to use the electrostatic constant K. However, in later chapters we will switch to the second version with ε_0.

There are three important observations regarding Coulomb's law:

1. Coulomb's law applies only to *point charges*. A point charge is an idealized material object with charge but with no size or extension. For practical purposes, however, Coulomb's law is appropriate if the size of two charged objects is much less than the separation between them.

2. Coulomb's law applies only to *static charges*—those that are at rest. For moving charges we must use the concept of the *electric field*, which we will introduce in the next section. In addition, moving charges experience additional forces that we will later identify as magnetic forces. So for now we will limit ourselves to the study of **electrostatics**.

3. Electric forces, like other forces, can be *superimposed*. That is, if multiple charges 1, 2, 3, ..., j, ... are present, the *net* electric force on charge j due to all other charges is given by

$$\vec{F}_{\text{on } j} = \vec{F}_{1 \text{ on } j} + \vec{F}_{2 \text{ on } j} + \vec{F}_{3 \text{ on } j} + \cdots, \quad (24\text{-}4)$$

where each of the $\vec{F}_{i \text{ on } j}$ is given by Eq. 24-3. This statement does *not* follow from Coulomb's law. It is an independent statement about the electric force that must be, and has been, experimentally confirmed.

EXAMPLE 24-3 Three charges—$q_1 = -50$ nC, $q_2 = +50$ nC, and $q_3 = +30$ nC—are placed on the corners of a 5 cm by 10 cm rectangle as shown in Fig. 24-18. What is the net force $\vec{F}_3$ on charge q_3 due to the other two charges?

SOLUTION The question asks for a *force*, so our answer will be a *vector*, the vector sum $\vec{F}_3 = \vec{F}_{1 \text{ on } 3} + \vec{F}_{2 \text{ on } 3}$. We can supply the answer in either magnitude and direction form or in component form. Our method will be to evaluate $\vec{F}_{1 \text{ on } 3}$ and $\vec{F}_{2 \text{ on } 3}$ separately and then to add them, using vector addition. We have drawn the figure to include a) a coordinate system, b) the individual force vectors, c) the vector sum $\vec{F}_3$, and d) the relevant geometry for finding distances. These should be a standard part of your electrostatics problem-solving strategy.

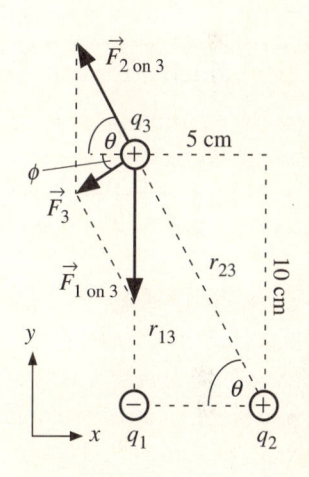

FIGURE 24-18 The charges and forces of Example 24-3.

Because q_1 and q_3 are opposite charges, $\vec{F}_{1 \text{ on } 3}$ is an attractive force in the $-y$-direction. Its magnitude is

$$|\vec{F}_{1 \text{ on } 3}| = \frac{K|q_1||q_3|}{r_{13}^2} = \frac{(9 \times 10^9 \text{ N m}^2/\text{C}^2)(50 \times 10^{-9} \text{ C})(30 \times 10^{-9} \text{ C})}{(.10 \text{ m})^2} = 1.35 \times 10^{-3} \text{ N},$$

where we have used $r_{13} = 10$ cm. But this is only its magnitude. By reference to the figure we see that its direction is down, in the $-y$ direction. So as a vector,

$$\vec{F}_{1 \text{ on } 3} = -1.35 \times 10^{-3} \hat{j} \text{ N}.$$

Charges q_2 and q_3 are both positive, so $\vec{F}_{2 \text{ on } 3}$ is a repulsive force in a direction away from q_2. The distance between these charges is given by the Pythagorean theorem as

$$r_{23} = \sqrt{(5 \text{ cm})^2 + (10 \text{ cm})^2} = 11.18 \text{ cm}.$$

Thus the magnitude of $\vec{F}_{2\,\text{on}\,3}$ is

$$|\vec{F}_{2\,\text{on}\,3}| = \frac{K|q_2||q_3|}{r_{23}^2} = \frac{(9\times10^9\,\text{Nm}^2/\text{C}^2)(50\times10^{-9}\,\text{C})(30\times10^{-9}\,\text{C})}{(0.1118\,\text{m})^2} = 1.08\times10^{-3}\,\text{N},$$

Again, this is only a magnitude. The *vector* $\vec{F}_{2\,\text{on}\,3}$ is

$$\vec{F}_{2\,\text{on}\,3} = -|\vec{F}_{2\,\text{on}\,3}|\cos\theta\,\hat{i} + |\vec{F}_{2\,\text{on}\,3}|\sin\theta\,\hat{j},$$

where angle θ is defined in the figure and the signs (negative *x*-component, positive *y*-component) were determined from the figure. We can evaluate θ from the geometry of the rectangle, finding

$$\theta = \tan^{-1}\left(\frac{10\,\text{cm}}{5\,\text{cm}}\right) = \tan^{-1}(2) = 63.4°.$$

Thus $\vec{F}_{2\,\text{on}\,3}$ is

$$\vec{F}_{2\,\text{on}\,3} = (-0.483\,\hat{i} + 0.966\,\hat{j})\times 10^{-3}\,\text{N}.$$

Now, finally, we can add $\vec{F}_{1\,\text{on}\,3}$ and $\vec{F}_{2\,\text{on}\,3}$ to find

$$\vec{F}_3 = \vec{F}_{1\,\text{on}\,3} + \vec{F}_{2\,\text{on}\,3} = (-4.83\,\hat{i} - 3.84\,\hat{j})\times 10^{-4}\,\text{N}.$$

This is an acceptable answer for most problems, although we sometimes need the net force as a magnitude and direction. With angle ϕ as defined in the figure, we can find these as

$$|\vec{F}| = \sqrt{F_{3x}^2 + F_{3y}^2} = 6.17\times10^{-4}\,\text{N}$$

$$\phi = \tan^{-1}\left|\frac{F_{3y}}{F_{3x}}\right| = 38.5°.$$

EXAMPLE 24-4 Two charges, $q_1 = +1\,\mu\text{C}$ and $q_2 = +3\,\mu\text{C}$, are 10 cm apart. Where (other than at infinity) could a third charge q_3 be placed so as to experience no net force?

SOLUTION Let the position of q_1 define the origin of a coordinate system and let q_2 be at $x = d = 10$ cm, as shown in Fig. 24-19. First we need to think about the region of space in which q_3 needs to be located to have no net force ($\vec{F}_3 = 0$). We have no information about the sign of q_3, so apparently the position for which we are looking will work for either sign. Suppose we place a positive q_3 at position A above the axis. Then the two forces $\vec{F}_{1\,\text{on}\,3}$ and $\vec{F}_{2\,\text{on}\,3}$ cannot possibly add to give zero. Similarly, suppose q_3 is located at position B on the *x*-axis but outside the charges. Then $\vec{F}_{1\,\text{on}\,3}$ and $\vec{F}_{2\,\text{on}\,3}$ are both going to point in the same direction. Once again, they cannot add to give zero. If, however, q_3 is placed at point C on the *x*-axis *between* the charges, then the two forces on it will be oppositely directed. This is true whether

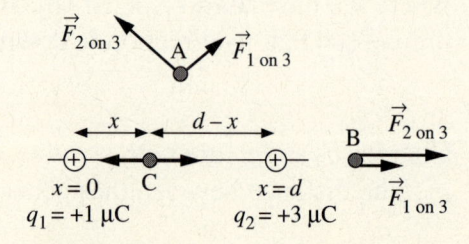

FIGURE 24-19 Geometry of Example 24-4.

q_3 is positive or negative. Thus the mathematical issue is find the specific location x for which the two forces $\vec{F}_{1 \text{ on } 3}$ and $\vec{F}_{2 \text{ on } 3}$ are equal in magnitude because, at that location, their sum will be zero.

If q_3 is distance x from q_1, it must be distance $d - x$ from q_2. The magnitudes of the forces, which are all we need, are

$$|\vec{F}_{1 \text{ on } 3}| = \frac{Kq_1|q_3|}{r_{13}^2} = \frac{Kq_1|q_3|}{x^2}$$

$$|\vec{F}_{2 \text{ on } 3}| = \frac{Kq_2|q_3|}{r_{23}^2} = \frac{Kq_2|q_3|}{(d-x)^2}.$$

Charges q_1 and q_2 are positive and do not need absolute value signs. Equating the two forces:

$$\frac{Kq_1|q_3|}{x^2} = \frac{Kq_2|q_3|}{(d-x)^2}.$$

The term $|q_3|$ cancels, and a rearrangement gives

$$q_1(d-x)^2 = q_2 x^2$$
$$\Rightarrow (q_2 - q_1)x^2 + (2q_1 d)x - q_1 d^2 = 0.$$

We see that, indeed, the sign and value of q_3 do not matter. Solving the quadratic equation with the known values for the charges and distance gives

$$x = +3.66 \text{ cm} \quad \text{or} \quad -13.66 \text{ cm}.$$

Both are legitimate points where the magnitudes of the two forces are equal. But $x = -13.66$ cm is to the left of q_1, where the magnitudes are equal but the directions are the same. So $\vec{F}_3 \neq 0$ at this point. The solution we want, which is between the charges, is $x = 3.66$ cm. So the point where we wish to place q_3 is 3.66 cm from q_1 along the line joining q_1 and q_2.

EXAMPLE 24-5 A small plastic sphere is charged to -10 nC. It is held 1 cm above a small glass sphere, which is at rest on a table. The glass sphere has a mass of 15 mg and a charge of $+10$ nC. Will the glass sphere "leap up" to the plastic sphere?

Figure 24-20 shows the geometry of the problem, with the plastic sphere designated as q_1 and the glass sphere as q_2. If $|\vec{F}_{1 \text{ on } 2}| < |\vec{W}| = m_2 g$, then the glass sphere will remain at rest on the table with $\vec{F}_{\text{net}} = \vec{F}_{1 \text{ on } 2} + \vec{W} + \vec{N} = 0$. If, however, $|\vec{F}_{1 \text{ on } 2}| > |\vec{W}| = m_2 g$, then the glass sphere will accelerate upward from the table. With the values provided,

$$|\vec{F}_{1 \text{ on } 2}| = \frac{K|q_1||q_2|}{r^2} = 9.0 \times 10^{-3} \text{ N}$$

$$|\vec{W}| = m_2 g = 1.5 \times 10^{-4} \text{ N}.$$

$|\vec{F}_{1 \text{ on } 2}|$ exceeds $|\vec{W}|$ by a factor of 60, so the glass sphere will leap upward!

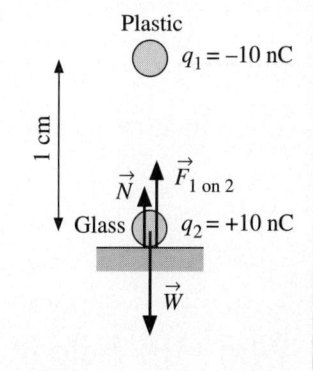

FIGURE 24-20 Example 24-5.

The values used in Example 24-5 are quite realistic for the amount of charge that can be generated through friction and for the mass of spheres ≈2 mm in diameter. What we learn from this example is that electric forces are, in general, *significantly* larger than gravitational forces. Consequently, we will usually *neglect* gravity when working electric force problems. Only when the charged particles are fairly massive will we include gravitational effects.

The fundamental unit of charge e has been measured to have the value

$$e = 1.60 \times 10^{-19} \text{ C}.$$

This is a very small amount of charge. An object that is missing 10^{10} electrons has a charge of 1.6×10^{-9} C = 1.6 nC. This is a fairly typical charge for an object charged by rubbing.

EXAMPLE 24-6 A hydrogen atom consists of a single electron orbiting a single proton at a distance of 0.053 nm. a) What is the force on the electron? b) What is the electron's orbital frequency, assuming the orbit to be circular?

SOLUTION a) Although the electron and the proton are subatomic particles, the force between them is given by Coulomb's law. Using $q_1 = e$ and $q_2 = -e$, the magnitude of the force is

$$|\vec{F}| = \frac{1}{4\pi\varepsilon_0} \cdot \frac{|q_1||q_2|}{r^2} = \frac{1}{4\pi\varepsilon_0} \cdot \frac{e^2}{r^2} = \frac{(9.0 \times 10^{-9} \text{ N m}^2/\text{C}^2)(1.60 \times 10^{-19} \text{ C})^2}{(5.3 \times 10^{-11} \text{ m})^2}$$
$$= 8.20 \times 10^{-8} \text{ N}.$$

b) The force on the electron is centrally directed toward the nucleus, so it acts as a *centripetal force* and moves the electron in a circular orbit. We can find the orbital speed from Newton's second law for circular motion:

$$|\vec{F}| = ma_c = \frac{mv^2}{r}$$

$$\Rightarrow v = \sqrt{\frac{r|\vec{F}|}{m}} = \sqrt{\frac{(5.3 \times 10^{-11} \text{ m})(8.20 \times 10^{-8} \text{ N})}{9.11 \times 10^{-31} \text{ kg}}} = 2.18 \times 10^6 \text{ m/s}.$$

The orbital frequency is related to the speed and the radius by

$$f = \frac{v}{2\pi r} = 6.56 \times 10^{15} \text{ s}^{-1}.$$

24.6 The Field Model

Electric and magnetic forces, like gravity, are *long-range forces*. No contact is required for one object to exert a force on another. Somehow, the force is transmitted through empty space—what scientists call *action at a distance*. The concept of action at a distance greatly troubled many of the leading thinkers of Newton's day, following the publication of his theory of gravity. Force, they believed, should have some *mechanism* by which it is exerted, and the idea of action at a distance—with no apparent mechanism—was more than most scientists could accept. Nonetheless, they could not dispute the great success of Newton's theory.

As long as it was "only" gravity that exhibited this action at a distance, the great prestige and success of Newton was able to keep scientists' doubts and reservations in check. But toward the end of the eighteenth century, as investigations of electric and magnetic phenomena came to the forefront of physics, the whole issue of action at a distance was reopened.

Consider the two charges A and B of Fig. 24-21 that exert attractive electric forces on each other. If the objects have been at rest for a long period of time, then we can confidently use Coulomb's law to determine the force being exerted on B by A. But suppose that A suddenly moves to a new location, as shown by the arrow. In response, the force vector on B must pivot to point toward the new location of A. Does this happen *instantly*, or is there some *delay* between when A moves and when the force $\vec{F}_{A \text{ on } B}$ responds? This troublesome issue also arises if A and B are masses that exert gravitational forces on each other.

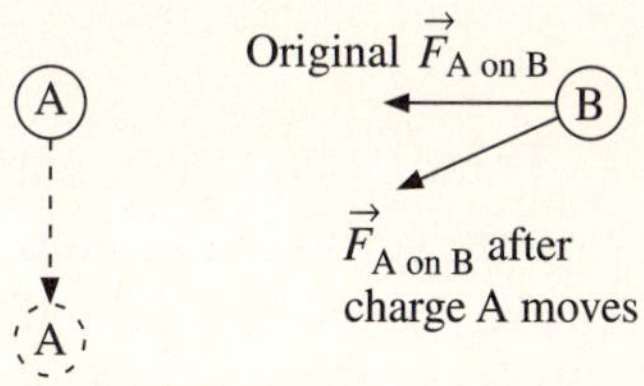

FIGURE 24-21 If charge A moves, how long does it take the force vector on B to respond?

Neither Coulomb's law nor Newton's law of gravity contains a time variable, so the answer from the perspective of Newtonian physics has to be "instantly." Yet most scientists found this troubling. What if A is a star on the other side of the galaxy, 100,000 light years away, while B is the sun? Will the sun respond *instantly* to an event that happens 100,000 light years away? The idea of instantaneous transmission of forces was becoming unbelievable to most scientists by the beginning of the nineteenth century. But if there is a delay, how long is it? How does the information to "change force" get sent from A to B? While the scientists did not like action at a distance, they did not have the answer to these vital questions. This was a major issue as a young Michael Faraday appeared on the scene.

Michael Faraday is certainly one of the most interesting figures in the history of science. The son of a poor blacksmith near London, Faraday was sent out to work at an early age with almost no formal education. As a teenager he found employment with a printer and bookbinder, and he began to read the books that came through the shop. By happenstance, a customer brought in a copy of the *Encyclopedia Britannica* to be rebound, and there Faraday discovered a lengthy article about electricity. It was all the spark he needed to set him on a course that, by his death in 1867, would make him one of the most esteemed scientists in Europe.

You will learn more about Faraday in coming chapters. For now, suffice it to say that Faraday was never able to become fluent in mathematics. Apparently the late age at which he started his studies was too much of a detriment for mathematical learning. In place of mathematics, Faraday's brilliant and insightful mind developed many ingenious *pictorial* methods for thinking about and describing physical phenomena. By far the most important of these was the field.

188 CHAPTER 24 ELECTRIC FORCES AND FIELDS

FIGURE 24-22 Iron filings sprinkled around the ends of a magnet are suggestive that the influence of the magnet extends into the space around it.

Faraday had been particularly impressed with the pattern that is made when iron filings are sprinkled around a magnet, as seen in the photograph of Fig. 24-22. The pattern's regularity and the curved lines suggested to Faraday that the *space itself* around the magnet was "filled up" with some kind of magnetic influence. He speculated that perhaps the magnet was in some unknown fashion *altering* the space around itself. If this were the case, then a piece of iron placed near the magnet would not respond to the magnet directly but, instead, to the alteration of the space. This space alteration—whatever it is—would then be the *mechanism* by which a long-range force is exerted.

In other words, rather than A and B interacting directly with each other, Faraday had the idea that A first alters or modifies the space around it. Object B then comes along and experiences a force due to this altered space. This idea is illustrated in Fig. 24-23. The alteration of space becomes an *agent* by which A and B interact. Furthermore, this alteration could easily be imagined to take a finite time to propagate outward from A—in, perhaps, a wavelike fashion—so B need not respond instantly to any changes in A. At this point we have no idea what the alteration would look like or how it would behave, so the wavy lines of Fig. 24-23 are meant to convey the idea with poetic license rather than to show the altered space literally.

Faraday's idea came to be called a **field**. The term *field* comes from mathematics. A function $f(x, y, z)$ that assigns a value to every point in space is sometimes called a *field* by mathematicians. It conveys the idea that the function exists everywhere in space. That is, indeed, what Faraday was suggesting about how long-range forces operate. The charge, or the mass, makes an alteration *everywhere* in space and other charges, or other masses, then respond to the alteration at their position. The specific alteration of the space around a mass is called the **gravitational field**. Similarly, the space around charges is altered to create the **electric field** and the **magnetic field**.

Faraday's proposal was an entirely novel way to think about how forces are exerted by one object on another. The field concept goes well beyond the boundary of Newtonian mechanics, where material particles and

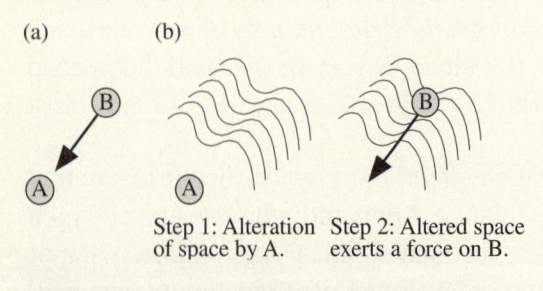

(a) (b)

Step 1: Alteration of space by A.   Step 2: Altered space exerts a force on B.

FIGURE 24-23 a) In Newton's view, A acts directly on B. b) Faraday suggested that for a long-range force A alters the space around it, then B experiences a force due to this altered space. The altered space is called a *field*.

forces are all that have significance. The field idea was not taken seriously at first—it seemed too vague and nonmathematical to scientists steeped in the Newtonian tradition. But the significance of the field grew as electromagnetic theory developed during the first half of the nineteenth century. What seemed at first a pictorial "gimmick" came to be seen as more and more essential for understanding how electric and magnetic forces behaved. Many of Faraday's own experiments provided solid evidence that the field is a reality, that it is more than just a convenient picture.

Einstein provided an excellent summary of the change in thinking:

> The old mechanical view attempted to reduce all events in nature to forces acting between material particles. Upon this mechanical view was based the first naive theory of the electric fluids [an early, incorrect theory of electricity]. The field did not exist for the physicist of the early years of the nineteenth century. For him only substance and its changes were real. He tried to describe the action of two electric charges only by concepts referring directly to the two charges.
>
> In the beginning, the field concept was no more than a means of facilitating the understanding of phenomena from the mechanical point of view. In the new field language it is the description of the field between the two charges, and not the charges themselves, which is essential for an understanding of their action. The recognition of the new concepts grew steadily, until substance was overshadowed by the field. It was realized that something of great importance had happened in physics. A new reality was created, a new concept for which there was no place in the mechanical description. Slowly and by a struggle the field concept established for itself a leading place in physics and has remained one of the basic physical concepts. The electromagnetic field is, for the modern physicist, as real as the chair on which he sits.

Faraday's field ideas were finally placed on a firm mathematical foundation in 1865 by James Clerk Maxwell, a young English physicist possessing both great physical insight and mathematical ability. Maxwell was able to describe completely all the known behaviors of electric and magnetic fields in four equations. Today, they are known as Maxwell's equations. You will not see the full Maxwell's equations in this text, because they require advanced mathematics for their use and understanding. However, the ideas that we will develop about electric and magnetic fields are, in essence, a subset of Maxwell's equations.

Maxwell made the following comments in his famous paper of 1865:

> I have preferred to seek an explanation [of electric and magnetic phenomena] by supposing them to be produced by actions which go on in the surrounding medium as well as in the excited bodies, and endeavoring to explain the action between distant bodies without assuming the existence of forces capable of acting directly at sensible [i.e., finite] distances.
>
> The theory I propose may therefore be called a theory of the 'Electromagnetic Field' because it has to do with the space in the neighborhood of the electric and magnetic bodies.

Had Maxwell's equations "merely" summarized Faraday's knowledge, they might be interesting but hardly radical. But the equations turned out to contain a remarkably wide range of predictions, including predictions that the fields could exist in the *absence* of charges, in the form of an *electromagnetic wave*. After the discovery of electromagnetic waves in 1887 by Heinrich Hertz, showing conclusively that Faraday's fields have a real existence, the success of the new field perspective was guaranteed. It was, of course, subsequently shown that light is simply an electromagnetic wave within a frequency range to which the retina is sensitive.

Maxwell's theory became the new standard of physics. All of the major new theories of the twentieth century, including Einstein's relativity and theories of the weak and strong nuclear forces, have all been field theories. Faraday, if he could return, could not begin to fathom the complex mathematics of present-day field theories. But once a present-day physicist began to sketch a few pictures at the blackboard, Faraday would certainly respond, "Aha! That's just the kind of idea I had in mind."

An Example of a Field: The Gravitational Field

Before we dive into investigating the electric field, we will take a slight detour to examine the gravitational field. The field concept is not as necessary for understanding the force of gravity as it is for electric and magnetic forces. Nonetheless, it will be useful to introduce the mathematical statement of the field concept in a familiar context before venturing into new and unexplored territory.

Recall that Newton's law of gravity says that two masses A and B exert equal but opposite attractive forces on each other of magnitude

$$|\vec{F}_{\text{A on B}}| = |\vec{F}_{\text{B on A}}| = \frac{Gm_A m_B}{r^2}, \qquad (24\text{-}5)$$

where G is the gravitational constant and r is the distance between the centers of the masses.

Suppose we were to represent the "action" of the force in the pictorial fashion shown in Fig. 24-24. Here we see a radial pattern of inward-directed arrows centered on mass A, with the length, or magnitude, of the arrows decreasing as they get farther away from the mass. Suppose, further, we make the outlandish claim that these arrows represent an actual *change in the space* around mass A, with the change depending both on the value of m_A and on the distance r. Admittedly, we have no evidence at all that such an alteration of the space is real—we are simply making a hypothesis. But if it should turn out that this hypothesis leads to predictions of new phenomena, that such phenomena are *not* predicted by the standard Newtonian theory, and that the predictions are confirmed, *then* we would have to consider seriously that this alteration of space really exists. Anticipating that such events will come to pass, let us call this alteration of space—which we see pictorially in Fig. 24-24—the *gravitational field*.

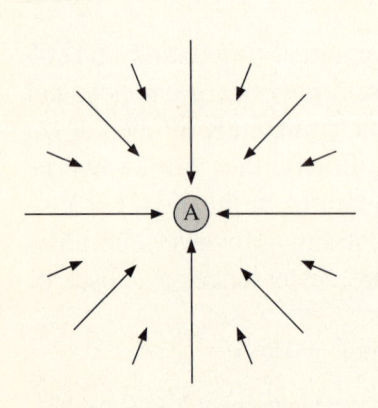

FIGURE 24-24 The "alteration" of the space around mass A.

Our hypothesis is that it is the gravitational field of mass A that exerts the force on mass B. Because force is a vector, we can write the force on B, due to A, as

$$\vec{F}_{\text{A on B}} = \left(\frac{Gm_A m_B}{r^2}, \text{ toward A}\right) = \left(\left(\frac{Gm_A}{r^2}\right) \cdot m_B, \text{ toward A}\right), \qquad (24\text{-}6)$$

where we have written the vector as a separate magnitude and direction. On the right-hand side we have separated out the mass m_B, on which the force is exerted, from the other factors that depend on the *source*—the mass m_A and the distance r. This suggests that the quantity Gm_A/r^2 must, somehow, characterize the gravitational field at a distance r from mass A.

With that in mind, let us try defining the gravitational field, which we will give the symbol $\vec{g}$, at any point in space as

$$\vec{g} \equiv \frac{\vec{F}_{\text{on } m}}{m} \quad \Rightarrow \quad \vec{F}_{\text{on } m} = m \cdot \vec{g}, \tag{24-7}$$

where $\vec{F}_{\text{on } m}$ is the gravitational force exerted on a mass m placed at that point. We are, with Eq. 24-7, defining the gravitational field to be a *vector* whose magnitude is a ratio: the force on mass m divided by the mass itself.

Note that the gravitational field is a *vector field* because we have defined, through Eq. 24-7, a vector at every point in space. Figure 24-24 can then be thought of as showing a representative sample of the vector $\vec{g}$ at just a *few* points in space near mass A. It is clearly impossible to draw a vector at every point, but you need to keep in mind that such a vector *is* at every point, whether shown or not.

Rearranging Eq. 24-7 tells us that if we place a mass m at some point in space where there is a gravitational field $\vec{g}$, the mass will experience a force given by $\vec{F}_{\text{on } m} = m\vec{g}$. This is the essence of the field idea—that *the field is the agent that exerts the force*. Figure 24-25 shows a mass placed in an arbitrary gravitational field $\vec{g}$, which is symbolized by vector arrows. Note that this is *not* the gravitational field of a single mass. In fact, we have no idea what masses are the source of this *field*—they are not shown. Nonetheless, simply knowing what the field is allows us to determine the gravitational force exerted on this mass to be $\vec{F}_{\text{on } m} = m\vec{g}$—a force of magnitude $|\vec{F}_{\text{on } m}| = m|\vec{g}|$ and in the same direction as $\vec{g}$. Note that the field exists at the point where the mass is placed, even though there is no field vector drawn at that exact point. The field exists at *all* points in space.

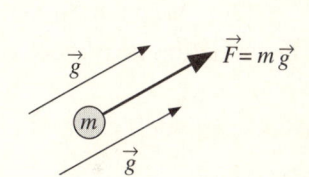

FIGURE 24-25 Whenever a mass is in a gravitational field $\vec{g}$, the force on it is $\vec{F} = m\vec{g}$.

We can find an explicit expression for the gravitational field of particle A, which has mass m_A. Let particle B be a *test mass* m_B that will *probe* the gravitational field of A. If B is placed at distance r from A, it experiences the gravitational force $\vec{F}_{\text{A on B}}$ given by Eq. 24-6, Newton's law of gravity. If we use this force in the Eq. 24-7 definition of the gravitational field, we find that the gravitational field a distance r away from mass A is

$$\vec{g} = \frac{\vec{F}_{\text{A on B}}}{m_B} = \left(\frac{Gm_A}{r^2}, \text{ toward A} \right). \tag{24-8}$$

Notice that the field is determined by the mass of A, the object that *creates* the field. The value m_B of the test mass that *experiences* the field does *not* enter into Eq. 24-8. Also notice that the gravitational field is a *vector*. The field at any point in space points directly at A. This is the field that was shown pictorially in Fig. 24-24.

EXAMPLE 24-7 a) What is the sun's gravitational field at the position of the earth? b) Use the gravitational field to find to force on the earth.

SOLUTION a) The sun's gravitational field is given by Eq. 24-8:

$$\vec{g} = \left(\frac{Gm_{sun}}{r_e^2}, \text{ toward sun}\right),$$

where r_e is the radius of the earth's orbit about the sun. From Table 12-2 we find that $m_{sun} = 1.99 \times 10^{30}$ kg and $r_e = 1.50 \times 10^{11}$ m. A calculation gives

$$\vec{g} = \left(0.0059 \text{ m/s}^2, \text{ toward sun}\right).$$

Note that the units of $\vec{g}$ are those of acceleration, m/s².

b) According to Eq. 24-7, the force that this gravitational field exerts on the earth is

$$\vec{F}_{on\ earth} = m_e \vec{g} = (5.98 \times 10^{24} \text{ kg})\left(0.0059 \text{ m/s}^2, \text{ toward sun}\right)$$

$$= \left(3.53 \times 10^{22} \text{ N}, \text{ toward sun}\right).$$

Our choice of the symbol $\vec{g}$ to represent the gravitational field was not arbitrary. Consider the situation near the surface of the earth. The gravitational force on mass m is then a simpler

$$\vec{F}_{on\ m} = -mg\hat{j}, \qquad (24\text{-}9)$$

where the minus sign indicates a force vector pointing in the downward direction. This g is the familiar acceleration due to gravity, $g = 9.80$ m/s². The gravitational field just above the earth's surface, according to Eq. 24-8, is thus

$$\vec{g} = \frac{\vec{F}_{on\ m}}{m} = -g\hat{j}. \qquad (24\text{-}10)$$

In other words, the gravitational field near the earth's surface is a vector of magnitude $|\vec{g}| = g$ pointing straight down. Our old friend g is nothing other than the magnitude of the gravitational field!

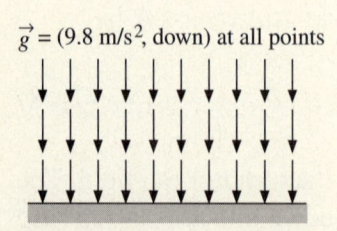

Figure 24-26 shows the gravitational field near the surface of the earth. All the vectors are of equal length $|\vec{g}| = g = 9.8$ m/s², and they are all pointing straight down. A mass m then *responds* to the field by experiencing a force $\vec{F} = m\vec{g}$. It is important to keep in mind that mass m does not *cause* the field. The field is created by the earth and is there whether or not mass m is present.

The field of Fig. 24-26 is what we will call a *uniform field*—one that is the same at all points in space. Contrast this picture with Fig. 24-24, which shows the gravitational field of a point mass. The uniform gravitational field of Fig.

$\vec{g} = (9.8$ m/s², down) at all points

FIGURE 24-26 The gravitational field $\vec{g} = -g\hat{j}$ near the earth's surface.

24-26 is valid, of course, only if we stay very near the earth's surface. Figure 24-24 is how the earth's gravitational field would appear from the distance of the moon.

24.7 The Electric Field of a Point Charge

Much of the electric phenomena you have seen in this chapter are considerably more complex than gravitational phenomena. The reason is twofold: first, the existence of two kinds of charge, as opposed to only one kind of mass, and thus of both attractive and repulsive forces; and second, the existence of two classes of materials, conductors and insulators, that have very different mobility for charge.

Although Coulomb's law is the basic force law for electric charges, it would be very difficult to use Coulomb's law directly to perform quantitative calculations about the phenomena we have observed—many of which involve quite large numbers of charges. This is where Faraday's concept of the field is going to start being very useful. We can begin our investigation of electric fields by invoking a two-level model of the interaction between charges:

1. Some charges, which we will call the **source charges**, alter the space around them by creating an *electric field*.

2. A separate charge *in* the electric field experiences a force exerted *by the field*. This charge is often called a **test charge**.

We must accomplish two tasks for this to be a useful model of electric interactions. First, we must learn how to calculate the electric field for a configuration of charges. Second, we must determine the forces on and the motion of the test charge in the electric field.

Suppose that a test charge q experiences an electric force $\vec{F}_{\text{on } q}$ due to the presence of other charges. (Remember that a charge does not exert a force on itself.) It is likely that the force varies from point to point in space, so that $\vec{F}_{\text{on } q}(x, y, z)$ is some continuous function of the test charge's position coordinates. This suggests that "something" is present at each point in space to "cause" the force that charge q experiences. With that in mind, let us define the electric field $\vec{E}$ at a point (x, y, z) in space as

$$\vec{E}(x, y, z) \equiv \frac{\vec{F}_{\text{on } q}(x, y, z)}{q}. \qquad (24\text{-}11)$$

The units of the electric field are "newtons per coulomb," or N/C.

Charge q can be used to determine whether an electric field is present at a point in space, which is why it is called a test charge. If charge q experiences an electric force at a particular point, then we *define* the electric field at that point to be the vector given by Eq. 24-11. You can imagine moving the test charge all through space and thereby "mapping out" the electric field. Notice that the definition assigns a *vector* to *every* point in space.

If this definition of the electric field seems a bit arbitrary—it is! We can define anything we want, but only subsequent testing and use will show whether the definition has any significance. Equation 24-11 will, indeed, turn out to be a "good" definition. Make sure you notice the very strong parallel between our definition of $\vec{E}$ and the definition, in the previous section, of the gravitational field $\vec{g}$.

The basic idea of the field model is that the *field* is the agent that exerts a force on test charge q. Although Eq. 24-11 is the definition of the field in terms of the force, in practice we will usually want to find the force exerted by a known field. A simple rearrangement of Eq. 24-11 tells us that a charge q at point (x, y, z) in space, where the electric field is $\vec{E}(x, y, z)$, experiences an electric force

$$\vec{F}_{\text{on } q} = q\vec{E}(x, y, z). \tag{24-12}$$

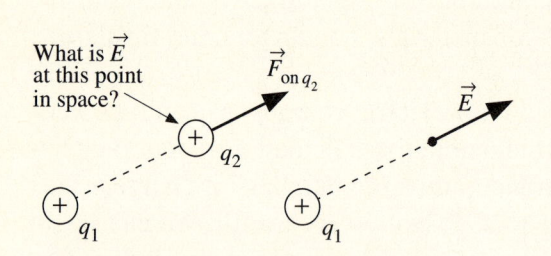

FIGURE 24-27 Charge q_2 acts as a test charge to determine the electric field of charge q_1.

We will begin to put the definition of the electric field to full use in the next chapter. For now, just to develop the ideas, we will determine the electric field of a single point charge. Consider two charges q_1 and q_2, as shown in Fig. 24-27. For the moment, let both charges be positive. We want to use q_2 as the test charge to determine the electric field of q_1 *at the point* where q_2 is located. So q_1 is the source charge. Because both charges are positive, the force on q_2 is repulsive and directed straight away from q_1. In particular, Coulomb's law gives us

$$\vec{F}_{\text{on } q_2} = \left(\frac{1}{4\pi\varepsilon_0} \cdot \frac{q_1 q_2}{r^2}, \text{ away from } q_1\right). \tag{24-13}$$

Using the definition of Eq. 24-11, the electric field at the point where q_2 is located, a distance r from q_1, is

$$\vec{E}_{\text{of } q_1} = \frac{\vec{F}_{\text{on } q_2}}{q_2} = \left(\frac{1}{4\pi\varepsilon_0} \cdot \frac{q_1}{r^2}, \text{ away from } q_1\right). \tag{24-14}$$

Notice that the subscripts 1 and 2 no longer serve a useful purpose. We needed them in Eq. 24-13 to distinguish the source charge q_1 from the test charge q_2, but the field of Eq. 24-14 doesn't depend on q_2. It is simply "the electric field of a point charge," so we can drop the subscripts and call the charge q. Thus the expression for the electric field at a distance r away from a point charge q is

$$\vec{E} = \left(\frac{1}{4\pi\varepsilon_0} \cdot \frac{q}{r^2}, \text{ away from } q\right) \quad \text{(electric field of point charge } q\text{)}. \tag{24-15}$$

It is convenient to refer to Eq. 24-15 as Coulomb's law for electric field of a point charge.

Figure 24-28 shows a charge q, which we still assume to be positive, and two points in the space near q. To determine the electric fields $\vec{E}_1$ and $\vec{E}_2$ at these two points, we first evaluate the *magnitude* of the field at each point from

$$|\vec{E}| = \frac{1}{4\pi\varepsilon_0} \cdot \frac{|q|}{r^2}. \tag{24-16}$$

Because $r_2 > r_1$, the field $\vec{E}_2$ at point 2 will have a smaller magnitude than the field $\vec{E}_1$ at point 1. Then, we draw the field vectors pointing straight away from charge q, as

Fig. 24-28 shows. Keep in mind that this is just the field at those two points, but the field exists at *all* points in space.

If we calculate the field at a sufficient number of points, we can draw a **field diagram** such as shown in Fig. 24-29a for a positive charge. The arrow attached to each point indicates both the direction *and* the strength of the field at that point. Again, remember that this is just a representative sample of points in space but that the field exists at all the other points as well. Such a diagram is sufficient, however, that you can tell fairly well what the field would be like at a neighboring point.

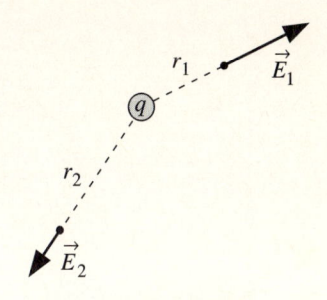

FIGURE 24-28 The electric field at two points in space near a positive charge.

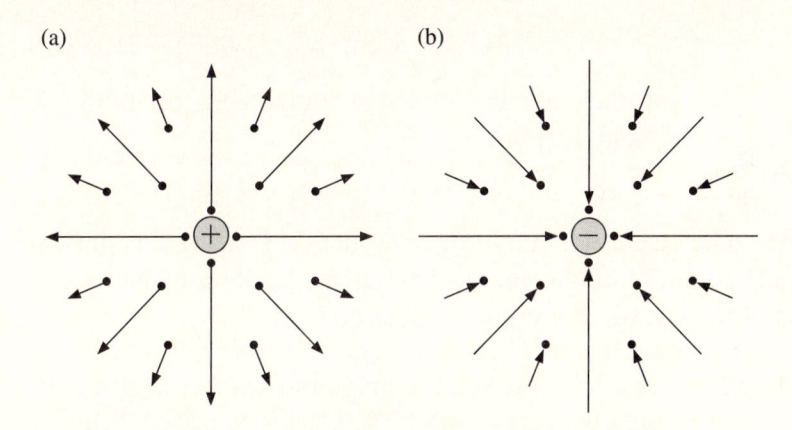

FIGURE 24-29 The electric field in the space surrounding a) a positive charge and b) a negative charge.

What happens if the source charge q is negative? If the test charge is still positive, the force vector on the test charge points *inward*, toward q. Thus the electric field of a negative charge points *inward*. Our expression for the field, Eq. 24-15, already takes this possibility into account by allowing the charge q to have a sign. If q is a negative number, then $\vec{E}$ is a negative number multiplying a vector that point away from q. But multiplying a vector by −1 simply reverses its direction, so the field of a negative charge has a magnitude given by Eq. 24-16 and points *toward q*. The field diagram of a negative charge is exactly the same as that of a positive charge *except* that the arrows are all reversed. Fig. 24-29b shows the field diagram of a negative q.

EXAMPLE 24-8 A charge of −1 nC is located at the origin. a) What is the electric field $\vec{E}$ at the positions $(x, y) = (1 \text{ cm}, 0 \text{ cm}), (0 \text{ cm}, 1 \text{ cm})$, and $(1 \text{ cm}, 1 \text{ cm})$? b) Draw a field diagram showing the electric field vectors at these points.

SOLUTION a) The magnitude of the electric field is

$$|\vec{E}| = \frac{1}{4\pi\varepsilon_0} \cdot \frac{|q|}{r^2},$$

196 CHAPTER 24 ELECTRIC FORCES AND FIELDS

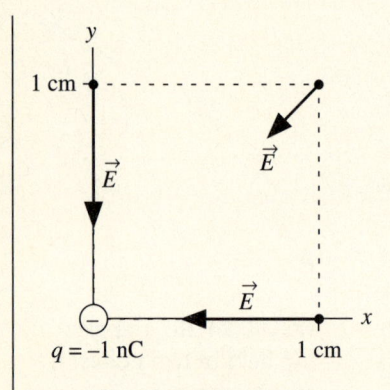

FIGURE 24-30 Electric field diagram for Example 24-8.

where $|q| = 1$ nC $= 10^{-9}$ C. The distance r is 1 cm $= 0.01$ m for the first two points and 1 cm $\times \sqrt{2} = 0.0141$ m for the third point. This gives

$|\vec{E}| = 90,000$ N/C at (1 cm, 0 cm) and (0 cm, 1 cm)

$|\vec{E}| = 45,000$ N/C at (1 cm, 1 cm).

Because q is negative, the field direction at each of these positions points *inward*, directly at charge q. In vector form we can write

$\vec{E}(1 \text{ cm}, 0 \text{ cm}) = -90,000 \, \hat{i}$ N/C

$\vec{E}(0 \text{ cm}, 1 \text{ cm}) = -90,000 \, \hat{j}$ N/C

$\vec{E}(1 \text{ cm}, 1 \text{ cm}) = (-31,800 \, \hat{i} - 31,800 \, \hat{j})$ N/C.

The numbers in the last entry are the *x*- and *y*-components of the 45,000 N/C vector. Figure 24-30 shows the field diagram.

EXAMPLE 24-9 The electron in a hydrogen atom orbits at a radius of 0.053 nm. a) What is the magnitude of the proton's electric field at the position of the electron? b) What is the magnitude of the electric force on the electron?

SOLUTION a) Because $q = e$ for a proton, the magnitude of the electric field is

$$|\vec{E}| = \frac{1}{4\pi\varepsilon_0} \cdot \frac{e}{r^2} = \frac{1}{4\pi\varepsilon_0} \cdot \frac{1.60 \times 10^{-19} \text{ C}}{(5.3 \times 10^{-11} \text{ m})^2} = 5.12 \times 10^{11} \text{ N/C}.$$

Notice how large this field is in comparison to the "typical" field that was calculated in the previous example.

b) In Example 24-6, we used Coulomb's law to find the force on the electron. But the whole point of knowing the electric field is that we can use it directly to find the force on a charge in the field. Because the electron has $|q| = e$, the magnitude of the force on it is

$$|\vec{F}| = |q||\vec{E}| = (1.60 \times 10^{-19} \text{ C}) \cdot (5.12 \times 10^{11} \text{ N/C}) = 8.20 \times 10^{-8} \text{ N}.$$

This agrees with the result of Example 24-6.

Summary

Important Concepts and Terms

neutral
charging
charge
like charges
opposite charges
discharging
conductor

insulator
electron cloud
ionization
ion
law of conservation of charge
sea of electrons
current

charge carriers	coulomb
electrostatic equilibrium	permittivity constant
grounded	electrostatics
polarization	field
polarization force	gravitational field
electric dipole	electric field
charging by induction	magnetic field
Coulomb's law	source charge
electrostatic constant	test charge
point charge	field diagram

In this chapter we introduced the "two-charge theory" of electric phenomena and its interpretation in terms of our contemporary knowledge of atomic structure. We have introduced a large number of new terms and concepts, including charge, charging and discharging, insulator, conductor, charge carrier, ion, dipole, electrostatic equilibrium, polarization, and ground. These concepts are very important for you to understand because they form the "language" of electricity. By using these ideas, you should be able to understand and explain a wide variety of electric phenomena. Quantitative calculations, which we will emphasize more in upcoming chapters, are of value only if you first understand the phenomena to which they pertain.

The essential ideas of the two-charge theory of electricity are:

1. There are two and only two types of charge, which we call positive and negative.
2. Charge is an inherent property of matter. Charges do not exist separate from matter. An object that is neutral has no *net* charge, which is not the same as "no charges."
3. At the atomic level, protons and electrons are the basic units of positive and negative charge. The net charge of an object is

$$q_{net} = \left(N_{protons} - N_{electrons}\right) \cdot e.$$

Charging occurs as electrons are added or removed from an object.

4. Like charges repel each other, opposite charges attract.
5. The strength of the electric force increases as the quantity of charge increases. It decreases as the distance between the charges increases.
6. There are two classes of materials with very different electric properties. Conductors are materials through or along which charges move easily. Insulators are materials in or on which charges stay fixed in place.

Coulomb's law for the electric force between two charges states:

1. If two charged objects having charges q_1 and q_2 are a distance r apart, the objects exert forces on each other of magnitude

$$|\vec{F}_{1 \text{ on } 2}| = |\vec{F}_{2 \text{ on } 1}| = \frac{K|q_1||q_2|}{r^2}.$$

2. The forces are directed along the line joining the two charges. The forces are *repulsive* for two like charges and *attractive* for two opposite charges.

The electric force is a long-range force. Faraday introduced the field concept as a way to understand how long-range forces occur. An electric field is created by one or more source charges. Their field at position (x, y, z) is measured by the force exerted on a test charge q at that point:

$$\vec{E}(x,y,z) \equiv \frac{\vec{F}_{\text{on } q}(x,y,z)}{q}.$$

Although this is the definition of $\vec{E}$, we use the field concept to *find* the force on a charge.

If a charge q is placed at position (x, y, z) in space where the electric field is $\vec{E}(x, y, z)$, then the charge experiences an electric force

$$\vec{F}_{\text{on } q} = q\vec{E}(x, y, z).$$

The electric field at a distance r from a point charge q is

$$\vec{E} = \left(\frac{1}{4\pi\varepsilon_0} \cdot \frac{q}{r^2}, \text{ away from } q\right).$$

Exercises and Problems

Exercises

1. Figure 24-10 showed how an electroscope becomes negatively charged. The leaves will also repel each other if you touch the electroscope with a positively charged glass rod. Use a series of pictures, similar to Fig. 24-10, to show how the electroscope becomes charged in this case. Keep in mind that the charge carriers in a metal are negative electrons.

2. Figure 24-16 and the discussion at the end of Section 24-4 described an electroscope being charged by induction. The final charge on the electroscope is *positive*.
   a. Use a series of pictures showing the charges to explain how charging by induction works.
   b. Write a paragraph explaining how charging by induction works. Your explanation should be based on the properties of charges, forces, insulators, and conductors.

3. A plastic balloon that has been rubbed with wool will stick to a wall.
   a. Can you conclude that the wall is charged? If so, where does the charge come from? If not, why not?
   b. Draw a series of charge pictures showing how the balloon is held to the wall.

4. What is the earth's gravitational field $\vec{g}$ (both magnitude and direction):
   a. At the surface of the earth?
   b. At the radius of the moon's orbit?

5. A glass rod that has been rubbed with silk has a charge of +5 nC.
   a. Has the rod had electrons removed or protons added? Explain.
   b. How many electrons have been removed or protons added?

6. Two small plastic spheres each have a mass of 2 g and a charge of –50 nC. They are placed 2 cm apart.
   a. What is the magnitude of the electric force on each sphere?
   b. By what factor is the electric force on a sphere larger than its weight?
   c. By what factor is the electric force larger than the gravitational force between the spheres?

7. A +10 nC charge is located at the origin. A –20 nC charge is located at $(x, y) = (0, 2 \text{ cm})$. Determine the force on each charge. Give each force in component form.

8. A +10 nC charge is located at the origin.
   a. What is the electric field at the positions $(x, y) = (5 \text{ cm}, 0 \text{ cm})$, $(-5 \text{ cm}, 5 \text{ cm})$, and $(-5 \text{ cm}, -5 \text{ cm})$? Give your answers in component form.
   b. Draw a field diagram showing the electric field vectors at these points.

9. A –10 nC charge is located at the origin.
   a. What is the electric field at the positions $(x, y) = (5 \text{ cm}, 0 \text{ cm})$, $(-5 \text{ cm}, 5 \text{ cm})$, and $(-5 \text{ cm}, -5 \text{ cm})$? Give your answers in component form.
   b. Draw a field diagram showing the electric field vectors at these points.

Problems

10. For each of the two diagrams in Fig. 24-31, find the force vector $\vec{F}$ on the 1 nC charge at the top due to the two lower charges.

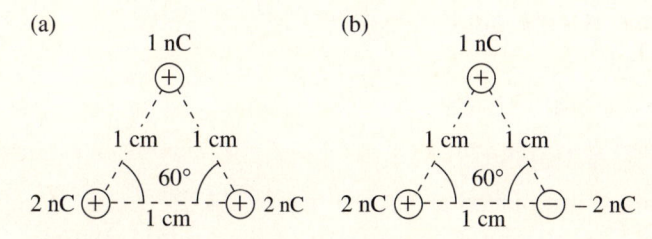

    FIGURE 24-31

11. What is the force vector $\vec{F}$ on the –10 nC charge in Fig. 24-32?

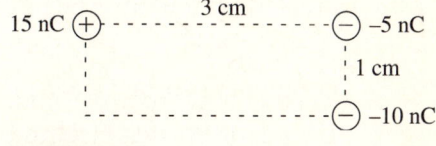

    FIGURE 24-32

12. For each of the two diagrams in Fig. 24-33, find the force vector $\vec{F}$ on the 1 nC charge in the middle due to the four other charges.

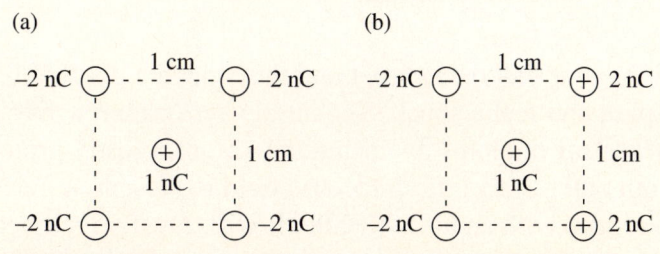

    FIGURE 24-33

13. +2 nC and –4 nC charges are spaced 1 cm apart, as shown in Fig. 24-34.
    a. Where could you place a proton so that it would experience no net force?
    b. Would $\vec{F}_{net} = 0$ for an electron placed at the same place? Explain.

    FIGURE 24-34

14. Objects A and B are both positively charged. Both have a mass of 100 g, but A has twice the charge of B. When A and B are placed 10 cm apart, B experiences an electric force of 0.45 N.
    a. How large is the force on A?
    b. What are the charges q_A and q_B?
    c. If the objects are released, what is the initial acceleration of A?

15. The electric field at a point in space is $\vec{E} = (200\,\hat{i} + 400\,\hat{j})$ N/m.
    a. What is the electric force on a proton at this point? Give your answer in component form.
    b. What is the electric force on a electron at this point? Give your answer in component form.
    c. In what ways are your answers to parts a) and b) similar and in what ways are they different?
    d. What is the magnitude of the proton's acceleration?
    e. What is the magnitude of the electron's acceleration?

16. A –10 nC charge is located at position $(x, y) = (2\text{ cm}, 1\text{ cm})$. At what (x, y) position or positions is the electric field:
    a. $-225{,}000\,\hat{i}$ N/C?
    b. $(161{,}000\,\hat{i} - 80{,}500\,\hat{j})$ N/C?
    c. $(28{,}800\,\hat{i} + 21{,}600\,\hat{j})$ N/C?

17. Three 1 nC charges are placed as shown in Fig. 24-35. Each of these charges creates an electric field $\vec{E}$ at a point 3 cm in front of the middle charge.
    a. What are the three fields $\vec{E}_1$, $\vec{E}_2$, and $\vec{E}_3$ created by the three charges? Write your answer for each as a vector in component form: $\vec{E} = E_x\,\hat{i} + E_y\,\hat{j}$.
    b. Do you think that electric fields obey a principle of superposition? That is, is there a "net field" at this point given by $\vec{E}_{net} = \vec{E}_1 + \vec{E}_2 + \vec{E}_3$? Based on what you learned in this chapter and previously in our study of forces, make an argument as to why this is or is not true.
    c. If it is true, what is $\vec{E}_{net}$?

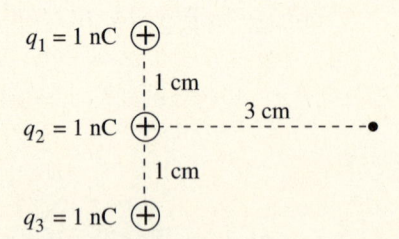

    FIGURE 24-35

18. When you shuffle your feet on a carpet and then touch a doorknob, you sometimes create a little spark and feel a shock. The air always has a few free electrons that have been knocked off of atoms by cosmic rays (ions, mostly protons, that impinge on the earth from outer space). If an electric field is present, a free electron is accelerated until it collides with an air molecule. It will transfer its kinetic energy to the

molecule, then accelerate, collide, accelerate, collide, and so on. The average distance traveled by an electron between collisions in air is 1 μm. If the electron has a kinetic energy of 2×10^{-18} J just before a collision, it will have sufficient energy to knock an electron out of the molecule it hits. Where once there was one free electron, now there are two! Each of these can then accelerate, hit a molecule, and knock out another electron—so now there are four electrons. So if the electric field is sufficiently strong, there is a "chain reaction" of electron production. This is called a *breakdown* of the air. The net result, shown in Fig. 24-36, is a current of moving electrons (that is what gives you the shock) and a burst of light (the spark) when the electrons eventually recombine with the positive ions and give off excess energy as light. The creation of lightning bolts occurs in much the same way on a larger scale.

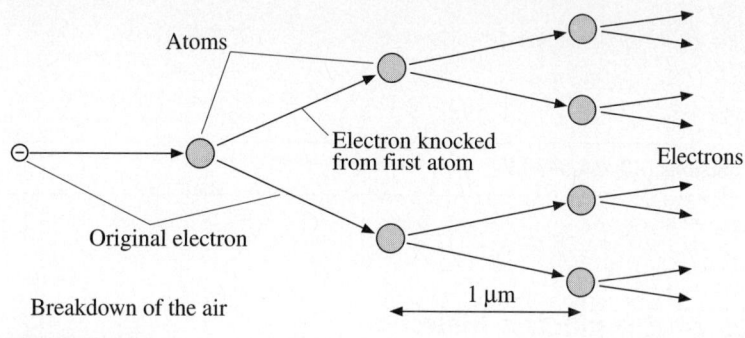

FIGURE 24-36

a. What acceleration must an electron have to gain 2×10^{-18} J of kinetic energy in a distance of 1 μm?
b. What force must act on the electron to give it the acceleration found in part a)?
c. A free electron in air is 1 cm away from a point charge. What minimum charge must this point charge have to cause a breakdown of the air and create a spark?
d. What is the electric field at the point of the initial electron?
e. Does it make any difference whether the point charge is positive or negative? Why or why not? (**Hint:** Think about whether the chain reaction of electron production can be maintained as the electrons move either toward or away from the point charge.)

[**Estimated 10 additional problems for the final edition.**]

Chapter 25

Calculating the Electric Field

LOOKING BACK Sections 6.3; 24.5–24.7

25.1 Sources of the Electric Field

We made a distinction in the last chapter between those charged particles that are the *sources* of the electric field and other charged particles that *experience* and move in the electric field. This distinction is very important. Most of this chapter will be concerned with the *sources* of the electric field. Only at the end, once we know how to calculate the electric field, will we look at how a charged particles moves in this field.

There is a lot of material in this chapter. So that you don't get confused, we will outline where we are going. We will first introduce the idea of *superposition*—that the net electric field due to several charged particles is the sum of the fields of each individual charge. This is easy to say, but it can be tricky to carry out because the sums are *vector* sums. We will practice with several examples. Then we will consider finding the electric field of a *continuous* distribution of charge, looking very carefully at the example of a line of charge. To find the field, we will divide the line up into small segments of charge Δq and use superposition to find the net electric field of these segments. Then, as in calculus, we will let $\Delta q \rightarrow dq$ and let the superposition sum become an integral. Once we know the field of a line of charge, we can proceed to find the field of a sheet of charge—like a charged flat electrode—by considering it to be a bundle of line charges. Finally, we will calculate the electric field between two parallel, oppositely charged electrodes. Such a device is called a *parallel-plate capacitor*, and we will find its electric field to be a *uniform field*. Every step we take in this chapter will move us a bit closer to this goal. Try to notice, as you study, the *method* by which we can find, step by step, the electric field of a fairly complex distribution of charge. This may seem difficult at first, but it is very important in practice to be able to do calculations like these. The homework problems at the end will give you a chance to practice.

25.2 The Field of Multiple Point Charges

You learned in the last chapter that electric forces obey the principle of superposition. That is, the net force on a charged particle due to several other charges is the *vector* sum of the forces due to each individual charge. It follows directly from this, and from the definition of the electric field, that the field itself also obeys the principle of superposition. The electric field $\vec{E}_{net}$ created at a point in space by *many* charges is given by

$$\vec{E}_{net} = \vec{E}_1 + \vec{E}_2 + \ldots + \vec{E}_n = \sum_i \vec{E}_i, \tag{25-1}$$

where $\vec{E}_i$ is the field from a single point charge. Superposition is the primary idea that allows us to calculate the electric field due to a distribution of charge.

It is especially important to notice that Eq. 25-1 is a *vector* sum. Thus Eq. 25-1 is, in reality, *three* simultaneous sums:

$$(E_{net})_x = (E_1)_x + (E_2)_x + \ldots = \sum (E_i)_x$$
$$(E_{net})_y = (E_1)_y + (E_2)_y + \ldots = \sum (E_i)_y \tag{25-2}$$
$$(E_{net})_z = (E_1)_z + (E_2)_z + \ldots = \sum (E_i)_z.$$

The key step in many of the calculations we will be doing is to choose an appropriate coordinate system for the problem and to break the Eq. 25-1 sum into the three component sums of Eq. 25-2. Once we have done that successfully, we are left with "only" the mathematical problem of evaluating the sums. One technique we will develop to simplify the evaluations of the sums is that of using geometrical *symmetry* in the problem. Let us look at an example.

EXAMPLE 25-1 Three equal, positive point charges q are located on the y-axis at $y = 0$ and at $y = \pm d$. What is the net electric field of these three charged particles at a point on the x-axis?

SOLUTION Figure 25-1 shows the geometry, with the charges numbered 1, 2, and 3. The question does not ask about any specific point, just "a point" on the axis. This point, as shown in Fig. 25-1, is distance x from charge q_2 and distance $(x^2 + d^2)^{1/2}$ from q_1 and q_3. We will be looking for a symbolic expression for the field, given in terms of the arbitrary distance x, rather than for a numerical value.

The three charged particles each create an electric field at this point. Call them $\vec{E}_1$, $\vec{E}_2$, and $\vec{E}_3$. Because the charges are positive, each of these fields points *away from* its source charge. The net field is the vector sum

$$\vec{E}_{net} = \vec{E}_1 + \vec{E}_2 + \vec{E}_3.$$

Before rushing in to calculate $\vec{E}_{net}$, we can make our task *much* easier by first thinking qualitatively about the situation and using the symmetry in the

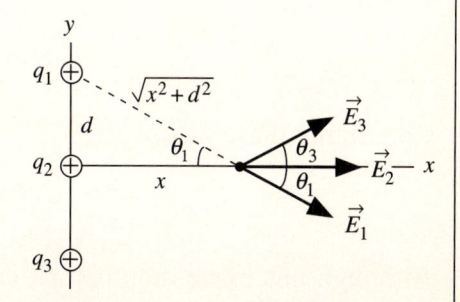

FIGURE 25-1 Calculating the electric field on the x-axis of three equal charges along the y-axis.

problem. First, we note that the three point-charge fields $\vec{E}_1$, $\vec{E}_2$, and $\vec{E}_3$ all lie in the xy-plane and have no z-components. Therefore, we can conclude that $(E_{net})_z = 0$. Next let's look at the y-component of the net field. Because charges q_1 and q_3 are located *symmetrically* on either side of the x-axis, are of equal value, and are equally distant from the position x, their fields must be of equal magnitude: $|\vec{E}_1| = |\vec{E}_3|$. Equally important, the symmetry of the problem tells us that the angle θ_1 by which $\vec{E}_1$ is tilted *below* the axis is *identical* to the angle θ_3 by which $\vec{E}_3$ is tilted *above* the axis. As a consequence, the y-components of $\vec{E}_1$ and $\vec{E}_3$ will *cancel* when added. Because $\vec{E}_2$ has no y-component, we can conclude that $(E_{net})_y = 0$ at the point of the calculation!

Lastly, we can see that $\vec{E}_1$ and $\vec{E}_3$ have *equal* x-components. So using Eq. 25-2, we find that the electric field is:

$$\vec{E}_{net} = \begin{cases} (E_{net})_x = (E_1)_x + (E_2)_x + (E_3)_x = 2(E_1)_x + (E_2)_x \\ (E_{net})_y = 0 \\ (E_{net})_z = 0. \end{cases} \quad (25\text{-}3)$$

We have saved ourselves a lot of calculation, and thus avoided a large number of potential calculational errors, by first *thinking* carefully about the calculation. Now we can proceed to the mathematics.

Because $\vec{E}_2$ has *only* an x-component, we find

$$(E_2)_x = |\vec{E}_2| = \frac{1}{4\pi\varepsilon_0} \cdot \frac{q_2}{r_2^2} = \frac{1}{4\pi\varepsilon_0} \cdot \frac{q_2}{x^2}, \quad (25\text{-}4)$$

where $r_2 = x$ is the distance from q_2 to the point at which we are calculating the field. Vector $\vec{E}_1$ points at angle θ_1 from the x-axis, so its x-component is given by

$$(E_1)_x = |\vec{E}_1|\cos\theta_1 = \frac{1}{4\pi\varepsilon_0} \cdot \frac{q_1}{r_1^2} \cdot \cos\theta_1,$$

where r_1 is the distance from q_1 to our point.

While this equation is correct, it is not sufficient. Both the distance r_1 and the angle θ_1 vary with the position x, and we need to do some geometry to express them as functions of x. From the Pythagorean theorem we have $r_1^2 = x^2 + d^2$. In addition, the right triangle of Fig. 25-1 gives

$$\cos\theta_1 = \frac{x}{r_1} = \frac{x}{(x^2+d^2)^{1/2}}.$$

Combining these pieces gives

$$(E_1)_x = \frac{1}{4\pi\varepsilon_0} \cdot \frac{q_1}{x^2+d^2} \cdot \frac{x}{(x^2+d^2)^{1/2}} = \frac{1}{4\pi\varepsilon_0} \cdot \frac{xq_1}{(x^2+d^2)^{3/2}}. \quad (25\text{-}5)$$

Although this expression is a bit cumbersome, it accurately describes how the x-component of $\vec{E}_1$ varies at all points along the x-axis.

Now, to finish the calculation, we need to combine Eqs. 25-4 and 25-5 into Eq. 25-3. Because the charges are all equal, we can drop the subscripts and use $q_1 = q_2 = q$. The

x-component of $\vec{E}_{net}$ is then

$$(E_{net})_x = 2(E_1)_x + (E_2)_x = \frac{q}{4\pi\varepsilon_0} \cdot \left[\frac{1}{x^2} + \frac{2x}{(x^2+d^2)^{3/2}}\right]. \quad (25\text{-}6)$$

Because the other two components are zero, the final result can be written as a vector:

$$\vec{E}_{net} = \frac{q}{4\pi\varepsilon_0} \cdot \left[\frac{1}{x^2} + \frac{2x}{(x^2+d^2)^{3/2}}\right]\hat{i}. \quad (25\text{-}7)$$

EXAMPLE 25-2 Three +1 nC charged particles are placed 1 cm apart on the y-axis, as shown in Fig. 25-1. a) What is the electric field at the point $x = 2$ cm? b) What is the force on an electron placed at $x = 2$ cm? c) What is the force on a Na$^+$ ion at the same location?

SOLUTION We found $\vec{E}_{net}$ for three charges in Example 25-1. We can use that result here.
a) With $q = 1 \times 10^{-9}$ C, $d = 0.01$ m, and $x = 0.02$ m, we can use Eq. 25-7 to find

$$\vec{E}_{net} = 38{,}600\,\hat{i}\text{ N/C}.$$

b) The force on a charge q in an electric field is $\vec{F} = q\vec{E}$. This is a vector equation, so signs are important. For an electron, with $q = -e$, the force is

$$\vec{F} = -e\vec{E}_{net} = -6.18 \times 10^{-16}\,\hat{i}\text{ N}.$$

Because of the minus sign, the force is in the *negative* x-direction. The electron will accelerate *toward* the positive source charges.

c) The Na$^+$ ion has $q = +e$, so $\vec{F} = +6.18 \times 10^{-16}\,\hat{i}$ N, in the positive x-direction. The electron and the Na$^+$ ion experience forces of the same magnitude. However, they will have very *different* accelerations $a = F/m$ in the electric field because their masses are quite different.

Can we check that our result for the electric field of three charged particles, as given by Eq. 25-7, makes sense? A useful skill to develop as you become a more sophisticated problem solver is to check your solution by looking at various limits or extreme cases where you know what the result "should" be. Suppose, in this example, that we let x become really, really small: $x \to 0$. The point where we calculated the field would then be right on top of q_2. The two fields $\vec{E}_1$ and $\vec{E}_3$ would be back-to-back (see Fig. 25-1), pointing exactly opposite each other, and would cancel. The field as $x \to 0$ *should* be that of the single charge q_2—a field we already know from Coulomb's law. Is it? While the initial $1/x^2$ term in Eq. 25-7 becomes very large as $x \to 0$, notice that

$$\lim_{x \to 0} \frac{2x}{(x^2+d^2)^{3/2}} = 0. \quad (25\text{-}8)$$

Thus $E_{net} \to q/(4\pi\varepsilon_0)x^2$ as $x \to 0$, exactly the expected field of a single point charge.

Now consider the opposite situation, where x becomes really, really large: $x \gg d$. We do not want to go quite to the limit $x \to \infty$, because that would make the field zero, but simply very far away in comparison to the spacing between the source charges. If the point is very far away, the three source charges will seem to merge into a single charge of size $3q$—just as three very distant light bulbs appear to be a single light. So the field for $x \gg d$ *should* be that of a single $3q$ point charge. Is it? If $x \gg d$, then the denominator of the second term in Eq. 25-7 becomes $(x^2 + d^2)^{3/2} \approx (x^2)^{3/2} = x^3$. The whole second term is then

$$\lim_{x \gg d} \frac{2x}{(x^2 + d^2)^{3/2}} = \frac{2}{x^2}, \tag{25-9}$$

giving the expected result for a point charge $3q$:

$$\vec{E}_{net}(x \gg d) = \frac{1}{4\pi\varepsilon_0} \cdot \frac{(3q)}{x^2} \hat{i}. \tag{25-10}$$

We will make similar checks of future calculations to see whether they agree with the expected results in the limit where the charge distribution appears "simple." Such checks help to provide confidence in the results of the calculation. Some of the homework problems will ask you to do the same.

It is useful to summarize the steps we have followed in these examples. They are a strategy for finding the electric field of multiple point charges:

Strategy for Calculating the Electric Field of Multiple Point Charges

1. Establish a coordinate system.
2. Determine the direction of each charge's electric field $\vec{E}_i$ at the point of interest.
3. Use geometric symmetry to determine where any components of $\vec{E}_{net}$ are zero.
4. Use geometry to determine distances and angles.
5. Calculate the magnitude $|\vec{E}_i|$ of each charge's electric field.
6. Write each vector $\vec{E}_i$ in component form.
7. Sum the components of each vector to determine $\vec{E}_{net}$.
8. Find the magnitude $|\vec{E}_{net}|$ and direction ϕ of $\vec{E}_{net}$.

25.3 The Field of a Dipole

We introduced the idea of a *dipole* in the last chapter when we described how an atom can be polarized, creating a small separation between the centers of positive and negative charge. That was an example of an **induced electric dipole** because the polarization was caused, or induced, by an external charge. There are also **permanent electric dipoles** in which two equal but oppositely charged particles maintain a small permanent separation. Many molecules, including the water molecule, are permanent dipoles due to the way in which the electrons are arranged.

Figure 25-2 shows two equal but oppositely charged particles, $+q$ and $-q$, permanently separated by a distance s. These form a permanent electric dipole. Although the dipole has zero net charge, it *does* have a net electric field. A point on the y-axis, for example, is slightly closer to one charge than to the other, so the individual fields from the two charges do *not* cancel.

Let's find the electric field of a dipole at a point on the y-axis, as shown in Fig. 25-2. The field $\vec{E}_+$ from the $+q$ charge points in the $+y$-direction while the field $\vec{E}_-$ from the $-q$ charge is in the $-y$-direction. $|\vec{E}_+|$ is slightly larger than $|\vec{E}_-|$, because the positive charge is slightly closer, so superposition of these two fields will give a small but nonzero net electric field in the upward $(+y)$ direction.

The point at which we wish to calculate the field is distance $y - s/2$ from the positive charge and $y + s/2$ from the negative. Superposition of the fields of the two point charges, each given by Coulomb's law, gives

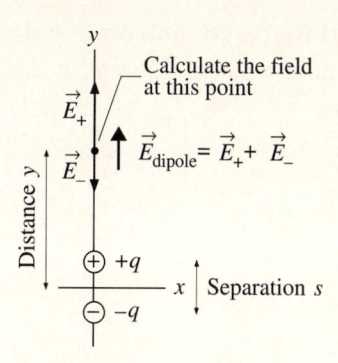

$$(E_{\text{dipole}})_y = (E_+)_y + (E_-)_y = \frac{1}{4\pi\varepsilon_0} \cdot \frac{q}{(y - \frac{1}{2}s)^2} + \frac{1}{4\pi\varepsilon_0} \cdot \frac{(-q)}{(y + \frac{1}{2}s)^2}$$

$$= \frac{q}{4\pi\varepsilon_0} \cdot \left[\frac{1}{(y - \frac{1}{2}s)^2} - \frac{1}{(y + \frac{1}{2}s)^2} \right]. \qquad (25\text{-}11)$$

FIGURE 25-2 An electric dipole. $\vec{E}_+$ is larger than $\vec{E}_-$ due to the charge separation s.

It will be useful to factor the y out of each denominator, leading to

$$(E_{\text{dipole}})_y = \frac{q}{4\pi\varepsilon_0} \cdot \left[\frac{1}{y^2\left(1 - \frac{s}{2y}\right)^2} - \frac{1}{y^2\left(1 + \frac{s}{2y}\right)^2} \right]$$

$$= \frac{q}{4\pi\varepsilon_0 y^2} \cdot \left[\left(1 - \frac{s}{2y}\right)^{-2} - \left(1 + \frac{s}{2y}\right)^{-2} \right]. \qquad (25\text{-}12)$$

Equation 25-12 is exact for a point on the y-axis. It is usually the case, however, that we are interested in the field of a dipole only at distances $y \gg s$. That is, we are going to observe the electric field of a dipole only for distances much larger than the separation s between the two charges. Thus we seek an approximation for Eq. 25-12 that is valid when $y \gg s$, or when $s/2y \ll 1$.

We have to be careful. A hasty effort would say that the $s/2y$ in each denominator can be neglected if $s/2y \ll 1$, but then we end up with $(E_{\text{dipole}})_y = 0$! We do not want to be *that* far away, so we need a better approximation technique.

We have previously used the *binomial approximation* to simplify expressions. The binomial approximation says that a term of the form $(1 + a)^n$ can be approximated, if $a \ll 1$, as:

$$(1 + a)^n \approx 1 + na \quad \text{if} \quad a \ll 1. \qquad (25\text{-}13)$$

First, apply the binomial approximation to each term in the square brackets of Eq. 25-12, using $a = \pm s/2y$ and $n = -2$. This gives:

$$\left[\left(1-\frac{s}{2y}\right)^{-2} - \left(1+\frac{s}{2y}\right)^{-2}\right] = \left[\left(1+\left(\frac{-s}{2y}\right)\right)^{-2} - \left(1+\frac{s}{2y}\right)^{-2}\right]$$

$$\approx \left[\left(1+(-2)\cdot\left(\frac{-s}{2y}\right)\right) - \left(1+(-2)\cdot\left(\frac{+s}{2y}\right)\right)\right] \quad (25\text{-}14)$$

$$= +\frac{2s}{y}.$$

You have to be careful with signs and the algebra, but we are left with a very simple and *very accurate* approximation. Next, put this result back in Eq. 25-12. This gives the electric field of a dipole at a point on the axis of the dipole:

$$(E_{\text{dipole}})_y \approx \frac{q}{4\pi\varepsilon_0 y^2} \cdot \frac{2s}{y}$$

$$\Rightarrow \vec{E}_{\text{dipole}}(\text{on axis}) = \frac{1}{4\pi\varepsilon_0} \cdot \frac{2qs}{y^3}\hat{j} \quad \text{if } y \gg s. \quad (25\text{-}15)$$

This is the result we were after. It tells us that the electric field of a dipole, along the axis of the dipole, decreases as the inverse *cube* of the distance (for $y \gg s$).

Does this inverse-cube equation violate Coulomb's law? Not at all. Coulomb's law does not say that *all* electric forces and fields decrease as the inverse square of the distance, only those of point charges. A dipole is not a point charge. The field of a dipole decreases much more rapidly than that of a single point charge, which is to be expected because the dipole is, after all, electrically neutral. Now that we know the dipole field (at least along the axis) we could find the attractive force between a dipole and a charged particle—a topic to which we will return.

Equation 25-15 gives the electric field only along the axis of the dipole, not at any other point. Now let's find the electric field along the line *perpendicular* to the dipole. This is equivalent to finding the field at a point on the x-axis. Figure 25-3 shows the geometry, with fields $\vec{E}_+$ and $\vec{E}_-$ from the two charges. Here we have a situation, similar to that of the last section, where we can use *symmetry* to simplify the calculation. In this case, because the charge magnitudes and distances are equal while the signs are opposite, the x-components of $\vec{E}_+$ and $\vec{E}_-$ will cancel while the y-components are equal and will add. We will leave it as a homework problem for you to complete the calculation and to show that

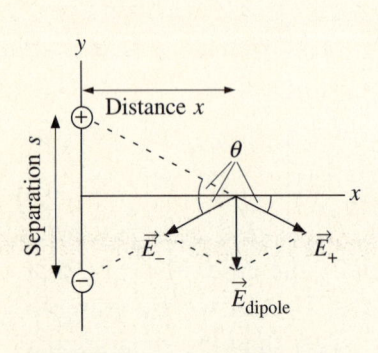

FIGURE 25-3 The electric field along the axis bisecting a dipole.

$$\vec{E}_{\text{dipole}}(\text{on bisecting axis}) = -\frac{1}{4\pi\varepsilon_0} \cdot \frac{qs}{x^3}\hat{j} \quad \text{if } x \gg s. \quad (25\text{-}16)$$

Notice that the field on the bisecting *x*-axis is *opposite* in direction to the field we found on the *y*-axis. If we calculate the field at a large number of points, which is not especially difficult now that you see the idea, we would end up with the electric field diagram of Fig. 25-4. We call this field the *electric field of a dipole*. Notice how it almost looks as if something is "flowing" from the positive charge to the negative charge. We will see this same field pattern again when we later talk about *magnetic dipoles*.

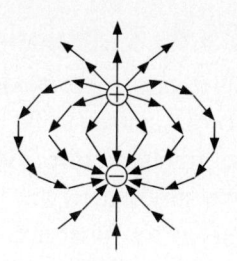

FIGURE 25-4 The full electric field of a dipole.

EXAMPLE 25-3 An electron and a proton are 0.1 nm apart. What is the force on a Na$^+$ ion 2 nm from the dipole if a) the ion is on the axis of the dipole, nearest the proton, and b) the ion is on a line perpendicular to the center of the dipole?

SOLUTION a) The force on a Na$^+$ ion, with $q = +e$, is $\vec{F} = e\vec{E}$. We need to find the field of the dipole at distance $y = 2$ nm. From Eq. 25-15, with $q = e$ for the dipole charges,

$$(E_{dipole})_y \approx \frac{1}{4\pi\varepsilon_0} \cdot \frac{2qs}{y^3}$$

$$= 9.0 \times 10^9 \text{ Nm}^2/\text{C}^2 \cdot \frac{2(1.60 \times 10^{-19} \text{ C})(1.0 \times 10^{-10} \text{ m})}{(2 \times 10^{-9} \text{ m})^3}$$

$$= 3.6 \times 10^7 \text{ N/C}.$$

Because the ion is nearest the proton, the field points *away from* the dipole. The force on the ion is thus

$$\vec{F} = e\vec{E} = (5.76 \times 10^{-12} \text{ N, straight away from dipole}).$$

b) From Eq. 25-16 you can see that the field strength at $x = 2$ nm is half of the field strength at $y = 2$ nm, or 1.8×10^7 N/C. The field direction, you can see from Fig. 25-4, is parallel to the dipole and pointing toward the electron's end. The force on a Na$^+$ ion at this point is

$$\vec{F} = e\vec{E} = (2.88 \times 10^{-12} \text{ N, parallel to and toward electron's end of dipole}).$$

A similar and very important electric field is that of just two like-charges *q* separated by distance *d*. This is a situation you can explore as a homework problem. Figure 25-5 shows the electric field diagram of two positive charges. Notice the symmetrical aspects of this field.

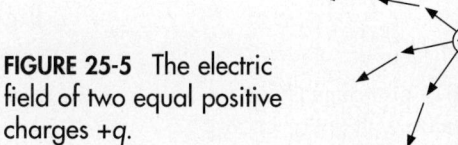

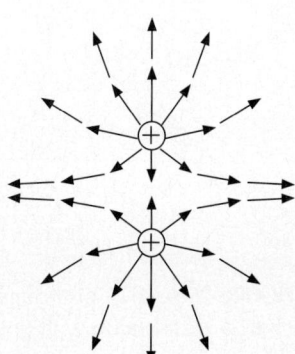

FIGURE 25-5 The electric field of two equal positive charges +*q*.

25.4 Calculating the Field of a Continuous Charge

Ordinary macroscopic objects—tables, chairs, beakers of water, flowing streams—seem to our senses to be *continuous* distributions of matter. There is no obvious evidence for an atomic structure, even though we have good reasons to believe that we would find atoms if we subdivided the matter sufficiently far. Rather than deal directly with atoms, it is much easier for most practical purposes to consider matter to be *continuous* and to talk about the *density* of matter—the number of grams per cubic centimeter or the number of kilograms per cubic meter. Then we can say that some parts of an object have "high density" and that others have "low density." In more precise terms, the idea of density allows us to describe the *distribution* of matter *as if* the matter were continuous rather than atomic.

Much the same situation occurs with charge. If we place a very large number of charged particles on an object—say, for example, 10^{12} electrons on a metal rod—then it is no longer practical to keep track of where every individual charge is located. It makes more sense to consider the charge to be *continuous* and to describe how it is *distributed* over the object. The basic problem we want to solve is to find the electric field of a continuous distribution of charge. What, for example, is the electric field of a charged rod or a charged sphere or a charged circular electrode? These are the "real world" objects whose fields we need to know for practical application. To calculate such a field, we need a step-by-step method.

Let's consider an extended, macroscopic object, such as a long rod or a flat disk, with total charge Q. We will use an upper case Q for the total charge of an object, reserving lowercase q for individual point charges. As you know, charge Q is really an excess or deficit of electrons. That is, $Q = Ne$ where N is the number of extra (or missing) electrons. But now we will assume N to be so enormous that, to a very good approximation, we can consider the charge to be continuously distributed throughout the object.

For a one-dimensional object of length L, such as a line with charge spread along it, the linear charge density λ is defined to be

$$\lambda = \frac{Q}{L}. \qquad (25\text{-}17)$$

The **linear charge density**, which has units of C/m, is the amount of charge *per meter* of length. Note that the linear charge density is analogous to the linear mass density that you used in Chapter 14 to find the speed of a wave on a string.

Figure 25-6a shows a linear distribution of charge. A small piece of the line, having length ΔL, contains the small amount of charge

$$\Delta q = \lambda \Delta L. \qquad (25\text{-}18)$$

Equation 25-18 will prove to be a very useful relationship between charge and length.

Figure 25-6b shows a charged two-dimensional surface of area A. We define the surface charge density η to be

$$\eta = \frac{Q}{A}. \qquad (25\text{-}19)$$

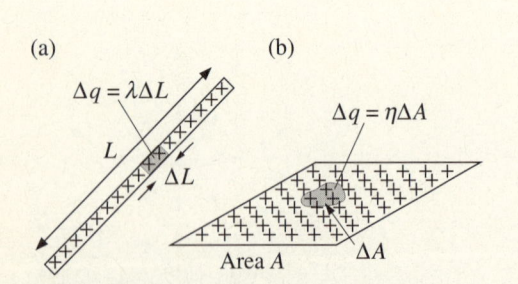

FIGURE 25-6 a) A one-dimensional charge distribution with linear charge density λ. b) A two-dimensional charge distribution with surface charge density η.

Surface charge density, with units of C/m², measures the amount of charge *per square meter*. A small piece of the surface with area ΔA contains the small quantity of charge
$$\Delta q = \eta \Delta A. \tag{25-20}$$
Figure 25-6 and the definitions of Eqs. 25-17 and 25-19 assume that the object is **uniformly charged**, meaning that the charges are evenly spaced over the object. Thus the charge density is constant along the entire length or across the entire surface of the object. We will consider only uniformly charged objects in this text.

Our knowledge of electric fields consists of two basic facts: the field of a point charge and the principle of superposition. To utilize this knowledge for a continuous distribution of charge, we will apply a three-part strategy: First, divide the total charge Q into many small point-like charges Δq; then, find the electric field of each Δq; finally, sum the fields of all the Δq to obtain the net field $\vec{E}_{net}$. In practice, as you may have guessed, we will let the sum become an integral. The difficulty with electric field calculations is not doing the integration itself, which is the last step, but setting up the calculation and knowing *what* to integrate. We will go step-by-step through several examples to illustrate the procedures. First, however, let's list the steps to be followed in the strategy:

Strategy for Calculating the Electric Field of a Continuous Charge

1. Draw a picture and establish a coordinate system.
2. Divide the total charge Q into small, regularly shaped pieces of charge Δq, using shapes for which you *already know* how to determine $\vec{E}$. This is often, but not always, a division into point charges.
3. Look for any possible symmetry of the charge distribution. A properly chosen coordinate system or a clever "pairing" of the Δq may allow you to conclude that some components of $\vec{E}$ are zero. Setting up and evaluating integrals can be a lot of work, with many chances for error, so you want to avoid doing all that work only to get a result of zero.
4. Use superposition to form an algebraic expression for *each* of the three components of $\vec{E}$ (unless you are sure one or more is zero) at an *arbitrary* point in the field. Keep in mind that $\vec{E}_{net} = \Sigma \vec{E}_i$ is really three separate equations, one for each component. By an "arbitrary point" we mean to let the (x, y, z) coordinates of the point remain as variables. Do not try to substitute specific numbers.
5. Replace the small charge Δq with an equivalent expression involving a *charge density* and the small *geometric quantity* that describes the shape of charge Δq. This is an essential step in making the transition from a sum to an integral because you will need a coordinate to serve as the integration variable.
6. Let the sum become an integral. The integration will be over the coordinate variable that is related to Δq by the charge density. All angles and distances must be expressed in terms of the integration variable. Think carefully about the integration limits for this variable; they will depend on the coordinate system you have chosen to use. Then carry out the integration and any subsequent simplification of the result.
7. Check that your result is consistent with any limits (such as $z \to \infty$) for which you know what the field should be.

These guidelines will help you to break down a difficult problem into smaller steps that are individually manageable. In the following sections you will see examples of this strategy being applied.

25.5 The Field of a Line of Charge

The first field of a continuous charge we will calculate is the electric field of a thin, uniformly charged rod. In particular, let the rod be of length L, and have total charge Q (either positive or negative), and let us find the field at a point that is a distance r straight away from the center of the rod. The geometry is shown in Fig. 25-7a.

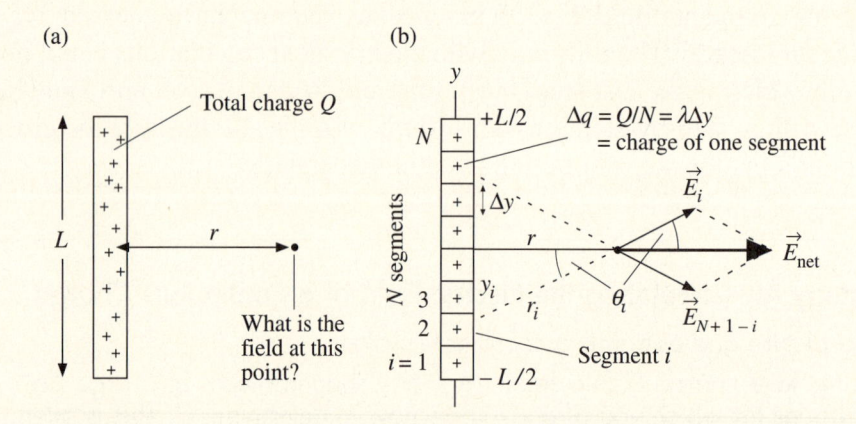

FIGURE 25-7 a) A charged rod, of length L and charge Q. b) Finding $\vec{E}$ at a point on the axis by dividing the rod into N segments of charge Δq each.

How shall we proceed? First let us assume, because the rod is "thin," that we can consider the charge Q to lie along a line of length L from $y = -L/2$ to $y = +L/2$. Note that we have, by this assumption, imposed a coordinate system on the problem. Our task, then, is to find the electric field of this **line of charge** at a point on the x-axis that is distance r away from the center of the rod.

Next, as we follow the strategy outlined in the previous section, let's divide the rod up into N small segments of length $\Delta y = L/N$, as shown in Fig. 25-7b. Each segment contains the small amount of charge $\Delta q = Q/N$ and, if N is sufficiently large, acts like a point charge. The field at a point on the x-axis is the sum of the fields of these N point charges.

Now that we have divided the line of charge into point-charge-like segments, we need to sum (like vectors!) the electric fields that each segment contributes at the point. This is exactly the problem that we did in Example 25-1 for the specific case that $N = 3$, so most of our work is already done. We need only to extend the result of that calculation to an arbitrary value of N. Recall, from our earlier calculation, that the y-components of the top and bottom charges canceled. Because of the symmetry of the charge placement, the field along the midline had *only* an x-component and pointed directly outward. Exactly the same reasoning is valid now. Because we are finding the field along the line that bisects the line of charge, the symmetry of the charge placement tells us that the net electric field will be $\vec{E} = E_x \hat{i}$ with no y- or z-components.

Let N be an even number, for convenience, and number the segments 1, 2, 3, ..., N starting from the bottom, as Fig. 25-7b shows. Select both segment i and its symmetrically located partner, which you should convince yourself is numbered $N + 1 - i$, then draw their electric fields $\vec{E}_i$ and $\vec{E}_{N+1-i}$. You can see from the figure that their y-components cancel, while their x-components add. As we proceed to sum over all the segments, we will get pairwise cancellation of the y-components and a net sum of x-components of the field. We will end up with $E_y = 0$, so the only *explicit* calculation we need to do is that for $E_x = \Sigma(E_i)_x$.

Segment i has charge Δq and is distance r_i from our point. The electric field of this point-charge-like segment is given by Coulomb's law, but we must then multiply by $\cos\theta_i$ to get just the x-component. The angle θ_i is defined in the figure. It is very important to note that each segment is at a different distance r_i and angle θ_i. So $(E_i)_x$, the x-component of the field of segment i, is given by

$$(E_i)_x = |\vec{E}_i|\cos\theta_i = \frac{1}{4\pi\varepsilon_0} \cdot \frac{\Delta q}{r_i^2} \cdot \cos\theta_i. \tag{25-21}$$

From the geometry of the figure we can see that $r_i = (y_i^2 + r^2)^{1/2}$, where y_i is the distance from the center to segment i, and $\cos\theta_i = r/r_i$. Thus

$$\begin{aligned}(E_i)_x &= \frac{1}{4\pi\varepsilon_0} \cdot \frac{\Delta q}{y_i^2 + r^2} \cdot \frac{r}{\sqrt{y_i^2 + r^2}} \\ &= \frac{1}{4\pi\varepsilon_0} \cdot \frac{r\Delta q}{(y_i^2 + r^2)^{3/2}}.\end{aligned} \tag{25-22}$$

This result is identical to Eq. 25-5 with the distance now called r.

We want to sum fields of all the Δq along the length of the rod and then replace the sum with an integral. But we can't integrate over q—it's not a geometric quantity. This is where the linear charge density enters. Because each segment has length Δy, the quantity of charge in each segment is given by Eq. 25-18 as

$$\Delta q = \lambda \Delta y, \tag{25-23}$$

where the linear charge density is $\lambda = Q/L$. Using Eq. 25-23 in Eq. 25-22, the field of segment i is

$$(E_i)_x = \frac{\lambda}{4\pi\varepsilon_0} \cdot \frac{r\Delta y}{(y_i^2 + r^2)^{3/2}}. \tag{25-24}$$

To finish the calculation, we need to sum $(E_i)_x$ from all N segments. This gives

$$\begin{aligned}E_x(r) &= \sum_{i=1}^{N}(E_i)_x = \sum_{i=1}^{N} \frac{\lambda}{4\pi\varepsilon_0} \cdot \frac{r\Delta y}{(y_i^2 + r^2)^{3/2}} \\ &= \frac{\lambda}{4\pi\varepsilon_0} \cdot \sum_{i=1}^{N} \frac{r\Delta y}{(y_i^2 + r^2)^{3/2}}.\end{aligned} \tag{25-25}$$

This is exactly the superposition we did for the $N = 3$ case. The only difference is that we have now written the result as an explicit summation so that N can have any value. (Note that $E_x(r)$ means E_x as a function of the distance r, *not* E_x multiplied by r.)

214 CHAPTER 25 CALCULATING THE ELECTRIC FIELD

Now we're ready to let the sum become an integral. If we let $N \to \infty$, then each segment becomes an infinitesimal length $\Delta y \to dy$ while the discrete position variable y_i becomes the continuous integration variable y. Because we are summing the electric field contribution from $i = 1$ at the bottom end of the line of charge to $i = N$ at the top end, the sum from 1 to N will be replaced with an integral from $y = -L/2$ to $y = +L/2$. Thus in the limit $N \to \infty$, Eq. 25-25 becomes

$$E_x(r) = \frac{\lambda}{4\pi\varepsilon_0} \cdot \int_{-L/2}^{L/2} \frac{r \, dy}{(y^2 + r^2)^{3/2}}. \tag{25-26}$$

This is a standard integral that you have learned to do in calculus and that can be found in any table of integrals. Integrating gives

$$E_x(r) = \frac{\lambda}{4\pi\varepsilon_0} \cdot \frac{y}{r\sqrt{y^2 + r^2}} \bigg|_{-L/2}^{L/2}$$

$$= \frac{Q/L}{4\pi\varepsilon_0} \cdot \left[\frac{L/2}{r\sqrt{(L/2)^2 + r^2}} - \frac{-L/2}{r\sqrt{(-L/2)^2 + r^2}} \right] \tag{25-27}$$

$$= \frac{1}{4\pi\varepsilon_0} \cdot \frac{Q}{r\sqrt{r^2 + (L/2)^2}}.$$

Integration gives us a result in closed form that is valid for *any* value of the distance r away from the charged rod. The only restriction is to remember that this is the electric field at point directly away from the *center* of the rod.

Finally, let us write the result of the integration in vector form, giving the full electric field at distance r away from the center of a charged rod of length L:

$$\vec{E}(r) = \frac{1}{4\pi\varepsilon_0} \cdot \frac{Q}{r\sqrt{r^2 + (L/2)^2}} \hat{i}. \tag{25-28}$$

This completes our first calculation of the electric field of a continuous charge. We have gone into great detail to make the procedure clear. Subsequent calculations will go more quickly. A more complete calculation of the field, using the same type of calculation we have done to arrive at Eq. 25-28, shows that the field diagram of a finite-length charged rod or wire has the pattern shown in Fig. 25-8. You need to imagine this figure rotated about the rod to "see" the three-dimensional structure of the electric field.

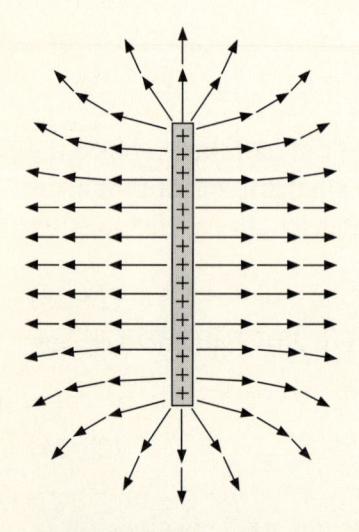

FIGURE 25-8 The electric field of a positively charged rod or wire. Visualize this in three dimensions as rotated about the rod.

EXAMPLE 25-4 What is the electric field 1 cm from a metal rod 8 cm long that has a charge of 10 nC?

SOLUTION This is a straightforward calculation to illustrate the use of Eq. 25-28. Using $L = 0.08$ m, $r = 0.01$ cm, and $Q = 10^{-8}$ C, we calculate

$$\vec{E} = 2.18 \times 10^5 \, \hat{i} \text{ N/C}.$$

For comparison, the field 1 cm from a 10 nC *point* charge would be a somewhat larger $9.0 \times 10^5 \, \hat{i}$ N/C.

Before moving on, let us check that Eqs. 25-27 and 25-28 agree with our expectations for limiting cases. Does this result "make sense"? One observation we can make is that the vector result, Eq. 25-28, is valid whether Q is positive or negative. The field for a positively charged rod will point in the direction $\hat{i}$ away from the rod, while a negative value for Q will cause $\vec{E}$ to point in the $-\hat{i}$ direction toward the rod. A second check is to see whether the field has the proper limit as r becomes very large (or, equivalently, if L becomes very small). If $r \gg L$, the rod "appears" as a point charge Q. In the limit $r \gg L$ we see that Eq. 25-27 becomes $E = Q/4\pi\varepsilon_0 r^2$—exactly as expected for a point charge.

Finally, what happens if the rod becomes very long, $L \to \infty$, while the linear charge density λ remains constant? That is, more charge is added so that the ratio $\lambda = Q/L$ stays constant as L increases. Then $L \gg r$ so we can neglect the r^2 term in the square root. Thus

$$|\vec{E}_{\text{very long rod}}| = E_{\text{wire}} = \frac{1}{4\pi\varepsilon_0} \cdot \frac{2(Q/L)}{r} = \frac{1}{4\pi\varepsilon_0} \cdot \frac{2\lambda}{r}. \quad (25\text{-}29)$$

This is the field of an infinitely long line of charge having linear charge density λ. The linear charge density measures how close together or far apart the charges are on the rod, and this did not change as the rod's length L was increased.

This result for a very long rod is actually of significant practical interest. A long, straight wire is essentially a very long, thin rod; so we can interpret Eq. 25-29 as being E_{wire}, the electric field at distance r from a charged wire. Unlike a charged point, for which the field decreases as $1/r^2$, the field of a charged wire decreases more slowly—as only $1/r$.

Figure 25-9 shows the electric field of an infinitely long charged wire. Notice that the field points straight away from the wire at all points, but the field gets weaker as the distance increases. By comparing Figs. 25-8 and 25-9 you can see the field fairly near the center of a finite-length rod can be well approximated by the simpler result for an infinite wire.

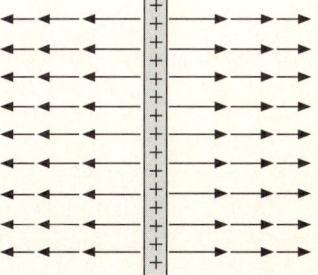

FIGURE 25-9 The electric field of a very long charged wire.

EXAMPLE 25-5 Consider a thin, circular ring of charge, having radius R and total charge Q uniformly distributed around the ring. Find an expression for the electric field at a point on the axis of the ring.

SOLUTION This problem asks us to find the on-axis field of a *ring of charge*. Although the charge is around a circle rather than on a straight line, this is still a "line of charge." We will briefly outline the seven steps of the problem-solving strategy.

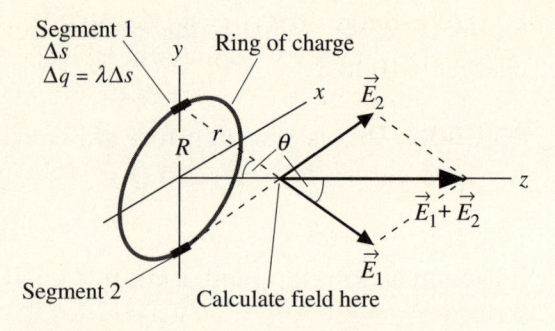

FIGURE 25-10 Calculating the on-axis field of a ring of radius R and charge Q.

Step 1: Figure 25-10 shows a picture of the situation. This is a three-dimensional perspective, showing the ring in the xy-plane with the z-axis perpendicular to the center of the ring. Let z measure the distance along the axis from the center of the ring to the point at which we want to calculate the electric field.

Step 2: We can divide the ring into N segments of length $\Delta s = 2\pi R/N$. Two such segments are shown in the figure. These will become point charges as $N \to \infty$.

Step 3: For every pair of segments that are diametrically opposed, such as segments 1 and 2 in the figure, the components of their fields that are perpendicular to the axis will cancel. So by symmetry, the field at a point on the axis will point straight outward along the z-axis. Only the z-component $E_z = \Sigma(E_i)_z$ of the net field will be nonzero.

Step 4: The z-component of $\vec{E}_i$, the field due to segment i, is given by

$$(E_i)_z = |\vec{E}_i|\cos\theta_i = \frac{\Delta q}{4\pi\varepsilon_0 r_i^2} \cdot \cos\theta_i.$$

From the geometry we see that every segment of the ring, independent of i, has

$$r_i = \sqrt{z^2 + R^2}$$

$$\cos\theta_i = \frac{z}{r_i} = \frac{z}{\sqrt{z^2 + R^2}}.$$

Inserting these into the expression for $(E_i)_z$ and summing over i gives

$$E_z = \sum_{i=1}^{N} \frac{1}{4\pi\varepsilon_0} \cdot \frac{z\Delta q}{(z^2 + R^2)^{3/2}}.$$

Step 5: The ring has circumference $L = 2\pi R$, giving it a linear charge density $\lambda = Q/L = Q/2\pi R$. This idea is still valid, even through the line in this case is curved, because the charge is spread out in only one dimension. Thus a segment of length Δs has charge

$$\Delta q = \lambda \Delta s = \frac{Q\Delta s}{2\pi r},$$

giving

$$E_z = \frac{1}{4\pi\varepsilon_0} \cdot \frac{Q}{2\pi R} \frac{z}{(z^2 + R^2)^{3/2}} \sum_{i=1}^{N} \Delta s.$$

Step 6: Let $\Delta s \to ds$ and the sum become an integral over s. But what are the limits? We are integrating along the "s-axis," which goes around the circumference of a circle, so the value of s increases from 0 at the start to $2\pi R$ at the end. Thus

$$\sum_{i=1}^{N} \Delta s \to \int_0^{2\pi R} ds = 2\pi R,$$

which finally yields

$$E_z(z) = \frac{1}{4\pi\varepsilon_0} \cdot \frac{zQ}{(z^2 + R^2)^{3/2}}. \qquad (25\text{-}30)$$

Step 7: It will be left to the homework problem to show that this result gives the expected limits when $z \to 0$ and when $z \gg R$. •

25.6 The Field of a Plane of Charge

[**Photo suggestion: A vacuum tube and a microphotograph of a field-effect transistor.**]

The two examples of continuous charge we looked at in the last section were both charges spread along a line—straight in one case and circular in the second. Most situations of practical interest involve charge distributed over some planar surface. For example, many electronic devices used charged metal surfaces—disks, squares, rectangles, and so on—to steer electrons along the proper paths. These charged surfaces are called **electrodes**. In this section we will look at how to calculate the electric field of a two-dimensional distribution of charge.

Consider an entire **plane of charge** that is infinite in all directions. The charge distributed across this two-dimensional surface is characterized by its *surface charge density* η. Surface charge density measures the quantity of charge per unit *area* of the surface. What is the electric field a distance z from this plane?

Although no physical object can be infinite in extent, an infinite plane is a reasonable approximation if we want to find the electric field of any finite plane of charge at a point that is relatively close to the center. As long as the distance z to the plane is small in comparison to the distance to the edges, we can reasonably treat the edges *as if* they are infinitely far away. This approximation will greatly simplify the mathematics. Our result will be valid for any finite planar surface of charge—circular as well as rectangular—as long as we do not get too close to an edge.

We can follow the step-by-step procedure of Section 25.4 to find the electric field of a plane of charge. First, we draw the picture that shows the charge as the xy-plane and a point at distance z above the plane (Fig. 25-11). Because the plane extends to infinity, there is no physical significance to the particular point we have chosen for calculation. It could be *any* point that is distance z from the plane.

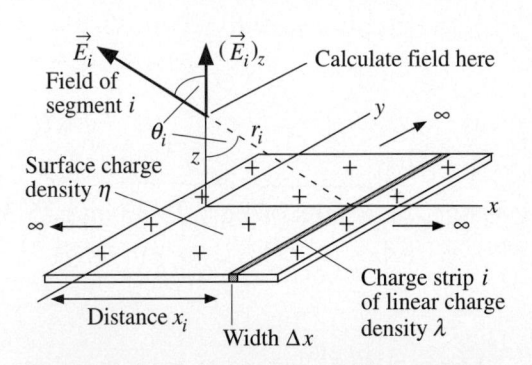

FIGURE 25-11 We can calculate the field of a plane of charge by dividing it into strips of charge, each having width Δx and field $\vec{E}_i$.

Next we divide the plane into small segments of charge whose field is already known. While we could divide it into small point-like squares, we will save ourselves lots of effort by utilizing another field that we now know—that of an infinite *line* of charge. To do this, divide the plane into many "strips" of charge parallel to the y-axis and of width Δx. One such strip, segment i, is shown in Fig. 25-11. It is distance x_i to the right of the point where we wish to calculate the field.

The field of a line of charge points straight away from the line, as we found in Section 25.5. Thus field $\vec{E}_i$ due to charged strip i points upward and to the left in the figure. But—this is where the use of symmetry comes in—there is another charge strip a distance x_i to the *left* of our point. Its field (not shown) will be equal in magnitude to $\vec{E}_i$ but will point upward and to the right. When added to $\vec{E}_i$, their horizontal components will *cancel* while their vertical components will add together. The net result is a field with a z-component only, pointing straight out from the plane. Because $E_x = E_y = 0$, our mathematical task is to find $E_z = \Sigma(E_i)_z$.

The magnitude of field $\vec{E}_i$ is that of an infinite line of charge, found in Eq. 25-29 to be

$$E_i = \frac{1}{4\pi\varepsilon_0} \cdot \frac{2\lambda}{r_i}. \tag{25-31}$$

Here r_i is the distance from charge strip i to our point. But what is λ, the linear charge density of the strip? Consider a length L of the strip. It has a surface area $A = L\Delta x$ and thus a charge $Q = \eta A = \eta L \Delta x$. The linear charge density, by definition, is

$$\lambda = \frac{Q}{L} = \frac{\eta L \Delta x}{L} = \eta \Delta x. \tag{25-32}$$

This is where the small width Δx enters, relating the linear charge density λ to the surface charge density η. Thus the magnitude of the field of strip i is

$$E_i = \frac{1}{4\pi\varepsilon_0} \cdot \frac{2\eta \Delta x}{r_i}. \tag{25-33}$$

The z-component of $\vec{E}_i$ is

$$(E_i)_z = E_i \cos\theta_i. \tag{25-34}$$

From the geometry of Fig. 25-11, r_i and $\cos\theta_i$ are related to x_i by

$$r_i = \sqrt{x_i^2 + z^2}$$

$$\cos\theta_i = \frac{z}{r_i} = \frac{z}{\sqrt{x_i^2 + z^2}}. \tag{25-35}$$

Using the results of Eqs. 25-33 and 25-35 in Eq. 25-34 gives

$$(E_i)_z = \frac{2\eta\Delta x}{4\pi\varepsilon_0 \sqrt{x_i^2 + z^2}} \cdot \frac{z}{\sqrt{x_i^2 + z^2}} = \frac{2\eta z \Delta x}{4\pi\varepsilon_0 (x_i^2 + z^2)}. \tag{25-36}$$

The net field at distance z from the plane of charge is found by summing over all the strips:

$$E_z = \frac{2\eta z}{4\pi\varepsilon_0} \cdot \sum_i \frac{\Delta x}{x_i^2 + z^2}. \tag{25-37}$$

The initial terms $2\eta z/4\pi\varepsilon_0$ are the same for each of the strips, making them constants as far as the summation is concerned, so they are brought outside the summation sign.

Finally, we let the summation become an integral by letting $\Delta x \to dx$ and replacing the discrete x_i with the continuous integration variable x. The integration limits run from the smallest to the largest values of x, which in this case are from $-\infty$ to $+\infty$. Thus

$$E_z = \frac{2\eta z}{4\pi\varepsilon_0} \cdot \int_{-\infty}^{\infty} \frac{dx}{x^2 + z^2}. \tag{25-38}$$

Having reached this point, you can recognize that performing the integration itself is not an especially difficult task. The difficulty was in finding and setting up the quantity to be integrated—that is, in doing the physics, not the mathematics. The integral can be looked up in a table of integrals, giving

$$E_z = \frac{2\eta z}{4\pi\varepsilon_0} \cdot \int_{-\infty}^{\infty} \frac{dx}{x^2 + z^2} = \frac{\eta z}{2\pi\varepsilon_0} \cdot \left[\frac{1}{z}\tan^{-1}\left(\frac{x}{z}\right)\right]_{-\infty}^{\infty}$$

$$= \frac{\eta}{2\pi\varepsilon_0} \cdot \left[\tan^{-1}(\infty) - \tan^{-1}(-\infty)\right]$$

$$= \frac{\eta}{2\pi\varepsilon_0} \cdot \left[\left(\frac{\pi}{2}\right) - \left(-\frac{\pi}{2}\right)\right]$$

$$= \frac{\eta}{2\varepsilon_0}. \tag{25-39}$$

Thus the electric field of a plane of charge having surface charge density η is

$$E_{\text{plane}} = E_z = \frac{\eta}{2\varepsilon_0} = \text{constant}. \tag{25-40}$$

This is a very simple result, but what does it tell us? First, the field strength is directly proportional to the charge density η. More charge, bigger field. Second, and more interesting, the field is the same at *all* points in space, independent of the distance z. You would measure the same field strength 1000 m from the plane as you would 1 mm from the plane.

How can this be? It seems seem like the field should get weaker as you move away from the plane of charge. But remember that we are dealing with an *infinite* plane of charge. What does it mean to be "close to" or "far from" an infinite object? For a square of finite edge length L, whether a point at distance z is "close to" and "far from" the square is a comparison of z to L. If $z \ll L$, then the point is close to the square, and if $z \gg L$, the point is far from the square. But as $L \to \infty$, we have no *scale* for distinguishing near and far. In essence, *every* point in space is "close to" a square having infinite size. Because no real plane is infinite in extent, we can interpret Eq. 25-40 as saying that the field of a planar surface of charge is a constant $\eta/2\varepsilon_0$ for those points whose distance z to the surface is small in comparison to their distance to the edge. But eventually, when $z \gg L$, any finite surface will begin to appear as a point charge Q and the field will have to decrease as $1/z^2$.

We do need to point out that the geometry of Fig. 25-11 considered only $z > 0$—that is, the $+z$-side of the plane. A positively charged plane has $\eta > 0$, and thus $E_z > 0$, indicating a field that points away from the plane. But the field must also point away from a positively charged plane if the observation point is on the other side, the $-z$-side, of the plane. This requires $E_z < 0$ on the $-z$-side. Thus the complete description of the electric field, valid for either side of the plane, is

$$(E_{\text{plane}})_z = \begin{cases} +\dfrac{\eta}{2\varepsilon_0} & z > 0 \\ -\dfrac{\eta}{2\varepsilon_0} & z < 0. \end{cases} \quad (25\text{-}41)$$

Figure 25-12 shows both a perspective view and an edge view of the electric field diagram for a positively charged plane. All the arrows would be reversed, pointing inward, for a negatively charged plane. On either side of the plane, the field is the same at *every* point in space and the field vectors $\vec{E}$ are thus parallel, equally spaced, and of equal length. A field such as this, which is the same at all points in space, is called a **uniform electric field**. The magnitude of the field $E = |\vec{E}|$ is often called the **electric field strength**. We have found that a plane of charge generates a uniform field of field strength $\eta/2\varepsilon_0$ on each side of the plane.

Perspective view   Edge view

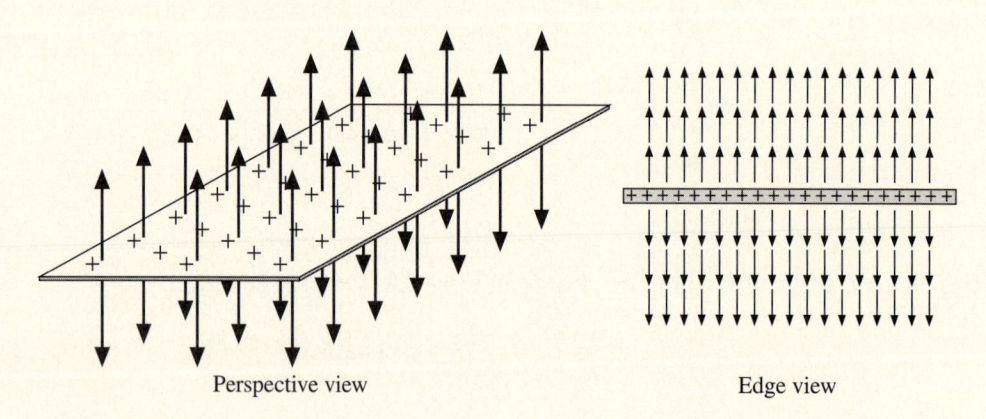

FIGURE 25-12 Perspective and edge views of the electric field of a positive plane of charge.

A uniform electric field is analogous to the uniform gravitational field near the surface of the earth. We see that such a field is produced by a plane of charge. This would have been very difficult to anticipate from Coulomb's law, but step by step we have been able to use the concept of the electric field to look at more and more complex distributions of charge. A uniform electric field would be very useful for applications, but for this to be practical we need a way to produce a uniform electric field using only *finite*, and thus realizable, electrodes.

EXAMPLE 25-6 A 10 cm × 10 cm plastic square is charged uniformly with an extra 10^{12} electrons. What is the electric field 1 mm above the surface?

SOLUTION The total charge of the plastic square is $Q = N(-e) = -1.6 \times 10^{-7}$ C. Because the charge is negative, due to excess electrons, the field will point straight *in* toward the square—the opposite of that shown in Fig. 25-12. The surface charge density is

$$\eta = \frac{Q}{A} = \frac{Q}{L \times L} = \frac{-1.6 \times 10^{-7} \text{ C}}{0.1 \text{ m} \times 0.1 \text{ m}} = -1.6 \times 10^{-5} \text{ C/m}^2.$$

The electric field, given by Eq. 25-41, is independent of z as long as we are not near the edge:

$$E_z = \frac{\eta}{2\varepsilon_0} = \frac{-1.6 \times 10^{-5} \text{ C/m}^2}{2 \times 8.85 \times 10^{-12} \text{ C}^2/\text{N m}^2} = -9.04 \times 10^5 \text{ N/m}.$$

This is the z-component of $\vec{E}$, so the minus sign indicates that the field points *toward*, rather than away from, the plane. •

25.7 The Field of a Capacitor

Consider two squares of edge length L, one charged to $+Q$ and the other to $-Q$, placed face-to-face a distance d apart as shown in Fig. 25-13a. Such an arrangement of two *electrodes*, charged equally but oppositely, is called a **parallel-plate capacitor**. Capacitors, which play important roles in many electronic circuits, are devices consisting of two conducting surfaces separated by an insulator. The air between the plates in Fig. 25-13a is the insulator of the parallel-plate capacitor.

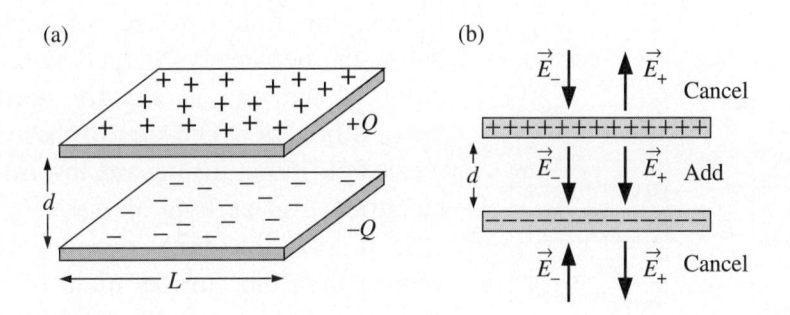

FIGURE 25-13 a) A parallel-plate capacitor with charges $+Q$ and $-Q$. b) The electric fields add between the electrodes but cancel outside. This is an *edge view* of the plates, which is how they are usually shown.

We want to find the electric field both inside the capacitor, between the plates, and outside the capacitor. We will assume that the spacing d is much less than the size of the squares: $d \ll L$. This is a reasonable assumption in practice because of how capacitors are constructed. The electric field of either plate, close to the plate, is very well approximated by the electric field of a plane of charge: $|\vec{E}| = \eta/2\varepsilon_0$, where $\eta = Q/A$ is the surface charge density on the plates. The field $\vec{E}_+$ of the positively charged electrode points away from the electrode. $\vec{E}_-$ of the negatively charged electrode has the same magnitude but points toward the electrode. When the two electrodes are placed close together, the net electric field is given by the superposition of their individual electric fields. The consequences are illustrated in Fig. 25-13b.

Between the electrodes, $\vec{E}_+$ from the positively charged square is of magnitude $\eta/2\varepsilon_0$ and points from the positive toward the negative side. The field $\vec{E}_-$ from the negatively charged disk is *also* of magnitude $\eta/2\varepsilon_0$ and *also* points from positive to negative. The net field inside the capacitor is thus

$$\vec{E}_{\text{capacitor}} = \vec{E}_+ + \vec{E}_- = \left(\frac{\eta}{\varepsilon_0}, \text{ from positive to negative}\right)$$
$$= \left(\frac{Q}{\varepsilon_0 A}, \text{ from positive to negative}\right). \quad (25\text{-}42)$$

Outside the squares, on either side, $\vec{E}_+$ and $\vec{E}_-$ are of equal magnitudes (nearly) but *opposite* directions and thus (nearly) cancel each other, giving $\vec{E} \approx 0$. We conclude that there is a strong *constant* field between the electrodes and essentially no field outside the electrodes.

Notice that the shape of the electrodes does not enter into the final result. Although we considered two squares, Eq. 25-42 refers only to their surface charge density, not to their geometry. In fact, Eq. 25-42 is the electric field between *any* two parallel electrodes—rectangular, circular, or other shape—as long as the spacing d between them is small in comparison to their lateral dimensions. In all cases, the field inside is uniform, of magnitude $|\vec{E}| = \eta/\varepsilon_0$, and points *from* the positive *to* the negative electrode.

Strictly speaking, the field of each electrode does decrease slightly with distance, because the squares are finite in size, not infinite. So just below the negatively charged plate, the field $\vec{E}_+$ points down and is slightly weaker than the upward field $\vec{E}_-$. The sum of these two is a very weak field pointing inward toward the negative electrode. But as long as $d \ll L$, which is the case in practice, this "outside" field is *much* weaker than the "inside" field of strength η/ε_0. The outer field is called the **fringe field**.

Figure 25-14a shows the actual field of a parallel-plate capacitor, including the weak fringe field. We will, however, adopt the idealization of Fig. 25-14b, in which the field is strong and *uniform* between the plates and *zero* elsewhere. Because $d \ll L$ in real capacitors, this idealization is a very good approximation for all but the most precise calculations. Note that the figure shows an *edge view* of the plates, which is how they are usually drawn, so keep in mind that these electrodes extend above and below the plane of the paper.

A parallel-plate capacitor is a very good and very practical way to produce a uniform electric field. In practice, the parallel electrodes are attached to the positive and negative terminals of a battery, as shown in Fig. 25-15. You will learn in Chapter 28 how to relate the electric field to

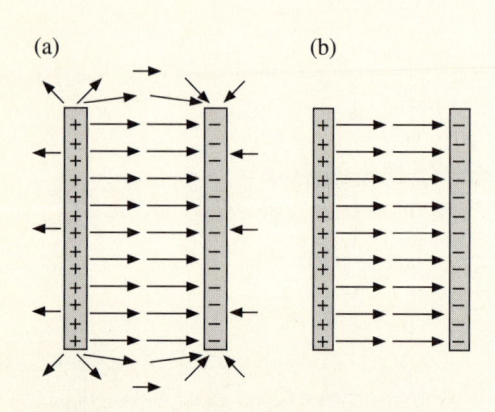

FIGURE 25-14 a) The electric field of a capacitor. b) An idealized field that is uniform inside the capacitor and zero outside.

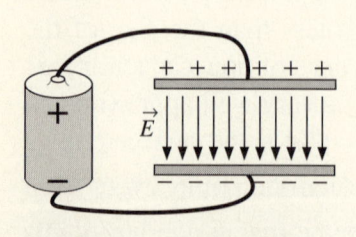

FIGURE 25-15 A capacitor is charged by connecting it to a battery, thus creating a uniform electric field inside.

the voltage of the battery. Many electric field problems utilize a uniform electric field. Such problems carry the implicit assumption that the action is taking place *inside* a parallel-plate capacitor, such as the Fig. 25-15 arrangement, even though the capacitor plates themselves are not "seen" in the problem.

EXAMPLE 25-7 Two rectangular electrodes of size 1 cm × 2 cm are 1 mm apart. a) What charge must be placed on each electrode to create an electric field of strength 2×10^6 N/C? b) How many electrons must be moved from one electrode to the other?

SOLUTION a) The field strength inside the capacitor is given by Eq. 25-42 as $E = Q/\varepsilon_0 A$. The needed charge is thus

$$Q = \varepsilon_0 AE = (8.85 \times 10^{-12} \text{ C}^2/\text{N m}^2)(2 \times 10^{-4} \text{ m}^2)(2 \times 10^6 \text{ N/C})$$

$$= 3.5 \times 10^{-9} \text{ C} = 3.5 \text{ nC}.$$

The positive plate must be charged to +3.5 nC and the negative plate to −3.5 nC. Notice that the plate spacing does not enter the result. As long as the spacing is much less than the plate dimensions, which is true in this example, the field is independent of the spacing.

b) The capacitor as a whole is neutral. The plates are charged by moving electrons from one plate, leaving it positive, to the other, making it negative. The number of electrons in 3.5 nC is

$$N = \frac{Q}{e} = \frac{3.5 \times 10^{-9} \text{ C}}{1.6 \times 10^{-19} \text{ C/electron}} = 2.2 \times 10^{10}.$$

25.8 The Field of a Sphere of Charge

There is one last charge distribution for which we need to know the electric field, and that is of a sphere of charge. This problem is exactly analogous to wanting to know the gravitational field of a spherical planet or star. Although the procedure for calculating the field of a sphere of charge is no different than we have used for lines and planes, the integrations are significantly more difficult. We will skip the details of the calculations and simply assert the results without proof.

A **sphere of charge** Q and radius R—be it a uniformly charged sphere or just a spherical shell—has an electric field *outside* the sphere ($r \geq R$) that is exactly the same as that of a point charge Q located at the center of the sphere:

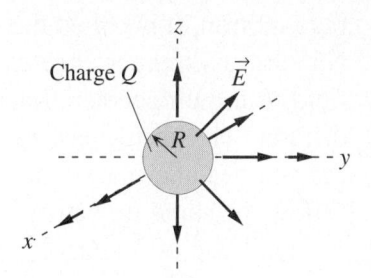

FIGURE 25-16 The electric field of a sphere of charge is the same as a point charge at the center of the sphere.

$$\vec{E}_{\text{sphere}} = \left(\frac{Q}{4\pi\varepsilon_0 r^2}, \text{ away from center} \right) \quad \text{for } r \geq \text{radius } R. \quad (25\text{-}43)$$

This is shown in Fig. 25-16. A sphere of charge acts as if all the charge is located at the center. This is, of course, exactly the same as our earlier assertion that the gravitational force between stars and planets can be computed as if all the mass is at the center. The field *inside* the sphere ($r < R$), be it a sphere of charge or a sphere of mass, is *not* given by Eq. 25-42, but we will not need to use the inside field in this text.

25.9 Motion of a Charged Particle in a Uniform Field

In the first eight sections of this chapter you have learned how to calculate the electric field of a variety of different charge distributions. An important goal of these calculations has been to find out how a charged object—be it a dipole, a charged rod, or a charged capacitor—interacts with other charges. After all, our original motivation for the concept of the electric field was to be able to analyze these complex electrical interactions. So in the remainder of this chapter we will turn our attention to how a charged particle moves in the electric field created by some configuration of source charges.

Consider a particle of charge q and mass m in an electric field $\vec{E}$ that has been produced by *other* charges, the source charges. The charged particle experiences a force

$$\vec{F}_{\text{on } q} = q\vec{E} \tag{25-44}$$

and hence an acceleration

$$\vec{a} = \frac{\vec{F}}{m} = \frac{q}{m} \cdot \vec{E}. \tag{25-45}$$

Thus the electric field vector at any point in space determines the force, and the acceleration, of a charged particle at that point. The acceleration given by Eq. 25-45 is the *response* of the charged particle to the source charges that created the field. Note that the force on a *negatively* charged particle is *opposite* in direction to the electric field vector. This follows from the fact that Eq. 25-44 is a vector equation, so signs are important.

The ratio q/m is especially important for understanding the dynamics of charged particle motion. It is called the **charge-to-mass ratio**. Two *equal* charges, say a proton and a Na$^+$ ion, will experience *equal* forces $\vec{F} = q\vec{E}$ if placed at the same point in an electric field. But their accelerations will be *different* because they have different masses and thus different charge-to-mass ratios. Two particles with the same charge-to-mass ratio, even if the charges and masses are themselves different, will undergo the same acceleration and follow the same trajectory.

The case of a charge particle moving in a *uniform* electric field is especially important, not only because of its basic simplicity but also due to many important applications. A uniform field is *constant* at all points—constant in both magnitude and direction—within the region of space where the charged particle is moving. It follows, from Eq. 25-45, that a charged particle in a uniform electric field will undergo *constant acceleration motion* with

$$|\vec{a}| = \frac{q|\vec{E}|}{m} = \frac{qE}{m} = \text{constant}, \tag{25-46}$$

where $E = |\vec{E}|$ is the electric field strength. The direction of $\vec{a}$ will be parallel or antiparallel to $\vec{E}$, depending on the sign of q.

Having identified the motion of a charged particle in a uniform field as being one of constant acceleration, we find that all the kinematic machinery that we developed in Chapters 3 and 6 for constant acceleration motion comes into play. The trajectories of charged particles will, in general, be parabolic—exactly analogous to the projectile motion of masses in the near-earth uniform gravitational field. The motion is one-dimensional in the special case of a charge moving parallel to the electric field vectors. This, of course, corresponds to the one-dimensional vertical motion of a mass. The one major difficulty for which we need to watch is that the electric field acceleration $\vec{a}_{elec}$ can point in *any* direction, whereas the gravitational acceleration $\vec{a}_{grav}$ always points straight down.

EXAMPLE 25-8 Example 25-6 found that the electric field 1 mm above a charged 10 cm × 10 cm plastic square is $\vec{E} = -9.04 \times 10^5 \hat{k}$ N/C. If a proton is released from rest at this point, how long does it take the proton to collide with the plastic square?

SOLUTION The electric field points down, toward the plastic square, so a positive proton will accelerate in the downward direction until it reaches the square. The field above the plane of charge is uniform, so the proton has constant acceleration

$$a = a_z = \frac{eE_z}{m} = \frac{(1.60 \times 10^{-19} \text{ C})(-9.04 \times 10^5 \text{ N/C})}{(1.67 \times 10^{-27} \text{ kg})} = -8.66 \times 10^{13} \text{ m/s}^2.$$

The proton is released from rest ($v_0 = 0$) at $z_0 = 1$ mm = 0.001 m and collides with the plastic square at $z_1 = 0$. The kinematics of constant acceleration motion are

$$z_1 = z_0 + v_0 \Delta t + \tfrac{1}{2} a (\Delta t)^2$$

$$\Rightarrow \Delta t = \sqrt{\frac{-2z_0}{a}} = 4.80 \times 10^{-9} \text{ s} = 4.80 \text{ ns}.$$

EXAMPLE 25-9 A capacitor is built of two circular disks 6 cm in diameter and spaced 5 mm apart. The capacitor is charged by transferring 1×10^{11} electrons from one electrode to the other. An electron is released from rest next to the negative electrode. What is the electron's speed as it collides with the positive electrode?

SOLUTION Figure 25-17 shows the capacitor and the electron. The electron will be repelled by the negative electrode and attracted to the positive electrode, accelerating across the distance d. But the electrodes are *not* point charges, so we *cannot* use Coulomb's law to find the force on the electron. Our analysis of the electron's motion must be in terms of the electric field inside the capacitor. The field is the agent that exerts the force on the electron, causing it to accelerate. The acceleration is $a = eE/m$. From one-dimensional kinematics,

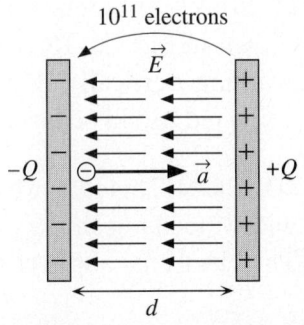

FIGURE 25-17 An electron accelerates in the field of a capacitor.

using $v_i = 0$, we can find the final velocity:

$$v_f^2 = v_i^2 + 2a\Delta x = 2ad$$

$$\Rightarrow v_f = \sqrt{2ad} = \sqrt{\frac{2eEd}{m}}.$$

The electric field inside a capacitor is

$$E = \frac{\eta}{\varepsilon_0} = \frac{Q}{\varepsilon_0 A} = \frac{Q}{\varepsilon_0 \pi R^2} = \frac{Ne}{\varepsilon_0 \pi R^2},$$

where Q/A is the surface charge density and $Q = Ne$ if N electrons are transferred from one plate to the other. Using the data supplied, we can calculate first $E = 6.39 \times 10^5$ N/m and then $v_f = 3.35 \times 10^7$ m/s.

[**Photo suggestion: A cathode ray tube.**]

Parallel electrodes such as the ones in Example 25-9 are often used to accelerate charged particles. If the positive plate has a small hole in the center, then a *beam* of electrons will pass through this hole and emerge with a speed of 3.35×10^7 m/s. This is the basic idea of the *electron gun* used in televisions, oscilloscopes, computer display terminals, and other *cathode-ray tube* (CRT) devices. A negatively charged electrode is often called a *cathode*, so the physicists who first learned to produce electron beams in the late nineteenth century called them *cathode rays*. The name has persisted, even though we have long since learned that cathode rays are simply high-velocity electrons.

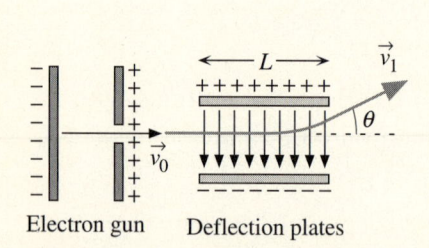

FIGURE 25-18 Deflection of an electron beam.

EXAMPLE 25-10 The capacitor of Example 25-9 is used as an electron gun to create a beam of electrons moving with a speed of 3.35×10^7 m/s. An electron emerging from the electron gun enters a horizontally oriented parallel-plate capacitor of length 2 cm, as shown in Fig. 25-18. The electric field in this capacitor is of strength 5×10^4 N/C and is pointing vertically downward. In which direction, and by what angle, is the electron deflected by this capacitor?

SOLUTION This is a two-dimensional projectile problem. The electron enters the capacitor with velocity *vector* $\vec{v}_0 = v_{x0} \hat{i} = 3.35 \times 10^7 \hat{i}$ m/s and leaves with velocity $\vec{v}_1 = v_{x1} \hat{i} + v_{y1} \hat{j}$. The electron's angle of travel upon leaving the capacitor is

$$\theta = \tan^{-1}\left(\frac{v_{y1}}{v_{x1}}\right),$$

This is the *deflection angle*. To find θ we must compute the final velocity vector $\vec{v}_1$.

The electron's acceleration $\vec{a}$ while inside the capacitor is given by Eq. 25-45 as $\vec{a} = (q/m)\vec{E} = -(e/m)\vec{E}$, because $q_{elec} = -e$. The field $\vec{E}$ is downward, in the $-\hat{j}$ direction, so $\vec{E} = -E\hat{j}$, where $E = |\vec{E}|$ is the scalar field strength of 5×10^4 N/C. The electron's acceleration inside the capacitor is thus

$$\vec{a} = -\frac{e}{m}\vec{E} = -\frac{e}{m} \cdot (-E\hat{j}) = +\frac{eE}{m}\hat{j} = 8.78 \times 10^{15}\hat{j} \text{ m/s}^2.$$

The electron is deflected *upward* because the acceleration is in the *positive* y-direction. Because there is no x-component to the acceleration, the electron's x-component of velocity will be unaffected: $v_{x1} = v_{x0} = 3.35 \times 10^7$ m/s. This is exactly analogous to gravitational trajectory problems in which the horizontal velocity remained constant. We can use this fact to determine the time interval Δt spent inside the capacitor of length L:

$$\Delta t = \frac{L}{v_{x0}} = 5.97 \times 10^{-10} \text{ s}.$$

Vertical acceleration will occur during this time interval, resulting in a final vertical velocity

$$v_{y1} = v_{y0} + a\Delta t = 0 + (8.78 \times 10^{15} \text{ m/s}^2)(5.97 \times 10^{-10} \text{ s}) = 5.24 \times 10^6 \text{ m/s}.$$

The electron's velocity as it leaves the capacitor is thus

$$\vec{v}_1 = (3.35 \times 10^7 \hat{i} + 5.24 \times 10^6 \hat{j}) \text{ m/s}$$

and the deflection angle θ is

$$\theta = \tan^{-1}\left(\frac{v_{y1}}{v_{x1}}\right) = 8.89°.$$

Example 25-10 demonstrates how an electron beam is "steered" to a point on the screen of a cathode ray tube. The capacitor plates in a CRT are called the *deflection plates*. A high-speed electron beam is created by an electron gun, such as that of Example 25-9. The beam then passes first through a set of *vertical deflection plates*, as in Example 25-10, then through a second set of *horizontal deflection plates*. Upon leaving the deflection plates it travels in a straight line to the screen of the CRT, where it strikes a phosphor coating on the inside surface and makes a dot of light. (The air has been removed from the inside of a CRT so that the electrons travel without collisions with air molecules.) The two sets of deflection plates independently control the vertical and horizontal deflection of the electrons. By properly choosing the electric fields within the deflection plates, which is done by controlling their voltage, the electron beam can be steered to any point on the screen of the CRT. By rapidly modulating the voltages, and thus the electric fields, the electron beam can be swept back and forth across the entire face of the CRT many times each second, giving the illusion of an image on the screen.

It is worth noting the enormous accelerations and velocities as well as the incredibly small time intervals involved when we work with electrons. These are far outside our range of common experience, and they arise as a consequence of the very tiny mass of the electron. Nonetheless, the motions are still just those you studied in single-particle dynamics.

25.10 Motion of a Dipole in an Electric Field

Let us conclude this chapter by returning to one of the more striking puzzles we faced when making the observations at the beginning of Chapter 24. There you discovered that charged objects of *either* sign exert forces on neutral objects—as when a comb picks up pieces of paper after being used to brush your hair. We reached a qualitative understanding that these phenomena occur as a consequence of a *polarization force*: The charged object polarizes the neutral object, creates an induced dipole, and then exerts an attractive force on the near end of the dipole that is slightly stronger than the repulsive force on the far end. We are now in a position to make that understanding more quantitative. In this section, we will focus our attention on *permanent* dipoles and will let you, as a homework problem, think about *induced* dipoles.

Dipoles in a Uniform Field

First consider a permanent electric dipole—two charges of $\pm q$ separated by a small distance s—placed in a *uniform* electric field, as shown in Fig. 25-19a. It is assumed, in a figure such as this, that the uniform field $\vec{E}$ is being created by source charges that we do not see. That is $\vec{E}$ is *not* the field of the dipole but, instead, is an "externally generated" field to which the dipoles are responding. In this case, because the field is uniform, the dipole is presumably inside an unseen parallel-plate capacitor.

The two charges forming the dipole each experience a force in the electric field. Because the charges $\pm q$ are equal in magnitude, while opposite in sign, the two forces $\vec{F}_+ = +q\vec{E}$ and $\vec{F}_- = -q\vec{E}$, are also equal but opposite. Thus the net force on the dipole is

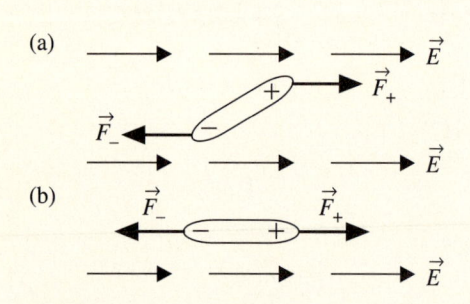

$$\vec{F}_{net} = \vec{F}_+ + \vec{F}_- = 0. \qquad (25\text{-}47)$$

FIGURE 25-19 a) A dipole in a uniform electric field experiences a torque but no net force. b) The equilibrium position, showing the dipole aligned with the field.

There is *no* net force on the dipole—at least in a uniform field—and so it will not undergo any translational motion.

That is not to say, however, that the electric field has no effect on the dipole. Because the two forces on the ends of the dipole are in opposite directions, the dipole is going to be *rotated* by the electric field—much as if two people grabbed the end of a bar and pulled in opposite directions. The rotating or twisting force that the field exerts on the dipole is called a **torque**.

For our purposes, you will not need to know how to compute a numerical value for the torque. The important thing for you to note is that an electric field *does* cause a torque to be exerted on a dipole and that the dipole, as a result, rotates until it is aligned with the electric field, as shown in Fig. 25-19b. In this position, the dipole experiences not only no net force but also no torque. Thus Fig. 25-19b represents the *equilibrium position* for a dipole in a uniform electric field. Notice that the positive end of the dipole points in the *same* direction as the field vectors. This is always a dipole's equilibrium orientation.

If permanent dipoles, such as water molecules, are placed in an electric field, the torque will cause the dipoles to rotate very quickly until they are aligned with the field. This is, in fact, the *mechanism* by which the sample becomes *polarized* because, once aligned, there will be a positive surface charge at one end of the sample and a negative surface charge at the other end. This is illustrated in Fig. 25-20. You will meet this same dipole behavior again in our study of magnetism, where it plays a critical role in understanding why some materials are magnetic but others are not.

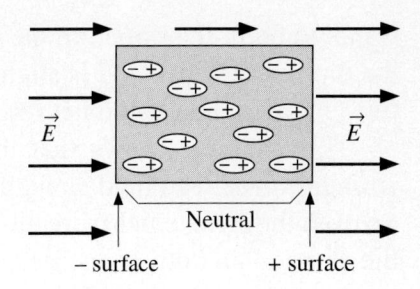

FIGURE 25-20 A neutral sample of molecules is polarized in an electric field.

Dipoles in a Nonuniform Field

Now let's suppose that a dipole is placed in a *nonuniform* electric field, one in which the field strength changes with position. Nearly all the fields we have examined in this chapter—point charges, rods, and loops—have been nonuniform. Figure 25-21 shows a dipole in the nonuniform field of a point charge. The first response of the dipole is to rotate until it is aligned with the field—the dipole's positive end pointing in the same direction as the field. Now, however, there is a slight *difference* between the forces acting on the two ends of the dipole. This difference occurs because the electric field, which depends on the distance from the point charge, is stronger at the near end of the dipole. This causes a *net force* to be exerted on the dipole.

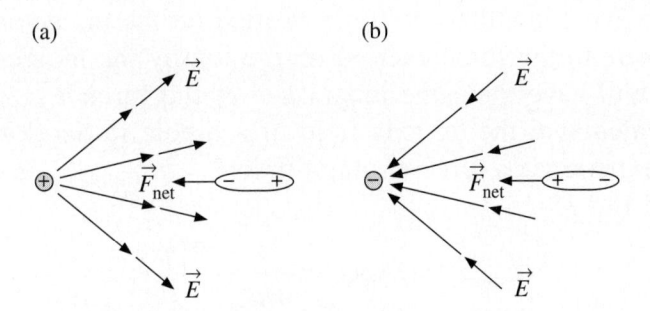

FIGURE 25-21 a) An aligned dipole is drawn toward a positive point charge because the electric field is slightly stronger at the negative end of the dipole. b) It is also attracted to a negative charge because the torque causes the dipole to flip over.

Which way does the force point? Figure 25-21a shows the nonuniform field of a positive point charge. Once the dipole is aligned, the leftward attractive force on its negative end will be slightly stronger than the rightward repulsive force on its positive end. This causes a net force to the *left*, toward the point charge. Now consider Fig. 25-21b, where the dipole is in the field of a negative point charge. The field vectors now point in the opposite direction. However, the field is still stronger at the near end of the dipole and weaker at the far end. The dipole now aligns in the opposite orientation, because its positive end still wants to be in the direction the electric field is pointing. The leftward attractive force on the dipole's positive end is slightly greater than the rightward repulsive force on its negative end. As a result, the net force is *still* to the left.

This example shows that the net force on a dipole is toward the direction of the *strongest field*. The field strength is larger on the left in both Fig. 25-21a and Fig. 25-21b. So even though the field direction is different, the force on the dipole is to the left—toward the charge—in both cases. Because any finite-size charged object, such as a point charge or charged rod, has a field strength that increases as you get closer to the object, we can conclude that a dipole will experience a net force toward *any* charged object.

EXAMPLE 25-11 The electric field of a dipole, formed by two charges $\pm q$ separated by distance s, was found in Eq. 25-15 to be proportional to the product qs. This product, with SI units of C m, is called the **dipole moment**. The dipole moment measures the "strength" of the dipole. The water molecule H_2O is known to have a dipole moment of 6.2×10^{-30} C m. Suppose that a water molecule is located 10 nm from a Na^+ ion in a salt water solution. What force does the ion exert on the water molecule?

SOLUTION A Na^+ ion is missing one electron and has charge $q = +e$. The electric field of this ion diverges, like that shown in Fig. 25-21a, so the water molecule will experience a torque that aligns it in the ion's field. The negative side of the molecule will then experience a slightly larger force than the positive side, drawing the water molecule toward the ion.

There are two ways that we can determine the magnitude of the force. First, from Newton's third law we know that the force exerted on the molecule by the ion must be equal but opposite to the force exerted on the ion by the molecule. So if we find $\vec{F}_{\text{molecule on ion}}$, it will have the same magnitude as the force $\vec{F}_{\text{ion on molecule}}$ that we are seeking. We calculated the on-axis field of a dipole in Section 25.3. An ion of charge $q = e$ will experience a force of magnitude $F = qE_{\text{dipole}} = eE_{\text{dipole}}$ when placed in that field. From Eq. 25-15,

$$|\vec{E}_{\text{dipole}}| = E_{\text{dipole}} = \frac{1}{4\pi\varepsilon_0} \cdot \frac{2(qs)}{y^3},$$

where $qs = 6.2 \times 10^{-30}$ C m is the molecule's dipole moment and $y = 10^{-8}$ m is the distance to the ion, The force on the ion in this field is

$$F = eE_{\text{dipole}} = \frac{1}{4\pi\varepsilon_0} \cdot \frac{2e(qs)}{y^3} = 1.79 \times 10^{-14} \text{ N}.$$

The force exerted on the water molecule by the ion will have the same magnitude. While 1.79×10^{-14} N may seem like a very small force, it is $\approx 10^{11}$ times larger than the size of the gravitational force on these atomic particles.

As a second method, we can directly compute the ion's force on the two charges in the dipole. Consider an aligned dipole of charges $\pm q$, separated by s, at distance y from an ion of charge $+e$. The negative end of the dipole, when aligned, is distance $y - (s/2)$ from the ion, while the positive end is slightly farther away at $y + (s/2)$. So there is a net force on the dipole, directed toward the ion, of magnitude

$$F = F_- - F_+ = qE_{\text{point}}(r = y - \tfrac{1}{2}s) - qE_{\text{point}}(r = y + \tfrac{1}{2}s)$$

$$= \frac{q}{4\pi\varepsilon_0} \cdot \frac{e}{(y - \tfrac{1}{2}s)^2} - \frac{q}{4\pi\varepsilon_0} \cdot \frac{e}{(y + \tfrac{1}{2}s)^2}$$

$$= \frac{eq}{4\pi\varepsilon_0 y^2} \cdot \left[\left(1 - \frac{s}{2y}\right)^2 - \left(1 + \frac{s}{2y}\right)^2 \right].$$

The term in brackets is exactly the same as the term in Eq. 25-12 that we previously simplified with the binomial approximation. Because $s \ll y$ in this example, we can use the earlier result and approximate the term in brackets as $2s/y$. This gives

$$F = \frac{1}{4\pi\varepsilon_0} \cdot \frac{2e(qs)}{y^3} = 1.79 \times 10^{-14} \text{ N},$$

exactly the same as our first method of solution.

The forces found in Example 25-11 cause water molecules to cluster around any ions that are in solution, and this clustering plays an important role in the microscopic physics of solutions. These are issues studied in a physical chemistry course. This example could, of course, be modified to find the force exerted on a dipole by a charged rod or any other charge distribution. The major conclusions are that (a) any charged object exerts an attractive force on a dipole, and (b) we now know how to calculate the magnitude of that force.

Summary

Important Concepts and Terms

induced electric dipole
permanent electric dipole
linear charge density
surface charge density
uniformly charged
line of charge
electrode
plane of charge

uniform electric field
electric field strength
parallel-plate capacitor
fringe field
sphere of charge
charge-to-mass ratio
torque
dipole moment

This has been a very dense chapter with many detailed calculations, and it will require careful study. Although we have derived quite a number of useful results, such as the field of a charged rod or that of a parallel-plate capacitor, by far the most important result has been the development of a *method* for calculating the electric field of a distribution of charges. This method was summarized in Section 25.4, and you should note how we followed the steps of the method in all our field calculations. The homework problems will let you try it for yet other charge distributions.

Our specific examples were chosen because they represent commonly occurring fields for which it is useful to have an explicit expression. The fields of a dipole:

$$\vec{E}_{\text{dipole}}(\text{on dipole axis}) = \frac{1}{4\pi\varepsilon_0} \cdot \frac{2qs}{y^3} \hat{j} \quad \text{if } y \gg s,$$

$$\vec{E}_{\text{dipole}}(\text{on bisecting axis}) = -\frac{1}{4\pi\varepsilon_0} \cdot \frac{qs}{x^3} \hat{j} \quad \text{if } x \gg s,$$

of a long charged wire:

$$\vec{E}_{\text{wire}} = \left(\frac{1}{4\pi\varepsilon_0} \cdot \frac{2\lambda}{r}, \text{ away from wire} \right),$$

of a plane of charge:

$$(E_{\text{plane}})_z = \begin{cases} +\dfrac{\eta}{2\varepsilon_0} & z > 0, \\ -\dfrac{\eta}{2\varepsilon_0} & z < 0, \end{cases}$$

of a parallel-plate capacitor:

$$\vec{E}_{\text{capacitor}} = \left(\frac{\eta}{\varepsilon_0}, \text{ from positive to negative} \right),$$

and of a sphere of charge:

$$\vec{E}_{\text{sphere}} = \left(\frac{Q}{4\pi\varepsilon_0 r^2}, \text{ away from center} \right) \quad \text{for } r \geq \text{radius } R,$$

are especially important and will be used again in upcoming chapters.

An important concept that we have introduced in this chapter has been that of a *uniform* electric field—a field that is constant throughout some region of space. A very large number of practical applications of electric fields utilize a uniform or very nearly uniform field. The region between two parallel electrodes having equal but opposite charges $\pm Q$, called a parallel-plate capacitor, is the most common practical method of realizing a uniform field.

Last, we looked at the dynamics of charged particles and dipoles in electric fields. The force on a charge q is $\vec{F} = q\vec{E}$, producing an acceleration $\vec{a} = (q/m)\vec{E}$. The ratio q/m is called the charge-to-mass ratio. The motion of charges in a uniform electric field is especially simple because it is constant acceleration motion. The charged particle trajectories are parabolas, exactly equivalent to the motion of masses near the earth's surface. Charged particle motion in nonuniform fields is more complicated, because the acceleration is not constant, and this type of motion will be easier to treat using the energy methods that will be introduced in Chapter 27. An electric dipole experiences a *torque* in an electric field that rotates it to align with the field. In addition, there is a net force on a dipole in a nonuniform field that draws it *into* regions of greater field strength.

Exercises and Problems

Exercises

1. A thin glass rod 10 cm long is uniformly charged to +100 nC. What is the force (magnitude and direction) on a small plastic sphere, charged to −10 nC, that is 5 cm from the center of the rod?

2. Two positive charges q are spaced distance s apart on the y-axis.
   a. Find an expression for the electric field at distance x on the axis that bisects the two charges.
   b. For $q = 1$ nC and $s = 8$ mm, evaluate $\vec{E}$ at $x = 0, 1, 2, 3, 4, 6,$ and 10 mm.
   c. Draw a graph of E versus x for $0 \leq x \leq \infty$.

3. Two spheres 2 cm in diameter are placed with a 2 cm space between them. One sphere is charged to +10 nC, the other to −15 nC. What is the field strength at the midpoint between the two spheres?

4. Two disks 2 cm in diameter are facing each other 1 mm apart. They are charged to ±10 nC.
   a. What is the electric field strength between the disks?
   b. A proton is shot from the negative disk toward the positive disk. What launch speed must it have to just barely reach the positive disk?

5. Example 25-5 calculated the on-axis field of a ring of charge.
   a. Draw a graph of E_z versus z for $-\infty < z < \infty$.
   b. Explain *why* the field should be zero at $z = 0$.
   c. Show that the field becomes that of a point charge Q in the limit $z \gg R$.

6. Two charged rings are facing each other 20 cm apart. Both rings are 10 cm in diameter. The left ring has a −50 nC charge, while the right ring is charged to +50 nC.
   a. What is the electric field $\vec{E}$, both magnitude and direction, at the midpoint between the two rings?
   b. What is the force $\vec{F}$ on a −1 nC charge placed at the midpoint?
   c. Suppose the −50-nC ring is replaced with a +50-nC ring. What will be the field $\vec{E}$ at the midpoint?

Problems

7. Figure 25-22 shows three charges placed at the corners of a rectangle.
   a. What is the electric field at point A? Give your answer both in component form *and* as a magnitude and direction.
   b. What is the force on a proton that is placed at point A?

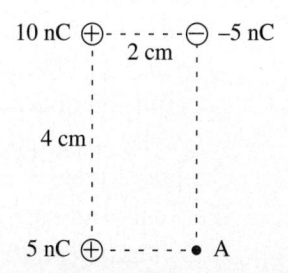

FIGURE 25-22

8. Figure 25-23 shows three charges placed at the corners of a rectangle.
   a. What is the electric field at point B? Give your answer both in component form *and* as a magnitude and direction.
   b. What is the force on an electron that is placed at point B?

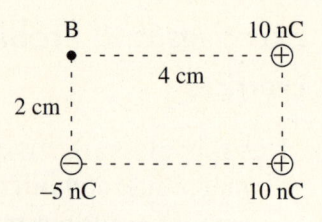

   FIGURE 25-23

9. Three charges are placed along the y-axis. Charges $-q$ are placed at $y = \pm d$ and charge $+2q$ is placed at $y = 0$.
   a. Find the field at a point on the x-axis.
   b. Verify that your answer to part a) has the expected behavior as x becomes very small and very large.
   c. Sketch a graph of E versus x for $0 \leq x \leq \infty$.

10. a. Find a general expression for the field $\vec{E}_{dipole}$ on the bisecting axis of a dipole, as shown in Fig. 25-3.
    b. Show that your result for part a) reduces to Eq. 25-16 in the limit $x \gg s$.

11. a. Find an expression for the electric field $\vec{E}$ a distance d from the left end of a metal rod of length L on which total charge Q has been uniformly distributed (Fig. 25-24).
    b. Evaluate $|\vec{E}|$ at $d = 3$ cm if $L = 5$ cm and $Q = 30$ nC.

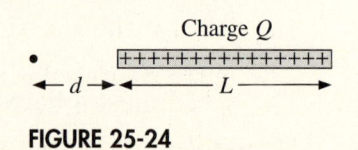

    FIGURE 25-24

12. A thin plastic rod 10 cm long is uniformly charged to $Q = 30$ nC. It is then bent into a semicircle, as shown in Fig. 25-25. What is the electric field $\vec{E}$ at the "center" of the semicircle? (**Hint:** A small piece of arc length Δs spans a small angle $\Delta \theta = \Delta s/R$ where R is the radius.)

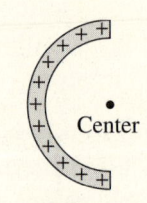

    FIGURE 25-25

13. A very long sheet of charge has width L and surface charge density η. Assume that the sheet lies in the xy-plane between $x = -L/2$ and $x = L/2$.
    a. What is the electric field at distance z above the center of the sheet?
    b. What is the electric field at a point in the xy-plane, but outside the sheet, at distance x from the center line?

14. a. Calculate the on-axis electric field, at distance z, of a uniformly charged circular *disk* having radius R and total charge Q. (**Hint:** Divide the disk into N concentric *rings* of charge, each of width $\Delta r = R/N$. Use geometry and charge density to find the charge Δq_i on ring i. Then use the on-axis field of a ring of charge to write the field of the disk as a sum of the fields of N rings. Finally, let the sum become an integral and integrate.)
    b. Test your result from part a) by showing that it reduces to the field of a point charge Q in the limit $z \gg R$.
    c. Let $R \to \infty$, giving you a plane of charge. Show that you get $E = \eta/2\varepsilon_0$ for a plane of charge. (This is an alternative derivation for the field of a plane of charge.)

15. Two parallel plates 1 cm apart are equally and oppositely charged. An electron is released from rest at the surface of the negative plate, while simultaneously a proton is released from rest at the surface of the positive plate. How far from the negative plate is the point where the electron and proton pass each other?

16. A problem of practical interest is to make a beam of electrons turn a 90° corner. This can be done with a parallel-plate capacitor oriented at 45°, as shown in Fig. 25-26. Consider an electron with kinetic energy 3×10^{-17} J moving vertically upward through a small hole in the bottom plate of the capacitor.
    a. Should the bottom plate be charged positive or negative, relative to the top plate, if you want the electron to turn to the right? Explain.
    b. What strength electric field is needed if the electron is to emerge from an exit hole 1 cm away from the entrance hole, traveling at right angles to its original direction? (**Hint:** The difficulty of this problem depends on how you choose your coordinate system. Can you find a coordinate system that makes this look like a projectile problem?)
    c. If the plates each have an area of 3 cm × 3 cm, what is the charge on the upper plate?
    d. What minimum separation s_{min} must the capacitor plates have?

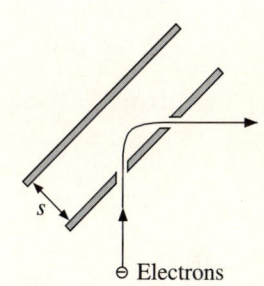

FIGURE 25-26

17. A 2 mm diameter glass sphere has a charge of +1 nC. What speed would an electron need to orbit the sphere 1 mm above the surface?

18. We want to understand just how a charged object can pick up a neutral piece of metal. Consider a small circular disk of aluminum foil lying flat on a table. The foil disk has radius R_F and thickness t. A glass ball of radius R_B is held at height h above the foil, as shown in Fig. 25-27a. The ball has been given a positive charge Q_B. Assume, to make the calculation easier, that $R_F \ll R_B$, $R_F \ll h$, and $t \ll R_F$. These assumptions imply that the electric field of the ball is very nearly constant over the volume of the foil disk.
    a. What is the ball's electric field $\vec{E}_{ball}$ (both magnitude and direction) at the position of the foil? Your answer will be an expression involving Q_B and h.
    b. As you learned in Chapter 24, the foil becomes polarized with + and − surfaces. The foil surfaces, with charges $+Q_F$ and $-Q_F$, act as the plates of a parallel-plate capacitor of separation t (Fig. 25-27b). Because the foil is a conductor in electrostatic equilibrium, the electric field $\vec{E}_{in}$ inside the foil *must* be zero. But a zero field seems inconsistent with the claim that the polarized foil is a parallel-plate capacitor, for which we know the interior field to be of magnitude η/ε_0, where η is the surface charge density. How are

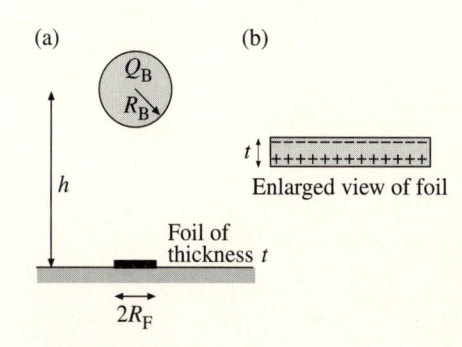

FIGURE 25-27

these two ideas to be reconciled? Use words and pictures to explain how $\vec{E}_{in} = 0$ even though the surfaces of the foil are charged. (**Hint:** Superposition!)

c. Now write the condition that $\vec{E}_{in} = 0$ as a mathematical statement and use it to find an expression for the charge Q_F on the upper surface of the foil. Your expression should be in terms of the quantities Q_B, R_F, and h.

d. Write an explanation for how the charged ball can lift the neutral foil. (**Hint:** The charged surfaces of the foil look very much like a dipole from the perspective of the ball.)

e. Suppose $R_B = 5$ mm, $Q_B = 100$ nC, $R_F = 1$ mm, and $t = 0.01$ mm. The density of aluminum is $\rho = 2700$ kg/m^3. How close must the ball be to lift the foil?

f. What is the foil's induced charge Q_F for the value of h you found in part e)?

[**Estimated 10 additional problems for the final edition.**]

Chapter 26

Current and Conductivity

LOOKING BACK Sections 24.3–24.5; 25.7, 25.9

26.1 Current and Charge Carriers

Devices that run on electricity, such as lights, stereos, and computers, are an important part of our lives. Such devices are connected, by wires, to a battery or an electrical outlet. What is happening *inside* the wire that makes the light come on or the stereo play? And *why* is it happening? We say that "electricity flows through the wire," but what does that statement mean? And more importantly, *how do we know*? If you simply *look* at a wire between a battery and a light bulb, you do not see anything at all that is moving or flowing. As far as visual appearance is concerned, the wire is absolutely the same whether it is "conducting electricity" or not.

When we say that "current flows through a wire":

- What does this statement actually say?
- How do we *know* that anything is really moving?
- How do we know *what* it is that moves?
- What keeps it (whatever it is) moving?
- What is the connection between "electric current" and the charging and electrostatic processes we have studied in the last two chapters?

These are the basic questions with which this chapter will be concerned. Our goal is to learn *how* and *why* charge moves through matter as what we call a *current*.

In Chapter 24 we defined a current as the motion of charges through a conductor. The charges that physically move are called the **charge carriers**. Different conducting media have different charge carriers: Solutions have both positive and negative ions, ionized gases have both positive ions and electrons, and semiconductors have negative electrons and positive "holes." Our emphasis in this chapter will be on metals, such as copper wires, for which the charge carriers are electrons. But it is important to keep in mind that electron motion in metals is only one of many important types of conduction.

237

Suppose that we have a charged capacitor. We can verify that the plates are charged—one positively and the other negatively—by holding them, one at a time, near hanging charged glass and plastic rods. However, if we now take a metal wire and connect the two capacitor plates together, as shown in Fig. 26-1a, both capacitor plates become neutral. "Something" has allowed the excess charge on one capacitor plate to move back to the plate from which it was originally removed, neutralizing both plates. In other words, the capacitor has been *discharged*.

This phenomenon is easy to understand. The excess charges on each plate repel each other, but they are unable to move. The wire, however, provides a path. Charges can move through a conductor, so the repulsive forces between like charges push some of the charges through the wire. The capacitor becomes discharged as excess charge moves from one capacitor plate *through the wire* to the other plate. This process is the essentially same as charging an electroscope, where charge moves from the top of the electroscope through the metal post to the leaves.

If the capacitor has sufficient charge, the wire that connects them is found to be hot after the discharge. If this wire includes a light bulb, the bulb glows briefly when the connection is first made but then fades quickly. If the wire passes over a compass lying flat on the table, the compass needle deflects when the connection to the capacitor is first made but then returns to its initial position. The generation of heat, the lighting of bulbs, and the deflection of compass needles are all indicators of a *current* of charge moving through the wire.

The current that discharges a capacitor quickly "runs down" and is not sustained. Suppose that we take the same wire and connect it to the two ends of a battery, as shown in Fig. 26-1b. After doing so, we find that the wire *continues* to generate heat, a light bulb *continues* to light, and a compass needle *continues* to be deflected. In other words, there must be a *sustained* current in the wire—that is, a sustained motion of charges—when it is connected to the two ends of a battery. This is an important observation. It is evidence that the current supplied by a battery is identical with the charge motion that we observed previously in charging and discharging processes. The only difference is that the battery can keep the motion going, whereas the plastic rod and capacitor plates quickly run out of charge. Thus we should be able to understand how currents move through wires and how circuits "work" on the basis of Coulomb's law and the electric field. You should think carefully about this because it is not obvious that batteries and circuits have anything at all to do with rubbed plastic rods or charged disks.

If we discharge a capacitor, or hook a wire to the ends of a battery, does positive charge move toward the negative side or does negative charge move toward the positive side? *Either* motion would explain the observations we have made. Figure 26-2 shows that the capacitor can be discharged if positive charge moves to the right *or* if negative charge moves to the left. So which charges are the charge carriers? In Chapter 24 we simply *asserted* that it is the negative electrons that move, but we offered no evidence. In fact, it is not easy to determine the charge carriers in metals.

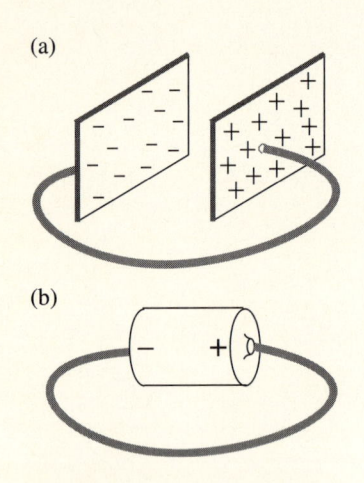

FIGURE 26-1 a) A current flows through a wire connecting two capacitor plates. b) A battery sustains the current in the wire.

One of the first clues was found by J. J. Thompson, the discoverer of the electron. In the late 1890s, Thompson discovered that metals heated until they are "white hot" emit electrons. (Today this *thermal emission* from hot tungsten filaments is how electrons are generated in electron guns.) This result suggested that the electrons are somehow "loose" in metals and can escape if they have sufficient thermal energy.

However, the first direct evidence the electrons are the charge carriers in metals did not appear until the *Tolman-Stewart experiment* of 1916. Tolman and Stewart caused a metallic rod to undergo a very sudden and very large acceleration, as illustrated in Fig. 26-3. If the charge carriers within the metal are free to move about, in a quasi-liquid fashion, their inertia (and Newton's first law) will cause them to be "thrown" to the back end of the metal (although we know that it is really the rigid part of the metal trying to leave the charge carriers behind as it accelerates away). If the charge carriers are positive, their displacement will cause the rear surface of the metal to become positively charged and leave the front surface negatively charged, much as if the metal were polarized by an electric field. Negative charge carriers are just the opposite. They will be "thrown" to the back, giving the rear surface a negative charge while leaving the front surface positive. The Tolman-Stewart experiment showed that the rear surface of a metallic rod becomes negatively charged as it accelerates. Thus the charge carriers in a metal are negative, and the only negatively charged particles are electrons. Later experiments with magnetic fields confirmed the Tolman-Stewart experiment: The charge carriers in metals are electrons.

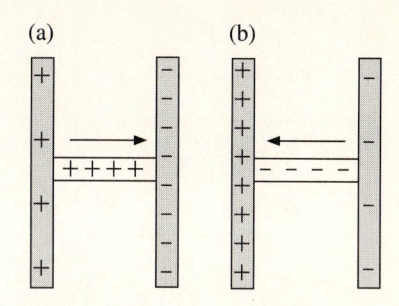

FIGURE 26-2 The capacitor is discharged if positive charge moves right or if negative charge moves left.

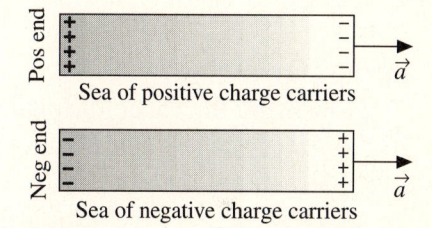

FIGURE 26-3 The Tolman-Stewart experiment to determine the sign of the charge carriers. Inertia causes the mobile charges to accumulate at the rear surface as the metal rod accelerates, causing a measurable distinction between positive and negative charge carriers.

26.2 The Electron Current

We now have evidence that the charge carriers in a metal are electrons. But what *sustains* the motion of electrons when a current flows in a circuit? What pushes them forward? Why don't they stop after a very brief interval, such as when you charge an electroscope by touching it with a negative rod?

Consider a conducting wire placed between the plates of a charged capacitor but not touching, as shown in Fig. 26-4a. The capacitor's electric field polarizes the wire, causing an excess of electrons on the left end, near the capacitor's positive plate, and a deficit of electrons on the right end next to the capacitor's negative plate. The electrons at the left end would "like" to leap over to the positive capacitor plate, where there is a deficit of electrons, but the restoring force from the positive ions in the wire prevents them from doing so.

240 CHAPTER 26 CURRENT AND CONDUCTIVITY

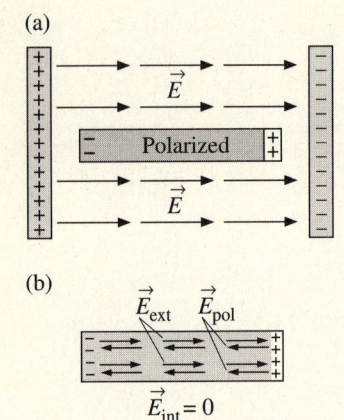

FIGURE 26-4 a) A wire is polarized in the field of a capacitor. b) The electric field inside the wire is zero.

The positive and negative surface charges at the ends of the polarized wire create an electric field $\vec{E}_{pol}$ inside the wire. The surface charges, in effect, act like the positive and negative plates of a capacitor. But the *net* field inside the wire has to be *zero* because the wire is in electrostatic equilibrium. If there were a field inside, it would exert a force on the electrons and cause them to move. But they aren't moving, so there cannot be a net field.

We can reconcile these two seemingly contradictory ideas by realizing that each point inside the wire experiences a *superposition* of the polarization field $\vec{E}_{pol}$ and the external field $\vec{E}_{ext}$ of the parallel-plate capacitor. In fact, the external field causes the wire to polarize until $\vec{E}_{pol}$ exactly balances $\vec{E}_{ext}$. After that, the wire is in electrostatic equilibrium and the external field of the capacitor has no more influence inside the wire. So in electrostatic equilibrium, the internal field $\vec{E}_{int}$ is

$$\vec{E}_{int} = \vec{E}_{ext} + \vec{E}_{pol} = 0. \qquad (26\text{-}1)$$

and no current flows

Now bring the capacitor plates forward to touch the ends of the wire, as shown in Fig. 26-5a. As contact is made, two things happen simultaneously. First, the excess electrons on the left end of the wire move onto the positive capacitor plate. Second, excess electrons from the negative capacitor plate are attracted to the positive end of the wire and move onto the wire's right end. The wire remains neutral, because the number of electrons leaving the left end is balanced by the number entering the right end, but the wire is *no longer polarized*.

The external electric field of the capacitor is still present in the wire, but the polarization field $\vec{E}_{pol}$ vanishes. Thus there is a *nonzero electric field* $\vec{E}_{int} = \vec{E}_{ext}$ inside the wire, as shown in Fig. 26-5b. This internal field exerts forces on the electrons in the wire and pushes the charge carriers through the wire as the capacitor discharges. Because the electrons are negative, they flow from right to left, opposite the field direction.

A conductor in electrostatic equilibrium has $\vec{E}_{int} = 0$. But a current-carrying wire is *not* in electrostatic equilibrium. The presence of an electric field *inside* the wire causes the electrons to move and a current to flow. As a result, the capacitor plates are discharged as charge is transferred *through* the wire from one capacitor plate to the other. No charges are created or destroyed; they are simply moved about under the influence of the electric field. Although we illustrated the ideas here with a capacitor, the physics is exactly the same with a battery that can sustain the current flow.

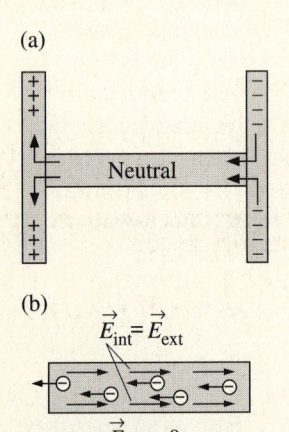

FIGURE 26-5 a) After contact is made, electrons can move between the capacitor plates and the wire. b) An electric field inside the wire causes the motion of the electrons through the wire.

A Model of Conduction

The basic question at the beginning of this section was, "What *sustains* the motion of electrons when a current flows in a circuit?" The answer, it appears, is that an electric field inside the conductor sustains the current. But that can't be the entire story. Why don't the electrons move faster and faster as they accelerate in the electric field? Why, once the electrons are moving, don't they obey Newton's first law and continue to move *without* a field?

The difficulty is that the electrons in a metal are not free particles. Although electrons can move through a conductor, they frequently collide with the positive ions in the lattice. These collisions provide a drag force, much like friction. A block moves in a straight line at constant speed *if* there are no forces on it, but you have to *push* the block for it to continue moving at constant speed along a surface with friction. Likewise, the electric field has to *continue* to push the electrons against the "friction" caused by collisions. We will later call this atomic-level drag force the *resistance* of the wire.

Current is a macroscopic quantity, but the process by which current flows is the microscopic motion of electrons through a forest of positive ions. The role of electrons in conduction is, in many ways, analogous to that of atoms in gases. Although we could characterize gases by their macroscopic parameters of temperature and pressure, we needed a microscopic atomic perspective to understand what temperature and pressure really are. The micro/macro link was the kinetic theory of gases. We can better understand conduction with a similar "kinetic theory" of current flow.

We noted earlier that each of the atoms in a conductor donates a valence electron to the sea of electrons. These *conduction electrons* act rather like a gas. We will treat the electrons as particles moving through the lattice structure of the metal under the influence of an electric field. The electrons, like the atoms in a gas, are moving randomly in all directions with a distribution of speeds. If we assume that the average thermal energy of the electrons is given by the same $\frac{3}{2}kT$ that applies to an ideal gas, then we can find that the average electron speed is $\approx 10^5$ m/s. This estimate turns out, for quantum physics reasons, to be not quite right, but it correctly indicates that the conduction electrons are moving very rapidly. However, an individual electron does not travel far before colliding with an ion and being "scattered" to a new direction.

If there is no electric field, then an electron moves in straight lines between collisions. Its *average* velocity is zero, as it bounces back and forth between collisions, and it undergoes no *net* displacement. This is similar to atoms in a container of gas.

If an electric field is present, the electric force causes the electrons to move along parabolic trajectories between collisions, as shown in Fig. 26-6. Because of the curvature, the negatively charged electrons begin to slowly "drift" in the direction opposite the field. The motion is similar to a ball moving in a pinball machine that has a slight downward tilt. An individual electron ricochets back and

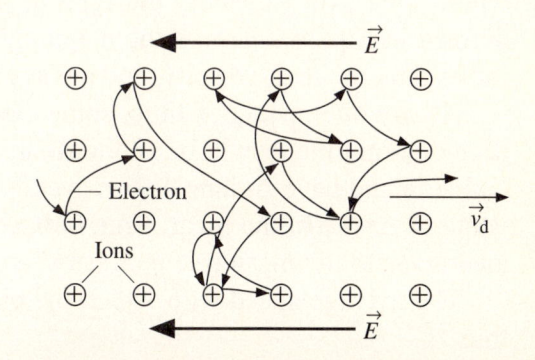

FIGURE 26-6 A microscopic view of a conduction electron moving through a metal. The drift velocity, opposite to $\vec{E}$, is superimposed upon the random collisions.

forth between the ions at a high rate of speed, but now there is a slow *net* motion in the "downhill" direction. But keep in mind that this net displacement is a very small effect superimposed on top of the much larger thermal motion.

Between collisions, the acceleration of each electron is

$$a = |\vec{a}| = \frac{|\vec{F}|}{m} = \frac{|q\vec{E}|}{m} = \frac{eE}{m}, \qquad (26\text{-}2)$$

where $E = |\vec{E}|$ is the field strength inside the wire. Because of the field, the electron's velocity increases linearly with time, $v = at$, until the electron collides with an ion. We will assume, as part of our model, that any velocity gained during the acceleration phase is lost upon collision with an ion, transferring the electron's kinetic energy to the ion and thus to the thermal energy of the metal. The electron then accelerates from $v = 0$ until the next collision. Figure 26-7 shows the field-induced electron velocity as a function of time.

All of the slopes in Fig. 26-7 are equal because the electron accelerates, after each collision, with the constant acceleration of Eq. 26-2. Each collision "resets" the velocity back to zero, so the electron doesn't continue to accelerate as it would in the absence of collisions. Our primary observation from Fig. 26-7 is that the electron has a nonzero *average* velocity for this repeating process of speeding up and colliding. This average velocity, due to the electric field, is called the **drift speed** v_d of the electrons.

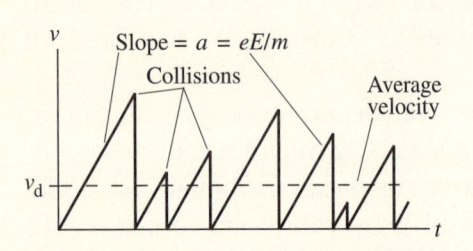

FIGURE 26-7 The electron accelerates between collisions, starting from rest after each collision. This causes the drift speed v_d.

If we observe all the electrons in the metal at one instant of time, we find that electron j has been accelerating for a time interval Δt_j since its last collision and consequently has speed $v_j = a\Delta t_j$. The average speed of all electrons at this instant is

$$v_d = \text{drift speed} = \overline{v_j} = a\overline{\Delta t_j}, \qquad (26\text{-}3)$$

where a bar over a quantity indicates an average value. Notice that the acceleration a was defined in Eq. 26-2 as $|\vec{a}|$, the magnitude of the acceleration. Thus v_d is the drift speed, rather than the drift velocity, and always has a positive value.

At any one instant of time, some electrons will have just recently collided and their acceleration time Δt will be shorter than average. Other electrons will be "overdue" for a collision and have Δt longer than average. When averaged over all electrons, the average value of Δt is just the **mean time between collisions**, which we will designate τ. This is analogous to the mean free path between collisions in the kinetic theory of gases.

The average speed with which the electrons are pushed along by the electric field is thus

$$v_d = a\tau = \frac{e\tau E}{m}, \qquad (26\text{-}4)$$

where the acceleration a from Eq. 26-2 was used. The major conclusion of this model of conduction is that electrons move through a wire, thus forming a current, with a drift speed proportional to the electric field strength. The field is needed to keep the electrons moving against the resistance caused by collisions with the positive ions in the metal's lattice.

EXAMPLE 26-1 Electrons in room-temperature copper have a mean time between collisions of 2.5×10^{-14} s. a) What is the rms velocity of an electron in copper if the electrons are treated as particles in a classical gas? b) How far, on average, do electrons move between collisions? c) What is the electron drift speed in copper for the "typical" electric field strength of 0.1 N/C?

SOLUTION a) The rms velocity of a particle was found in Chapter 22 to be

$$v_{\text{rms}} = \sqrt{\frac{3kT}{m}} = \sqrt{\frac{3(1.38 \times 10^{-23} \text{ J/K})(293 \text{ K})}{9.11 \times 10^{-31} \text{ kg}}} = 1.15 \times 10^5 \text{ m/s}.$$

b) During a time interval $\Delta t = \tau$, an electron moves

$$\Delta x = v_{\text{rms}} \tau = 2.9 \times 10^{-9} \text{ m} = 2.9 \text{ nm}.$$

c) The electron drift speed is found from Eq. 26-4:

$$v_d = \frac{e\tau E}{m} = \frac{(1.60 \times 10^{-19} \text{ C})(2.5 \times 10^{-14} \text{ s})(0.1 \text{ N/C})}{9.11 \times 10^{-31} \text{ kg}} = 4.4 \times 10^{-4} \text{ m/s}.$$

Example 26-1 has several interesting implications. First, electrons only move very small distances between collisions—roughly 10 times the spacing between atoms in the solid. Second, the drift speed is very small in comparison to the speed of the electron's thermal motion. Third, the drift speed is *extremely slow*. It would take an electron nearly 40 minutes to move 1 m at this speed!

The Electron Current

The *rate* at which electrons flow through a wire is called the **electron current**. The electron current is measured in units of "electrons per second." Consider a wire of cross-sectional area A, as shown in Fig. 26-8. If ΔN electrons move past the observation point during a time interval Δt, then the electron current in the wire is

$$\text{electron current} = \frac{\Delta N}{\Delta t}. \quad (26\text{-}5)$$

The electrons are moving through the wire with average speed v_d, so they all move distance $\Delta L = v_d \Delta t$ to the right during interval Δt. Thus the number of electrons that will pass the observation point during interval Δt is just the number of conduction electrons in length ΔL of the wire.

This section of wire is a cylinder with volume $V = A\Delta L$. If the number density of conduction electrons is n electrons per cubic meter, then the total

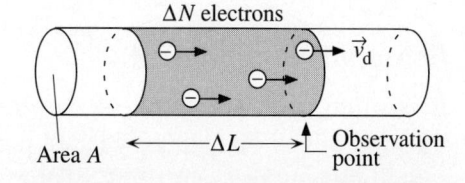

FIGURE 26-8 The ΔN electrons in the segment of wire of length ΔL move forward past during time interval Δt, causing an electron current.

number of electrons in the cylinder is
$$N = nV = nA\Delta L. \tag{26-6}$$
Using Eq. 26-6 in Eq. 26-5, the electron current in the wire is
$$\text{electron current} = \frac{\Delta N}{\Delta t} = \frac{nA\Delta L}{\Delta t} = \frac{nAv_d \Delta t}{\Delta t} = nAv_d. \tag{26-7}$$

It makes sense that the electron current should depend on the three quantities n, A, and v_d. We can increase the flow of electrons if we increase their density, if we increase the size of the "pipe" through which they flow, or if we increase the speed with which they flow. If we use Eq. 26-4 for the drift speed v_d, then we can write the electron current directly in terms of the electric field that causes the current:

$$\text{electron current} = \frac{\Delta N}{\Delta t} = \frac{ne\tau}{m} AE. \tag{26-8}$$

Table 26-1 gives the measured values of the conduction electron density n for several common metals. You can see that there is not a great deal of variation from one metal to another. This reflects the fact that atomic spacings do not vary greatly from one element to another.

TABLE 26–1 Conduction electron densities in some typical metals.

Metal	Density (m^{-3})
Aluminum	6.0×10^{28}
Copper	8.5×10^{28}
Iron	8.5×10^{28}
Gold	5.9×10^{28}
Silver	5.8×10^{28}

• **EXAMPLE 26-2** What is the electron current in a 2 mm diameter copper wire with an internal electric field of 0.1 N/C?

SOLUTION Copper has an electron density $n = 8.5 \times 10^{28}$ m^{-3}, and a 2 mm diameter wire has a cross-sectional area $A = \pi r^2 = 3.14 \times 10^{-6}$ m^2. Example 26-1 found the drift speed in copper for this field strength to be 4.4×10^{-4} m/s. Thus the electron current is

$$\frac{\Delta N}{\Delta t} = nAv_d = 1.2 \times 10^{20} \text{ electrons/s}.$$

Note that this is *a lot* of electrons going past each second!
•

Example 26-2 illustrates an important point. A wire can carry a very large current not because the electrons are moving through the wire quickly (they're hardly moving at all!) but because the wire contains an *enormous* number of electrons.

Conservation of Current

It is often useful to think of current flow as analogous to the flow of water in rivers or through pipes. The analogy is not perfect, but it provides a good mental image of how currents behave. You can think of the electron current, the number of electrons per second flowing past a point, as being the equivalent of a fluid current of some number of water molecules per second passing through a pipe.

Consider a wire, perhaps with some bends like the one shown in Fig. 26-9a, through which a current flows. How does the electron current flowing into one end of this wire compare with the electron current flowing out of the other end? Suppose the wire were, instead, a water hose. Because we cannot create or destroy water molecules, and assuming that there are no holes in the hose, every water molecule entering the hose at one end must leave at the other end. That part is obvious. It is a bit more subtle to realize that the *rate* at which water molecules enter one end must be matched by the *rate* at which they leave the other.

Suppose this was not true. If 100 molecules per second enter one end of the hose but only 80 molecules per second leave the other, then 20 molecules are getting "lost" within the hose each second. But molecules cannot be destroyed, and the hose has no means of "storing" molecules on a continuing basis. It is just not physically possible to lose 20 molecules per second, so our assumption of a difference in rates must be false. It is *not possible* for the rate at which molecules enter one end of a pipe to differ from the rate at which they emerge on the other end.

This same reasoning applies to electrons in the wire. Charge is conserved; we cannot create or destroy electrons. Neither can the wire "store" electrons. So the rate at which charge enters one end of the wire must be identical with the rate at which it leaves the other end. But the rate of electrons entering or leaving is simply the electron current, so it must be that the electron current entering one end equals the electron current leaving the other end. In other words, electron current is *conserved* as it flows through a wire.

Suppose, however, that the wire passes through the light bulb of Fig. 26-9b, causing it to glow. Now what happens when we compare the input electron current to the output electron current? Your initial reaction may be that the output current will be less than the input because the light bulb is using up electricity. But return to a fluid analogy. Consider a river on either side of a hydroelectric dam. The water is forced to spin the giant turbines of the hydroelectric plant, generating electricity, before it can get by. Is the river flow downstream different than the river flow upstream? No! If it were less, either water would have to be destroyed (impossible) or the water level behind the dam would have to keep getting higher and higher and higher. In a steady-state situation such as this, the water level is constant and the flow rate out of the dam is exactly the same as the flow rate into the dam.

Likewise with the wire. Because electrons cannot be destroyed, and because the wire "upstream" from the bulb is not getting increasingly negative, it *must* be that the electron current "downstream" is *exactly the same* as the electron current "upstream" from the bulb! This is, again, conservation of current. What, then, is being "used" by the bulb? That is a fair question. It is not the electrons themselves that are being "used up." Instead, *energy* is being transferred from the electrons to the light bulb filament, but *without* altering the flow of the electrons. This is an issue we will examine carefully later.

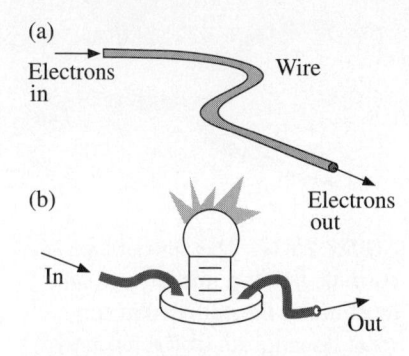

FIGURE 26-9 a) The current flowing out of a wire is the same as the current flowing in. b) This remains true even if the current lights a bulb.

246 CHAPTER 26 CURRENT AND CONDUCTIVITY

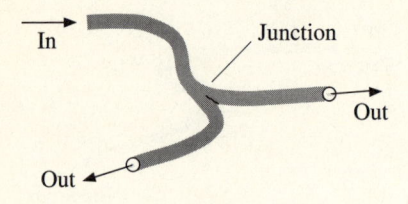

FIGURE 26-10 The sum of the currents flowing into a junction must equal the sum of the currents flowing out of the junction.

Figure 26-10 shows a more complicated situation in which one wire splits into two, with current flowing out of *both* ends. But our reasoning still holds. We cannot create or destroy electrons in the wire, and neither can we store them up. The rate at which electrons flow into one *or many* wires must be exactly balanced by the rate at which they flow out of others. When we examine a *junction* such as the one of Fig. 26-10, conservation of charge requires that

$$\Sigma(\text{input electron currents}) = \Sigma(\text{output electron currents}),$$

where, as usual, the Σ symbol means "summation." This basic conservation statement, that the sum of the currents flowing into a junction equals the sum of the currents flowing out, is called **Kirchhoff's junction law**. The junction law, which plays an important role in circuit analysis, is a very basic consequence of the conservation of charge and the conservation of current.

26.3 The Field Inside the Wire

As we have seen, an electron current in a wire requires the presence of an electric field *inside* the wire to push the electrons along. It is reasonable to guess that a constant electron current through the wire requires an electric field of constant magnitude, although the field's direction will change as the wire bends. Figure 26-11 illustrates the situation. (Notice that the electron flow is *opposite* the field direction because the electron charge is negative.) Where does this internal electric field come from? What causes it?

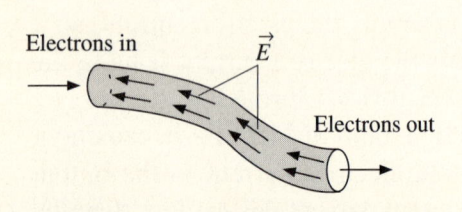

FIGURE 26-11 The electric field inside a wire is of constant magnitude and follows the shape of the wire.

Consider two wires attached to a battery, as shown in Fig. 26-12a. The two free, or unconnected, ends of the wires are close but *not* touching. Both wires are in electrostatic equilibrium, so the electric field inside the wires is zero and no current is flowing. The wires *do*, however, have a surface charge. They are, in effect, extensions of the ends of the battery, so the battery's charge spreads out as far as possible due to the repulsive electrostatic forces between the charges.

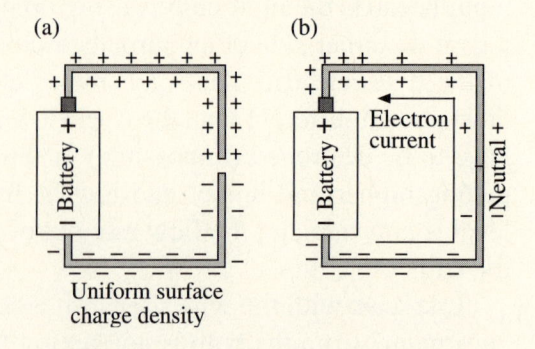

FIGURE 26-12 a) Before the circuit is completed, the wires have a uniform surface charge and $\vec{E}_{\text{int}} = 0$. b) After the wires make contact, the surface charge varies smoothly from positive to negative and creates the internal field $\vec{E}_{\text{int}}$.

Now let's connect the ends of the wires together, as in Fig. 26-12b. What happens? Within a *very* brief interval of time (~10^{-9} s), the surface charge rearranges itself into a new pattern that varies smoothly from positive at the battery's positive end through zero (neutral) at the midpoint and then to negative at the negative end of the battery. This new distribution of surface charge is shown in Fig. 26-12b.

Although this new surface charge distribution may seem fairly obvious for a wire connected to the positive and negative ends of a battery, it has an *extremely* important consequence. Figure 26-13 shows a section from a wire on which the surface charge density becomes more positive toward the left and more negative toward the right. Three circular "rings" of charge are illustrated. They act like the ring of charge that we analyzed in the last chapter. There we found that the on-axis field of a positive ring of charge points away from the ring and gradually weakens with distance.

Looking inside the wire in region 1 of Fig. 26-13, we see a field pointing to the right from charge ring A and a field pointing to the *left* from charge ring B. Ring A is more highly charged than ring B, so these fields do *not* cancel. There is a *net* field pointing to the right in region 1. The situation is similar in region 2. There we find fields in opposite directions due to charge ring B and charge ring C, but ring B is more highly charged than ring C, so the net field is to the right. Although a complete analysis would have to add (integrate) the fields due to all the rings along the edge of the wire, this three-ring model gives us the basic idea of what happens: The *nonuniform* surface charges on the wire create a net *internal* electric field inside the wire that points from the more positive end of the wire toward the more negative end of the wire. This is the internal field $\vec{E}_{\text{int}}$ that pushes the electron current through the wire.

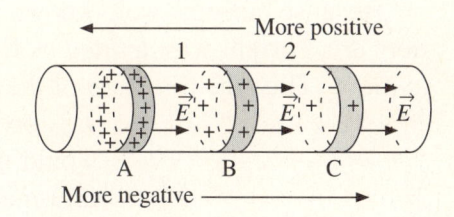

FIGURE 26-13 The varying surface charge distribution of Fig. 26-12b creates an internal electric field $\vec{E}_{\text{int}}$ inside the wire.

Why didn't this happen before the wires made contact? Before contact, the wires on either side have a *uniform* surface charge distribution. In that case, the electric fields from the different rings *are* of equal magnitude and cancel, giving $\vec{E}_{\text{int}} = 0$. It is the *rearrangement* of the surface charge after the wire becomes a continuous path between the two sides of the battery, or of a capacitor, that creates the internal electric field that, in turn, drives the current through the wire. Think about this carefully. It is important to understand that: a) there is an internal electric field in a current-carrying wire, b) this field arises from a nonuniform distribution of surface charge on the wire, and c) this field exerts a force on the charge carriers and causes a current to flow.

26.4 Current and Current Density

Many of the properties of currents were known long before the discovery that electrons are the charge carriers in metals. Because the coulomb is the unit of charge, and because currents are charges in motion, it seemed quite natural in the nineteenth century to define **current** as the *rate*, in coulombs per second, at which charge moves through a wire.

If an amount of charge ΔQ flows past an observation point on a wire during time interval Δt, we can define the current $\vec{I}$ in the wire to be

$$\vec{I} = \left(\frac{|\Delta Q|}{\Delta t}, \text{ in direction of } \vec{E}_{\text{int}} \right). \tag{26-9}$$

Notice that, strictly speaking, current is a vector. It has both a magnitude and a direction, and we will draw current arrows on diagrams to indicate the direction. For calculations, however, we are usually interested only in the magnitude $I = |\vec{I}|$.

Because currents were known and studied before it was known what the charge carriers are, current was *defined* as flowing in the direction in which positive charges *seem* to move. Thus the direction of the current is the same as that of the internal electric field $\vec{E}_{\text{int}}$ because that is the direction that a positive charge carrier would move. If the charge carriers are negative, as in the case of a metal, then the charge carrier's actual motion is *opposite* to $\vec{I}$.

If this choice of direction seems arbitrary, it is! On the other hand, it makes no real difference. Figure 26-14 shows a capacitor being discharged by a wire. There is an electric field from left to right, and its direction is unambiguous. The capacitor will be discharged regardless of whether positive charges are pushed to the right by $\vec{E}_{\text{int}}$ or whether, instead, negative charges are pushed to the left. A *macroscopic* analysis will simply conclude that a current flows from the positive side to the negative side.

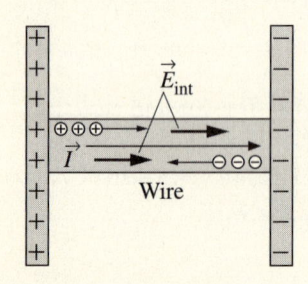

FIGURE 26-14 The current direction is left to right whether the charge carriers are positive or negative.

It is important to distinguish the macroscopic current I from the electron current $\Delta N/\Delta t$. Current I is the measurable quantity, but we can relate I to $\Delta N/\Delta t$. Because the *quantity* of charge ΔQ is related to the *number* of charge carriers ΔN by $\Delta Q = e\Delta N$, we have

$$I = \frac{|\Delta Q|}{\Delta t} = e \cdot \frac{\Delta N}{\Delta t} = e \cdot (nAv_d) = (nev_d) \cdot A, \tag{26-10}$$

where $\Delta N/\Delta t$ is the electron current, which we know from Eq. 26-7 to be nAv_d.

The SI unit for current is the coulomb per second, which is called the **ampere** A:

1 ampere = 1 A = 1 coulomb per second = 1 C/s.

The current unit is named after the French scientist André Marie Ampère, who made major contributions to the study of electricity and magnetism in the early nineteenth century. The unit of an *amp* is an informal abbreviation of ampere. Household currents are typically ~1 A. A 100 watt light bulb, for example, "draws" a current of 0.85 A and an electric hair dryer's current is ≈10 A. Currents in consumer electronics, such as stereos and computers, are much less. They are typically measured in milliamps (1 mA = 10^{-3} A) or even microamps (1 μA = 10^{-6} A).

EXAMPLE 26-3 If the electron flow through the wire shown in Fig. 26-14 is 1×10^{19} electrons per second, what is the current?

SOLUTION The quantity given is the electron current $\Delta N/\Delta t$. Each electron has charge $|q| = e$, so the current is

$$I = e \cdot \frac{\Delta N}{\Delta t} = (1.60 \times 10^{-19} \text{ C}) \cdot (1 \times 10^{19} \text{ s}^{-1}) = 1.6 \text{ C/s} = 1.6 \text{ A}.$$

As you can see in Eq. 26-10, the current in a wire of cross-sectional area A is given by the product of the geometric quantity A with the quantity nev_d. Note that nev_d depends on the charge carriers and on the internal electric field that determines the drift speed, whereas A is simply a physical dimension of the wire. It will be useful to separate these quantities by defining the **current density** J in a wire as the current per square meter of cross section:

$$J = \text{current density} = \frac{I}{A}. \tag{26-11}$$

The current density has units of A/m². From Eq. 26-10 we see that

$$J = nev_d. \tag{26-12}$$

As with any density, J is independent of any particular piece of conductor. The current density describes how charge flows through *any* piece of a particular kind of metal in response to an electric field. A *specific* piece of metal, shaped into a wire with cross-sectional area A, then has the specific current

$$I = JA. \tag{26-13}$$

EXAMPLE 26-4 A 2 A current flows through a 1 mm diameter aluminum wire. a) What is the current density? b) What is the electron's drift speed through the wire?

SOLUTION a) The current density is

$$J = \frac{I}{A} = \frac{I}{\pi r^2} = \frac{2 \text{ A}}{\pi (0.0005 \text{ m})^2} = 2.5 \times 10^6 \text{ A/m}^2.$$

This is a typical current density in a wire of typical size.

b) The drift speed is given in terms of the current density in Eq. 26-13. The conduction electron density for aluminum is found from Table 26-1. Thus

$$v_d = \frac{J}{ne} = 2.6 \times 10^{-4} \text{ m/s} = 0.26 \text{ mm/s}.$$

Wires can carry large currents not because the electrons move especially fast (as we see, they move excruciatingly slowly) but because the electron density n is so high. This, however, raises a puzzle. If the electrons are moving so slowly, why is it that the light overhead, many meters away, seems to come on instantly when you turn on the light switch? You can think about this as a homework question.

26.5 Conductivity and Resistivity

We know, from our microscopic model of conduction, how to express the drift speed v_d in terms of the microscopic collision parameter τ. Using Eq. 26-4 for v_d, we can write the current density as

$$J = nev_d = ne\left(\frac{e\tau E}{m}\right) = \frac{ne^2\tau}{m} \cdot E. \tag{26-14}$$

The quantity $ne^2\tau/m$ in Eq. 26-14 depends *only* on the conducting material. It contains the electron density n and the mean time between collisions τ. Materials with larger values of these numbers will conduct current better than materials with smaller values. That is, a given electric field will generate a larger current density in the material that has the larger values of n and τ. It makes sense, then, to define the **conductivity** σ of a material as

$$\sigma = \text{conductivity} = \frac{ne^2\tau}{m}. \tag{26-15}$$

With this definition, Eq. 26-14 becomes

$$J = \sigma E. \tag{26-16}$$

This is a result of fundamental importance. Equation 26-16 tells us three things:

1. Current is caused by an electric field exerting forces on the charge carriers.
2. The current density, and hence the current $I = JA$, depends linearly on the strength of the electric field.
3. The current density also depends on a material-dependent quantity called the *conductivity*. Different conducting materials have different conductivities because they have different values of the electron density n and, especially, different values of the mean time between electron collisions with the lattice of ions.

The value of the conductivity is affected by the crystalline structure, by any impurities in the metal, and by the temperature. As the temperature increases, so do the thermal vibrations of the lattice atoms. This makes them "bigger targets," causes collisions to be more frequent, lowers τ, and thus decreases the conductivity. Metals conduct better at low temperatures than they do at high temperatures.

Conductivity, like density, characterizes a material as a whole. All pieces of copper have the same value of σ (at a common temperature). But the conductivity of copper is different than from that of aluminum. Various experimental techniques exist for measuring the conductivities of metals. The mean time between collisions τ can then be inferred from the measured conductivity.

Although the conductivity is used to characterize different conductors, in practical applications we will often find it more convenient to use the inverse of the conductivity, $1/\sigma$. This is called the **resistivity** and has the symbol ρ. Thus

$$\rho = \text{resistivity} = \frac{1}{\sigma} = \frac{m}{ne^2\tau}. \tag{26-17}$$

Note that this is not the same as "resistance," a related quantity that you will meet in Chapter 29.

Table 26-2 gives representative values of the conductivity and resistivity for several metals. You can see that they do vary quite a bit, with copper and silver being the best two conductors. The units of conductivity, from Eq. 26-15, are those of J/E, namely A C/N m^2. These are clearly awkward. In Chapter 29 we will introduce a new unit called the ohm, symbolized by Ω (uppercase Greek "omega"). It will then turn out that resistivity has units of Ω m while conductivity has Ω^{-1} m^{-1}.

TABLE 26–2 Resistivity and conductivity of some typical conducting materials.

Metal	Resistivity (Ωm)	Conductivity (Ω^{-1}m^{-1})
Aluminum	2.8×10^{-8}	3.5×10^{7}
Copper	1.7×10^{-8}	6.0×10^{7}
Gold	2.4×10^{-8}	4.1×10^{7}
Iron	9.7×10^{-8}	1.0×10^{7}
Silver	1.6×10^{-8}	6.2×10^{7}
Tungsten	5.6×10^{-8}	1.8×10^{7}
Nichrome*	1.5×10^{-6}	6.7×10^{5}
Carbon	3.5×10^{-5}	2.9×10^{4}

*Nickel-chromium alloy used for heating wires

EXAMPLE 26-5 What is the mean time between collisions in copper?

SOLUTION The mean time between collisions is given by Eq. 26-15 as

$$\tau = \frac{m\sigma}{ne^2}.$$

The measured electron density is found in Table 26-1 and the measured conductivity is found in Table 26-2. A quick calculation gives $\tau = 2.5 \times 10^{-14}$ s. This is the value that was given in Example 26-1.

EXAMPLE 26-6 A 2 mm diameter aluminum wire carries a current of 300 mA. What is the electric field inside the wire?

SOLUTION The electric field is

$$E = \frac{J}{\sigma} = \frac{I}{\sigma A} = \frac{I}{\sigma \pi r^2} = \frac{0.3 \text{ A}}{(3.5 \times 10^7 \ \Omega^{-1}\text{m}^{-1})\pi(0.001 \text{ m})^2} = 0.0027 \text{ N/C},$$

where the conductivity of aluminum was taken from Table 26-2. This is a *very* small field in comparison with those we calculated in Chapters 24 and 25 for point charges and charged objects. It is, in fact, roughly the size of the electric field 1 mm away from a *single* electron. The lesson to be learned from this example is that it takes *very few* surface charges on a wire to create the internal electric field necessary for the wire to carry considerable current. Just a few excess electrons every centimeter is sufficient. The reason, once again, is the enormous value of the carrier density n. Even through the electric field is very tiny and the drift speed is agonizingly slow, substantial current can flow in the wire due to the vast number of charge carriers able to move.

Superconductivity

In 1911, the Dutch physicists Kammerlingh Onnes was studying the conductivity of metals at very low temperatures. Scientists had just recently discovered how to liquefy helium, and this opened a whole new field of *low-temperature physics*. As we noted earlier, metals become better conductors—higher conductivity, lower resistivity—at lower temperatures. But the effect is gradual. Onnes, however, found that mercury suddenly and dramatically loses *all* resistance to current flow when cooled below a temperature of 4.2 K. This complete loss of resistance at low temperatures is called **superconductivity**.

Later experiments established that the resistivity of a superconducting metal is not just small, it is truly zero. The electrons are moving in a frictionless environment, and current will continue to flow through a superconductor *without an electric field*. Superconductivity was not understood until the 1950s, when it was explained as a specifically quantum effect.

Superconducting wires can carry enormous currents because the wires are not heated by electrons colliding with the ions. Very strong magnet fields can be created with superconducting electromagnets. However, the number of applications remained limited because all known superconductors required temperatures of less than 20 K. This situation changed dramatically in 1986 with the discovery of *high-temperature superconductors*. These ceramic-like materials are superconductors at temperatures as "high" as 125 K. Although −150°C may not seem like a high temperature to you, the technology for producing such temperatures is simple and inexpensive. It is now likely that a wide range of superconductor applications will appear in the coming years.

Summary

Important Concepts and Terms

charge carriers   ampere
drift speed   current density
mean time between collisions   conductivity
electron current   resistivity
Kirchhoff's junction law superconductivity
current

Our focus in the chapter has been on the *mechanisms* by which current flows. We first examined the experimental evidence by which we know that the charge carriers in metals are electrons and that currents are not fundamentally different from electrostatic charging and discharging processes. You should review this line of reasoning. We then developed a microscopic model of conduction in which electrons are "pushed along" by an internal electric field. Their collisions with the lattice atoms cause the electrons to move with a steady drift speed

$$v_d = \frac{e\tau E}{m},$$

where τ is the mean time between collisions. The electric field in the wire is created by surface charges on the wire when it is connected to a battery or other source of charges.

We then defined the electron current

$$\text{electron current} = \frac{\Delta N}{\Delta t} = nAv_d$$

as the rate at which electrons move through a conducting wire of cross-sectional area A. This provided us with a means of measuring the flow of current. The electron current is conserved as it flows through wires or into and out of junctions.

The macroscopic current I is the rate, in coulombs per second, at which charge flows through a wire:

$$\vec{I} = \left(\frac{|\Delta Q|}{\Delta t}, \text{ in direction of } \vec{E}_{int}\right).$$

The current direction is *defined* as the direction in which positive charge carriers would move.

The current density is

$$J = \frac{I}{A} = nev_d = \sigma E,$$

where σ is the conductivity of the material. This is an important result because it ties the macroscopic current I to the field in the wire and to the microscopic drift speed. The current in a wire of cross-sectional area A is then $I = JA$, where I is measured in amperes.

Exercises and Problems

Exercises

1. A 100 watt light bulb draws a current of 0.85 A. The filament inside the bulb is 0.25 mm in diameter.
   a. What is the current density in the filament?
   b. What is the electron current through the filament?

2. A 10 A current flows through a hollow copper wire that has an inner diameter of 1 mm and an outer diameter of 2 mm. What is the current density in the wire?

3. A cube of metal 1 cm on each side is sandwiched between two electrodes. The electrodes create an electric field of 5.0×10^{-3} N/C in the metal. A current of 9.0 A flows through the cube, from the positive electrode to the negative electrode. Identify the metal.

4. A 20 mA current flows through a silver wire that is 0.5 mm in diameter.
   a. What is the electron current in the wire?
   b. What is the electric field in the wire?
   c. What is the electron drift speed in the wire?

5. Figure 26-15 shows a wire connected to a battery. Redraw this figure on a sheet of paper.
   a. Show, by using plusses and minuses, how surface charge is distributed along the wire. Place your symbols closer together in regions where the charge density is higher, farther apart in regions where the charge density is lower.

b. Show the electric field vector $\vec{E}$ at each of the seven points in the wire marked with a dot. The length of each vector should be proportional to $|\vec{E}|$ at that point.
c. Explain the reasoning you used in answering part b).

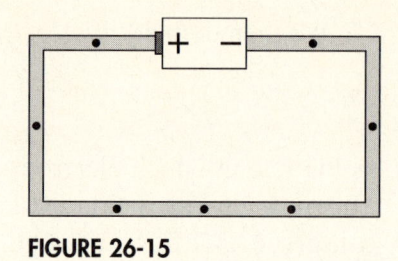

FIGURE 26-15

◣ Problems

6. A television picture tube has an electron beam that is 0.5 mm in diameter and that carries a current of 50 μA. This electron beam impinges on the inside of the picture tube screen.
   a. How many electrons strike the screen each second?
   b. What is the current density of the electron beam?
   c. The electrons move with a constant velocity of 5×10^7 m/s. What electric field strength is needed to accelerate electrons from rest to this velocity in a distance of 5 mm?
   c. Each electron transfers its kinetic energy to the picture tube screen upon impact. What is the *power* delivered to the screen by the electron beam?

7. The starter motor of a car engine draws a current of 100 A from the battery. The copper wire to the motor is 5 mm in diameter and 1.2 m long. The starter motor runs for 0.5 s until the car engine starts.
   a. How much charge passes through the starter motor?
   b. How far does an electron travel down the wire while the starter motor is on?

8. A wire is made of two materials having equal diameters but different conductivities σ_1 and σ_2, as shown in Fig. 26-16. Current I flows through this wire. If the conductivities have the ratio $\sigma_2/\sigma_1 = 2$, what is the ratio E_2/E_1 of the electric field strengths in the two segments of the wire?

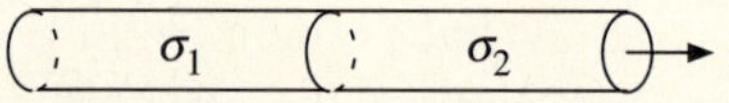

FIGURE 26-16

9. An aluminum wire consists of three segments: a 2 mm diameter top segment, a 1 mm diameter middle segment, and finally a 2 mm diameter bottom segment (Fig. 26-17). A 10 A current flows into the top segment. For each of these three segments, find the
   a. Current I.   d. Drift velocity v_d.
   b. Current density J.   e. Mean time between collisions τ.
   c. Electric field E.   f. Electron current $\Delta N/\Delta t$.

See Tables 26-1 and 26-2 for the necessary data. Place your results in a table for easy viewing.

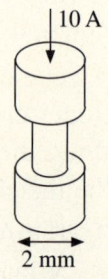

FIGURE 26-17

[**Estimated 10 additional problems for the final edition.**]

Chapter 27

The Electric Potential

LOOKING BACK | Sections 10.1–10.5; 12.5; 24.6–24.7; 25.5–25.8

27.1 An Alternative View of Electrical Interactions

In Part II of this text, on interacting particles and conservation laws, we extended the tangible concepts of force and motion to the more abstract idea of energy. Energy and its conservation gave us a different, but very useful, perspective on motion. Now we are going to extend the tangible concepts of field and current to a new, and more abstract, idea that we will call the *electric potential*. Like energy, the electric potential will provide us with an alternative, and ultimately quite practical, view of electric interactions.

In this chapter and the next we will introduce the electric potential and develop its properties. Then, in Chapter 29, we will see how the potential becomes an essential idea for the understanding of circuits. In fact, the familiar unit of *volts* are the units of the electric potential.

The electric potential is an important concept, but it is also a fairly subtle and abstract concept. We will begin our study by using analogies with the more familiar gravitational force and potential energy. We will also use many examples to illustrate how the electric potential is used in practical situations. The Looking Back sections, especially those from Chapters 10 and 12, are highly recommended for review.

27.2 Electric Potential Energy

There is a close analogy between gravitational forces and electrical forces. The gravitational force between two masses depends inversely on the square of the distance between them, as does the electrical force between two point charges. Similarly, the uniform gravitational field near the earth's surface looks very much like the uniform electric field inside a parallel-plate capacitor. Consequently, it will be useful to briefly review gravitational potential energy before defining the electric potential energy.

You will recall that the mechanical energy $E_{mech} = K + U$ is conserved for particles that interact with each other via conservative forces, where K and U are the kinetic and potential energy. That is,

$$\Delta E_{mech} = \Delta K + \Delta U = 0. \tag{27-1}$$

We need to be careful with notation. Because we are now using E to represent the electric field strength, we will usually represent energy as $K + U$, rather than as E, to avoid confusion.

The potential energy U, you will recall, measures the *interaction energy* of a system of particles. In particular, we defined the *change* in potential energy as

$$\Delta U = U_f - U_i = -W_{\text{interaction force}}(\text{position i} \rightarrow \text{position f})$$
$$= -\int_{s_i}^{s_f} F_s(s)\,ds, \tag{27-2}$$

where W is the work done *on* the particle by force $\vec{F}$ as the particle moves along the s-axis from position i to position f. $F_s(s)$ is the s-component of the force (which may be a function of the position variable s).

Using this definition, we found the gravitational potential energy near the earth to be

$$U_{grav}(y) = U_0 + mgy, \tag{27-3}$$

where U_0 is the value of U at $y = 0$. We often choose $U_0 = 0$, in which case $U_{grav}(y) = mgy$, but such a choice is *not* necessary. The zero point of potential energy is an arbitrary choice, because we have defined ΔU rather than U.

The gravitational *field* $\vec{g}$ near the surface is a *uniform* field, $\vec{g} = (g, \text{down}) = \text{constant}$. A particle of mass m loses potential energy as it moves in the direction of the field, as shown in Fig. 27-1a. This is because the gravitational force is in the same direction as the particle's displacement, causing the gravitational field to do a *positive* amount of work as the particle falls. Thus, because of the minus sign in the Eq. 27-2 definition of ΔU, the potential energy *decreases*. You also know that a particle released from rest falls down, not up, so it must lose potential energy while gaining kinetic energy as it moves in the direction of the field.

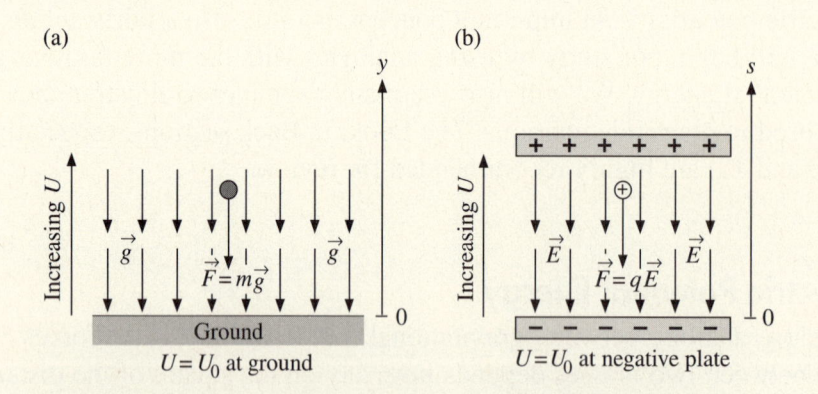

FIGURE 27-1 a) When a mass moves in the direction of the gravitational field, the field does positive work and the gravitational potential energy decreases. b) When a positive charge moves in the direction of the electric field, the field does positive work and the electric potential energy decreases.

The uniform gravitational field near the earth's surface is *exactly analogous* to the uniform electric field between the plates of a parallel-plate capacitor. Let's establish a coordinate axis s that points from the negative plate of the capacitor toward the positive plate, as shown in Fig. 27-1b. We'll let the origin $s = 0$ be at the negative plate. This is a valid coordinate axis no matter what the physical orientation of the capacitor. A *positive* charge q released from rest inside the capacitor will "fall" to the negative plate, exactly like an earthly projectile. Placing $s = 0$ at the negative plate is thus like placing $y = 0$ at the ground for a projectile.

The electric field exerts a *constant* force $F = qE$ on the charge, just as the earth's gravitational field exerts a constant force $F = mg$ on a mass. Thus we can repeat the derivation for the gravitational potential energy, with qE replacing mg. The electric field does work $W = qEs$ on the charge as it moves from position s to the negative plate at $s = 0$. The particle's change in potential energy is

$$\Delta U = U_f - U_i = U_0 - U(s) = -W(s \to 0) = -qEs.$$

So the **electric potential energy** of charge q in a uniform electric field is

$$U_{\text{elec}}(s) = U_0 + qEs \quad \text{(uniform field } E\text{)}, \tag{27-4}$$

where s is the distance from the negative plate and $U_0 = U(s = 0)$ is the potential energy right at the negative plate, $s = 0$. Note that the potential energy of a positive charge *decreases* as it moves in the direction of the field, exactly the same as for a mass moving down. This observation—that the potential energy of a positive charge decreases along the field direction—is one we will use many times.

Although Eq. 27-4 was derived assuming q to be positive, it is valid for either sign of q. A negative charge released from rest moves *away from* the negative plate. As it speeds up, it needs to lose potential energy, $\Delta U < 0$, in order to gain kinetic energy. A negative value for q in Eq. 27-4 causes the potential energy $U(s)$ to become *more negative* as s increases, which is exactly what we need to conserve energy.

EXAMPLE 27-1 A 2 cm × 2 cm capacitor with a 2 mm spacing is charged to ±1 nC. A proton and electron are released from rest at the midpoint of the capacitor, as shown in Fig. 27-2. a) Which way does each move, and why? b) What is each particle's change in potential energy from the release point until it strikes one of the plates? c) What is each particle's impact speed as it strikes one of the plates?

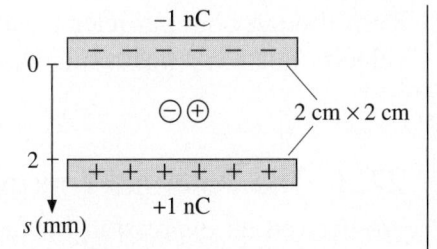

FIGURE 27-2 Charges in the capacitor of Example 27-1.

SOLUTION a) The electric field points upward, from the positive toward the negative plate, so the s-axis points *downward*. A proton in the center will experience an upward force, in the direction of field $\vec{E}$ and away from the positive plate. This is the direction of *decreasing* potential energy for a positive charge, and we know that a particle speeds up (increasing K) in the direction of decreasing U. All of this tells us that the proton will move upward. Similarly, the force on the negative electron is downward, opposite the field, which is also the direction of decreasing U for the electron. The electron will move downward.

b) If d is the distance between the two plates, then both charges start at $s_i = d/2$. The change of potential energy is $\Delta U = U(s_f) - U(s_i)$. The proton has $q = +e$ and ends at $s_f = 0$. Using the potential energy function of Eq. 27-4 gives

$$\Delta U_{proton} = U(s_f) - U(s_i) = (U_0 + 0) - \left(U_0 + eE\frac{d}{2}\right) = -\tfrac{1}{2}eEd.$$

ΔU_{proton} is negative—a loss of potential energy—as expected. The electron has $q = -e$ and ends at $s_f = d$. Thus

$$\Delta U_{electron} = U(s_f) - U(s_i) = (U_0 - eEd) - \left(U_0 - eE\frac{d}{2}\right) = -\tfrac{1}{2}eEd,$$

exactly the same as for the proton! The capacitor's electric field is

$$E = \frac{\eta}{\varepsilon_0} = \frac{Q}{\varepsilon_0 A} = 2.82 \times 10^5 \text{ N/C}.$$

Using $d = 0.002$ m, we find

$$\Delta U_{proton} = \Delta U_{elec} = -4.52 \times 10^{-17} \text{ J}.$$

c) From conservation of energy, $\Delta K + \Delta U = 0$. Because the particles are released from rest, $\Delta K = K_f - 0 = \tfrac{1}{2}mv_f^2$. Thus

$$\Delta K + \Delta U = \tfrac{1}{2}mv_f^2 + \Delta U = 0$$

$$\Rightarrow v_f = \sqrt{\frac{-2\Delta U}{m}}.$$

Using the masses of the proton and the electron, we find

$$v_{proton} = 2.33 \times 10^5 \text{ m/s}$$

$$v_{electron} = 9.96 \times 10^6 \text{ m/s}.$$

Even though both particles have the same ΔU, the electron achieves a much larger final velocity due to its much smaller mass.

27.3 The Potential Energy of Point Charges

We derived an expression for the potential energy of two point masses in Section 12.5, which you should review. The potential energy of two point charges is derived in the same way, so we will not repeat all the steps here. The key feature of both interactions is that the force depends on the inverse square of the separation.

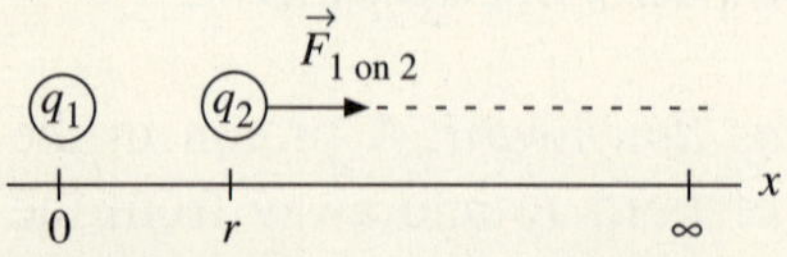

Figure 27-3 shows two *like* charges q_1 and q_2, separated by distance r, exerting repulsive interaction forces on each other. The potential energy of their interaction can be found by calculating the work done by q_1 on q_2 as q_2 moves from $x = r$ to $x = \infty$. An important choice we made in Chapter 12, which

FIGURE 27-3 Two point charges have a potential energy because of the interaction forces they exert on each other.

applies to both gravitational and electric potential energy, is to set the zero point at infinity. That is, we will choose $U(r = \infty) = 0$. This is a reasonable choice because that is the distance at which the two charges cease to interact with each other.

Having made this choice for $U = 0$, the definition of Eq. 27-2 gives

$$\Delta U = U(\infty) - U(r) = 0 - U(r) = -\int_r^\infty (F_{1 \text{ on } 2})_x \, dx$$
$$= -\int_r^\infty \frac{Kq_1q_2}{x^2} \, dx \qquad (27\text{-}5)$$
$$= -\frac{Kq_1q_2}{r},$$

from which we conclude

$$U_{\text{elec}}(r) = \frac{Kq_1q_2}{r} = \frac{1}{4\pi\varepsilon_0} \cdot \frac{q_1q_2}{r} \quad \text{(two point charges)}. \qquad (27\text{-}6)$$

Note that the potential energy of two point charges looks *almost* the same as the coulomb force between the charges—the difference being r in the denominator here but r^2 in Coulomb's law. Make sure you remember which is which!

Although we derived Eq. 27-6 for two like charges, it is equally valid for two opposite charges. Thus two like charges have a positive potential energy while the potential energy of two opposite charges is *negative*. Figure 27-4 shows the potential energy function $U(r)$ for two like charges and two unlike charges.

A charge released from rest always moves in the direction of a *decreasing* potential energy. This is because U must decrease in order for the particle to move and gain kinetic energy. So you see from Fig. 27-4 that two like charges will move apart, because U decreases as r increases, and that two opposite charges will move together. This is the expected behavior, based on Coulomb's law, but energy conservation provides an alternative perspective on the process.

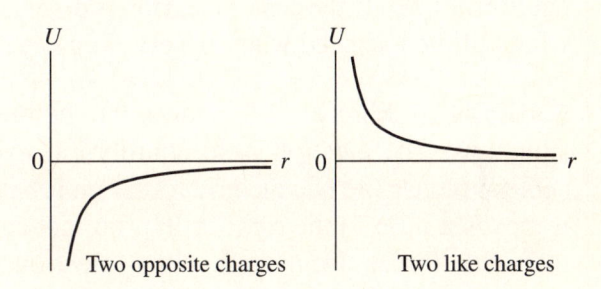

FIGURE 27-4 The potential energy function $U(r)$ for two like charges and two opposite charges.

Because the electric field outside a *sphere of charge* is the same as that of a point charge, it comes as no surprise that Eq. 27-6 is also the electric potential energy of two charged spheres. Distance r is the distance between their centers.

If more than two charges are present, the potential energy of charge i is just the sum of its potential energies due to all the other charges:

$$U_i = \sum_{j \neq i} \frac{Kq_iq_j}{r_{ij}}, \qquad (27\text{-}7)$$

where r_{ij} is the distance between q_i and q_j. This addition follows from the fact that the electric force on q_i is the sum of the forces due to each individual charge.

EXAMPLE 27-2 A proton is fired from far away at a 1 mm diameter glass sphere that has been charged to +100 nC. What speed must the proton have to just reach the surface of the glass?

SOLUTION You learned in gravitational potential energy problems that "far away" is interpreted as meaning $r \approx \infty$ and $U = 0$. To "just reach" the glass, the proton will end with speed $v_f = 0$ as it reaches $r_f = 0.5$ mm (the *radius* of the sphere). Conservation of energy gives

$$K_i + U_i = K_f + U_f$$

$$\Rightarrow \tfrac{1}{2}mv_i^2 + 0 = 0 + \frac{Kq_{proton}q_{sphere}}{r_f}.$$

Using $q_{proton} = e$, we can solve for the proton's initial speed:

$$v_i = \sqrt{\frac{2Keq_{sphere}}{mr_f}} = 1.86 \times 10^7 \text{ m/s}.$$

EXAMPLE 27-3 Three electrons are spaced 1 mm apart along a line. The outer two electrons are fixed in position. a) Is the center electron at a point of stable or unstable equilibrium? b) If the center electron is displaced horizontally by an infinitesimal distance, what will be its speed when is very far away?

SOLUTION a) Figure 27-5 shows the before and after situations. The electron is in equilibrium *exactly* in the center because the two electric forces on it balance. But if it moves a little to the right or left, no matter how small, then the forces from both outer electrons will have horizontal components that will push the electron farther away. This is an unstable equilibrium, like being on the top of a hill.

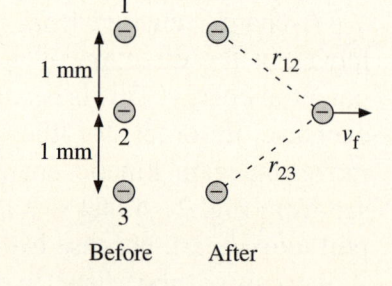

FIGURE 27-5 The electrons of Example 27-3.

b) With even an infinitesimal displacement the electron will begin to move away. If the displacement is only infinitesimal, the initial conditions are $(r_{12})_i = (r_{23})_i = 1$ mm and $v_i = 0$. "Very far away" is interpreted as $r \to \infty$, although in practice the electron merely has to move far enough that $U \approx 0$. Conservation of energy gives

$$K_i + U_i = K_f + U_f$$

$$\Rightarrow \tfrac{1}{2}mv_i^2 + \left[\frac{Kq_1q_2}{(r_{12})_i} + \frac{Kq_2q_3}{(r_{23})_i}\right] = \tfrac{1}{2}mv_f^2 + \left[\frac{Kq_1q_2}{\infty} + \frac{Kq_2q_3}{\infty}\right].$$

This is easily solved, using $v_i = 0$ and $q_1 = q_2 = q_3 = -e$, to give

$$v_f = \sqrt{\frac{2}{m} \cdot \left[\frac{Ke^2}{(r_{12})_i} + \frac{Ke^2}{(r_{23})_i}\right]} = 1006 \text{ m/s}.$$

Refer to Section 12.5 for examples of particles colliding and for calculations of the escape speed. Although those examples use the gravitational potential energy of two point masses, the calculations for two point charges follow an identical procedure.

27.4 The Electric Potential

In Chapter 24 we introduced the concept of the *electric field* because of concerns and difficulties with action at a distance. The field provides an intermediary through which two charges exert forces on each other: Charge q_1 somehow alters the space around it by creating an electric field $\vec{E}_1$, and charge q_2 then responds to the field, experiencing force $\vec{F} = q_2 \vec{E}_1$, rather than responding directly to q_1.

We chose to define the electric field so as to separate the *source* of the field from the charge *in* the field. The definition $\vec{E} = \vec{F}/q$ did that, giving us essentially the force *per unit charge*. Then $\vec{F} = q\vec{E}$ gives us the force—a tangible quantity—as

force on q = [charge q] × [alteration of space by source charges].

Suppose we were to try the same procedure with the potential energy. We want to separate the *source* of the potential energy from the charge which *has* the potential energy. After all, a charge does not create its own potential energy. The potential energy is a consequence of its interaction with *other* charges. Can we divide these so that

potential energy of q = [charge q] × [*potential* for interaction with source charges]?

From Eqs. 27-4 and 27-6, you see that the charge q can easily be separated out from the expressions for the potential energy of a charge in a parallel-plate capacitor and for the potential energy of a charge with another point charge. This leads us to propose defining a new quantity V as

$$V = \frac{U_{\text{of } q}}{q}. \tag{27-8}$$

This implies, of course, that the potential energy of charge q is

$$U_{\text{of } q} = qV, \tag{27-9}$$

where V depends on the *source charges* but is independent of q.

What is this V? Does this make any sense? How is it useful? By its definition, V is the potential energy *per unit charge*. Because the potential energy U depends on the *position* of charge q, it must be that the quantity V is defined throughout *all* of the space around a source charge or charges. The space has somehow been altered by the presence of the source charges and it now has the *potential* to endow charge q, if it is present, with a potential energy. But this ability, or potential, is present throughout space *regardless* of whether charge q is there to experience it. For this reason, we call V the **electric potential**, or sometimes, for brevity, just the *potential*. The electric potential depends only on the source charges and their geometry.

The unit of electric potential is the joule per coulomb, which is called the **volt** V:

1 volt = 1 V = 1 J/C.

This unit is named in honor of Alessandro Volta, who invented the electric battery in the year 1800. Microvolts (μV), millivolts (mV), and kilovolts (kV) are all commonly used units because the electric potentials used in practice differ significantly in magnitude. Make sure you don't confuse the symbol V for the *value* of the potential with the symbol V for the *unit* of potential.

The field potential and the electric field are two different ways of describing the *same* alteration of space. To see this, return to the definition of ΔU in Eq. 27-2. The difference in potential energy between any two points s_1 and s_2 is

$$\Delta U = U(s_2) - U(s_1) = -\int_{s_1}^{s_2} F_s(s)\,ds. \tag{27-10}$$

We are specifically interested in the electric potential energy, so the force is $F_s = qE_s$, where the subscript s indicates the s-components of the force and electric field vectors. Thus

$$\Delta U_{\text{of } q} = -q\int_{s_1}^{s_2} E_s(s)\,ds. \tag{27-11}$$

From the definition of the electric potential V, in Eq. 27-8, we have

$$\Delta V(s) = V(s_2) - V(s_1) = \frac{\Delta U_{\text{of } q}(s)}{q} = -\int_{s_1}^{s_2} E_s(s)\,ds, \tag{27-12}$$

where ΔV is the **potential difference** between points 1 and 2. Because the potential difference, like the potential, is measured in volts, ΔV is often called the **voltage** between the two points.

Equation 27-12 is an important result, not so much for doing calculations as for understanding what it all means. Equation 27-12 tells us that the electric potential is a function of the electric field, and that if we know $\vec{E}$ we can determine ΔV. In the next chapter, you will learn how to do the reverse: determine $\vec{E}$ from a knowledge of ΔV. The point to be made is that the electric potential *and* the electric field are two different ways to describe how a source charge has altered the space around it. The potential cannot exist without the field, and vice versa.

When working with potential energy, only the *change* in potential energy ΔU is precisely defined. Energy conservation problems always use ΔU, not U itself. The point in space we choose to call $U = 0$ has no observable or measurable results. Likewise only ΔV, the difference in the electric potential between two points in space, is physically meaningful. We can place the zero of potential anywhere that is convenient for doing calculations, but only ΔV can be measured. The potential difference is an extremely important quantity in applications of electricity, as you will see.

The terminology of electricity, which has been established by long historical usage, is potentially confusing. The quantity V is the electric potential, while the quantity U is the electric potential energy. They are two completely and totally different quantities and concepts, yet the nearly identical names make it very easy (but wrong!!) to think that they are interchangeable. In hindsight, it would have been far better for the pioneers of electricity to have given V an entirely different name—but, alas, they did not. One of your highest priorities in studying this chapter is to make sure you understand which is which, what each means and measures, and how each is used.

EXAMPLE 27-4 A proton and an electron are traveling side by side with speeds of 2×10^5 m/s. The two charges enter a region of space in which source charges have created an electric potential, and they undergo a potential difference of 100 V. What are their final speeds?

SOLUTION A charge q that moves through a potential difference ΔV experiences a change in potential *energy* $\Delta U = q\Delta V$. Using energy conservation, we can find the final speed v_f:

$$\Delta K + \Delta U = \Delta K + q\Delta V = 0$$
$$\Rightarrow K_f = \tfrac{1}{2}mv_f^2 = K_i - q\Delta V = \tfrac{1}{2}mv_i^2 - q\Delta V$$
$$\Rightarrow v_f = \sqrt{v_i^2 - \frac{2q}{m}\Delta V}.$$

Because $q_{\text{proton}} = e$, we can calculate that the proton slows to

$$v_{f\,\text{proton}} = \sqrt{(2\times 10^5 \text{ m/s})^2 - \frac{2(1.6\times 10^{-19}\text{ C})(100\text{ V})}{(1.67\times 10^{-27}\text{ kg})}} = 1.44\times 10^5 \text{ m/s}.$$

An electron has $q_{\text{electron}} = -e$, so it *speeds up* to $v_{f\,\text{electron}} = 5.93 \times 10^6$ m/s.

It is important to note in Example 27-4 that the electric potential *already existed* in space because of some other charges, perhaps on a capacitor, that are not explicitly seen in the problem. The electron and proton have nothing to do with creating the potential. Instead, they *respond* to the potential by having a potential energy $U = qV$. When a charge moves through a region of space in which there is an electric potential, we say that the charge "moves through a potential difference ΔV." As you can see from this example, charges speed up or slow down as they move through a potential difference.

Also notice, in Example 27-4, that the proton slows down while the electron speeds up. Because $\Delta V > 0$, the charges are moving through a region of space in which the electric potential *of the space* is increasing. This causes the proton's potential energy to increase, as if it were going uphill, so it slows. But the electron, with $q < 0$, has $\Delta U < 0$. It *loses* potential energy as it "falls" through a region of space where the potential increases. This loss of potential energy is balanced by an increase of kinetic energy. Had we used the electric field and electric forces, rather than the potential and potential energy, we would not have been surprised to see these results because the forces, and thus the accelerations, on the two opposite charges would point in opposite directions. Electric potential is more abstract and harder to visualize than the field, so it will take some time (and practice at interpreting results) before it all begins to "make sense."

27.5 The Electric Potential of a Point Charge

Our first order of business, on the way to making the electric potential a useful and effective tool, is to determine the potential of some standard charge distributions.

We have already found, in Eq. 27-6, the potential energy of two point charges. Let q_1 be considered the source charge and q_2 a test charge that experiences a force and a potential energy due to q_1. Then, by definition, the electric potential of charge q_1 is

$$V_{\text{of } q_1}(r) = \frac{U_{\text{of } q_2}}{q_2} = \frac{1}{4\pi\varepsilon_0} \cdot \frac{q_1}{r}. \quad (27\text{-}13)$$

There is, of course, nothing special about the subscript. We needed subscripts 1 and 2 when we calculated the potential *energy* because we had to distinguish the two charges, but now we are describing the electric potential of a single charge. Thus we can drop the subscript and state

$$V(r) = \frac{1}{4\pi\varepsilon_0} \cdot \frac{q}{r} \quad \text{(electric potential of a point charge).} \quad (27\text{-}14)$$

This is a potential that extends through all of space, showing the influence of charge q, but weakening with distance as $1/r$.

This result for the potential of charge q is similar to that for the electric field of charge q. The difference most quickly seen is that V depends on $1/r$ while $\vec{E}$ depends on $1/r^2$. But it is also especially important to notice that the potential is a *scalar* while the field is a *vector*. This will make the mathematics of using the potential much easier than the vector mathematics required to use the electric field.

Equation 27-14 includes the assumption that we have chosen $V = 0$ to be at $r = \infty$. This is the most logical choice for a point charge because the influence of charge q ends at infinity. Regardless of the zero point location, however, the potential *difference* between two points at distance r_1 and r_2 from the charge is

$$\Delta V_{12} = V(r_2) - V(r_1) = \frac{q}{4\pi\varepsilon_0} \cdot \left(\frac{1}{r_2} - \frac{1}{r_1}\right). \quad (27\text{-}15)$$

Equation 27-15 is interesting because it depends only on distances from the source charge, *not* on directions. Figure 27-6 shows several point in space around charge q. Because points 1, 2, and 3 are all the *same* distance r_1 from the charge, the potential differences between each point and point 4 are all equal:

$$\Delta V_{14} = \Delta V_{24} = \Delta V_{34}.$$

This is true despite the fact that the distances r_{14}, r_{24}, and r_{34} are *not* equal. Keep in mind that ΔV describes the potential difference between two points *in space* as a consequence of the presence of charge q. There need *not* be, and usually is not, anything physically located at those points.

What is the potential difference ΔV_{12} between points 1 and 2 in Fig. 27-6? Because they are an equal distance from the charge, $r_1 = r_2$ and $\Delta V_{12} = 0$. Each of these two points in space has the same ability, or potential, to provide a test charge with a potential energy. There is *no difference* in their ability, hence *no potential difference*. This seems quite strange, after our experiences with the electric field, but it follows from the scalar nature of the electric potential.

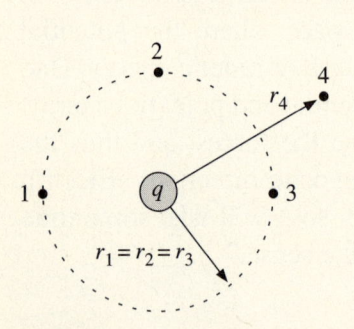

FIGURE 27-6 Points 1, 2, and 3 all have the same potential, but the potential at point 4 is different.

EXAMPLE 27-5 a) What is the electric potential 1 cm from a +1 nC charge? b) What is the potential difference between a point 1 cm away and a second point 3 cm away?

SOLUTION a) The potential is calculated from Eq. 27-14 as

$$V(r) = \frac{1}{4\pi\varepsilon_0} \cdot \frac{q}{r} = (9.0 \times 10^9 \text{ N} \cdot \text{m}^2/\text{C}^2) \cdot \frac{10^{-9} \text{ C}}{0.01 \text{ m}} = 900 \text{ V}.$$

b) We can similarly calculate $V(r = 3 \text{ cm}) = 300$ V. Thus $\Delta V = 600$ V. Because 1 nC is typical of the electrostatic charge produced by rubbing, we see that such a charge creates a fairly large potential nearby. Why aren't we shocked and injured when working with the "high voltages" of such charges? The sensation of being shocked is a result of current, not potential. Some high-potential sources simply do not have the ability to generate much current. We will look at this issue in Chapter 29.

In practice, we are more likely to be working with a charged sphere, of radius R, than a perfect point charge. Because the potential energy of two charged spheres is the same as that of two point charges, the electric potential of a sphere of charge is indistinguishable from that of a point charge. Thus we can conclude that the electric potential of a sphere of charge Q and radius R, for positions *outside* the sphere ($r \geq R$), is

$$V(r \geq R) = \frac{1}{4\pi\varepsilon_0} \cdot \frac{Q}{r} \quad \text{(sphere of charge)}. \quad (27\text{-}16)$$

We can cast this in a more useful form. It is customary to speak of charging an electrode, such as a sphere, "to" a certain potential—as in "Bob charged the sphere to a potential of 3000 volts." This is the potential right on the surface of the sphere, $r = R$, which we will call V_0. From Eq. 27-16 we can see that

$$V_0 = V(r = R) = \frac{Q}{4\pi\varepsilon_0 R}.$$

Consequently, a sphere of radius R that is charged to potential V_0 has charge

$$Q = 4\pi\varepsilon_0 R V_0. \quad (27\text{-}17)$$

If we use this expression for Q in Eq. 27-16, we can write the potential outside a sphere charged to potential V_0 as

$$V(r) = \frac{R V_0}{r}. \quad \text{(sphere charged to potential } V_0\text{)}. \quad (27\text{-}18)$$

EXAMPLE 27-6 A proton is released from rest at the surface of a 1 cm diameter sphere that has been charged to +1000 V. a) What is the proton's speed when it is 1 cm from the sphere? b) How much charge is on the sphere?

SOLUTION a) A sphere charged to $V_0 = +1000$ V is *positively* charged. The proton will be repelled by this charge and fly away. Using $U = qV = eV$ for the proton's potential energy, we have

$$\Delta K + \Delta U = \Delta K + e\Delta V = 0$$
$$\Rightarrow \tfrac{1}{2}mv_f^2 = \tfrac{1}{2}mv_i^2 - e[V(r_f) - V(r_i)]$$
$$\Rightarrow \tfrac{1}{2}mv_f^2 = \tfrac{1}{2}mv_i^2 - e\left(\frac{RV_0}{r_f} - V_0\right).$$

To arrive at the last line, we used Eq. 27-18 for the potential of a sphere. We also noted that $V_i = V_0$ because the proton starts from the surface of the sphere ($r_i = R$). When the proton is 1 cm from the *surface* of the sphere its value of r_f is $r_f = 1$ cm $+ R = 1.5$ cm. Using this, along with $v_i = 0$, we can solve for v_f:

$$v_f = \sqrt{\frac{2eV_0}{m}\left(1 - \frac{R}{r_f}\right)} = 3.57 \times 10^5 \text{ m/s}.$$

This example illustrates how the ideas of potential and potential energy work together, yet they are *not* the same thing.
b) From Eq. 27-17,

$$Q = 4\pi\varepsilon_0 R V_0 = 1.11 \times 10^{-9} \text{ C} = 1.11 \text{ nC}.$$

27.6 The Electric Potential Inside a Capacitor

Having been successful at finding the potential of a point charge, let us do the same for a parallel-plate capacitor. We found the potential energy of charge q inside a parallel-plate capacitor, in Eq. 27-4, to be

$$U_{elec} = U_0 + qEs,$$

where s is measured from the *negative* plate and where U_0 is the charge's potential energy at the negative plate. Let us consider charge q to be a test charge that measures the pre-existing electric potential inside the capacitor. Then, from the definition of V, the electric potential inside the capacitor must be

$$V(s) = \frac{U_{elec}}{q} = V_0 + Es, \quad (27\text{-}19)$$

where $V_0 = U_0/q$ is the value of the potential at the negative plate. The notation $V(s)$ means that the potential V is a function of the distance s.

We can set V_0 to any value that is convenient because it will not matter when we find potential *differences*. You can see, in fact, that the potential difference ΔV_{12} between two points at distances s_1 and s_2 from the negative plate is

$$\Delta V_{12} = V_2 - V_1 = (V_0 + Es_2) - (V_0 - Es_1) = E \cdot (s_2 - s_1), \quad (27\text{-}20)$$

a very simple result that is independent of V_0.

Figure 27-7 illustrates these ideas, both pictorially and graphically, for a capacitor of spacing d. We've chosen $V_{net} = V_0 = 0$ at the negative plate, so the potential increases linearly with distance across the gap of the capacitor until it reaches the value $V_{pos} = Ed$ at the positive plate. The total potential difference between the two plates is

$$\Delta V_0 = V_{\text{pos}} - V_{\text{neg}} = Ed.$$

The quantity ΔV_0 is usually called the "voltage across the capacitor." In practice, this potential difference is established by connecting the capacitor plates to a battery of potential difference ΔV_0.

We can turn this around and arrive at the important conclusion that the electric field inside a capacitor that has been charged to a total potential difference ΔV_0 is

$$E = \frac{\Delta V_0}{d} \quad \text{(electric field inside a parallel-plate capacitor)}. \tag{27-21}$$

Notice that the electric field points in the direction of *decreasing* electric potential. Equations 27-19 and 27-21 allow us to write the electric potential inside as capacitor as

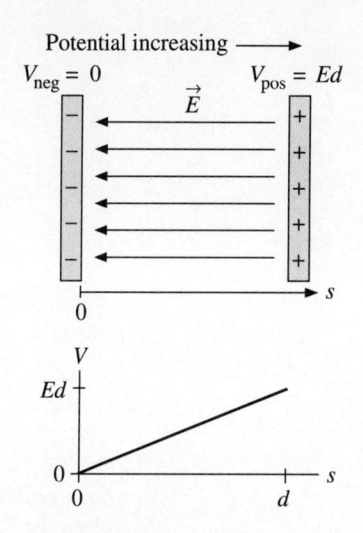

FIGURE 27-7 The electric potential increases linearly from the negative to the positive plate of a capacitor.

$$V(s) = V_0 + \frac{s}{d} \cdot \Delta V_0 \quad \text{(potential inside a parallel-plate capacitor)}. \tag{27-22}$$

Equation 27-21 implies that the units of electric field are "volts per meter," or V/m. We have been using electric field units of "newtons per coulomb." In fact, as you can show for a homework problem, these units are equivalent to each other. That is,

Units of the electric field: 1 N/C = 1 V/m.

Volts per meter are the *preferred* units that are customarily used by scientists and engineers. We will now adopt them as our standard electric field unit.

EXAMPLE 27-7 A 1.5 V battery is connected to a parallel-plate capacitor having a spacing of 3 mm. a) What is the electric field inside the capacitor? b) What is the potential 2 mm from the negative plate?

SOLUTION a) The battery charges to capacitor to $\Delta V_0 = 1.5$ V. Using $d = 0.003$ m we find

$$E = \frac{\Delta V_0}{d} = \frac{1.5 \text{ V}}{0.003 \text{ m}} = 500 \text{ V/m}.$$

Amazing! All that effort in Chapter 25 to calculate the field inside a capacitor, and it ends up being this simple! Notice the new units of the electric field.

b) The potential inside the capacitor is given by Eq. 27-22 as $V(s) = V_0 + (s/d) \cdot \Delta V_0$, where V_0 is the potential at the negative plate and s is measured from the negative plate. If we set $V_0 = 0$, which we are free to do, and use $s = 2$ mm, we find

$$v(s = 2 \text{ mm}) = 0 + \tfrac{2}{3} \cdot 1.5 \text{ V} = 1.0 \text{ V}.$$

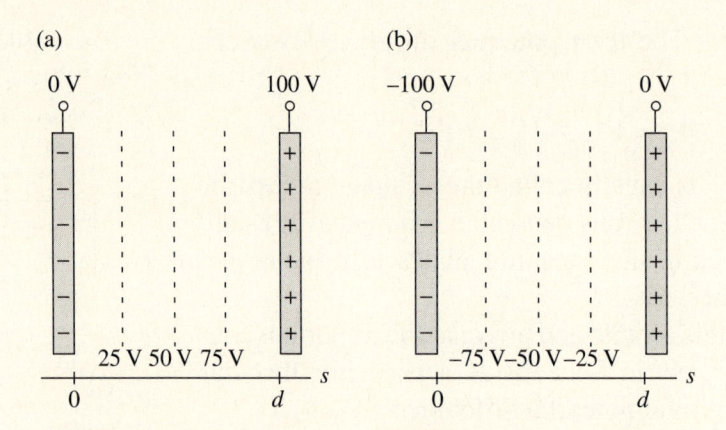

FIGURE 27-8 Two capacitors charged to a potential difference of 100 V. a) The zero of potential is at the negative plate. b) The zero of potential is at the positive plate. The *physical situation* is the same for both.

Example 27-7 raises the issue of the zero of the potential. In the example we chose to let $V = 0$ at the negative plate, but that is not the only possible choice. Figure 27-8 shows a parallel-plate capacitor that has been charged to a potential *difference* of 100 V. The lines with little circles coming off the capacitor plates are a standard symbol representing a *terminal* that is attached to some unseen but known voltage. In Fig. 27-8a we have chosen to let $V_{neg} = V_0 = 0$ V. In this case, the positive plate has potential

$$V_{pos} = V(s = d) = V_0 + 100 \text{ V} = 100 \text{ V}.$$

The electric potential exists at all points between the plates. *Any* point that is halfway between the plates, at $s = d/2$, has potential

$$V(s = \tfrac{1}{2}d) = V_0 + \tfrac{1}{2}\Delta V_0 = 50 \text{ V}.$$

The dotted line in the center of the capacitor of Fig. 27-8a represents an edge-view of a *plane* through the capacitor on which the potential is 50 V at all points. Similarly, we can locate the 25 V plane and the 75 V plane at one-quarter and three-quarters of the distance from the negative to the positive plate. These dotted lines begin to give us a *picture* of the electric potential between the plates.

Suppose, however, we chose the *positive* plate to have a potential of zero: $V_{pos} = 0$. This situation is shown in Fig. 27-8b. In this case, the negative plate has potential $V_{neg} = V_0 = -100$ V and the potential at distance s from the negative plate is

$$V(s) = -100 \text{ V} + \frac{s}{d} \cdot 100 \text{ V}.$$

Figure 27-8b shows the −75 V, the −50 V, and the −25 V planes.

The important point to be made is that Figs. 27-8a and 27-8b represent the same *physical* situation. The potential *difference* between any two points is the same whether we use Fig. 27-8a or Fig. 27-8b to find it. For example, the potential difference between $s = 0.75d$ and $s = 0.25d$ is $\Delta V = 75$ V − 25 V = +50 V in one case and an equal (−25 V) − (−75 V) = +50 V in the other. No matter which choice of zeros we use, a charge q moving between

$s = 0.25d$ and $s = 0.75d$ experiences a change of potential energy $\Delta U = 50q$ joules. Although we may *prefer* one of these figures over the other (or even a third, where the two plates are at –50 V and +50 V), there is no measurable, physical difference between them.

The electric potential inside a capacitor is caused by the charges on the positive and negative capacitor plates. These are the source charges. A charge q inside the capacitor has potential energy $U = qV$ as it interacts with the charges on the plates. Thus we can use all we know about energy conservation to learn how the charge moves in response to this interaction.

EXAMPLE 27-8 A parallel-plate capacitor is constructed of two 2 cm diameter disks spaced 2 mm apart. It is charged to a potential difference of 500 V. a) What is the electric field inside? b) How much charge is on each plate? c) A proton is shot through a small hole in the negative plate with a speed of 2×10^5 m/s. Does it reach the other side? If not, where is the turning point?

SOLUTION a) From Eq. 27-21, $E = \Delta V_0/d = (500 \text{ V})/(0.002 \text{ m}) = 2.50 \times 10^5$ V/m.
b) Because $E = \eta/\varepsilon_0$ for a parallel-plate capacitor, with $\eta = Q/A = Q/\pi R^2$ being the surface charge density, we find

$$Q = \pi R^2 \varepsilon_0 E = 6.95 \times 10^{-10} \text{ C} = 0.695 \text{ nC}.$$

c) Conservation of energy requires $\Delta K + \Delta U = \Delta K + e\Delta V = 0$. The electric potential increases as the proton moves from the negative side toward positive side. Thus the proton's potential energy increases ($\Delta U = e\Delta V > 0$) and it slows down ($\Delta K < 0$). To make it all the way across to the positive plate requires

$$K_i \geq e\Delta V_0.$$

We can calculate that $K_i = 3.34 \times 10^{-17}$ J and that $e\Delta V_0 = e \cdot 500 \text{ V} = 8.00 \times 10^{-17}$ J. The proton does *not* have sufficient initial kinetic energy to be able to gain 8.00×10^{-17} J of potential energy, so it will not make it across. Instead, the proton will reach a turning point and reverse direction.

The proton starts at $s_0 = 0$, the negative plate. Let the turning point be at s_1. The potential inside the capacitor is given by $V = V_0 + (s/d) \cdot \Delta V_0$ with $d = 0.002$ m and $\Delta V_0 = 500$ V. The value of V_0 is irrelevant, so let us set $V_0 = 0$ for convenience. Then

$$\Delta K + \Delta U = \Delta K + e\Delta V = (K_1 - K_0) + e(V_1 - V_0) = 0$$

$$\Rightarrow \left(0 - \tfrac{1}{2}mv_0^2\right) + e\left(\frac{s_1}{d}\Delta V_0 - 0\right) = 0.$$

Solving for the turning point s_1 gives

$$s_1 = \frac{mdv_0^2}{2e\Delta V_0} = 8.35 \times 10^{-4} \text{ m} = 0.835 \text{ mm}.$$

The proton does not make it quite halfway across before being turned back.

27.7 The Electric Potential of a Continuous Charge

Suppose many charges are located within a region of space. Number them 1, 2, ..., i, ..., j, ..., N. One particular charge is q_i. As we found in Section 27.3, in Eq. 27-7, the electric potential energy of q_i is the sum of its potential energies due to all the other charges:

$$U_{\text{of } q_i} = \sum_{j \neq i} \frac{Kq_iq_j}{r_{ij}} = q_i \cdot \sum_{j \neq i} \frac{1}{4\pi\varepsilon_0} \cdot \frac{q_j}{r_{ij}}, \qquad (27\text{-}23)$$

where r_{ij} is the distance between q_i and q_j. The summation is over all the *other* charges, excluding i because charge q_i does not interact with itself. Consequently, q_i can be brought out of the summation, as shown in the last step of Eq. 27-23.

We can find the electric potential V at the point in space where q_i is located by considering q_i to be a test charge. According to the definition of the electric potential:

$$V = \frac{U_{\text{of } q_i}}{q_i} = \sum_j \frac{1}{4\pi\varepsilon_0} \cdot \frac{q_j}{r_j}, \qquad (27\text{-}24)$$

where r_j is the distance from charge q_j to the point in space where the potential is being calculated. But $q_j/4\pi\varepsilon_0 r_j$ is simply the potential at this point due to the point charge q_j. Thus the electric potential—like the electric field—obeys a principle of superposition. In other words, the potential at any point in space is just the sum of the potentials contributed—*at that point*—by each and every source charge.

Equation 27-24 is important and useful because it gives us a way to determine the potential of a continuous distribution of charge, such as a charged rod or a charged disk. Consider the arbitrary but continuous distribution of charge shown in Fig. 27-9. To find the potential of this total charge Q at some point in space, imagine dividing it up into many small parcels of charge Δq_j. Each such parcel—two are shown—is small enough to be treated as a point charge. Notice that each parcel is a *different* distance r_j away from the point at which we wish to compute the potential. The potential is given by Eq. 27-24 as simply the sum of the potentials contributed by all the many parcels. In the limit that $\Delta q \to dq$ and $N \to \infty$, the summation can be replaced by an integral and we have

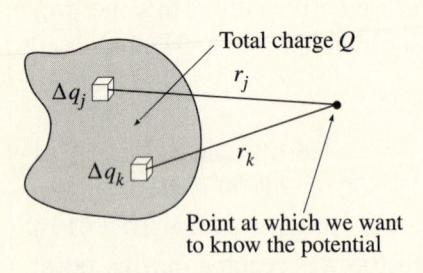

FIGURE 27-9 Finding the potential from a continuous distribution by dividing it into many small charges Δq_j.

$$V = \sum_{j=1}^{N} \frac{1}{4\pi\varepsilon_0} \cdot \frac{\Delta q_j}{r_j} \xrightarrow{N \to \infty} \frac{1}{4\pi\varepsilon_0} \cdot \int \frac{dq}{r}. \qquad (27\text{-}25)$$

This is analogous to how we found the electric field of a continuous distribution of charge, but here we have one big advantage—Eq. 27-25 is a *scalar* integral whereas we had to perform a vector integral (separate x, y, and z components) to find $\vec{E}$.

Note that Eq. 27-25 is still a somewhat formal statement of the potential and is *not*, in this form, ready to integrate. First, q is not a proper integration variable. We will need to use the idea of charge density to relate dq to some small piece of length or area, as we did with the electric field integrations, before we can "do" the integral. Second, we will have to perform some geometric analysis to express the varying distance r in terms of the integration variable. Let's look at some examples to help clarify how Eq. 27-25 is used, in practice, to find the potential of a continuous charge.

The Potential of a Ring of Charge

Suppose we have a ring of charge with radius R and total charge Q. Let's find the potential of the ring at a point on the axis of the ring. Let the ring be in the xy-plane, as shown in Fig. 27-10. Then the point where we want to find the potential is on the z-axis at distance z from the center of the ring.

Divide the ring into N segments of length $\Delta s = 2\pi R/N$. Each segment has charge $\Delta q = \lambda \Delta s$, where $\lambda = Q/L = Q/2\pi R$ is the linear charge density. The distance r_j between segment j and the point of observation is

$$r_j = \sqrt{R^2 + z^2}. \tag{27-26}$$

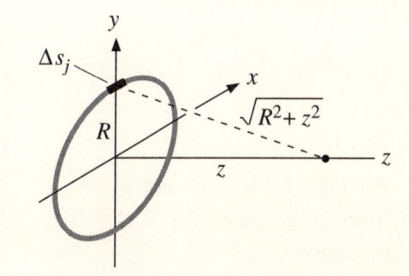

FIGURE 27-10 Geometry for finding the potential of a ring of charge.

Note that r_j is a constant distance, independent of j.

The potential V at distance z is then given by Eq. 27-25:

$$V = \sum_{j=1}^{N} \frac{1}{4\pi\varepsilon_0} \cdot \frac{\Delta q_j}{r_j} = \frac{1}{4\pi\varepsilon_0} \cdot \frac{Q/2\pi R}{\sqrt{r^2 + z^2}} \sum_{j=1}^{N} \Delta s_j. \tag{27-27}$$

The sum is easily evaluated to be

$$\sum_{j=1}^{N} \Delta s_j = s_{\text{total}} = \text{total length of ring} = 2\pi R,$$

so it is not necessary to "do" an integration in this case. Using this result in Eq. 27-27 gives

$$V_{\text{ring}} = \frac{1}{4\pi\varepsilon_0} \cdot \frac{Q}{\sqrt{R^2 + z^2}}. \tag{27-28}$$

Notice that the potential approaches that of a point charge, $Q/4\pi\varepsilon_0 z$, when $z \gg R$.

The Potential of a Disk of Charge

Now consider a thin charged *disk* of radius R and total charge Q. What is the potential of this disk at a point on the axis at distance z? This is a somewhat more complex situation, but we can take advantage of now knowing the on-axis potential of a ring of charge.

Orient the ring in the xy-plane, as shown in Fig. 27-11. Then divide the disk into N rings of equal width $\Delta r = R/N$. Ring i has radius r_i and charge Δq_i. We can find the potential V_i of this ring from Eq. 27-28 if we replace Q by Δq_i and R by r_i:

$$V_i = \frac{1}{4\pi\varepsilon_0} \cdot \frac{\Delta q_i}{\sqrt{r_i^2 + z^2}}. \tag{27-29}$$

If you "unrolled" ring i you would have a rectangle of length $2\pi r_i$ and width Δr. Thus ring i covers the small area

$$\Delta A_i = 2\pi r_i \Delta r. \quad (27\text{-}30)$$

The amount of charge on ring i is $\Delta q_i = \eta \Delta A_i$, where η is the surface charge density of the disk. But $\eta = Q/A = Q/\pi R^2$, so we have

$$\Delta q_i = \eta \Delta A_i = \frac{Q}{\pi R^2} \cdot 2\pi r_i \Delta r = \frac{2Q r_i \Delta r}{R^2}. \quad (27\text{-}31)$$

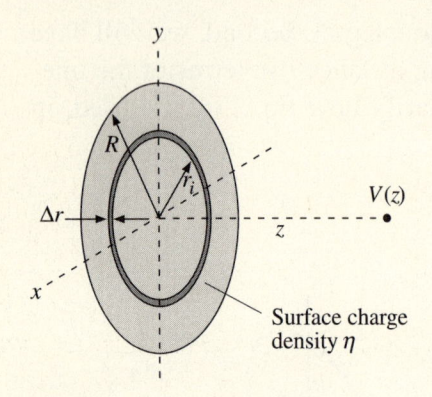

FIGURE 27-11 Geometry for finding the potential of a ring of charge.

Equation 27-31, which relates the charge of each segment to its dimensions, is a critical step in the solution. Notice that all the segments have equal widths but they do *not* have equal charges because the charge depends on the radius r_i.

Now we can add the contributions from all of the rings and let $N \to \infty$, giving

$$V = \sum_i V_i = \frac{1}{4\pi\varepsilon_0} \cdot \sum_{i=1}^{N} \frac{2Q}{R^2} \cdot \frac{r_i \Delta r_i}{\sqrt{r_i^2 + z^2}} \to \frac{Q}{2\pi\varepsilon_0 R^2} \int_0^R \frac{r\,dr}{\sqrt{r^2 + z^2}}. \quad (27\text{-}32)$$

This is an integral that we can find in integration tables to be

$$\int_0^R \frac{r\,dr}{\sqrt{r^2 + z^2}} = \sqrt{r^2 + z^2}\Big|_0^R = \sqrt{R^2 + z^2} - z, \quad (27\text{-}33)$$

which gives the potential of a charged disk as

$$V_{\text{disk}}(z) = \frac{Q}{2\pi\varepsilon_0 R^2}\left(\sqrt{R^2 + z^2} - z\right). \quad (27\text{-}34)$$

Although there were a number of steps to go through, this was *much* easier than evaluating the electric field, because we did not have to worry about components. This same procedure can be followed to find the potential for any continuous distribution of charge.

We can find the potential V_0 of the disk itself by setting $z = 0$, giving $V_0 = Q/2\pi\varepsilon_0 R$. In other words, placing charge Q on a disk of radius R "charges" it to a potential V_0. We can write Eq. 27-34 in terms of the disk's potential V_0:

$$V_{\text{disk}}(z) = V_0 \cdot \left[\frac{\sqrt{R^2 + z^2} - z}{R}\right]$$
$$= V_0 \cdot \left[\sqrt{1 + (z/R)^2} - (z/R)\right]. \quad (27\text{-}35)$$

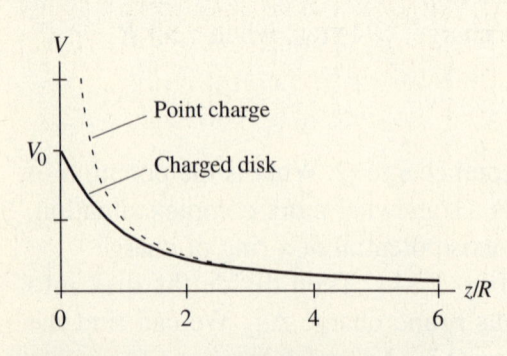

FIGURE 27-12 Potential of a charged disk. For comparison, the dotted line is the potential of a point charge having the same Q.

Figure 27-12 shows a graph of V_{disk} as a function of distance z along the axis and, for comparison, the potential of a point charge Q. You can see graphically that the charged disk begins to look like a point charge for $z > 4R$ but differs significantly from a point charge for $z \leq R$. We will leave it as a homework problem for you to demonstrate algebraically that V_{disk} reduces to the potential of a point charge when $z \gg R$.

EXAMPLE 27-9 A dime, which is 17.5 mm in diameter, is given a charge of +5 nC. a) What is the potential of the dime? b) What is the potential 1 cm above the dime?

SOLUTION a) The dime is a charged disk. The potential of the dime itself is given by the potential of a disk at $z = 0$:

$$V_0 = \frac{Q}{2\pi\varepsilon_0 R} = 10{,}300 \text{ V}.$$

b) A quick calculation at $z = 1$ cm shows that $z/R = 1.143$. Using Eq. 27-35 for the potential on the axis of the dime, we find

$$V = V_0\left[\sqrt{1+(z/R)^2} - (z/R)\right] = 3870 \text{ V}.$$

Summary
Important Concepts and Terms

electric potential energy   potential difference
electric potential   voltage
volt

This chapter has introduced the new and important concept of the electric potential V. The electric potential, like the electric field, is a property of the source charges and it fills the space around them. Because of the electric potential, any other charge q that visits this region of space will acquire a potential energy $U = qV$.

There were two major topics in this chapter:

1. Establishing the connection between the electric potential energy U and the electric potential V.
2. Learning the potential of several common and important charge configurations:

Point charge (or sphere): $V(r) = \frac{1}{4\pi\varepsilon_0} \cdot \frac{q}{r}.$

Parallel-plate capacitor: $V(s) = V_0 + Es = V_0 + \frac{s}{d} \cdot \Delta V_0.$

Charged disk: $V(z) = V_0\left[\sqrt{1+(z/R)^2} - (z/R)\right].$

An important corollary is that the electric field inside a parallel-plate capacitor is related to the potential difference ΔV_0 across the capacitor by $E = \Delta V_0/d$. As a consequence, we will adopt the revised units of "volts per meter" for the electric field.

This has been an *introduction* to the concept of the electric potential, but not the final work on the subject. We will continue our exploration of the potential in the next chapter by looking more closely at the connection between the potential V and the field $\vec{E}$, and looking at *sources* of potential difference. Then, in Chapter 29, we will see how these ideas have practical value in the analysis of circuits.

Exercises and Problems

Exercises

1. Show that 1 V/m = 1 N/C.
2. Demonstrate that the on-axis potential of a charged disk, as derived in Section 27.7, reduces to the potential of a point charge when $z \gg R$.
3. Figure 27-13 shows three charges placed at the corners of a rectangle.
   a. What is the electric potential at point A?
   b. What is the potential energy of a proton placed at point A?

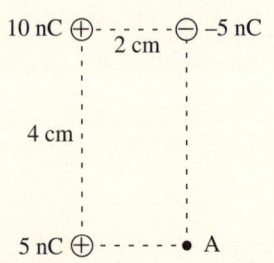

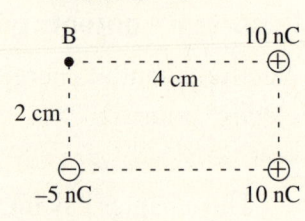

FIGURE 27-13

4. Figure 27-14 shows three charges placed at the corners of a rectangle.
   a. What is the electric potential at point B?
   b. What is the potential energy of an electron placed at point B?

FIGURE 27-14

Problems

5. Two small metal spheres, of 2 g and 4 g, are tied together by a 5 cm long massless string and are resting on a frictionless surface (Fig. 27-15). Each is charged to + 2 µC.
   a. What is the initial energy of this system?
   b. What is the initial momentum of this system?
   c. How much tension is in the string?
   d. The string is cut. After a long time has elapsed, what is the speed of each of the spheres? (**Hint:** There are *two* conserved quantities. Make use of both.)

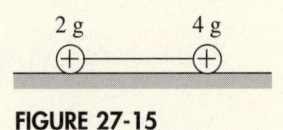

FIGURE 27-15

6. One form of nuclear radiation is *beta decay*. This occurs when a neutron changes into a proton, an electron, and a massless, chargeless particle called a *neutrino*: n → p⁺ + e⁻ + ν where ν is the symbol for a neutrino. When this change happens to a neutron within the nucleus of an atom, the proton remains behind in the nucleus while the electron and neutrino are ejected from the nucleus. The ejected electrons are called *beta radiation*. One nucleus that exhibits beta decay is the isotope of

hydrogen ^{3}H, called *tritium*, whose nucleus consists of one proton (making it hydrogen) and two neutrons (giving tritium an atomic mass $A = 3$). Tritium is radioactive, and it decays to the $A = 3$ isotope of helium, ^{3}He: ^{3}H $\rightarrow$ ^{3}He + e$^-$ + v. (This is an important nuclear reaction in the *nuclear fusion* that powers the sun.)
   a. Is charge conserved in the beta decay process? Explain.
   b. Why is the final product a helium atom? Explain.
   c. The nuclei of both ^{3}H and ^{3}He have radii of 1.5×10^{-15} m. With what minimum speed must the electron be ejected if it is to "escape" from the nucleus and not fall back?

7. The nucleus of a gold atom contains 79 protons and 118 neutrons. The diameter of the nucleus is 14 fm (1 fm = 1 femtometer = 10^{-15} m).
   a. With what speed must a proton be fired toward a gold nucleus if it is to just touch the surface?
   b. An electron is fired toward the gold nucleus with the speed you found in part a). What is its speed as it collides with the nucleus?

8. Two spherical drops of mercury each have a charge of 0.1 nC and a potential of 300 V at the surface. The two drops merge to form a single drop. What is the potential at the surface of the new drop?

9. A parallel-plate capacitor consists of two disks, each 2 cm in diameter, that are spaced 2 mm apart. The electric field between the disks is 5.0×10^5 V/m.
   a. What is the voltage across the capacitor?
   b. How much charge is on each disk?
   c. An electron is launched from the negative plate. It strikes the positive plate at a speed of 2.0×10^7 m/s. What was the electron's speed as it left the negative plate?

10. Figure 27-16 shows a charged rod of length L and total charge Q.
    a. Find an expression for the electric potential a distance x away from the center of the rod.
    b. Find an expression for the electric potential a distance d away from the end of the rod.
    c. Evaluate your answers to parts a) and b) for distances 2 cm from a rod 6 cm long with a charge of 50 nC.

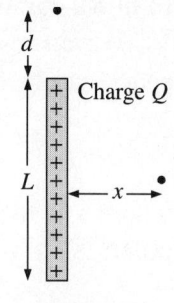

FIGURE 27-16

[**Estimated 10 additional problems for the final edition.**]

Chapter 28

Potential and Field

LOOKING BACK Sections 10.7; 25.4–25.7; 26.3–26.5; 27.4–27.7

28.1 Connecting Potential and Field

We introduced the concept of the *electric potential* in the last chapter. This chapter will continue our exploration of this important concept, focusing in particular on the relationship between the electric potential and the electric field. These are not two distinct entities but, instead, two different perspectives or two different mathematical representations of how source charges alter the space around them.

We discovered in the last chapter that the potential difference between two points in space can be found from the electric field via

$$\Delta V = V(s_2) - V(s_1) = -\int_{s_1}^{s_2} E_s(s)\,ds, \tag{28-1}$$

where s is the position along a line from point 1 to point 2. This method for finding the potential is useful if the electric field is already known.

EXAMPLE 28-1 The electric field along the axis of a ring of charge Q was found in Chapter 25 to be

$$E_z(z) = \frac{1}{4\pi\varepsilon_0} \cdot \frac{zQ}{(z^2 + R^2)^{3/2}},$$

where z is the distance from the center of ring and R is the radius of the ring. Find the potential $V(z)$ along the axis of the ring.

SOLUTION Let point 1 be at distance $s_1 = z$ from the ring and point 2 be at $s_2 = \infty$. If we choose to let the potential be zero at infinity, then Eq. 28-1 gives

$$\Delta V = V(\infty) - V(z) = -V(z) = -\int_z^\infty E_z(z)dz$$

$$\Rightarrow V(z) = \int_z^\infty E_z(z)dz = \frac{Q}{4\pi\varepsilon_0} \int_z^\infty \frac{zdz}{(z^2 + R^2)^{3/2}}.$$

The integral is easily evaluated, giving

$$V(z) = \frac{1}{4\pi\varepsilon_0} \cdot \frac{Q}{\sqrt{z^2 + R^2}}.$$

This is exactly the result we found for a ring of charge in Eq. 27-28, where we worked directly with the continuous distribution of charge. Here we have found the potential by explicitly recognizing the connection between the potential and the field. ●

Our main purpose in this section is to look at the reverse operation—finding the electric field when we know the potential. Consider two points separated by a very small distance Δs, so small that the electric field is essentially constant over this very short distance. The potential difference between these two points is given by Eq. 28-1 as

$$\Delta V = -\int_{s_1}^{s_2} E_s\, ds = -E_s \int_{s_1}^{s_2} ds = -E_s\, \Delta s$$

$$\Rightarrow E_s = -\frac{\Delta V}{\Delta s}. \tag{28-2}$$

We have used the constancy of E_s over this small distance to take it outside the integral. In the limit $\Delta s \to 0$ we have

$$E_s = -\frac{dV}{ds}. \tag{28-3}$$

Now we have reversed Eq. 28-1 and have a way to find the electric field from the potential.

A geometric interpretation of Eq. 28-3 is that the electric field is the negative of the *slope* of the potential-versus-position graph. This should be familiar. You learned in Chapter 10 that the force on a particle is the negative of the slope of the potential energy function: $F = -dU/ds$ (Eq. 10-24). Equation 28-3, in fact, is simply Eq. 10-24 with both sides divided by q to yield E and V. This geometric interpretation is an important step in developing an understanding of potential.

The s-axis is a generic axis, which is replaced in an actual computation by x, y, r, or whatever coordinate is most appropriate to the geometry. In Cartesian coordinates, Eq. 28-3 is

$$E_x = -\frac{dV}{dx}$$
$$E_y = -\frac{dV}{dy} \tag{28-4}$$
$$E_z = -\frac{dV}{dz}.$$

278 CHAPTER 28 POTENTIAL AND FIELD

(*Mathematical aside*: Strictly speaking, these should be partial derivatives. The field $\vec{E}$ is the negative of the *gradient* of the potential: $\vec{E} = -\nabla V = -[(\partial V/\partial x)\hat{i} + (\partial V/\partial y)\hat{j} + (\partial V/\partial z)\hat{k}]$. If you have reached gradients in your study of calculus, then you will recognize Eq. 28-4 as a simple way of writing the vector components. None of our applications will require a knowledge of gradients, but more advanced treatments of the electric field make extensive use of this mathematical relationship.)

Calculating the electric field $\vec{E}$ of a continuous distribution of charge, as you learned to do in Chapter 25, requires careful attention to evaluating the *components* of the field for each Δq and then, in general, performing three separate integrations to find E_x, E_y, and E_z. By contrast, evaluating the potential V takes only a single scalar integral, with no concern about components. Once V is known, $\vec{E}$ can be found by taking derivatives. Thus finding $\vec{E}$ by first calculating V, then using Eq. 28-4, is often *much* easier than calculating $\vec{E}$ directly.

• **EXAMPLE 28-2** What is the electric field inside a parallel-plate capacitor that has been charged to a potential difference ΔV_0?

SOLUTION Consider a capacitor of spacing d that is oriented along the x-axis. We found in the last chapter that the potential inside a parallel-plate capacitor, charged to a potential difference ΔV_0, is

$$V(x) = V_0 + \frac{x}{d}\Delta V_0.$$

Figure 28-1 shows a graph of this potential function, assuming the negative plate is located at $x = 0$ and the positive plate at $x = d$. It is easy to see that the slope of this graph is $dV/dx = \Delta V_0/d$, from which we conclude that the electric field inside this capacitor is

$$E_x = -\frac{dV}{dx} = -\frac{\Delta V_0}{d}.$$

Why the minus sign? Because this is the *x*-component of the vector $\vec{E}$. The arrangement we have chosen causes the field $\vec{E}$ to point in the $-x$-direction, so $E_x < 0$. The *magnitude* of the capacitor field is simply $\Delta V_0/d$, as we had already concluded in Eq. 27-21. The point of this first example is not to find something new but to see how easily the electric field can be found from the potential.

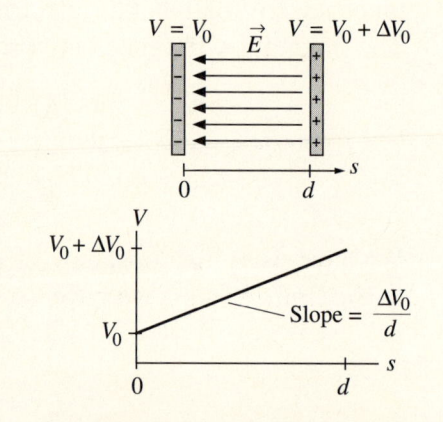

FIGURE 28-1 A graph of the potential inside a parallel-plate capacitor. E_x is the negative of the slope of this graph.

• **EXAMPLE 28-3** What is the electric field of a positive point charge?

SOLUTION A graph of $V(r)$ for a point charge is shown in Fig. 28-2. Notice that the slope is negative for all values of r. We can conclude, from the minus sign in Eq. 28-3, that the

CONNECTING POTENTIAL AND FIELD 279

field $\vec{E}$ will point in the positive, or outward, direction. With $s = r$ we can compute a specific value:

$$E_r = -\frac{dV}{dr} = -\frac{d}{dr}\left(\frac{1}{4\pi\varepsilon_0} \cdot \frac{q}{r}\right) = +\frac{1}{4\pi\varepsilon_0} \cdot \frac{q}{r^2}$$

$$\Rightarrow \vec{E} = \left(\frac{1}{4\pi\varepsilon_0} \cdot \frac{q}{r^2}, \text{ away from charge}\right).$$

Again nothing new; we are just confirming that Eqs. 28-3 and 28-4 allow us to calculate the field from a knowledge of the potential.

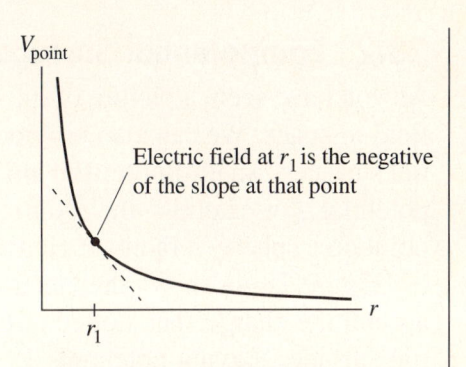

FIGURE 28-2 The potential of a point charge. The slope is negative at all points.

EXAMPLE 28-4 In Chapter 27 we calculated the on-axis potential of a charged disk to be

$$V_{\text{disk}}(z) = \frac{Q}{2\pi\varepsilon_0 R^2}\left(\sqrt{R^2 + z^2} - z\right).$$

Find the on-axis electric field of a charged disk.

SOLUTION The graph of $V(z)$ is shown in Fig. 28-3. The x- and y-components of $\vec{E}$ are zero on the axis because V has no dependence on x or y. E_z is directly from V as the negative of the slope of the graph at each point:

$$E_z = -\frac{dV}{dz} = -\frac{d}{dz}\left[\frac{Q}{2\pi\varepsilon_0 R^2}\left(\sqrt{R^2 + z^2} - z\right)\right]$$

$$= \frac{Q}{2\pi\varepsilon_0 R^2}\left[1 - \frac{z}{\sqrt{R^2 + z^2}}\right].$$

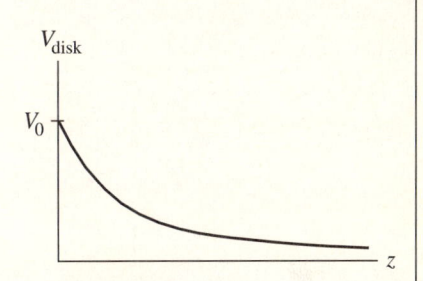

FIGURE 28-3 The potential of a disk of charge.

This is in perfect agreement with the result of Homework Problem 25.14.

The significance of Eqs. 28-3 and 28-4 is thus twofold:

1. They provide a direct relationship between the electric field and the electric potential. When combined with Eq. 28-1, they allow one quantity to be found if the other is known. This procedure establishes that the field and the potential are two aspects of how the source charges alter the space around them. The field and the potential are *not* separate, independent quantities.

2. They provide what is often the easiest and most direct means to find the electric field of a distribution of charge. These equations are widely used in practice.

28.2 Equipotential Surfaces

As you have seen, a picture of the electric field vectors can help you visualize the electric field in space. We can also provide a pictorial model of the potential by using equipotential surfaces. An **equipotential surface** is a surface for which all the points are at the same potential. For example, the electric potential of a point charge q has the same value at *any* point on a sphere of radius r. Hence a sphere is an equipotential surface for a point charge.

Figure 28-4a shows the electric field of a point charge and a three concentric circles around the charge. The dotted circles are slices through three of the spherical equipotential surfaces, having potentials V_1, V_2, and V_3. In a two-dimensional figure we see only *equipotential lines*. The value of the potential is the same at *any* two points on one of these lines. Notice that the electric field vectors are perpendicular to the equipotential surfaces. This is an important point to which we will return.

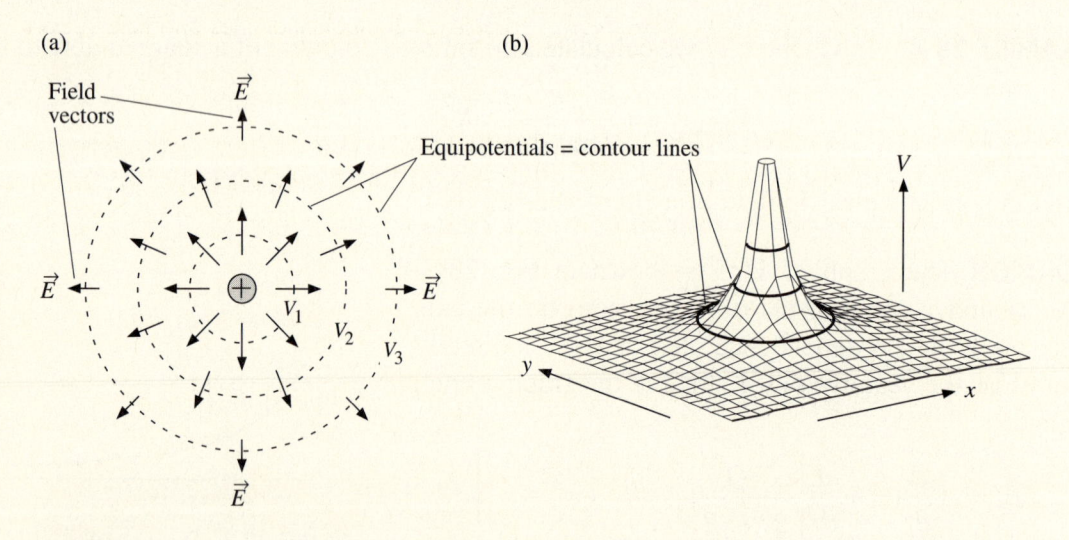

FIGURE 28-4 a) Equipotential lines and field vectors for a point charge. b) A surface plot of the potential. The equipotential lines represent equal-elevation contours on the surface plot.

Figure 28-4b shows a surface plot of the potential of a point charge, with the potential graphed vertically above an *xy*-plane. This gives us a much more vivid impression of a "potential mountain." In comparing Figs. 28-4a and 28-4b, you should notice that the equipotential lines on the left correspond to circular paths of constant elevation around the mountain. Thus the dotted lines in Fig. 28-4a are exactly like the *contour lines* on a topographic map—each line corresponds to the same "elevation" on the potential mountain.

Figure 28-5a shows the surface plot of the more complex potential of a positive *and* a negative charge—a *dipole*. Here we see both a "mountain" and a "valley." What does a topographic map look like for this terrain? If we compute the potential at many points in the *xy*-plane and then connect together all the points having the same values, we end up with the **potential map** of Fig. 28-5b. Notice that this is, indeed, just the contour map for the mountain and valley of Fig. 28-5a.

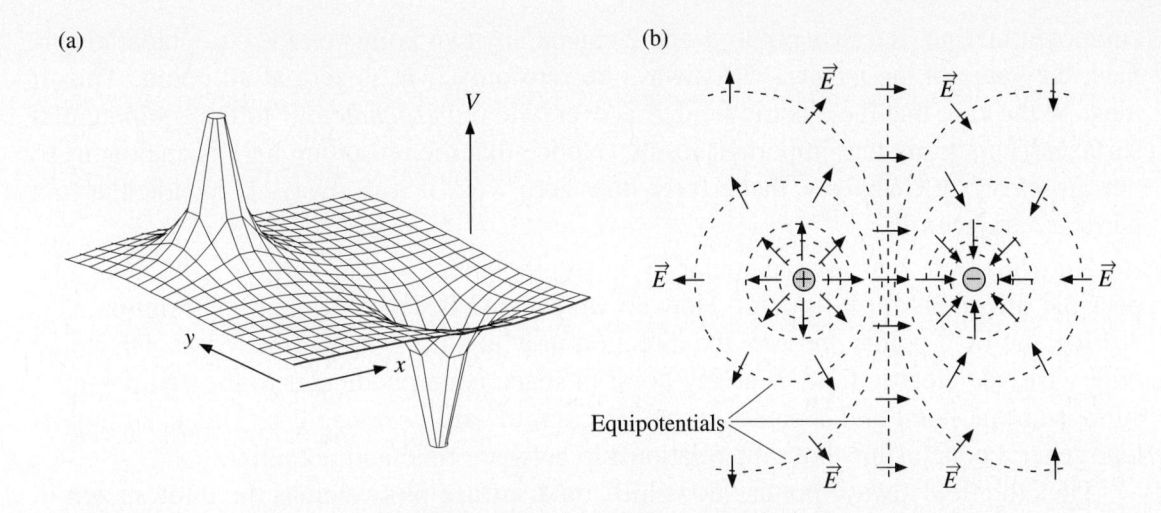

FIGURE 28-5 a) Surface plot of the potential of a dipole. b) Equipotential lines and field vectors.

Figure 28-6a shows a surface plot of the potential inside a capacitor—a two-dimensional version of Fig. 28-1, to which you should refer for comparison. You see that the contours of equal altitude are simply straight lines across the hillside. Compare this to the potential map of Fig. 28-6b, where we see the contours as straight lines between the capacitor plates.

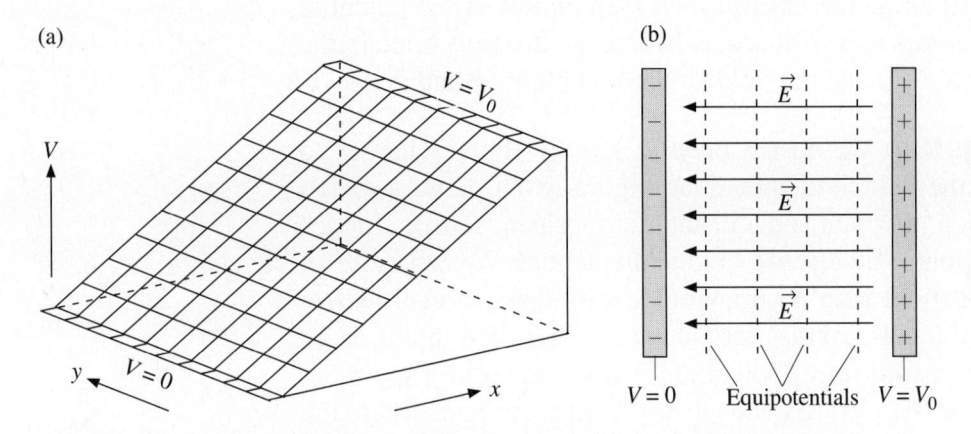

FIGURE 28-6 a) Surface plot of the potential inside a parallel-plate capacitor. b) Equipotential lines and field vectors.

Consider two points labeled 1 and 2 on an equipotential surface, as shown in Fig. 28-7. $\Delta V_{12} = 0$ between these two points because, by definition, they have the same potential. The relationship between the electric potential and the electric field in this region of space is

$$\Delta V_{12} = -\int_{s_1}^{s_2} E_s \, ds = 0,$$

where s is measured along the equipotential line from 1 to 2 and E_s is the component of $\vec{E}$ parallel to the

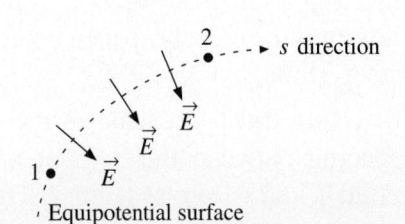

FIGURE 28-7 The electric field is perpendicular to the equipotential surface at each point in space.

equipotential line. Because points 1 and 2 can be *any* two points on the equipotential surface, the value of the integral will always be zero only if E_s is zero at all points. Thus it must be the case that the electric field $\vec{E}$ is everywhere *perpendicular* to the equipotential surfaces! This is another important result. (Notice that the reasoning here is analogous to our discovery, in Chapter 9, that a force does zero work if it is always perpendicular to a particle's trajectory.)

Knowing that $\vec{E}$ is perpendicular to the equipotential line or surface still leaves two possible directions for it to point. How do we decide? Because of the minus sign in $E_s = -dV/ds$, the field points *opposite* the direction in which the potential increases. Or, stated otherwise, the electric field $\vec{E}$ at any point in space is perpendicular to the equipotential surface at that point *and* it points in the direction toward *decreasing V*. This is an important general conclusion about the relationship between field and potential.

Thus the field always points "downhill" on a surface plot, such as the three shown in Figs. 28-4b, 28-5a, and 28-6a. This is shown by the relationship between the field vectors and the equipotential surfaces in Figs. 28-4a, 28-5b, and 28-6b. The field vector at each point is perpendicular to the equipotential surface and pointing downhill toward lower potential. Knowing that the field must point from higher toward lower potential will be an important piece of our understanding of how currents flow in wires.

EXERCISE 28-5 Figure 28-8a shows a set of equipotential lines. a) Draw the electric field map on top of the potential map for this region of space. b) Use the distance scale in the figure to estimate the field strength at points 1 and 2.

SOLUTION a) We do not need to see the source charges to relate the field to the potential. Figure 28-8a shows a region of space in which some distant, but unseen, source charges have created an electric field and potential. We can draw an electric field map by drawing vectors that are everywhere perpendicular to the equipotential lines and point from higher toward lower potential. This is shown in Fig. 28-8b. Because $|\vec{E}| = \Delta V/\Delta s$, the electric field is stronger where the equipotential lines are closer together (smaller Δs) and weaker where the equipotential lines are farther apart (larger Δs). Thus the field vectors are longer on the left, shorter on the right. If this were a topographic map, you would interpret the closely spaced contour lines on the left as a very steep slope.

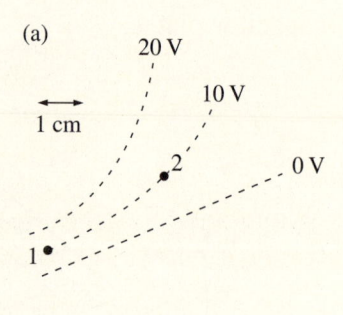

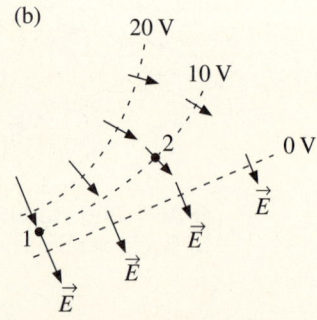

FIGURE 28-8 a) Equipotential lines of Example 28-5. b) The electric field map.

b) Using the 1 cm scale in the figure, we can measure the spacing between the equipotential lines to be ≈ 0.5 cm at point 1 and ≈ 1 cm at point 2. Thus

$$|\vec{E}_1| \approx \frac{\Delta V}{\Delta s_1} \approx \frac{10 \text{ V}}{0.005 \text{ m}} = 2000 \text{ V}/\text{m}$$

$$|\vec{E}_2| \approx \frac{\Delta V}{\Delta s_2} \approx \frac{10 \text{ V}}{0.010 \text{ m}} = 1000 \text{ V}/\text{m}.$$

Closed Loops

Figure 28-9 shows two points, 1 and 2, in a region of electric field and potential. The potential difference between these two points is the same along any two paths that join them. For example, the potential difference along path 1-a-b-c-2 is

$$\Delta V_{12} = V_2 - V_1 = \Delta V_{1a} + \Delta V_{ab} + \Delta V_{bc} + \Delta V_{c2}$$
$$= 0 \text{ V} + 10 \text{ V} + 0 \text{ V} + 10 \text{ V} = 20 \text{ V}.$$

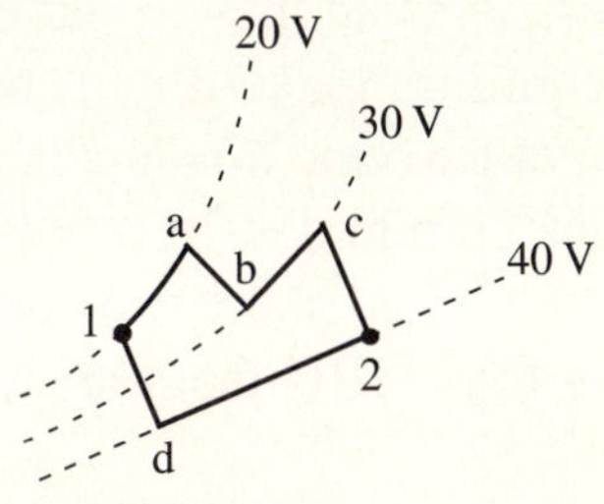

Here we have used the fact that $\Delta V = 0$ for any two points along an equipotential line, such as 1 and a. Similarly, path 1-d-2 gives the same value:

$$\Delta V_{12} = V_2 - V_1 = \Delta V_{1d} + \Delta V_{d2} = 20 \text{ V} + 0 \text{ V} = 20 \text{ V}.$$

FIGURE 28-9 The potential difference between points 1 and 2 is the same along either path.

This leads us to an important conclusion about the electric potential. Consider the *closed path* 1-a-b-c-2-d-1 that ends where it started. What is the potential difference "around" this closed path? We can compute:

$$\Delta V_{\text{loop}} = \sum_j (\Delta V)_j = \Delta V_{1a} + \Delta V_{ab} + \Delta V_{bc} + \Delta V_{c2} + \Delta V_{2d} + \Delta V_{d1}$$
$$= 0 \text{ V} + 10 \text{ V} + 0 \text{ V} + 10 \text{ V} + 0 \text{ V} + (-20) \text{ V}$$
$$= 0,$$

where we used $\Delta V_{d1} = -\Delta V_{1d} = -20$ V.

The numbers are specific to this example, but the idea applies to any **closed loop** path through an electric field. The situation is analogous to hiking on the side of a mountain. You may walk uphill during parts of your hike and downhill during other parts, but if you return to your starting point, your *net* change of elevation is zero. So for any path that starts and ends at the same point, we can conclude that

$$\Delta V_{\text{closed loop}} = \sum_j (\Delta V)_j = 0. \tag{28-5}$$

Stated in words, the sum of all the potential differences encountered while moving around a closed path is zero.

Equation 28-5 is really just a statement of energy conservation, because a charge that moves around a closed loop and returns to its starting point has $\Delta U = q\Delta V = 0$. This conservation statement will turn out to be one of the fundamental principles of circuit analysis.

28.3 The Potential and Field of a Conductor in Electrostatic Equilibrium

The basic relationships between potential and field allow us to draw some interesting and very important conclusions about conductors. Consider a conductor, such as a metal, that is in electrostatic equilibrium. The conductor may be charged but all the charges are at rest. You learned, in Chapter 24, that any excess charges on a conductor in electrostatic equilibrium are always located on the *surface* of the conductor. Using similar reasoning, we can conclude that the electric field is zero, $\vec{E} = 0$, at any interior point of a conductor.

Why? If the field were other than zero, then there would be a force $\vec{F} = q\vec{E}$ on the charge carriers and they would move, creating a current. But there are no currents in electrostatic equilibrium, so it must be that $\vec{E} = 0$ at all interior points. (This is in contrast with a current-carrying conductor, where there most definitely *is* an interior electric field to create the current density $J = \sigma E$.)

Consider any two points in a conductor. They can be connected by a line, such as shown in Fig. 28-10, that remains entirely inside the conductor, and we can let s measure the distance along this line from 1 to 2. We can then find the potential difference $\Delta V = V_2 - V_1$ between these points by using Eq. 28-1 to integrate E_s along the line from 1 to 2. But $\vec{E} = 0$ inside, so the component E_s must also equal zero and therefore the value of the integral is zero! Thus $\Delta V = 0$. That is, *any* two points in a conductor (because 1 and 2 were arbitrary) have the *same* potential.

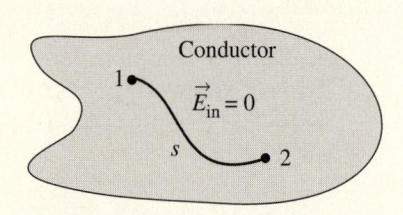

FIGURE 28-10 The interior electric field of a conductor is zero and the entire conductor is at the same potential.

Because $\Delta V = 0$ between any two points in a conductor, we can conclude that the *entire conductor* is at the same potential! If we charge a metal sphere, then the entire sphere is at a single potential. Similarly, a charged metal rod or wire is at a single potential *if* it is in electrostatic equilibrium. (A current, however, destroys our assumption that $\vec{E} = 0$, so this conclusion does *not* apply to a current-carrying conductor.)

If $\vec{E} = 0$ inside a charged conductor but $\vec{E} \neq 0$ outside, what happens right at the surface? Because electric fields start from charges, and because the excess charge is all on the surface, there must be an electric field right at the surface of a charged conductor. But if the entire conductor is at the same potential, then the surface of the conductor is an equipotential surface. You have already seen that the electric field is always perpendicular to an equipotential surface, hence the electric field $\vec{E}$ must be *perpendicular* to the surface of a charged conductor at all points!

Thus we know five very important properties about conductors in electrostatic equilibrium:

1. All excess charge is located on the surface.
2. The interior electric field is zero: $\vec{E}_{in} = 0$.
3. The exterior electric field is always perpendicular to the surface: $\vec{E}_{out} \perp$ surface.
4. The entire conductor has a single value of the electric potential.
5. The surface is an equipotential surface.

These are important as well as practical conclusions because conductors are primary the components of electrical devices.

Figure 28-11 summarizes the situation. It shows a conductor that has received a net positive charge. You can see that all the charge is on the

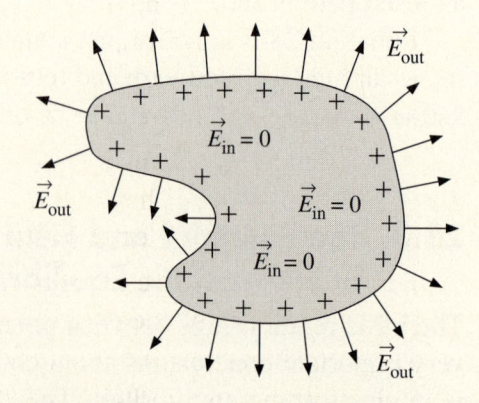

FIGURE 28-11 A charged conductor has $\vec{E}_{in} = 0$ and $\vec{E}_{out}$ perpendicular to the surface.

surface, that the interior field is zero, that the exterior field is perpendicular to the surface, and that all points on the surface have potential V_0. Finding the field at points farther away would require some calculations, but we have learned a lot about the field near the conductor without having to do *any* calculations. We have learned this from *reasoning* about the properties of fields, potentials, and conductors.

EXAMPLE 28-6 Figure 28-12a shows two L-shaped electrodes charged to a potential difference of 100 V. Sketch the equipotential lines and the electric field vectors.

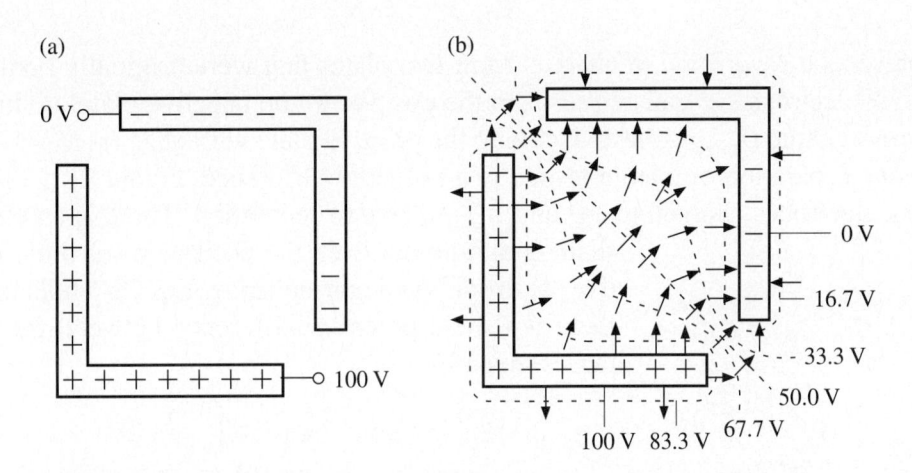

FIGURE 28-12 a) Two L-shaped electrodes with a 100 V potential difference. b) The equipotential lines and electric field vectors for these electrodes.

SOLUTION Because the electrodes are metals in electrostatic equilibrium, they are equipotential surfaces. You can envision what a surface plot of the potential would look like. The 100 V electrode is a high L-shaped plateau whereas the 0 V electrode is a flat L-shaped area on the "plains" below. The surface has to flow smoothly from the plains up to the plateau. The equipotential lines will go around the potential mountain, hugging close to the electrodes at first and then gradually blending together. These equipotential lines are shown at 13.3 V intervals in Fig. 28-12b. Notice that the diagonal line from upper left to lower right is equidistant between the electrodes. From the symmetry, we can conclude that this will be the 50 V equipotential.

The electric field vectors must be perpendicular to the metal surface *and* perpendicular to the equipotential lines. They will point from the higher-potential 100 V electrode toward the lower-potential 0 V electrode. These field vectors are also shown in Fig. 28-12b.

Electrodes shaped like these are common in integrated circuits and elsewhere. Now that we know how the field and potential look, we could begin to think about how an electron would move in this region of space. While quantitative calculations would be necessary to understand the details, a qualitative analysis, such as this example, provides a good overall picture of the field and potential.

28.4 Sources of Potential

We've now established the relationship between the potential and the field. But what *causes* a potential difference? This is a question we must answer before making practical applications of electricity.

One source of a potential difference, as we have seen, is a charged capacitor. The potential difference ΔV_{cap} between the positive and the negative plates is directly proportional to the charge on each plate:

$$\Delta V_{cap} = Ed = \frac{\eta}{\varepsilon_0} \cdot d = \left(\frac{d}{\varepsilon_0 A}\right) \cdot Q. \tag{28-6}$$

Here Q represents a *separation* of charges from two plates that were originally both neutral. That is, N electrons are transferred from the positive to the negative plate, giving the negative plate a charge $Q_{neg} = -Ne$ and leaving the positive plate with $Q_{pos} = +Ne = -Q_{neg}$.

In fact, *any* separation of charge results in a potential difference. Figure 28-13 shows two arbitrary electrodes charged to $+Q$ and $-Q$—a charge separation. There is an electric field $\vec{E}$ that points from the positive toward the negative electrode. As a consequence, Eq. 28-1 tells us that there will be a potential difference between the electrodes given by

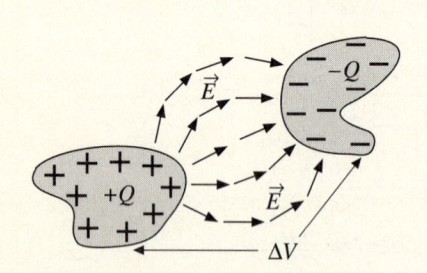

$$\Delta V = V_{pos} - V_{neg} = -\int_{neg}^{pos} E_s \, ds, \tag{28-7}$$

FIGURE 28-13 A potential difference occurs as a result of charge separation.

where the integral runs from a point on the negative electrode to a point on the positive electrode. The key idea is that *we can create a potential difference by creating a charge separation*.

A charged capacitor is a source of potential, but you saw in Chapter 26 that the charge cannot be maintained if we try to "use" the capacitor. Figure 28-14a shows a wire connected to a capacitor that has been charged to potential difference ΔV_{cap}. A charge q at the positive plate has potential energy $U_{top} = q\Delta V_{cap}$. This charge, along with many others, "falls downhill" through the wire and forms the current that discharges the capacitor.

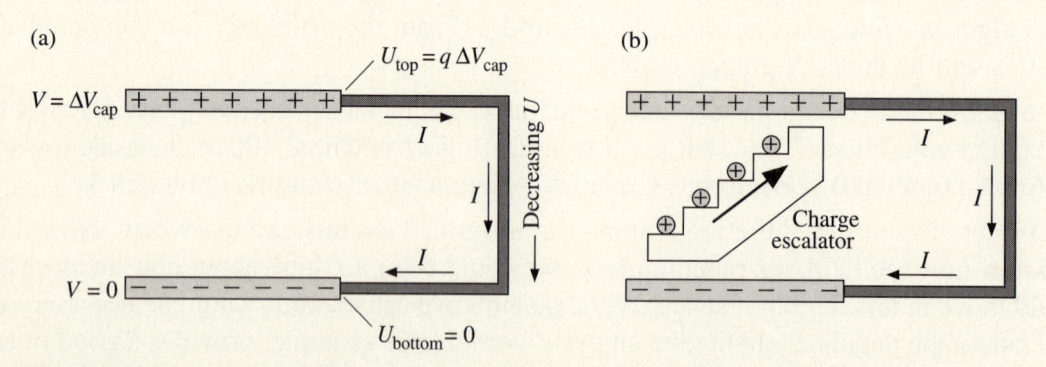

FIGURE 28-14 a) Current flows through a wire connecting the separated charge on two capacitor plates, but it cannot be sustained. b) A "charge escalator" lifts the charge back to the positive plate and sustains the current. This is what a battery does.

Each charge loses potential energy as it falls, reaching the negative plate with $U_{bottom} = 0$. Once the charge reaches the negative plate it has no means to return to the other side.

Suppose, however, that there existed a "charge escalator" between the capacitor plates that could "lift" the charges from the negative side back to the positive, as shown in Fig. 28-14b. Once there, they could again "fall downhill" through the wire and sustain the current. The escalator, by moving charge across, is able to *sustain* the potential difference of the capacitor. Such an escalator would, however, have to be powered from the outside—it is doing work on the charges to "lift" them, and that work is an energy transfer from the outside.

This is exactly what a **battery** does. Rather complex chemical reactions cause a separation of ions within the battery. For our purposes, you simply need to know that these reactions separate charge by moving positive ions to one end of the battery and negative ions to the other end. The ends of the battery are called the positive and negative **terminals**. As a consequence of the charge separation, a potential difference ΔV_{bat} is established between the terminals. The value of ΔV_{bat} is determined by the battery's construction and by the specific chemical reactions employed.

Figure 28-15 shows a wire connected to the terminals of a battery. A charge starts with energy $U_{top} = q\Delta V_{bat}$ at the positive terminal, "falls downhill" through the wire as a current, and reaches the negative terminal with $U = 0$. This much of the process is identical to the current that discharges a capacitor. But now the charge is "lifted" back to the positive terminal by the charge escalator of chemical reactions, regaining energy $q\Delta V_{bat}$ in the process.

The battery does work $W = q\Delta V_{bat}$ on the charge to transport it from the negative to the positive terminal. The energy to perform this work is provided by the chemical reactions within the battery. Without these reactions the battery would not be able to sustain its potential difference or the current in the wire. When the chemicals are used up and the reactions cease, the battery is "dead" or "run down."

The work performed per charge, W/q, is called the **emf** of the battery—pronounced as the sequence of three letters "e-m-f." The symbol for emf is $\mathcal{E}$, a "script E." The units of emf are those of the electric potential—joules per coulomb, or volts. The "rating" of a battery, such as 1.5 V or 9 V, is a measure of the battery's *emf*. Originally the term emf was an abbreviation of "electromotive force."

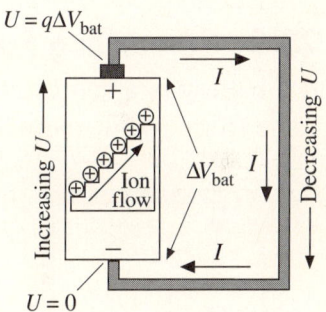

FIGURE 28-15 A battery is a "charge escalator," doing work to "lift" charges to the opposite terminal.

That is an outdated term (work per charge is not even a force!), so today we just call it "emf" and it is not an abbreviation of anything! (*Caution*: The term *emf*, often capitalized as "EMF," is widely used in popular science articles in newspapers and magazines to mean "electromagnetic field." It is used in the debate over whether transmission lines, which generate electromagnetic fields, are a health hazard. This is *not* how we will use the term *emf*.)

For an **ideal battery**, the potential difference between the terminals is simply

$$\Delta V_{bat} = \mathcal{E}. \tag{28-8}$$

That is, the potential difference maintained between the terminals of an ideal battery is numerically equal to the battery's emf. In practice, the potential difference ΔV_{bat} between the terminals of a real battery (often called the *terminal voltage*) is usually slightly less than $\mathcal{E}$. You will learn the reason for this in the next chapter. The difference is often quite small, so we will usually adopt the *ideal battery approximation* and assume that Eq. 28-8 is valid.

Generators, photocells, and other sources of potential difference have different means for separating charges but otherwise function exactly the same as a battery. The common feature of all such devices is that they use a *nonelectrical* means to separate charge and thus to create a potential difference. The emf $\mathcal{E}$ of any device is the work performed per charge as the charge is separated, and its terminal voltage is, to a good approximation, equal to the emf. Thus we will use the term *battery* in a generic sense to mean any potential difference that is sustained by an outside source of energy.

What is especially noteworthy is that a battery is a source of potential difference, *not* a source of current. It is true that a current flows if a wire is attached to the battery, but that is a *consequence* of the battery supplying a potential difference. The battery's emf exists whether or not it is connected to a wire or a circuit, and whether or not a current flows out of the battery. You can think of the battery's emf as being the "causative agent" in a circuit. Current, heat, light, sound, and so on are all *consequences* or *effects* that happen when the battery is used in certain ways. Distinguishing between *cause* and *effect* is extremely important for understanding the practical applications of batteries, such as in circuits, and one of the most common reasoning errors is to forget which is which.

28.5 Connecting Potential and Current

If a battery is a source of potential difference, not current, then where does the current come from? While current does come "from" the battery, the point to be made is that this is a consequence of the battery being a source of potential difference, not because the battery is a source of current. The distinction is subtle but important. Consider a battery of emf $\mathcal{E}$ generating a steady current I in a wire that connects the two battery terminals, as shown in Fig. 28-16. You learned in Section 28.2 that the potential difference between any two points is independent of the path between them. Consequently, the potential difference between the two ends of the wire, along a path through the wire, is equal to the potential difference between the two terminals of the battery:

$$\Delta V_{\text{wire}} = \Delta V_{\text{bat}}. \quad (28\text{-}9)$$

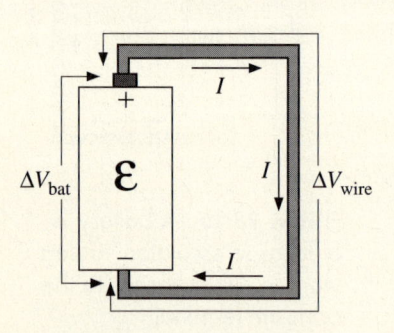

FIGURE 28-16 A wire connected between the terminals of a battery has a potential difference $\Delta V_{\text{wire}} = \Delta V_{\text{bat}}$.

So the battery is the *source* of the potential difference between the ends of the wire.

But if there is a potential difference between the ends of the wire, then there must be an electric field *inside* the wire. And if $\vec{E}_{\text{wire}} \neq 0$, a current will flow! The wire's electric field is related to the potential difference ΔV_{wire} by Eq. 28-1:

$$\Delta V_{\text{wire}} = -\int_0^L E_s \, ds, \quad (28\text{-}10)$$

where L is the length of the wire and s is measured along the wire. Because $\Delta V_{\text{wire}} = V_{\text{pos}} - V_{\text{neg}}$, the s-axis runs from $s = 0$ at the battery's negative end to $s = L$ at the positive end.

But the field vector $\vec{E}_{\text{wire}}$ inside the wire points "downhill," from the positive toward the negative end of the wire. Thus the s-component of $\vec{E}_{\text{wire}}$, along the line of integration, is

$$E_s = -E_{\text{wire}}.$$

With this in mind, Eq. 28-10 gives

$$\Delta V_{\text{wire}} = -\int_0^L (-E_{\text{wire}}) ds = E_{\text{wire}} \cdot \int_0^L ds = E_{\text{wire}} L$$

$$\Rightarrow E_{\text{wire}} = \frac{\Delta V_{\text{wire}}}{L}.$$

(28-11)

This is an important result: The electric field inside a wire—the field that drives the current forward—is simply the potential difference between the ends of the wire divided by the length of the wire. And ΔV_{wire} can be established by connecting the wire to a battery.

Figure 28-17a shows a straightened wire with a potential difference of 5 V between the ends. A series of equally spaced equipotentials between the two ends of the wire is indicated as dotted lines. The electric field $\vec{E}_{\text{wire}}$ is perpendicular to the equipotentials and thus parallel to the wire. This is the field that drives the current I. Notice that the field points in the direction of *decreasing* potential, implying that the current flows in the direction of decreasing potential. This will be an important result for circuit analysis in the next chapter.

Figure 28-17b shows the same wire actually connected to a 5 V battery. The battery is the *source* of the 5 V potential difference between the ends of the wire. Note that the electric field $\vec{E}$ follows the curvature of the wire, always pointing in the direction of decreasing potential, while the field strength $|\vec{E}| = \Delta V_{\text{wire}}/L$ remains constant.

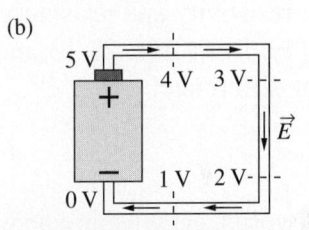

FIGURE 28-17 a) Equipotentials of a current-carrying wire, with $\vec{E}$ pointing in the direction of decreasing potential. b) The same wire connected to a 5 V battery.

Now that we know E_{wire}, from Eq. 28-11, we can find the current I in the wire. As you learned in Chapter 26, the electric field $\vec{E}_{\text{wire}}$ determines the current density

$$J = \sigma E,$$

where $J = I/A$. These are combined to give

$$I = AJ = \sigma A E_{\text{wire}} = \frac{A}{\rho} \cdot E_{\text{wire}},$$

(28-12)

where $\rho = 1/\sigma$ is the *resistivity*, which we tabulated for several metals in Table 26-2. Using Eq. 28-11 for E_{wire} in terms of ΔV gives

$$I = \frac{A}{\rho L} \cdot \Delta V.$$

(28-13)

We have, for convenience, dropped the subscript "wire" because the meaning of ΔV is clear from the context.

It will be useful to define a quantity called the **resistance** R:

$$\text{resistance} = R = \frac{\rho L}{A}. \qquad (28\text{-}14)$$

The resistance is a property of a *specific* wire or conductor because it depends on the conductor's length and thickness as well as on the resistivity of the material.

The SI unit of resistance is the **ohm**, defined as one volt per ampere:

$$1 \text{ ohm} = 1 \, \Omega = 1 \text{ V/A}.$$

The ohm is the basic unit of resistance, although kilohms (kΩ) and megohms (MΩ) are widely used in practice. Now you can see, from Eq. 28-14, why resistivity has units of Ω m, as we introduced in Chapter 26, and conductivity $\sigma = 1/\rho$ has units of $\Omega^{-1} \, \text{m}^{-1}$.

The resistance of a wire increases as the length increases. This seems reasonable, because it should be harder to push electrons through a long wire than a short one. Decreasing the cross-sectional area also increases the resistance. This again seems reasonable, because the same electric field can push more electrons through a fat wire than a skinny one.

It is important to distinguish clearly between resistivity and resistance. Resistivity describes just the *material*, not any particular piece of it. Resistance, on the other hand, characterizes a very specific piece of the conductor having a specific geometry. The relationship between resistivity and resistance is analogous to that between mass density and mass.

Finally, from Eqs. 28-13 and 28-14 we can find the connection between current and potential difference:

$$I = \frac{\Delta V}{R} \quad \text{(current in a wire)}. \qquad (28\text{-}15)$$

In other words, establishing a potential difference ΔV between the ends of a wire of resistance R creates an electric field that, in turn, causes a current $I = \Delta V/R$ to flow through the wire. The smallest the resistance, the larger the current that flows.

• **EXAMPLE 28-7** A nichrome wire 20 cm long and 1 mm in diameter is connected to the ends of a 1.5 V battery. a) What is the wire's resistance? b) What is the electric field inside the wire? c) How much current flows through the wire?

SOLUTION a) The resistance is

$$R = \frac{\rho L}{A} = \frac{\rho L}{\pi r^2}.$$

From Table 26-2 we find that $\rho = 150 \times 10^{-8} \, \Omega$ m for nichrome. After converting L and r to meters, we find $R = 0.382 \, \Omega$.

b) Connecting the wire to the battery implies $\Delta V_{\text{wire}} = \Delta V_{\text{bat}} = \mathcal{E} = 1.5$ V. The electric field, from Eq. 28-11, is

$$E_{\text{wire}} = \frac{\Delta V_{\text{wire}}}{L} = \frac{1.5 \text{ V}}{0.2 \text{ m}} = 7.5 \text{ V/m}.$$

This is not a very large field.

c) The current is $I = \Delta V/R = 3.93$ A.
•

In the next chapter we will look more carefully at how currents flow in circuits. Our purpose here was to establish the connection between a source of potential difference and the current that flows in a wire. The cause-and-effect sequence is the main idea:

1. The battery is a source of potential difference $\Delta V_{bat} = \mathcal{E}$.
2. The battery *causes* a potential difference $\Delta V_{wire} = \Delta V_{bat}$ between the ends of the wire.
3. The potential difference ΔV_{wire} *causes* an electric field $E = \Delta V/L$ inside the wire.
4. The electric field *causes* a current $I = JA = \sigma AE$ to flow.
5. The magnitude of the current flow is determined *jointly* by the battery and the wire's resistance R to be $I = \Delta V/R$.

28.6 Capacitance

We have made frequent use of the idea of a charged parallel-plate capacitor. A capacitor can be charged by attaching the plates, with wires, to a battery. Figure 28-18 shows a capacitor connected to a battery. Once the wires are attached, there is a *brief* flow of current that carries charges to the plates, but then the current stops flowing—it cannot be sustained, unlike the case of a continuous wire connecting the terminals.

How does the current "know" when to stop? We have indicated four potential differences in the figure—ΔV_{bat} of the battery, ΔV_{cap} of the capacitor, ΔV_1 between the two ends of the top wire, and ΔV_2 between the ends of the bottom wire. Because the potential difference between the two battery terminals is independent of the path between them, it must be that

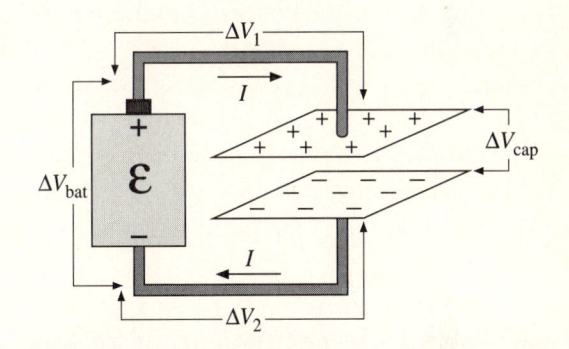

FIGURE 28-18 A parallel-plate capacitor being charged by a battery.

$$\Delta V_{bat} = \Delta V_1 + \Delta V_{cap} + \Delta V_2. \qquad (28\text{-}16)$$

The capacitor plates are not charged when the wires are first connected, so initially $\Delta V_{cap} = 0$. From the symmetry of the set-up there is no reason for ΔV_1 to differ from ΔV_2. So initially, a brief instant after the wires are connected, we must have $\Delta V_1 = \Delta V_2 = \frac{1}{2}\Delta V_{bat}$.

Because there is a potential difference along the wires, currents will begin to flow through them. The current flowing in the bottom wire removes positive charge from the lower plate, leaving it with a negative charge. The "charge escalator" in the battery "lifts" this charge up and send it out as a current through the top wire to the upper plate of the capacitor. The net effect is that the battery removes charge from one plate of the capacitor and transfers it to the other plate. This is what we mean by "charging the capacitor."

As current flows and the charge on the capacitor plates increases, the potential difference ΔV_{cap} increases. Because ΔV_{bat} is fixed, an increasing ΔV_{cap} must result in *decreasing* values of ΔV_1 and ΔV_2. But as they decrease, so does the current. When ΔV_{cap} equals ΔV_{bat}, then $\Delta V_1 = \Delta V_2 = 0$ and the current stops flowing. Thus the capacitor charges until its potential difference equals the battery's potential difference, then it stops charging because there is no longer a potential difference along the wires to cause any further current.

The capacitor's potential difference is related to its electric field by $\Delta V_{cap} = Ed$, where d is the separation of the plates. But

$$E = \frac{Q}{\varepsilon_0 A},$$

where A is the surface area of the plates. Combining these gives

$$\Delta V_{cap} = Ed = \frac{d}{\varepsilon_0 A} \cdot Q$$

$$\Rightarrow Q = \frac{\varepsilon_0 A}{d} \cdot \Delta V_{cap}.$$

(28-17)

In other words, the charge on the capacitor plates is directly proportional to the potential difference between the plates.

The ratio of the charge Q to the potential difference ΔV_{cap} is called the **capacitance** C:

$$capacitance = C = \frac{\varepsilon_0 A}{d}.$$

(28-18)

Capacitance is a purely *geometric* property of two electrodes because it depends only on their surface area and spacing. The SI unit of capacitance is called the **farad**, in honor of Michael Faraday. One farad is equal to 1 coulomb per volt:

1 farad = 1 f = 1 C/V.

One farad turns out to be an enormous amount of capacitance, and practical capacitors are usually measured in units of microfarads (µf) or picofarads (1 pf = 10^{-12} f). (*Note*: Textbooks and practitioners are not consistent with the symbol for farads. Some, as we will do, use a lowercase f, but others prefer an uppercase F. This is one of the rare cases where either is acceptable.)

With this definition of capacitance, Eq. 28-17 can be written as

$$Q = C\Delta V \quad \text{(charge of a capacitor)}.$$

(28-19)

This result is analogous to Eq. 28-15 for the current of a wire. There we found $I = \Delta V/R$—that is, the current in a wire that is attached to a battery is determined jointly by the potential difference supplied by the battery *and* a property of the wire called resistance. Now we see that the charge on a capacitor attached to a battery is determined jointly by the potential difference supplied by the battery *and* a property of the electrodes called capacitance. Equations 28-15 and 28-19 are two of the fundamental relationships governing the behavior of circuits.

• **EXAMPLE 28-8** a) What is the surface area of a 1 µf capacitor having a spacing of 0.05 mm? b) How much charge is on the plates if this capacitor is attached to a 1.5 V battery?

SOLUTION a) From Eq. 28-18, the definition of capacitance, $A = dC/\varepsilon_0 = 5.65$ m^2.
b) The charge is $Q = C\Delta V = 1.5 \times 10^{-6}$ C = 1.5 µC.

The capacitance of Example 28-8 is a quite typical value, but the required surface area is enormous—2.4 m × 2.4 m. In practice, two very large sheets of thin metal foil are separated by a thin insulator ($d \approx 0.05$ mm). The sheets are then rolled up to form a cylinder that is typically 1 cm in diameter and 3 cm long. The insulating sheets and the rolling do have some effect on the capacitance, which you can learn about in more advanced courses, but the main thing to remember is that real capacitors are close cousins to the parallel-plate capacitors that we have studied.

EXAMPLE 28-9 A 0.01 µf capacitor and an aluminum wire that is 50 cm long and 0.1 mm in diameter are both connected to a battery. The capacitor plates are found to have charges of ±90 nC. What is the current through the wire?

SOLUTION Because $\Delta V_{cap} = \Delta V_{bat}$ and $\Delta V_{wire} = \Delta V_{bat}$, connecting both the capacitor and the wire to the battery makes

$$\Delta V_{wire} = \Delta V_{cap} = \frac{Q}{C} = \frac{90 \times 10^{-9} \text{ C}}{0.01 \times 10^{-6} \text{ f}} = 9.0 \text{ V}.$$

The wire's resistance is

$$R = \frac{\rho L}{A} = \frac{\rho L}{\pi r^2} = \frac{(2.8 \times 10^{-8} \, \Omega \text{m})(0.5 \text{ m})}{\pi (5 \times 10^{-5} \text{ m})^2} = 1.8 \, \Omega,$$

where the resistivity of aluminum was taken from Table 26-2. Thus the current through the wire is

$$I = \frac{\Delta V_{wire}}{R} = \frac{9.0 \text{ V}}{1.8 \, \Omega} = 5.0 \text{ A}.$$

Summary

Important Concepts and Terms

equipotential surface
potential map
closed loop
battery
terminals
emf

ideal battery
resistance
ohm
capacitance
farad

We have now explored, in some depth, the connection between the electric field and the electric potential. These are *not* two independent quantities but, instead, two different aspects of how source charges alter the space around them. The field and the potential are closely related to each other, and knowing one allows us to determine the other via

$$\Delta V = -\int_{s_1}^{s_2} E_s \, ds$$

$$E_s = -\frac{dV}{ds}.$$

Only the *change* in potential ΔV is meaningful, not the absolute value, and we can choose the zero of potential to be anywhere that is convenient.

We now have several means of calculating the field and the potential, either directly from the charge distribution or indirectly through the field/potential relationships, and we have looked at several examples.

The importance of the field/potential connection becomes more apparent and important when we look at charges in two and three dimensions. We introduced the concept of an *equipotential surface* and a *potential map*, and you should be able to interpret such a diagram. The electric field vectors are perpendicular to the equipotential surfaces and point in the direction of decreasing potential, allowing us to easily move back and forth between field maps and potential maps.

We have also made a number of practical discoveries about potential and its connection to other ideas and quantities. These include:

1. Batteries are a *source* of potential difference, *not* a current source.
2. A conductor in electrostatic equilibrium has $\vec{E}_{in} = 0$ and $\vec{E}_{out}$ perpendicular to the surface. The conductor is an equipotential.
3. A current-carrying wire is *not* in electrostatic equilibrium. It has a potential difference ΔV_{wire} between the two ends and an internal electric field

$$E = \frac{\Delta V_{wire}}{L}.$$

   As a consequence, the wire carries a current

$$I = \frac{\Delta V_{wire}}{R},$$

   where

$$R = \frac{\rho L}{A}$$

   is the wire's resistance. Resistance is related both to the wire's geometry and to the resistivity ρ of the material.

4. A capacitor connected to a battery becomes charged to the same potential difference as the battery: $\Delta V_{cap} = \Delta V_{bat}$. The charge on the two capacitor plates is

$$Q = \pm C \Delta V_{cap},$$

   where

$$C = \frac{\varepsilon_0 A}{d}$$

   is the capacitance. Capacitance is a *geometric* property of the electrodes.

Exercises and Problems

Exercises

1. Wires 1 and 2 are made of the same metal. Wire 2 is twice as long and twice as wide as wire 1.
   a. What is the ratio ρ_2/ρ_1 of the resistivities of the two wires?
   b. What is the ratio R_2/R_1 of the resistances of the two wires?

2. What is the resistance of
   a. A copper wire that is 1 m long and 0.5 mm in diameter?
   b. A piece of carbon that is 10 cm long and has a 1 mm × 1 mm square cross section?

3. A wire that is 10 m long with a diameter of 0.8 mm has a resistance of 1.1 Ω. What is the material of this wire?

4. Two 2 cm × 2 cm square aluminum electrodes are spaced 0.5 mm apart. The electrodes are attached to a 100 V battery.
   a. What is the capacitance?
   b. What is the charge on each electrode?

5. Two metal plates 2 cm × 2 cm in size are spaced 1 mm apart and connected by wires to the terminals of a 9 V battery. The wires are disconnected, and the plates are then pulled apart to a new spacing of 3 mm. What is the final potential difference between the two plates?

6. Figure 28-19 is a graph of V versus x in a region of space. The potential is independent of y and z.
   a. Draw a graph of E_x versus x in this same region.
   b. Draw a potential map in the xy-plane in the square-shaped region $-3 \text{ m} \leq x \leq 3 \text{ m}$ and $-3 \text{ m} \leq y \leq 3 \text{ m}$. Show and label the equipotential lines for −10 V, −5 V, 0 V, +5 V, and +10 V.
   c. Draw the electric field vectors on your potential map of part b).

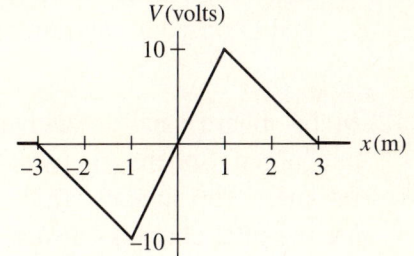

FIGURE 28-19

Problems

7. A 15 cm long nichrome wire is connected across the terminals of a 1.5 V battery.
   a. What is the electric field inside the wire?
   b. What is the current density inside the wire?
   c. If the current in the wire is 2 A, what is the wire's resistance?
   d. What is the wire's diameter?

8. A hollow nichrome tube of inner diameter 2.8 mm, outer diameter 3.0 mm, and length 20 cm is connected to a 3 V battery. What is the current in the tube?

296   CHAPTER 28 · POTENTIAL AND FIELD

9. Figure 28-20 shows a set of equipotential lines and five labeled points.
   a. From measurements made on this figure with a ruler, determine the electric field $\vec{E}$ at the five points indicated. Express your answer as a magnitude and a direction.
   b. Redraw the figure a separate piece of paper, then add to it the electric field vectors $\vec{E}$ at the five points. For each vector, identify and label the angle you used in part a) to specify its direction.

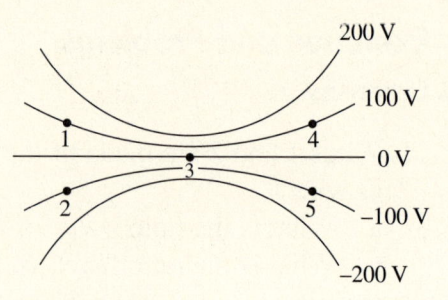

FIGURE 28-20

10. The electric potential in space is given by $V(x, y) = 100(x^2 - y^2)$ volts.
    a. Draw a potential map, showing and labeling the equipotentials lines for –400 V, –100 V, 0 V, +100 V, and +400 V.
    b. Find an expression for the electric field $\vec{E}(x, y)$ at position (x, y).
    c. Show the electric field vectors on your diagram of part a).

11. Figure 28-21 shows the electric potential at points on a 4 cm × 4 cm grid.
    a. Reproduce this figure on a separate sheet of paper. Then draw the 50 V, 75 V, and 100 V equipotential surfaces.
    b. Determine the electric field (magnitude and direction) at points A, B, C, and D.
    c. Draw the electric field vectors at points A, B, C, and D on your diagram of part a).

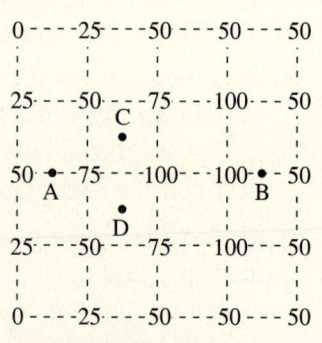

FIGURE 28-21

12. Metal sphere 1 has a positive charge of 6 nC, while metal sphere 2 is initially uncharged. Sphere 2 is twice the diameter of sphere 1, and the spheres are located far apart. The spheres are then connected together by a long, thin metal wire. What are the final charges on each sphere?

13. Consider a dipole of two charges ±q spaced distance s apart along the y-axis and centered at the origin, as shown in Fig. 28-22.
    a. Find an expression for the potential $V(x, y)$ at an arbitrary point in the xy-plane. Your answer will be in terms of q, s, x, and y.
    b. Use the binomial approximations with your result of part a) to find an approximate expression for the potential of a dipole that is valid when $s \ll x$ and $s \ll y$.
    c. Assuming $s \ll x, y$, find expressions for E_x and E_y, the components of $\vec{E}$ for a dipole.
    d. What is the on-axis field $\vec{E}$? Does your result agree with Eq. 25-15?
    e. What is the field $\vec{E}$ on the bisecting axis? Does your result agree with Eq. 25-16?

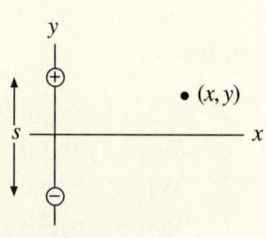

FIGURE 28-22

[**Estimated 12 additional problems for the final edition.**]

Chapter 29

Fundamentals of Circuits

LOOKING BACK | Sections 26.2–26.5; 28.4–28.5

29.1 Practical Electricity

[**Photo suggestion: A circuit board.**]

In this chapter we will bring together many of the ideas of field and potential and apply them to a most practical endeavor—the analysis of electrical circuits. This single chapter will not pretend to be a full course on circuit analysis. You will learn nothing about the design of circuits, nor about fancy things like diodes or transistors. Instead, as the title implies, our more modest goal is to describe the *fundamental physical principles* by which circuits operate. An understanding of these basic principles will prepare you to undertake a course in circuit design and analysis at a later time.

We have been developing a theoretical understanding of charges and their interactions. Several ideas are of particular relevance to electrical circuits:

1. Charge is able to move through conductors but not through insulators. The charge carriers in wires are negative electrons, but this "detail" is not relevant to circuit analysis. We have adopted the convention that a current I is the motion of positive charge carriers.

2. A battery is a source of potential difference that is generated by the chemical separation of charges. Although a battery does supply current to a circuit, it is *not* a "current source." The distinction is subtle but important.

3. When a conducting path, such as a wire, connects the two battery terminals, the battery's potential difference causes an electric field $E = \Delta V_{bat}/L$ in the wire, where L is the wire's length. The electric field in turn generates a current density $J = \sigma E$ and thus a current $I = JA$. The cause and effect sequence is important. The current depends both on the battery's potential difference *and* on the properties (resistance) of the conducting material.

4. The electric field $\vec{E}$ in the wire points in the direction of decreasing potential V, from the higher potential end of the wire toward the lower potential end. As a consequence, the current flows in the direction that the potential decreases. Stated the other way around, the potential of a current-carrying wire decreases along the direction of current flow.

Now we are ready to see how this information can be put to use.

29.2 Resistors and Ohm's Law

[**Photo suggestion: A group of resistors.**]

Why does the filament of a light bulb glow, but not the wires leading to and from it? This may seem to be a silly question, but it contains the essence of circuit analysis. The primary idea is one that we introduced in the last chapter—resistance.

Figure 29-1 shows a conducting material, such as a wire or filament, with a potential difference ΔV between the ends. As a consequence, there is an electric field $\vec{E}$ *inside* the wire and a current I flows through it. As you learned in the last chapter, there is a simple relationship between I and ΔV.

Let the conductor have length L and cross-sectional area A. The electric field strength is $E = \Delta V/L$, causing a current density

$$J = \sigma E = \frac{E}{\rho},$$

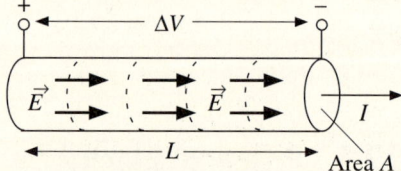

FIGURE 29-1 Current I is related to the potential difference ΔV through the field $\vec{E}$.

where σ is the material's conductivity and $\rho = 1/\sigma$ is its resistivity. These were tabulated for several conductors in Table 26-2. Because the wire's current is $I = JA$, we have

$$I = JA = \frac{E}{\rho}A = \frac{\Delta V/L}{\rho}A = \frac{\Delta V}{(\rho L/A)} = \frac{\Delta V}{R}, \quad (29\text{-}1)$$

where

$$R = \frac{\rho L}{A} \quad (29\text{-}2)$$

is the *resistance* of the conductor. Resistance is a number that characterizes the conducting material; it depends on both the material's properties and its geometry. The units of resistance were defined as 1 ohm = 1 Ω = 1 V/A.

• **EXAMPLE 29-1** a) What is the resistance of a copper wire that is 10 cm long and 1 mm in diameter? b) What potential difference between the wire's ends will cause a current of 3 A to flow through it?

SOLUTION a) Using $\rho = 1.7 \times 10^{-8}$ Ω m for copper and $A = \pi r^2$,

$$R = \frac{\rho L}{A} = \frac{\rho L}{\pi r^2} = 2.2 \times 10^{-3} \text{ Ω}.$$

b) The needed potential difference is
$$\Delta V = IR = 6.6 \times 10^{-3} \text{ V} = 6.6 \text{ mV}.$$

This is a very small potential difference because the resistance of a copper wire is so small.

Because the resistance of metals is so small, a circuit made exclusively of copper wires would have enormous currents and would quickly deplete the battery. It is useful to control and limit the flow of current by using circuit elements which, although they are conductors, have a resistance that is significantly larger than the wires. Such devices are called **resistors**. Resistors are made either from poorly conducting materials, such as carbon, or by depositing extremely thin metal films on an insulating substrate. The cross-sectional area A of a thin film is so small that a film's resistance can be many times larger than that of a solid wire. Circuits typically use resistors ranging from 10 Ω to 1 MΩ. Thus the resistance of the connecting wires, for all practical purposes, is zero.

Light bulbs are common elements in electrical circuits. Figure 29-2 shows the components of a light bulb. The light itself comes from the *filament*, which is a very thin tungsten wire heated by the passage of electric current until it glows white hot. The filament is sealed in a glass envelope with an inert gas (no oxygen) so that it doesn't immediately burn out. Connecting wires go from the filament to both the *tip* and the *side* of the metal base. When the bulb is screwed into a socket, current flows *in* the tip, through the filament, and back *out* the side. So a light bulb, like a wire or a resistor, is a device *through* which a current flows. It is often useful to think of a light bulb as simply a resistor that happens to give off light when current flows through. The resistance of a glowing light bulb is typically in the range from 10 Ω to 500 Ω.

If the resistance R of a device remains constant over a wide range of conditions, the device is said to be **ohmic**. In this case there is a direct proportionality between current and potential difference, as given by Eq. 29-1 and as seen in the graph of Fig. 29-3a. Doubling the potential difference results in a doubling of the current. The slope of the I-versus-ΔV graph is $1/R$.

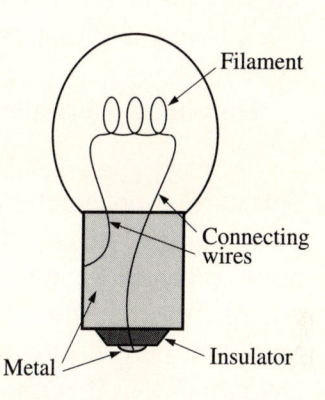

FIGURE 29-2 The anatomy of a light bulb.

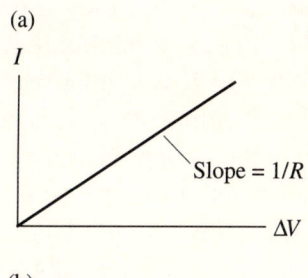

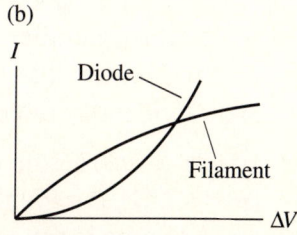

FIGURE 29-3 a) Current is directly proportional to ΔV for an ohmic material. b) Diodes and filaments are examples of nonohmic materials.

Some materials, however, are **nonohmic**, meaning that their resistance is *not* a constant value. We can list several important examples of nonohmic materials:

1. Batteries, where $\Delta V = \mathcal{E}$ is determined by chemical reactions, independent of I.
2. Semiconductors, where R depends on the value of ΔV.
3. Filaments, where R changes significantly as the filament heats up.

Figure 29-3b shows the I versus ΔV graphs of a filament and of a semiconductor device called a *diode*. You can see that the current is *not* directly proportional to the potential difference. The lack of proportionality is why these are nonohmic devices. Resistance is not a very useful number for nonohmic materials, because it does not have a unique value.

For ohmic materials, where R does remain constant, Eq. 29-1 is often written as

$$\Delta V = IR \quad \text{(Ohm's law)}. \tag{29-3}$$

This current-potential difference relationship is called **Ohm's law**.

Despite its name, Ohm's law is *not* a "law" of nature. It has a very limited applicability to just those materials whose resistance remains constant during use. These materials include resistors, but not batteries, semiconductors, or filaments. Ohm's law is an important part of circuit analysis because resistors are essential components of almost any circuit, but it is important that you apply Ohm's law *only* to the resistors and not to anything else.

Circuit textbooks usually write Ohm's law as $V = IR$ rather than $\Delta V = IR$. This can be very misleading until you have sufficient experience to interpret this statement properly. Ohm's law relates the current flow through an ohmic material to the potential *difference* between the ends. As we have seen, only ΔV is well defined and we cannot give a unique value to "the potential" at one point. Engineers and circuit designers *mean* "potential difference" when they use the symbol V, but this use of the symbol is often forgotten by beginners who think that it means "the potential." We will consistently use ΔV rather than V in this text so that the meaning is clear.

Writing $\Delta V = IR$ is perhaps aesthetically more pleasing than $I = \Delta V/R$, but it can convey the wrong message if you are not careful. $\Delta V = IR$ suggests that a current flow I causes a potential difference ΔV. As you have seen, the proper cause-and-effect sequence is the other way around: Current is a *consequence* of a potential difference. Thus $I = \Delta V/R$ is a better description of cause and effect, so we will usually give Ohm's law in this form. (Note that this is similar to our use, in Part I of this text, of $\vec{a} = \vec{F}/m$ rather than $\vec{F} = m\vec{a}$ in order to indicate that force causes acceleration rather than the other way around.)

We can identify three basic classes of circuit materials, based on their resistance:

1. *Wires*: These are metals with very small resistivities ρ and thus very small resistances ($R \ll 1 \, \Omega$). An **ideal wire** has $R = 0$, so the potential difference between the ends of an ideal wire is *zero* even if a current is flowing through it. For most practical purposes you can assume that any wires in a circuit are ideal.

2. *Resistors*: These are "poor" conductors that have resistances typically in the range 10–$10^6 \, \Omega$. They are used to control the current flow in a circuit. A resistor has a constant value of R and thus obeys Ohm's law. Light bulbs function as resistors as long as they are glowing, but the filament is really nonohmic and the value of its resistance

when hot is quite different (much larger) than its room-temperature value. Even though the filament is a metal wire (tungsten), its resistance is high because it has an extremely small cross-sectional area A (a typical diameter is 0.1 mm).

3. *Insulators*: These are nonconductors such as glass, plastic, or air. The ideal insulator has $R = \infty$, so no current ($I = \Delta V/R = 0$) will flow through an insulator even if there is a potential difference. This is why insulators can be used to hold apart two conductors that have different potentials. All practical insulators have $R \gg 10^9 \, \Omega$ and can be treated, for our purposes, as ideal.

Consider a light bulb filament connected to two copper wires, as shown in Fig. 29-4. The filament's resistance when hot is *much* larger than the resistance of the wires. Recall that current is conserved as it flows through a circuit. Therefore the current I flowing through the filament is the same as the current flowing through each wire. However, their potential differences are quite different. Because of its much higher resistance, the potential difference across the filament $\Delta V_{fil} = IR_{fil}$ is *much* larger than the potential difference $\Delta V_{wire} = IR_{wire}$ between the two ends of each wire.

Figure 29-5 shows the potential along the wire/filament combination, from left to right, where we see a large *voltage drop*, or potential difference, across the filament. The voltage drops across the two wires are very small. For ideal wires, having $R_{wire} = 0$, we would have $\Delta V_{wire} = 0$ and $\Delta V_{fil} = \Delta V_{total}$. In that case, the two segments of the Fig. 29-5 graph corresponding to the wires would be perfectly horizontal.

Figure 29-5 is part of the answer to our question of why the filament glows and the wires do not. Because of the large difference in the resistances of the filament and wires, essentially all of the potential difference supplied by a battery occurs across the filament. The wires convey current from the battery to the filament, but there is essentially no potential difference along the wires. We still have to learn how potential differences are related to the heating of the conductors, and that we will do in Section 29.5.

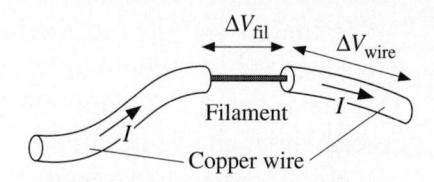

FIGURE 29-4 When a filament is connected to copper wires, ΔV_{fil} is *much* larger than ΔV_{wire}.

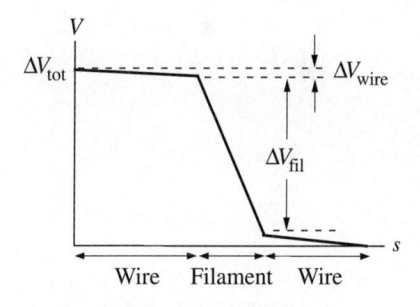

FIGURE 29-5 The potential along the wire/filament combination. Nearly all of the voltage drop occurs across the filament.

29.3 The Basic Circuit

The most basic electrical circuit, from which all others are derived, is that of a single resistor connected to the two terminals of a battery. This is shown in Fig. 29-6. Note that this is a *complete circuit*, forming a continuous path between the battery terminals. The resistor might be a labeled resistor, such as a "10 Ω resistor," or it might be some other resistive device, such as a light bulb. Regardless of what the resistor is, it is called the **load**.

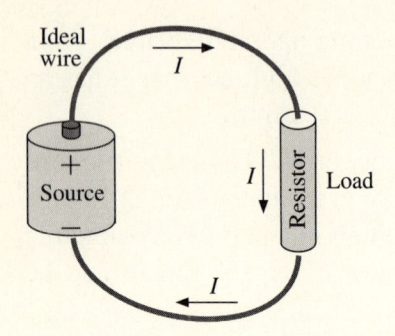

FIGURE 29-6 The "basic circuit" of a resistor connected to a battery.

The battery, or other emf, is called the **source**. We will assume, in the rest of this chapter, that any wires used to connect components are ideal, with $R_{\text{wire}} = 0$.

Our interest is with circuits in which the battery's potential difference is unchanging and all the currents flowing in the circuit are constant. These form what are called **direct current** (DC) circuits. There is a whole class of circuits in which the emf oscillates, in a sinusoidal manner, from positive to negative and back again. This, in turn, causes oscillatory currents called **alternating current** (AC). The wall outlets in your home or apartment provide an oscillating emf, and home appliances are AC circuits. The principles are the same, but the mathematics is more complicated. We will leave AC circuit analysis to more advanced texts.

To understand the functioning of this circuit, it is not necessary to know whether the wires are bent or straight or whether the battery is to the right or to the left of the resistor. The literal "picture" of Fig. 29-6 provides many irrelevant details. It is customary when describing or analyzing circuits to use a more abstract picture called a **schematic diagram** or a *circuit diagram*. A schematic diagram is really just a "logical" picture of what is connected to what. The actual circuit, once built, may *look* quite different from the schematic diagram, but it will have the same logic and connections. A schematic diagram also replaces pictures of the objects with symbols. Figure 29-7 shows the basic symbols that we will need: a battery, a wire, a resistor, a light bulb, and a junction where wires connect together. Notice that the *longer* line on the battery symbol represents the positive terminal of the battery.

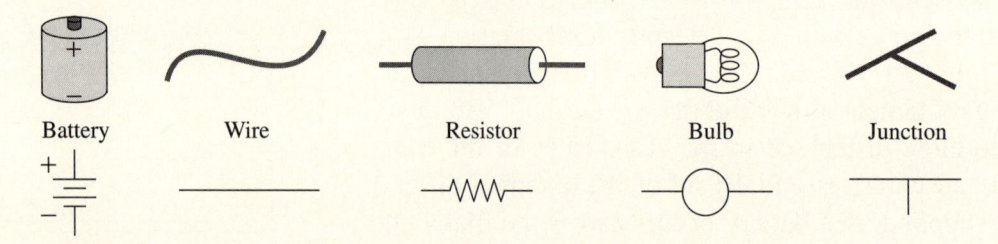

FIGURE 29-7 A library of basic symbols used for electrical circuit drawings.

With these symbols, we can draw a schematic diagram of the basic circuit—which we have done in Fig. 29-8. Notice how things are labeled. The battery's emf $\mathcal{E}$ is shown beside the battery, and + and − symbols, even though somewhat redundant, are shown beside the terminals. The resistance R of the resistor is written beside it. If we knew specific values for $\mathcal{E}$ and R, we would use the numbers on the diagram rather than the algebraic symbols. The wires, which in practice may bend and curve, are shown as straight-line connections between the battery and the resistor. You should get into the habit of drawing your own schematic diagrams to be similar.

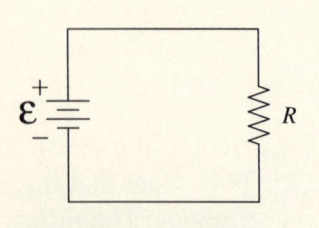

FIGURE 29-8 A schematic diagram of the basic circuit of Fig. 29-6

29.4 Kirchhoff's Laws and Simple Circuits

We are now ready to begin analyzing circuits. To analyze a circuit means a) to find the potential difference across each component in the circuit and b) to find the current flowing through each component of the circuit. The laws of circuit analysis, which are conservation laws, are called *Kirchhoff's laws*.

One of the important properties of potential, which you learned about in Section 28.2, is that the sum of the potential differences around any *closed* path is zero. This is a statement of energy conservation, because a charge that moves around a closed path and returns to its starting point has $\Delta U = 0$. If we apply this idea to a circuit, we can add all of the potential differences around any *closed loop* of the circuit. Doing so gives

$$\sum_{\text{closed loop}} \Delta V_i = 0, \tag{29-4}$$

where ΔV_i is the potential difference of the *i*th component in the loop. This statement is known as **Kirchhoff's loop law**.

You also learned, in Section 26.2, that current is conserved. When currents flow into and out of a junction, where two or more wires join, the total current flowing into the junction must equal the total current leaving the junction. We can state this as

$$\Sigma(\text{input currents}) = \Sigma(\text{output currents}). \tag{29-5}$$

This statement is called **Kirchhoff's junction law**. Kirchhoff's laws are simple statements about conservation, but they form very powerful tools for circuit analysis.

Let's begin by analyzing the basic circuit that was shown in Fig. 29-8. We will go through the steps carefully so that the procedure will be easier to follow for more complicated circuits. As a first step in the analysis, we will redraw the schematic diagram and show the symbols for each potential difference and each current. We have done this in Fig. 29-9.

Because there are no junctions in this circuit, and because current is not used up, the current *I* must be the same through all four sides of the circuit. Figure 29-9 shows this. There are four potential differences to consider: the battery's emf $\mathcal{E}$, the potential difference ΔV_R across the resistor, and the potential differences ΔV_{top} and ΔV_{bottom} along the top and bottom connecting wires. You have learned that the potential *decreases* along the direction in which current flows, so each of the three nonbattery sections has little + and − signs to indicate the more positive and more negative ends of these segments. If you are not sure why the "top" end of the resistor is more positive than the "bottom" end, then go back and review Section 28.5.

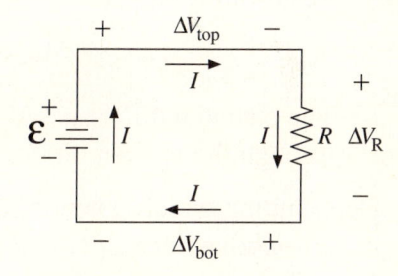

FIGURE 29-9 Analysis of the basic circuit, showing current and potential differences.

To use Kirchhoff's loop law, we want to pick a starting point in the circuit and then sum all of the potential differences around a closed loop. It is customary to "travel" around the loop in the direction that the current flows. Let us choose the lower left corner of the circuit in Fig. 29-9 as our starting point and then add the potential differences in a clockwise fashion. Kirchhoff's loop law then tells us that

$$\Sigma \Delta V_i = \Delta V_{\text{bat}} + \Delta V_{\text{top}} + \Delta V_R + \Delta V_{\text{bot}} = 0. \tag{29-6}$$

Note that all the terms are *added* because this is a statement about the *net* potential difference around a closed loop. Some of the terms may be *negative quantities*, but that information doesn't appear in Eq. 29-6.

We need to be explicit as to what the Δ means in each ΔV. Because $\Delta V = V_f - V_i$, and because we are "traveling" through each component in the direction that the current flows, each ΔV is interpreted to mean

$$\Delta V = V_{\text{downstream}} - V_{\text{upstream}}. \tag{29-7}$$

This is an important piece of information, because without it we would not be sure which of the potential differences are positive and which negative. At least one of the ΔV *must* be negative if the sum of Eq. 29-6 has to be zero.

Now let's look more closely at each of the four terms in Eq. 29-6:

1. The potential *increases* as we travel through the battery on our journey around the loop. We enter the negative terminal and, further downstream, exit the positive terminal after having gained potential $\mathcal{E}$. Thus $\Delta V_{\text{bat}} = +\mathcal{E}$.

2. We are assuming the connection wires between the battery and the resistor to be ideal, with $R_{\text{wire}} = 0$. Thus $\Delta V_{\text{top}} = \Delta V_{\text{bot}} = IR_{\text{wire}} = 0$. No potential difference is required for a current to flow through a material having no resistance.

3. The *magnitude* of the potential difference across the resistor is given by Ohm's law as $\Delta V = IR$, but Ohm's law does not tell us whether this should be positive or negative—and the difference is crucial. The potential of a conductor decreases in the direction that current flows, and that is the direction we are traveling around the loop. Thus the "upstream" end of the resistor is more positive than the "downstream" end, as indicated by the + and − in Fig. 29-9. For the purpose of evaluating Kirchhoff's loop law, this means that

$$\Delta V_R = V_{\text{downstream}} - V_{\text{upstream}} = -IR < 0. \tag{29-8}$$

The potential difference ΔV_R is *negative* because the potential *decreases* as we "travel" through the resistor on our journey around the loop.

Determining which potential differences are positive and which negative is perhaps *the* most important step in circuit analysis, so make sure you understand the "why" of steps 1 and 3.

If we return to Eq. 29-6 with this new information, it becomes

$$\Delta V_{\text{bat}} + \Delta V_{\text{top}} + \Delta V_R + \Delta V_{\text{bot}} = \mathcal{E} - IR = 0. \tag{29-9}$$

We can now solve the loop equation to find the current in the circuit:

$$I = \frac{\mathcal{E}}{R}. \tag{29-10}$$

Finally, we can use the current of Eq. 29-10 in Eq. 29-8 to find the resistor's potential difference:

$$\Delta V_R = -IR = -\mathcal{E}. \tag{29-11}$$

This result should come as no surprise, because the potential "gained" in the battery is subsequently "lost" in the single resistance of the circuit.

Equations 29-10 and 29-11 are the primary result of our circuit analysis. Both $\mathcal{E}$ and R are usually known pieces of information, so Eq. 29-10 can be used to find how much current flows in the circuit. It is worth reiterating that the battery is a voltage source, not a current source, so that the potential difference $\mathcal{E}$ supplied by the battery remains constant. The current, however, depends on the size of the resistance.

EXAMPLE 29-2 A 15 Ω resistor is connected to a 1.5 V battery. a) How much current flows in the circuit? b) Draw a graph showing the potential as a function of distance traveled through the circuit.

SOLUTION a) This is the basic circuit of a single resistor connected to a single battery, as shown in Fig. 29-9. The current is given by Eq. 29-10:

$$I = \frac{\mathcal{E}}{R} = \frac{1.5 \text{ V}}{15 \text{ Ω}} = 0.1 \text{ A}.$$

b) We already know three of the four potential differences: $\Delta V_{bat} = \mathcal{E} = 1.5$ V and $\Delta V_{top} = \Delta V_{bot} = 0$. The resistor's potential difference is found from Eq. 29-10: $\Delta V_R = -\mathcal{E} = -1.5$ V. Figure 29-10 is a graphical means of showing how the potential changes as the current flows around the circuit. The distance s is measured from our starting point at the lower left corner, and we have chosen to let $V = 0$ at that point. The potential increases by 1.5 V as the current passes through the battery, does not change as it flows through the ideal wires, and drops by 1.5 V through the resistor. The potential ends back at the value from which it started.

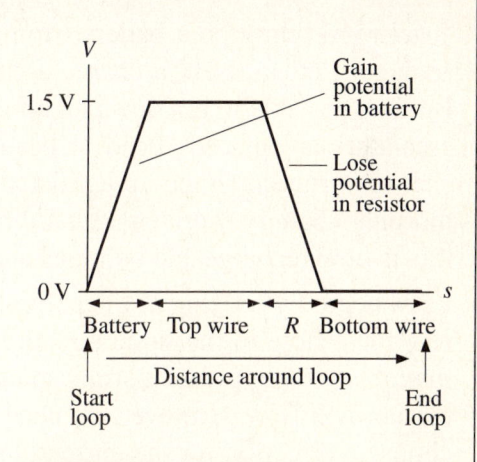

FIGURE 29-10 A graphical presentation of how the potential changes around the loop of the circuit.

This completes the analysis of the basic circuit. We have determined the current through and the potential difference across every component of the circuit. The four basic steps of the analysis can be summarized as follows:

1. Write Kirchhoff's loop law for all the potential differences around a closed loop.
2. Evaluate each ΔV by using Ohm's law (for resistors) or the inherent properties of the component (for batteries). Pay very careful attention to signs!
3. Solve Kirchhoff's loop law for the current I.
4. Use the current and Ohm's law to find values of ΔV for each resistor.

A somewhat more complex example will help clarify the procedure.

EXAMPLE 29-3 Analyze the circuit shown in Fig. 29-11a. a) Find the magnitude and direction of the current flow and the potential difference across each component. b) Draw a graph showing how the potential changes throughout the circuit.

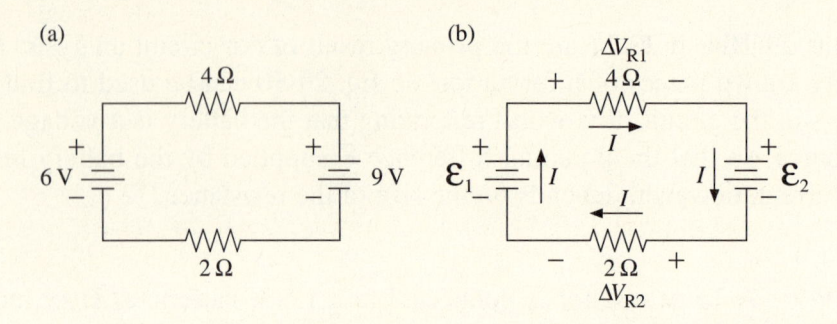

FIGURE 29-11 a) Circuit for Example 29-3. b) Current and potential differences established for the analysis of this circuit.

SOLUTION a) This circuit presents several new challenges. How do we deal with *two* batteries? How do we know which way the current will flow? Can current flow "backward" through a battery, from positive to negative? Consider the charge escalator analogy. Left to itself, a charge escalator will lift charge from lower to higher potential. But it *is* possible to run down an up-escalator—as many of you have probably tried. If two escalators are placed "head to head," whichever is the stronger will, indeed, force the charge to run down the up-escalator of the other battery. So current *can* run "backward" through a battery *if* driven in that direction by a larger emf from a second battery. Indeed, this is how rechargeable batteries are recharged.

Let's begin by looking at the current flow. Because there are no junctions, the same current must flow in the same direction through *each* component in the circuit. With some thought we might successfully guess which way—clockwise or counterclockwise—the current will flow. However, we need not know in advance of our analysis. We will simply choose a direction for the current, call it I, and solve for the value of I. If our solution is positive, then the current really does flow in the chosen direction, but if the solution should turn out to be negative, we will know that the flow direction is opposite the one we chose. Figure 29-11b shows that we have chosen the clockwise direction for current I, and we have labeled the battery emf's as $\mathcal{E}_1$ and $\mathcal{E}_2$. The resistors' potential differences are ΔV_{R1} and ΔV_{R2}. You have seen that $\Delta V_{wire} = 0$ for ideal wires, so we do not need to include them. The + and − beside each resistor are based on the assumed current direction.

Kirchhoff's loop law for this circuit is

$$\sum \Delta V = \Delta V_{bat\ 1} + \Delta V_{R1} + \Delta V_{bat\ 2} + \Delta V_{R2} = 0,$$

where we started from the lower left corner and proceeded in the direction of I. As before, *all* the signs are + because this is a formal statement of *adding* potential differences around the loop. Next we can evaluate each ΔV. As we go in the clockwise direction the current *gains* potential in battery 1 but *loses* potential in battery 2. Thus $\Delta V_{bat\ 1} = +\mathcal{E}_1$ while $\Delta V_{bat\ 2} = -\mathcal{E}_2$. There is a *loss* of potential in traveling through each resistor, so $\Delta V_{R1} = -IR_1$ and $\Delta V_{R2} = -IR_2$. Thus Kirchhoff's loop law becomes

$$\sum \Delta V = \mathcal{E}_1 - IR_1 - \mathcal{E}_2 - IR_2 = \mathcal{E}_1 - \mathcal{E}_2 - I(R_1 + R_2) = 0$$

$$\Rightarrow I = \frac{\mathcal{E}_1 - \mathcal{E}_2}{R_1 + R_2}.$$

Using the given values of $\mathcal{E}_1$, $\mathcal{E}_2$, R_1, and R_2, we can calculate

$$I = \frac{6\text{ V} - 9\text{ V}}{4\text{ }\Omega + 2\text{ }\Omega} = -0.5\text{ A}.$$

Because the current I turned out to be negative, we conclude that the current in this circuit is 0.5 A flowing *counterclockwise*. We could have anticipated this from the orientation of the larger 9 V battery.

b) The potential difference across the 4 Ω resistor is

$$\Delta V_{R1} = -IR_1 = +2.0\text{ V}.$$

Because the current circulates counterclockwise, the resistor's potential *increases* in the clockwise direction of our travel around the loop. Similarly, the potential difference across the 2 Ω resistor is $\Delta V_2 = 1.0$ V. Figure 29-12 is a graph of potential versus position, following a clockwise path around the loop starting from the lower left-hand corner. Notice how the potential *drops* 9 V upon passing through battery 2 in this direction, but then gains 2 V upon passing through R_2 and ends at the starting potential.

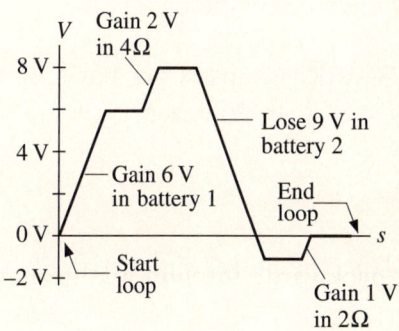

FIGURE 29-12 Graph of potential differences for the circuit of Example 29-3.

29.5 Energy and Power

A battery not only supplies a potential difference, it also supplies the *energy* to a circuit. The charge escalator is an energy transfer process, transferring chemical energy E_{chem} stored in the battery to the potential energy U of the charges. A battery can operate only so long as it has chemicals to keep this energy transfer going.

Because the battery "lifts" each charge q through a potential difference ΔV_{bat}, it must provide to each charge an increase of potential energy $\Delta U = q\Delta V_{\text{bat}}$. But as we found in Chapter 28, an ideal battery's potential difference is just $\Delta V_{\text{bat}} = \mathcal{E}$, the emf. So the battery supplies energy $\Delta U = q\mathcal{E}$ to charge q as the charge moves from the negative to the positive terminal.

It is useful to know the *rate* at which the battery supplies energy to the charges. The rate at which energy is supplied, you will recall from Chapter 9, is *power*, measured in joules per second or *watts*. If energy $\Delta U = q\mathcal{E}$ is transferred to charge q, then the *rate* at which energy is transferred from the battery to the charges is

$$P_{\text{bat}} = \text{rate of energy transfer} = \frac{dU}{dt} = \frac{dq}{dt}\mathcal{E}. \qquad (29\text{-}12)$$

But dq/dt, the rate at which charge moves through the battery, is none other than the current I. So we are led to the conclusion that the power supplied by a battery, or the rate at which the battery transfers energy to a current I passing through it, is

$$P_{\text{bat}} = I\mathcal{E}. \qquad (29\text{-}13)$$

Because $q\mathcal{E}$ is a potential energy, with units of joules, $I\mathcal{E}$ has units of joules per second, or watts.

EXAMPLE 29-4 A 3 V battery delivers 2 A of current to a light bulb. What is the power delivered by this battery?

SOLUTION This is a straightforward calculation of $P_{bat} = I\mathcal{E} = (2\text{ A})(3\text{ V}) = 6\text{ W}$.

EXAMPLE 29-5 A 90 Ω load is connected to a 120 V battery. How much power is delivered by the battery?

SOLUTION This is our basic battery-and-resistor circuit that we analyzed above, finding $I = \mathcal{E}/R$. In this case

$$I = \frac{\mathcal{E}}{R} = \frac{120\text{ V}}{90\text{ Ω}} = 1.33\text{ A}.$$

Thus the power supplied is $P_{bat} = I\mathcal{E} = 160\text{ W}$.

P_{bat} is the energy transferred per second from the battery's store of chemicals to the moving charges that make up the current. But what happens to this energy? Where does it end up? Figure 29-13 shows a section of a current-carrying resistor. Our microscopic understanding of conductivity, you will recall from Section 26.2, is that the charge carriers alternately accelerate in the direction of $\vec{E}$, then collide with atoms in the lattice, coming to rest and starting over again. The acceleration phase is a conversion of potential to kinetic energy, while the collisions are a *heat transfer* of the charge carrier's microscopic kinetic energy to the thermal energy of the lattice. The potential energy originates in the battery, from the conversion of chemical energy, so the entire energy transfer process looks like

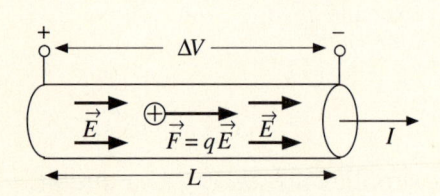

FIGURE 29-13 A current-carrying resistor dissipates power because the electric force does work on the charges.

$$E_{chem} \rightarrow U \rightarrow K \rightarrow E_{therm}.$$

The *net* result is that the battery's chemical energy is transferred to the thermal energy of the resistors, raising their temperature.

Now let's look at the transfer process in the resistor. The electric force $\vec{F}$ exerted on each charge q by the electric field, $\vec{F} = q\vec{E}$, does work as it pushes the charge through length L of wire. The field is constant inside the resistor, so the force must also be constant. The work done on charge q is then a simple product:

$$W = F\Delta s = FL = qEL. \tag{29-14}$$

According to the work-kinetic energy theorem, this work increases the kinetic energy of each charge q by

$$\Delta K = W = qEL. \tag{29-15}$$

This kinetic energy is transferred to the lattice when charge q collides with a lattice atom, causing the thermal energy of the lattice to increase:

$$\Delta E_{therm} = \Delta K = qEL. \tag{29-16}$$

But EL is just the potential difference ΔV_R between the two ends of the resistor, so we are left with the conclusion that *each* charge q, as it travels the length of the resistor and undergoes a potential decrease ΔV_R, transfers energy to the atomic lattice in the amount

$$\Delta E_{therm} = q \Delta V_R. \tag{29-17}$$

The *rate* at which energy is transferred from the current to the resistor is thus

$$P_R = \frac{dE_{therm}}{dt} = \frac{dq}{dt} \cdot \Delta V_R = I \Delta V_R. \tag{29-18}$$

We say that this power—so many joules per second—is *dissipated* by the resistor as current flows through it. The resistor, in turn, transfers this energy to the air and to the circuit board on which it is mounted, causing the circuit and all its surroundings to heat up.

A primary result of our analysis of the basic circuit, in which a battery is connected to a single resistor, was that $\Delta V_R = \mathcal{E}$—that is, the potential difference across the resistor is exactly the emf supplied by the battery. But then Eqs. 29-13 and 29-18, for P_{bat} and P_R, are numerically equal and we find that

$$P_R = P_{bat}.$$

This answers the question of where the energy supplied by the battery goes: It is transformed into the thermal energy of the resistor. The *rate* at which the battery supplies energy is exactly equal to the *rate* at which the resistor dissipates energy. This is, of course, exactly what we would have expected from energy conservation.

EXAMPLE 29-6 How much current is "drawn" by a 100 W light bulb connected to a 120 V outlet?

SOLUTION Many household appliances have a "power rating," such as a 100 W light bulb or a 1500 W hair dryer. The rating does *not* mean that these appliances *always* dissipate that much power. These appliances are intended for use with a "standard" household voltage of 120 V, and their rating is the power they will dissipate *if* operated with a potential difference of 120 V. Their power consumption will differ from the rating if they are operated at any other potential difference. Because the light bulb in this example is operating as intended, it will dissipate 100 W of power. Thus

$$I = \frac{P}{\Delta V} = \frac{100 \text{ W}}{120 \text{ V}} = 0.833 \text{ A},$$

In other words, a current of 0.833 A flowing through this light bulb will transfer 100 joules per second to the thermal energy of the filament, which, in turn, will dissipate 100 joules per second as heat and light to its surroundings.

EXAMPLE 29-7 Most household circuits contain a 20 A fuse which "blows" if the current through it exceeds 20 A. This is an important safety feature to prevent fires or electrocution from faulty electrical devices. What is the maximum power that can be delivered by a circuit that contains a 20 A fuse?

SOLUTION If I_{max} = 20 A, the maximum power that a 120 V household circuit can deliver is

$$P_{max} = I_{max} \Delta V = (20 \text{ A})(120 \text{ V}) = 2400 \text{ W} = 2.4 \text{ kW}.$$

Many common appliances that involve heat, such as a hair dryer, toaster, or popcorn maker, are rated at $P \approx 1500$ W. If you try to use two such appliances simultaneously, you will probably blow the fuse.

Most of the loads in which we are interested are resistors obeying Ohm's law, $\Delta V = IR$. This gives us two alternative ways of writing the power dissipated by the load. We can either substitute IR for ΔV_R or substitute $\Delta V/R$ for I, giving

$$P_R = I\Delta V_R = I^2 R$$
$$P_R = I\Delta V_R = \frac{(\Delta V_R)^2}{R}. \tag{29-19}$$

If the same current I flows through several conductors in series, then $P = I^2 R$ tells us that most of the power will be dissipated by the largest resistance. This is the last piece of information we need to understand why the light bulb filament glows but the wires do not. Because $R_{filament} \gg R_{wires}$ and because the current through the filament is the same as the current through the wires, we see from Eq. 29-19 that $P_{filament} \gg P_{wires}$. In fact, an ideal wire with $R_{wire} = 0$ would dissipate no power at all! Essentially *all* of the power supplied by the battery is dissipated by the high-resistance filament and essentially no power is dissipated by the wires. That is why the filament gets very hot, until it glows, but the wires do not.

EXAMPLE 29-8 Most loudspeakers have a resistance of 8 Ω. If an 8 Ω loudspeaker is connected to a stereo amplifier with a rating of 100 W, what is the maximum current that might flow through the loudspeaker?

SOLUTION The "rating" of an amplifier is the *maximum* power it can deliver. Most of the time it delivers far less, but the maximum might be reached for brief, intense sounds like cymbal clashes. Because the loudspeaker is a resistive load, $P_{max} = (I_{max})^2 R$ and thus

$$I_{max} = \sqrt{\frac{P_{max}}{R}} = \sqrt{\frac{100 \text{ W}}{8 \text{ Ω}}} = 3.5 \text{ A}.$$

EXAMPLE 29-9 How much power is dissipated by a 60 W light bulb when operated, using a dimmer switch, at 100 V?

SOLUTION The 60 W rating is for "normal" operation at 120 V. We can use this rating to find the filament's resistance:

$$R = \frac{(\Delta V)^2}{P} = \frac{(120 \text{ V})^2}{60 \text{ W}} = 240 \text{ Ω}.$$

When operated at $\Delta V = 100$ V the power dissipation will be

$$P = \frac{(\Delta V)^2}{R} = \frac{(100 \text{ V})^2}{240 \text{ Ω}} = 42 \text{ W}.$$

Note: This really is not quite correct. A filament, as we noted previously, is not an ohmic material, because the resistance changes with temperature. The filament's resistance at 100 V, where it glows less brightly and the temperature is lower, will not be quite the same as its resistance at 120 V. The voltage in this example is still near 120 V, so the temperature of the filament will decrease only slightly and our answer should be fairly good. However, this calculation would not give a good result for the power dissipation at 20 V, where the filament's temperature would be much less than at 120 V. •

Now we know what is "used up" by a light bulb or other load attached to a battery. It is not the current that is used up. Instead, it is the stored chemical energy of the battery that is used up. It is used by being converted first to the potential energy of the charges, from there to the thermal energy of the filament, thus heating it, and lastly to the heat and light energy that we feel and see coming from the bulb. The law of conservation of energy is of paramount importance for understanding electrical circuits.

29.6 Combinations of Resistors

Most of our attention thus far has been focused on circuits with a single resistor. But many practical circuits contain two or more resistors connected to each other in various ways. Thus much of circuit analysis consists of analyzing different *combinations* of resistors.

For example, Fig. 29-14a shows a combination of three resistors placed end-to-end between points a and b. Resistors that are aligned end-to-end, with no junctions between them, are said to be connected in **series**. Because there are no junctions, and because current is conserved, the current I must be the same through each of these resistors. That is, the current flowing out of the last resistor in a series is equal to the current flowing into the first resistor. This is the primary fact about series resistances.

FIGURE 29-14 a) Three resistors in series. b) The same current flows through all three resistors. c) An equivalent resistor.

Each resistor in the series has a potential difference $\Delta V_i = IR_i$. The total potential difference ΔV_{ab} between points a and b, as shown in Fig. 29-14b, is the sum of the individual potential differences:

$$\Delta V_{ab} = \Delta V_1 + \Delta V_2 + \Delta V_3$$
$$= IR_1 + IR_2 + IR_3 \qquad (29\text{-}20)$$
$$= I(R_1 + R_2 + R_3).$$

But the resistance R_{ab} between points a and b, by definition, is

$$R_{ab} = \frac{\Delta V_{ab}}{I}. \qquad (29\text{-}21)$$

312 CHAPTER 29 FUNDAMENTALS OF CIRCUITS

Using Eq. 29-20 for *I*, we find the resistance between points a and b to be
$$R_{ab} = R_1 + R_2 + R_3. \quad (29\text{-}22)$$

In other words, the three resistors R_1, R_2, and R_3 act exactly the same as a *single* resistor of value $R_1 + R_2 + R_3$. Neither the current nor the potential difference between points a and b will change if the three separate resistors are replaced by the single resistor R_{ab}. We can say that the single resistor R_{ab} is *equivalent* to the three resistors in series.

There was nothing special about having chosen three resistors to be in series. If we have *N* resistors in series, the **equivalent resistance** is
$$R_{eq} = R_1 + R_2 + \cdots + R_N \quad \text{(series resistors)}. \quad (29\text{-}23)$$

In other words, the behavior of the circuit will be unchanged if the *N* series resistors are replaced by the single resistor R_{eq}.

EXAMPLE 29-10 a) What is the current in the circuit of Figure 29-15a? b) Find the potential difference of each resistor. c) Draw a graph of potential versus position in the circuit.

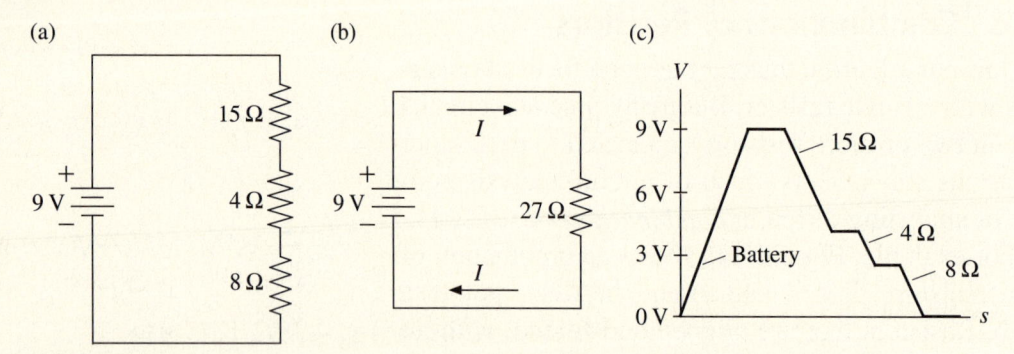

FIGURE 29-15 a) Series resistor circuit of Example 29-10. b) Equivalent circuit. c) Graph of potential in circuit.

SOLUTION a) The circuit contains three series resistors. Nothing about the circuit's behavior will change if we replace them by their equivalent resistance:
$$R_{eq} = 15\ \Omega + 4\ \Omega + 8\ \Omega = 27\ \Omega.$$

This is shown as an equivalent circuit in Fig. 29-15b. Now we have a circuit with a single battery and a single resistor, for which we know the current to be
$$I = \frac{\mathcal{E}}{R_{eq}} = \frac{9\ \text{V}}{27\ \Omega} = 0.333\ \text{A}.$$

Because the current is the same in the original circuit of Fig. 29-15a and the equivalent circuit of Fig. 29-15b circuit, the current flowing through each of the three resistors in the original circuit is identical to the 0.333 A current flowing through R_{eq}.

b) The potential difference across resistor R_i is $\Delta V_i = IR_i = 0.333 R_i$. This gives potential differences of 5 V, 1.33 V, and 2.67 V for the 15 Ω, the 4 Ω, and the 8 Ω resistors, respectively.

c) Figure 29-15c shows that the potential increases by 9 V due to the battery's emf, then decreases by 9 V in three steps.

Figure 29-16a shows a different combination of three resistors between the two points c and d. Resistors that are aligned side by side, with their ends joined at a junction, are said to be connected in **parallel**. Because the left ends of the three resistors are all connected, the left ends are all at the same potential V_c. Likewise, the right ends are all at the same potential V_d. Thus the potential *differences* ΔV_1, ΔV_2, and ΔV_3 are the *same* for all three resistors and are simply ΔV_{cd}, as shown in Fig. 29-16b. This is key to the analysis of the parallel resistors.

Kirchhoff's junction law applies at the junctions. The input current I splits into three separate currents I_1, I_2, and I_3 at the left junction. On the right, the three currents are recombined into current I. According to the junction law,

$$I = I_1 + I_2 + I_3. \qquad (29\text{-}24)$$

Using Ohm's law for each resistor, along with the fact the $\Delta V_1 = \Delta V_2 = \Delta V_3 = \Delta V_{cd}$, the current is

$$\begin{aligned}I &= \frac{\Delta V_1}{R_1} + \frac{\Delta V_2}{R_2} + \frac{\Delta V_3}{R_3} \\ &= \frac{\Delta V_{cd}}{R_1} + \frac{\Delta V_{cd}}{R_2} + \frac{\Delta V_{cd}}{R_3} \\ &= \Delta V_{cd} \cdot \left[\frac{1}{R_1} + \frac{1}{R_2} + \frac{1}{R_3}\right].\end{aligned} \qquad (29\text{-}25)$$

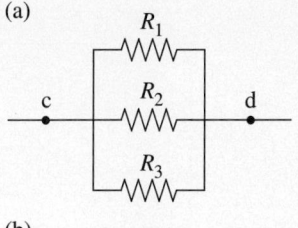

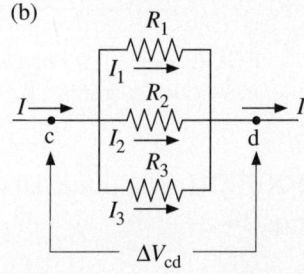

But the resistance between points c and d is $R_{cd} = \Delta V_{cd}/I$. Thus

$$R_{cd} = \left[\frac{1}{R_1} + \frac{1}{R_2} + \frac{1}{R_2}\right]^{-1}, \qquad (29\text{-}26)$$

which is more easily written as

$$\frac{1}{R_{cd}} = \frac{1}{R_1} + \frac{1}{R_2} + \frac{1}{R_2}.$$

We can again say that R_{cd} is the *equivalent* of the three resistors R_1, R_2, and R_3. Neither the current I nor the potential difference ΔV_{cd} will change if these three resistors are replaced by the single resistor R_{cd}, as shown in Fig. 29-16c.

In general, for N resistors in parallel, the equivalent resistance is

$$\frac{1}{R_{eq}} = \frac{1}{R_1} + \frac{1}{R_2} + \cdots + \frac{1}{R_N} \quad \text{(parallel resistors)}. \qquad (29\text{-}27)$$

FIGURE 29-16
a) Resistors in parallel.
b) All three resistors have the same potential difference. c) An equivalent resistor.

EXAMPLE 29-11 The three resistors of Example 29-10 are connected in parallel to a 9 V battery, as shown in Fig. 29-17a. Find the potential difference across and current through each resistor.

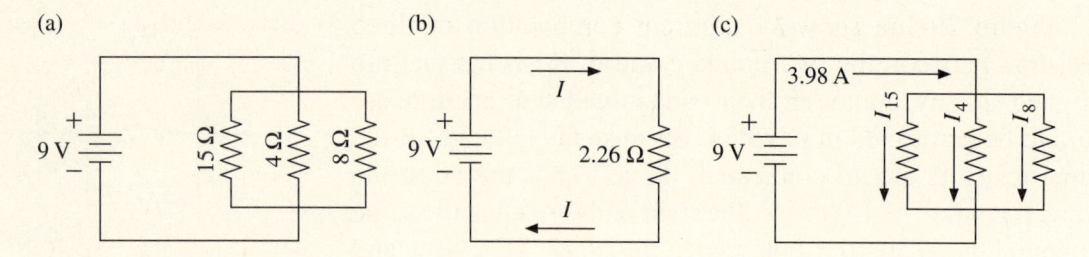

FIGURE 29-17 a) Parallel resistor circuit of Example 29-11. b) Equivalent circuit. c) The individual currents.

SOLUTION The three parallel resistors can be replaced by a single equivalent resistor. From Eq. 29-27,

$$\frac{1}{R_{eq}} = \frac{1}{15\ \Omega} + \frac{1}{4\ \Omega} + \frac{1}{8\ \Omega} = 0.4417\ \Omega^{-1}$$

$$\Rightarrow R_{eq} = \left[0.4417\ \Omega^{-1}\right]^{-1} = 2.26\ \Omega.$$

The equivalent circuit is shown in Fig. 29-17b, from which we find the current to be

$$I = \frac{\mathcal{E}}{R_{eq}} = \frac{9\ \text{V}}{2.26\ \Omega} = 3.98\ \text{A}.$$

The potential difference across R_{eq} is $\Delta V_{eq} = \mathcal{E} = 9.0$ V. Now we have to be careful. The current through the three original resistors is *not* 3.98 A. Current *I divides* at the junction into the smaller currents I_{15}, I_4, and I_8. Furthermore, the division is *not* into three equal currents, as you might expect. According to Ohm's law, resistor *i* has current $I_i = \Delta V_i / R_i$. Because the resistors are in parallel, their potential differences are equal:

$$\Delta V_{15} = \Delta V_4 = \Delta V_8 = \Delta V_{eq} = 9.0\ \text{V}.$$

So we can calculate

$$I_{15} = 0.60\ \text{A} \quad I_4 = 2.25\ \text{A} \quad I_8 = 1.13\ \text{A}.$$

Notice that the *sum* of the three currents is 3.98 A, as required by Kirchhoff's junction law.

The result of Example 29-11 seems surprising. The equivalent of a parallel combination of 15 Ω, 4 Ω, and 8 Ω was calculated to be 2.26 Ω. How can the equivalent of a group of resistors be *less* than any single resistor in the group? Shouldn't more resistors imply more resistance? That is true for resistors in series, but not for resistors in parallel. Even though a resistor is an obstacle to the flow of current, parallel resistors provide *more pathways* for current to get through. As an analogy, several parallel water pipes can carry more water than a single pipe, so they are *equivalent* to a larger-diameter pipe with less resistance. The equivalent of *N* resistors in parallel is always *less* than any single resistor in the group.

If only two resistors R_1 and R_2 are in parallel, we can solve explicitly for R_{eq}. From Eq. 29-27,

$$\frac{1}{R_{eq}} = \frac{1}{R_1} + \frac{1}{R_2} = \frac{R_1 + R_2}{R_1 R_2}$$

$$\Rightarrow R_{eq} = \frac{R_1 R_2}{R_1 + R_2}.$$

(29-28)

If the two parallel resistors are identical, with $R_1 = R_2 = R$, then

$$R_{eq} = \frac{R}{2}.$$

(29-29)

That is, two equal resistors in parallel have an equivalent resistance that is *half* the value of either. Contrast this with two equal resistors in series, which have an equivalent resistance $R_{eq} = 2R$.

Quite complex combinations of resistors can be reduced to a single equivalent resistance through a step-by-step application of the series and parallel rules. The examples in the remainder of this chapter will illustrate these ideas.

EXAMPLE 29-12 What is the equivalent resistance for the group of resistors shown in Fig. 29-18a?

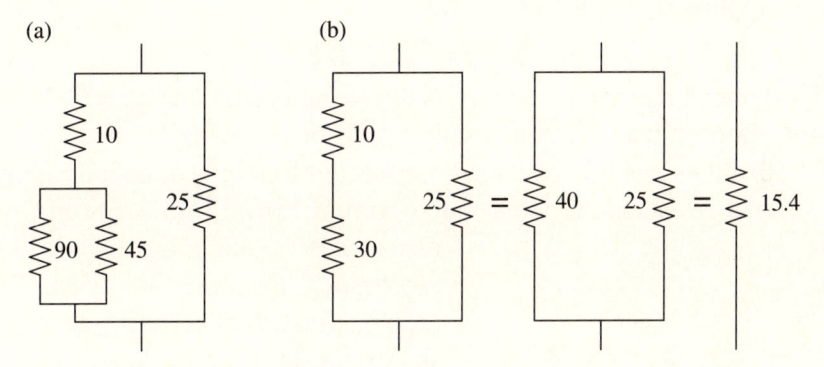

FIGURE 29-18 a) A combination of resistors. b) Step-by-step reduction to a single equivalent resistor.

SOLUTION Reduction to a single equivalent resistance is best done in a series of steps, redrawing the circuit after each step. We have shown this in Fig. 29-18b. First, the 45 Ω and 90 Ω resistors are in parallel, so they can be replaced by an equivalent resistor that is calculated from Eq. 29-28 to be 30 Ω. (Note that the 10 Ω and 25 Ω resistors are *not* in parallel. They are connected together at their top ends but not at their bottom ends. Resistors must be *directly* connected at *both* ends to be in parallel. Similarly, the 10 Ω and 45 Ω resistors are *not* in series because of the junction between them.) For the next step, the new 30 Ω resistor is in series with the 10 Ω resistor, having an equivalent resistance of 40 Ω. Last, the 40 Ω resistor is in parallel with the 25 Ω resistor, giving a final equivalent resistance of 15.4 Ω. If the original group of four resistors occurred within a larger circuit, they could be replaced with a single 15.4 Ω resistor without having any effect on the rest of the circuit.

29.7 Real Batteries

In previous sections we've assumed that the source of the potential in a circuit is an ideal battery. Now let's look at how real batteries differ from the ideal we have been assuming. Like ideal batteries, real batteries separate charge and create a potential difference. However, real batteries also provide a slight resistance to any current flowing through them. They have what is called an **internal resistance**, which is symbolized by r. Figure 29-19 shows both an ideal and a real battery, with the real battery consisting of both an emf $\mathcal{E}$ *and* an internal resistance r.

From our vantage point outside a battery we cannot see $\mathcal{E}$ and r separately—they are both inside the battery. The battery simply provides to us, the user, a potential difference ΔV_{bat} between its positive and negative terminals. But the presence of the internal resistance does affect ΔV_{bat}. Suppose current I is flowing through the battery. As current travels from the negative to the positive terminal, it first gains potential $\mathcal{E}$ but then *loses* potential Ir due to the internal resistance. The potential difference supplied by the battery is thus

$$\Delta V_{bat} = \mathcal{E} - Ir \leq \mathcal{E}. \qquad (29\text{-}30)$$

Only when $I = 0$, meaning that the battery is not being used, is $\Delta V_{bat} = \mathcal{E}$. It is worthwhile to see how this affects a basic circuit having a load of one resistor.

Figure 29-20 shows a single resistor R connected to the terminals of a battery having emf $\mathcal{E}$ and internal resistance r. How much current flows in this circuit? Although the resistances R and r are physically separated, *electrically* they are in series. Thus we can replace them, for the purpose of circuit analysis, with a single equivalent resistor $R_{eq} = R + r$, as shown in the figure. The current in the circuit is thus

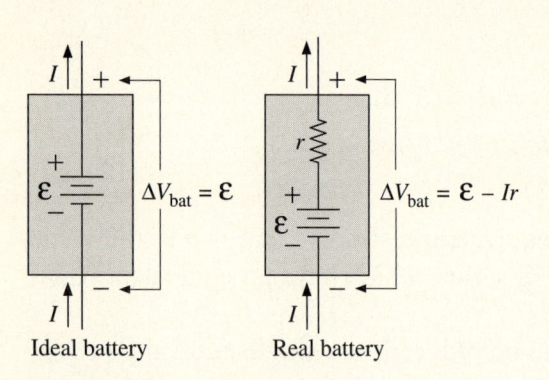

FIGURE 29-19 An ideal battery and a real battery.

$$I = \frac{\mathcal{E}}{R_{eq}} = \frac{\mathcal{E}}{R+r}. \qquad (29\text{-}31)$$

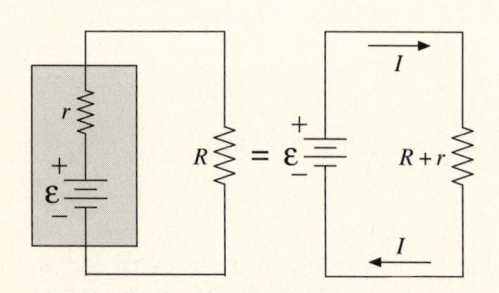

FIGURE 29-20 A single resistor connected to a real battery is in series with the battery's internal resistance, giving $R_{eq} = R + r$.

If $r \ll R$, so that the internal resistance of the battery is negligible, then $I = \mathcal{E}/R$ —exactly as we found in Eq. 29-10. But if $r \approx R$, the current is significantly decreased due to the internal resistance of the battery.

This current causes the potential difference across the load resistor R to be

$$\Delta V_R = IR = \frac{R}{R+r} \cdot \mathcal{E}. \qquad (29\text{-}32)$$

Similarly, the potential difference across the terminals of the battery is

$$\Delta V_{\text{bat}} = \mathcal{E} - Ir = \mathcal{E} - \frac{r}{R+r} \cdot \mathcal{E} = \frac{R}{R+r} \cdot \mathcal{E}. \quad (29\text{-}33)$$

We see that the potential difference across the resistor is equal to the potential difference between the *terminals* of the battery, which is where the resistor is attached, but not to the emf of the battery.

EXAMPLE 29-13 A 6 Ω flashlight bulb is powered by a 3 V battery having an internal resistance of 1 Ω. a) What is the power dissipation of the bulb? b) What is the terminal voltage of the battery?

SOLUTION a) Figure 29-20 is a schematic diagram of the flashlight circuit. R is the resistance of the bulb's filament. The current, from Eq. 29-31, is

$$I = \frac{\mathcal{E}}{R+r} = \frac{3 \text{ V}}{6\,\Omega + 1\,\Omega} = 0.43 \text{ A}.$$

This is 15% less than the 0.5 A an ideal battery would supply. The potential difference across the bulb is

$$\Delta V_R = IR = 2.6 \text{ V}.$$

The power dissipation is thus $P = I\Delta V_R = 1.1$ W.

b) The battery's terminal voltage is given by Eq. 29-33 as

$$\Delta V_{\text{bat}} = \frac{R}{R+r} \cdot \mathcal{E} = \frac{6\,\Omega}{6\,\Omega + 1\,\Omega} \cdot 3 \text{ V} = 2.6 \text{ V}.$$

1 Ω is a typical internal resistance for a flashlight battery. You can see that the internal resistance causes the battery's voltage to be 0.4 V less than the battery's emf.

Suppose that we replace the resistor in the circuit of Fig. 29-20 with an ideal wire having $R_{\text{wire}} = 0$, as shown in Fig. 29-21. When a connection of low or zero resistance is made between two points in a circuit that are normally separated by a higher resistance, then we have what is called a **short circuit**. We say that the wire in Fig. 29-21 is *shorting out* the battery. If the battery were ideal, shorting it with an ideal wire ($R = 0$) would cause the current to be $I = \mathcal{E}/0 = \infty$. The current, of course, cannot really become infinite. Instead, the battery's internal resistance r becomes the only resistance in the circuit. If we use Eq. 29-31 with $R = 0$, we find that the *short circuit current* is

$$I_{\text{short}} = \frac{\mathcal{E}}{r}. \quad (29\text{-}34)$$

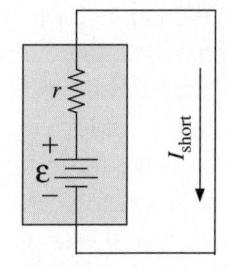

FIGURE 29-21
The short circuit current of a battery.

A 3 V battery with 1 Ω internal resistance generates a short circuit current of 3 A. This is the maximum possible current that this battery can deliver. Adding any external resistance R will decrease the current to a value less than 3 A.

EXAMPLE 29-14 a) What is the short circuit current of a 12 V car battery with an internal resistance of 0.02 Ω? b) What happens to the power generated by this battery?

SOLUTION a) From Eq. 29-34 we find that

$$I_{short} = \frac{\mathcal{E}}{r} = \frac{12 \text{ V}}{0.02 \text{ Ω}} = 600 \text{ A}.$$

This is very realistic. Car batteries are designed to drive the starter motor, which has a very small resistance and can draw a current of a few hundred amps. That is why the battery cables are so thick. A shorted car battery can produce an enormous amount of current.

b) Power is generated by chemical reactions in the battery and dissipated by the load resistance as $P = I^2R$. But with a short-circuited battery, the load resistance is *inside* the battery! The shorted battery has to dissipate power $P = I^2r = 7200$ W *internally*. The normal response of a shorted car battery is to explode—it simply cannot dissipate this much power any other way. Shorting flashlight batteries can make them rather hot, but your life is not in danger. Car batteries, by contrast, are truly dangerous, even though the voltages are relatively small, and should be treated with great respect if you work on cars or need to "jump start" a dead battery with a working battery.

Most of the time a battery is used under conditions in which $r \ll R$ and the internal resistance is negligible. The ideal battery assumption is fully justified in that case. We will assume, in problems, that batteries are ideal *unless stated otherwise*. You should realize, though, that batteries (or any other source of emf) do have an internal resistance, that the internal resistance can become important when the load has a very small resistance, and that it is the internal resistance that controls the maximum possible current of the battery.

29.8 Resistive Circuits

We can use the information of this chapter to analyze a variety of more complex but also more realistic circuits. This will give us a chance to bring together the many ideas of this chapter and to see how they are used in practice. We will look at examples of both qualitative reasoning and quantitative calculations.

One important factor to keep in mind when analyzing circuits is that you need to maintain a "global perspective" in which you think about the circuit as a whole. If you alter the circuit at one point, perhaps by altering the value of a resistor, it is likely that *all* the currents and potential differences in the circuit will change as a result. The change is not limited to just the altered component.

EXAMPLE 29-15 Figure 29-22 shows three identical light bulbs attached to a battery. All of the bulbs are glowing. a) What is the "order of brightness" of the bulbs? That is, which is brightest, second brightest, and so on? b) Suppose bulb B is removed from its socket. What happens to the brightness of bulbs A and C? c) When bulb B is removed, does the potential difference ΔV_{12} between points 1 and 2 increase, decrease, or stay the same?

SOLUTION a) We assume, because not informed otherwise, that the battery and wires are ideal. Bulbs A and B are resistors in series, the A + B combination is in parallel with bulb C, and both the A + B combination and C are connected to the battery terminals with ideal wires. The potential difference ΔV_C must be equal to the potential difference ΔV_{A+B} across the A + B combination, with both being equal to ΔV_{bat}. Because A and B are identical, it follows that

$$\Delta V_A = \Delta V_B = \tfrac{1}{2}\Delta V_{bat} = \tfrac{1}{2}\Delta V_C$$

$$\Rightarrow I_A = I_B = \frac{\tfrac{1}{2}\Delta V_{bat}}{R} = \tfrac{1}{2}I_C.$$

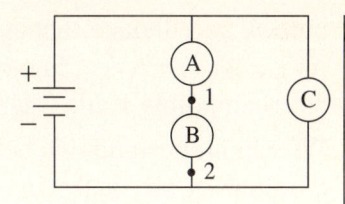

FIGURE 29-22 A three-bulb circuit.

In other words, sides A + B and C have the same potential difference, because they are in parallel, but side A + B has twice the resistance and therefore only half the current as side C. Because the current determines the bulb brightness, we conclude that A and B are equally bright but dimmer than C. We can state this as C > (A = B). Note two errors to avoid. First, current is not used up as it passes through bulbs, so A = B rather than A > B. Second, the current from the battery does *not* divide equally at the junction, so the brightness of C is *not* equal to that of A and B. The current that travels down each side depends on the resistance of that side. Side C has less resistance than A + B, so more current goes through bulb C.

b) When bulb B is removed, there is a "break" in the A + B side and *no* current can flow through. Without current, bulb A goes dark. A naive guess would be that the current gets sent instead to C, making it brighter, but this is *not* what happens. The current through C is determined by $I = \Delta V_C/R$, and ΔV_C is always equal to ΔV_{bat} because C is connected directly to the battery with ideal wires. Thus the current I_C is completely unaffected by anything that happens on side A + B. Removing B has no effect on the brightness of C. (The naive guess would be based on an assumption that a battery is a *current* source, always pumping out the same current that has to go somewhere. This is not the case. The battery is a source of potential difference, which remains constant, and the current is determined by the amount of resistance present in the circuit.)

c) Care is called for in evaluating the potential difference ΔV_{12} when bulb B is removed. Before B was removed, $\Delta V_{A+B} = \Delta V_{bat}$ and this difference was equally divided between the two bulbs to give $\Delta V_B = \Delta V_{12} = \tfrac{1}{2}\Delta V_{bat}$. With the bulb gone, it is tempting to reason "there is no resistance and there is no current, so it must be that $\Delta V_{12} = 0$." *No!* "No resistor" is *not* the same thing as "no resistance." Removing the bulb leaves an air gap, which is an insulator, so the *resistance* between points 1 and 2 is $R_{12} = \infty$, not zero. Furthermore, Ohm's law applies only to current flowing through *resistors*, not to insulators without current flow.

The proper way to understand the situation is to use Kirchhoff's loop law for the closed loop consisting of the battery and side A + B. This tells us that $\Delta V_{bat} - \Delta V_A - \Delta V_B = 0$, where we have used minus signs because the potential *decreases* as we "travel" through the bulbs. With both bulbs in place, $\Delta V_A = \Delta V_B$ and Kirchhoff's law gives $\Delta V_A = \Delta V_B = \tfrac{1}{2}\Delta V_{bat}$, as we had already surmised. With bulb B removed, we *can* apply Ohm's law to bulb A and deduce that $\Delta V_A = 0$ because $I_A = 0$. But then Kirchhoff's law implies that $\Delta V_B = \Delta V_{12} = \Delta V_{bat}$. In other words, the full potential difference of the battery will appear between points 1 and 2 when bulb B is removed.

We can understand this better by noting that point 2 is connected to the battery's negative terminal by an ideal wire, so point 2 is at the same potential as the negative terminal. If bulb A is *not* glowing, so that there is no potential difference across the filament, then point 1 is connected to the battery's positive terminal by a conducting path along which $\Delta V = 0$. Thus point 1 is at the same potential as the positive terminal. In this case, the potential difference between 1 and 2 is *identical* to the potential difference between the positive and negative battery terminals, ΔV_{bat}. We conclude that ΔV_{12} will *increase* when bulb B is removed.

EXAMPLE 29-16 The circuit of Fig. 29-23 has three identical light bulbs. Initially the switch is open so that only bulbs A and B are glowing. Then the switch is closed. What happens to the brightness of bulbs A and B?

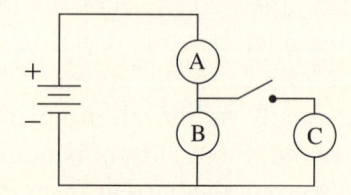

FIGURE 29-23 A circuit with a switch and three bulbs.

SOLUTION A *switch* is simply two ideal wires that are brought into contact as the switch closes. Once closed, it functions identically to an ideal wire. It is tempting here to look only at bulb B and think that nothing will happen to bulb A, but that is thinking "locally" rather than "globally."

Initially A and B are in series and are equally bright. Closing the switch puts C in parallel with B, and the B + C combination in series with A. The equivalent resistance of B + C is *less* than the resistance of B alone, because that is the nature of parallel resistances, so the *total* resistance of A + (B + C) is *less* than the original resistance of A + B. Consequently, *more* current will flow from the battery after the switch closes. All of that current passes through A, so closing the switch will *increase* the brightness of A. Most people are surprised by this because they have not realized the "global" implications of adding bulb C to the circuit.

Because B and C are identical, the current will divide equally at the junction. I_B and I_C will each be half of I_A, so B and C will be dimmer than A. But simply knowing that B is dimmer than A doesn't tell us how B *changes* when the switch is closed. We need to compare the current I_B without bulb C to the current I_B with bulb C. Initially the total resistance of A + B is $2R$, so current $I_B = \frac{1}{2}\mathcal{E}/R$ passes through A and B. With C added, $R_{B+C} = \frac{1}{2}R$, because they are in parallel, and the equivalent resistance of B + C in series with A is $R + \frac{1}{2}R = \frac{3}{2}R$. The total current delivered by the battery is then $I_{bat} = \mathcal{E}/R_{eq} = \frac{2}{3}\mathcal{E}/R$. This current increase from the battery makes A brighter. But the current through B is only *half* of the current supplied by the battery, so $I_B = \frac{1}{3}\mathcal{E}/R$. This is *less* current than before C was added. The conclusion: B will grow dimmer when the switch is closed.

The qualitative reasoning demonstrated in the last two examples carries over into quantitative circuit analysis. While calculating specific numbers is important in practice, it is equally important to *understand* the behavior of a circuit and to realize, without calculations, the implications of changing part of the circuit.

EXAMPLE 29-17 Find the current through and the potential difference across each of the four resistors in the circuit shown in Fig. 29-24.

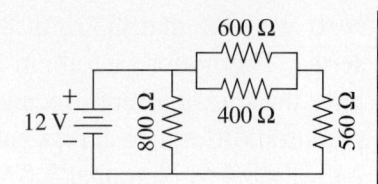

FIGURE 29-24 A complex multiresistor circuit for analysis.

SOLUTION There are two aspects to the analysis of a more complex circuit such as this. First, break it down step by step into a one-battery, one-resistor circuit. This is a matter of using series and parallel resistor combinations. Second, build the circuit back up, step by step, finding the currents and potential differences at each step. Let's begin by looking at the resistor combinations. In this example, the 600 Ω and 400 Ω resistors are in parallel and combine to an equivalent 240 Ω. The 240 Ω is then in series with 560 Ω to give an equivalent 800 Ω. Last, the two 800 Ω resistors are in parallel and combine to give 400 Ω. These replacements are shown in Fig. 29-25a. The equal signs imply that each version of the circuit is equivalent to the circuit in the preceding step. It is highly recommended that you follow the same step-by-step strategy in your own circuit analysis.

FIGURE 29-25 a) Decomposing the circuit of Fig. 29-24 into a single equivalent resistor. b) Rebuilding the circuit to find the currents and potential differences. Currents are given in mA.

The final circuit of Fig. 29-25a, with a battery and single resistor, is our basic circuit, which we know how to analyze. The current is

$$I = \frac{\mathcal{E}}{R} = \frac{12 \text{ V}}{400 \text{ Ω}} = 0.030 \text{ A} = 30 \text{ mA}.$$

The potential difference across the resistor is $\Delta V_{400} = \Delta V_{bat} = \mathcal{E} = 12$ V. We can now use the potential difference as the starting point for "rebuilding" the circuit, as shown in Fig. 29-25b.

The steps of Fig. 29-25b repeat the steps of Fig. 29-25a exactly, but in reverse order. The 400 Ω resistor came from two 800 Ω resistors in parallel. The major fact about parallel resistances is that all the individual resistors *and* R_{eq} have the *same* potential difference. Because $\Delta V_{400} = 12$ V, it must be true that each $\Delta V_{800} = 12$ V. The current through each 800 Ω resistor is then given by $I = \Delta V/R = 15$ mA. Note that 15 mA + 15 mA = 30 mA, so Kirchhoff's junction law is satisfied at the junction.

Next we note that the right 800 Ω resistor was formed by 240 Ω and 560 Ω resistors in series. The major fact about series resistances is that all the individual resistors *and* R_{eq} have the *same* current. Because I_{800} = 15 mA, it must be true that $I_{240} = I_{560}$ = 15 mA. The potential difference across each is given by Ohm's law as $\Delta V = IR$, so ΔV_{240} = 3.6 V and ΔV_{560} = 8.4 V. Note that 3.6 V + 8.4 V = 12 V = ΔV_{800}, so the potential differences add as they should.

Finally, the 240 Ω resistor came from 600 Ω and 400 Ω resistors in parallel. They will each have the same 3.6 V potential difference as their 240 Ω equivalent, and we can find each of their currents from $I = \Delta V/R$. This gives I_{600} = 6 mA and I_{400} = 9 mA. Note that 6 mA + 9 mA = 15 mA, so the junction law is again satisfied. We now know all currents and potential differences, the original circuit has been "rebuilt," so the analysis is complete.

•

It is important to notice two things about Example 29-17. First, the use of a very systematic procedure. Second, we *checked our work* at each step by verifying that currents summed properly at junctions and that potential differences summed properly along a series of resistances. This "check as you go" procedure is extremely important. It provides you, the problem solver, with a built-in error finder. It is nearly impossible to do an incorrect circuit analysis if you always check your current and voltage sums after each step. They will immediately inform you if a mistake has been made.

29.9 Getting Grounded

[**Photo suggestion: Three-prong plug.**]

People who work with electronics are often heard to talk about things being "grounded." It always sounds quite serious, perhaps somewhat mysterious. What is it? Why do it?

The circuit analysis procedures we have been discussing so far deal only with potential *differences*. We talk about the potential difference across a resistor or the potential difference across a battery. This goes back to our basic definition of potential, where all that we really defined was ΔV rather than V itself. As noted there, we are free to choose the zero-point of potential anywhere that is convenient. Our analysis of circuits has not revealed any need to establish a zero-point; potential differences are all we need.

Difficulties can begin to rise, however, if we want to connect two *different* circuits together. Perhaps you would like to connect your CD player to your amplifier or your computer monitor to the computer itself. Although we need not get into any details, incompatibilities can arise unless all the circuits to be connected have a *common* reference point for potential.

You learned previously that the earth itself is a conductor. Suppose we have two circuits. If we connect *one* point of each circuit to the earth by an ideal wire, and we also agree to call the potential of the earth V_{earth} = 0, then both circuits have a common reference point. But notice something very important: *One* wire connects the circuit to the earth, but there is no second wire returning to the circuit. That is, the wire connecting the circuit to the earth is not part of a *complete circuit*, so there is *no current* flowing through this wire! The presence of the connecting wire, because it is an equipotential, makes one point in the circuit have the same potential as the earth, but it does *not* in any way change how the circuit functions. A circuit connected to the earth in this way is said to be **grounded**, and the wire is called the *ground wire*.

Figure 29-26a shows a fairly simple circuit with a 10 V battery and two resistors in series. The symbol beneath the circuit is the *ground symbol*. In this circuit, the symbol indicates that a wire has been connected between the earth and the wire that goes from the 12 Ω resistor to the negative battery terminal. This ground wire does not make a complete circuit, so no current flows through it. Consequently, the presence of the ground wire does not affect the circuit's behavior. The total resistance is 8 Ω + 12 Ω = 20 Ω, so a current $I = (10\text{ V})/(20\text{ Ω}) = 0.5$ A flows around the loop. The potential differences across the two resistors are found, using Ohm's law, to be $\Delta V_8 = 4$ V and $\Delta V_{12} = 6$ V. These are the same values of the current and the potential differences that we would find if the ground wire were *not* present. So what has grounding the circuit accomplished?

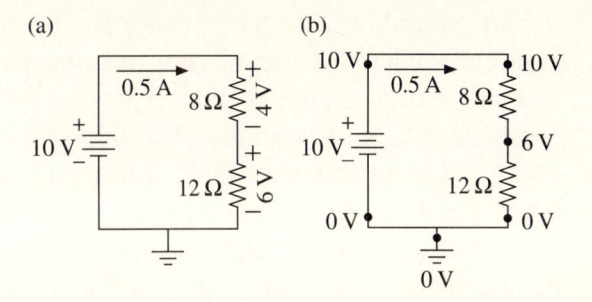

FIGURE 29-26 a) A circuit that is grounded at one point. b) The potential at various points in the circuit, with $V_{earth} = 0$.

The potential is shown at several points in Fig. 29-26b. By definition, $V_{earth} = 0$. Now, because of the ground wire, both the negative battery terminal and the bottom of the 12 Ω resistor are connected by ideal wires to the earth, so the potential at these two points must also be zero. The positive terminal of the battery is 10 V more positive than the negative terminal, so $V_{neg} = 0$ implies $V_{pos} = +10$ V. Similarly, the fact that the potential *decreases* by 6 V as current flows through the 12 Ω resistor now implies that the potential at the junction of the resistors must be +6 V. The potential difference across the 8 Ω resistor is 4 V, so the top has to be at +10 V if the bottom is at +6 V. This agrees with the potential at the positive battery terminal, as it must because these two points are connected by an ideal wire.

Grounding the circuit has not changed the current nor has it changed any of the potential differences. All that grounding the circuit does is allow us to have *specific values* for the potential at each point in the circuit, rather than always having to use potential differences. Now we can say, "The voltage at the resistor junction is 6 V," whereas before all we could say was that "there is a 6 V potential difference across the 12 Ω resistor." There is one particularly important lesson from this: Nothing happens in a circuit "because" it is grounded. You cannot use "because it is grounded" to *explain* anything about a circuit's behavior. Being grounded *does not affect* the circuit's behavior under normal conditions!

There is one exception, which is why we added "under normal conditions." Most circuits have a case of some sort around them. This case is not connected to the circuit; the circuit is held apart with insulators. Sometimes, however, a circuit breaks or malfunctions in some way such that the case comes into electrical contact with the circuit. If the circuit uses high voltage, or even ordinary 120 V household voltage, anyone touching the case could be injured or killed by electrocution. To prevent this, many appliances or electrical instruments have the case itself grounded. This ensures that the potential of the case will always remain at 0 V and be safe. If a malfunction occurs that connects the case to the circuit, a large current will flow through the ground wire to the earth and cause a fuse to blow. This is the *only* time a current would ever flow through the ground wire, and it is *not* a normal operation of the circuit.

324 CHAPTER 29 FUNDAMENTALS OF CIRCUITS

So grounding a circuit serves two functions. First, it provides a common reference potential so that different circuits or instruments can be correctly interconnected. Second, it is an important safety feature to prevent injury or death from a defective circuit. For this reason you should *never* tamper with or try to defeat the ground connection (the third prong) on an instrument's plug. If it has a ground connection, then it *needs* a ground connection and you should not try to plug it in to a two-prong ungrounded outlet. Grounding the instrument does not affect its operation *under normal conditions*, but the abnormal and the unexpected are always with us. Play it safe.

EXAMPLE 29-18 Suppose the circuit of Fig. 29-26 were grounded at the junction between the two resistors instead of at the bottom. Find the potential at each corner of the circuit.

SOLUTION Figure 29-27 shows the new circuit. (It is customary to draw the ground symbol so that its "point" is always down.) Changing the ground point does not affect the circuit's behavior. There is still a current of 0.5 A, and the potential differences across the two resistors are still 4 V and 6 V. All that has happened is that we have moved the $V = 0$ reference point. Because the earth has $V_{earth} = 0$ and we have connected the earth through a wire to the resistors' junction, the junction itself now has a potential of zero. The potential decreases by 4 V as current flows through the 8 Ω resistor. Because it *ends* at 0 V, the potential at the top of the 8 Ω must be +4 V. Similarly, the potential decreases by 6 V through the 12 Ω resistor. Because it *starts* at 0 V, the bottom of the resistor must be at −6 V.

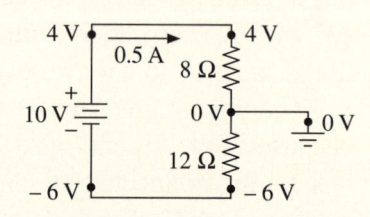

FIGURE 29-27 Circuit of Fig. 29-26 if grounded at the point between the resistors.

The negative battery terminal is at the same potential as the bottom of the 12 Ω resistor, because they are connected by a wire, so $V_{neg} = -6$ V. Finally, the potential increases by 10 V as the current flows through the battery, so $V_{pos} = +4$ V, in agreement, as it should be, with the potential at the top of the 8 Ω.

You may wonder about the negative voltages that are now in the circuit. A negative voltage means only that the potential *at that point* is less than the potential at some other point which we have decided to call $V = 0$. Only potential *differences* are physically meaningful, and only potential *differences* enter into Ohm's law: $I = \Delta V/R$. The potential *difference* across the 12 Ω resistor in this example is 6 V, decreasing from top to bottom, regardless of which point we choose to call $V = 0$.

Summary
Important Concepts and Terms

resistor
ohmic
nonohmic

Ohm's law
ideal wire
load

source	series resistors
direct current	equivalent resistance
alternating current	parallel resistors
schematic diagram	internal resistance
Kirchhoff's loop law	short circuit
Kirchhoff's junction law	grounded

This chapter concludes our study of electricity and prepares the way for a look at magnetism. We have tried to tie together theory and observation, and we hope to leave you with an understanding that circuits obey the same fundamental physical laws about charge that you first met in electrostatics. The difference is one of technique rather than physics—namely, that batteries and wires allow us to move charge about under controlled conditions.

The basic quantities that characterize circuits are potential differences ΔV, currents I, and resistances R. Resistance is a property of a conducting material, one that determines the amount of current that can flow through the material. Given a potential difference, usually supplied by a battery, the current through a conductor of resistance R is given by Ohm's law:

$$I = \frac{\Delta V}{R}.$$

However, it is important to remember that Ohm's law has strict limitations as to its applicability. Only ohmic materials—those whose resistance remains constant—are correctly described by Ohm's law.

The behavior of the circuit is governed by Kirchhoff's two laws:

$$\sum_{\text{closed loop}} \Delta V_i = 0 \qquad \text{Kirchhoff's loop law}$$

$$\sum(\text{input currents}) = \sum(\text{output currents}) \quad \text{Kirchhoff's junction law.}$$

These are basic statements about conservation of charge and conservation of energy. Conservation of energy is also relevant when we consider energy and power in circuits. The battery is the source of energy, and its chemical energy is transferred first to the potential energy of charges and then, ultimately, to the thermal energy of the resistors. The rate at which the battery supplies energy, what we call the power delivered to the circuit, is

$$P_{\text{bat}} = I\mathcal{E}$$

and the power consumed or dissipated by each resistor is

$$P_{\text{res}} = I\Delta V = I^2 R = \frac{(\Delta V)^2}{R}.$$

Resistors often occur in various combinations. Two standard patterns can be reduced to a single equivalent resistance:

Series resistors:
$$R_{\text{eq}} = R_1 + R_2 + R_3 + \cdots$$
$$I_{\text{eq}} = I_1 = I_2 = I_3 = \cdots$$

Parallel resistors:
$$1/R_{\text{eq}} = 1/R_1 + 1/R_2 + 1/R_3 + \cdots$$
$$\Delta V_{\text{eq}} = \Delta V_1 = \Delta V_2 = \Delta V_3 = \cdots$$

326 CHAPTER 29 FUNDAMENTALS OF CIRCUITS

Quite complex circuits can be analyzed with a careful step-by-step application of resistor combinations and Kirchhoff's laws. In performing such an analysis, Kirchhoff's laws should be used at every step to ensure that no mistakes have been made.

Exercises and Problems

Exercises

1. What is the equivalent resistance between points a and b for the circuit of Fig. 29-28?

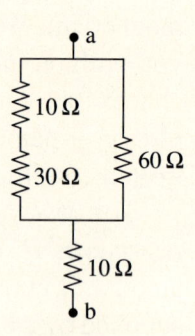

   FIGURE 29-28

2. Consider the circuit shown in Fig. 29-29. Find the current, the potential difference, and the power dissipation of each resistor.

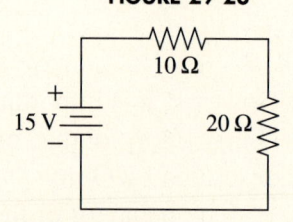

   FIGURE 29-29

3. Consider the circuit shown in Fig. 29-30.
   a. How much current flows through the 30 Ω resistor, and in which direction?
   b. Draw a graph of potential versus position, starting at the lower left corner of the circuit.

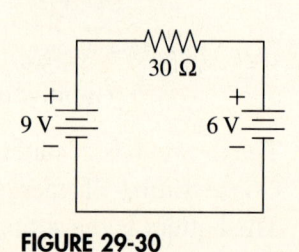

   FIGURE 29-30

4. The 10 Ω resistor in Fig. 29-31 is dissipating 40 W of power. How much power is being dissipated by the other two resistors?

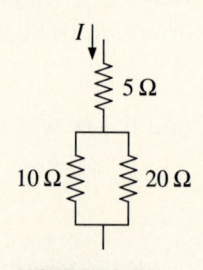

   FIGURE 29-31

5. A standard 100 W light bulb contains a tungsten filament 7 cm long. The high-temperature resistivity of tungsten is 9.0×10^{-7} Ω m. What is the diameter of the filament?

6. In the circuit shown in Fig. 29-32, the battery has an emf $\mathcal{E}$ and no internal resistance. Initially light bulbs A and B are both lit. Bulb B is then removed. Consequently, the potential difference ΔV_{12} between points 1 and 2
   a. increases.
   b. stays the same.
   c. decreases.
   d. becomes zero.

   Explain your choice.

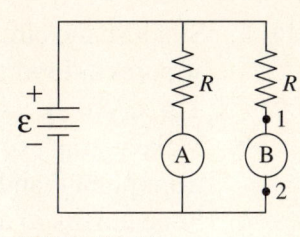

FIGURE 29-32

◢ Problems

7. Figure 29-33 shows five identical bulbs connected to an ideal battery. All the bulbs are glowing. What is the order of brightness of the bulbs?

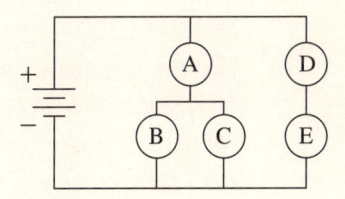

FIGURE 29-33

8. Six identical light bulbs, lettered A through F, are powered by a single battery in the circuit shown in Fig. 29-34. Assume that the battery's emf is large enough for all bulbs to be glowing. What is the order of brightness, from most bright to least bright, of the bulbs in the circuit? Explain your reasoning.

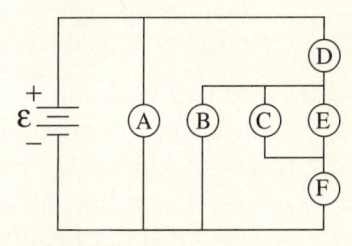

FIGURE 29-34

9. A current of 0.5 A flows in the circuit of Fig. 29-35.
   a. In which direction does it flow? Explain.
   b. What is the value of the resistance R?
   c. What is the power dissipated by R?
   d. Make a graph of potential versus position, starting from the lower left corner and proceeding clockwise.

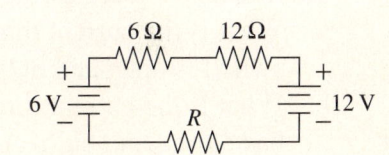

FIGURE 29-35

10. In the circuit shown in Fig. 29-36, what is the equivalent resistance between points a and b?

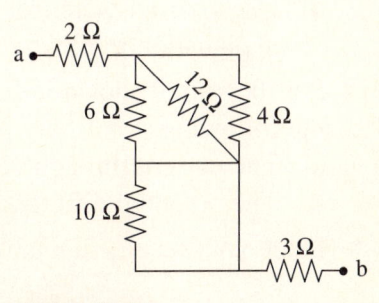

FIGURE 29-36

11. a. Suppose the circuit shown in Fig. 29-37 is grounded at point d. Find the potential at each of the four points a, b, c, and d.
    b. Make a graph of potential versus position, starting from point d and proceeding clockwise.
    c. Repeat parts a) and b) if the circuit is grounded at point a instead of d.

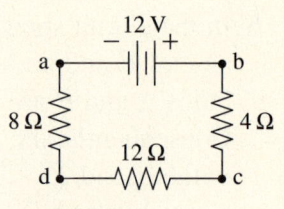

FIGURE 29-37

12. a. A battery of emf $\mathcal{E}$ and internal resistance r is attached to a load resistor R. For what value of the load resistance R, in terms of $\mathcal{E}$ and r, will the power dissipated *by the load resistor* be a maximum? (**Hint**: This is a calculus problem.)
    b. What is the maximum power that the load can dissipate if the battery has $\mathcal{E} = 9$ V and $r = 1$ Ω?
    c. For the battery of part b), calculate the power dissipated by the load for at least 10 values of R between $R = 0$ and $R = 3R_{max}$, where R_{max} is the load resistance for which power is maximized. Then draw a graph of power P dissipated versus load resistance R. Verify that your graph has a maximum at R_{max}.
    d. Why should the power dissipated by the load have a maximum value? Give a qualitative explanation. (**Hint**: Consider what happens to the power dissipation when R is either very small or very large.)

13. Consider the circuit shown in Fig. 29-38. For an ideal battery ($r = 0$), closing the switch does not affect the brightness of bulb A. You may have seen in lab, however, that bulb A dims *just a little* when bulb B is added to the circuit by closing the switch. The following series of questions will help you determine the reason. Assume that the 1.5 V flashlight battery has an internal resistance $r = 0.5$ Ω and that the resistance of a glowing bulb is $R = 6$ Ω.
    a. What is the current through bulb A when the switch is open and only A is glowing?
    b. What is the current through bulb A after the switch has closed and bulbs A and B are in parallel?
    c. By what percent does the current through A change when the switch is closed?
    d. Would the current through A change if $r = 0$?
    d. Use your answers from a) through d) to write an explanation as to why bulb A dims slightly when the switch is closed?

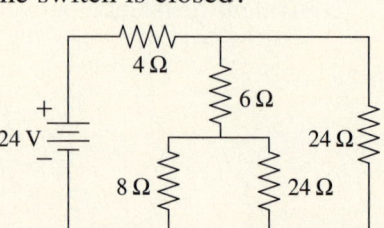

FIGURE 29-38

14. For the circuit shown in Fig. 29-39, determine the following:
    a. The current through each resistor.
    b. The potential difference across each resistor.

    Place your results in a table for ease of reading.

FIGURE 29-39

15. Dear Poly Engineering:

Our prison has recently had several "incidents" (i.e., prisoner escapes) when the exterior floodlights burned out during the night. The guards in the control room were not aware that the floodlights had gone out until it was too late. We would like your firm to design for us a new circuit to power the floodlights. This circuit should cause a red indicator light to come on in the control room if a floodlight bulb burns out or goes off for any other reason. Each floodlight bulb will have its own indicator so that guards will know immediately where the problem is. Here are the specifications:

Floodlight bulbs: There are 75 of these bulbs, each rated at 120V/240W. They currently operate in a standard 120 V parallel circuit. Each bulb *must* operate at the full output power of 240 W.

Indicator lights: 300 Ω resistance per bulb. The indicator bulb lights when the current through it exceeds 600 mA; it is dark if the current is less than 600 mA.

Power supplies: Voltages of 60 V, 120 V, 180 V, and 240 V are available at the prison. These cannot be combined with each other.

Please design a circuit to meet our needs. You may, of course, specify additional resistors, capacitors, or other elements as needed. Your design should be accompanied by a full analysis so that our in-house engineering staff can evaluate it.

Sincerely,
R. Knight, Warden

[**Estimated 12 additional problems for the final edition.**]

Chapter 30

The Magnetic Field

30.1 Observing Magnetism

[**Photo suggestion: Paper clips hanging from magnet.**]

You use magnets every day. In addition to holding shopping lists and cartoons on refrigerators, magnets allow us to run electric motors, produce pictures on television and computer screens, store information on cassettes and computer disks, cook food in microwave ovens, find directions, and listen to music over loudspeakers. Magnets are also used in magnetic resonance imaging to produce images of the interior of the human body, in high-energy physics experiments to identify subatomic particles, and in magnetic levitation trains. Understanding these applications, and countless others, starts with a basic understanding of magnets and **magnetism**—the properties of magnets.

In Chapter 24 we began our investigation of electricity by looking at the results of some simple experiments. Let's do the same for magnetism, using a few small magnets like the type you might use on your refrigerator. A typical refrigerator magnet exerts a force that is easily sensed by your fingers when the magnet gets within about a half-inch of the refrigerator door. You have also probably used a magnet to pick up small objects, such as paper clips, and seen them leap upward a half-inch or more to the magnet. The force exerted by a magnet is clearly a long-range force.

But is this a *new* force, or is this an electrical or a gravitational force? What is the *evidence* for it being a new force? And if it is a new force, what are its basic properties? In what ways does it differ from the electrical and gravitational forces?

To answer these questions, we need to experiment with magnets much as we experimented with charged rods at the start of Chapter 24. A careful observation of how magnets behave shows the following properties:

1. Magnets have two sides or two ends that are different. The ends sometimes attract each other, and they sometimes repel. These forces are clearly long-range because they are easily sensed without the magnets being in contact. The forces can be quite strong—certainly much stronger than the gravitational force, because one magnet can hold up another against the downward force of its weight.

2. If a magnet is taped to a piece of cork, so that it floats in a dish of water, it always turns to align itself in an approximate north–south direction. No matter how you orient the

magnet when you first place it in the dish, the same end of the magnet always turns to point to the geographic north. The same behavior is observed if you suspend the center of the magnet from a thread, so that it is free to rotate. The end of a magnet that points to the north is called the *north-seeking pole*, or simply, **north pole**. The other end is the **south pole**. At the moment these names are merely descriptive. They tell us nothing about *why* this effect happens or what a pole is.

3. If the north pole of one magnet is brought near the north pole of another magnet, they exert repulsive forces on each other. The same is true if two south poles are brought together. However, if the north pole of one magnet is brought near the south pole of another, the force between them is attractive. Thus like poles repel and unlike poles attract, as shown in Fig. 30-1. (*Note:* We will indicate the north pole of a magnet by shading.)

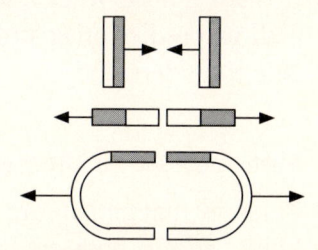

FIGURE 30-1 Forces between magnets. The shaded end of each magnet is its north pole.

4. If a magnet is brought near a charged electroscope, the leaves do not move. If a charged rod is brought near a magnet, it exerts a weak *attractive* force on *both* ends of the magnet. This force is exactly the same as the force exerted by the charged rod on a metal bar that isn't a magnet, so this is simply a polarization force like the ones we studied in Chapter 24. Other than polarization forces, charged objects are found to have *no effects* on magnets.

These are very important observations, because from them we can conclude that the force exerted by a magnet is *not* an electrical force. Despite the similar behavior of "likes repel" and "opposites attract," magnet poles are *not* the same as electric charges.

We can also take a bar magnet and drop it first with one pole down, then the other. It falls the same in either case, allowing us to conclude that magnetic forces are *not* gravitational. (This may seem a little silly, but without doing such experiments we might postulate that magnets exert very strong gravity and antigravity forces!) So it seems safe to conclude that the force exerted by a magnet is not a known force but, instead, is a new force that we call the **magnetic force.**

5. Magnets can pick up some objects, but not all. If an object is attracted to one end of a magnet, it is also attracted to the other end. (The objects themselves are *not* magnets and exert no forces on each other.) An important but frequently overlooked observation is that no objects are repelled by a magnet. The materials that are attracted to a magnet are called **magnetic materials**. The most common magnetic material is iron. Other magnetic materials include nickel, cobalt, and gadolinium, but these are not common household items. Most materials, including copper, aluminum, glass, and plastic, are not magnetic and experience no force from a magnet.

6. If we bring a compass near a magnet, the compass needle swings around so that one end points directly toward the magnet. The other end of the compass needle is attracted if the magnet is reversed. Apparently the compass needle itself is a little bar magnet with a north pole and a south pole.

7. Cutting a bar magnet in half produces two weaker but still complete magnets, each with a north and a south pole. The two poles of a magnet cannot be separated. No matter how small the magnets are cut, even down to microscopic sizes, the pieces remain complete magnets with all the properties described in the previous six items.

Our goal is to develop a theory of magnetism that will explain these phenomena, as well as predict other properties of magnetism. At the end of this chapter, you should be able to understand *why* all the preceding phenomena happen.

Monopoles and Dipoles

Poles in magnetism are analogous to charges in electricity. In both cases there are two kinds, and in both cases likes repel but opposites attract. Despite the analogy, magnetic poles and electrical charges are very different entities, so make sure you do not think of them as the same thing. Notice that giving *names* to the ends of magnets has not *explained* anything. The names *north* and *south* are arbitrary—we could have equally well called the ends *red* and *blue*. It is simply by convention that we agree to call a magnetic pole that is attracted geographically northward a *north pole*.

It is a strange observation that cutting a magnet in half yields two weaker but still complete magnets, each with a north and a south pole. The two poles of a magnet are permanently wedded and cannot be separated, thus forming a **magnetic dipole**. A magnetic dipole is analogous to an electric dipole, which is formed from a positive and a negative charge. But the two charges in an electric dipole can be separated and used individually. Apparently this is *not* true for a magnetic dipole. Every magnet that has ever been observed has both a north pole and south pole.

An isolated magnetic pole, such as a north pole in the absence of a south pole, would be called a **magnetic monopole**. As far as we know, magnetic monopoles do not exist in nature. No one knows why this is. In fact, many theories of subatomic particles predict the existence of magnetic monopoles. Some of the most talented physicists of the twentieth century have conducted very ingenious and sensitive experiments to find a magnetic monopole, but no one has ever succeeded. On the other hand, no one has ever given a convincing explanation of why isolated magnetic poles should not exist. Their existence remains an open question.

A compass needle is a weak magnetic dipole, with a north and a south pole. The north pole of a compass is attracted toward the geographic north pole of the earth and repelled by the earth's geometric south pole. Apparently the earth itself is a large magnet! The reasons for the earth's magnetism are complex and still actively researched by geophysicists, but it is generally agreed that the earth's magnetic poles arise from currents that flow in its molten iron core. One interesting fact to note is that the geographical north pole is actually a *south* magnetic pole! As an exercise, you should use what you have learned thus far to convince yourself that this is the case.

An introductory physics course often gives the impression that our understanding of nature is complete. Nothing could be further from the truth. Whether magnetic monopoles exist is an unanswered question at the most fundamental level of physics. Why the earth has a magnetic field is only a partially answered question. Perhaps some of you will one day find the answers to these questions.

30.2 Oersted's Discovery

Magnetism has been known since antiquity. The ancient Greeks recorded that certain minerals called lodestones could attract iron objects. Compasses using lodestones were commonly used in China by 1000 A.D., but they were not described in the West until nearly 1200. In 1600 William Gilbert realized that the earth itself acts as a magnet. He also wrote about the similarities and differences between electricity and magnetism, as they were known at that time.

As electricity began to be studied seriously in the eighteenth century, some scientists speculated about a connection between electricity and magnetism. Benjamin Franklin, for example, tried to magnetize a needle with an electrical discharge. A pair of French scientists, in 1805, tried suspending a battery from threads to see if it would orient itself like a compass needle. None of these early experiments were successful.

Then in 1820 the Danish scientist Hans Oersted discovered that there was indeed a relationship between magnetism and electricity. In the midst of a classroom lecture demonstration, Oersted passed a large current through a wire. By chance, a compass was sitting next to the wire, as shown in Fig. 30-2, and Oersted noticed that the compass needle turned when the current was flowing. In other words, the compass acted as if a magnet had been brought near. Oersted had previously been interested in a possible connection between electricity and magnetism, so the significance of this serendipitous discovery was immediately apparent to him—he realized that magnetism can be caused by an electric current.

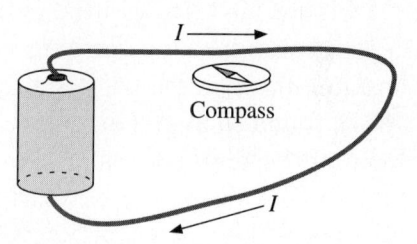

FIGURE 30-2 Oersted's experiment.

This was the long-sought link between electricity and magnetism. You will learn later that the connection between electricity and magnetism is, in fact, much more profound than Oersted's discovery, but a careful study of how a current influences a compass needle will be our starting point for developing a theory of magnetism.

The Effect of Current on a Compass

We will find in our study of magnetism that it will be necessary to use all three dimensions. All the physics we have considered up to this point could be described in either one or two dimensions, which has been convenient for making drawings on a page. Now, however, we will need to indicate an object or a vector going into or coming out of the plane of the page. The standard convention for drawing these is shown in Fig. 30-3. A dot (•) represents an object coming *out of* the page. You can think of it as the tip of a vector arrow pointing toward you. An × is used to represent an object directed *into* the page. This convention can represent either vectors or currents that are perpendicular to the plane of the paper.

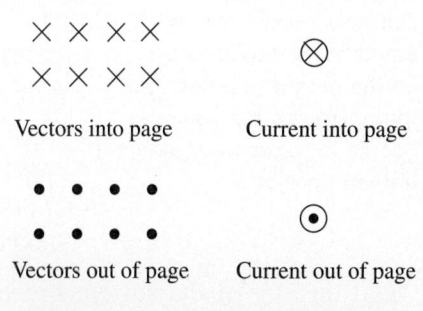

FIGURE 30-3 The notation for showing vectors and currents that are perpendicular to the page.

334 CHAPTER 30 THE MAGNETIC FIELD

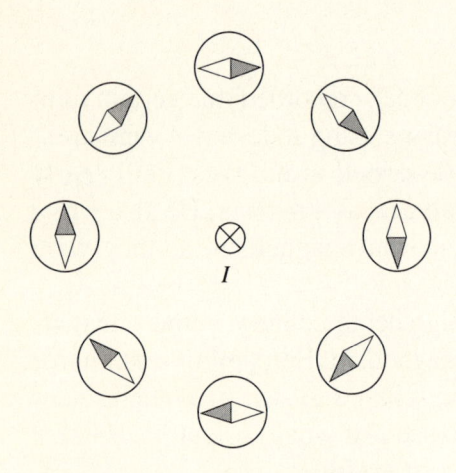

FIGURE 30-4 Response of compass needles to a current directed into the page.

[**Photo suggestion: Compass next to a current-carrying wire.**]

Let us use a compass as a probe of the magnetism created when an electric current I passes through a long, straight wire. We will assume that the current is large enough that the magnetic force exerted on the compass by the current is much larger than the force exerted on it by the earth's magnetism. (This is easy to do today, but it was difficult to produce such large currents in Oersted's era. This made his task more difficult, because the compass was responding to both the wire and the earth. In fact, the difficulty of producing large currents is one likely reason that no one prior to Oersted had seen the effect on a compass.)

Figure 30-4 shows what is observed if we place a compass at various points around a long, straight wire carrying current I *into* the page. (Imagine the wire that has been inserted through the page, starting above it and coming out below it.) The compass needle is found to align tangent to a circle around the wire. The relation between the direction of current flow and the orientation of the compass needles is given by the **right-hand rule**: If you point your *right* thumb along the wire in the direction that the current flows and then curl your fingers as if wrapping them around the wire, your fingers will be oriented in the same direction as the north poles of the compass needles. (You should verify that this is the case in Fig. 30-4.)

Because charges do not affect magnets, it must be that the current is exerting a magnetic force or forces on the compass needle. The *net* force on a needle must be zero because the compass does not undergo an acceleration or displacement. Instead, the needle is *rotated* into alignment. This is what we would expect if the two magnetic poles of the needle experience forces of equal strength but in opposite directions—one attractive and one repulsive—as shown in Fig. 30-5. This twisting force on a magnetic dipole is a *torque*, and it is analogous to the torque experienced by an electric dipole in an electric field.

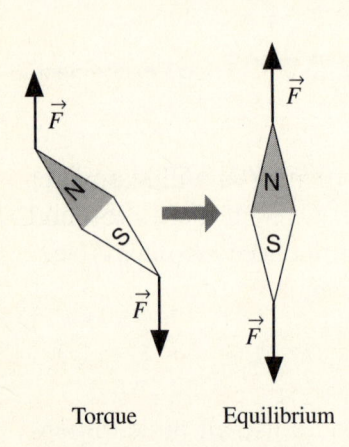

FIGURE 30-5 Torque on a compass needle due to equal but opposite forces on the magnetic poles. The torque causes the compass needle to rotate to an equilibrium position.

The size of these forces can be found by attaching springs to the compass needles and measuring how far the springs stretch when the current is turned on. Doing such experiments yields the following results:

1. If the *size* of the current I is varied, the strength of the forces on the compass needles is found to be directly proportional to the current: $F \propto I$.

2. If the *distance* from the wire to the compass d is varied, the strength of the forces is found to be inversely proportional to the distance: $F \propto 1/d$.

3. If the *direction* of the current is reversed, coming out of the page rather than flowing into the page, then the direction of all the compass needles also reverses.

These are observations. We will begin to develop a *theory* of magnetism in the next section, and one of our first goals will be to use the theory to *explain* Oersted's observations.

You might be concerned that we have introduced two kinds of magnetism. We opened this chapter discussing permanent magnets and their forces. Then, without warning, we switched to the apparent magnetic forces caused by a current. Even if this is a magnetic force, it is not at all obvious that it is the same kind of magnetism exhibited by stationary chunks of metal called "magnets." Perhaps there are two different types of magnetic forces, one having to do with currents and the other responsible for permanent magnets. One of the major goals for our study of magnetism is to see that these two apparently quite different ways of producing magnetic effects are really just two different aspects of a *single* magnetic force.

30.3 The Magnetic Field

Recall that in our study of electrical forces we introduced the idea of a *field* as a way to understand long-range forces. A charge alters the space around it by creating an electric field. If a second charge is nearby, it experiences a force due to the presence of the electric field. The electric field is thus the agent by which charges interact with each other. Although this idea appeared rather farfetched at first, it turned out to be very useful. It seems as though we need a similar idea here to understand the long-range forces exerted by a current on a compass needle.

Let us define the **magnetic field** $\vec{B}$ as having the following properties:

1. A magnetic field is created at *all* points in space surrounding a current-carrying wire.
2. At each point, the magnetic field is a vector and has both magnitude and direction.
3. The magnetic field exerts forces on magnetic poles, with the magnitude of the force proportional to the field strength at that point. The force on a north magnetic pole is parallel to the magnetic field direction, and the force on a south magnetic pole is opposite to the magnetic field direction.

Figure 30-6 shows a compass needle in a magnetic field. The vectors labeled $\vec{B}$ show the magnitude and direction of the field at a few specific points, but keep in mind that the field is present at *all* points in space. The two poles of the compass each have a magnetic force exerted on them, parallel to $\vec{B}$ for the north pole and opposite to $\vec{B}$ for the south pole. This pair of opposite forces produces a torque on the needle, thus rotating the needle to an equilibrium position that is parallel to the magnetic field at that point. Further, the north pole of the compass needle, when it reaches the equilibrium position, is pointing in the direction of the magnetic field.

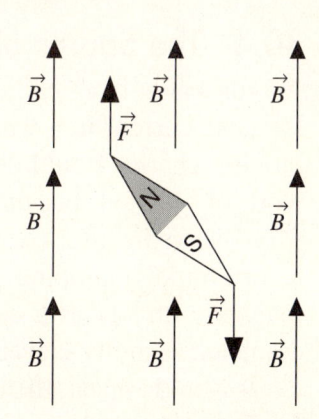

FIGURE 30-6 The magnetic field exerts forces on the poles of a compass, creating a torque and causing the needle to align with the field.

FIGURE 30-7 The magnetic field around a current that is directed into the page.

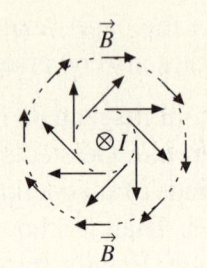

Thus a compass needle can be used as a probe of the magnetic field, just as a test charge was a probe of the electric field. The magnetic force causes the compass needle to be aligned parallel to the field, with the north pole of the compass showing the direction in which the magnetic field vector points.

From this definition of the magnetic field, we can use the compass alignments of Fig. 30-4 to deduce the magnetic field of a current-carrying wire. The field is shown in Fig. 30-7.

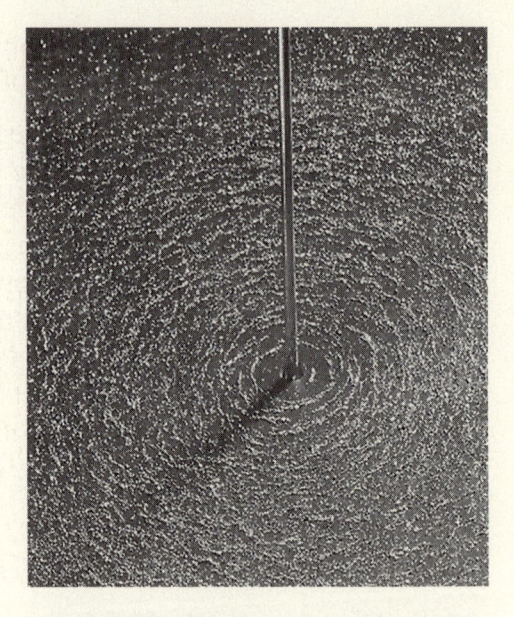

FIGURE 30-8 Iron filings reveal the magnetic field around a current-carrying wire.

Notice that the field is weaker (shorter vectors) at greater distances from the wire. The geometric structure of the magnetic field differs greatly from that of an electric field. Electric fields diverge *outward* from positive charges and converge *inward* to negative charges. The magnetic field, by contrast, circulates *around* a current-carrying wire. Figure 30-8 shows a photograph of iron filings sprinkled around a current-carrying wire. You can easily see the circular patterns of the filings. It was patterns such as this that suggested the field concept to Faraday.

30.4 The Source of Magnetic Fields: Moving Charges

As you know from earlier chapters, an electric current is a collection of moving charges. Because current in a wire generates a magnetic field, it's natural to wonder whether *any* moving charge would do the same. This was generally assumed to be the case after the work of Oersted, but this assumption was not confirmed until 1875, by H. A. Rowland, fifty-five years after Oersted's discovery. Rowland's proof was to charge the perimeter of a very rapidly spinning, nonconducting disk and to see that it produced exactly the same magnetic effects as a current flowing through a circular loop of wire having the same diameter. A pretty clever demonstration!

It would appear, then, that the *source* of a magnetic field is *moving charge*. At the most fundamental level, we can specify the magnetic field of a *single* charged particle q moving with speed v. Consider a point in space at distance r from the charge and angle θ away from the its direction of motion. The magnetic field at this point is found to be

$$\vec{B} = \left(\frac{\mu_0}{4\pi} \frac{qv \sin \theta}{r^2}, \quad \text{direction given by the right-hand rule} \right). \tag{30-1}$$

Equation 30-1 is called the **law of Biot and Savart** for a point charge. It is analogous to Coulomb's law for the electric field of a point charge. Because the magnetic field is a vector, Eq. 30-1 specifies both the magnitude and the direction of $\vec{B}$.

The direction of $\vec{B}$ is given by the right-hand rule, using the charge's velocity as the reference direction: Place your right thumb along the direction of $\vec{v}$, and your fingers will curl in the direction that $\vec{B}$ points. The $\vec{B}$ vectors are all tangent to circles drawn about the charge's line of motion, similar to the field of a current-carrying wire. Figure 30-9 shows the magnetic field of a positive moving charge. Notice, because of the $\sin\theta$ term in Eq. 30-1, that $\vec{B}$ is zero along the line of motion (where $\theta = 0$), but nonzero at all other points. All the vector arrows in Fig. 30-9 would have the same magnitude but be reversed in direction for a negative charge.

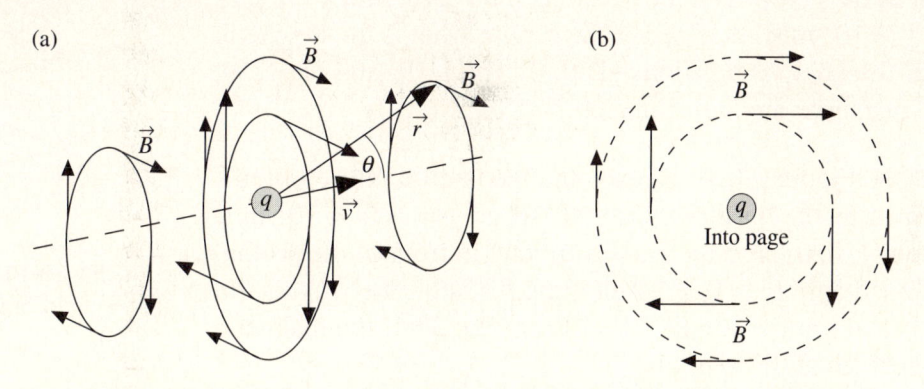

FIGURE 30-9 The magnetic field of a charge moving with velocity $\vec{v}$. The field strength depends both on distance and on angle θ. a) Perspective view. b) In the plane of the charge.

The requirement that the charge be moving to generate a magnetic field is explicit in Eq. 30-1. If the speed v of the particle is zero, the magnetic field (but not the electric field!) will be zero. This property helps to emphasize a fundamental distinction between electric and magnetic fields: Charges cause electric fields, but only *moving* charges cause magnetic fields.

The magnetic field is measured in the SI unit called the **tesla**, abbreviated as T. You will see later in the chapter, when we discuss magnetic forces on current-carrying wires, that the definition of the tesla is

1 tesla = 1 T = 1 N/Am.

One tesla is quite a large field. Table 30-1 shows some typical magnetic field strengths. You can see that magnetic fields are usually a small fraction of a tesla.

The constant μ_0 in Eq. 30-1 is called the **permeability constant**. Its value is

TABLE 30-1 Some typical magnetic field strengths.

At the surface of the earth	5×10^{-5} T
Typical refrigerator magnet	5×10^{-3} T
Laboratory magnet	10^{-1} T
Large superconducting magnet	10 T

$$\mu_0 = 4\pi \times 10^{-7}\,\mathrm{T\,m/A} = 1.257 \times 10^{-6}\,\mathrm{T\,m/A}.$$

This constant plays a role in magnetism similar to that played by the permittivity constant ε_0 in electricity.

Equation 30-1 is the starting point for generating all magnetic fields, just as our earlier expression for the electric field of a point charge was the start for generating all electric fields. (Note that the charge we're considering here still has its associated electric field *in addition* to the magnetic field generated by virtue of its motion.) You can see that the magnetic field of a moving point charge is given by an inverse square law, as was the case for the electric field of a point charge. However, things are somewhat more complex for the magnetic field because the expression also contains directional information—the angle θ between the charge's velocity vector $\vec{v}$ and the vector $\vec{r}$ from the charge to the point where the field is evaluated.

EXAMPLE 30-1 A proton moves along the *x*-axis with a velocity $v_x = 1.0 \times 10^7$ m/s. As it passes the origin, what is the magnetic field at the (*x*, *y*, *z*) positions (1, 0, 0), (0, 1, 0), and (1, 1, 0), where distances are in millimeters?

SOLUTION Figure 30-10 shows the geometry. The proton's motion is in the +*x*-direction with speed $v = 1.0 \times 10^7$ m/s. Position (1, 0, 0) is on the *x*-axis, directly in front of the proton. Thus $\theta = 0$ and $\vec{B} = 0$. Position (0, 1, 0) is on the *y*-axis with $\theta = 90°$ and $r = 1$ mm $= 0.001$ m. From Eq. 30-1, the magnetic field strength at this point is

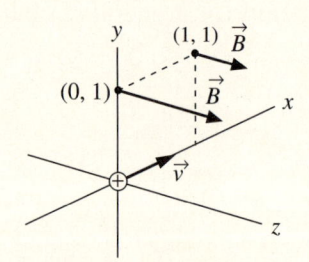

FIGURE 30-10 The magnetic field of Example 30-1.

$$|\vec{B}| = \frac{\mu_0}{4\pi} \frac{qv\sin\theta}{r^2} = 10^{-7} \cdot \frac{(1.60 \times 10^{-19} \text{ C})(1.0 \times 10^7 \text{ m/s})\sin 90°}{(0.001 \text{ m})^2}$$

$$= 1.60 \times 10^{-13} \text{ T.}$$

From the right-hand rule, the field at a point on the +*y*-axis points in the +*z*-direction. So

$$\vec{B}(0, 1 \text{ mm}, 0) = 1.60 \times 10^{-13} \hat{k} \text{ T.}$$

The field at position (1, 1, 0) also points in the +*z*-direction, but it is weaker than at (0, 1, 0) both because *r* is larger *and* because θ is smaller. From geometry, $r = \sqrt{2}$ mm $= 0.00141$ m and $\theta = 45°$. A similar calculation using Eq. 30-1 gives

$$\vec{B}(1 \text{ mm}, 1 \text{ mm}, 0) = 0.57 \times 10^{-13} \hat{k} \text{ T.}$$

Superposition

You learned previously that the total electric field caused by several charges $q_1, q_2, \ldots, q_n$ is given by the superposition of the fields of each charge calculated separately:

$$\vec{E}_{\text{total}} = \vec{E}_1 + \vec{E}_2 + \ldots + \vec{E}_n.$$

Does the magnetic field of several moving charges also obey the principle of superposition? This is an experimental question.

Figure 30-11 shows a wire, carrying current *I*, that has been looped and then doubled back on itself. A compass can be used to probe the magnetic field at points 1, 2, and 3. At point 2 there are two wires carrying current *I* in the same direction, but at point 3 the two wires carry current in opposite directions. Measurements show that the field strength at point 2 is twice that at 1 but the field at point 3 is zero. These results are exactly what we would expect if the magnetic fields of each wire superimpose. The two fields at point 3 are equal in magnitude, but because of the right-hand rule they point in opposite directions—one into the page and one out.

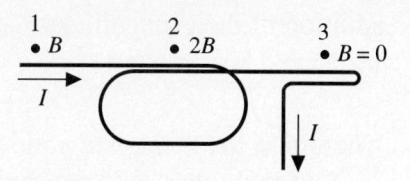

FIGURE 30-11 Superposition of the magnetic fields of a wire.

Magnetic fields thus obey the principle of superposition. If there are n moving point charges, the total magnetic field is given by the vector sum

$$\vec{B}_{total} = \vec{B}_1 + \vec{B}_2 + \cdots + \vec{B}_n, \tag{30-2}$$

where each individual $\vec{B}_i$ is calculated with Eq. 30-1. By utilizing the principle of superposition, we can now determine the magnetic fields of several important current distributions.

30.5 The Magnetic Field of a Long, Straight Wire

We can use the law of Biot and Savart and the principle of superposition to determine the magnetic field of a long, straight wire carrying current I. Consider a wire stretching along the x-axis from $-\infty$ to $+\infty$ carrying current I in the positive direction, as shown in Fig. 30-12. We want to calculate the field at a point that is distance d from the wire.

First, let's establish a coordinate system so that the point of interest is located at $x = 0$. Then divide the wire into infinitesimal lengths dx, each of which contains a very small amount dq of *moving charge* that will generate a small magnetic field $d\vec{B}$ at the point of interest. Charge dq is effectively a point charge, so $d\vec{B}$ is given by Eq. 30-1, the law of Biot and Savart. According to Eq. 30-2, the net magnetic field is the superposition of all the fields of all the little segments of moving charge. That is,

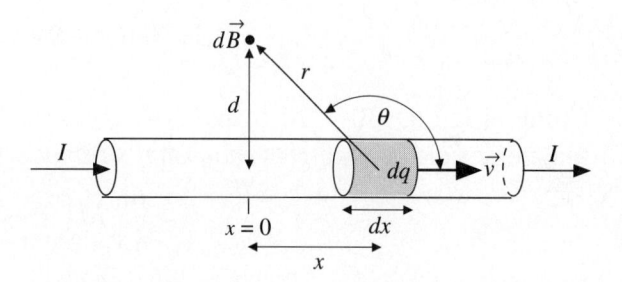

FIGURE 30-12 Calculating the magnetic field of a long, straight wire carrying current I.

$$\vec{B} = d\vec{B}_1 + d\vec{B}_2 + \cdots + d\vec{B}_n \xrightarrow[n \to \infty]{} \int_{-\infty}^{\infty} d\vec{B}(x), \tag{30-3}$$

where the notation in the last step indicates that $d\vec{B}$ is really a function of x and that we will need to integrate over the entire length of the wire.

Evaluating such an integral can, in general, be very difficult because we're adding *vectors* that do not necessarily all point in the same direction. Fortunately for us, in this case they do. If you look back at Fig. 30-9 and think about the right-hand rule, you will see that every single small section dx of the wire generates a small $d\vec{B}$ at the point of interest that is directed *out of* the page. Adding vectors that all point in the same direction is a simple addition of their magnitudes, so we can simplify Eq. 30-3 to be a scalar equation:

$$B = \int_{-\infty}^{\infty} dB(x), \qquad (30\text{-}4)$$

where $B = |\vec{B}|$ is the *magnetic field strength*.

Our main task is to determine the small field $dB(x)$ generated by a small length dx of wire. The small amount of moving charge in dx is dq, and it is moving with speed v. The time it takes this charge to move distance dx is simply $dt = dx/v$. Because current is defined as $I = dq/dt$, the current in the wire is related to dq and v by

$$I = \frac{dq}{dt} = \frac{dq}{dx/v} = \frac{v\,dq}{dx}.$$

Thus we find the important relationship

$$v\,dq = I\,dx. \qquad (30\text{-}5)$$

Now we can use Eq. 30-5 in the law of Biot and Savart to determine the small field strength dB generated by the moving charge dq in dx:

$$dB = \frac{\mu_0}{4\pi}\frac{dq\,v\sin\theta}{r^2} = \frac{\mu_0}{4\pi}\frac{I\,dx\sin\theta}{r^2} = \frac{\mu_0}{4\pi}\frac{I\sin\theta}{r^2}dx. \qquad (30\text{-}6)$$

Completing the calculation is now a calculus exercise. Both the variables r and θ are functions of x and are easily determined to be

$$r = \sqrt{x^2 + d^2}$$

$$\sin\theta = \sin(\pi - \theta) = \frac{d}{r} = \frac{d}{\sqrt{x^2 + d^2}}. \qquad (30\text{-}7)$$

Combining Eqs. 30-4, 30-6, and 30-7 gives the final integral that we must do to obtain the magnetic field of a current in a long, straight wire:

$$B = \frac{\mu_0}{4\pi}Id\int_{-\infty}^{\infty}\frac{dx}{(x^2 + d^2)^{3/2}}. \qquad (30\text{-}8)$$

This is a standard integral that we can evaluate to find

$$B_{\text{wire}} = \frac{\mu_0}{4\pi}Id\frac{x}{d^2(x^2+d^2)^{1/2}}\bigg|_{-\infty}^{\infty} = \frac{\mu_0}{2\pi}\frac{I}{d}. \qquad (30\text{-}9)$$

This result gives only the magnitude of the field. The direction is determined at each point by using the right-hand rule.

Figure 30-13 shows the magnetic field of a current in a long, straight wire. The perspective view shows the three-dimensional nature of the field as it circulates around the wire. Usually, however, it will be easier to draw the side view of Fig. 30-13b, using the notation of Fig. 30-3 to show that the field above the wire is coming out of the page and the field beneath the wire is pointing into the page. You should convince yourself that the field above the plane of the page is pointing down and the field behind the plane of the page is pointing up.

The results of this calculation confirm Oersted's discovery that the field strength is proportional to the ratio I/d and that the field circles the wire. You should note the similarities and differences between the magnetic field of a current in a long, straight wire and the electric field of a charged long, straight wire (Fig. 25-9). The strength of both depends

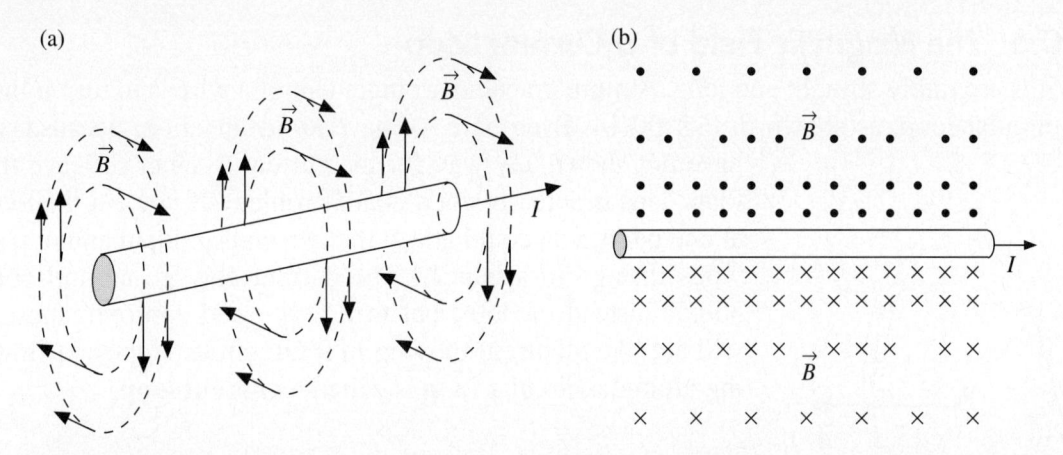

FIGURE 30-13 The magnetic field of a long, straight wire carrying current I. a) Perspective view. b) Side view.

inversely on the distance d, but $\vec{E}$ points radially outward from the wire whereas $\vec{B}$ goes around the wire.

This calculation was pretty difficult—in fact, the most difficult that you will need to do in your study of magnetism. Many of the steps in it, however, recur frequently in other problems. Our basic procedure was to divide the wire into many small pieces, use the principle of superposition to set up an integral for $\vec{B}$, then relate $d\vec{B}$ to dx because it is ultimately x that we must integrate over. This last step—converting the terms of the integrand to functions of x—is often the key step and, equally often, the step that causes the most difficulty. Careful study of this example will make it easier for you to attack similar problems on your own.

EXAMPLE 30-2 A nichrome heater wire 1 m long and 1 mm in diameter is connected to a 12 V battery. What is the magnetic field 1 cm away from the wire?

SOLUTION The current through the wire is

$$I = \frac{\Delta V_{\text{bat}}}{R},$$

where the wire's resistance R is given by

$$R = \frac{\rho L}{A} = \frac{\rho L}{\pi r^2} = 1.91 \ \Omega.$$

The nichrome resistivity $\rho = 150 \times 10^{-8} \ \Omega \ \text{m}$ was taken from Table 26-2. Thus the current in the wire is $I = 12 \ \text{V}/1.91 \ \Omega = 6.28 \ \text{A}$. The magnetic field at distance $d = 1 \ \text{cm} = 0.01 \ \text{m}$ from the wire is found from Eq. 30-9 to be

$$B_{\text{wire}} = \frac{\mu_0}{2\pi} \frac{I}{d} = (2 \times 10^{-7} \ \text{T m / A}) \frac{6.28 \ \text{A}}{0.01 \ \text{m}} = 1.26 \times 10^{-4} \ \text{T}.$$

The magnetic field of the wire is slightly more than twice the strength of the earth's magnetic field.

30.6 The Magnetic Field of a Current Loop

Wires are rarely straight and long. A more practical arrangement of a wire is to turn it into a circular loop, as shown in Fig. 30-14. Even here we have an idealization, because we have not shown any way for the current to enter or leave the loop. This is actually not a major problem. If you cut the loop at one point, you could attach the two ends to input and output wires that go to a battery. The current may go around only 359.5° instead of 360°, but to a very good approximation it will act like a current moving in a full circle. A current flowing around a circular loop is called a **current loop**.

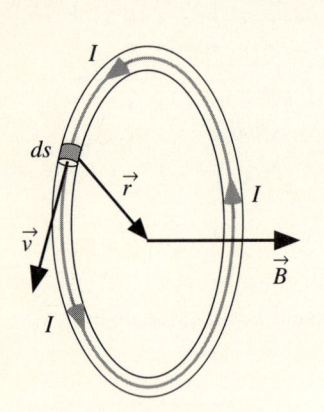

FIGURE 30-14 The magnetic field at the center of a current loop.

The Field at the Center of the Loop

As another example of calculating magnetic fields, let's calculate the field at the *center* of a current loop of radius R that carries current I. We will proceed in the same manner as we did for the straight wire by dividing the loop into little pieces of length ds (not dx because they are not along a coordinate axis), finding $d\vec{B}$ for each, and adding them up by integrating. Once again, the *direction* of each $d\vec{B}$ is identical. If you apply the right-hand rule, you will see that the field at the center of the loop due to the current in any little ds, no matter where around the loop it is located, is to the right—*perpendicular* to the loop. So again we can perform a scalar integral to get the magnetic field strength:

$$B = \int_{\text{around loop}} dB(s), \qquad (30\text{-}10)$$

where

$$dB(s) = \frac{\mu_0}{4\pi} \frac{I \sin\theta}{r^2} ds. \qquad (30\text{-}11)$$

Notice that we've used Eq. 30-6 for dB in terms of the current I passing through a small segment of length ds. The limits of integration in Eq. 30-11 will be determined when we get to a final integral, but for now we will indicate that the integration must add up the contribution from every ds "around the loop."

Now we must consider the specific geometry of the loop, which is quite different from that of the straight wire. For *every* element ds, the distance from the wire to the point of interest, namely the center of the loop, is $r = R$. Also, for *every* element the angle θ between $\vec{v}$, which is tangent to the loop, and $\vec{r}$, which points to the center, is 90°. Therefore $\sin\theta = 1$ for each ds segment of the loop. Equation 30-11 simplifies to

$$dB = \frac{\mu_0 I}{4\pi R^2} ds,$$

which then gives

$$B = \frac{\mu_0 I}{4\pi R^2} \int_{\text{around loop}} ds. \qquad (30\text{-}12)$$

There are two ways we can evaluate the integral. First, the small length ds is geometrically related to the small angle $d\phi$ by $ds = Rd\phi$, as shown in Fig. 30-15. This is a variable change to make ϕ the integration variable. Going "around the loop" then requires that we integrate ϕ from 0 to 2π. Evaluating the integral in this way gives

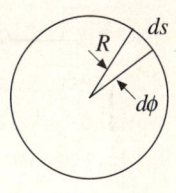

$$B_{\text{loop}} = \frac{\mu_0 I}{4\pi R^2} \int_0^{2\pi} R\, d\phi = \frac{\mu_0 I}{4\pi R^2}(2\pi R) = \frac{\mu_0 I}{2R}. \quad (30\text{-}13)$$

FIGURE 30-15 The small arc length ds is equal to $Rd\phi$.

An alternative means of evaluating the integral is to stick with s as the integration variable. This is not a coordinate axis like you are used to integrating along (e.g., the x-axis), but the meaning of integration as "adding up" all the pieces is still valid. The variable s measures distance along the loop, and integrating around the loop requires s to increase from 0 to $2\pi R$. We can then evaluate the integral as

$$\int_{\text{around loop}} ds = \int_0^{2\pi R} ds = 2\pi R. \quad (30\text{-}14)$$

An integral like this along a curve, rather than along a coordinate axis, is called a *line integral*. Using the result of Eq. 30-14 in Eq. 30-12 gives exactly the same result for B_{loop} that we found in Eq. 30-13.

If you compare this result to that of a long, straight wire, you see that it really differs only by a factor of π in the denominator—a factor that depends on the geometry being considered. The significant result, namely that the field depends on the ratio of current to distance, is the same in both cases. The field at the center of a current loop is a factor of π larger than the field an equal distance from a straight wire carrying an equal current because the wire in the loop "stays close" to the point of observation. Each element of the loop contributes equally, whereas for the straight wire only a small number of elements nearest the observation point make much of a contribution.

EXAMPLE 30-3 What current is needed in a 10 cm diameter loop to cancel the earth's magnetic field at the center of the loop?

SOLUTION The earth's magnetic field, from Table 30-1, is 5×10^{-5} T. Scientists frequently need to carry out a measurement in zero magnetic field. One way to do this is to generate a magnetic field of magnitude equal to the earth's field but pointing in the opposite direction. The vector sum of the two fields is zero, so the scientist has created a "field-free" region of space! Although there are better ways to do this than at the center of a single current loop, this illustrates the idea. The necessary current, from Eq. 30-13, is

$$I = \frac{2RB}{\mu_0} = \frac{2(0.05 \text{ m})(5 \times 10^{-5} \text{ T})}{4\pi \times 10^{-7} \text{ T m/A}} = 4.0 \text{ A}.$$

This is easily accomplished.

The Field Near the Loop

The very center of a current loop is a unique and special place that is equal distance and equal angle from every ds. This made our evaluation of Eq. 30-10 straightforward. In principle, we could apply this procedure to evaluate $\vec{B}$ at any point in space. We would quickly find, though, that the mathematics of evaluating the integral for $\vec{B}$ at an arbitrary point is far more than we want to want to tackle. Even if we did, the resulting equations (which are very complicated) would not tell us much physics. Detailed calculations are appropriate in some circumstances, but we have a somewhat different goal of trying to understand *how* magnetic fields are generated rather than trying to calculate precise numbers. For our purposes, pictures rather than equations will tell us more about the physics.

Thus we would like to know, in an approximate way, what the magnetic field looks like in the space near a current loop. The principle of superposition and the right-hand rule are sufficient to answer this question. Figure 30-16 shows a cross section through the center of a current loop that is perpendicular to the page. The current in the top of the loop is coming out of the page, curving around above the page, going back into the page at the bottom of the loop, then completing the circle below the page. The magnetic field at the center of the loop, which we can calculate from Eq. 30-13, is shown at point A. To verify its direction, place your right thumb pointing against the × where the current enters the page at the bottom of the loop and circle your fingers toward your palm. Note that your fingers circle clockwise, thus pointing toward the right at the center of the circle. If you point your right thumb away from the page at the •, where the current exits, your fingers again point toward the right at the center of the circle.

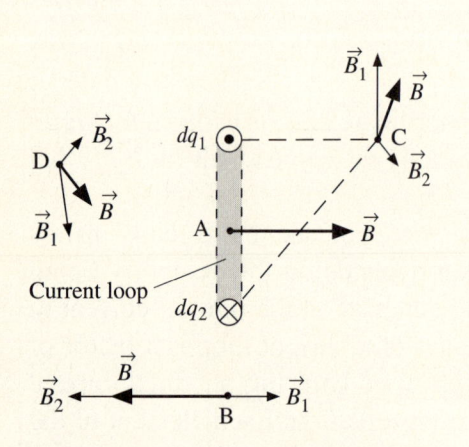

FIGURE 30-16 The magnetic field at several points near a current loop that is perpendicular to the page.

What is the magnetic field at point B, directly beneath the loop? Let's consider the two fields $\vec{B}_1$ and $\vec{B}_2$ generated by moving charges dq_1 and dq_2 in small segments at the top and bottom of the loop, respectively. The right-hand rule applied to the current exiting the page at the top of the loop (•) shows that $\vec{B}_1$ points to the right. Similarly, the current entering the page at the bottom of the loop (×) generates $\vec{B}_2$ pointing to the left. Point B is significantly closer to the bottom of the loop than to the top, so the inverse-square nature of $\vec{B}$ indicates that the magnitude B_2 will be much larger than B_1. Adding the vectors, using the principle of superposition, gives a net field pointing to the left.

Now you might object, and rightly so, that we have considered only two current elements in the loop. What about all the others, those above and below the page? That is a good question, and answering it takes some pretty good visualization skills. We will leave it to you as an exercise to think about it, but the net result of considering all the other elements gives the same general conclusion as considering only the top and bottom current elements. After all, the inverse-square nature of the law means that the *closest* element will have by far the largest effect, with more distant elements making only minor contributions to the field. So considering only the top and bottom elements gives a fairly decent indication of what the magnetic field looks like near the loop.

Next let's look at the field at point C in Fig. 30-16. This point is not in the plane of the loop as points A and B were. We again use the right-hand rule to determine the directions of fields $\vec{B}_1$ and $\vec{B}_2$ generated by current elements at the top and bottom of the loop. Recall that the field of a current element is tangent to a circle about the velocity vector $\vec{v}$, which points out of the page at the top of the loop and into the page at the bottom. Further, magnitude B_1 will in this case be larger than B_2 because of the proximity to the top of the loop. Adding $\vec{B}_1$ and $\vec{B}_2$ as vectors gives the net field. Considering all the other current elements would slightly alter the details but not the basic conclusion that the field at point C is directed upward and somewhat to the right. As an exercise, confirm for yourself that the field at point D points to the right and somewhat downward.

By continuing this process we could map the appearance of the magnetic field at many points near the loop. The result of doing so is shown in Fig. 30-17a. Note that this figure has *rotational symmetry*. You are looking at a slice through the field. To picture the full three-dimensional field, you need to imagine this slice rotated about the axis of the loop. There is a clear sense of a circulation of the magnetic field about the loop, with the field leaving the loop on the right, flowing around the outside, and returning to the loop on the left. These directions would reverse, of course, if the current in the loop were to flow the other way. Figure 30-17b shows the magnetic field in the plane of the loop as seen from the right side of Fig. 30-17a.

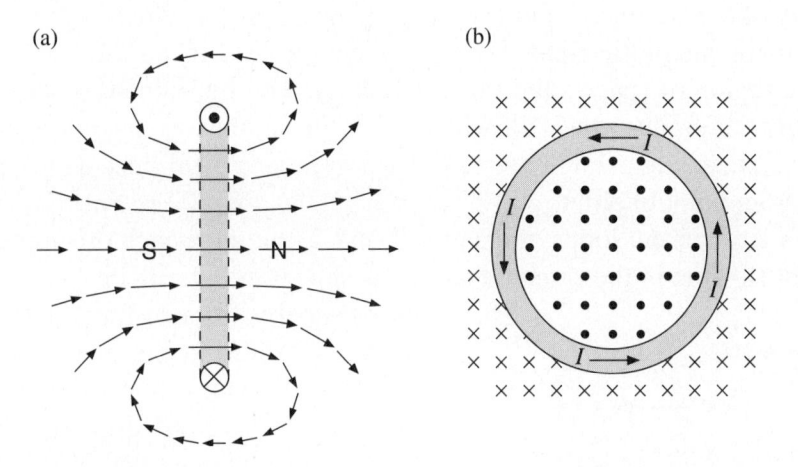

FIGURE 30-17 a) Side and b) front views of the magnetic field of a current loop.

A current loop has two distinct sides—one with the magnetic field pointing outward and the other with the field pointing inward. You might wonder whether there is a connection to the magnetic field of permanent magnets, which also have two distinct sides or ends. Indeed, if a current loop is suspended vertically by a thread, it aligns itself so that the side of the loop where the field points outward is facing geographic north. Furthermore, the side of the current loop where the field points outward is repelled by the north pole of a bar magnet but attracted toward the bar magnet's south pole. The opposite side, where the field points inward, is attracted to a north pole.

Thus a current loop is a magnetic dipole! It has its own north pole and south pole, just like a permanent bar magnet. Using our definition of magnetic poles, the side of the current loop where the field points outward is its north magnetic pole. Obviously, then, the side where the field points inward is the south magnetic pole. These poles are indicated on Fig. 30-17. Evidently there is a strong connection between current loops and permanent magnets, and this is a topic to which we will return to shortly.

Because a current loop is magnetic dipole, Fig. 30-17 is the field of a magnetic dipole. A permanent bar magnet, which is also a magnetic dipole, creates a field *outside the magnet* that looks essentially the same. The field of a permanent magnet is shown in Fig. 30-18.

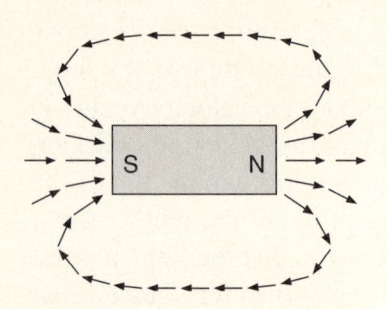

FIGURE 30-18 The magnetic field of a permanent magnetic dipole.

30.7 Uniform Magnetic Fields

In our study of electricity, we made extensive use of the idea of a uniform electric field. This was a field that was the same at every point in space. We found that two parallel charged plates generate a uniform electric field between them, and for this reason we focused much attention on learning about the electric field of parallel plates.

Similarly, there are many applications of magnetism for which we would like to generate a **uniform magnetic field**. We define such a field to be the same at every point within some region of space—the same magnitude and the same direction. The problem we face is how to produce one. Clearly none of the sources we have looked at thus far—a moving charge, a long, straight wire, and a current loop—produce a uniform field. Their fields vary from point to point.

Consider two current loops, numbered 1 and 2, that face each other and that have the same current flowing in the same direction, as shown in Fig. 30-19. We can use the principle of superposition to find the magnetic field at point A, which is in the plane midway between the loops. We know that each loop produces a dipole field, so we can draw the two vectors $\vec{B}_1$ and $\vec{B}_2$ showing the field at point A from each loop. Because of the symmetry of the dipole field, $\vec{B}_1$ points upward at the same angle that $\vec{B}_2$ points downward. Consequently, the vector sum at A is a purely horizontal field pointing to the right. The same is true at any point in the midplane, such as points B and C. Within this plane, the magnetic field vectors are all parallel to the axis, which is fairly suggestive of a uniform field. Unfortunately, the field at other points is not parallel to the axis because the contribution from one loop is larger than from the other loop, so perfect cancellation of the vertical components does not occur. Nonetheless, we see that combining two loops has taken us at least part way toward

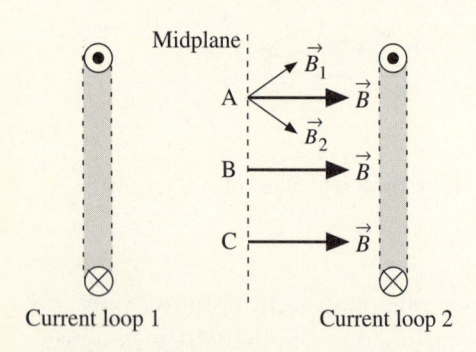

FIGURE 30-19 The magnetic field in the midplane between the two loops.

producing a uniform field. We simply need to extend this idea yet further.

Figure 30-20 shows a large number of current loops side by side with the same current flowing the same direction in each. The analysis of this problem gets too mathematical for us, but the results are quite plausible if you understand the previous example of just two current loops. The net field of many current loops close together is a *uniform* magnetic field inside the loops (at least if you do not get too close to the end). The same horizontal magnetic field exists at each point inside the loops, as shown in Fig. 30-20, and that is what we mean by a uniform field.

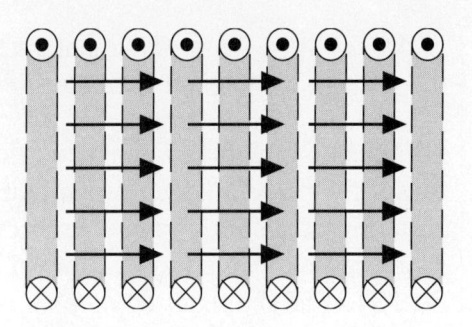

FIGURE 30-20 A uniform magnetic field is generated inside a group of side-by-side current loops.

As a practical matter, uniform fields are generated with a **solenoid**. A solenoid is a coil of wire made by winding the wire as a helix around a cylindrical form, as shown in Fig. 30-21a. When the solenoid is connected to a battery, the same current flows around each loop in the coil and the net effect is equivalent to a large stack of separate current loops. If the coil is wound on a hollow cylinder, the interior is open for use. Figure 30-21b shows that the magnetic field *inside* a solenoid is a uniform magnetic field.

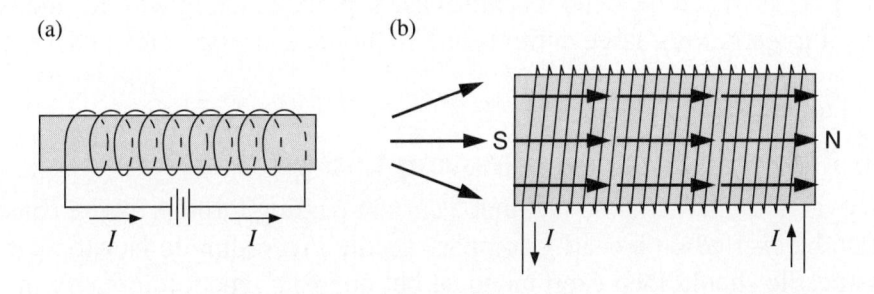

FIGURE 30-21 a) A solenoid. b) The magnetic field *inside* a solenoid is a uniform field. The ends of the solenoid are magnetic poles, creating a dipole field *outside* the solenoid.

A more detailed analysis shows that the strength of the magnetic field inside a solenoid depends on the current I and also on the length L of the solenoid and on the number of turns of wire N. The result of that analysis gives

$$B_{\text{solenoid}} = \frac{\mu_0 N I}{L}. \qquad (30\text{-}15)$$

The field strength can be made very high by using a coil with a large number of turns of wire.

Note that the field is uniform only *inside* the solenoid. As with a current loop, we can identify the north magnetic pole and the south magnetic pole of a solenoid. The north pole is the end where the magnetic field points away from the coil. So from the *outside* a solenoid looks and acts very much like a bar magnet with a north and a south pole. It is what we call an **electromagnet**, and its external field looks like that of Fig. 30-18.

EXAMPLE 30-4 We wish to generate a field of 0.1 T inside a coil that is 10 cm long. How many turns of wire are necessary if the wire can carry a maximum current of 10 A?

SOLUTION Generating a magnetic field with a solenoid is a trade-off between current and turns of wire. A larger current requires fewer turns. However, wires do have resistance and so too large a current can overheat the solenoid. Maximum safe currents are established based on the wire's cross-sectional area. For a wire that can carry 10 A, we can use Eq. 30-15 to find the required number of turns:

$$N = \frac{LB}{\mu_0 I} = \frac{(0.1 \text{ m})(0.1 \text{ T})}{(4\pi \times 10^{-7} \text{ T m/A})(10 \text{ A})} = 800 \text{ turns.}$$

A wire that can carry 10 A without overheating is about 1 mm in diameter, so only 100 turns can be placed in a 10 cm length. Thus it takes eight layers to reach the required number of turns.

Objects inserted into the center of a solenoid are in a uniform magnetic field. Many physics experiments that need a uniform magnetic field are conducted inside a solenoid, which can be quite large in many cases. You have probably seen pictures of hospital patients undergoing magnetic resonance imaging (MRI), where they are positioned inside a large white cylinder. The cylinder is a solenoid, with the coil of wire hidden from view by the housing of the machine. MRI machines use superconducting wire for the solenoid, which allows it to carry very large currents and to produce strong fields of several tesla.

30.8 The Magnetic Force on a Moving Charge

Oersted discovered the remarkable fact that a current passing through a wire causes magnetic forces to be exerted on a nearby compass needle. According to Newton's third law, the compass needle should then exert an equal but opposite reaction force on the current in the wire! Does it?

Ampère's Experiment

Experimental verification of this assertion was made by the French scientist André-Marie Ampère within a week of the news of Oersted's discovery reaching Paris. The reaction force on the wire was much too small to observe in Oersted's original procedure, so Ampère faced the difficulty of increasing the size of the forces. He reasoned that the current was acting like a magnet and thereby exerting a force on the compass needle, which he knew to be a weak magnet. If he were to replace the compass with a second current-carrying wire, which would also act like a magnet, would *it* experience a force due to the first current? And would the first current then experience a reaction force? Ampère used this argument to predict that two current-carrying wires would exert equal but opposite forces on each other.

Ampère's experiment measured the forces between two parallel wires that could carry large currents in either the same direction or in opposite directions. He found that the two parallel wires *attract* each other when the currents are in the *same* direction and that they *repel* each other when the currents flow in *opposite* directions, as shown in Fig. 30-22. (Notice in this experiment that "likes" attract and "opposites" repel, the reverse of forces between electric charges or magnetic poles.) As expected, *each* wire experiences an equal force, but in opposite directions. Before reporting his results, Ampère was careful to verify that these were magnetic forces, not electrical forces—after all, he was using batteries and currents, so charges were present. He noted, for one thing, that current-carrying wires are electrically neutral. Even if the wires did manage somehow to become charged, the arrangement in which the currents flow in the same direction should charge the wires equally and cause a repulsion of like charges. But he found an attractive force under these circumstances. His conclusion was that current-carrying wires exert *magnetic* forces on each other.

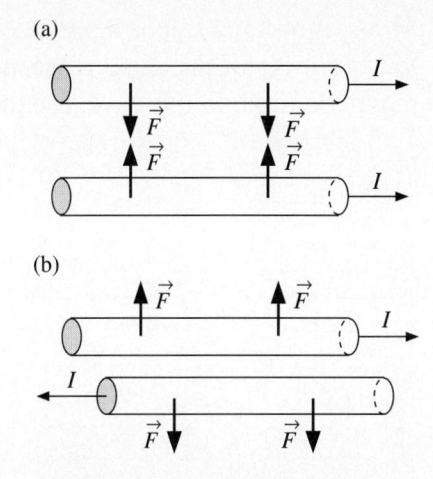

FIGURE 30-22 Ampere's experiment to observe the forces between parallel current-carrying wires. a) Parallel currents attract each other. b) Opposite currents repel each other.

The Force on a Moving Charge

Ampère's experiment showed that *a magnet exerts a force on a current*. But a current is simply a moving charge. This fact strongly suggests that a magnetic field will exert a force on a moving charge—and indeed it does, although the exact form of the force law was not discovered until later in the nineteenth century.

The magnetic force on a moving charge turns out to be rather tricky. It depends not only on the charge and the charge's velocity but also on how the velocity vector is oriented relative to the magnetic field. Experiments on moving charges in a magnetic field reveal the following behavior:

1. A charge moving *parallel* to the magnetic field experiences *zero* force.
2. A charge q moving with a velocity $\vec{v}$ that is *perpendicular* to $\vec{B}$ experiences a force of magnitude

$$F = |\vec{F}| = qvB$$

in a direction given by the right-hand rule.

We can summarize these by saying that the force on a charged particle moving in a magnetic field is given by

$$\vec{F} = \begin{cases} 0 & \vec{v} \text{ parallel to } \vec{B} \\ (qvB, \text{ right-hand rule}) & \vec{v} \text{ perpendicular to } \vec{B}. \end{cases} \quad (30\text{-}16)$$

More advanced courses will consider the case of an arbitrary angle between $\vec{v}$ and $\vec{B}$, but we will limit ourselves to just the cases of parallel and perpendicular motion.

A right-hand rule is again needed to determine a direction, because the force is a vector, but it is *not* the same right-hand rule used previously to determine the direction of a magnetic field. In this case, use the thumb and first two fingers of your right hand to make three mutually perpendicular directions. Orient your hand so that your thumb points in the direction of $\vec{v}$ and your index finger in the direction of $\vec{B}$. Your middle finger is then pointing in the direction of the force $\vec{F}$. Note carefully that the force is perpendicular to *both* $\vec{v}$ and $\vec{B}$.

The geometrical relationship among $\vec{v}$, $\vec{B}$, and $\vec{F}$ is illustrated in Fig. 30-23 for several moving charges. (Notice that the *source* of the magnetic field isn't shown in pictures like this, just the field itself.) For a negative charge, such as an electron, the negative value of qvB in Eq. 30-16 indicates a force pointing in the *opposite* direction to that given by the right-hand rule. Again we see the inherent three-dimensionality of magnetism, with the force perpendicular to the magnetic field. This is very different from the case for electric forces, which were parallel to the electric field.

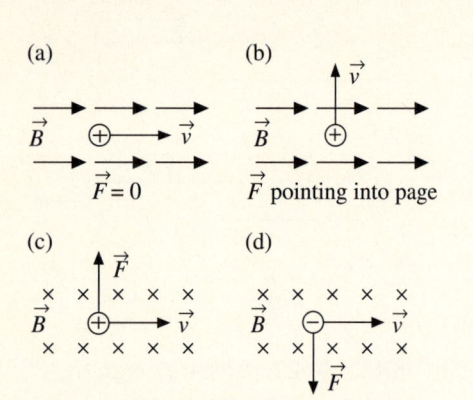

FIGURE 30-23 Magnetic forces on moving charges.

- **EXAMPLE 30-5** An electron is moving parallel to a wire carrying a current of 10 A. The electron is 1 cm above the wire and traveling in the same direction as the current. What are the magnitude and the direction of the magnetic force on the electron?

    SOLUTION Figure 30-24 shows a current and an electron moving to the right. The field above the wire is out of the page, so the electron is moving perpendicular to the field and the force on it is given by Eq. 30-16. The right-hand rule with $\vec{v}$ and $\vec{B}$ gives a downward-pointing vector, but the electron is a negative charge so the force on it is upward, away from the wire. The force is of magnitude qvB, and the field is that of a long, straight wire. From Eq. 30-9,

    $$B = \frac{\mu_0 I}{2\pi d} = 2.0 \times 10^{-4} \text{ T}.$$

    The magnitude of the force on the electron is thus

    $$F = evB = 3.2 \times 10^{-16} \text{ N}.$$

    The electron will curve away from the wire because of this force.

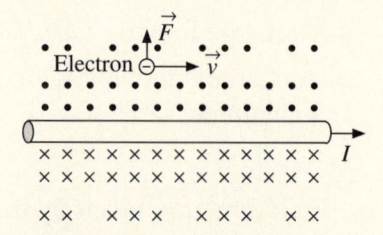

    FIGURE 30-24 An electron moving parallel to a current-carrying wire.

We can draw an interesting and important conclusion at this point. You have seen that the magnetic field is created *by* moving charges. Now you also see that magnetic forces are exerted *on* moving charges. It thus appears that this thing we call *magnetism* is an *interaction between moving charges*. Any two charges, whether moving or stationary, interact with each other through the electric field. In addition, two *moving* charges also interact with each other through the magnetic field. This fundamental observation is easy to lose sight of when we talk about currents, magnets, torques, and all the other phenomena of magnetism. But the most basic underlying feature of all these events is an interaction between moving charges. Your understanding of magnetism will be greatly enhanced if you think carefully about how the many large-scale magnetic effects occur because of how the individual charges are moving.

Cyclotron Motion

Many important applications of magnetism involve the motion of charged particles in a magnetic field. Your television picture tube functions by using magnetic fields to steer electrons as they move through a vacuum from the electron gun to the screen. Microwave generators, which are used in applications ranging from ovens to communications, use a device called a *magnetron* in which electrons oscillate rapidly in a magnetic field. And charged particles moving in the earth's magnetic field are responsible for the *aurora*, or northern lights.

Let's look at the simplest case of what happens when a charged particle moves in a *uniform* magnetic field, such as that generated inside a solenoid. In Fig. 30-23a we saw a charge with velocity $\vec{v}$ in a direction *parallel* to the field $\vec{B}$. In this case, according to Eq. 30-16, the force on the charge is $\vec{F} = 0$. Consequently, by Newton's first law, the charge will continue to move in a straight line at *constant* velocity. In other words, a magnetic field has *no effect* on a charge moving parallel to the field. This effect is quite different from a charge moving in an electric field, where it is accelerated parallel to the field.

Of much more importance and interest, consider a charge q moving with a velocity $\vec{v}$ that is *perpendicular* to a uniform magnetic field $\vec{B}$, as shown in Fig. 30-25a. Equation 30-16 and the right-hand rule tell us that a force of constant magnitude $F = qvB$ is always directed *perpendicular* to the direction in which the charge is moving. As the charge changes direction, the force vector turns so as to remain perpendicular to $\vec{v}$. This should seem familiar—it is exactly the same situation as a mass on a string, which experiences an always-perpendicular tension force $\vec{T}$ as it moves in a circle. The same motion occurs here, with the magnetic force acting as a *centripetal force* that accelerates the charge around a circular path at constant speed v. Stated more directly, a charged particle moving in a plane *perpendicular* to a uniform magnetic field will travel in a circular path around the magnetic field vectors. This particular type of motion, shown in Fig. 30-25b, is called the **cyclotron motion** of a charged particle in a magnetic field. Note that a negative charge will orbit in the direction opposite that shown in Fig. 30-25 for a positive charge.

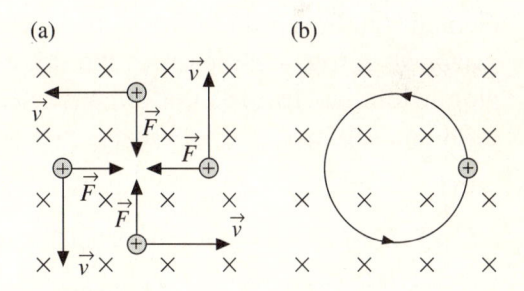

FIGURE 30-25 Cyclotron motion of a charged particle moving in a magnetic field. a) The magnetic force acts as a centripetal force. b) The circular trajectory of the particle.

352 CHAPTER 30 THE MAGNETIC FIELD

Because the magnetic force $F_{mag} = qvB$ acts as a centripetal force $F_c = mv^2/r$, we can equate the two:

$$\left[F_{mag} = qvB\right] = \left[F_c = \frac{mv^2}{r}\right]. \quad (30\text{-}17)$$

This is easily solved to give the radius of the cyclotron orbit:

$$r = \frac{mv}{qB}. \quad (30\text{-}18)$$

The inverse dependence on B indicates that the size of the orbits can be decreased by increasing the magnetic field strength.

We can also determine the frequency of the circular cyclotron motion. Recall from your earlier study of circular motion that the frequency of revolution f is related to the speed and radius by $f = v/2\pi r$. A rearrangement of Eq. 30-18 gives the **cyclotron frequency**:

$$f_{cyclotron} = \frac{qB}{2\pi m}, \quad (30\text{-}19)$$

where the ratio q/m is the *charge-to-mass ratio* of the charged particle.

Equation 30-19 indicates that the cyclotron frequency depends on the charge-to-mass ratio and the magnetic field, but *not* on the charge's velocity. This result has important applications. Positive atomic ions (e.g., atoms with one electron removed) all have the same charge $q = e$. But because every element has a unique atomic mass m, the charge-to-mass ratio e/m for each element is unique. The ratio e/m is thus a "fingerprint" that identifies the element (and even the isotope) precisely. A technique developed in the 1970s called *ion cyclotron resonance* measures the different cyclotron frequencies present in a sample of atomic ions and from that information deduces which elements are present. It is a very powerful and sophisticated tool that is now widely used in science and engineering.

EXAMPLE 30-6 An electron is accelerated from rest through a potential difference of 1000 V, then fired into a uniform magnetic field. The electron is observed to orbit perpendicularly to the field at a frequency of 300 MHz. What is the radius of its orbit?

SOLUTION We can determine the magnetic field strength from the charge-to-mass ratio and the cyclotron frequency of the electron:

$$B = \frac{2\pi m f}{e} = \frac{2\pi (9.11 \times 10^{-31} \text{ kg})(300 \times 10^6 \text{ Hz})}{1.60 \times 10^{-19} \text{ C}} = 0.0107 \text{ T}.$$

The radius of the orbit is given by Eq. 30-18, but first we need to find the electron's speed. Because it was accelerated from rest ($v_i = 0$) through a known potential difference, conservation of energy gives

$$\Delta K + \Delta U = \Delta K - e\Delta V = 0$$
$$\Rightarrow \tfrac{1}{2} m v_f^2 = e\Delta V.$$

This is easily solved to give the speed upon entering the magnetic field:

$$v = \sqrt{\frac{2e\Delta V}{m}} = \sqrt{\frac{2(1.60 \times 10^{-19} \text{ C})(1000 \text{ V})}{9.11 \times 10^{-31} \text{ kg}}} = 1.87 \times 10^7 \text{ m/s}.$$

The cyclotron radius is thus

$$r = \frac{mv}{qB} = 9.95 \times 10^{-3} \text{ m} = 9.95 \text{ mm}.$$

Note that Eq. 30-18 can also be used to measure the momentum $mv = rqB$ of a charged particle by measuring the radius of its orbit in a known field. This technique is used in high-energy physics experiments to determine the momentum of subatomic particles produced when two particles (such as electrons or protons) collide at speeds nearly the speed of light. The collision occurs inside a device called a *bubble chamber*, where the particles leave a map of their trajectories in the form of a string of tiny bubbles in liquid hydrogen. This bubble pattern is photographed, and the radius of the particle's orbit is measured from the photograph. Because the magnetic field strength inside the bubble chamber is known, the particle's momentum can be determined.

Figure 30-26 shows photographs of the circular motion of subatomic particles in a bubble chamber. The magnetic field is perpendicular to the page. The particle on the left is an electron. Its radius slowly decreases, and it spirals inward, as it loses speed due to collisions with the hydrogen atoms.

FIGURE 30-26 Subatomic particles moving in a bubble chamber. The magnetic field is perpendicular to the page.

30.9 Magnetic Forces on Current-Carrying Wires

Having determined the magnetic force on an individual moving charge, we can now use that knowledge to analyze Ampère's experiment. As a first step, let us find the force that a uniform magnetic field exerts on a straight wire that carries a current I.

Force on a Single Wire in a Magnetic Field

If a current-carrying wire is *parallel* to a magnetic field, the force on it is zero. This follows from Eq. 30-16. Because the charges in the wire are moving parallel to the field, the force on each of them is zero. Thus we will restrict our attention to the case of a wire that is *perpendicular* to the magnetic field, as shown in Fig. 30-27. Note that the field shown here is an *external* magnetic field, created by *other* currents, and is *not* the field of the current I.

The direction of the force on the current in Fig. 30-27 is determined by considering the force on each charge in the current. By the right-hand rule, each charge is seen to have a force of magnitude qvB directed to the right. Consequently, the entire length of wire experiences a sideways force that is perpendicular to both the current direction and the field direction. In fact, you can easily see that the same right-hand rule can be applied directly to a current-carrying wire if you place your thumb along the direction in which the current is flowing.

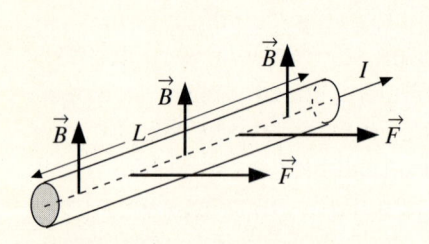

FIGURE 30-27 Magnetic force on a current-carrying wire.

To find the magnitude of the force, we must relate qv of the charge to the current I in the wire. This is essentially the same procedure as we followed in Section 30.5 to determine the magnetic field of a straight current-carrying wire. Consider a section of wire of length L that contains total charge q moving with speed v. The current I, by definition, is the charge q divided by the time T it takes the charge to flow out of this section of the wire: $I = q/T$. The time required is simply $T = L/v$, giving

$$I = \frac{q}{L/v}.$$

Thus

$$qv = IL. \qquad (30\text{-}20)$$

Inserting this into the force equation $F = qvB$ gives the magnetic force on a current-carrying wire of length L:

$$F = ILB \quad \text{(force on a current-carrying wire)}. \qquad (30\text{-}21)$$

Equation 30-21 is a very simple result, but remember the two assumptions behind it: The wire is perpendicular to the field, and the field is constant over the length of the wire. As an aside, you can see from Eq. 30-21 that the magnetic field must have units of N/A m. This is why we defined 1 T = 1 N/A m in Section 30.4.

Force Between Two Parallel Wires

Now consider Ampère's experimental arrangement of two parallel wires of length L, each carrying current I. The wires are a distance d apart. Figure 30-28a shows the case of the currents traveling in the same direction; Fig. 30-28b shows the currents in opposite directions. We will assume that the wires are sufficiently long to allow us to use the earlier result for the magnetic field of a long, straight wire: $B = \mu_0 I/2\pi d$. You will recall that this is not valid near the ends of the wire, but if the wires are sufficiently long a slight deviation near the ends will not have much effect on our results.

As Fig. 30-28a shows, the current in the lower wire produces a magnetic field at the position of the upper wire that is directed out of the page, perpendicular to the current direction. This field, due to the lower wire, exerts a magnetic force on the upper wire. Using the right-hand rule, we see that the force on the upper wire is downward, attracting it toward the lower wire. Although the field of the lower current is not a uniform field, it is nonetheless the *same* at all points along the upper wire because the two wires are parallel. Consequently, we can use the field of a long straight wire for B in Eq. 30-21 to determine the magnetic force exerted by the lower wire on the upper wire:

$$F = \frac{\mu_0 L I^2}{2\pi d} \qquad \text{(force between two parallel wires).} \qquad (30\text{-}22)$$

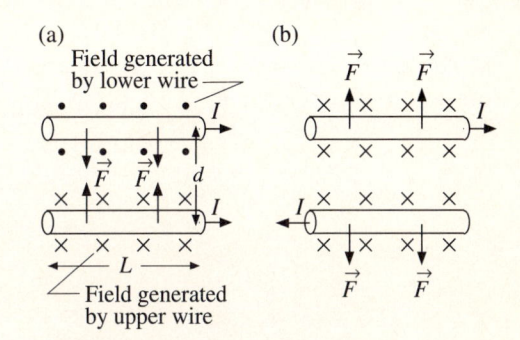

FIGURE 30-28 Magnetic forces between parallel current-carrying wires.
a) Currents in the same direction.
b) Currents in opposite directions.

As an exercise, you should convince yourself that the upper wire exerts an upward-directed force on the lower wire of exactly the same magnitude.

So as Ampère originally surmised, two parallel wires do indeed exert equal but opposite forces on each other as required by Newton's third law. If the two currents are in the same direction, the forces are attractive and tend to pull the wires together. You should convince yourself, using the right-hand rule, that the forces are repulsive and tend to push the wires apart if the two currents are in opposite directions. This is shown in Fig. 30-28b.

EXAMPLE 30-7 Two parallel wires 50 cm in length are connected at the ends by metal springs. Each spring has an unstretched length of 5 cm and a spring constant of 0.02 N/m. The wires push each other apart when a current flows around the loop. How much current is required to stretch the springs to lengths of 6 cm?

SOLUTION The springs are conductors, so a current can flow around the loop. The currents in the two wires are in opposite directions, so the wires exert repulsive forces on each other. In equilibrium, the repulsive force between the wires will be balanced by the

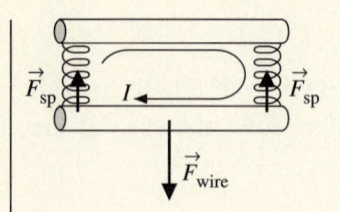

FIGURE 30-29 Current-carrying wires of Example 30-7.

restoring forces $F_{sp} = k\Delta s$ of the springs. Figure 30-29 shows the forces on the lower wire. Because the net force must be zero,

$$F_{wire} = \frac{\mu_0 L I^2}{2\pi d} = 2 \cdot k\Delta s,$$

where k is the spring constant and $\Delta s = 1$ cm is the amount by which each spring stretches. Solving for the current at $d = 6$ cm gives

$$I = \sqrt{\frac{4\pi k d \Delta s}{\mu_0 L}} = 15.5 \text{ A}.$$

30.10 Forces and Torques on Current Loops

Our goal is to understand not only electromagnets but also permanent magnets. You have already seen that a current loop is a magnetic dipole, very much like a bar magnet. We will now look at some important features of how current loops behave in magnetic fields. This discussion will be qualitative, but it will highlight some of the important properties of magnets and magnetic fields. We will then use many of these ideas in the next section, where we will make the connection between electromagnets and permanent magnets.

Consider two current loops facing each other, as shown in Fig. 30-30, with the current circulating in the same direction in each. Do these two loops exert forces on each other? We might suspect, by analogy with parallel straight wires, that the loops will attract each other if the currents circulate in the same direction and repel each other if the currents circulate in opposite directions. It is not hard to confirm this guess because we know the magnetic field of a current loop (see Fig. 30-17). Figure 30-30 shows the magnetic field at the top and bottom of the right loop due to the current in the left loop. The current of the right loop at these points, which is flowing into or out of the page, is perpendicular to the field. We can use the right-hand rule to determine the direction of the magnetic forces $\vec{F}_1$ and $\vec{F}_2$ on the top and bottom of the right loop. Forces $\vec{F}_1$ and $\vec{F}_2$ are of equal magnitude, each perpendicular to both $\vec{B}$ and the current, and $\vec{F}_1$ is tilted upward by exactly the same angle that $\vec{F}_2$ is tilted downward. The vertical components of these forces cancel when they are added, but their horizontal components reinforce each other to give a net force to the left. That is, the left current loop exerts a net magnetic force on the right current loop.

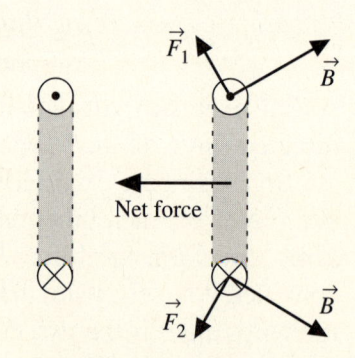

FIGURE 30-30 Magnetic forces between two parallel current loops.

By Newton's third law, the right current loop then exerts a force on the left current loop that is equal in magnitude but directed toward the right. (You should confirm this by considering the magnetic field due to the right loop at the position of the left loop.) Thus we find that parallel current loops with current circulating in the same direction do attract each other. Similar analysis shows that the loops repel each other when the currents circulate in opposite directions.

Figure 30-30 analyzed the situation from the perspective of magnetic forces on currents, but we can also look at it from the standpoint of magnetic poles. Figure 30-31 shows the north and south magnetic poles of the current loops. If the currents circulate in the same direction, a north and a south pole face each other and exert attractive forces on each other. For currents circulating in opposite directions, the two north poles repel each other.

Here, at last, we have a real connection to the behavior of magnets that opened our discussion of magnetism—namely that like poles repel and opposite poles attract. Two current loops exert forces on each other exactly as if they were two bar magnets. Two solenoids, which also have north and south poles, exhibit the same behavior. Our tour through interacting moving charges is finally starting to show some practical results! We still have to relate this behavior of current loops to the properties of permanent magnets, and we will do so in the next section.

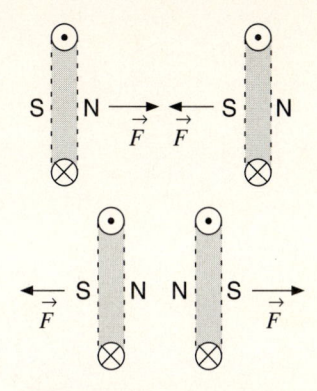

FIGURE 30-31 The forces between current loops in terms of their north and south poles.

Now let's consider the forces on a current loop in a *uniform* magnetic field, as shown in Fig. 30-32a. Because the field in this case is the same at the top and bottom of the loop, the forces on these two sections of the loop are equal in magnitude. However, application of the right-hand rule shows that the directions are not the same: The force on the top of the loop is directed upward, but that on the bottom of the loop is downward. The net force is zero, but because these forces are not collinear they will *turn* the loop by exerting a torque on it.

Both the torque and the net force are zero when the plane of the loop is perpendicular to the magnetic field, so this is the equilibrium position. If a current loop is pivoted so that it is free to rotate, the net effect of placing it in a magnetic field is for it to rotate until it reaches the equilibrium position perpendicular to the field.

Figure 30-32b also shows the current loop's magnetic poles. The torque on the loop rotates the magnetic dipole until the north pole is in the direction that the magnetic field points. But this is exactly the behavior of a compass needle! A compass needle also rotates until its north pole is in the direction that the field points.

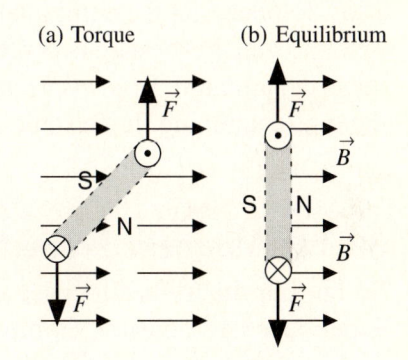

FIGURE 30-32 a) Torque on a current loop in a uniform external magnetic field. b) In equilibrium the north pole of the current loop points in the direction of the field.

Here is another situation in which a current loop acts just like a permanent magnet, but we now have a *physical mechanism* for understanding that the torque occurs due to magnetic forces on the charges moving in the wire. We continue to make progress toward an actual *understanding* of magnetic phenomena.

The torque on a current loop in a magnetic field is the basis for how an electric motor works, as shown in Fig. 30-33. The *armature* of a motor is a coil of wire, rather than a single loop, that is wound on an axle and is free to rotate. A magnetic field exerts a torque on the armature when a current flows through the coil, causing it to spin. If the current were

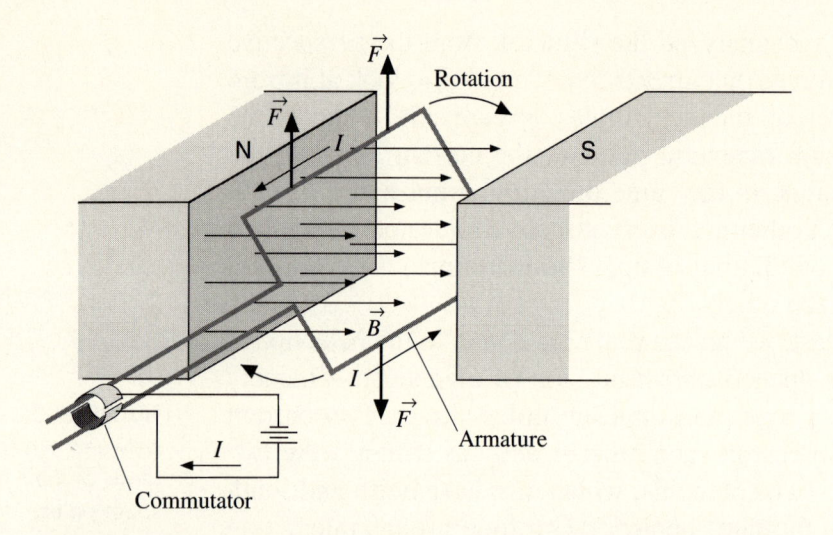

FIGURE 30-33 A simple electric motor. The magnetic torque causes the coil to rotate. The commutator directs the current flow through the coil so as to prevent the coil from reaching an equilibrium position.

steady, the armature would rotate to the equilibrium position and stop. To keep the motor turning, a device called a *commutator* reverses the direction of current flow in the coils every 180°. (Notice that the commutator is split, so the positive terminal of the battery sends current into whichever wire is contacting the bottom half of the commutator.) This reversal prevents the armature from ever reaching an equilibrium position, so the magnetic torque keeps the motor spinning in the same direction as long as current flows through the coil.

30.11 Magnetic Properties of Matter

So far, our theory has focused only on the magnetic properties of currents. Our everyday experience, by contrast, is mostly with permanent magnets. We have seen that current loops and other circulating currents have magnetic poles and exhibit behaviors that are like those of permanent magnets, but we still lack a specific connection between electromagnets and permanent magnets. The goal of this section is to complete our understanding of magnetism by developing an atomic-level view of the magnetic properties of matter.

Atomic Magnets

We like to think of an atom as an object with little negatively charged electrons orbiting in solar-system fashion about a positive nucleus. However, this "classical atom," which obeys the laws of Newtonian mechanics, fails to behave in the expected ways when subjected to experimental tests. It was the failure of this simple theory of atoms to predict experimental results that eventually led to the discovery, in the 1920s, of quantum mechanics. Atoms are properly understood only through quantum theory, and any "classical" picture of what is happening on the atomic level has to be viewed with extreme suspicion. We will nonetheless use a classical point of view in this chapter to explore the magnetic properties of atoms. Such an approach does give a *qualitatively* correct understanding, and that is all we seek at this time.

A hydrogen atom has a single electron orbiting a proton—the simplest possible atom. In this classical picture of the atom, shown in Fig. 30-34a, the electron's motion is exactly that of a current loop! It is a microscopic current loop, to be sure, but a current loop nonetheless. Consequently, the theory of magnetism we have developed predicts that a hydrogen atom will act as a tiny magnetic dipole, with a north and a south pole. The current direction is opposite the electron's motion, because it is a negative charge, and we can use this fact along with the right-hand rule to locate the two poles. Thus our theory of magnetism, applied at the atomic level, strongly suggests that the moving electrons within atoms will both generate magnetic fields and experience magnetic forces.

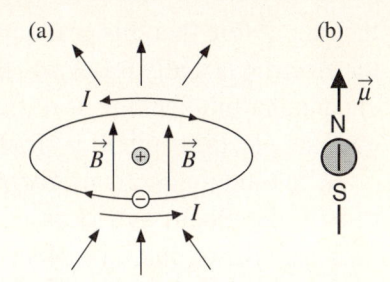

FIGURE 30-34 a) A hydrogen atom. b) The magnetic moment of the electron's current loop.

When dealing with atoms, it is customary to talk about **magnetic moments** rather than current loops. The magnetic moment is a vector $\vec{\mu}$ that is centered on the atom and points in the direction of the atom's north magnetic pole, as shown in Fig. 30-34b. The magnetic moment is perpendicular to the plane of the electron's orbit. The strength of the magnetic field at points near the atom is proportional to the size of the magnetic moment: $|\vec{B}| \propto |\vec{\mu}|$.

We cannot give a precise definition of $\vec{\mu}$ in our classical atomic picture, but we do not need one because we will not be doing any calculations. Still, the concept of the magnetic moment as representing the *strength* and *orientation* of the atomic magnet will be useful. If you wish, think of the magnetic moment vector as an atomic-size bar magnet.

The atoms of most elements, you recall from chemistry, contain many electrons. Unlike the solar system, where all of the planets orbit in the same direction, electron orbits are usually arranged to oppose each other: One electron moves counterclockwise for every electron that moves clockwise. Thus the magnetic moments of individual orbits tend to cancel each other. Even if the cancellation is not complete, an atom's *net* magnetic moment is nonetheless small.

The cancellation continues as atoms are joined into molecules and the molecules into solids. The atomic arrangement required by quantum mechanics is such that any magnetic moment of one atom tends to cancel that of neighboring atoms. When all is said and done, the net magnetic moment of any bulk matter due to the orbiting electrons is so small as to be negligible. There are various subtle magnetic effects that can observed under laboratory conditions, but orbiting electrons cannot explain the very strong magnetic effects of a piece of iron. The magnetic effects of a single orbiting electron, which are very real, fail to account for the magnetic properties of bulk matter.

The missing key to atomic magnetism was the 1922 discovery that electrons have the property of *spin*. In a classical picture, you can think of the electron as a very small spherical charged particle spinning on its axis like a top. The electron itself looks like a small current loop, as shown in Fig. 30-35, because each piece of the electron's charge is moving in a circular path. Consequently, each and every electron has a magnetic moment and acts as a small

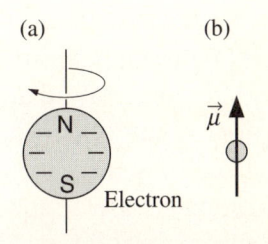

FIGURE 30-35 a) Electron spin. b) The electron's spin magnetic moment.

360 CHAPTER 30 THE MAGNETIC FIELD

magnet. (Note that this crude picture of a spinning electron is not a realistic portrayal of how the electron is really behaving. But quantum physics does require an electron to have an inherent angular momentum, as if it were spinning, and an inherent magnetic moment. We will discuss the electron spin in more detail in Chapter 37.)

The total magnetic moment of an atom is the vector sum of the magnetic moments due to both the electrons' orbital motions *and* the electrons' spins. Here again we must appeal to the results of quantum physics to find out what happens in an atom with many electrons. The spin magnetic moments, like the orbital moments, tend to oppose each other as the electrons are placed into their shells: For each electron spinning clockwise, another is spinning counterclockwise. So the net magnetic moment for all *filled* shells is zero. However, *unfilled* shells contain valence electrons, which may not have paired spins. Thus an atom can end up with a net magnetic moment if several valence electrons have their spins oriented in the same direction.

For most elements, the net magnetic moment of each atom is equally likely to point up as to point down when the atoms join together to form a solid. Figure 30-36 illustrates how this *random* arrangement of the magnetic moments might look for the atoms in a typical solid. Clearly, the solid as a whole will have a net magnetic moment that is very close to zero. This agrees with our common experience that most materials are not magnetic—you cannot pick them up with a magnet or make a magnet from them. Our atomic-level picture explains this lack of magnetic properties as a cancellation of the atomic magnetic moments when the atoms join together in a solid, even though each individual electron does have a magnetic moment. On the other hand, there are those materials such as iron that do exhibit strong magnetic properties, so we need to discover why these magnetic materials are different.

FIGURE 30-36 Random magnetic moments of the atoms in a typical solid. The net magnetic moment is zero.

Ferromagnetism

It happens that in iron, and a few other substances, the magnetic moments due to the electrons' spins tend to line up in the *same* direction, rather than in the random directions of Fig. 30-36. The reasons are purely quantum mechanical, so we cannot give a classical explanation. Materials that behave in this fashion are called **ferromagnetic**, with the prefix *ferro* meaning "iron-like." Figure 30-37 shows how the spin magnetic moments are aligned for the atoms making up a ferromagnetic solid.

In ferromagnetic materials, the individual magnetic moments do not cancel. Instead, they all add together to create a *macroscopic* magnetic dipole. Keep in mind that a magnetic moment is just a convenient way to talk

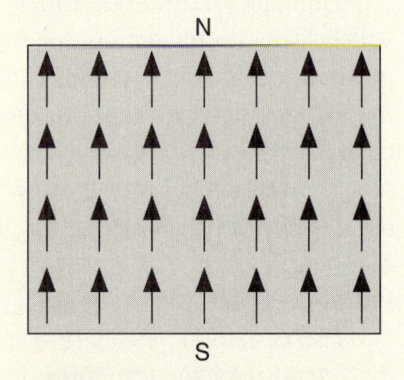

FIGURE 30-37 Aligned spin magnetic moments in a ferromagnetic material. This arrangement creates a macroscopic magnetic dipole, with a north and a south pole.

about the microscopic current loops in an atom. So when an object such as iron has a macroscopic magnetic dipole, the many atomic-level currents are cooperating to produce the equivalent of a macroscopic current loop. A material in which the atomic-level magnetic moments are all aligned thus has a north and a south magnetic pole, generates a magnetic field, and aligns itself with an external magnetic field—in other words, it *is* a magnet! But what appears to us on the bulk matter scale as "a magnet" is a *consequence* of moving charges at the atomic level.

Although iron is a magnetic material, a typical piece of iron is not a strong permanent magnet. You need not worry that a steel nail, which is mostly iron and is easily lifted with a magnet, will leap from your hands and pin itself against the hammer because of its own magnetism. The reason becomes clear from an examination of a piece of iron. It turns out that a crystal of iron is divided into small regions called **magnetic domains**, as shown in Fig. 30-38. A typical domain size is roughly 0.1 mm—small, but not unreasonably so. The spin magnetic moments of all of the iron atoms within each domain are perfectly aligned, causing each domain to be a magnetic dipole. The arrows in Fig. 30-38 represent the magnetic dipoles of each domain. An individual domain is, indeed, a strong permanent magnet. This is easily confirmed with grains of iron sufficiently small (<0.1 mm) to be a single domain. Figure 30-39 shows the magnetic field of a single cylindrically shaped domain of iron. Notice that the field looks very much like that of a solenoid.

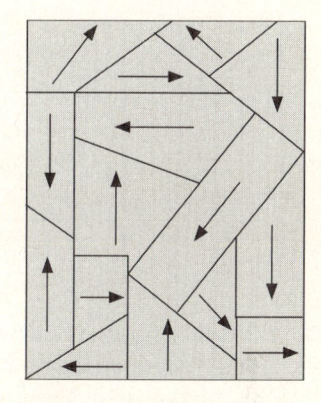

FIGURE 30-38 Magnetic domains in a ferromagnetic material. The net magnetic dipole is near zero.

However, the various magnetic domains that form a larger solid, such as you might hold in your hand, are randomly arranged. Their magnetic dipoles thus cancel, much like the cancellation that occurs on the atomic scale for nonferromagnetic substances, so the solid as a whole does not have a magnetic dipole. That is why the nail is not a strong permanent magnet.

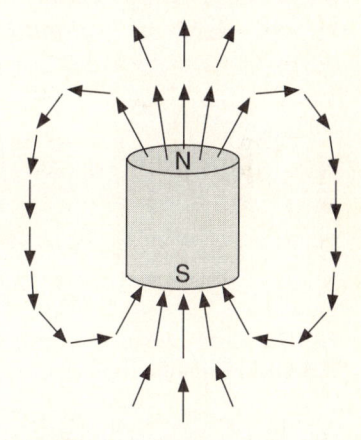

FIGURE 30-39 The magnetic field of a single magnetic domain.

What happens to a ferromagnetic substance if it is subjected to an *external* magnetic field, as would be the case if a magnet is brought close to a piece of iron? The magnetic dipole of each domain experiences a torque—just like a current loop—due to the external field. The torque causes many of the domains to rotate and become aligned with the external field, although internal forces between the domains generally prevent the alignment from being perfect. In addition, atomic-level forces between the spins can cause the *domain boundaries* to move. Domains that are aligned along the external field become larger, at the expense of domains that are opposed to the field. As a result of these changes in the size and orientation of the domains, the material develops a *net magnetic dipole* that is aligned with the external field. This magnetic dipole has been *induced* by the external field, so it is called an **induced magnetic dipole**.

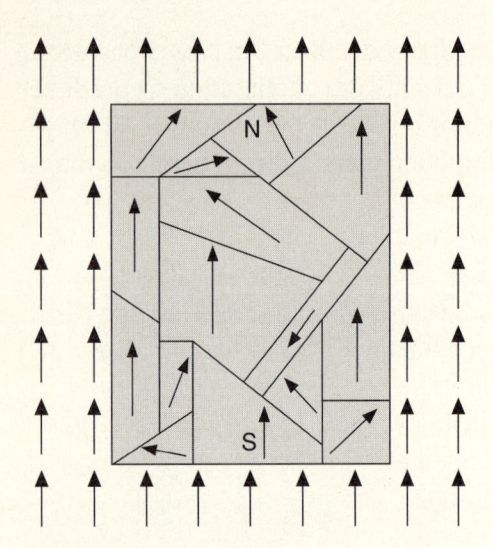

FIGURE 30-40 Alignment of the magnetic domains in an external magnetic field. This creates an induced magnetic dipole in the material.

Figure 30-40 shows the same ferromagnetic material we looked at in Fig. 30-38, only now the material is in an external magnetic field. Compare the two figures carefully to see how the domains' magnetic dipoles have been rotated by the field and how some domain boundaries have shifted. The solid as a whole now has a magnetic dipole, with a north and a south pole.

Suppose the external field is generated by a solenoid, as shown in Fig. 30-41. Because the induced magnetic dipole of the object is aligned with the external field, its south pole faces the solenoid's north pole. The magnetic force between the poles pulls the object to the electromagnet. If the solenoid were reversed so that its south pole faced the ferromagnetic object, the induced magnetic dipole would also reverse to produce a north pole facing the solenoid. Again, the opposite poles would attract. In other words, either end of an electromagnet will attract a piece of ferromagnetic material! This is, of course, one of the basic observations about magnetism that we started with at the beginning of the chapter. Now we have an *explanation* of how it works, based on three ideas: 1) the existence of a magnetic moment of each electron due to its spin, 2) the organization of spins into magnetic domains in a ferromagnetic material, and 3) the alignment of the domains in an external magnetic field to produce an induced magnetic dipole for the entire object.

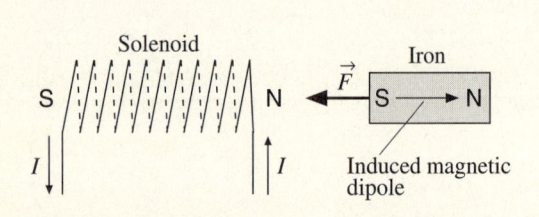

FIGURE 30-41 Attractive force on the induced magnetic dipole of a piece of iron.

When the external field is removed, internal atomic-level forces tend to randomize the domain magnetic dipoles. However, the object's net magnetic dipole will not return all the way to zero. Some domains become "frozen" in the alignment they had in the external field. Others, which grew in size when the external field was applied, manage to keep their larger size intact when the external field is removed. Through these mechanisms, a ferromagnetic object that has been in an external field may be left with a net, nonzero magnetic dipole after the field is removed. In other words, the object is left as a **permanent magnet**. Whether this happens depends both on the strength of the external field and on the internal crystalline structure of the material. *Steel* is an alloy of iron with other elements. An alloy of mostly iron with the right percentages of chromium and nickel produces *stainless steel*, which has virtually no magnetic properties at all because its particular crystalline structure is not conducive to the formation of domains. A very different steel alloy called "Alnico V" is made with 51% iron, 24% cobalt, 14% nickel, 8% aluminum, and 3% copper. It has extremely prominent magnetic properties and is used to make high-quality permanent magnets. You can see from the complex formula that developing good magnetic materials requires a lot of engineering skill as well as a lot of patience!

A permanent bar magnet has a magnetic dipole aligned along the axis, with magnetic poles at the ends of the cylinder. Its magnetic field is essentially the same as that of a solenoid. Because we have already seen how a solenoid can exert forces on an unmagnetized ferromagnetic object by inducing a magnetic dipole, that explanation carries over directly to understanding how a permanent magnet can exert a force on a ferromagnetic object.

So we've come full circle. Our initial observation about magnetism was that a permanent magnet can exert forces on some materials but not others. The *theory* of magnetism that we then proceeded to develop concerned the interactions between moving charges. It was not obvious what moving charges had to do with permanent magnets. But finally, by considering magnetic effects at the atomic level, we find that properties of permanent magnets and magnetic materials *can* be traced to the interactions of vast numbers of spinning electrons.

Summary

Important Concepts and Terms

magnetism	current loop
north pole	uniform magnetic field
south pole	solenoid
magnetic force	electromagnet
magnetic material	cyclotron motion
magnetic dipole	cyclotron frequency
magnetic monopole	magnetic moment
right-hand rule	ferromagnetic
magnetic field	magnetic domain
law of Biot and Savart	induced magnetic dipole
tesla	permanent magnet
permeability constant	

During the course of this chapter we have developed a theory of magnetism that allows us to *explain* the many observations with which we started the chapter. The underlying idea is that of an interaction between moving charges—two moving charges exert forces on each other in addition to any forces resulting from gravity or electricity. *Why* they should do so is a question that physics cannot answer, any more than it can explain *why* the earth and the sun exert gravitational forces on each other. They simply do. Nonetheless, we recognized the existence of this interaction between moving charges, and we were able to express mathematically how it works.

All else followed from this basic starting point. First we defined the magnetic field as being the agent through which moving charges interact. Second, we used the properties of individual moving charges to learn about the magnetic fields generated by currents and about the magnetic forces experienced by currents. This allowed us to understand how currents can exert forces on each other and how current loops rotate to align themselves with an external magnetic field. Lastly, we looked at the moving charges within atoms and developed an atomic-level view of the magnetic properties of matter. This led us to an explanation of permanent magnets in terms of the collective behavior of vast number of spinning electrons in the iron atoms.

The magnetic field of a moving charge is given by the law of Biot and Savart:

$$\vec{B} = \left(\frac{\mu_0}{4\pi} \frac{qv\sin\theta}{r^2}, \quad \text{direction given by the right-hand rule}\right).$$

A charge moving in a magnetic field experiences a force

$$\vec{F} = \begin{cases} 0 & \vec{v} \text{ parallel to } \vec{B} \\ (qvB, \text{ right-hand rule}) & \vec{v} \text{ perpendicular to } \vec{B}. \end{cases}$$

Thus magnetism, at its most fundamental level, is an interaction between two moving charges.

In practice, we are usually more interested in the magnetic fields of currents:

$$B_{\text{wire}} = \frac{\mu_0 I}{2\pi d} \quad B_{\text{loop}} = \frac{\mu_0 I}{2R} \quad B_{\text{solenoid}} = \frac{\mu_0 N I}{L}.$$

The magnetic field *inside* a solenoid is a uniform field. Charged particles undergo circular cyclotron motion in a uniform magnetic field.

Current-carrying wires experience magnetic forces:

$$F = ILB \qquad \text{(force on a current-carrying wire)}$$

$$F = \frac{\mu_0 L I^2}{2\pi d} \qquad \text{(force between two parallel wires)}.$$

A current loop is a magnetic dipole, with the north pole being the side from which the field emerges. Forces between current loops and torques on current loops are most easily visualized as interactions between the poles.

Exercises and Problems

Exercises

1. Summarize the experimental evidence that the magnetic force is a *new* kind of force, different from other forces we have studied.

2. a. What currents are needed to generate the magnetic field strengths of Table 30-1 at a point 1 cm from a long, straight wire?
   b. At what distances from a long, straight wire carrying a 10 A current would the magnetic field strengths of Table 30-1 be generated?

3. The magnetic field is 0.0025 T at the center of a current loop 1 cm in diameter.
   a. How much current flows in the loop?
   b. A long, straight wire carries the same current you found in part a). At what distance from the wire is the magnetic field 0.0025 T?

4. An electron moves along the y-axis with a velocity of $+1.0 \times 10^7$ m/s. As it passes the origin, what is $\vec{B}$ (both magnitude *and* direction) at the following (x, y) points?
   a. (0 cm, +1 cm)   c. (1 cm, 0 cm)
   b. (−1 cm, 0 cm)   d. (1 cm, 1 cm)

5. Determine the *initial* direction of deflection for the charged particles entering each of the magnetic fields shown in Fig. 30-42.

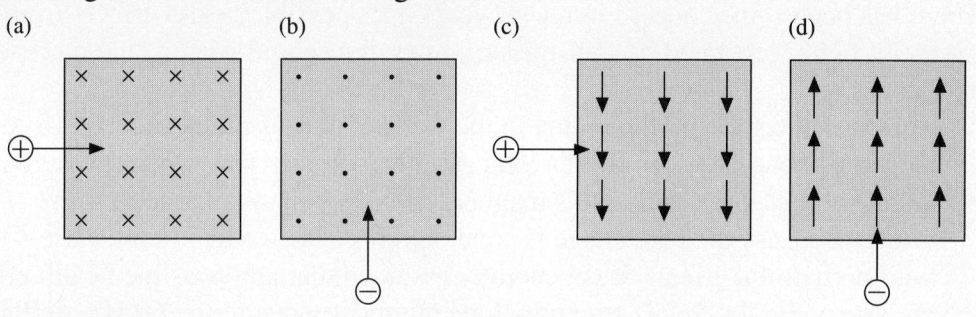

(a) (b) (c) (d)

FIGURE 30-42

6. Determine the magnetic field direction that causes each of the charged particles shown in Fig. 30-43 to experience the indicated magnetic force.

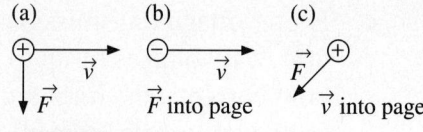

FIGURE 30-43

7. The aurora is caused by electrons and protons from the *solar wind* when these charged particles are captured by the earth's magnetic field of $\approx 5 \times 10^{-5}$ T. As the particles execute cyclotron orbits high above the earth, they collide with molecules of the tenuous atmospheric gases and cause them to glow. What is the radius of the cyclotron orbit for each of the following particles?
   a. An electron with speed 1×10^6 m/s
   b. A proton with speed 5×10^4 m/s

8. The atomic masses of several atoms are shown in the following table. Calculate the cyclotron frequency in a 3.00 T magnetic field for the ions: a) N_2^+, b) O_2^+, and c) CO^+. Express your answers in MHz. The accuracy of your answers should reflect the accuracy of the data. (3 T is typical for ion cyclotron resonance machines. Note that although both N_2 and CO have a *nominal* molecular mass of 28, they are easily distinguished by virtue of their different cyclotron resonance frequencies.)

Atomic masses

^{12}C	12.0000 u
^{14}N	14.0031 u
^{16}O	15.9949 u

9. Consider the two current loops in the uniform magnetic field shown in Fig. 30-44.
   a. Use force diagrams to show that both loops are in equilibrium, having a net force of zero and no torque. (Assume that each loop interacts only with the external field, not with the other loop.)
   b. One of the loop positions is stable, so the forces will return it to equilibrium if it is rotated slightly. The other position is unstable, like an upside-down pendulum, so the forces will cause it to flip over if it is rotated ever so slightly away from equilibrium. Which position is which?

FIGURE 30-44

Problems

10. There has been some concern in recent years over possible health effects from the magnetic fields generated by transmission lines, household wiring, and electrical appliances.
    a. The current carried by the wiring in the walls of houses rarely exceeds 10 A. What is the magnetic field 2 m from a long straight, wire carrying a current of 10 A?
    b. What percentage of the earth's magnetic field is your answer to part a)?
    (**Note:** Because the percentage is small and because we live in the earth's field with no harmful effects, it is generally assumed that any possible health effects are due not to the field's strength alone but mostly due to the 60 Hz oscillatory nature of fields from power lines.)
    c. High-voltage transmission lines, which you see on tall towers, typically carry 200 A at voltages of up to 500,000 V. Although this is much larger than household currents, the lines are roughly 20 m overhead. Estimate the magnetic field on the ground underneath such lines. Would you expect transmission lines to be a *significantly* different factor, either for better or worse, than household wiring?
    d. One common electrical appliance is an electric blanket. Some consumer groups urge pregnant women not to use electric blankets just in case there is a health risk. The current through the heater wires is approximately 1 A. Estimate, stating clearly any assumptions you make, the magnetic field a fetus might experience. How does this compare to your answer to part a)?

11. a. A current loop made of N turns of wire, each carrying current I, generates a magnetic field at the center of the loop given by $B = \mu_0 NI/2R$ *if* the width of the bundle of wires is much less than the loop radius R. Make an argument why this should be so. While doing so, make sure you explain the restriction on the size of the bundle of wires.
    b. A small wire can safely carry a current of 1 A. How many turns are needed to produce a field of 10^{-3} T at the center of a loop 1 cm in diameter?

12. You are asked to build a 20 cm long solenoid with an interior field of 5×10^{-3} T. The specifications call for a single layer of wire, wound with the coils as close together as possible. Two spools of wire are available: Wire of #18 gauge has a diameter of 1.02 mm and has a maximum current rating of 6 A, and #26 gauge wire is 0.41 mm in diameter and can carry up to 1 A. Which wire should you use, and what current will you need?

13. An Alnico magnet 2 cm in diameter by 8 cm long has a magnetic field strength of 0.10 T. To produce the same field with a solenoid of the same size, carrying a current of 2 A, how many turns of wire would you need? Does this seem feasible? (See Problem 12 for data about wire sizes and maximum current.)

14. a. Derive an expression for the magnetic field strength at distance d from the center of a straight wire of finite length L that carries current I.
    b. Determine the field strength at the center of a current-carrying *square* loop having sides of length $2R$.
    c. Compare your answer to part b) to the field at the center of a *circular* loop of diameter $2R$. Do so by computing the ratio B_{square}/B_{circle}.

15. a. Derive an expression for the magnetic field strength $B(z)$ a distance z along the axis of a circular current loop of radius R and current I. (**Hint:** The procedure is similar to that of Section 30.6. However, the $d\vec{B}$ due to the different lengths ds are now *not* parallel to the axis or to one another. Therefore, consider what happens when you add the two $d\vec{B}$ from two sections ds that are diametrically opposite each other.)
    b. Confirm that your answer reduces to Eq. 30-13 for the center of a current loop when $z \to 0$.
    c. Find an *approximate* expression for $B(z)$ when $z \gg R$.
    d. Graph $B(z)$ over the interval $-\infty < z < +\infty$. Where is B a maximum?

16. An electron of speed 1×10^7 m/s travels between two parallel charged plates, as shown in Fig. 30-45. The plates are separated by 1 cm and are attached to a 200 V battery. What magnetic field $\vec{B}$ (both strength *and* direction) would allow the electron to pass through without being deflected?

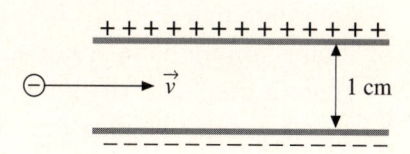

    FIGURE 30-45

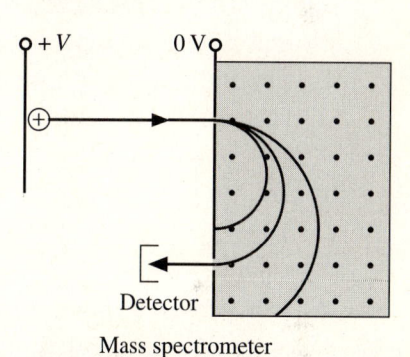

17. Figure 30-46 shows a *mass spectrometer*, which is an analytical instrument used to identify the various molecules in a sample by measuring their charge-to-mass ratio e/m. The sample is ionized, the positive ions are accelerated (starting essentially from rest) through a potential difference ΔV, and they then enter a region of uniform magnetic field. The field bends the ions into circular trajectories, but after just half a circle they either strike the wall or pass through a small opening to a detector. As the accelerating voltage is slowly increased, different ions are focused into the detector and measured. Typical design values are a magnetic field $B = 0.200$ T and a spacing between the entrance and exit holes $d = 8$ cm. What accelerating voltages V are required to detect: a) N_2^+, b) O_2^+, and c) CO^+. (See Exercise 8 for atomic data, and note there the comment about accuracy.)

    Mass spectrometer
    FIGURE 30-46

18. A long, straight wire having a linear mass density of 50 g/m is suspended by threads, as shown in Fig. 30-47. When a 10 A current flows through the wire, it experiences a horizontal magnetic force that deflects it to an equilibrium angle of 10°. What is the magnetic field $\vec{B}$ (strength *and* direction)?

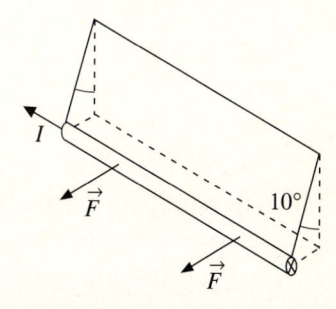

    FIGURE 30-47

19. Figure 30-48 shows three long, straight wires of linear mass density 50 g/m directed into the page. They each carry equal currents in the directions shown. The lower two wires are 4 cm apart and are attached to a table. What current I will allow the upper wire to "float" so as to form an equilateral triangle with the lower wires?

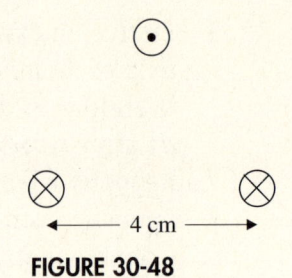

FIGURE 30-48

20. You have a horizontal cathode ray tube (CRT) for which the controls have been adjusted such that the electron beam *should* make a single spot of light exactly in the center of the screen. You observe, however, that the spot is deflected to the right. It is possible that the CRT is broken. But as a clever scientist, you realize that your laboratory might be in either an electric or a magnetic field. Assuming that you do not have a compass, any magnets, or any charged rods, how can you use the CRT itself to determine whether the CRT is broken, is in an electric field, or is in a magnetic field? You cannot remove the CRT from the room.

21. Figure 30-49 shows two small current loops next to a long, straight wire carrying current into the page.
    a. Show the forces on the top and bottom elements of each loop due to the current in the straight wire (similar to Figs. 30-30 and 30-32).
    b. Do the loops experience torques?
    c. Draw a diagram showing the equilibrium position for each loop.
    d. Label, on your diagram, the north and south magnetic poles of each loop.
    e. Compare your diagram to Fig. 30-4. What conclusions can you draw?

FIGURE 30-49

22. A computer diskette is a plastic disk coated with a ferromagnetic paint. A single magnetic domain can have its magnetic moment oriented to point either up or down, and these two orientations can be interpreted as 0s (up) or 1s (down). Each 0 or 1 is called a *bit* of information. Hence binary data can be stored on a disk as a series of magnetic domains that are aligned in the proper up/down orientations. A diskette stores roughly 500,000 *bytes* of data on one side, and each byte contains 8 bits. Estimate the width of a magnetic domain, and compare your answer to the typical domain size given in the text. Include a list of all the assumptions you use in your estimate.

23. In the semiclassical Bohr model of the hydrogen atom, the electron moves in a circular orbit of radius 5.3×10^{-11} m with speed 2.2×10^6 m/s. What is the magnetic field at the center of a hydrogen atom? (**Hint:** You will need to determine the *average* current of the orbiting electron.)

24. The ends of two permanent bar magnets can either attract or repel each other. We have asserted that magnetism is an interaction between moving charges. Give a step-by-step description of how these magnetic forces—for both the attractive and the repulsive cases—really do result from the interaction between moving charges, in this case spinning electrons.

25. Give a step-by-step description, in terms of moving charges, of how a permanent magnet can pick up a piece of nonmagnetized iron.

[**Estimated 5 additional problems for the final edition.**]

Chapter 31

Electromagnetic Induction

LOOKING BACK Sections 28.4, 28.5; 30.6–30.9

31.1 Faraday and Henry

Oersted's 1820 discovery that an electric current creates a magnetic field generated enormous excitement throughout scientific circles. Dozens of scientists in Europe and America immediately began experiments to explore the implications of this discovery. One of the questions scientists hoped to answer was whether the converse of Oersted's discovery was true—that is, whether a magnet could be used to create an electric current. There was not yet a good understanding of the origins or properties of electricity and magnetism, so scientists hoping to generate a current from magnetism had little to guide them. Many experiments were reported in which wires and coils had been placed in or around magnets of various sizes and shapes, but no one was able to generate a current from a magnet.

A few months later, in 1821, Michael Faraday (Fig. 31.1) repeated Oersted's experiment. You met Faraday in Chapter 24 as the originator of the field concept. The field idea came to him as he observed how a compass needle, as it is moved around a current-carrying wire, stays tangent to a circle around the wire. Faraday initially ascribed the compass needle's behavior to "circular lines of force"; only later were these lines of force called the *magnetic field*. The lines of force provided Faraday with a pictorial model of how magnetic effects are related to currents. Based on this idea, he invented the first electric motor. His motor was a simple device, not at all like a contemporary electric motor, but it demonstrated that

FIGURE 31-1 Michael Faraday, one of the greatest scientists of the nineteenth century.

the magnetic effects of currents could produce a continuous mechanical motion. But Faraday's motor raised another question: If a current can produce mechanical motion, can mechanical motion cause a current?

Faraday was one of the many scientists who joined the search for new ways to generate a current from a magnet. His journal throughout the 1820s describes many different experiments, but they all end with the phrase "no effect." At the same time, on the other side of the Atlantic, the American physicist Joseph Henry was following this research with great interest. Henry taught mathematics and philosophy at the Albany Academy, in Albany, New York. American teachers at the time were supposed to devote all of their time to teaching, so Henry had little opportunity for research. But during a one-month vacation in 1831, Henry became the first to discover how to produce electricity from magnetism, a process that we now call **electromagnetic induction**. But Henry had no time for follow-up studies, and he was not able to publish his discovery until the following year.

Meanwhile, and entirely independently, Faraday made the same discovery and immediately published his findings. Credit in science usually goes to the first to publish, so today we study *Faraday's law* rather than *Henry's law*. The situation, however, is not entirely unjust. Even if Faraday did not have priority of discovery, it was Faraday who studied the new phenomenon of electromagnetic induction exhaustively, established its properties, and realized that he had discovered a new law of nature. Henry had discovered an *effect*, but he was not able to do the research needed to understand the implications of his discovery.

Faraday's Discovery

Faraday's 1831 discovery, like Oersted's, was a happy combination of an unplanned event and a mind that was prepared to recognize its significance immediately. Faraday was experimenting with two coils of wire wrapped around an iron ring, as shown in Fig. 31-2. He had hoped that the magnetic field generated by a current in the coil on the left would induce a magnetic field in the iron, which in turn might somehow create a current in the circuit on the right. As with all his previous attempts, this experiment had failed to generate a current. But Faraday happened to notice that the needle of the current meter jumped ever so slightly at the instant he closed the switch in the circuit on the left. Once the switch was closed, however, the needle immediately returned to zero. When he later opened the switch, the needle again jumped, but this time in the opposite direction. Faraday recognized that the motion of the needle indicated a very slight current in the circuit on the right. But the effect happened only during the very brief interval that the current on the left was starting or stopping, not while it was flowing continuously.

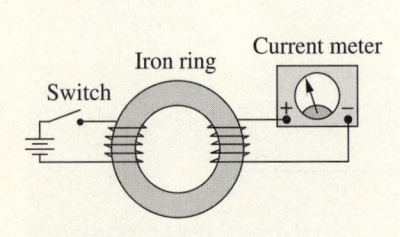

FIGURE 31-2 Faraday discovered that a current flows through the circuit on the right for a brief interval as the switch on the left is opened or closed.

Faraday applied his mental picture of lines of force to this discovery. When the current on the left magnetizes the iron ring, the magnetic field of the iron ring passes through the coil on the right. Faraday's observation that the current needle jumped only when the switch was opened and closed suggested to him that a current was generated only if the magnetic field was *changing* as it passed through the coil. This also would explain why all the previous attempts to generate a current were unsuccessful: They had used only steady, unchanging magnetic fields.

Faraday set out at once to test his hypothesis. If the critical issue was *changing* the magnetic field through the loop, then the iron ring should not be a necessary component. That is, any method that created a changing magnetic field should work. Faraday began a series of experiments to find out whether this was true. First, he placed one coil directly above the other without an iron ring, as shown in Fig. 31-3a. When he closed the switch in the upper circuit, creating a magnetic field, the momentary current again appeared in the lower circuit. As before, *no* current flowed in the lower circuit during the interval in which the switch remained closed. The momentary current returned, in the opposite direction, when the switch was opened and the magnetic field disappeared.

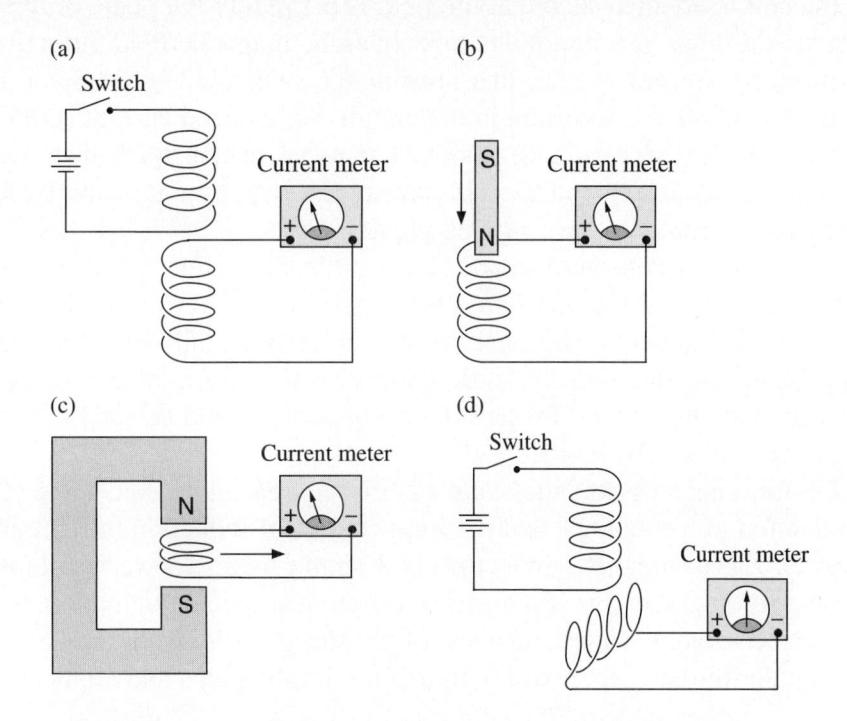

FIGURE 31-3 Faraday was able to induce a current in a circuit by a) opening or closing a switch in another circuit, b) inserting a bar magnet, or c) pulling the circuit out of a magnetic field. d) No current is induced if the coils are at right angles.

Next, Faraday tried pushing a bar magnet into a coil of wire, as shown in Fig. 31-3b. This action caused the magnetic field through the loop to increase quickly. Faraday found that this also caused a momentary deflection of the current meter needle, although *holding* the magnet inside the coil had no effect. A quick withdrawal of the magnet deflected the needle in the other direction.

Must it be the magnet that moves? Figure 31-3c shows a coil of wire being pulled rapidly out of the field of a permanent magnet. Faraday found that this action also deflected the needle, although no current flowed if the coil was stationary in the magnetic field. Pushing the coil *into* the magnet caused the needle to deflect in the opposite direction.

Finally, as shown in Fig. 31-3d, Faraday considered the orientation between the magnetic field and the coil of wire. If the test coil was turned 90°, perpendicular to the coil of the current loop, the current meter showed *no* activity as the switch opened or closed. Similarly, no momentary current flowed in the coil of Fig. 31-3c if it was rotated 90° before being pulled out of the magnet.

To summarize Faraday's discovery, he found that a *momentary* current flows in a coil of wire if, and only if, the magnetic field passing *through* the coil is *changing*. It makes no difference what causes the changing magnetic field: current stopping or starting in another circuit, moving a magnet through the coil, or moving the coil near a magnet. The effect is the same in all cases. No current flows if the field is not changing, so it is not the field itself that is responsible for the current flow but, instead, it is the *changing of the field*. Finally, the changing magnetic field must pass *through* the coil. No effect is observed if the coil is oriented such that the field is parallel to the plane of the coil.

The current that flows in a circuit due to a changing magnetic field through the circuit is called an **induced current**. We say that opening the switch in Fig. 31-3a or moving the magnet in Fig. 31-3b *induces* a current in the circuit. An induced current is *not* caused by a battery or by a physical separation of positive and negative charges. This is a completely new way to generate a current, and we will have to discover how it is similar to and how it is different from currents we have studied previously.

The first induced currents were small, barely noticeable effects. Neither Faraday nor Henry could have answered the question, "What good is it?" Yet in the decades that followed, electromagnetic induction became the basis of commercial electrical generation, of radio and television broadcasting, of audio and video recording, of computer memories and data storage, and much more. Present-day technology would be vastly different without the discoveries of Faraday and Henry.

Our goal in this chapter is to understand electromagnetic induction, some of the applications of induction to technology, and the implications of induction to light and electromagnetic waves. Electromagnetic induction is a subtle topic, so we will build up to it gradually. The next three sections will examine different aspects of induction. Section 31.5 will then introduce Faraday's law, a new law of physics governing electromagnetic induction. The remainder of the chapter will then explore implications and applications.

31.2 Motional emf

As you've just seen, an induced current can be created either by changing the magnetic field through a stationary coil, as in Figs. 31-3a and 31-3b, or by removing a coil from a stationary magnetic field, as in Fig. 31-3c. Although the effects are the same, the causes turn out to be different. In this section, we'll take a closer look at situations in which the magnetic field is fixed while the circuit moves or changes.

To begin, let's consider a metal bar or metal wire of length L, as shown in Fig. 31-4a, that is moving with speed v through a uniform magnetic field of strength $B = |\vec{B}|$. The charge carriers inside the wire are also moving with speed v at right angles to a magnetic field, so they each experience a magnetic force of magnitude $|\vec{F}| = qvB$. The charge carriers in a metal are free to move, so this force will cause a current to begin flowing parallel to the wire. The right-hand rule establishes that the force on a positive charge is upward, so the current will flow in the upward direction. Conversely, if you prefer to think of electrons as the actual charge carriers in a metal, the force on a negative charge is downward. In either case, motion of the wire through the magnetic field creates a current that makes the top end of the wire positive and the bottom end negative.

However, this current cannot be sustained. As the magnetic force separates the positive and negative charges, a downward electric field is created within the wire. The electric field points from the positive charge at the top to the negative charge at the bottom. The current in the wire will flow, and the charge separation and electric field will increase, only until the upward magnetic force on a positive charge in the wire is exactly balanced by a downward electric force. As soon as the electric force becomes big enough to balance the magnetic force, the *net force* on a charge in the wire becomes zero and the current ceases.

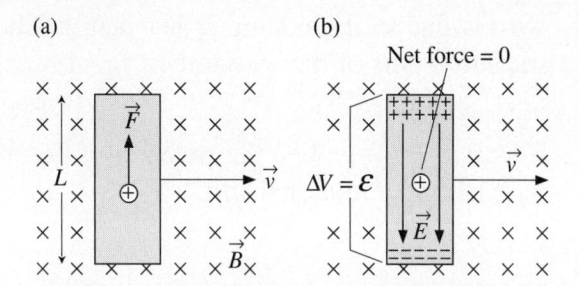

FIGURE 31-4 a) The magnetic field exerts a force on charge carriers in a moving wire. b) The charge separation causes an electric field in the wire and a potential difference between the ends.

The electric field whose force exactly balances the magnetic force is given by

$$\left[|\vec{F}_{\text{elec}}| = qE \right] = \left[|\vec{F}_{\text{mag}}| = qvB \right]$$
$$\Rightarrow E = vB.$$
(31-1)

Because the magnetic field is uniform, the electric field will have this strength at all points in the wire.

The electric field causes a potential difference between the top and the bottom of the wire. For an electric field of constant strength E, the potential difference between two points separated by distance L is simply

$$\Delta V_{\text{wire}} = V_{\text{top}} - V_{\text{bottom}} = EL = vLB,$$
(31-2)

where we've used $E = vB$ from Eq. 31-1. Thus the motion of the wire through a magnetic field *induces* a potential difference between the ends of the wire. The potential difference depends on the strength of the magnetic field and on the wire's speed through the field.

There's a strong analogy between this potential difference due to the motion of a wire through a magnetic field and the potential difference of a battery. You saw in Section 28.4 that a battery uses a *non*electric force to separate charge and that the work performed per charge (W/q) was defined as the emf $\mathcal{E}$ of the battery. An isolated battery, with no current flowing, has a potential difference $\Delta V_{\text{bat}} = \mathcal{E}$ *because* of the work done by the nonelectric force to separate the charge. We could refer to a battery, where the charges are separated by chemical reactions, as a source of *chemical emf*.

The moving wire develops a potential difference *because* of the work done by magnetic forces to separate the charge. You can think of the moving wire as a "battery" that stays charged only as long as it keeps moving but "runs down" if it stops. The emf of the wire is due to its motion, rather than to chemical reactions inside, so we can define the **motional emf** of the wire to be

$$\mathcal{E} = vLB. \tag{31-3}$$

Equation 31-2 is then the potential difference that we measure as a *consequence* of the wire having a motional emf.

EXAMPLE 31-1 A flashlight battery is 6 cm long and has an emf of 1.5 V. With what speed must a wire 6 cm long move through a 0.1 T magnetic field to create a motional emf of 1.5 V?

SOLUTION This is a straightforward computation. Using Eq. 31-3, we find

$$v = \frac{\mathcal{E}}{LB} = \frac{1.5 \text{ V}}{(0.06 \text{ m})(0.1 \text{ T})} = 250 \text{ m/s} \approx 500 \text{ mph.}$$

This might not be a very practical substitute for a battery, but it would work as long as the wire continued to move through the field with this speed.

Induced Current in a Circuit

Figure 31-5a shows a circuit made by a wire sliding, with velocity $\vec{v}$, along a U-shaped conducting rail in a magnetic field $\vec{B}$. The moving wire alone was not able to sustain a current, because the charges had nowhere to go. But now charges pushed to the top of the moving wire by the magnetic force can continue to flow around the circuit. The moving wire acts like a battery, with an emf given by Eq. 31-3, and an *induced current* flows counterclockwise around the circuit. If the total resistance of the circuit is R, the induced current is given by Ohm's law as

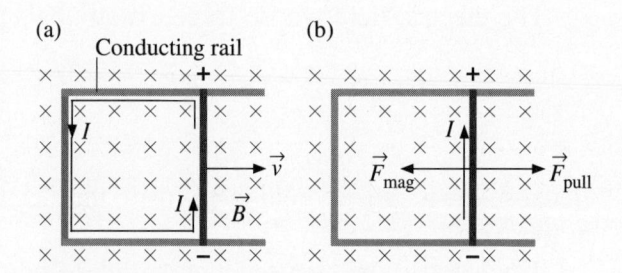

FIGURE 31-5 a) A wire moving through a magnetic field induces a current in the circuit. b) The magnetic force on the current in the moving wire must be balanced by a pulling force.

$$I = \frac{\mathcal{E}}{R} = \frac{vLB}{R}. \tag{31-4}$$

The induced current in this case is due to magnetic forces on moving charges.

Because the induced current in the moving wire is in a magnetic field, it experiences a magnetic force $\vec{F}_{mag}$. Application of the right-hand rule, as described in Section 30.9, reveals that the magnetic force on the current-carrying wire is to the left. This magnetic force will cause the wire to slow down and stop *unless* we exert an equal but opposite pulling force $\vec{F}_{pull}$ to the right, as shown in Fig. 31-5b. In other words, our assumption that the wire is moving with a *constant* speed v can be true only if we exert a pulling force to balance the magnetic "drag force."

The magnetic force on a current-carrying wire was given in Eq. 30.21. Using that result, we find that the force required to pull the wire with a *constant* speed v is

$$F_{\text{pull}} = |\vec{F}_{\text{pull}}| = |\vec{F}_{\text{mag}}| = ILB = \left(\frac{vLB}{R}\right)LB = \frac{vL^2B^2}{R}. \qquad (31\text{-}5)$$

The environment is doing work on the wire to pull it: $W = F_{\text{pull}}\Delta x$. So what is happening to the energy that is being transferred to the wire by this work? It will be easier to answer this question by thinking about power rather than work. Power is the *rate* at which work is done on the wire, and you learned in Chapter 9 that the power exerted by a constant force pushing or pulling an object with velocity v is

$$P_{\text{input}} = F_{\text{pull}} v = \frac{v^2 L^2 B^2}{R}. \qquad (31\text{-}6)$$

This is the rate at which energy is being added to the circuit by the pulling force. But the circuit also dissipates energy by transforming electrical energy into the thermal energy of the wires and components, heating them up. In fact, you learned in Chapter 29 that the power dissipated by current I flowing through resistance R is $P = I^2R$. Using Eq. 31-4 for the current I, the power dissipated by the circuit of Fig. 31-5 is

$$P_{\text{dissipated}} = I^2 R = \frac{v^2 L^2 B^2}{R}. \qquad (31\text{-}7)$$

You can see that Eqs. 31-6 and 31-7 are identical. Thus *energy is conserved*, with the work done by the pulling force being transformed into an increased thermal energy of the wires.

If you have to *pull* on the wire to get it to move to the right, you might think that it would move to the left on its own. Figure 31-6 shows the same circuit with the wire moving to the left. You should convince yourself that reversing the direction of motion reverses the emf, which reverses the current direction, which reverses the direction of the force $\vec{F}_{\text{mag}}$ on the wire. In this case, you have to *push* the wire to keep it moving. The magnetic force is always opposite to the wire's direction of motion, so we will call this a *retarding force*.

In both Fig. 31-5, where the wire is pulled, and Fig. 31-6, where it is pushed, a mechanical force is being used to create an electrical current. In other words, we have a conversion of *mechanical* energy to *electrical* energy. A device that converts mechanical energy to electrical energy is called a **generator**. The slide-wire circuit of Fig. 31-5 is a simple example of a generator. We will look at more practical examples of generators later in the chapter.

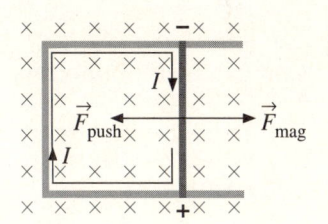

FIGURE 31-6 A pushing force is required to move the wire to the left.

We can summarize our analysis of the circuits of Figs. 31-5 and 31-6 as follows:

1. Pulling or pushing the wire through the magnetic field at speed v creates a motional emf $\mathcal{E}$ in the wire and induces a current $I = \mathcal{E}/R$ to flow in the circuit.

2. To keep the wire moving at constant speed, a pulling or pushing force is needed to balance the magnetic force on the wire. This force is a source of energy that is doing work on the circuit.

3. The energy input of the pulling or pushing force exactly balances the energy dissipated by the current as it flows through the resistance of the circuit.

EXAMPLE 31-2 Figure 31-7 shows a circuit consisting of flashlight bulb, rated 3.0 V/1.5 W, and ideal ($R = 0$) wires. The right wire of the circuit, which is 10 cm long, is being pulled at constant speed v through a magnetic field of strength 0.1 T. a) What speed must the wire have to light the bulb to full brightness? b) With what force must the wire be pulled?

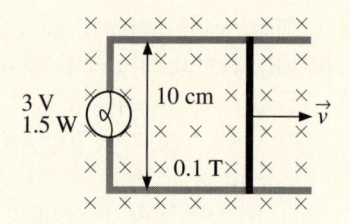

FIGURE 31-7 Circuit of Example 31-2.

SOLUTION a) The bulb's rating of 3.0 V/1.5 W means that at full brightness it will dissipate 1.5 W at a potential difference of 3.0 V. Because the power is related to the voltage and current by $P = IV$, the current causing full brightness is

$$I = \frac{P}{V} = \frac{1.5 \text{ W}}{3.0 \text{ V}} = 0.5 \text{ A}.$$

The bulb's resistance, which is the total resistance of the circuit, is

$$R = \frac{V}{I} = \frac{3.0 \text{ V}}{0.5 \text{ A}} = 6 \, \Omega.$$

Using Eq. 31-4, the speed needed to induce this current is

$$v = \frac{IR}{LB} = \frac{(0.5 \text{ A})(6 \, \Omega)}{(0.1 \text{ m})(0.1 \text{ T})} = 300 \text{ m/s}.$$

You can confirm from Eq. 31-6 that the input power at this speed is 1.5 W.

b) From Eq. 31-5, the pulling force must be

$$F_{\text{pull}} = \frac{vL^2B^2}{R} = 5.0 \times 10^{-3} \text{ N}.$$

You can also obtain this result from $F_{\text{pull}} = P/v$.

Eddy Currents

Figure 31-8 shows a *rigid* square loop of wire being pulled between the poles of a magnet. The upper pole is a north pole, so the magnetic field points downward and is confined

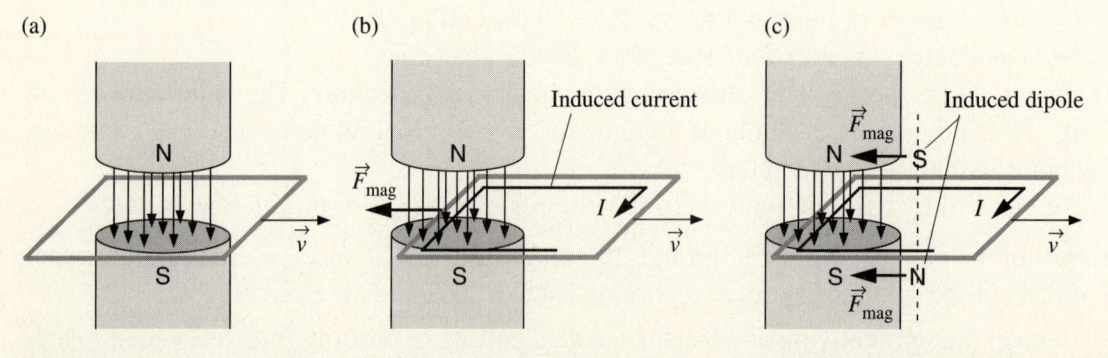

FIGURE 31-8 a) A loop that surrounds a magnetic field, but is not in the field, can move freely. b) The loop experiences a retarding force as one edge passes through the field. c) The retarding force can be thought of as an interaction between the magnet and the induced magnetic dipole of the loop.

to the region between the poles. In Fig. 31-8a, the magnetic field passes through the loop, but the wires of the loop are not in the field. Thus none of the charge carriers experience a magnetic force, there is no induced current, and it takes no force to pull the loop to the right. The magnetic field has no effect on the loop.

But when the left edge of the loop enters the field, as shown in Fig. 31-8b, the magnetic force induces a current to flow around the loop. Now there is a retarding magnetic force on the current, pointing toward the left, so *a pulling force must be exerted to pull the loop out of the magnetic field*. Note that the wire—typically copper—is *not* a magnetic material. A piece of the wire held near the magnet would feel no force. Nor would a force be required to pull the wire out if there were a gap in the loop, breaking the circuit and preventing a current from flowing. It is the *induced current* in the complete loop that experiences the retarding force of the magnetic field.

Figure 31-8c offers an alternative way of viewing the situation. As the left edge of the loop passes through the field, an induced current flows around the loop in a clockwise direction. This current loop is a magnetic dipole. The rules for current loops given in Chapter 30 tell us that the north pole is beneath the loop and the south pole is above the loop. Opposite poles attract, so the induced south pole is attracted to the permanent magnet's north pole and the induced north pole is attracted to the magnet's south pole. Thus pulling the loop out of the field is like pulling apart two magnets that are stuck together or pulling a magnet off the refrigerator door. A net force must be applied that equals or exceeds the magnetic force trying to pull the loop back in.

These ideas have interesting implications. Consider pulling a *sheet* of metal through a magnetic field, as shown in Fig. 31-9a. The metal, we will assume, is not a magnetic material, so it experiences no magnetic force if it is at rest. The figure shows a view looking down from the north pole of the magnet toward the south pole. As the metal moves, the charge carriers in the magnetic field experience a sideways force in the plane of the metal. Thus a current is induced, with charge carriers flowing into the bottom and out of the top. These induced currents flow because of the same magnetic force that induced a current in the loop of Fig. 31-8, but here the currents do not have wires to define their path. As a consequence, two "whirlpools" of current begin to circulate in the metal. These spread-out current whirlpools in a solid metal are called **eddy currents**.

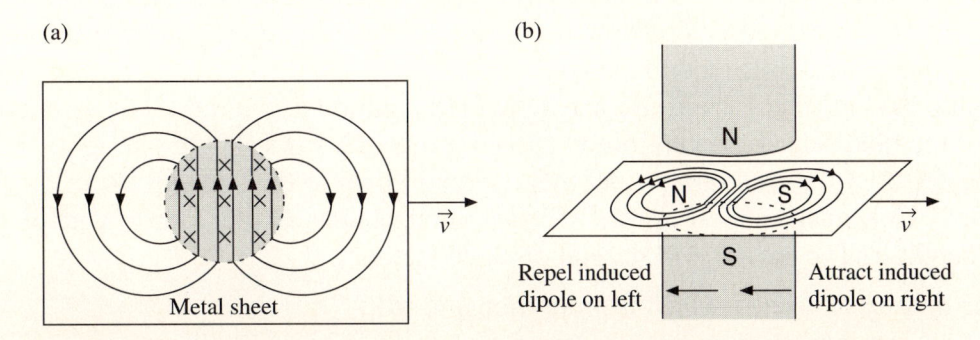

FIGURE 31-9 a) Eddy currents are induced if a solid metal sheet is pulled through a magnetic field. b) The magnetic forces on the eddy currents are retarding forces, opposite the direction of motion.

As Fig. 31-9b shows, the eddy current on the right rotates clockwise, forming a magnetic dipole with the south pole above the plane. The eddy current on the left rotates in a counterclockwise direction and forms a magnetic dipole with the north pole above the plane. The forces between the poles are such that the permanent magnet attracts the induced magnetic dipole on the right but *repels* the induced magnetic dipole on the left. *Both* pairs of forces cause a retarding magnetic force that opposes the motion of the metal through the field. Thus a net force is required to pull the metal through the field. If the pulling force ceases, the retarding magnetic forces will quickly decelerate the metal until it stops.

Eddy currents are often undesirable. Extra energy must be expended to move metals in magnetic fields, and the power dissipation of the eddy currents, due to the material's resistance, can cause significant unwanted heating. But eddy currents also have important useful applications. A good example is magnetic braking, which is used in some trains and transit-system vehicles. The moving car has an electromagnet that straddles the rail, as shown in Fig. 31-10. During normal travel, no current flows through the electromagnet and there is no field. To stop the car, a current is switched into the electromagnet. The current creates a strong magnetic field that passes *through* the rail, and the motion of the rail relative to the magnet induces eddy currents in the rail. The induced magnetic dipoles in the rail exert a braking force on the magnet and thus, by Newton's third law, on the car. Magnetic braking systems are very efficient, and they have the added advantage that they heat the rail rather than the brakes.

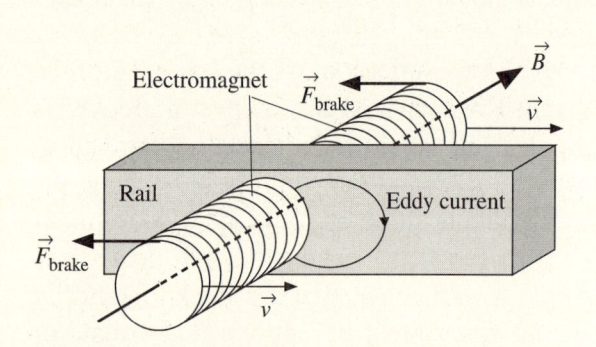

FIGURE 31-10 Magnetic braking systems are an application of eddy currents.

31.3 Magnetic Flux

Faraday showed that the induced current depends on changing the amount of magnetic field passing *through* a coil or a loop of wire. No induced current was seen in the coil of Fig. 31-3d, which was turned 90° so that the magnetic field vectors of the upper coil pass across but not through the loop. Now we need to make a more precise determination of "the amount of field passing through a loop."

Figure 31-11a shows a uniform magnetic field pointing to the right. For the moment, think of the field vectors as a uniformly spaced group of arrows shot to the right. Assume that the density of arrows (arrows per m²) is proportional to the strength $|\vec{B}|$ of the magnetic field. How many of these arrows will pass through a rectangular loop of wire of width a and height b? The answer will depend on three factors:

1. The density of arrows, which is proportional to $|\vec{B}|$,
2. The area $A = ab$ of the loop, and
3. The angle θ between the loop and the direction of the field.

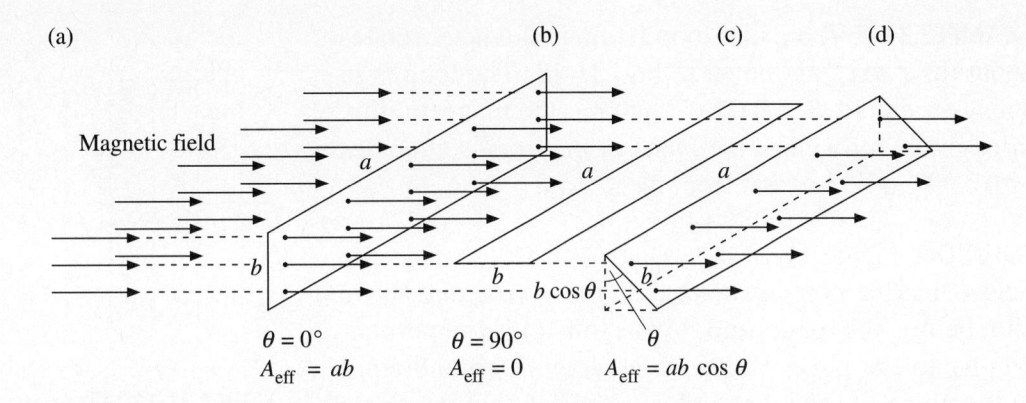

FIGURE 31-11 a) A uniform magnetic field. b) A loop with maximum flux. c) A loop with zero flux. d) A loop with an intermediate flux.

The maximum number of arrows can pass through when the loop is perpendicular to the arrows of the magnetic field ($\theta = 0°$), as in Fig. 31-11b. If the loop is tilted 90° to the field, as in Fig. 31-11c, then no arrows pass through the loop.

As seen from the arrows' perspective, the loop of Fig. 31-11d, which is tilted at angle θ, presents a "window" of width a and height $b \cos \theta$. This window is the *effective area* through which the arrows must pass. Thus

$$A_{\text{eff}} = ab \cos \theta = A \cos \theta, \tag{31-8}$$

where $A = ab$ is the loop's actual area. The effective area is equal to A when $\theta = 0°$ (Fig. 31-11b) and is equal to zero when $\theta = 90°$ (Fig. 31-11c).

With this in mind, let's define a quantity called the **magnetic flux** Φ as

$$\Phi = A_{\text{eff}} |\vec{B}| = A |\vec{B}| \cos \theta. \tag{31-9}$$

The magnetic flux measures, as we wanted, the amount of magnetic field passing through a loop of area A if the loop is tilted at angle θ from the field. The SI unit of magnetic flux is the **weber**. From Eq. 31-9 you can see that

$$1 \text{ weber} = 1 \text{ Wb} = 1 \text{ T m}^2.$$

Equation 31-9 is reminiscent of the vector dot product: $\vec{A} \cdot \vec{B} = |\vec{A}||\vec{B}| \cos \theta$. Let's define an **area vector** $\vec{A}$ to be a vector that is *perpendicular* to a loop of area A and whose magnitude is equal to the area of the loop: $|\vec{A}| = A$. Vector $\vec{A}$ has units of m². Figure 31-12 shows the area vector $\vec{A}$ for a circular loop of area A. If the area vector is tilted at angle θ to a magnetic field, then the magnetic flux passing through the loop is

$$\Phi = \vec{A} \cdot \vec{B}. \tag{31-10}$$

Equation 31-10 is the same definition of magnetic flux as Eq. 31-9. Our purpose in writing it as a dot product is to make clear how angle θ is defined: θ is the angle between the magnetic field and a line *perpendicular* to the plane of the loop.

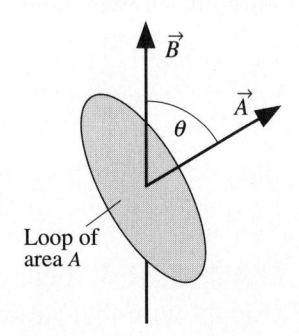

FIGURE 31-12 The area vector $\vec{A}$ is perpendicular to a circular loop of area A.

EXAMPLE 31-3 A circular loop 10 cm in diameter rotates about the z-axis, as shown in Fig. 31-13. The loop is in a magnetic field $\vec{B} = 0.05\,\hat{j}$ T. What is the magnetic flux through the loop when the plane of the loop is at angles of 0°, 30°, 60°, and 90° from the x-axis?

SOLUTION Figure 31-13 shows the xy-plane, the magnetic field $\vec{B}$, and the loop's area vector $\vec{A}$. The loop extends above and below the page and pivots on an axis perpendicular to the page. Vector $\vec{A}$ is drawn perpendicular to the plane of the loop, and you can see that the angle θ between $\vec{A}$ and $\vec{B}$ is the same as the angle between the plane of the loop and the x-axis. Vector $\vec{A}$ has magnitude $|\vec{A}| = \pi r^2 = 7.85 \times 10^{-3}$ m². Using Eq. 31-10, the flux is found to be

$$\Phi = \vec{A}\cdot\vec{B} = |\vec{A}||\vec{B}|\cos\theta = \begin{cases} 3.93\times 10^{-4} \text{ Wb} & \theta = 0° \\ 3.40\times 10^{-4} \text{ Wb} & \theta = 30° \\ 1.96\times 10^{-4} \text{ Wb} & \theta = 60° \\ 0 \text{ Wb} & \theta = 90°. \end{cases}$$

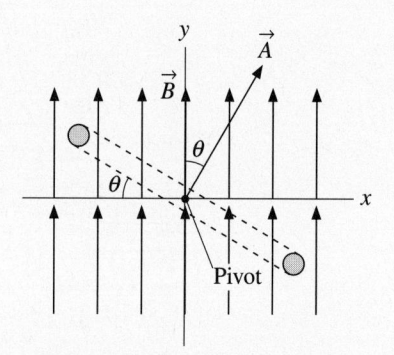

FIGURE 31-13 Geometry of Example 31-3.

Equation 31-10 for the magnetic flux assumes that the field is uniform over the area of the loop. If the field changes from one side of the loop to the other, we need to divide the loop into many small pieces of area dA, find the flux $d\Phi$ through each little piece, then integrate to add up all the small individual fluxes. That is, define an infinitesimal flux $d\Phi$ through area dA as

$$d\Phi = \vec{B}\cdot d\vec{A},$$

then integrate to find the total magnetic flux through the loop:

$$\Phi = \int_{\text{area of loop}} \vec{B}\cdot d\vec{A}. \qquad (31\text{-}11)$$

Equation 31-11 may look rather formidable, so we'll illustrate its use with an example.

EXAMPLE 31-4 Figure 31-14 shows a rectangular loop 1 cm × 4 cm parallel to a long straight wire that carries a current of 1.0 A. What is the magnetic flux through the loop?

SOLUTION The magnetic field of the wire varies with distance from the wire, so the field is *not* uniform over the area of the loop. Using the right-hand rule, we see that the field, as it circles the wire, is perpendicular to the plane of the loop. Let the loop have dimensions a and b, as shown, with the near edge distance c from the wire. Divide the loop into many narrow rectangles of area $dA = b\,dx$. For each of these rectangles, the area vector $d\vec{A}$

is perpendicular to the loop and thus parallel to $\vec{B}$. So $\theta = 0°$ for each rectangle. The infinitesimal flux through one of these rectangles is:

$$d\Phi = \vec{B} \cdot d\vec{A} = B\, dA = bB\, dx = \frac{\mu_0 I b}{2\pi x} dx,$$

where we've used $B = \mu_0 I/2\pi x$ from Eq. 30-9 as the magnetic field at distance x from a long, straight wire. Integrating "over the area of the loop" means to integrate from the near edge of the loop at $x = c$ to the far edge at $x = c + a$. Thus

$$\Phi = \frac{\mu_0 I b}{2\pi} \int_c^{c+a} \frac{dx}{x} = \frac{\mu_0 I b}{2\pi} \ln x \Big|_c^{c+a} = \frac{\mu_0 I b}{2\pi} \ln\left(\frac{c+a}{c}\right).$$

Evaluating for $a = c = 0.01$ m, $b = 0.04$ m, and $I = 1.0$ A gives

$$\Phi = 5.55 \times 10^{-9} \text{ Wb}.$$

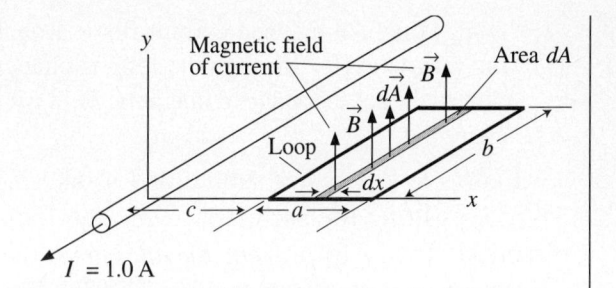

FIGURE 31-14 A rectangular loop parallel to a current-carrying wire.

The flux measures how much of the wire's magnetic field passes through the loop, but we had to integrate, rather than simply using Eq. 31-10, because the field is stronger at the near edge of the loop than at the far edge.

31.4 Lenz's Law

Figure 31-15a shows a loop of wire, a current meter, and a bar magnet. Faraday discovered that a momentary current is induced in the loop as the magnet is pushed into loop, causing the meter's needle to jump. No current flows if the magnet is stationary, even through there is a magnetic flux through the loop.

The cause of the induced current in a loop of wire is a *change* in the magnetic flux through the loop. Pushing a magnet into the loop increases the flux through the loop and, as the flux is changing, induces a current to flow. Pulling the magnet back out of the loop decreases the flux, and an induced current flows in the opposite direction. But *which way* does the induced current flow in each situation? This is the issue we need to examine.

German physicist Heinrich Lenz began to study electromagnetic induction after learning of Faraday's discovery. Three years later, in 1834, Lenz announced a rule for determining the direction of the induced current. We now call his rule **Lenz's law**, and it can be stated as follows:

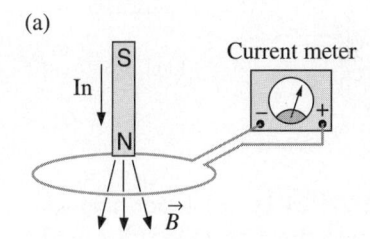

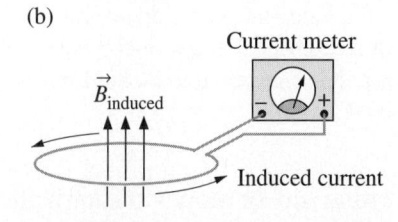

FIGURE 31-15 a) A bar magnet pushed into a loop increases the flux through the loop and induces a current to flow. b) The current flows in a direction that generates an upward magnetic field in order to try to keep the flux from changing.

382 CHAPTER 31 ELECTROMAGNETIC INDUCTION

Lenz's law: An induced current flows around a closed, conducting loop if and only if the magnetic flux through the loop is changing. The induced current flows in a direction such that the induced magnetic field tries to prevent the flux through the loop from changing.

Lenz's law is rather subtle, and it takes some practice to see how to apply it. In Fig. 31-15a, pushing the bar magnet into the loop increases the flux pointing in a downward direction. To try to *prevent the increase*, the loop itself needs to generate the *upward*-pointing magnetic field of Fig. 31-15b. From the right-hand rule, you know that the induced magnetic field at the center of the loop will point upward if the current flows counterclockwise. Thus pushing the north end of a bar magnet toward the loop induces a current that flows in a *counterclockwise* direction. Once the magnet stops moving, the induced current stops flowing.

Now suppose the bar magnet of Fig. 31-15 is pulled back out of the loop, as shown in Fig. 31-16a. There is a downward magnetic flux through the loop at the start, but the flux will decrease as the magnet moves away. According to Lenz's law, the induced magnetic field of the loop will try to prevent this decrease. To do so, the induced field needs to point in the *downward* direction, as shown in Fig. 31-16b. Thus as the magnet is withdrawn, the induced current flows clockwise, opposite to the induced current of Fig. 31-15b.

Notice that the magnetic field of the bar magnet is pointing downward in both Figs. 31-15 and 31-16. It is not the *field* of the magnet that the induced current opposes, but the *change* in the field. This is a subtle but critical distinction. If the induced current opposed the field itself, the current in both Figs. 31-15 and 31-16 would flow counterclockwise to generate an upward magnetic field. But that's not what happens. When the field of the magnet is down and increasing, the induced current opposes the *increase* by generating an upward field. When the field of the magnet is down but decreasing, the induced current opposes the *decrease* by generating a downward field.

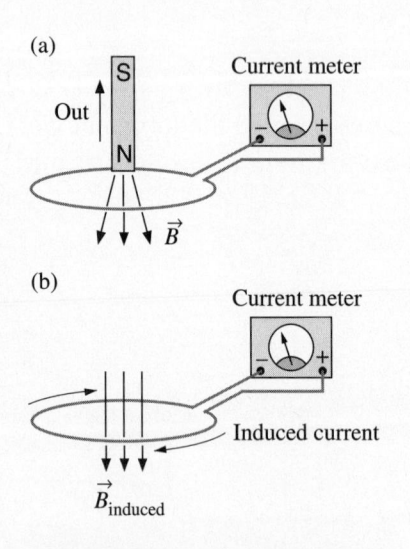

FIGURE 31-16 a) A bar magnet is pulled out of a loop. The magnetic field still points down, but is decreasing. b) The induced current flows opposite to that in Fig. 31-15b.

Figure 31-17 shows four basic situations in which a current is induced. The magnetic field can point either up or down through the loop, and it can be either increasing or decreasing in strength. In each case, the induced current flows in a direction such that the induced field tries to keep the flux from changing. A good strategy is to think about in what direction the induced *field* needs to point, then use that information and the right-hand rule to determine the direction of the induced current. Let's look at some examples.

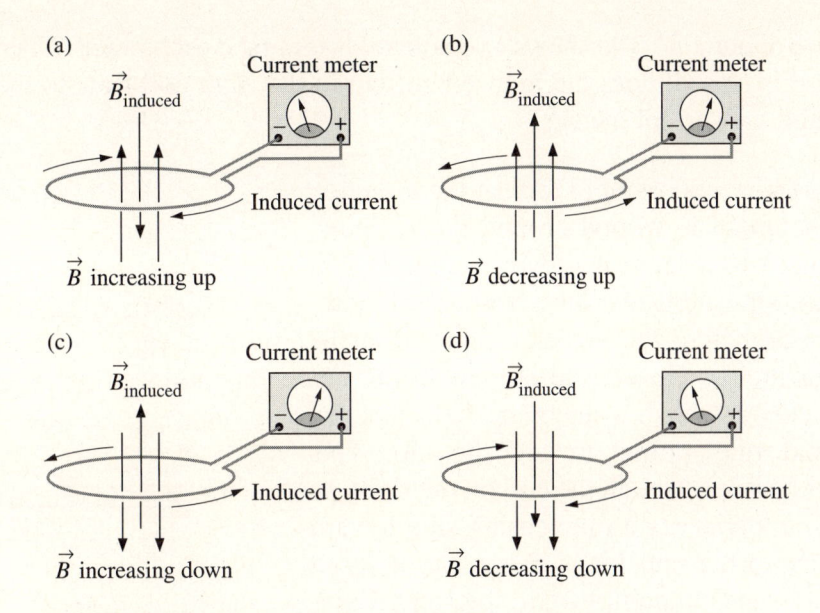

FIGURE 31-17 The induced current for four different situations: a) Field up and increasing. b) Field up and decreasing. c) Field down and increasing. d) Field down and decreasing.

EXAMPLE 31-5 The switch in the upper current loop of Fig. 31-18a has been closed for a long time. At the moment it is opened, what happens in the lower loop?

SOLUTION While the switch is closed, the current in the upper loop flows counterclockwise and generates an upward-pointing magnet field through the lower loop. As long as the switch stays closed, the flux is not changing and *no* current flows in the lower loop. But when the switch is opened, as shown in Fig. 31-18b, the upward flux rapidly decreases. In an effort to prevent this decrease, the induced magnetic field needs to point upward. Thus a current is induced in the counterclockwise direction in the lower loop. This current lasts only the very brief time interval until the magnetic field of the upper loop drops to zero.

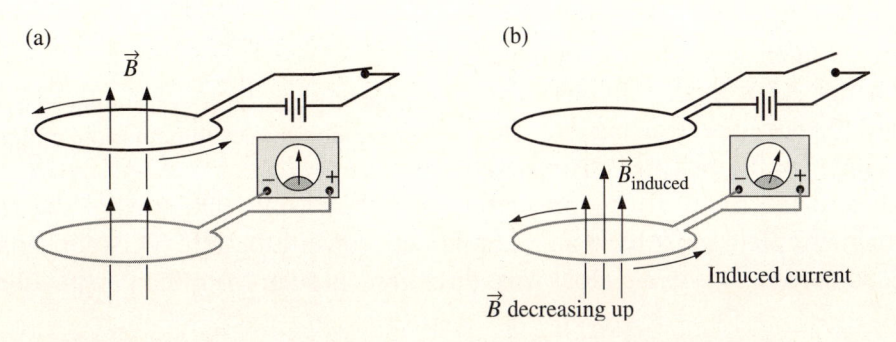

FIGURE 31-18 a) Loops of Example 31-5. b) The induced current flows counterclockwise to oppose the decreasing upward flux.

384 CHAPTER 31 ELECTROMAGNETIC INDUCTION

EXAMPLE 31-6 Figure 31-19a shows two solenoids that face each other. When the switch of the left coil is closed, does the induced current in the right coil flow to the right or to the left through the current meter?

SOLUTION It is very important to look at the *direction* in which a solenoid is wound around the cylinder. Notice that the two solenoids shown in Fig. 31-19a are wound in opposite directions. No current flows in the right coil before the switch is closed on the left. As the switch is closed, the current in the left coil will quickly establish a magnetic field that, by the right-hand rule, points toward the left. This magnetic field passes through the coil on the right, so the right coil experiences an increasing flux toward the left. To try to prevent this increase, the induced magnetic field needs to point toward the right. Careful attention to the direction in which the coil is wound, and use of the right-hand rule, establishes that the induced current must move to the left as it passes through the current meter. The induced current is only momentary. It lasts only until the field from the coil on the left reaches full strength and is no longer changing.

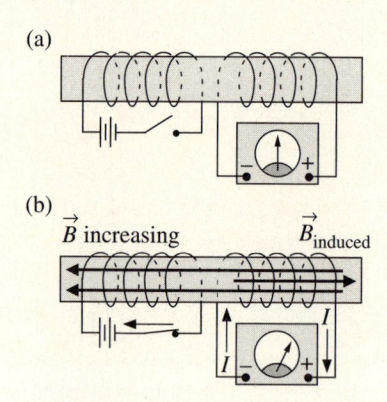

FIGURE 31-19 a) Solenoids of Example 31-6. b) The induced current in the right coil flows to the left through the meter to oppose the increasing magnetic flux toward the left.

EXAMPLE 31-7 The loop of wire in Fig. 31-20 was initially in the *xy*-plane, parallel to the magnetic field. It is suddenly rotated 90° about the *y*-axis until it is in the *yz*-plane, perpendicular to the magnetic field. As the loop rotates, in what direction does the induced current flow?

SOLUTION Unlike the previous examples, the magnetic field is constant and unchanging. Nonetheless, the *flux* through the loop changes as it rotates. Initially the flux is $\Phi = 0$, but after rotating 90° (to where the angle $\theta = 0°$) the flux is $\Phi = A|\vec{B}|$

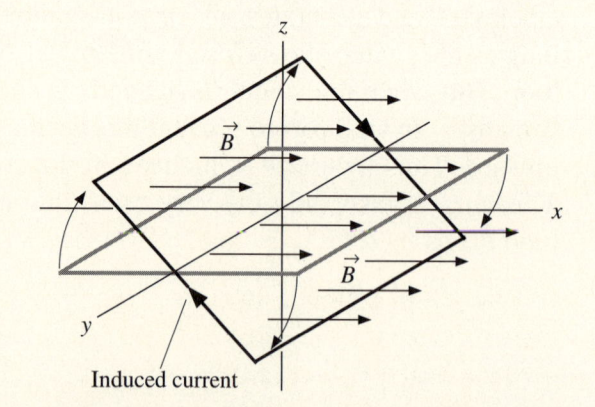

FIGURE 31-20 A current is induced in a loop as the loop rotates in a constant magnetic field.

and pointing to the right. To try preventing this increasing flux toward the right, the induced magnetic field must have an *x*-component toward the left. This will be the case if an induced current flows in a clockwise direction, as seen from the perspective of Fig. 31-20.

31.5 Faraday's Law

As Faraday and Henry discovered, a *change* in the magnetic flux through a loop of wire induces a current to flow in the wire. But a current requires an emf, such as a battery, to provide the energy. Currents don't spontaneously start flowing in circuits that have no source of energy. So if a current is induced to flow in a loop of wire, it must be because the changing flux generates an **induced emf** $\mathcal{E}$. Then, if there is a complete circuit having resistance R, a current

$$I_{\text{induced}} = \frac{\mathcal{E}}{R} \qquad (31\text{-}12)$$

flows in the wire as a *consequence* of the induced emf. The direction of the current flow is given by Lenz's law. The last piece of information we need is the size of the induced emf $\mathcal{E}$.

The research of Faraday, Lenz, and others eventually led to the discovery of the basic law of electromagnetic induction, which we now call **Faraday's law**. Faraday's law is a new law of physics, not derivable from any previous laws you have studied. It states:

Faraday's Law: An emf $\mathcal{E}$ is induced in a conducting loop if the magnetic flux through the loop changes. The magnitude of the emf is

$$|\mathcal{E}_{\text{loop}}| = \left|\frac{d\Phi}{dt}\right|, \qquad (31\text{-}13)$$

and the direction of the emf is such as to drive an induced current in the direction given by Lenz's law.

In other words, the induced emf is the *rate of change* of the magnetic flux through the loop. As a corollary to Faraday's law, a coil of wire consisting of N turns in a changing magnetic field acts like N batteries in series. The induced emf's in all the coils add, so the induced emf of the entire coil is

$$|\mathcal{E}_{\text{coil}}| = N \left|\frac{d\Phi}{dt}\right|. \qquad (31\text{-}14)$$

As a first example of using Faraday's law, let's return to the situation of Fig. 31-5 where a wire moved through a magnetic field by sliding on a U-shaped conducting rail. Figure 31-21 shows the wire again. The magnetic field is perpendicular to the plane of the conducting loop, so $\theta = 0°$ and the magnetic flux is $\Phi = A|\vec{B}| = AB$. If the slide wire is at distance x from the end, the area of the loop is $A = xL$ and the flux at that instant of time is

$$\Phi = AB = xLB.$$

As the wire moves, the flux through the loop increases. The changing flux creates an induced emf given by Eq. 31-13,

$$|\mathcal{E}_{\text{loop}}| = \left|\frac{d\Phi}{dt}\right| = \frac{d}{dt}(xLB) = \frac{dx}{dt}LB = vLB, \qquad (31\text{-}15)$$

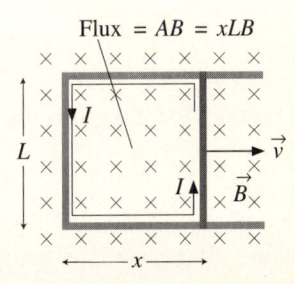

FIGURE 31-21 The magnetic flux increases as the slide wire moves.

where we've noted that the wire's velocity is related to its position by $v = dx/dt$. Thus, using Eq. 31-12, the induced current is

$$I = \frac{\mathcal{E}_{\text{loop}}}{R} = \frac{vLB}{R}. \tag{31-16}$$

Because the flux is increasing into the loop, the induced magnetic field will try to prevent this increase by pointing out of the loop. This requires the induced current to flow in a counterclockwise direction. The net result of our Faraday's law analysis is that an induced current given by Eq. 31-16 will flow around the loop in a counterclockwise direction. This is exactly the conclusion we reached in Section 31.2, where we analyzed the situation from the perspective of magnetic forces on moving charge carriers. Thus Faraday's law confirms what we already knew but, at least in this case, doesn't seem to offer anything new.

EXAMPLE 31-8 A loop of wire, which has a diameter of 2 cm and a resistance of 0.01 Ω, is placed inside a solenoid. The solenoid is 4 cm in diameter, 20 cm long, and wrapped with 1000 turns of wire, as shown in Fig. 31-22a. The current through the solenoid as a function of time is shown by the graph of Fig. 31-22b. Determine the current in the loop as a function of time. Present the results of the calculation as a graph. Interpret a positive current as one that flows into the page at the top of the loop.

SOLUTION The magnetic field of the solenoid creates a flux through the loop inside the solenoid. The flux will change with time as the solenoid current changes, and this change will induce a current in the loop. Because the field is uniform inside the solenoid and perpendicular to the loop ($\theta = 0°$), the flux is

$$\Phi = AB,$$

where $A = \pi r^2 = 3.14 \times 10^{-4}$ m² is the area of the loop. The field of a solenoid was found in Eq. 30.15 to be

$$B = \frac{\mu_0 N I_{\text{sol}}}{L},$$

so the flux through the loop when the solenoid current is I_{sol} is

$$\Phi = \frac{\mu_0 A N I_{\text{sol}}}{L}.$$

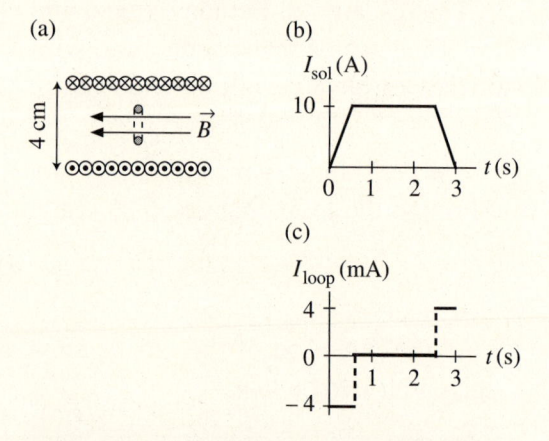

FIGURE 31-22 a) A loop inside a solenoid. b) The current through the solenoid as a function of time. c) The induced current in the loop as a function of time.

The changing flux creates an induced emf $\mathcal{E}$, given by Eq. 31-13, and the induced current is then found from Eq. 31-12 to be

$$I_{\text{loop}} = \frac{1}{R}\left|\frac{d\Phi}{dt}\right| = \frac{\mu_0 A N}{LR}\left|\frac{dI_{\text{sol}}}{dt}\right| = 1.97 \times 10^{-4}\left|\frac{dI_{\text{sol}}}{dt}\right|.$$

From the graph, we see that

$$\left|\frac{dI_{\text{sol}}}{dt}\right| = \begin{cases} 20 \text{ A/s} & 0.0 \text{ s} < t < 0.5 \text{ s} \\ 0 & 0.5 \text{ s} < t < 2.5 \text{ s} \\ 20 \text{ A/s} & 2.5 \text{ s} < t < 3.0 \text{ s}. \end{cases}$$

Thus the induced current has *magnitude*

$$I_{\text{loop}} = \begin{cases} 3.94 \text{ mA} & 0.0 \text{ s} < t < 0.5 \text{ s} \\ 0 & 0.5 \text{ s} < t < 2.5 \text{ s} \\ 3.94 \text{ mA} & 2.5 \text{ s} < t < 3.0 \text{ s}. \end{cases}$$

Although the induced current has the same magnitude during the first and the third intervals, it does not have the same direction. The magnetic field of the solenoid points to the left. The flux through the loop is increasing toward the left during the first half-second, so to try preventing this change of flux the induced magnetic field must point toward the right. The induced current will come *out* of the page at the top of the loop, which, according to the sign convention for this problem, is a negative current. The flux is to the left but decreasing during the last half-second. In an attempt to prevent the decrease, the induced magnetic field of the current loop must point to the left. This requires a positive current *into* the page at the top of the loop. These results are shown in the graph of Fig. 31-22c. ●

The induced emf of the sliding wire can be understood as a motional emf due to magnetic forces on moving charges. But not the induced emf of Example 31-8. There is no motion in this situation. The induced emf in Example 31-8 is new physics, nothing we can explain or predict on the basis of previous laws or principles. It might be helpful to write Eq. 31-13 as

$$|\mathcal{E}_{\text{loop}}| = \left|\frac{d\Phi}{dt}\right| = \left|\vec{B} \cdot \frac{d\vec{A}}{dt} + \vec{A} \cdot \frac{d\vec{B}}{dt}\right|. \tag{31-17}$$

The first term on the right side of Eq. 31-17 represents a motional emf. The flux changes because the loop itself is changing in some fashion, and that motion causes magnetic forces on the charge carriers in the loop. We had not anticipated this kind of current in Chapter 30, but it takes no new laws of physics to understand it. The second term on the right of Eq. 31-17 is new, however. It says that simply changing a magnetic field, even if nothing is moving, *also* creates an emf. This was the source of the emf in Example 31-8. So we have the interesting situation in which there are two distinct ways to create an induced emf. Either the circuit changes *or* the magnetic field changes. Regardless of which it is, the induced emf is simply the rate of change of the magnetic flux through the loop.

Unlike the emf of a battery, which is localized within a well-defined region of a circuit, an emf induced by changing $\vec{B}$ is *distributed* throughout the entire circuit. In other words, there is an induced emf at all points in the loop of Example 31-8, not just at a single location. This has interesting implications that we will explore in Section 31.7.

388 CHAPTER 31 ELECTROMAGNETIC INDUCTION

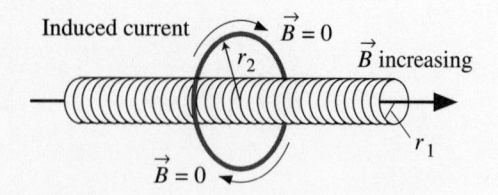

FIGURE 31-23 A changing current in the solenoid induces a current in the loop, even though the loop is not in a magnetic field at all!

As a final example in this section, consider the loop shown in Fig. 31-23. A very long, tightly wound solenoid of radius r_1 passes through the center of a conducting loop having a larger radius r_2. As you saw in Chapter 30, the field is intense inside the solenoid but, if the solenoid is sufficiently long, essentially zero outside the solenoid. Thus $\vec{B} = 0$ at all points on the conducting loop. That is, the loop doesn't know there's a magnetic field passing through its center. Now suppose the current in the solenoid is increased, as it was in Example 31-8. Will there be an induced current in the loop?

Despite the fact that $\vec{B} = 0$ (and thus $dB/dt = 0$) at all points on the loop itself, there *is* an induced current because the flux *through* the loop changes. The flux is

$$\Phi = \pi r_1^2 B_{\text{sol}}. \tag{31-18}$$

A changing solenoid current will cause a changing B_{sol}. This will cause a changing flux that, in turn, will cause an induced emf and a current in the loop. Note that the radius in Eq. 31-18 is r_1, not r_2, because the field is zero outside of the solenoid.

Faraday's law allows for some strange and nonintuitive situations, and this is one of them. There really is a current induced in the loop, despite the fact that the loop itself neither moves nor experiences a magnetic field. Quantum physics is not the only area of physics where seemingly bizarre things can happen!

31.6 Induced Currents: Four Applications

There are many applications of Faraday's law and induced currents in modern technology. In this section we will look at four: generators, transformers, metal detectors, and magnetic recording devices.

Generators

[**Photo suggestion: Large electrical generator.**]

We noted in Section 31.2 that the slide-wire on a U-shaped track is a simple generator because it transforms mechanical energy into electrical energy. Figure 31-24a shows a more practical generator. Here you see a coil of wire rotating in a magnetic field. Although both the field and the area of the loop are constant, the magnetic flux through the loop is continuously changing as the loop rotates. Thus an emf and a current are induced in the coil. The current is removed from the rotating loop by *brushes* that press up against rotating *slip rings*.

The flux through the coil is given by $\Phi = AB\cos\theta = AB\cos\omega t$, where ω is the angular frequency of rotation ($\omega = 2\pi f$) of the coil. If the coil consists of N turns, the induced emf is given by Eq. 31-14:

$$\mathcal{E}_{\text{coil}} = N\frac{d\Phi}{dt} = ABN\frac{d}{dt}(\cos\omega t) = -\omega ABN\sin\omega t. \tag{31-19}$$

(a)

(b)

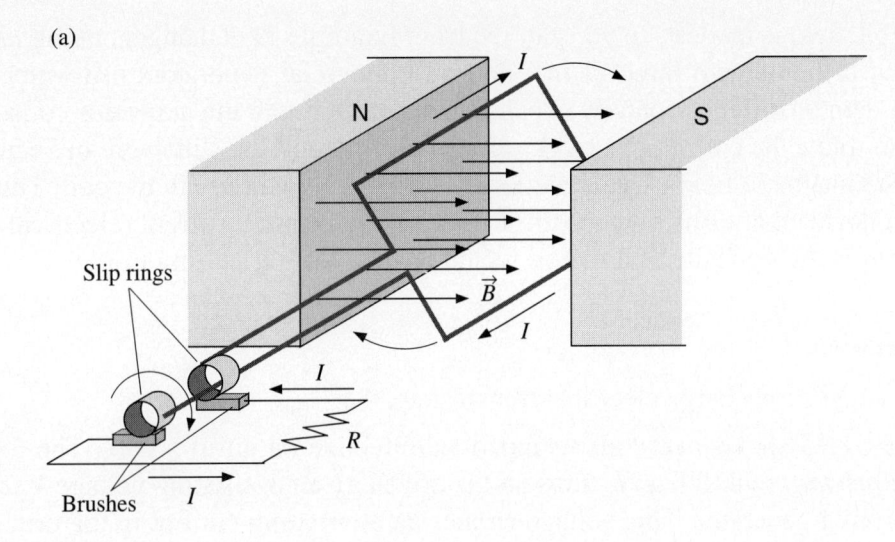

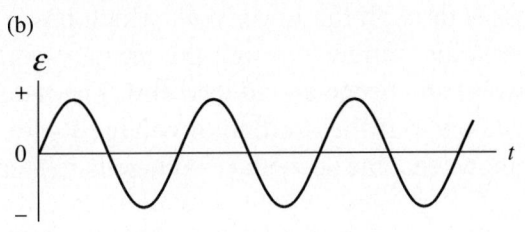

FIGURE 31-24 a) An alternating-current generator. b) The induced emf as a function of time.

We've dropped the absolute value signs to demonstrate that the sign of $\mathcal{E}_{coil}$ alternates between positive and negative. Figure 31-24b shows a graph of the induced emf as a function of time. Because the emf alternates in sign, the current through resistor R alternates back and forth in direction. Hence the generator of Fig. 31-24a is an *alternating-current generator*, producing what we usually call an *AC voltage*. If the external resistance R is much larger than the resistance of the coil itself, then essentially all of the induced AC voltage is delivered to the resistance.

EXAMPLE 31-9 A coil with area 2 m² rotates in a 0.01 T magnetic field at a frequency of 60 Hz. How many turns are needed to generate a peak voltage of 160 V?

SOLUTION The coil's maximum voltage is given by Eq. 31-19 as

$$\mathcal{E}_{max} = \omega ABN = 2\pi fABN.$$

The number of turns on the coil to generate $\mathcal{E}_{max} = 160$ V is

$$N = \frac{\mathcal{E}_{max}}{2\pi fAB} = \frac{160 \text{ V}}{2\pi (60 \text{ s}^{-1})(2 \text{ m}^2)(0.01 \text{ T})} = 21 \text{ turns.}$$

A 0.01 T field is modest, so you can see from Example 31-9 that generating large voltages is not difficult with large (2 m^2) coils. Commercial generators use water flowing through a dam or turbines spun by expanding steam to rotate the generator coils. Work is required to rotate the coil, just as work was required to pull the slide wire in Section 31.2, because the magnetic field exerts retarding forces on the currents in the coil. Thus a generator is a device that turns motion (mechanical energy) into a current (electrical energy). A generator is the opposite of a motor, which turns a current into motion.

Transformers

[**Photo suggestion: Large electrical transformer.**]

Figure 31-25 shows two coils wrapped side by side on an iron core. The top coil is called the *primary coil*. It has N_1 turns and is driven by an oscillating voltage $V_1 \cos \omega t$—perhaps from a generator. This voltage creates an alternating current in the primary coil. The magnetic field of the primary passes through the lower coil, which has N_2 turns and is called the *secondary coil*. The alternating current through the primary coil causes an oscillating flux through the secondary coil and hence an induced emf. The induced emf of the secondary coil is delivered to resistance R as the oscillating voltage $V_2 \cos \omega t$.

A detailed analysis of this device shows that the secondary voltage is related to the primary voltage by

$$V_2 = \frac{N_2}{N_1} V_1. \qquad (31\text{-}20)$$

Depending on the ratio N_2/N_1, the voltage V_2 delivered to the resistance can be *transformed* to a higher or a lower voltage than V_1. Consequently, this device is called a **transformer**. Transformers are widely used in the commercial generation and transmission of electricity. A *step-up transformer*, with $N_2 > N_1$, boosts the voltage of a generator up to several hundred thousand volts. Power can be delivered with smaller currents at higher voltages, so losses due to resistance in the wires are lessened. High-voltage transmission lines carry electrical power to urban areas, where *step-down transformers* ($N_2 < N_1$) lower the voltage to 120 V. An AC/DC adapter, such as used for cassette recorders, modems, and other consumer electronics, is a step-down transformer that lowers the voltage even further, typically to 12 V (and also converts it from alternating current to direct current).

Note that the induced emf of a transformer is created by a changing magnetic field, whereas the induced emf of a generator is a motional emf. These two applications illustrate the two types of induced emf's, but both are governed by Faraday's law.

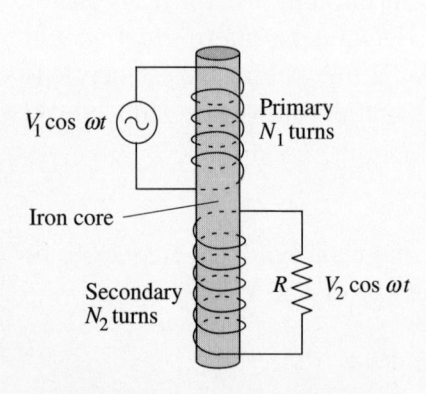

FIGURE 31-25 A transformer consists of two coils linked by a magnetic field. An alternating current through the primary coil induces an alternating current in the secondary coil.

Metal Detectors

Metal detectors, such as used in airports for security, seem fairly mysterious. How can they detect the presence of *any* metal—not just magnetic materials such as iron—but not detect plastic or other materials? Metal detectors work because of induced currents.

A metal detector, shown in Fig. 31-26, consist of two coils: a *transmitter coil* and a *receiver coil*. A high-frequency alternating current flows through the transmitter coil, generating an alternating magnetic field along the axis. This magnetic field creates a changing flux through the receiver coil and causes an alternating induced current. The transmitter and receiver are very similar to a transformer.

Now suppose a piece of metal is placed between the transmitter and the receiver. The alternating magnetic field through the metal induces eddy currents to flow in a plane parallel to the transmitter and receiver coils. The receiver coil then responds to the *superposition* of the transmitter's magnetic field and the magnetic field of the eddy currents. Because the eddy currents attempt to prevent the flux from changing, in accordance with Lenz's law, the net field at the receiver *decreases* when a piece of metal is inserted between the coils. Electronic circuits detect the current decrease in the receiver coil and set off an alarm. Eddy currents can't flow in an insulator, so this device detects only metals.

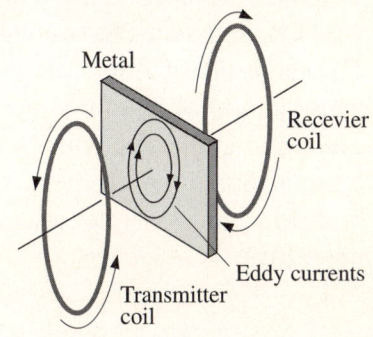

FIGURE 31-26 A metal detector. Eddy currents induced in a piece of metal reduce the induced current in the receiver coil.

Magnetic Recording Devices

[**Photo suggestion: Microphotograph of magnetic domains on a computer hard disk.**]

Audio cassettes, video cassettes, and computer disks store information magnetically. Figure 31-27a shows a highly magnified view of a computer disk. The disk has a thin ferromagnetic coating on the surface, and magnetic domains in the coating can be magnetized with their magnetic moments pointing either up or down. As the disk spins, these domains move rapidly past a very small *pick-up coil*.

Each domain is a small magnet. As a domain passes beneath the pick-up coil, its field extends upward through the coil. The coil would not respond at all if the entire disk surface were magnetized in the same way, because the flux through the coil would not change. But when a *transition* from an up-domain to a down-domain passes the coil, the *change* in the flux through the coil induces a voltage. A domain change

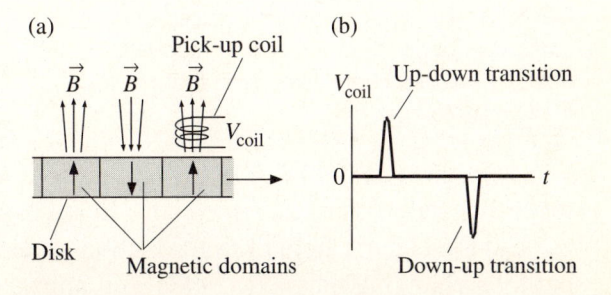

FIGURE 31-27 A pick-up coil detects the *change* from an up-domain to a down-domain as a ferromagnetic surface passes beneath it.

from down to up induces a current in the opposite direction or, equivalently, and induced emf of opposite sign. Figure 31-27b shows an example of how the coil's output voltage would look as the up-down-up domains of Fig. 31-27a pass by. Notice that the coil responds to *changes* in the domains, not to the domains themselves.

Different schemes are used to encode information on the disk as binary 0's and 1's. The large increase in storage capacities of magnetic disks over the last few decades is a consequence of learning how to produce ever-smaller domains *and* learning how to make ever-smaller pick-up coils. Audio and video recording on magnetic tapes utilizes the same basic idea, but the magnetic field strengths vary continuously rather than simply being up or down. This generates a variable *analog output* rather than the *digitized output* of a computer disk.

31.7 Induced Fields and Electromagnetic Waves

In Section 31.5 you saw that it is possible to induce a current in a stationary loop. But what *causes* the current to flow in this situation? That is, what *force* pushes the charges around the loop against the resistive forces of the metal? The only agents that exert forces on charges are electric fields and magnetic fields. Magnetic forces are responsible for motional emf's, such as that in the slide wire of Section 31.2. But magnetic forces cannot explain the current induced in a *stationary* loop by a changing magnetic field.

Figure 31-28a shows a conducting loop in an increasing magnetic field. A current is induced, according to Lenz's law, in the counterclockwise direction. Consequently, there must be an *electric* field, tangent to the loop at all points. It is this electric field that drives the current, creating a current density $J = \sigma E$. This electric field is *caused* by the changing magnetic field and is called an **induced electric field**.

Because the induced electric field is caused by the changing magnetic field, its existence does not depend on the loop. If the loop is removed, as in Fig. 31-28b, we find that the space is filled with a pinwheel-pattern of induced electric fields! But this is a rather peculiar electric field. All the electric fields that we have examined before this were created by *charges*, where electric field vectors pointed away from positive charges toward negative charges. An electric field created by charges is called a *Coulomb field*.

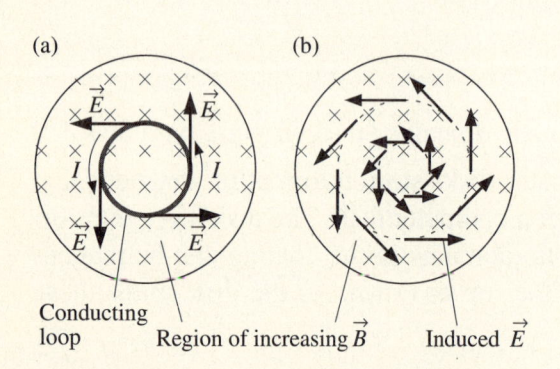

FIGURE 31-28 a) A current is induced in a conducting loop in an increasing magnetic field. b) The induced electric field that causes the current to flow.

But the electric field of Fig. 31-28b is a *non-Coulomb field*. It is caused not by charges but by a changing magnetic field. So it appears that there are *two* different ways to create an electric field.

1. A Coulomb electric field is created by positive and negative charges.

2. A non-Coulomb electric field is created by a changing magnetic field.

Both kinds of field exert a force $\vec{F} = q\vec{E}$ on a charge, and both cause a current to flow in a conductor. However, the origins of the fields are very different.

We introduced the idea of a field as a way of thinking about how two charges exert long-range forces on each other through the emptiness of space. The field may have seemed to you like a useful pictorial model of charge interactions, but we really had no evidence that fields are *real*, that they actually exist. Now we do. The electric field has shown up in a completely different context, independent of charges, as an idea that we need in order to understand the phenomena of nature. The electric field is *not* just a pictorial model; it has a real existence.

Faraday's field concept was capable of explaining the phenomena of electricity and magnetism as they were known in the 1830s and 1840s. But Faraday, despite his intuitive genius, lacked the mathematical skills to develop a true *theory* of electric and magnetic fields. It was not easy to predict new phenomena or develop applications without a theory.

In 1855, less than two years after receiving his undergraduate degree, the English physicist James Clerk Maxwell presented a paper titled "On Faraday's Lines of Force." In this paper, he began to sketch out how Faraday's pictorial model could be given a rigorous mathematical basis. Maxwell then spent the next ten years developing the mathematical theory of electromagnetism.

One item that troubled Maxwell early in the process was a certain lack of symmetry. Oersted had found a link between electricity and magnetism by showing that currents create magnetic fields. Faraday had completed the symmetry by discovering that magnetic fields can, under the proper circumstances, create currents. Then Faraday found that a changing magnetic field creates an induced electric field, a non-Coulomb electric field not tied to charges. But what, Maxwell begin to wonder, about a changing *electric* field? Does it create a magnetic field?

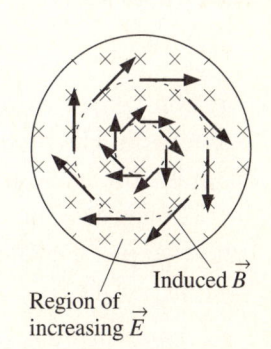

To complete the symmetry, Maxwell proposed that a changing electric field creates an **induced magnetic field**, a new kind of magnetic field not tied to the existence of currents. Figure 31-29 shows a region of space in which the *electric* field is increasing. This region of space, according to Maxwell, is also filled with a pinwheel pattern of induced magnetic fields. The situation looks the same as Fig. 31-28b, with $\vec{E}$ and $\vec{B}$ interchanged, except—for technical reasons—that the induced $\vec{B}$ point the opposite way from the. induced $\vec{E}$. Although there was no experimental evidence at the time to think that induced magnetic fields existed, Maxwell's went ahead and included them in his electromagnetic field theory. This was an inspired hunch, soon to be vindicated, because it allowed Maxwell to be the first to recognize that light is an *electromagnetic wave*.

In Maxwell's theory, a changing magnetic field creates an induced electric field and a changing electric field creates an induced magnetic field. Maxwell soon realized that it might be possible, under the proper circumstances, to establish self-sustaining electric and magnetic fields that would be entirely independent of any charges or currents. The idea is shown

Region of increasing $\vec{E}$ / Induced $\vec{B}$

FIGURE 31-29 Maxwell proposed that a changing electric field creates an induced magnetic field.

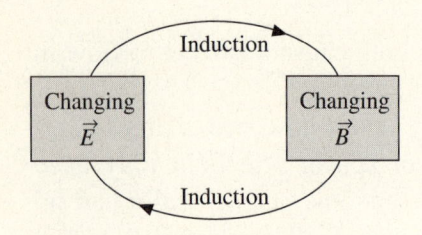

FIGURE 31-30 Maxwell suggested that the proper configuration of changing electric and magnetic fields could be self-sustaining because of electromagnetic induction.

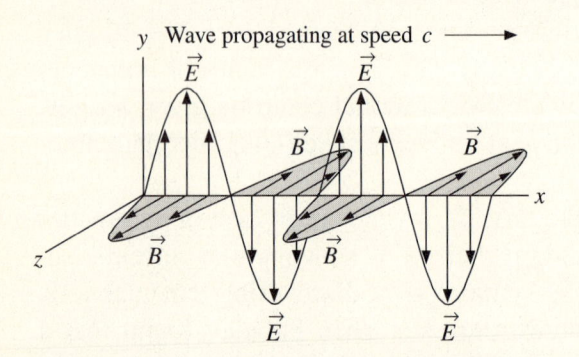

FIGURE 31-31 A self-sustaining electromagnetic wave consists of mutually perpendicular electric and magnetic fields traveling with speed c.

schematically in Fig. 31-30. There you see that a properly changing electric field $\vec{E}$ creates a magnetic field $\vec{B}$, which changes in just the right way to re-create the electric field, which then changes in just the right way to again re-create the magnetic field, and so on. The fields are continually re-created through electromagnetic induction.

The mathematics of Maxwell's theory are difficult, and it took him quite some time to find a self-sustaining solution of his field equations. But he did eventually succeed. Maxwell predicted that electric and magnetic fields would be able to sustain themselves, free from charges and currents, if they took the form of an **electromagnetic wave**. The wave would have to have a very specific geometry, shown in Fig. 31-31, in which $\vec{E}$ and $\vec{B}$ are perpendicular to each other as well as perpendicular to the direction of travel. That is, an electromagnetic wave would be a *transverse* wave. Furthermore, Maxwell predicted, the wave would have to travel with the very specific speed

$$v_{\text{em wave}} = \frac{1}{\sqrt{\varepsilon_0 \mu_0}}, \qquad (31\text{-}21)$$

where ε_0 is the permittivity constant from Coulomb's law and μ_0 is the permeability constant from the law of Biot and Savart. Maxwell quickly computed that an electromagnetic wave, if it existed, would have to travel with speed

$$v_{\text{em wave}} = 3.00 \times 10^8 \text{ m/s}.$$

We don't know Maxwell's immediate reaction, but it must have been both shock and excitement. His predicted speed for electromagnetic waves, a prediction that came directly from his theory, was none other than the well-known speed of light!

The agreement between the prediction of the theory and the measured speed of light could be just a coincidence, but Maxwell didn't think so. Making a bold leap of imagination, he wrote:

> The velocity of transverse undulations in our hypothetical medium, calculated from the electromagnetic experiments of Kohlrausch and Weber [who had measured ε_0 and μ_0], agrees so exactly with the velocity of light calculated from the optical experiments of Fizeau that we can scarcely avoid the inference that *light consists in the transverse undulations of the same medium which is the cause of electric and magnetic phenomena.*

In other words, light is an electromagnetic wave.

Maxwell was convinced, but most physicists remained skeptical. To persuade them, Maxwell made specific predictions that could be tested. The most important was that electromagnetic waves could exist at any frequency, not just the high frequencies of light waves, and that all of these waves would travel with the same speed. It took roughly 25 years until Maxwell's predictions could be tested. In 1886 the German physicist Heinrich Hertz discovered by chance—like Oersted and Faraday—how to generate and transmit radio waves. Two years later, in 1888, he was able to measure the speed of the new radio waves. He found it to be the speed of light.

Maxwell did not live to see his full triumph; he died in 1879 at the age of 48. But the closing years of the nineteenth century were a second golden age for physics, a period of successes unparalleled since the time of Newton two centuries earlier. Much of the credit goes to Faraday, for his study of electromagnetic induction, and to Maxwell. Electricity, magnetism, and optics were now unified, thermodynamics was well understood, and scientific applications were growing at a rapid pace. There were some who thought that physics was nearing an end, that everything worth discovering was already known. But trouble was just over the horizon, as you will see when we begin Part VI.

Summary

Important Concepts and Terms

electromagnetic induction Lenz's law
induced current induced emf
motional emf Faraday's law
generator transformer
eddy current induced electric field
magnetic flux induced magnetic field
weber electromagnetic wave
area vector

Faraday and Henry independently discovered that a current is induced to flow in a conducting loop if the magnetic flux $\Phi = \vec{A} \cdot \vec{B}$ through the loop *changes*. This phenomenon is called electromagnetic induction. The induced current has magnitude $I = \mathcal{E}/R$, where the induced emf $\mathcal{E}$ is given by Faraday's law:

$$|\mathcal{E}_{\text{loop}}| = \left|\frac{d\Phi}{dt}\right|.$$

The direction of the induced current is given by Lenz's law, which states that the induced current flows in a direction such that the induced magnetic field tries to prevent the flux through the loop from changing.

A closer look at induction shows that there are two distinct causes of an induced current.

1. The magnetic flux can change as a consequence of the motion of part of a circuit through the magnetic field. In this case, the emf and the induced current are due to magnetic forces on the moving charge carriers. Motional emf is the basis of applications such as generators.

2. The magnetic flux can change because the magnetic field itself changes. In this case, the changing magnetic field creates an induced electric field. The induced electric field, which is a non-Coulomb field, exerts forces on the charge carriers and causes a current to flow. Induced fields are the basis of applications such as transformers and magnetic recording.

Faraday's law correctly describes both of these different situations.

An induced electric field is independent of point charges. It is a completely new way of creating an electric field. Maxwell hypothesized that induced magnetic fields should also exist, created by a changing electric field. This led to his prediction of self-sustaining electromagnetic waves, a prediction that was confirmed by Hertz's discovery of radio waves.

Exercises and Problems

Exercises

1. Figure 31-32 shows an equilateral triangle, 20 cm on each side, halfway into a magnetic field of strength 0.1 T.
   a. What is the magnetic flux through the loop?
   b. If the magnetic field decreases, in which direction will current flow in the loop?

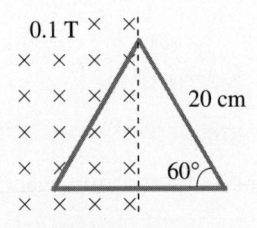

   FIGURE 31-32

2. Figure 31-33 shows a loop of wire 10 cm in diameter in three different magnetic fields. The loop's resistance is 0.1 Ω. For each situation, determine the induced emf, the induced current, and the direction of current flow.

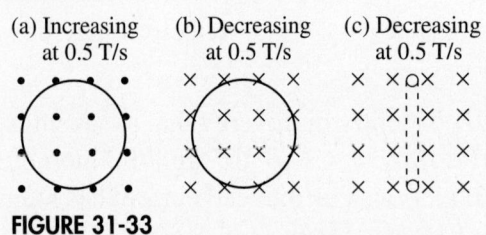

   FIGURE 31-33

3. The 20 cm × 20 cm square loop of Fig. 31-34 has a resistance of 0.1 Ω. The field strength is described by $B = 4t - 2t^2$.
   a. Determine B, $\mathcal{E}$, and I at half-second intervals from 0 s to 2 s.
   b. Use your results of part a) to graph B and I versus time.

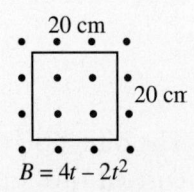

   FIGURE 31-34

4. A 1000-turn coil of wire that is 2 cm in diameter is in a magnetic field that drops from 0.1 T to zero in 10 ms. The axis of the coil is parallel to the field. What is the emf of the coil?

5. Figure 31-35 shows a 10 cm × 10 cm square bent in a 90° angle. A magnetic field of strength 0.05 T is pointed downward at a 45° angle. What is the magnetic flux through the loop?

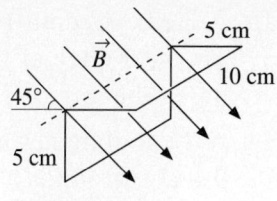

FIGURE 31-35

Problems

6. A slide wire 4 cm wide moves outward with a speed of 100 m/s in a 1.0 T magnetic field. (See Fig. 31-32.) At the instant the circuit forms a 4 cm × 4 cm square, with $R = 0.04 \, \Omega$, determine:
   a. The induced emf.
   b. The induced current.
   c. The potential difference between the two ends of the moving wire.

7. A zero-resistance slide wire of length 20 cm moves outward, on zero-resistance rails, at a steady speed of 10 m/s in a 0.1 T magnetic field. On the far side, a 1 Ω carbon resistor completes the circuit by connecting the two rails. The resistor has a mass of 50 mg.
   a. What is the induced current in the circuit?
   b. How much force is needed to pull the wire at this speed?
   c. If the wire is pulled for 10 s, what is the temperature increase of the carbon? (The specific heat of carbon is 0.170 cal/g °C.)

8. Figure 31-36 shows a vertically oriented loop in a 1 T horizontal magnetic field. The loop is 20 cm × 20 cm, has a mass of 10 g, and a resistance of 0.01 Ω, and is halfway in the field. The loop is released from rest.
   a. Will the loop fall with constant speed or constant acceleration?
   b. How long will it take the loop to leave the field? How does this compare to the time it would take the loop to fall the same distance in the absence of a field?

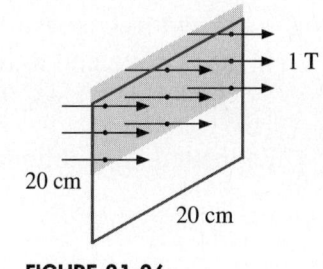

FIGURE 31-36

9. Figure 31-37 shows an L-shaped conductor moving at 10 m/s across a fixed L-shaped conductor in a 0.1 T magnetic field. The two vertices overlapped, so that the enclosed area was zero, at $t = 0$. The conductor has a resistance of 0.01 ohms *per meter*.
   a. In which direction does the induced current flow?
   b. Find expressions for the induced emf and the induced currents as functions of time. (**Hint:** It is a calculus problem to find dA/dt in terms of v. Remember that the enclosed area is always a square.)
   c. Evaluate $\mathcal{E}$ and I at $t = 0.1$ s.

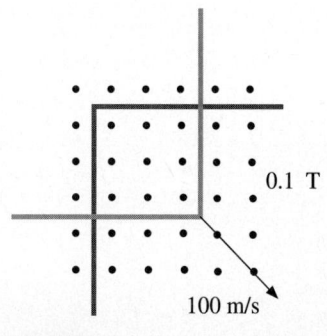

FIGURE 31-37

10. A 2 cm × 2 cm square loop of resistance 0.01 Ω is parallel to a long, straight wire. The near edge of the loop is 1 cm from the wire. The current in the wire is increasing at the rate of 100 A/s. What is the current in the loop?

11. A 4 cm diameter loop of resistance 0.1 Ω surrounds a 2 cm diameter solenoid. The solenoid is 10 cm long, has 100 turns, and carries the current shown in the graph of Fig. 31-38. A positive current is interpreted as a current into the page at the top of the loop. Determine the current in the loop as a function of time. Give your answer as a graph of current-versus-time for 0 s ≤ t ≤ 3 s.

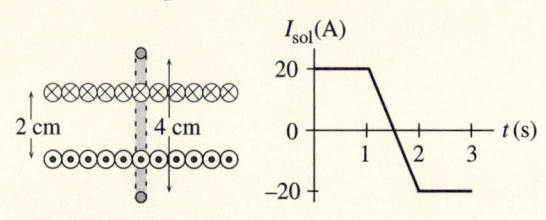

FIGURE 31-38

12. A 10-turn coil of wire having a diameter of 1 cm and a resistance of 0.2 Ω is in a 0.001 T magnetic field, with the coil oriented for maximum flux. The coil is connected to an uncharged 1 μf capacitor rather than to a current meter. The coil is quickly pulled out of the magnetic field. Afterward, what is the voltage across the capacitor? (**Hint:** Use $I = dq/dt$ to relate the *net* change of flux to the amount of charge that flows to the capacitor.)

13. A square loop, length L on each side, is shot with velocity v_0 into a uniform magnetic field B. The plane of the loop is perpendicular to the field. The loop has mass m and resistance R, and it enters the field at $t = 0$.
  a. Find an expression for the loop's velocity as a function of time as it travels through the magnetic field. You can ignore gravity. (**Hint:** Use calculus to solve Newton's second law for a nonconstant force.)
  b. Calculate and draw a graph of v over the interval 0 s ≤ t ≤ 0.04 s for the case that v_0 = 10 m/s, L = 10 cm, m = 1 g, R = 0.001 Ω, and B = 0.10 T.

[**Estimated 12 additional problems for the final edition.**]

PART V Scenic Vista

The Global Village

> What hath God wrought?
> First message sent by telegraph.

Mass and charge are the two most fundamental properties of matter. The first four parts of this text were about the properties and interactions of masses. Part V has been a study of the physics of charge—what charge is, how charges interact, and how charged objects move. The scientific understanding of the interactions and motions of charges has provided the basis for many of the technical innovations of the past two centuries. The terms *electricity* and *magnetism* denote the large collection of phenomena—from electric circuits to magnetic resonance imaging—that spring from charged-particle interactions.

To understand how the long-range forces of electricity and magnetism work, it was necessary to recognize the existence of electric and magnetic *fields*. The field concept is subtle, but it is an essential part of our modern understanding of the physical universe. One charge—the source charge—alters the space around it by creating an electric field and, if it is moving, a magnetic field. Other charges experience forces exerted *by the fields*. Thus the electric and magnetic fields are the agents by which two charges interact.

Faraday first introduced the field concept to describe the long-range electric interactions between charges and the magnetic interactions between currents and compasses. And it was Faraday who later discovered that fields can also be created by other fields: A changing magnetic field induces an electric field and a changing electric field induces a magnetic field. Faraday's and Henry's study of electromagnetic induction paved the way for the eventual discovery of electromagnetic waves—the quintessential electromagnetic phenomenon.

Part V has introduced many new concepts and laws. Table SV V-1 draws these many ideas together with a knowledge structure for electricity and magnetism. The major division of Table SV V-1 is between the laws governing the sources of the field (the laws of Coulomb, Biot and Savart, and Faraday) and the laws governing the response of charges in the field. A secondary theme is the distinction between the macroscopic electric and magnetic fields and the microscopic motions of atomic-level charges and magnetic moments. The micro/macro connection that we introduced in thermodynamics is equally important for understanding electric and magnetic phenomena.

TABLE SV V-1 The knowledge structure of electricity and magnetism.

Sources of fields

	Source charges		Changing fields
	Coulomb's law	Biot and Savart law	Faraday's law
	$\vec{E}_{pt} = \left(\dfrac{q}{4\pi\varepsilon_0 r^2}, \text{away from } q\right)$ $\quad E_s = -\dfrac{dV}{ds}\quad$ $V_{pt} = \dfrac{q}{4\pi\varepsilon_0 r}$ $\vec{E} = \Sigma \vec{E}_{pt} \to \int d\vec{E} \qquad \Delta V = -\int_{s_1}^{s_2} E_s\, ds \quad V = \Sigma V_{pt} \to \int \dfrac{dq}{4\pi\varepsilon_0 r}$	$\vec{B}_{pt} = \left(\dfrac{\mu_0 qv \sin\theta}{4\pi r^2}, \text{RHR}\right)$ $\vec{B} = \Sigma \vec{B}_{pt} \to \int d\vec{B}$	$\mathcal{E} = \left|\dfrac{d\Phi}{dt}\right|$ Direction so as to oppose the *change* in the flux.
	Special fields line of charge sphere of charge plane of charge parallel-plate capacitor	*Special fields* long straight current current loop solenoid	

Charges in fields

Particle:	$\vec{F} = q\vec{E}$	$U = qV$	$\vec{F} = \begin{cases} 0 & \vec{v} \parallel \vec{B} \\ (qvB, \text{RHR}) & \vec{v} \perp \vec{B} \end{cases}$	
Current:	$J = \sigma E$	$I = \dfrac{\Delta V}{R}$	$\vec{F} = \begin{cases} 0 & \vec{I} \parallel \vec{B} \\ (ILB, \text{RHR}) & \vec{I} \perp \vec{B} \end{cases}$	$I_{\text{induced}} = \dfrac{\mathcal{E}}{R}$
Applications:	circuits		magnetic materials	electromagnetic waves
Microscopic model:	charge carriers conductors and insulators		magnetic moments magnetic domains	

Although electric and magnetic phenomena have been known since antiquity, they were difficult to reproduce and nearly impossible to control. The year 1800 marked a turning point, for in that year Alessandro Volta invented the battery. The two centuries since Volta have seen the most profound and far-reaching technological changes in the history of the human species. Many of these changes have been direct applications of the physics of electricity and magnetism. The remainder of this Scenic Vista will look more closely at one particularly important application of electricity and magnetism—telecommunications.

The Telegraph: Dawn of a New Age

In 1800, the year that Volta invented the battery and that Thomas Jefferson was elected the third U.S. president, the fastest a message could travel was the speed of a man or woman on horseback. It took at least three days for news to travel from New York to Boston and well over a month for it to reach the frontier outpost of Cincinnati. It had been so since the earliest civilizations, and not even the most progressive thinkers of the age—Jefferson and Benjamin Franklin—anticipated any change.

But Oersted's 1820 discovery that a current creates a magnetic field soon introduced revolutionary changes to communications. The American scientist Joseph Henry, who with Faraday shares credit for the discovery of electromagnetic induction, saw a simple electromagnet in 1825. Inspired, he set about improving the device by using a soft iron core and many turns of wire. By 1831, Henry had an electromagnet, powered by a single battery, that could lift 750 pounds of iron.

In 1830 Henry strung more than a mile of wire around his classroom, then used a current through the wire to activate an electromagnet and strike a bell. This device, in essence, was the first telegraph. He soon had wires running between his house and his laboratory so he could send messages to his wife by ringing bells.

Henry was a scientist and an inventor with little concern about the use of his ideas. In 1835, however, he met an entrepreneur interested in the commercial development of electric technology—Samual F. B. Morse. Morse was one of the most prominent American artists of the early nineteenth century, and his paintings hang in museums around the world. But Morse also had an abiding interest in technology. In the 1830s he invented the famous code that bears his name—Morse code—and began to experiment with electromagnets. With advice and encouragement from Henry, Morse went on to develop the first practical telegraph.

Morse patented the telegraph system in 1840. Figure SV V-1 shows the basic features of the telegraph. When the telegraph key is pushed, the circuit is completed and the distant electromagnet is activated. A small piece of iron is pulled against the electromagnet, making an audible "click," then pulled back by a spring (not shown) when the key is released. Skilled operators could listen to the pattern of clicks and decode messages at a very high speed. Other inventors, including Thomas Edison and Alexander Graham Bell, soon made devices in which the telegraph made marks on paper, providing a permanent record.

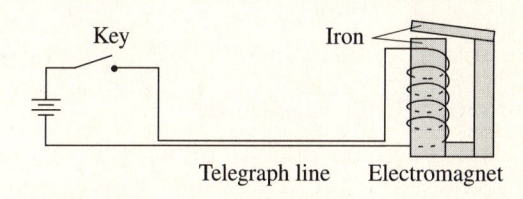

FIGURE SVV-1 The telegraph.

By 1843, Morse had persuaded Congress to fund a telegraph line between Washington, D.C., and Baltimore. The telegraph first operated on May 24, 1844, sending the message, "What hath God wrought?" For the first time, long-distance communication could take place essentially instantaneously.

Telegraph communication advanced as quickly as wire could be strung. England was connected to Europe in 1850, and the first transatlantic telegraph cable was laid in 1858. By 1875, only 50 years after Henry began experimenting with magnetism, a worldwide network of cables had been laid. The United States was connected to Europe, India, Australia, and the Far East.

The telegraph was the first successful device to transmit electric signals over long distances, but it didn't hold its monopoly for long. Inventors immediately began to think about using electromagnetic devices to transmit the more complex signals associated with speech. The first to succeed was Alexander Graham Bell, who invented the telephone in 1876. The first practical telephone, shown in Fig. SV V-2, was a modification of the telegraph. Bell replaced the on-off telegraph key with a *transmitter* consisting of an electromagnet and a thin

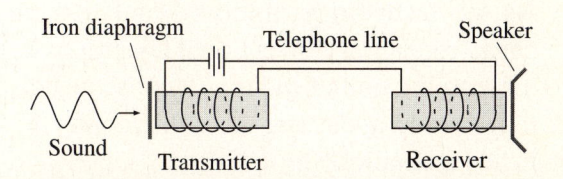

FIGURE SVV-2 An early telephone system. Sound waves vibrate the iron diaphragm, causing an induced current that, in turn, creates a sound wave in the loudspeaker on the right.

iron diaphragm. The *receiver* was a similar electromagnet next to a loudspeaker with an iron diaphragm.

An incident sound wave causes the iron diaphragm to vibrate. Because iron is magnetic, these oscillations cause the magnetic flux through the transmitter coil to oscillate and an oscillatory current to be induced in the circuit. The oscillating current in the receiver electromagnet then exerts an oscillating force on the loudspeaker diaphragm, causing it to broadcast a sound wave.

Telephone companies were quickly established, led by Bell's own Bell Telephone Company, but the early service was often unimpressive. In 1880, apparently after having difficulty hearing over the telephone lines, Mark Twain wrote a letter to a New York newspaper:

> It is my heart-warm and world-embracing Christmas hope and aspiration that all of us—the high, the low, the poor, the rich, the admired, the despised—may eventually be gathered together in a heaven of everlasting rest and peace and bliss—except the inventor of the telephone!

Twain's objections notwithstanding, telephone technology has improved until you can now easily speak with another person anywhere in the world.

Radio and Television

The telegraph and its immediate descendant, the telephone, provided electromagnetic communication over wires. Hertz's 1886 discovery of electromagnetic waves opened up an entirely new possibility—wireless communication at the speed of light. There is no single inventor of radio because many individuals were working simultaneously to develop practical methods of transmitting and receiving radio waves. Nonetheless, the Italian inventor Guglielmo Marconi is the man most associated with the first successful use of radio waves.

The earliest radio was a radiotelegraph, sending Morse code rather than the now-familiar audio signals. Others had already attached a telegraph key to Hertz's transmitter and were sending on-off radio signals over short distances. Marconi made significant improvements in the technology, especially in antennas, and by 1896 he was transmitting radiotelegraph signals over distances of a mile. The distances increased quickly, and in 1899 Marconi established a commercial radiotelegraph link between England and France. In 1901 he sent and received the first transatlantic radio signal—the single letter "S" in Morse code.

The first major application of radiotelegraphs was communicating with ships at sea. As the technology improved and transmitters became more powerful, radiotelegraphs began replacing conventional telegraphs connected by wire. World War I greatly hastened the introduction of radio as a consequence of both the need to communicate with units as they moved about and the vulnerability of telegraph cables. Thus radio was a well-established technology by 1920.

[**Photo suggestion: Early vacuum tube triode.**]

Sound transmission by radio was not possible until the 1906 invention of the triode vacuum tube. The vacuum tube made possible both reliable oscillators and high-power amplifiers. Experimental long-distance broadcasts were made in 1915, and in 1919 the newly formed Radio Corporation of America (RCA) purchased the patent rights to the triode vacuum tube as well as to the technology for modulating radio waves to encode audio information. Commercial radio was underway, and by 1925 there were more than one thousand radio stations operating in the United States.

Interestingly, the history of television goes back further than the history of radio. Several inventors in the late nineteenth century devised schemes for using telegraph or telephone technology to transmit pictures that were scanned by mechanically moving a light-sensitive cell across an image. None of these were practical devices, but the idea of transmitting pictures was widely known to scientists and inventors. A patent for a television system had been granted in 1911 for a device based on the cathode-ray tube, but the inventor himself had noted that "it is an idea only, and the apparatus has never been constructed."

The first working device that we would call a television was developed in England in 1925. It transmitted a crude, barely recognizable black-and-*red* picture to a one-inch viewing screen! Nonetheless, a kit for home assembly was marketed in 1926. Short television transmissions were made by the BBC starting in 1929, and by the early 1930s it was clear that television was here to stay. Technological improvements throughout the 1930s led to black-and-white televisions of respectable quality by 1940.

A color television system was demonstrated in 1928, but it relied on mechanical scanning and was not practical. Not until 1947 was a color system developed that used separate red, green, and blue phosphor dots. The current standards for television broadcasts in the United States were adopted in 1953. They establish a 525-line picture and the three-phosphor scheme for color broadcasts.

Communication Satellites

Telecommunication systems spanned the globe by 1960, but receiving "live" information was not always possible. Telephones and shortwave radio broadcasts could reach anywhere in the world, but reception was often poor. Commercial radio stations reached a few hundred miles at best, and television transmission was pretty much limited to each city. National broadcasts within the United States required the signal to be transmitted via microwave relays to local stations for rebroadcast. This system made network television news possible, but not "live-from-the-scene" broadcasts. News journalists had to film events, then return the film to the studio for broadcast. Television images from overseas couldn't be seen until the next day, after film was flown back to the United States.

The world's first communication satellite was launched by NASA in 1960. It was a 100-foot-diameter balloon, inflated after reaching its 1000-mile-high orbit, with a metalized surface that reflected microwaves. Voice communication was established between California and New Jersey during each orbital pass of the satellite. But because this was a passive satellite—merely reflecting microwaves—reception required specialized high-sensitivity receivers. Although not practical for commercial use, the satellite allowed engineers to learn how microwaves propagate through the upper atmosphere.

Two years later, in 1962, NASA launched the first Telstar satellite. Telstar was an active satellite, using solar power to amplify signals received from earth and to beam them back down. The first live transatlantic television transmission was made on July 11, 1962, and was broadcast throughout the United States. Telephone links were established the next day.

Telstar could be used only during the short interval of each orbit that it was in sight of both Europe and the United States. Plans were being made for a system of roughly 100 satellites, so that one would always be available. But another idea soon proved to be more practical. In 1945, twelve years before the first satellite was launched, the scientist and science-fiction writer Arthur C. Clarke proposed placing satellites in circular orbits 22,300 miles above the earth. A satellite at this altitude orbits with the same 24-hour period that the earth rotates, so from the ground it appears to hang stationary in space. We now call this a *geosynchronous orbit*. One such satellite would allow microwave communication between two points one-third of a world apart, so just three geosynchronous satellites would span the entire earth. Another advantage of Clark's proposal was that ground antennas could remain pointed at the satellite at all times and wouldn't have to track it.

The energy required to reach geosynchronous orbit is much higher than to reach low earth orbit, as you learned in Chapter 12. Yet rocket technology was advancing faster than NASA could build Telstar satellites, and the first commercial communications satellite was placed in geosynchronous orbit in 1965. For the first time, television images could be broadcast live to anywhere in the world. Today all of the world's intercontinental television and much of the intercontinental telephone traffic travels via microwaves to a cluster of these manmade stars floating high above the earth.

Telecommunication Goes Digital—Again

We live in a digital age, but the idea is not new. The telegraph of the nineteenth century encoded information into binary form—dots and dashes, on and off, ones and zeros. Many of these ideas about the digital encoding of information were forgotten during the first half of the twentieth century, when the emphasis was on the *analog* transmission of telephone, radio, and television signals. Two technical innovations in the second half of the twentieth century have prompted a massive shift back to digital communication: the laser and the optical fiber.

The laser was invented in 1960, and it was immediately clear that it could be modulated off and on at a very high speed to communicate digital information via light waves. This was, in a sense, an updated version of Marconi's radiotelegraph. Even so, the laser could not be a practical communication device until it was possible to move light from point to point easily and cheaply.

Research in the 1960s showed that laser light was readily transmitted through very thin, flexible strands of glass called *optical fibers*. The fiber acts like a pipe for light traveling parallel to the fiber, so light injected at one end of a fiber follows a long, circuitous route and then emerges from the other end. It took a decade of work to learn how to manufacture optical fibers of high quality, but telephone transmission over optical fibers began in the late 1970s and then exploded during the 1980s.

All optical-fiber communication is performed by *digitizing* the signal, using a binary code. The signal is then transmitted by modulating the laser off and on. Modulation can be extremely rapid (more than 10^{10} times per second) because of the extremely high frequencies of light. Consequently, hundreds of thousands of telephone conversations or dozens of television signals can be transmitted simultaneously over a single fiber no wider than a human hair.

The Global Village

As we enter the twenty-first century we have truly become a global village. Information and images span the world as quickly—or quicker—than they once moved through a small village. You can pick up the phone and talk to friends or relatives anywhere in the world. Each evening's news brings live images from remote parts of the world. You can determine your exact location on earth to within a few meters with an inexpensive receiver that monitors the Global Positioning System satellites. And the Internet allows you instant access to information from computers all over the world.

The technology that unites the global village is invisible to the user. You cannot tell whether a phone call to another country is being transmitted by old-fashioned copper wire, by microwaves beamed into space and back, or by laser light gliding through a microscopic strand of glass. Voices, data, and images are retrieved from around the globe and displayed on your computer screen at the click of a mouse. Telecommunication unites our world, and the technologies of telecommunications are direct descendants of Coulomb, Ampère, Oersted, Henry, and—most of all—Michael Faraday.

PART VI
Quantum Physics and the Structure of Atoms

PART **VI** Overview

The Atomic Structure of Matter

Our journey into physics is nearing its end. We started our journey with a specific destination in mind—namely, learning about the atomic structure of matter. Much of the physics of the past century has focused on probing ever deeper into the structure of matter—first the structure of atoms, then later the structure of nuclei and subatomic particles. And much of today's technology, from semiconductor devices to nanostructures to lasers, depends upon a knowledge of atomic structure.

We noted at the very beginning of this text, in the Overview to Part I, that "atoms and their properties are described by quantum physics, but we cannot leap directly into that subject and expect that it would make any sense. To reach our destination, we are going to have to study many other topics along the way." Now, at last, you *have* visited those many other topics. You've learned how particles move under the influence of forces, how to describe a particle's motion in terms of its energy, about the physics of waves, and about the electromagnetic interactions that form the atomic glue. We can begin the last phase of our journey with confidence.

Atoms

We've shown you some glimpses of atoms during the first five parts of this text. You've seen how Rutherford used collisions between atomic particles to infer the size of the nucleus and to postulate a solar-system model of the atom (Chapter 8). You've seen how macroscopic measurements of density, pressure, and temperature allow us to learn about the sizes and speeds of atoms (Chapter 22). You've seen how atoms can be polarized (Chapter 24), how electrons and ions can transfer charge as an electric current (Chapter 26), and how atoms act as tiny magnets (Chapter 30). And very importantly, you've seen evidence that matter on the atomic scale sometimes acts like a wave (Chapter 18).

But when you get right down to it, what *is* an atom? What are the properties of atoms? What makes an atom of carbon different from an atom of gold? And what's inside an atom? As important as these questions are, there's an even deeper and more important question: How do we *know* about atoms?

One of the important goals of this part of the text will be to answer this question: *How do we know about atoms?* To reach this goal, we will emphasize both *experiments* that probe into the structure of matter and the *interpretation* of those experiments as support for an atomic theory of matter. As you learned in Part III, the wave-nature of light prevents us from seeing individual atoms through a microscope. Our knowledge of atoms must be inferred from the spectrum of the light they emit or from currents that flow through devices such as electron microscopes and scanning tunneling microscopes. These are *macroscopic* measurements. Because atomic phenomena lie beyond the immediate realm of our senses, learning to interpret macroscopic data in terms of atomic-level events is an essential aspect of learning about atomic physics. It is an interesting piece of detective work to analyze the structure and properties of microscopic atoms by piecing together the evidence from macroscopic measurements. To do so, we must enter the world of quantum physics.

Quantum Physics

We ended Part III with experimental evidence that light sometimes acts like a particle and that electrons and neutrons sometimes act like waves. These were observations only; we did not offer any theory or explanation at the time. Now we want to return to that thread of thought and see how it leads to the new ideas and theory of quantum physics. We'll begin by looking at *evidence* about the atomic world. *What* do we know about electrons and atoms, and *how* do we know it? We will approach these question from an historical perspective in order to gain some insight into how scientific theories develop and change.

Many scientists of the eighteenth and nineteenth centuries believed that atoms existed, but they had little evidence for that belief. To them, atoms were tiny, indivisible, fundamental pieces of matter that moved about in accordance with Newton's laws of motion. But it became increasingly clear toward the end of the nineteenth century that atoms *can* be divided into smaller pieces. Furthermore, the evidence was growing that Newton's mechanics and Maxwell's electrodynamics—what we call *classical physics*—were unable to explain the behavior of atoms or the light they emit.

We will focus our attention on one key experiment, called the *photoelectric effect*. The photoelectric effect was a very simple experiment, and the data were unambiguous. But difficulty will arise when we begin to *interpret* the data and try to find an understanding of the effect. We will find that the experimental evidence was totally at odds with the predictions of the classical theories. It was Albert Einstein who offered a fresh and original interpretation of the photoelectric effect in terms of the *quantization* of energy. Although not recognized at the time, Einstein's paper of 1905 on the photoelectric effect signified the end of classical physics and the start of a twenty-year period in which a new *quantum physics* was developed.

Next, we will look at Niels Bohr's efforts to apply the ideas of energy quantization to atoms. The *Bohr model* of the atom was the first to account for the discrete spectra emitted by atoms, but his model could not be extended beyond the simplest hydrogen atom. Bohr was on the right track, which is why we still study the Bohr model today, but his ideas were still too classical. The missing ingredient was de Broglie's postulate that matter should have wave-like properties—a postulate we looked at in Chapter 18.

The complete theory of quantum physics was developed by Erwin Schrödinger soon after he learned of de Broglie's postulate. Schrödinger's new theory described atomic particles in terms of an entirely new concept called a *wave function*. One of your most important tasks in Part VI will be to learn what a wave function is, how it is used, and how a mathematical description of the wave function can be connected to experimental measurements.

Finally, you will see how *quantum mechanics*—the quantum physics analysis of motion—succeeds, where classical physics failed, at providing an explanation for the structure of atoms that is in excellent agreement with experiments. You will see where the electron shell model of chemistry comes from, how molecular bonds are formed, how atoms emit and absorb light as a discrete spectrum, and how lasers work. We will concentrate on quantum mechanics in one dimension. This will allow us to focus on the physical phenomena we are trying to understand without becoming sidetracked by the difficult mathematics of quantum mechanics in three dimensions. You will see that there are quite a few situations for which a one-dimensional model is a reasonable, if not perfect, description. One-dimensional quantum mechanics will allow you to understand the essential features of radioactive decay, the scanning-tunneling microscopes, and various kinds of semiconductor devices.

Quantum physics will give you an entirely new perspective on the nature of the physical universe. The quantum world can seem strange and mysterious, with elusive wave functions giving only probabilities of where particles are to be found. Yet quantum mechanics gives the most definitive and accurate predictions of any physical theory. Perhaps it is fitting that the theory of wave-particle duality should have this two-sided personality.

Chapter 32

The End of Classical Physics

I must confess that I am jealous of the term atom; *for though it is easy to talk of atoms, it is very difficult to form a clear idea of their nature.*

Michael Faraday

LOOKING BACK Sections 8.7; 18.1, 18.2; 27.2–27.5

32.1 Physics in the 1800s

By the end of the late nineteenth century, many of the forces of nature were understood and could be controlled. The concepts of *classical physics*—Newtonian mechanics, thermodynamics, and Maxwell's theory of electromagnetism—had given rise to an impressive body of knowledge with immense explanatory power. Many scientists of the time felt that they could use these concepts to explain just about anything, and some even felt there was nothing significant left to discover. But within the span of just a few years, at the turn of the century, investigations into the structure of matter led to many astonishing new discoveries that classical physics was at a loss to explain. It came to be recognized that the laws of classical physics break down when applied to atomic systems. Physicists in the early years of the twentieth century had to reexamine their most basic assumptions about the nature of matter and light.

Our goal in this chapter is twofold. First, we will learn how scientists in the nineteenth and early twentieth centuries discovered the properties of atoms. Second, we will recognize that many of the newly found atomic properties could not be reconciled with classical physics. Scientists were faced with many new phenomena that they could not explain or understand. Before we launch into quantum physics, it is important to recognize where classical physics failed and why a new theory of light and matter was needed.

It is "easy to talk of atoms," as Faraday noted long ago, but quite another thing to have real knowledge of atoms. We have all read or been told that matter consists of atoms, but how do scientists *know* that atoms exist? We cannot see or touch atoms, so what is the evidence that leads us, with great certainty, to our current understanding of the atomic theory of matter? That is also what this chapter is about—establishing the *evidence* upon which our upcoming theories will be based.

We will join the historical progression in the early 1800s, then quickly speed along until the main action begins in the 1890s. Scientists in 1800 had three major realms of inquiry. They wanted to understand the nature of matter, the nature of electricity, and the nature of light.

Matter

The idea that matter consists of small, indivisible particles is due to Leucippus and his student Democritus, who flourished in ancient Greece around 440–420 B.C. They called these particles *atoms*, Greek for "not divisible." Atomism did not become a widespread belief due, in no little part, to the complete lack of evidence for atoms. But atomic ideas did manage to maintain a minority status throughout the Middle Ages until, at about the time of Newton and the beginnings of a mechanistic conception of the world, they experienced a revival of interest.

Robert Boyle began his study of gases in the 1660s, from which we today know Boyle's law as $pV = $ constant for an isothermal process. Newton himself noted that Boyle's law could be explained if a gas consisted of small particles. In 1738, Daniel Bernoulli advanced a more modern theory that gases are composed of small moving particles. The *evidence* for atoms, however, was still far too weak for Bernoulli's ideas to be more than a curiosity, and they did not gain acceptance at the time.

Things begin to change in the early years of the nineteenth century. The English chemist John Dalton published his *A New System of Chemical Philosophy* in 1808. He argued that much of what was known about chemical reactions, in particular the law of definite proportions, could be understood if elements consisted of identical indestructible atoms. The unique feature of Dalton's work, which made it more science than speculation, was his attempt to determine the relative masses of the atoms of different elements. These ideas were further extended in 1811 by the Italian chemist Amedeo Avogadro, who postulated that atoms could stick together to form more complex entities he called *molecules* and that equal volumes of gases at equal temperatures contain equal numbers of molecules.

Other evidence for atoms began to accumulate as thermodynamics and the kinetic theory of gases developed in the mid-nineteenth century. Two lines of inquiry led to rough estimates of atomic sizes. First, slight deviations from the ideal gas law at high pressures could be understood if the atoms were beginning to come into close proximity to one another. Second, the viscosity of a gas—which is readily measured—could be related to the mean free paths of the molecules and these, in turn, to the sizes of the atoms. By 1890, it was known that atoms exist and have diameters of $\approx 10^{-10}$ m.

Electricity

Although static electricity had been known since antiquity, electrical currents and conductivity were not discovered until the seventeenth century. This raised interesting new questions. Is the "electrical substance" a continuous fluid, or does it consist of granular particles of electricity? There was no direct evidence at the time, but the flow of current suggested a fluid of some sort. This was analogous to the prevailing belief that heat was also a fluid, called *caloric*.

Electrical studies in the eighteenth century relied on electrostatic generators—mechanical devices that could produce large, but uncontrolled, potential differences from friction. These were useful for studying electrical forces, but they were not suited to generating currents in wires. A major improvement came in 1800, with Volta's invention of the battery. This provided a controlled and reproducible potential difference. Further, it was perfect for causing currents to flow through conductors. The invention of the battery stimulated a wave of electrical research during the opening decades of the nineteenth century.

It only took two months from Volta's invention of the battery until the discovery that an electrical current through water decomposes the water into hydrogen and oxygen—the process that came to be called **electrolysis**. The basic experiment, as done today in chemistry classes, is shown in Fig. 32-1. The positive and negative terminals of a battery are connected to pieces of metal called *electrodes*. The negative electrode is called the **cathode** and the positive one is the **anode**. Bubbles of gas appear at the electrodes—hydrogen at one and oxygen at the other—as current flows, and they can be collected in tubes as they rise to the surface.

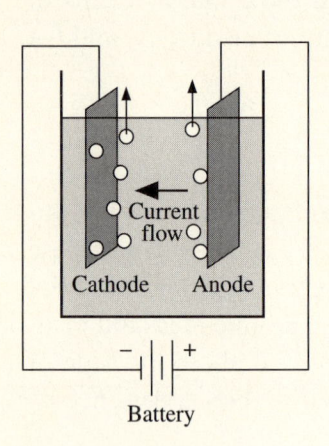

FIGURE 32-1 A current through water decomposes the water into hydrogen and oxygen gas bubbles, a process known as electrolysis.

This does not surprise us today, but water had long been regarded as one of the basic elements. The decomposition of water forced scientists to reconsider the basic building blocks of matter. These newly discovered effects of electrical currents suggested a previously unsuspected connection between electricity and matter.

Light

The question "What is light?" had long been debated. Newton, as we have noted previously, favored a *corpuscular* theory of small particles, or corpuscles, of light traveling in straight lines. His conclusion was based largely on the sharp shadows cast by sunlight, in contrast to the "bending" of water waves as they pass barriers. Although some of Newton's contemporaries had argued in favor of a wave theory, Newton's prestige carried the day and the corpuscular theory was dominant throughout the eighteenth century.

This view changed quickly as the nineteenth century opened. English linguist, physician, and scientist Thomas Young, in a series of experiments between 1802 and 1804, demonstrated the interference of light with his celebrated two-slit experiment. Young's wave hypothesis was given a more rigorous mathematical foundation in 1818 by the French physicist Fresnel. His theory predicted a number of diffraction effects that had not been previously observed. These were initially criticized as being contrary to common sense, but their subsequent experimental verification validated the wave theory of light.

But if light is a wave, what is waving? What is the medium? How can light travel through a vacuum? The corpuscular theory had not faced these difficulties, but they were brought to the forefront by the overwhelming evidence that light must be some kind of wave.

32.2 Faraday

Throughout the eighteenth century, studies of matter, electricity, and light had seemed to be separate and independent lines of inquiry. These three lines of inquiry—matter, electricity, and light—came together during the 1820s in the person of Michael Faraday, one of the most remarkable scientific geniuses in history. Faraday conducted three investigations of particular interest to us:

Electrical Conduction in Liquids

Others had already begun to study electrolysis, the conduction of electrical current through liquids, but it was Faraday's systematic and careful measurements that revealed the laws governing electrolysis. Faraday showed that electrolysis is most easily understood on the basis of an atomic theory of matter, and he found that there is a *charge* associated with each atom in the solution. Some charges are positive, others are negative. Today we call these positive and negative *ions*.

Faraday's discoveries strongly suggested that:

1. Atoms exist.
2. Electric charges are somehow—Faraday did not know how—associated with atoms.
3. Two separate kinds of charge, positive and negative, both exist.
4. Electricity is "granular" rather than a continuous fluid.
5. Electric charge is *quantized*. That is, it comes in discrete amounts with a basic "unit" of charge.

Electrical Conduction in Gases

Faraday also investigated whether electrical currents could pass through air. He sealed metal electrodes into a glass tube, lowered the pressure with a primitive vacuum pump, then attached an electrostatic generator. When he started the generator, the tube began to glow with a bright purple color! Faraday's device, called a **gas discharge tube**, is shown in Fig. 32-2.

Faraday's investigations demonstrated three properties:

1. Current flows through a low-pressure gas, creating an electrical discharge.
2. The color of the discharge depends on the type of gas in the tube.
3. Regardless of the type of gas, there is a separate, constant glow around the negative electrode—the cathode. This is called the **cathode glow**.

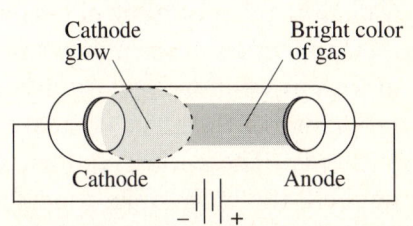

FIGURE 32-2 Faraday's gas discharge tube.

Today we know the purple color to be characteristic of nitrogen, the primary component of air. Most of us are more familiar with the reddish-orange color of the neon discharge tubes used for signs, but neon was not discovered until long after Faraday. His investigations, though, showed an unexpected connection between the color of the light and type of atoms in the tube.

Electromagnetic Fields

Perhaps Faraday's most important contributions to physics were in the realms of magnetism and light. Recall from our study of electricity and magnetism that Faraday was the originator of the concept of electric and magnetic *fields*. Although first devised as simply a way of envisioning electric and magnetic processes, Faraday's later studies of electromagnetic induction showed that these fields have a real existence and real properties. These investigations paved the way for the discovery, about thirty years later, that light is an electromagnetic wave. We will return to the implications of this discovery in Chapter 33.

Faraday's work was a major step toward providing real *evidence* for the existence of atoms. Altogether, Faraday established that atoms are associated with electricity, he demonstrated that different colors of light are associated with different kinds of atoms, and he prepared the way for showing that light is associated with electricity and magnetism. So matter, electricity, and light, previously three separate ideas, had been intertwined. As you will see, the connection of atoms with electric charges turned out to be of particular significance. Even so, Faraday recognized that he had barely scratched the surface, that far more research was needed before atoms could be understood. In his own words:

> It is impossible, perhaps, to speak on this point without committing oneself beyond what the present facts will sustain; and yet it is equally impossible, and perhaps would be impolitic, not to reason upon the subject. Although we know nothing of what an atom is, yet we cannot resist forming some idea of a small particle which represents it to the mind; and though we are in equal, if not greater, ignorance of electricity, so as to be unable to say whether it is a particular matter or matters, or mere motion of ordinary matter, or some third kind of power or agent, yet there is an immensity of facts which justify us in believing that the atoms of matter are in some way endowed or associated with electrical powers to which they owe their most striking qualities, and amongst them their mutual chemical affinity.

32.3 Cathode Rays

By 1850 it was widely accepted that chemicals are composed of various combinations of a small number of basic elements, that the atoms of each element are all alike, and that the relative masses of the atoms can be determined from chemical reactions. But the "charge" of ions in solution was still mysterious and unexplained, as was the repetition of chemical properties of the elements that was to lead, a short time later, to the periodic table of the elements. These observations suggested that atoms are *not*, as the classical picture implied, the indivisible, fundamental entities of nature. Instead, atoms appeared to have some form of internal structure that determines their electrical and chemical properties. This finding was unexpected, but it would take several decades of technological and scientific advances before much progress could be made at understanding atomic structure.

An important technological breakthrough came in the 1850s, with the development of much improved vacuum pumps. In 1858, the German scientist Plücker began a study of Faraday's gas discharge tube, using lower gas pressures. He found:

1. At lower pressures, the colored glow of the gas diminishes and the cathode glow becomes more extended.

2. The cathode glow tends to follow a magnetic field "as if it were constituted of flexible chains of iron filings attached at one end to the cathode."

3. At sufficiently low pressures the cathode glow extends to the glass wall and, significantly, the glass itself emits a greenish glow at that point.

In 1869, Plücker's student Hittorf sealed a solid object inside the tube, between the cathode and the glass wall. He had previously seen that the glass wall would glow where the cathode glow touched it. Now, however, he found that a solid object cast a *shadow* on the glass wall, as shown in Fig. 32-3. Hittorf's discovery suggested that the cathode emits *rays* of some form that travel in straight lines. The rays themselves are invisible, but they cause the glass to glow where they strike it. The gas slows or deflects these rays, because it takes a very low pressure for them to reach the glass walls at all, and the solid object blocks them entirely so as to cast a sharp shadow. These "rays" were quickly dubbed **cathode rays**, a name that lives on today in the *cathode-ray tube* that forms the picture tube in televisions and most computer displays. But naming the rays did nothing to explain them. What were they?

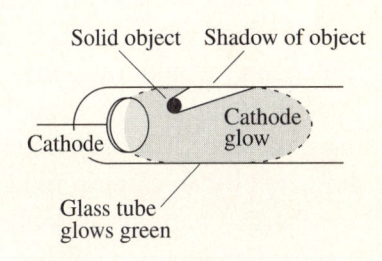

FIGURE 32-3 A solid object in the cathode glow casts a shadow, suggesting that some form of "rays" are leaving the cathode.

The most systematic studies on the new cathode rays were carried out during the 1870s by the English scientist Sir William Crookes. Whereas earlier studies had been primarily qualitative observations, Crookes' studies made measurements using a set of tubes he devised. Such tubes are today called **Crookes tubes** (see Fig. 32-4). His primary innovation was to elongate the tube, use yet lower pressures, and introduce one or more collimating holes for the rays to pass through. The net result was to generate a "beam" of cathode rays with a small glowing spot where they struck the end of the tube.

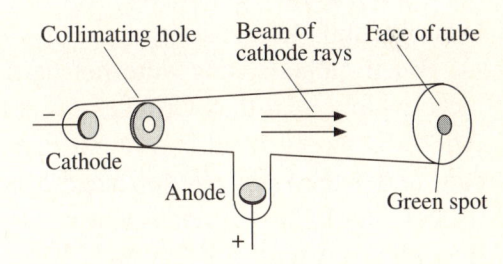

FIGURE 32-4 A Crookes tube.

The work of Crookes and others demonstrated:

1. Cathode rays are emitted from the cathode in an evacuated tube in which an electric current is flowing.

2. The rays are deflected by a magnetic field *as if* they were negative charges.

3. All metal cathodes produce cathode rays, and the ray properties are independent of the cathode material.

4. The rays can exert forces on objects and can transfer energy to objects, causing a thin foil in the cathode beam to glow red hot.

The results of Crookes' experiments led to more questions than they answered. Were the cathode rays some sort of particles? Or a wave? Were the rays themselves the carriers of the electric current, or were they something else that happened to be emitted when the current flowed? Note that item 2, although suggestive that the cathode rays are negative particles, is not by itself sufficient to demonstrate that they really are. Item 3 is worthy of note because it suggests that the cathode rays are a *fundamental* entity, not a part of the element from which they are emitted.

Although you can read the final answers in a book today, it is important to realize how difficult these questions were at the time and how experimental evidence was used to answer them. Crookes advanced a theory that molecules in the gas collided with the cathode, somehow acquired a negative charge (like Faraday's negative ions), and then "rebounded" with great speed as they were repelled by the negative cathode. The "charged molecules" would travel away in a straight line, carry energy and momentum, be deflected by a magnetic field, and could cause the tube to glow, or *fluoresce*, where it was struck by the molecules. Crookes' theory predicted, of course, that the negative ions would also be deflected by an electric field. This would be a definitive demonstration that cathode rays are charged particles. Crookes attempted to demonstrate it by sealing electrodes into the tube and creating an electric field, but his efforts were inconclusive. Other than this troublesome difficulty, Crookes' model seemed to explain the known observations.

Crookes' theory was immediately attacked, however. It was noted that the cathode rays could, at a pressure of 0.008 mm of mercury, travel the length of a 90-cm-long tube with no discernible deviation from a straight line. But the mean free path for molecules at this pressure, due to collisions with other molecules, is only about 6 mm. There was no chance at all that molecules could travel in a straight line for 150 times their mean free path! It was later discovered, in 1891, that the cathode rays could even penetrate very thin (≈ 2 μm thick) metal foils and, if these foils were at the end of the tube, then travel about 1 cm through the air. No atomic-size particle could penetrate these foils. Crookes' theory, seemingly adequate when it was proposed, was wildly inconsistent with subsequent experimental observations.

But if cathode rays were not particles, then what were they? An alternative theory, more widely held than Crookes', was that the cathode rays were electromagnetic waves. After all, waves travel in straight lines, cast shadows, carry energy and momentum, and can, under the right circumstances, cause materials to fluoresce. It was known that hot metals emit light—incandescence—so it seemed plausible that the cathode could be emitting waves. A long path through the gas would present no problem, and it was known by 1890 that long-wavelength electromagnetic waves—radio waves—could penetrate thin foils. The major obstacle for the wave theory was the deflection of cathode rays by a magnetic field. But the theory of electromagnetic waves was still quite new at the time, and many characteristics of these waves were still unknown. Although visible light was not deflected by a magnetic field, it was easy to think that some other form of electromagnetic waves might be so influenced.

The controversy over particles versus waves was intense. British scientists generally favored particles, but waves were preferred by their continental counterparts. Such controversies are an integral part of science, for they stimulate the best minds to come forward with new ideas and new experiments. In the end, as you will see in the next section, the attempts to understand the nature of cathode rays led to the discovery of the first subatomic particles.

32.4 J. J. Thompson and the Discovery of the Electron

Shortly after Wilhelm Röntgen's 1895 discovery of x rays, the young English physicist J. J. Thompson began using them to study electrical conduction in gases. He found that x rays could discharge an electroscope and concluded that they must be ionizing the air molecules, thereby making the air conductive. That is, the x rays were splitting the molecules into charged fragments—ions! This simple observation was of profound significance. Up until then, the only form of ionization known was the creation of positive and negative ions in solutions where, for example, a complex molecule such as NaCl splits into two smaller charged pieces. Although not yet understood, the fact that two atoms could acquire charge as a molecule splits apart did not jeopardize the idea that the atoms themselves were indivisible. But after observing that even monatomic gases, such as helium, could be ionized by x rays, Thompson realized that the *atom itself* must have charged constituents that could be split apart! Although there had been earlier hints, this was the first *direct* evidence that the atom is a complex structure made up, somehow, of charged pieces. The atom could no longer be thought of as the fundamental unit of matter.

Thompson was also conducting a series of experiments regarding the nature of cathode rays. He had previously leaned toward the charged-particle model of cathode rays, and this new insight about the atom strengthened his belief. Other scientists had placed an electrode in the cathode-ray beam, as shown in Fig. 32-5a, and measured a current flow. Although this seemed to demonstrate that the rays are charged particles, proponents of the wave model argued that the current might be a separate, independent event that just happened to be following the same straight line as the rays.

Thompson realized that he could use magnetic deflection of the cathode rays to settle the issue. He built a modified tube that had an electrode off to the side, as shown in Fig. 32-5b. Under normal operation, the cathode rays struck the center of the tube face and created a greenish spot on the glass. No current was measured by the electrode under these circumstances. Thompson then placed the tube in a magnetic field, which deflected the cathode rays to the side. He could determine their trajectory by the location of the green spot as it moved across the face of the tube. Just when the field was strong enough to deflect the cathode rays onto the electrode, it began to collect a current! At an even stronger field, when the cathode rays were deflected completely to the other side of the electrode, the current ceased.

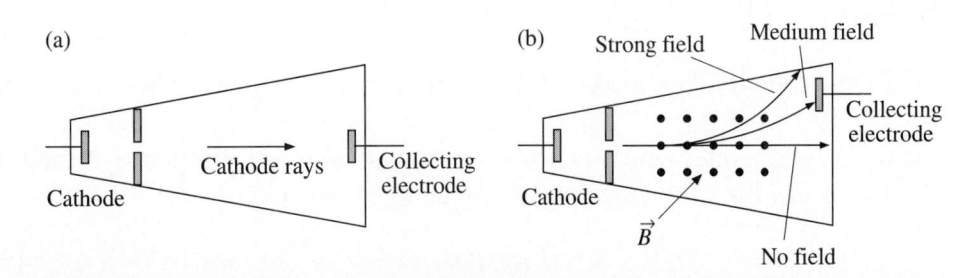

FIGURE 32-5 a) Earlier experiments had shown that an electrode in the cathode-ray beam would collect a current. b) Thompson's modified cathode-ray tube, which he used to steer the cathode rays onto the electrode with a magnetic field.

This was the first conclusive demonstration that cathode rays really are negatively charged particles. But why were they not deflected by an electric field? Thompson's first experiment to look for such a deflection met with the same inconclusive results that others had found. But it was Thompson's experience with the x-ray ionization of gases that led him to recognize the difficulty. He realized that the rapidly moving particles in the cathode-ray beam must be *ionizing* the molecules in the gas—which were still plentiful even at low pressures. The collision of a cathode-ray particle with a molecule could, just as with x rays, split the molecules into charged pieces. The electric field created by these charges effectively neutralized the field of the electrodes; hence there was no deflection.

Fortunately, vacuum technology was getting ever better. By using the most sophisticated techniques of his day, Thompson was eventually able to lower the pressure enough that ionization of the residual gas was not a problem. And then, just as he had expected, the cathode rays *were* deflected by an electric field! In Thompson's words:

> As the cathode rays carry a charge of negative electricity, are deflected by an electrostatic force as if they were negatively electrified, and are acted on by a magnetic force in just the way in which the force would act on a negatively electrified body moving along the path of these rays, I can see no escape from the conclusion that they are charges of electricity carried by particles of matter.

This was a decisive victory for the charged-particle model, but it still did not indicate anything about the nature of the particles. What were they? Thompson could measure their deflection for various strengths of the magnetic field, but the deflection depends on both the particle's charge-to-mass-ratio *q/m and* on its velocity. Finding the charge-to-mass-ratio, and thus learning something about the particles themselves, would require some means of accurately measuring their velocity. To do so, Thompson devised the experiment for which he is most remembered.

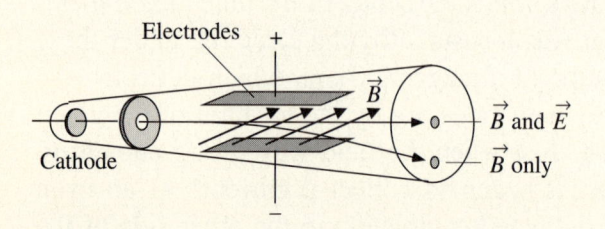

FIGURE 32-6 Thompson's crossed-field experiment to measure the velocity and the charge-to-mass-ratio of cathode rays.

Having demonstrated that he could deflect the charged particles with an electric field, Thompson built a tube containing sealed-in parallel-plate electrodes; then he placed the tube between the poles of a magnet. The tube was oriented, as shown in Fig. 32-6, such that the electric and magnetic fields were perpendicular to each other—thus creating what came to be known as a **crossed-field experiment**. Thompson's original tube is shown in the photograph of Fig. 32-7.

The magnetic field alone, which is always perpendicular to a particle's velocity vector, exerts a *downward* force on each negatively charged particle of magnitude

$$|\vec{F}| = qvB. \tag{32-1}$$

This force is at right angles to $\vec{v}$ and thus, as you learned in Chapter 30, acts as a centripetal force causing the particle to move along a circular arc of radius

$$r = \frac{mv}{qB}. \tag{32-2}$$

FIGURE 32-7 Photograph of Thompson's tube in which he discovered the electron. The coils with which he produced the magnetic field are seen below the tube. The cathode is on the right end.

The particle does *not*, however, move in a circular *orbit* such as we studied earlier. Because the velocity is large and because the magnetic field is limited in extent, a particle moves around only a small *arc* of the circle before it leaves the field and continues in a straight line. The net result of the magnetic field is to *deflect* the beam of particles downward. This deflection is easily observed by monitoring the green spot where the particles strike the glass at the end of the tube. It is a straightforward problem in geometry to determine the radius of curvature r the particle had while in the field by working backward from a measured deflection of the green spot.

Thompson's brilliant new idea was to establish an electric field between the parallel plate electrodes that would exert an *upward* force on the negative charges, pushing them back toward the center of the tube. The magnitude of the electric force on each particle is

$$|\vec{F}_{elec}| = qE. \tag{32-3}$$

The electric field strength is easily changed by varying the potential difference ΔV across the parallel plates. Thompson adjusted the field strength until the particle beam, in the presence of both electric and magnetic fields, was exactly in the center of tube—no deflection at all. Zero deflection requires that there is no net force on the particles as they move down the tube, so the magnetic and electric forces must, in this case, exactly cancel each other. Equating their magnitudes (the force vectors point in opposite directions) gives

$$|\vec{F}_{mag}| = |\vec{F}_{elec}|$$
$$\Rightarrow qvB = qE \tag{32-4}$$
$$\Rightarrow v = \frac{E}{B}.$$

By balancing the magnetic force against the electric force, Thompson could use the known magnetic and electric field strengths to measure the velocity of the charged-particle beam. (The ratio of volts per meter to tesla does, it turns out, give units of m/s.) Once v was known, he could then use Eq. 32-2 to determine the charge-to-mass-ratio

$$\frac{q}{m} = \frac{v}{rB}. \tag{32-5}$$

Thompson found, in a series of experiments under different conditions, a consistent value for the charge-to-mass ratio of cathode rays: $q/m \approx 1 \times 10^{11}$ C/kg. Although this seems not terribly accurate in comparison to a modern value of 1.76×10^{11} C/kg, keep in mind both the experimental limitations of his day *and* the fact that, prior to his work, no one had *any* idea of the charge-to-mass-ratio value.

• **EXAMPLE 32-1** An electron in a cathode-ray tube is fired between two parallel-plate electrodes that are 5 mm apart and 3 cm long. A potential difference ΔV between the electrodes establishes an electric field between them. A magnetic field of strength 0.001 T and width 3 cm is perpendicular to the electric field of the electrodes. When $\Delta V = 0$, the electron is deflected by 2 mm as it passes between the plates. What value of ΔV will allow the electron to pass between the plates without deflection?

SOLUTION We can find the needed electric field, and thus ΔV, if we know the electron's speed v. We can find the electron's speed from Eq. 32-5 by using the information provided to determine the radius of the electron's circular arc. Figure 32-8 shows an electron passing through the magnetic field between the plates when $\Delta V = 0$. (The curvature has been exaggerated to make the geometry clear.) Based on the right triangle, the electron's deflection Δy is related to the radius r and the length L of the plates by

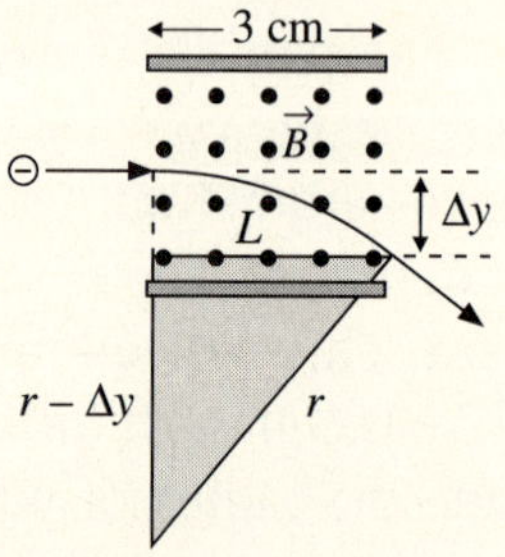

$$(r - \Delta y)^2 + L^2 = r^2.$$

FIGURE 32-8 Geometry of the electron's trajectory in Example 32-1.

This is easily solved to give the radius of the arc as

$$r = \frac{(\Delta y)^2 + L^2}{2\Delta y} = \frac{(0.002 \text{ m})^2 + (0.030 \text{ m})^2}{2(0.002 \text{ m})} = 0.226 \text{ m}.$$

From Eq. 32-5, the speed of an electron with an arc of this radius is

$$v = \frac{erB}{m} = 4.0 \times 10^7 \text{ m/s}.$$

From Eq. 32-4, the electric field that will allow the electron to pass through without deflection is

$$E = vB = 4000 \text{ V/m}.$$

The electric field of a parallel-plate capacitor of spacing d is related to the potential difference by $E = \Delta V/d$, so the needed potential difference is
• $\Delta V = Ed = (4000 \text{ V/m})(0.005 \text{ m}) = 20 \text{ V}.$

Notable an achievement as this was, Thompson did not stop there. It was noted previously that cathodes made of different metals generated cathode rays that seemed to behave the same. So one of the first things Thompson did was to measure q/m for different cathode materials. They were all the same. Whatever the charged particles were, they were identical for all the different elements. Next, Thompson compared his result to the charge-to-mass-ratio of the hydrogen ion—known from electrolysis to have a value of $\approx 1 \times 10^8$ C/kg, roughly 1000 times smaller than for the cathode-ray particles. This, he noted, could imply that the cathode-ray particles had much larger charges than the hydrogen ions, or much smaller masses, or some combination of these.

Electrolysis experiments suggested a basic unit of charge, so it was tempting to assume that the cathode-ray charge was the same. But the cathode rays were so different that such an assumption could not be justified without some other evidence. So Thompson called attention to the previous experiments showing that cathode rays penetrate thin metal foils and even up to 1 cm of air—evidence that was earlier used to argue that cathode rays could *not* be particles. Because it was now certain that they were, indeed, particles, Thompson concluded that they must be vastly smaller than atoms and thus much less massive than atoms.

In a paper published in 1897, J. J. Thompson assembled all of the evidence to announce the discovery that cathode rays are negatively charged particles, that they are much less massive ($\approx 0.1\%$) than atoms, and that they are identical when generated by different elements. Thompson, in other words, had discovered the existence of a **subatomic particle**, one of the constituents of which atoms themselves are constructed. In recognition of the role this particle plays in electricity, it was later named the **electron**.

Thompson's discovery of the electron, dramatic as it was, did not immediately convince everyone that the particles of cathode rays were a ubiquitous component of all atoms. But experiments by Thompson and others over the next few years showed that negative particles emitted from hot metal wires (discovered by Thomas Edison in his development of the light bulb) had the same q/m, that one type of radioactive decay (today called *beta radiation*) consisted of particles with the same q/m, and that certain changes in the spectra of atoms when placed in a magnetic field could be understood if the atoms had a charged constituent with the same q/m. By 1900 it was clear to all that electrons were a fundamental building block of atoms. J. J. Thompson was awarded the Nobel Prize in 1906.

32.5 Millikan and the Fundamental Unit of Charge

Thompson had measured the electron's charge-to-mass-ratio and had *surmised* that the mass must be much smaller than that of an atom, but clearly it was desirable to measure the charge q directly. Thompson and his students set out to do this, making use of their discovery that air can be ionized by x rays. They had subsequently found that the water vapor in moist air would easily condense to form small droplets around ions, thereby producing small charged water droplets. By collecting charged water drops, measuring the charge collected and the mass of the water, they were able to show that the unit of charge involved in the ionization of gases was roughly 1×10^{-19} C.

These experiments were not very accurate, but the value they obtained was close to the charge of the hydrogen ion as determined (also rather crudely) from electrolysis in liquids. Thompson interpreted this information, that the same unit of charge is responsible both for conduction through liquids and for conduction through gases, as implying the existence of a fundamental unit of charge. This fundamental unit of charge is designated e, so a hydrogen ion has a charge $q_{\text{H ion}} = +e$ and an electron has $q_{\text{elec}} = -e$.

In 1906, the American scientist Robert Millikan began making his own measurements of *e*. He started by duplicating the experiments of Thompson and his colleagues, but he soon realized the limitations of this approach. After a series of improvements, Millikan realized that he could "catch" a charged oil droplet and then accurately measure its motion using the combined influence of gravity and an electric field.

The **Millikan oil-drop experiment**, as we call it today, is illustrated in Fig. 32-9. A squeeze-bulb atomizer sprayed out a mist of oil droplets, some of which were charged because of frictional forces in the sprayer. The droplets slowly settled toward a horizontal pair of parallel-plate electrodes where a few of them passed through a small hole in the top plate. Millikan observed the drops by shining a bright light between the plates and using an eyepiece to see the droplets' reflections. He then quickly applied a voltage to the plates to establish an electric field between them. If the electric field exerted an upward force on a charged drop that exactly balanced the downward gravitational force, the drop would remain suspended between the plates for a lengthy period of time. In this situation,

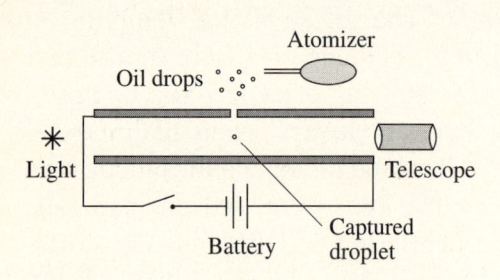

FIGURE 32-9 Millikan's oil-drop apparatus to measure the fundamental unit of charge.

$$m_{\text{drop}}g = q_{\text{drop}}E$$
$$\Rightarrow q_{\text{drop}} = \frac{m_{\text{drop}}g}{E}. \tag{32-6}$$

Note that *m* and *q* are the mass and charge of the oil droplet, not of an electron. But because the droplet is charged by acquiring (or losing) electrons or ions, the charge of the droplet should be related to the fundamental unit of charge.

Ideally, the mass of a droplet could be found by measuring its diameter and using the known density of the oil. However, the drops were too small (≈ 1 μm) to measure accurately by viewing through the eyepiece. Instead, Millikan devised an ingenious method of finding the size of the droplets. Objects this small are *not* in free fall. The air resistance forces are so large that they fall with a very small but constant velocity. The motion of a sphere through a viscous medium is a problem that had been solved in the nineteenth century, and it was known that the sphere's velocity depends on its radius and on the viscosity of air. So rather than holding the droplets motionless, Millikan caused them, by altering the field, to slowly move up and down through a known distance. By timing them with a stopwatch, he could determine their velocities. Then, using the known viscosity of air, he could calculate their radii, compute their masses, and, finally, arrive at a value for their charge. Although it was a somewhat round-about procedure, Millikan was able to measure their charge with an accuracy of ±0.1% (one part in a thousand).

Millikan found that some of his droplets were positively charged and some negatively charged, but all had charges that were some integer multiple of a certain *minimum* charge value. Whether the charges were due to electrons (either an excess or a deficiency) or to ions (either positive or negative) could not be determined—he likely observed all of these conditions. Millikan measured many hundreds of droplets, some for hours at a time, under a wide variety of conditions. In many cases a drop would suddenly change its charge—but always by some integer multiple of this minimum charge.

Millikan's conclusion was that "the electrical charges found on ions all have either exactly the same value or else some small exact multiple of that value." That value, the fundamental unit of charge we now call e, is measured to be

$$e = 1.602 \times 10^{-19} \text{ C}.$$

Using the value for the charge-to-mass ratio e/m, we then find the mass of the electron to be

$$m_{elec} = 9.11 \times 10^{-31} \text{ kg}.$$

EXAMPLE 32-2 Oil has a density of 860 kg/m³. An oil droplet 1 μm in diameter acquires 10 extra electrons as it is sprayed. What potential difference between two parallel plates 1 cm apart will cause the droplet to be suspended in air?

SOLUTION The magnitude of the charge on the drop is $q_{drop} = 10e$. The mass of the charge is related to its density ρ and volume V by

$$m_{drop} = \rho V = \tfrac{4}{3} \pi R^3 \rho = 4.50 \times 10^{-16} \text{ kg},$$

where the droplet's radius is $R = 5.0 \times 10^{-7}$ m. The electric field that will suspend this droplet against the force of gravity is

$$E = \frac{m_{drop} g}{q_{drop}} = 2760 \text{ V/m}.$$

To establish this electric field between two plates spaced by $d = 0.01$ m requires a potential difference

$$\Delta V = Ed = 27.6 \text{ V}.$$

It is often said that Millikan measured the charge of the electron. This is not strictly true because most, if not all, of the charges on the oil droplets are ions rather than electrons. Yet because it had become quite clear that atoms consisted of positive and negative constituents, the electron charge must be equal in magnitude to the fundamental ion charge if matter, as a whole, is to be electrically neutral. Taken together with the experiments of Thompson and others, Millikan's measurements provided overwhelming evidence that electric charge is particulate and, even more significantly, that *all* charges found in nature are multiples of a fundamental unit of charge we call e.

32.6 Rutherford and the Discovery of the Nucleus

It was clear, by the opening years of the twentieth century, that atoms are not indivisible but, instead, are constructed of charged particles. Atomic sizes were known to be $\approx 10^{-10}$ m, but the electrons common to all atoms are much smaller and much less massive than the smallest atom. So how do they "fit" into the larger atom? What is the positive charge of the atom? Where are the charges located inside the atoms, and what are the internal dynamics of the atom as these charges interact?

J. J. Thompson proposed the first model of an atom. Because the electrons are very small and light compared to the whole atom, it seemed reasonable to think that the positively charged part would take up most of the space. Thompson suggested that the atom consists of a spherical "cloud" of positive charge, roughly 10^{-10} m in diameter, in which the smaller negative electrons are embedded. The positive charge exactly balances the negative, so the atom as a whole has no net charge. This model of the atom has often been called the "plum-pudding model" or the "raisin-cake model" for reasons that should be clear from the picture of Fig. 32-10.

Thompson found, by careful analysis, that there are configurations of the electrons for which this atom is mechanically stable. He thought that small oscillations of the electrons around their equilibrium positions might account for the chemical and optical properties of atoms, but he was never successful at making any predictions that could be tested. The Thompson atom did not stand the tests of time, and his model is of interest today primarily to remind us that our current models of the atom are by no means "obvious." Science, as we have noted, has many side steps and dead ends as it progresses.

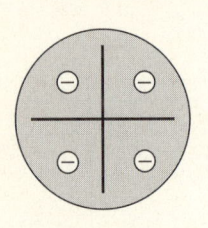

FIGURE 32-10 Thompson's raisin-cake model of the atom, with electrons embedded in a positive sphere of charge.

One of Thompson's students was a New Zealander named Ernest Rutherford. While Rutherford and Thompson were studying the ionizing effects of x rays, in 1896, the French physicist Becquerel announced the discovery that some new form of "rays" were emitted by crystals of uranium. Like x rays, these rays could expose film, pass through objects, and ionize the air. Yet they were emitted continuously from the uranium without having to "do" anything to it. This was the discovery of **radioactivity**.

In 1896, with x rays only a year old and cathode rays not yet completely understood, it was very difficult to know whether all these various kinds of rays were truly different or merely variations of a single type. Rutherford immediately began a study of these new rays. He quickly discovered that there are at least two *different* kinds of rays emitted by a uranium crystal. The first, which he called **alpha rays**, were easily absorbed by a piece of paper or a thin sheet of metal. The second, obviously named **beta rays**, could penetrate through at least 0.1 inch of metal and through much greater thicknesses of soft materials.

Thompson, as we have already noted, showed that beta rays have the same charge-to-mass-ratio as cathode rays. The beta rays thus turned out to be high-speed electrons emitted by the uranium crystal. Rutherford, using similar techniques, showed that alpha rays are *positively* charged particles. By 1906 he had measured their charge-to-mass-ratio to be

$$\frac{q}{m} = \frac{1}{2}\frac{e}{m_H},$$

where m_H is the mass of a hydrogen atom. This value could indicate either a singly ionized hydrogen molecule H_2^+ ($q = e$, $m = 2m_H$) *or* a doubly ionized helium atom He^{++} ($q = 2e$, $m = 4m_H$). How could he determine which it was?

In an ingenious experiment, Rutherford sealed a sample of radium—an emitter of alpha radiation—into a glass tube. Alpha rays cannot penetrate the glass, so the particles were contained within the tube. After waiting about a week, he used electrodes in the tube to create a discharge and observed the spectrum of the emitted light. He found the characteristic wavelengths of helium, but not those of hydrogen. Alpha rays—or alpha particles, as we now call them—consist of doubly ionized helium atoms (a helium nucleus) emitted at high speed ($\approx 3 \times 10^7$ m/s) from the sample.

It had been quite a shock to discover that atoms are not indivisible, that they have an inner structure. Now, with the discovery of radioactivity, it appeared that some atoms were not even stable but could spit out various kinds of charged particles! Physics had come a long way from the simple atomic idea of Democritus.

Rutherford was particularly interested in using these high-speed charged particles as a probe of other atoms. We discussed Rutherford's experiment previously (Section 8.7) as an example of momentum, and you should refer back to that section for a more complete description. In 1909, Rutherford and his students Geiger and Marsden used an apparatus like that of Fig. 32-11 to shoot alpha particles at *very* thin metal foils—about 2000 atoms thick. Some of the alpha particles would penetrate the foil, but in doing so the beam of particles would spread out a bit. This was not surprising. The alpha particle is charged, and it experiences forces from the positive and negative charges of the atoms as it passes through the foil. Thompson's model of the atom, which had only recently been proposed, had both the positive and the negative charges spread fairly equally throughout the volume of the atom. It was expected that the forces exerted on the alpha particle by the positive atomic charges would roughly cancel the forces from the negative charges. The alpha particle would be only slightly deflected, which is what Geiger and Marsden found.

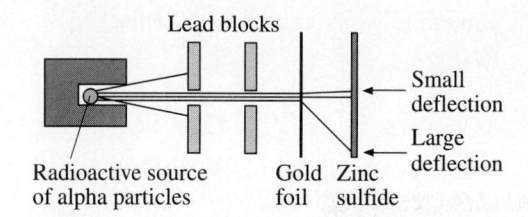

FIGURE 32-11 Rutherford's experiment to shoot high-speed alpha particles through a thin gold foil. They were detected by flashes of light on a zinc sulfide scintillation screen. Unexpectedly, a few showed large deflections.

But at Rutherford's suggestion, Marsden set up the apparatus to see whether any alpha particles were deflected at very large angles. As Rutherford later told the story:

> I may tell you in confidence that I did not believe that they would be, since we knew that the alpha particle was a very fast, massive particle, with a great deal of energy, and you could show that if the scattering was due to the accumulated effect of a number of small scatterings [as expected using the Thompson model], the chance of an alpha particle's being scattered backward was very small. Then I remember two or three days later Geiger coming to me in great excitement and saying, "We have been able to get some of the alpha particles coming backward." It was quite the most incredible event that has ever happened to me in my life. It was almost as if you fired a 15-inch shell at a piece of tissue paper and it came back and hit you. On consideration, I realized that this scattering backward must be the result of a single collision, and when I made calculations I saw that it was impossible to get anything of that order of magnitude unless you took a system in which the greater part of the mass of the atom was concentrated in a minute nucleus. It was then that I had the idea of an atom with a minute massive center, carrying a charge.

Rutherford's concept was a new model of the atom that we now call the **nuclear model**. He envisioned an atom in which negative electrons orbit, like a miniature solar system, an unbelievably small, massive, positive **nucleus** at the center. Nearly all of the atom is then merely empty space—the void!

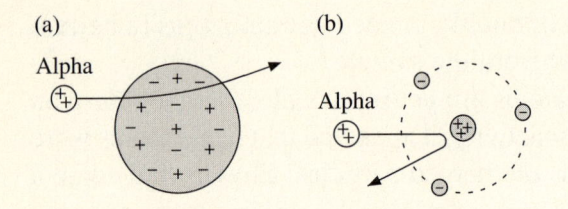

FIGURE 32-12 a) The spread-out charge of the Thompson atom cannot exert a big enough force to deflect the alpha particle by much. b) With a concentrated positive nucleus, the alpha particle can get close enough to experience a large deflecting force.

As Fig. 32-12a shows, the spread-out positive charge in Thompson's model of the atom only exerts enough force on the passing alpha particle to deflect it by a small amount. But if the positive charge is concentrated into a very small sphere, then, on at least a few occasions when the alpha particle is shot straight at the nucleus, the two will be able to come *very* close together. Because the electric force varies with the inverse square of the distance, the very large force of this very close approach can cause a backward deflection of the alpha particle, as shown in Fig. 32-12b. (Recall, from your study of electric fields, that a sphere of charge—as long as you stay outside the sphere—acts as if all the charge is concentrated at the center.)

EXAMPLE 32-3 An alpha particle is shot with a speed of 2×10^7 m/s directly toward a gold atom. What is the distance of closest approach to the nucleus?

SOLUTION Assume that the alpha particle is moving directly toward the center of the nucleus so that it will slow down, stop instantaneously, then reflect directly backward. We are not interested in how long it takes or any of the details of the trajectory, so a conservation of energy approach rather than a Newton's laws approach is called for. Initially, when the alpha particle is very far away, the system has only kinetic energy. At the moment of closest approach, just before the alpha recoils, the charges are at rest and the system has only potential energy. The gold nucleus is held essentially at rest the entire time by its surrounding electrons, and the electrons also keep the atom itself locked in place in the crystal lattice of the metal. So we are justified in thinking of the gold nucleus as remaining stationary throughout. Equating initial kinetic and final potential energies,

$$E_i = E_f$$
$$\Rightarrow \frac{1}{2}mv^2 = \frac{1}{4\pi\varepsilon_0}\frac{q_\alpha q_{Au}}{r},$$

where q_α and q_{Au} are the charges of the alpha particle and of the gold nucleus, m is the alpha particle mass, and r is the distance between them as the alpha particle reverses—the distance of closest approach. Solving for r gives

$$r = \frac{1}{4\pi\varepsilon_0}\frac{2q_\alpha q_{Au}}{mv^2}.$$

The alpha particle is a helium nucleus, so $m = 4$ u $= 6.64 \times 10^{-27}$ kg and $q_\alpha = 2e = 3.20 \times 10^{-19}$ C. Gold has atomic number 79, so $q_{Au} = 79e = 1.26 \times 10^{-17}$ C. We then calculate $r = 2.7 \times 10^{-14}$ m. This is only about 1/10,000 the size of the atom itself!

You may be concerned that we ignored the atom's electrons in this example. In fact, they make almost no contribution to the alpha particle's trajectory. First of all, the alpha particle is exceedingly massive compared to the electrons, so it easily pushes them aside without any noticeable change in its velocity. Once inside the atom, the alpha particle "sees" a spherical electron cloud surrounding it. The forces from opposite sides of the electron cloud are in opposite directions and cancel, so there is *no* electron force on the alpha particle once inside the atom.

Rutherford went on to make careful experiments of how the alpha particles scattered at different angles, and he compared his results to calculations based on point charges. When the experimental results began to differ from the calculations, he reasoned that the alpha particle must then be coming in contact with the nucleus. From such experiments he deduced that the atomic nucleus is $\approx 1 \times 10^{-14}$ m = 10 fm in diameter (1 fm = 1 femtometer = 10^{-15} m), increasing a little for elements of higher atomic number and atomic mass.

It may seem surprising to you today that the Rutherford model of the atom, with its solar system analogy, was not Thompson's original choice. But scientists at the time could not imagine matter having the extraordinarily high density implied by a small nucleus. Neither could they understand what holds the nucleus together, why the positive charges do not push each other apart. Thompson's model, in which the positive charge was spread out and balanced by the negative electrons, actually made more sense. It would be several decades before the forces holding the nucleus together began to be understood, but Rutherford's evidence for a very small nucleus was incontrovertible.

The Rutherford model makes it easier to understand and picture such processes as ionization. Because electrons orbit a positive nucleus, an x-ray photon or a rapidly moving particle, such as another electron, can "knock" one of the orbiting electrons away, creating a positive ion. Removing one electron makes a singly charged ion, with $q = +e$, and removing two creates a doubly charged ion with $q = +2e$. This is shown for lithium (atomic number 3) in Fig. 32-13. It is also possible to sneak an extra electron into orbit, creating a negative ion, but this is harder to do because the repulsive forces between the electrons quickly overpower the attractive force of the nucleus. It was known experimentally that most atoms have several stages of positive ionization—$q = +e, +2e, +3e, \ldots$—but only a single $q = -e$ stage of negative ionization. Rutherford's model makes it clear why this can happen. It also allows us to understand why it is that electrons are easily transferred during chemical reactions and during such processes as frictional charging, but protons are not. The protons are tightly bound in the nucleus, shielded by all the electrons, but outer electrons are easily stripped away. Rutherford's model has explanatory power that was lacking in Thompson's model.

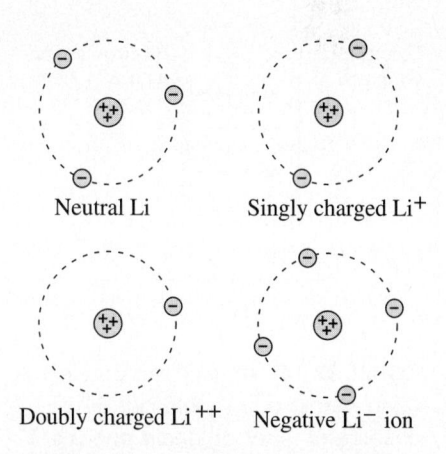

FIGURE 32-13 Different ionization stages of the lithium atom ($Z = 3$).

EXAMPLE 32-4 How much energy is required to ionize a hydrogen atom, for which the electron orbits in a radius of 5.29×10^{-11} m at a speed of 2.19×10^6 m/s?

SOLUTION A hydrogen atom has one electron orbiting a single proton. Removing the electron creates a hydrogen ion H⁺, which is just the proton alone. Initially the electron has both kinetic and potential energy. If it is moved very far away and is at rest, it will have neither kinetic nor potential energy. The energy needed to remove the electron is $\Delta E = E_f - E_i$. This is

$$\Delta E = E_f - E_i = (K_f + U_f) - (K_i + U_i)$$

$$= (0+0) - \left(\frac{1}{2}mv^2 + \frac{1}{4\pi\varepsilon_0}\frac{(e)(-e)}{r}\right)$$

$$= 2.17 \times 10^{-18} \text{ J}.$$

If the electron receives 2.17×10^{-18} J of energy—from a photon, in a collision with another electron, or by other means—it will be knocked out of the atom and leave a H⁺ ion behind.

The Electron Volt

You have probably noticed that the energies of charged particles are typically a tiny fraction of a joule. For atomic particles, such as electrons and protons, energies such as 10^{-18} J are common. The joule was a unit of appropriate size in mechanics and thermodynamics, where we dealt with macroscopic objects, but it is poorly matched to the needs of atomic physics. It will be very useful, as we continue our study of atoms, to have an energy unit appropriate to atomic events.

Consider an electron released from rest at the negative side of a parallel-plate capacitor, as shown in Fig. 32-14. There is a 1 V potential difference between the capacitor plates, creating an electric field that will accelerate the electron across the gap. What is the electron's kinetic energy when it reaches the positive plate? We know from energy conservation that

$$\Delta K + \Delta U = 0$$
$$\Rightarrow (K_f - K_i) + q\Delta V = 0,$$

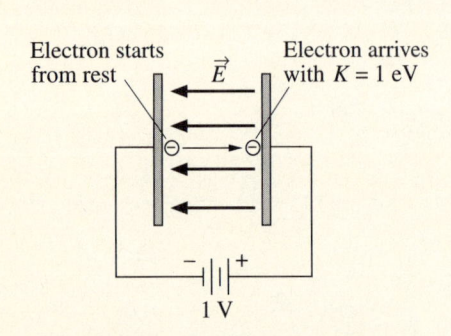

FIGURE 32-14 An electron accelerating across a 1 V potential difference gains 1 eV of kinetic energy.

where we have used $U = qV$, the relationship between the electric potential and the electric potential energy for a charge q. Now $K_i = 0$, because the electron starts from rest, and the electron's charge is $q = -e$. The potential *increases* as the electron moves across to the more positive side, so $\Delta V = 1$ V. Thus

$$K_f = -q\Delta V = e\Delta V = (1.60 \times 10^{-19} \text{ C})(1 \text{ V}) = 1.60 \times 10^{-19} \text{ J}.$$

Let us define a new unit of energy, called the **electron volt**, as

$$1 \text{ electron volt} = 1 \text{ eV} = 1.60 \times 10^{-19} \text{ J}.$$

With this definition, the final kinetic energy of the electron in our example is
$$K_f = 1 \text{ eV}.$$
In other words, 1 electron volt is the kinetic energy gained by an electron (or proton) if it accelerates through a potential difference of 1 volt. Note that the abbreviation eV uses a lowercase e but an uppercase V. Units of keV (10^3 eV), MeV (10^6 eV), and GeV (10^9 eV) are also common.

The electron volt is a troublesome unit for many students. This stems in part from its unusual name, which looks less like a unit than, say, "meter" or "second." A more significant difficulty is that the name suggests a relationship to volts. But *volts* are units of electric potential, whereas this new unit—with the admittedly confusing name of *electron volt*—is a unit of energy! It is crucial to distinguish between the potential V, measured in volts, and an energy—E or K or U—which can be measured either in joules or in electron volts. You can now use electron volts anywhere that you would previously have used joules. This is no different from converting back and forth between pressure units of pascals and atmospheres.

The main point, we reiterate, is that the electron volt is a unit of *energy*, convertible to joules, and *not* a unit of potential. Potential is always measured in volts. Note, however, that the joule remains the SI unit of energy. Although we will find it useful to express energies in eV, you *must* convert this energy to joules before doing most calculations.

EXAMPLE 32-5 What is the speed of a 8.30 MeV alpha particle?

SOLUTION Alpha particles are helium nuclei, having $m = 4 \text{ u} = 6.64 \times 10^{-27}$ kg. This alpha particle has a kinetic energy of 8.30×10^6 eV. First, convert the energy to joules:

$$K = 8.30 \times 10^6 \text{ eV} \cdot \frac{1.60 \times 10^{-19} \text{ J}}{1 \text{ eV}} = 1.33 \times 10^{-12} \text{ J}.$$

Now we can find the speed:

$$K = \frac{1}{2} mv^2 = 1.33 \times 10^{-12} \text{ J}$$

$$\Rightarrow v = \sqrt{\frac{2K}{m}} = 2.0 \times 10^7 \text{ m/s}.$$

This was the speed of the alpha particle in Example 32-3.

EXAMPLE 32-6 The electron in a hydrogen atom moves with a speed of 2.19×10^6 m/s in an orbit of radius 5.29×10^{-11} m. What is the electron's energy in eV?

SOLUTION The electron has both kinetic energy of motion *and* a potential energy due to its interaction with the proton. We can use the potential energy of two point charges, with $q_{proton} = +e$ and $q_{elec} = -e$. Then

$$E = K + U = \frac{1}{2} mv^2 + \frac{1}{4\pi\varepsilon_0} \frac{(e)(-e)}{r}$$

$$= -2.17 \times 10^{-18} \text{ J}.$$

Conversion to eV gives

$$E = -2.17 \times 10^{-18} \text{ J} \cdot \frac{1 \text{ eV}}{1.60 \times 10^{-19} \text{ J}} = -13.6 \text{ eV}.$$

The negative energy reflects the fact that the electron is *bound*. You would have to *add* energy to remove the electron—which is exactly what we calculated in Example 32-4. Make sure you understand how these two examples are related: If the electron has an energy of –13.6 eV when bound, it takes the addition of 13.6 eV to ionize the atom.

32.7 Into the Nucleus

We are not going to discuss nuclear physics in depth in this text, but it will be helpful to give a brief outline of what has been learned about the nucleus of the atom. The relative masses of many of the elements were known, from chemistry experiments, by the mid-nineteenth century. By arranging the elements in order of ascending mass, and noting recurring regularities in their chemical properties, Mendeleev first proposed the *periodic table* of the elements in 1872. But what did it mean to say that hydrogen was atomic number 1, helium number 2, lithium number 3, and so on? No one at the time suspected that the numbering might be connected with an inner structure of the atoms.

It soon became known that hydrogen atoms can only be singly ionized, producing H$^+$, but a doubly ionized H^{++} is never observed. Helium, by contrast, can be both singly and doubly ionized, with He$^+$ and He^{++} seen, but He^{+++} is not observed. Once Thompson discovered the electron and Millikan established the fundamental unit of charge, it seemed fairly clear that hydrogen atoms must consist of one electron and one unit of positive charge, helium of two electrons and two units of positive charge, and so on. That is, the **atomic number** of an element—which is always an integer—describes the number of electrons and the number of units of positive charge within the atom. The atomic number is symbolized by Z, so hydrogen has $Z = 1$, helium $Z = 2$, lithium $Z = 3$.

Because each atom has an integer number of electrons and an integer number of units of positive charge, a reasonable assumption to make is that the positive charge is associated with a charged particle. This particle must have charge $+e$. Further, because nearly all the atomic mass is associated with the positive charge, this particle must be much more massive than the electron. This positive subatomic particle is called the **proton**. With the advent of Rutherford's nuclear model, physicists recognized that atoms of atomic number Z consist of Z negative electrons, with net charge $-Ze$, orbiting a massive nucleus that contains protons and has net charge $+Ze$. The Rutherford atom went a large way toward explaining the periodic table.

But there was a problem. Helium has atomic number 2 and thus has twice as many electrons as hydrogen. Lithium, $Z = 3$, has three electrons. But from chemistry measurements it was known that helium is *four times* as massive as hydrogen and lithium is *seven times* as massive. Using the symbol A to represent the *atomic mass*, in atomic mass units, hydrogen has $A = 1$ u, helium has $A = 4$ u, and lithium $A = 7$ u. If a nucleus contains Z protons to balance the Z orbiting electrons, and because nearly all the atomic mass is contained in the nucleus, then helium should be simply twice as massive as hydrogen and lithium three times as massive. That is, the atomic mass should, in this simple model of the nucleus, be the same as the atomic number. Something else must be happening in the nucleus to make the atoms more massive than our simple nuclear model predicts.

Physicists immediately recognized how to solve this problem. Because protons have the same charge as electrons, with opposite sign, but are much more massive, a nucleus containing A protons *and* $A - Z$ electrons will have a mass of A but a net charge of only Ze. Lithium, for example, would have a nucleus containing 7 protons and $7 - 3 = 4$ electrons. This would give it a net charge of $+3e$, to balance the three orbiting electrons, but a mass seven times that of hydrogen. A nucleus constructed of protons *and* electrons seemed to explain all the known facts. Yet, as you may have surmised, this is another example of an "obvious" theory that was doomed to extinction.

In the early 1900s, J. J. Thompson and others turned their attention to studying the positive ions that are created when a gas is ionized. The ions, being much more massive than electrons, are much harder to deflect in electric and magnetic fields. But by 1913 Thompson had developed techniques for deflecting the ions sufficiently to measure their charge-to-mass ratio. These techniques were further developed by his student Aston into measuring devices called **mass spectrometers**. (We looked at an example of a mass spectrometer in Problem 17 of Chapter 30.)

As Aston and others began collecting data, they soon discovered that many elements consisted of atoms of *differing* mass! Neon, for example, had been assigned an atomic mass of 20. Aston found, as seen in the data of Fig. 32-15, that while most neon atoms (91%) have $A = 20$ u, a few (9%) have $A = 22$ u and a very small percentage have $A = 21$ u. Chlorine was found to contain a mix of 75% $A = 35$ u atoms and 25% $A = 37$ u atoms, both having $Z = 17$.

Atoms of the same element but of different mass are called **isotopes**. They are designated by using the atomic mass A as a leading superscript on the chemical symbol. Thus ^{20}Ne and ^{22}Ne are isotopes of neon, and ^{35}Cl and ^{37}Cl are isotopes of chlorine. All isotopes of an element have the same *chemical* behavior, because that is determined by the number of orbiting electrons. Their differences become apparent only in experiments that depend on the mass of the atoms. It is now known that every element in the periodic table has more than one isotope, although in most cases a single isotope accounts for nearly all of the atoms. The concept of isotopes explained a previous difficulty, namely, that chemical experiments had given a puzzling noninteger atomic mass of 35.4 to chlorine. It was now recognized that chemical measurements give only an *average* atomic mass.

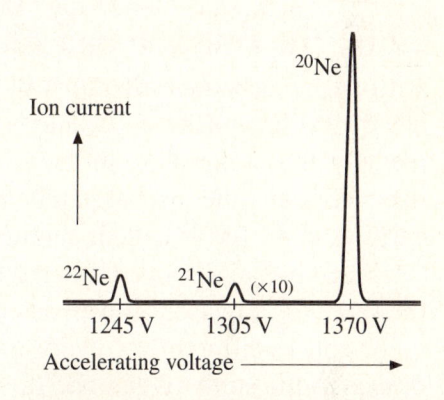

FIGURE 32-15 The mass spectrum of neon. Ions of different charge-to-mass-ratio are collected for different accelerating voltages. The major isotopes have $A = 20$ u and $A = 22$ u. An isotope with $A = 21$ u is also present ($\approx 0.3\%$ abundance) and has been magnified by 10.

These discoveries raised troubling new issues about the nature of the nucleus, difficulties that were not resolved until the discovery, in 1932, of a third subatomic particle. This particle has the same mass as a proton but has *no* electric charge. It is called the **neutron**. Neutrons reside in the nucleus with protons, contributing to the mass of the atom but

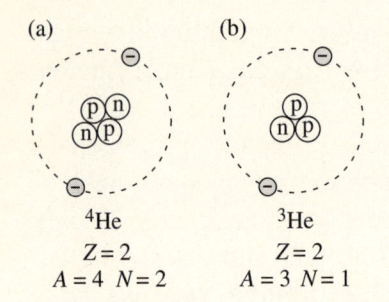

FIGURE 32-16 The two isotopes of helium. a) ⁴He. b) ³He, only 0.0001% abundant.

not to the charge. The number of neutrons is designated N. In our current understanding of the atom, an isotope of atomic number Z and atomic mass A consists of Z electrons orbiting a nucleus of Z protons and $N = A - Z$ neutrons. All the atoms of a particular element have the same number of protons in their nuclei. Differing numbers of neutrons, however, give rise to different isotopes having different masses. Thus ^{35}Cl has 17 protons and 18 neutrons in its nucleus, and ^{37}Cl has 17 protons and 20 neutrons. Both have 17 orbiting electrons, which determine the chemical properties of chlorine. Familiar helium also has two isotopes, shown in Fig. 32-16. The rare ^{3}He is only 0.0001% abundant, but it can be isolated and has important uses in scientific research.

32.8 The Emission and Absorption of Light

All of the many investigations of cathode rays leading to Thompson's discovery of the electron followed from Faraday's invention of the gas discharge tube. As these experiments were taking place in the mid- to late 1800s, a completely separate group of scientists was using the discharge tube for different purposes. Their findings would also, in the early years of the twentieth century, come to bear on the issue of atomic structure.

Faraday had discovered that the discharge tubes exhibit both a *cathode glow*, which is the same for all gases, as well as a *positive column*, as it is called, which glows with bright color and is different for every gas. The bright glow of the positive column was a hindrance to the study of cathode rays, and those investigators learned that they could eliminate the bright glow, and extend the cathode rays, by reducing the gas pressure sufficiently.

But other scientists were intrigued by the brightly colored light of the positive column. What causes it? Why does every gas emit a different color? Can these colors tell us anything about the nature of the atoms and molecules? As luck would have it, Faraday's discovery came just at the time that the interference and diffraction of light were first being understood. The production of diffraction gratings was well underway by mid-century, and these were the ideal tool to study the light emitted by a discharge tube.

Figure 32-17a shows a typical experimental arrangement for recording the *spectrum* of light emitted by a gas. The light is collected by a lens, focused through the entrance slit of a *spectrometer*, then diffracted by a grating—as you studied in Chapter 18. Different wavelengths in the light are diffracted at different angles, then brought to focus on a film or a photographic plate. A modern spectrometer, widely used today in physics, chemistry, and astronomy, is little changed except that the film is replaced with an electronic photodetector.

Examples of *emission spectra* are shown in Fig. 32-18. Each line represents one of the wavelengths of light coming from the discharge. These wavelengths can be measured with extremely high accuracy if the instrument is well calibrated. It was quickly learned that:

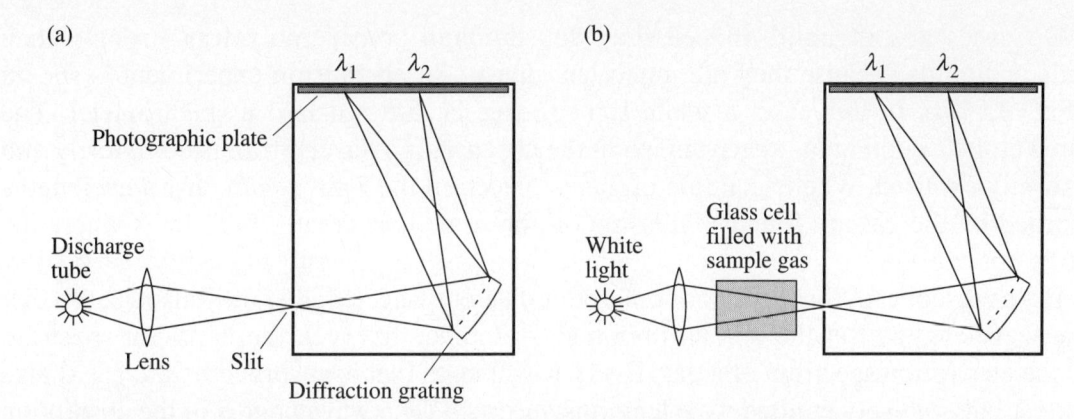

FIGURE 32-17 A grating spectrometer is used to study the spectra of light. a) Recording the emission spectrum from a gas discharge tube. b) Recording the absorption spectrum of a gas.

1. Gases emit a **discrete spectrum**, consisting of discrete, specific wavelengths of light. This is in contrast to the *continuous* rainbow-like spectrum of the sun or an incandescent light source.
2. Every element and every compound emits a unique spectrum.

But substances not only emit light, they can also absorb light. If you look at a light bulb through a piece of red glass, the bulb looks red because the glass *absorbs* the yellow, green, and blue wavelengths of the white light. Only the red wavelengths make it through to be seen. Similarly, grass and leaves appear green because they absorb both red and blue wavelengths (red and blue wavelengths are the ones that drive photosynthesis), reflecting only the green and yellow wavelengths in the center of the visible spectrum. Note, by the way, a peculiarity of the English language: The process of *absorbing* light is called *absorption*, not "absorbtion."

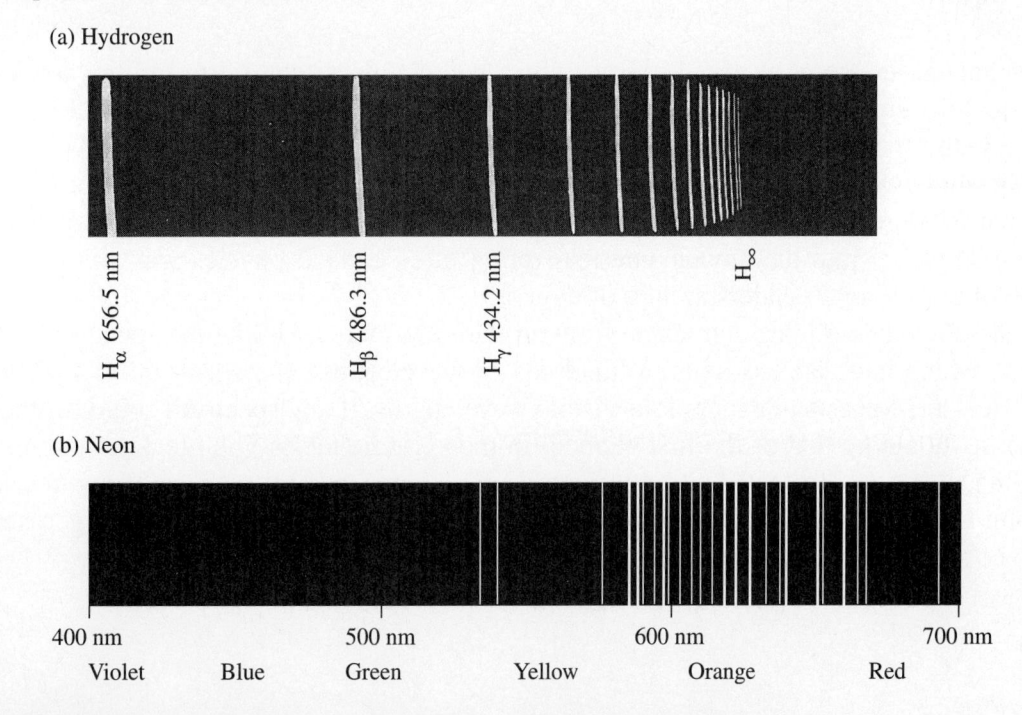

FIGURE 32-18 Sample emission spectra from a) hydrogen and b) neon.

Do gases absorb light? Indeed they do, although considerably less strongly than solids or liquids because they are much less dense. An absorption experiment is shown in Fig. 32-17b. In this case, a white light source is directed into a spectrometer. The source emits a continuous spectrum, so in the absence of a gas the film is completely and uniformly exposed. When a sample of gas is placed in the light's path, any wavelengths absorbed by the gas are found to missing on the film. This creates dark lines where the film is not exposed.

It was discovered that gases not only emit discrete wavelengths, they also absorb discrete wavelengths. But there is an important difference between the emission spectrum and the absorption spectrum of a gas: Every wavelength that is absorbed by the gas is also emitted, but *not* every emitted wavelength is absorbed. The wavelengths in the absorption spectrum appear as a subset of the wavelengths in the emission spectrum. As an example, Fig. 32-19 shows both the emission and the absorption spectra of sodium atoms. Notice that all of the absorption wavelengths are prominent in the emission spectrum but that there are many emission lines for which no absorption occurs.

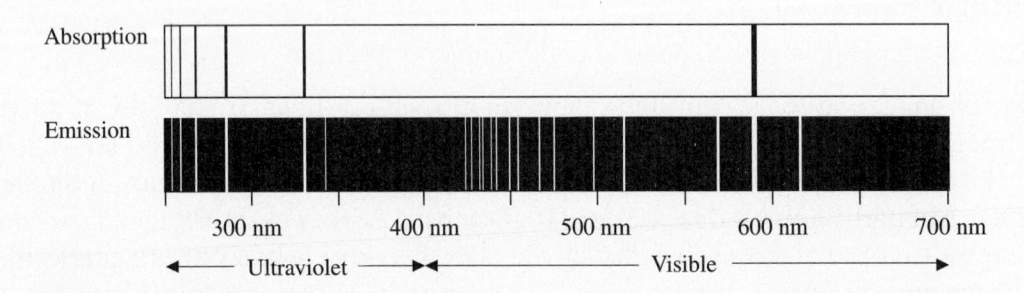

FIGURE 32-19 Absorption and emission spectra of sodium vapor. These are replicas of photographic plates.

What causes atoms to emit or absorb light? Why a discrete spectrum? Why are some wavelengths emitted but not absorbed? Why is every element and compound different? What do the spectra reveal about the nature or the structure of the atoms? These were vexing questions with which nineteenth-century physicists struggled but could not answer. Ultimately, as you will see in more detail later, their inability to answer these questions forced scientists into the unwelcome realization that classical physics was simply incapable of providing an understanding of atoms.

The only encouraging sign came from an unlikely source. Unlike the spectra of other atoms, which have dozens or even hundreds of wavelengths, the spectrum of hydrogen (Fig. 32-18a) consists of a mere four visible wavelengths. If any spectrum could be understood, it should be that of the first element in the periodic table. The breakthrough came in 1885, not by an established and recognized scientist but by a Swiss schoolteacher, Johann Balmer. Balmer showed that the four known wavelengths in the hydrogen spectrum could be represented by the simple formula

$$\lambda = \frac{91.18 \text{ nm}}{\left(\frac{1}{2^2} - \frac{1}{n^2}\right)}, \quad n = 3, 4, 5, 6. \tag{32-7}$$

Balmer's story was told more completely in Section 18.2, to which you should refer.

Later experimental evidence, as ultraviolet and infrared spectroscopy developed, showed that Balmer's result could be generalized to

$$\lambda = \frac{91.18 \text{ nm}}{\left(\frac{1}{m^2} - \frac{1}{n^2}\right)}, \quad m = 1, 2, 3, \ldots \quad n = m+1, m+2, \ldots . \tag{32-8}$$

We now refer to Eq. 32-8 as the **Balmer formula**, although Balmer himself only suggested the original version of Eq. 32-7. Other than at the very highest levels of resolution, where new details appear that need not concern us in this text, the Balmer formula accurately describes *every* wavelength in the emission spectrum of hydrogen.

The Balmer formula is what we call *empirical knowledge*. It is an accurate mathematical representation found empirically—that is, through experimental evidence—but it does not rest on any physical principles or physical laws. No one was able to *derive* Balmer's formula from Newtonian mechanics or the theory of electromagnetism. So although the Balmer formula was useful, no one *understood* where Balmer's formula came from or why it worked. Yet it was so simple that it must, everyone agreed, have a simple explanation. It would take thirty years to find it.

32.9 Classical Physics at the Limit

At the start of the nineteenth century, few scientists believed in the existence of atoms. By century's end, there was substantial evidence not only for atoms but for the existence of charged subatomic particles. The explorations into atomic structure culminated with Rutherford's nuclear model. This model matched the experimental evidence, but it had a serious shortcoming. Electrons, as they orbit the nucleus, are oscillating charged particles. According to Maxwell's theory of electricity and magnetism, these orbiting electrons should act as small antennas and, consequently, should radiate electromagnetic waves. This radiation would be at very high frequencies and would be perceived as visible light. But it was easy to calculate that such an atom would radiate a *continuous* spectrum rather than the observed discrete spectrum of atoms. In addition, the atoms would continuously lose energy as they radiated electromagnetic waves. This would cause the electrons to spiral into the nucleus rather than to move in a fixed orbit. Calculations showed that a Rutherford atom can last no more than about a microsecond before the electrons impact the nucleus!

This clearly does not happen. One of the most notable aspects of matter is its stability. But according to the imposing edifice of classical physics, the Rutherford model of the atom would be highly unstable and would immediately self-destruct. Neither, for that matter, could classical physics explain how the positive charges in the nucleus of a Rutherford atom hold themselves together against the overwhelming repulsive electrostatic force.

The experimental efforts of the nineteenth century had been impressive. There could be no doubt about the existence of electrons, about the small positive nucleus, and about the unique discrete spectrum emitted by each atom. But the theoretical framework for understanding such observations had lagged behind. As the new century dawned, physicists could not explain the structure of atoms, could not explain the stability of matter, could not explain the discrete spectra or why the absorption spectrum differs from the emission spectrum, and could not explain the origin of x rays or radioactivity.

Yet few physicists were willing to abandon the successful and long-cherished theories of classical physics. Despite the attention we have focused on the search for atomic structure, the large majority of scientists were working in other fields—electricity, acoustics, thermodynamics—for which classical physics remained completely satisfactory. Most considered these "problems" with atoms to be "minor discrepancies" that would soon be resolved. But classical physics had, indeed, reached its limit, and a whole new generation of brilliant young physicists, with new ideas, was about to take the stage. Among the first was an unassuming young man in Berne, Switzerland. His scholastic record had been mediocre, and the best job he could find upon graduation was as a clerk in the patent office, examining patent applications. He needed the job, having recently married a fellow student because of, at least in part, a child conceived out of wedlock. His name was Albert Einstein.

Summary

Important Concepts and Terms

electrolysis	alpha rays
cathode	beta rays
anode	nuclear model
gas discharge tube	nucleus
cathode glow	electron volt
cathode rays	atomic number
Crookes tube	proton
crossed-field experiment	mass spectrometer
subatomic particle	isotope
electron	neutron
Millikan oil-drop experiment	discrete spectrum
radioactivity	Balmer formula

It is important for engineers, as well as scientists, to know *why* we accept and use an atomic theory of matter. Understanding the structure and properties of matter in terms of atoms has been a triumph of modern science, and the application of this knowledge is an essential part of contemporary engineering. But unlike Newtonian mechanics, where we can directly experience forces and can measure kinematic quantities with a meter stick and stopwatch, the *evidence* for atoms and their properties is much more indirect. Hence the goal of this chapter has been to acquaint you with some of the evidence by which we know about atoms.

The information and evidence has been presented from a historical perspective. This approach was chosen, in part, because the evidence is easier to understand if you see how it grew out of a scientific framework—mechanics, thermodynamics, electromagnetism—with which you are now fairly familiar. But, in addition, the story reveals many facets of how science works and progresses. You should have noticed several examples of scientific models and theories that, reasonable as they seemed when first proposed, failed to match the subsequent experimental evidence. The route to our present understanding had many more detours and dead ends than is usually recognized.

Two figures dominate the story of the atom: Michael Faraday and J. J. Thompson. Faraday, in a remarkable career, provided convincing evidence for atoms, showed that atoms were somehow associated with electric charges, invented the gas discharge tube that led to the discovery of cathode rays and of the discrete spectra of atoms, and made fundamental discoveries about electricity and magnetism that paved the way to the realization that light is an electromagnetic wave. Thompson was one of the greatest experimentalists of all times. Not only did Thompson himself discover the electron, but many of the other major discoveries of the period were made either in his laboratory or by students he had trained.

In reviewing this chapter you want to be aware of three streams of investigation: searching for the nature of matter, the nature of electricity, and the nature of light. Although initially separate, Faraday showed that these three topics were closely related. Rutherford's nuclear model of the atom, fully uniting matter with electricity, was the culmination of several decades of intense research. But impressive as this was, there remained nagging difficulties. How did these atoms emit and absorb the discrete spectra of light? What was the significance of Balmer's formula for the hydrogen atom? How do the orbiting electrons avoid a death spiral into the nucleus, as predicted by electromagnetic theory? Although Thompson, Rutherford, and their colleagues felt confident that these difficulties could be overcome, new experimental evidence would soon exacerbate rather than ameliorate the situation. Classical physics, in an attempt to understand the atom, had reached its limits.

Exercises and Problems

Exercises

1. What was the significance of Thompson's experiment in which an off-center electrode was used to collect charge deflected by a magnetic field?
2. What is the evidence by which we know that an electron from an iron atom is identical to an electron from a copper atom?
3. Express in eV (or keV or MeV if more appropriate):
   a. The kinetic energy of an electron moving with a speed of 5×10^7 m/s.
   b. The potential energy of an electron and a proton 0.1 nm apart.
   c. The kinetic energy of a proton that has accelerated from rest through a potential difference of 5000 V.
   d. The kinetic energy of a doubly charged Li^{++} ion that has accelerated from rest through a potential difference of 5000 V.
   e. The kinetic energy, just before impact, of a 200 g ball dropped from a height of 1 m.
4. Determine:
   a. The velocity of a 100 eV electron.
   b. The velocity of a 20 MeV neutron.
   c. The specific type of particle that has 2.09 MeV of kinetic energy when moving with a velocity of 1.0×10^7 m/s.

5. Figure 32-18a identified the wavelengths of three lines in the spectrum of hydrogen.
   a. Determine the Balmer formula n and m values for these wavelengths.
   b. Predict the wavelength of the fourth line in the spectrum.
   c. Figure 32-18a also labels a feature called H_∞, although there is no line present at that point. This is called the *series limit* for the series of spectral lines that begins with the three wavelengths of part a). Determine the wavelength of the series limit.

6. Consider the gold isotope ^{197}Au.
   a. How many electrons, protons, and neutrons are in a neutral ^{197}Au atom?
   b. The gold nucleus has a radius of 7 fm. What is the density of matter in a gold nucleus?
   c. The density of lead is 11,400 kg/m^3. How many times the density of lead is your answer to part b)?

7. An electron in a cathode-ray beam passes between two parallel-plate electrodes that are 5 mm apart and 2.5 cm long. A magnetic field of strength 0.002 T and width 2.5 cm is perpendicular to the electric field between the plates. The electron passes through the plates without being deflected if the potential difference across the plates is 600 V.
   a. What is the electron's speed?
   b. If the potential difference across the plates is set to zero, through what angle is the electron deflected as it passes through the magnetic field?

▴ Problems

8. A lithium atom has three electrons. As you will soon discover, two of these electrons form an "inner core," but the third—the valence electron—orbits at much larger radius. From the valence electron's perspective, it is orbiting a spherical ball of charge having net charge $+1e$. The energy required to ionize a lithium atom is 8.22×10^{-19} J. According to the Rutherford model of the atom, what are the orbital radius and speed of the valence electron? (**Hint:** Consider both the energy needed to remove the electron *and* the force needed to give the electron a circular orbit.)

9. Two parallel electrodes are 5 cm long and spaced 1 cm apart. A proton enters the plates from one end and moves along the center line, as shown in Fig. 32-20. A potential difference ΔV is applied to the plates, and it is found that 500 V is just sufficient to cause the proton to strike the outer end of the lower electrode. What magnetic field, both magnitude *and* direction, will allow the proton to pass through undeflected when the 500 V potential is applied? (Assume that both the electric and magnetic fields are confined to the space between the electrodes.)

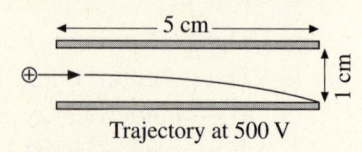

FIGURE 32-20

10. One of Thompson's earliest experiments involved placing a thin metal foil in the electron beam and measuring its temperature rise. Consider a cathode-ray tube in which electrons are accelerated through a 2000 V potential difference, then strike a copper foil of mass 10 mg.
    a. How many electrons strike the foil if the foil temperature rises 6°C in 10 s?
    b. What is the current of the electron beam?

11. An experimental nuclear physicist wants to shoot a proton *into* a ^{197}Au nucleus. In order to initiate the desired nuclear reaction, the proton must impact the nucleus with a kinetic energy of 6.3 MeV. The nuclear radius is 7 fm.
    a. With what speed must the proton be fired toward the target?
    b. Through what potential difference must the proton be accelerated from rest to acquire this speed?

12. Consider an oil droplet of mass m and charge q. We want to determine the charge on the droplet in a Millikan-type experiment. We will do this in several steps. Assume, for simplicity, that the charge is positive and that the electric field between the plates points upward.
    a. An electric field is established by applying a potential difference to the plates. It is found that a field of magnitude E_0 will cause the droplet to be suspended motionless. Write an expression for the droplet's charge in terms of the suspending field E_0 and the droplet's weight mg.
    b. The field E_0 is easily determined by knowing the plate spacing and measuring the potential difference applied to them. The larger problem is to determine the mass of a microscopic droplet. Consider a particle of mass m falling through viscous medium in which there is a retarding force or drag force. Suppose the retarding force is given by $F_{drag} = -bv$ where b is a constant and v the particle's velocity. The sign recognizes that the drag force vector points upward when the particle is falling (negative v). A falling particle quickly reaches a *constant* velocity, called the *terminal velocity*. Write an expression for the terminal velocity v_t in terms of m, g, and b. (**Hint:** What is the net force on a particle falling at constant speed?)
    c. A spherical object of radius r moving slowly through the air is known to experience a retarding force $F_{drag} = -6\pi \eta r v$ where η is the *viscosity* of the air. Use this and your answer to part b) to show that a spherical droplet of density ρ falling with a terminal velocity v_t has a radius

    $$r = \sqrt{\frac{9\eta |v_t|}{2\rho g}}.$$

    d. Oil has a density 860 kg/m^3. An oil droplet is suspended between two plates 1 cm apart by adjusting the potential difference between them to 1177 V. When the voltage is removed, the droplet falls and quickly reaches constant speed. It is timed, with a stopwatch, to fall 3.00 mm in 7.33 s. The viscosity of air is 1.83×10^{-5} kg/m s. What is the droplet's charge?
    e. How many units of the fundamental electric charge does this droplet possess?

[**Estimated 10 additional problems for the final edition.**]

Chapter 33

Waves, Particles, and Quanta

All the fifty years of conscious brooding have brought me no closer to the answer to the question, "What are light quanta?" Of course today every rascal thinks he knows the answer, but he is deluding himself.

Albert Einstein

LOOKING BACK   Sections 16.3–16.5; 17.2; 17.6; 18.3–18.5; 27.4–27.6

33.1 Particles and Waves

Rutherford's nuclear model of the atom, although not proposed until 1911, was the capstone of the nineteenth-century quest for the nature of matter. Although firmly grounded in experimental evidence, Rutherford's model, for reasons articulated in the previous chapter, brought the classical physics of Newton and Maxwell to its limits. The scientific giants of the nineteenth century thought that physics as they knew it, perhaps with slight modifications, would ultimately be able to explain the atom. But the early years of the twentieth century brought in new faces, new experiments, and new ideas. These younger scientists quickly pushed beyond the bounds of classical physics. Nature, at the microscopic level of atoms, turns out to be far too subtle and far too complex to describe with the macroscopic concepts of force and field.

One problem is that classical physics contains an implicit *dichotomy*—a division into two mutually exclusive categories—between particles and waves. The fundamental entities of nature must be one or the other, with no room for a middle ground. Particles are discrete, localized objects, but waves are continuous, extended, and subject to the principle of superposition. Two waves can occupy the same point in space and exhibit interference; two particles cannot occupy the same point in space and do not interfere. At the end of the nineteenth century it seemed clear that the electrical substance consisted of charged *particles*, that atoms were constructed of these same charged particles, and that larger objects—often modeled as a single particle in mechanics—were aggregates of atomic particles. Light and sound, by contrast, exhibited interference and clearly were *waves*.

As we discussed in Chapter 18, Einstein, de Broglie, and others showed that reality is not so simple. The wave–particle dichotomy, as obvious as it may seem, simply does not exist in the microscopic world of atoms. Instead, light and matter exhibit characteristics of *both* particles *and* waves. The new **wave–particle duality**, as it is called, completely defies our common-sense picture of how things "ought" to behave. The quotation by Einstein that opens this chapter shows how hard it was, and is, to grasp the meaning of the new concepts—even by those who invented them!

The experimental evidence for wave–particle duality is now overwhelming and cannot be doubted. Wave-like electrons and particle-like photons of light are no longer just scientific curiosities. Modern engineering devices, such as *quantum-well semiconductors*, make explicit use of wave–particle duality. Yet we still cannot provide a clear picture of just what an electron "is" or of what light "is."

In this chapter we will explore two critical ideas in depth: Einstein's introduction of a particle-like nature for light and de Broglie's suggestion of a wave-like nature for matter. We will begin to see how to think about and describe matter and light without resorting to overly simplified classical models. This chapter will rely heavily on topics that have been developed throughout the text, especially the phenomenon of interference, the concept of the electric potential, and the material in Chapter 18. A review of the sections indicated in the Looking Back section is highly recommended.

33.2 The Photoelectric Effect

In 1886 Heinrich Hertz was the first to demonstrate that electromagnetic waves can be artificially generated. By verifying the predictions of Maxwell's electromagnetic theory, Hertz cemented the last blocks of classical physics into place. Yet in one of those ever-present ironies of history, Hertz happened, quite by chance, to discover the very phenomenon that would bring classical physics down. He noticed, in the course of his investigations, that ultraviolet light could discharge a negatively charged metal electrode.

Hertz noted this observation in his published papers, but he did not pursue it. It did, however, catch the attention of J. J. Thompson. Thompson inferred that the metal, when illuminated with ultraviolet light, must emit negative charges and thus restore itself to electrical neutrality. In 1899, using magnetic deflection techniques similar to those with which he discovered the electron, Thompson showed that the emitted charges had exactly the same charge-to-mass ratio as electrons and, presumably, were electrons. The emission of electrons from a substance due to light striking its surface came to be called the **photoelectric effect**. The emitted electrons are often called *photoelectrons* to indicate their origin, but they are identical in every respect to all other electrons.

Although this discovery might seem to be a minor footnote in the history of science, it soon became a, or maybe *the*, pivotal event that sealed the fate of classical physics and opened the door to new ideas. We will look at the photoelectric effect in a fair bit of detail. Our goals in this section and the next are to understand how classical physics was unable to explain the details of such a simple experiment and to recognize the startling new concept introduced by Einstein.

Characteristics of the Photoelectric Effect

It was not the discovery itself that dealt the fatal blow to classical physics, but the specific characteristics of the photoelectric effect that one of Hertz's students, Philipp Lenard, found during experiments around 1900. Lenard built a glass tube, like that shown in Fig. 33-1, with two facing electrodes and a quartz window, to allow better transmission of ultraviolet wavelengths. He then pumped out as much air from the tube as possible. An adjustable potential difference ΔV was established between the two electrodes, and a meter recorded the current flow through the "circuit." Lenard allowed light to shine on the cathode; then he systematically studied how the photoelectron current varied as the potential difference and the light's wavelength and intensity were changed.

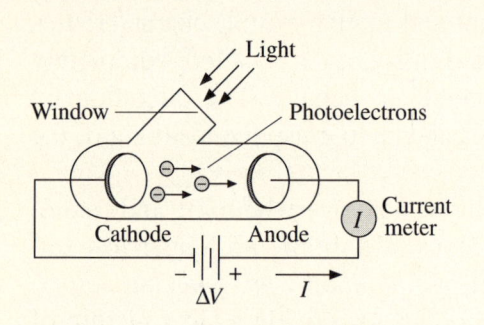

FIGURE 33-1 Lenard's experimental device to study the photoelectric effect.

Lenard found the photoelectric effect to have the following properties:

1. Short-wavelength radiation causes the cathode to emit electrons, but never positive ions. These electrons travel through the vacuum to the anode, where they are collected. The electron motion forms a current which, because there are no junctions, must also flow through the battery and through the meter, where it is measured.

2. The current is directly proportional to the light intensity. If the light intensity is doubled, the current also doubles.

3. Photoelectrons are emitted *only* if the light frequency f exceeds a **threshold frequency** f_0. The value of the threshold frequency depends on the type of metal from which the cathode is made. If $f > f_0$ (short wavelengths), current flows no matter how weak the light's intensity. But if $f < f_0$ (long wavelengths), even if the frequency is very close to f_0, no current flows no matter how intense the light. This is shown in the graph of Fig. 33-2.

4. The current appears without delay when the light is applied. To Lenard, this meant within the ≈ 0.1 s with which his equipment and reactions could respond. Later experiments showed that the current begins within 10^{-9} s of the light!

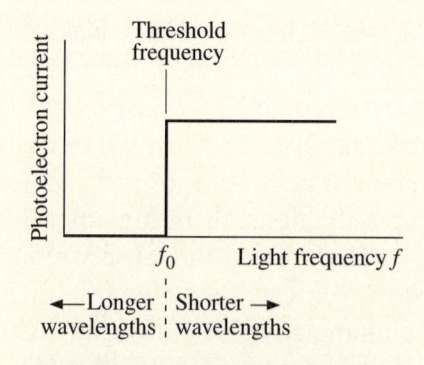

FIGURE 33-2 Photoelectrons are emitted only if the light frequency f exceeds some threshold frequency f_0. The value of f_0 depends on the type of metal from which the cathode is made.

5. If the potential difference ΔV is positive (anode positive with respect to the cathode), the current (for a fixed light intensity) is constant and does not change with ΔV. If ΔV is made negative (anode negative with respect to the cathode), by reversing the battery, the current decreases until, at some specific

value $\Delta V = -V_{\text{stop}}$ the current reaches zero. The value of V_{stop} is called the **stopping potential**. No current flows if $\Delta V < -V_{\text{stop}}$. This behavior is shown in Fig. 33-3.

6. Finally, the value of V_{stop} is the same for both weak light and intense light. More current flows if the light is more intense, as Fig. 33-3 shows, but in both cases the current ceases when $\Delta V = -V_{\text{stop}}$.

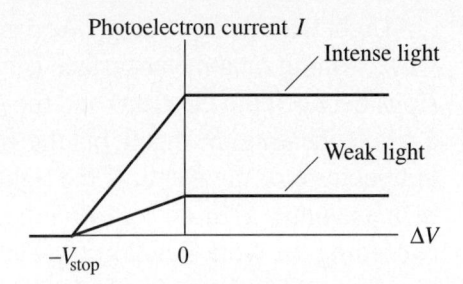

FIGURE 33-3 The current is constant for positive ΔV but stops at a specific negative voltage $\Delta V = -V_{\text{stop}}$.

Classical Interpretation of the Photoelectric Effect

To what extent can we understand and explain these observations? The mere existence of the photoelectric effect is not, as some assume, a difficulty for classical physics. Recall, from Chapter 26, that electrons are the charge carriers in metal and are somehow able to move around freely inside. They are, in some sense, a gas of negatively charged particles. As the electrons move about, they must have some *distribution* of speeds and energies. Some electrons have a speed that is higher than average; others are lower than average.

Electrons do not spontaneously spill out of a piece of metal. To free an electron from the metal, you would have to exert a force on it and pull it—a displacement—through the surface. That is, you would have to do *work* on the electron, increasing its speed until it has enough energy to escape. The *minimum* energy E_0 needed to free an electron that is right at the surface is called the **work function** of the metal. Some electrons, depending on their location, may require more energy than E_0 to escape, but all will require *at least* E_0. Different metals, due to their different densities and crystal structures, have different values of the work function E_0. These values can be found in tables of material properties, and a short list is given in Table 33-1. (Note that the work functions are given in electron volts.)

One way to increase the electron energies is to heat the metal. At a sufficiently high temperature, some electrons will have enough thermal energy to make it "over the wall," leaving the metal. This can happen if a particular electron's thermal energy E_{elec} exceeds the work function: $E_{\text{elec}} \geq E_0$. This **thermal emission** of electrons, which had been discovered by Thomas Edison in his research leading to the incandescent light bulb, requires a temperature $\geq 1500°C$ for most metals. As you can see from Table 33-1, there are only a few elements, such as tungsten, for which the thermal emission of electrons can become significant before the metal melts! (Thermal emission from a hot tungsten filament is the first stage in the *electron gun* of cathode ray tubes in televisions and computer display terminals.)

TABLE 33-1 The work function and melting temperature for some of the elements.

Element	E_0 (eV)*	T_{melt} (°C)
Potassium	2.3	64
Sodium	2.7	98
Aluminum	4.3	660
Tungsten	4.5	3380
Iron	4.7	1530
Copper	4.7	1080
Gold	5.1	1060

*1 eV = 1.60×10^{-19} J

Light waves carry energy. According to Maxwell's electrodynamics, the energy in a light wave shining on a metal surface is absorbed by the *electrons* in the metal. Subsequent collisions between the electrons and the positive ions transfer some of this energy to the metal as a whole, heating the metal, but the *electron temperature* may be significantly higher than the temperature of the metal. If the light is sufficiently intense, the electron temperature should increase until, after enough time has elapsed, a few of the electrons have a thermal energy exceeding the work function E_0. These electrons will escape, becoming photoelectrons. This is similar to the thermal emission of electrons, but there is one major difference: Light waves heat primarily the electrons, whereas the entire metal must be heated to cause thermal emission. In 1900 it was reasonable to think that an intense light source could heat the electrons enough to cause photoemission but without melting the metal.

Once photoelectrons leave the metal cathode, they will move out in all directions. Some of the electrons might strike the anode, creating a measurable current, but many would not. But if the anode is biased positively with respect to the cathode, the electric field will steer *all* of the photoelectrons to the anode. Because all of the electrons are collected, the current should be the same for any positive value of ΔV. This is what is seen on the right side of the Fig. 33-3 graph. It should also be no surprise that a more intense light wave will deposit more energy, free more electrons, and thus generate a larger current.

It is important to realize that photoelectrons leave the cathode with kinetic energy. An electron having thermal energy $E_{elec} > E_0$ inside the metal must lose an amount of energy ΔE to escape, so it emerges as a photoelectron with kinetic energy $K = E_{elec} - \Delta E$. Recall that it takes *at least* the work function energy E_0 to escape, so $\Delta E \geq E_0$. Thus an electron with thermal energy E_{elec} will leave the cathode as a photoelectron with kinetic energy

$$K \leq E_{elec} - E_0. \qquad (33\text{-}1)$$

Notice that there is a *range* of kinetic energies, depending on how much energy was required to free each photoelectron, but the *maximum* kinetic energy is $K_{max} = E_{elec} - E_0$.

A photoelectron with sufficient initial kinetic energy will reach the anode even if the anode is negative. A slightly negative anode voltage repels the slowest electrons but allows the faster ones to get through. The current is less than that obtained for a positive ΔV, where *all* electrons are collected, but some current continues to flow due to the faster electrons. The current will steadily decrease as the voltage becomes increasingly negative, because fewer and fewer photoelectrons will have enough kinetic energy to get through. Finally, at the stopping potential, the anode's voltage is so negative that *all* electrons are turned back and no current flows. This is the behavior observed on the left side of the Fig. 33-3 graph.

We can analyze this process more carefully. Let the cathode be the point of zero potential energy, as shown in Fig. 33-4. An electron emitted from the cathode with kinetic energy K_i has an initial total energy

$$E_i = K_i + U_i = K_i + 0 = K_i.$$

When the electron reaches the anode, which is at potential ΔV relative to the cathode, it has potential energy $U = q\Delta V = -e\Delta V$. So an electron reaching the anode has a final total energy

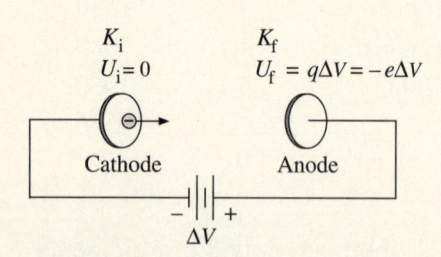

$$E_f = K_f + U_f = K_f - e\Delta V.$$

FIGURE 33-4 Energy transfer of a photoelectron as it moves from the cathode to the anode.

We can use conservation of energy to find the electron's final kinetic energy:
$$E_f = E_i$$
$$\Rightarrow K_f - e\Delta V = K_i \qquad (33\text{-}2)$$
$$\Rightarrow K_f = K_i + e\Delta V.$$

The electron speeds up ($K_f > K_i$) if ΔV is positive. The electron slows down if ΔV is negative, but it still reaches the anode ($K_f > 0$) if K_i is large enough. The specific potential difference $\Delta V = -K_i/e$ will stop an electron just as it reaches the anode.

EXAMPLE 33-1 A photoelectric effect experiment is performed with an aluminum cathode. An electron inside the cathode has a speed of 1.5×10^6 m/s. If the potential difference between the anode and cathode is -2.0 V, what is the fastest possible speed with which this electron could reach the anode?

SOLUTION If this electron succeeds in escaping as a photoelectron, its *maximum* possible kinetic energy is $K_{max} = E_{elec} - E_0$, where $E_0 = 4.3$ eV is the work function of aluminum. If we assume that the electron escapes with the maximum possible kinetic energy, its kinetic energy at the anode will be given by Eq. 33-2, with $\Delta V = -2.0$ V.
The electron's initial energy is
$$E_{elec} = \tfrac{1}{2}mv^2 = \tfrac{1}{2}(9.11 \times 10^{-31} \text{ kg})(1.5 \times 10^6 \text{ m/s})^2 = 1.03 \times 10^{-18} \text{ J} = 6.40 \text{ eV}.$$

Its maximum kinetic energy as it leaves the cathode is
$$K_i = K_{max} = E_{elec} - E_0 = 2.1 \text{ eV},$$
leading to a kinetic energy at the anode of
$$K_f = K_i + e\Delta V = 2.1 \text{ eV} - (e)(2.0 \text{ V}) = 0.1 \text{ eV}.$$
Notice that the electron loses 2.0 eV of *energy* as it moves through the potential difference of -2.0 V, so we can compute the final kinetic energy in eV without having to convert to joules. We must, though, convert K_f to joules to find the final speed:
$$K_f = \tfrac{1}{2}mv_f^2 = 0.1 \text{ eV} = 1.6 \times 10^{-20} \text{ J}$$
$$\Rightarrow v_f = \sqrt{\frac{2K_f}{m}} = 1.9 \times 10^5 \text{ m/s}.$$

Although the electrons are emitted with a distribution of kinetic energies, there is some maximum K_{max} that is not exceeded. That is, the photoelectrons all have $K_i \le K_{max}$. The potential difference that will succeed in turning back the fastest electrons, with $K = K_{max}$, is
$$\Delta V_{\text{stop fastest electrons}} = -\frac{K_{max}}{e}.$$

But, by definition, the potential difference that causes the photoelectron current to cease is $\Delta V = -V_{stop}$, where V_{stop} is the stopping potential. Thus
$$V_{stop} = \frac{K_{max}}{e}. \qquad (33\text{-}3)$$

This is the stopping potential seen in Fig. 33-3. It is the voltage that stops the very fastest of the electrons, and it tells us directly the maximum kinetic energy of the photoelectrons.

Limits of the Classical Interpretation

A purely classical analysis of the electrons has provided a satisfactory explanation of observations number 1, 2, and 5 above. But nothing in this explanation suggests that there should be a threshold frequency. If even a weak intensity at a frequency just slightly above f_0 can generate a current, then certainly a strong intensity at frequency just slightly below f_0 should be able to do so. There is no reason that a very slight change in frequency should matter, but Lenard found there to be a very sharp threshold at f_0.

And what about his observation that the current starts instantly? It should take some length of time for the light to heat the electrons sufficiently for some to escape. In fact, fairly straightforward calculations show that, for a light of modest intensity, it should take several minutes before current starts flowing! The experimental evidence was in sharp disagreement: If $f > f_0$, the current starts instantly for both weak light and intense light.

And lastly, more intense light would be expected to heat the electrons to a higher temperature. This should increase the maximum kinetic energy of the photoelectrons and thus should increase the stopping potential V_{stop}. But as Lenard found, the stopping potential is the same for strong light as it is for weak light.

Although the mere presence of photoelectrons did not seem surprising, classical physics was unable to explain the observed behavior of the photoelectrons. The threshold frequency and the instantaneous current seemed particularly anomalous. The photoelectric effect seemed to suggest, at least to a few inquisitive minds, that an entirely new approach was called for.

33.3 Einstein's Explanation

In 1905 Albert Einstein was a little-known young man of 26. A photograph from the time, shown in Fig. 33-5, bears little resemblance to the familiar picture of a white-haired older Einstein. He had recently graduated from the Polytechnic Institute in Zurich, Switzerland, with the Swiss equivalent of a Ph.D. in physics. Although his mathematical brilliance was recognized, his overall academic record was mediocre. Rather than pursue an academic career, Einstein took a job with the Swiss Patent Office in Berne. This was a fortuitous choice because it provided him with plenty of spare time to think about physics in his own unique way.

Einstein found certain aspects of Maxwell's electromagnetic theory puzzling, even paradoxical, when applied to the nature of light. Although other scientists of the day had brushed these aside as of little consequence, Einstein

FIGURE 33-5 A young Einstein.

recognized that they went right to the foundations of classical physics. In 1905, within the span of a single year, Einstein published three papers on three different topics, all three of which would revolutionize physics.

The first was his initial paper on the theory of relativity, the subject with which Einstein is most associated in the public mind. Interestingly, this paper received far less attention at the time than the other two. The second paper explained a phenomenon called *Brownian motion*. An English botanist, Robert Brown, had used a microscope to examine small pollen grains suspended in water. He had observed, in 1827, that the pollen grains jiggled about rather than remained at rest. Einstein, using the techniques of statistical mechanics, provided a convincing explanation that the jiggling resulted from the continual random collisions of water molecules with the pollen grain. His analysis provided one of the most definitive pieces of evidence for the reality of atoms and molecules.

But it is Einstein's third paper of 1905, on the nature of light, in which we are most interested. In it he offered an exceedingly simple, but amazingly bold, idea to explain the photoelectric effect data of Lenard. A few years earlier, in 1900, the German physicist Max Planck had been trying to understand the details of the rainbow-like spectrum of light emitted by a glowing, incandescent object. This problem didn't yield to a classical physics analysis, but Planck found that he could calculate the spectrum exactly if he made an unusual assumption. The atoms in a solid vibrate back and forth around their equilibrium positions with frequency f. An oscillator's energy, as you learned in Chapter 13, depends on its amplitude and can have *any* possible value. But to predict the spectrum correctly, Planck had to assume that the oscillating atoms are *not* free to have any possible energy. Instead, the energy of a vibrating atom has to be one of the specific energies $E = 0$, hf, $2hf$, $3hf$, etc., where h is a constant. That is, the vibrational energies are *quantized*, and in-between values of the energy cannot exist.

Planck was able to determine the value of the constant h by comparing his calculations of the spectrum to experimental measurements. The constant that he introduced into physics is now called **Planck's constant**. Its contemporary value is

$$h = 6.63 \times 10^{-34} \text{ J s} = 4.14 \times 10^{-15} \text{ eV s}.$$

The first value, with SI units, is the proper one for most calculations, but we will find the second to be useful when energies are expressed in eV.

Planck considered this to be a "trick." He was skeptical that the atoms *really* had quantized energy, and he felt sure that classical physics would soon reveal how this trick worked. Einstein was the first to take Planck's idea seriously and to suggest that the quantization is real. Einstein went even further and suggested that *electromagnetic radiation itself is quantized!* That is, light is not really a continuous wave but, instead, arrives in small packets or bundles of energy. Einstein called each packet of energy a **light quantum**, and he postulated that the energy of one light quantum is directly proportional to the frequency of the light. That is, each quantum of light has an energy

$$E = hf, \tag{33-4}$$

where h is Planck's constant and f is the frequency of the light.

The idea of light quanta is subtle, so let's look at an analogy with raindrops. Although water can exist as a continuous fluid in some circumstances, such as in a beaker, rain consists of water that falls in discrete packets called raindrops. Raindrops are analogous to quanta of light. A downpour has a torrent of raindrops, but in a light shower the drops are few. The difference between "intense" rain and "weak" rain is the *rate* at which the drops arrive. An intense rain makes a continuous noise on the roof, so you are not aware of the individual drops, but the individual drops become apparent during a light rain. Similarly, intense light consists of a great number of quanta arriving each second, but weak light consists of only a few quanta per second. Raindrops can also come in different sizes, and drops with larger mass fall with larger kinetic energy. Likewise, higher frequency light quanta carry a larger amount of energy. Although this analogy is not perfect, it does provide a useful mental picture of light quanta arriving at a surface.

Einstein framed three postulates about light quanta and their interaction with matter:

1. Light of frequency f consists of individual, discrete quanta, each of energy $E = hf$.
2. Light quanta are emitted or absorbed on an all-or-nothing basis. A substance can emit 1 or 2 or 3 quanta, but not 1.5. Similarly, the electrons in a metal cannot absorb half a quantum but, instead, only an integer number.
3. A light quantum, when absorbed by a metal, delivers its entire energy to *one* single electron. The light's energy is transformed into the electron's thermal energy.

These three postulates—that light comes in chunks, that the chunks cannot be divided, and that the energy of one chunk is delivered to one electron—are crucial for understanding the new ideas that will lead to quantum physics. They are completely at odds with the concepts of classical physics, where energy can be continuously divided, so they deserve careful thought.

Now let's look at how Einstein's postulates apply to the photoelectric effect. If Einstein is correct, the light shining on the metal is a torrent of light quanta, each of energy hf. Each quantum is absorbed by *one* electron, giving that electron an energy $E_{elec} = hf$. (The electron's thermal energy at room temperature is so much less than hf that we can neglect it.) This leads us to several interesting conclusions:

1. As the classical analysis showed, an electron with sufficient thermal energy can "escape" from the metal, becoming a photoelectron. To do so, its energy E_{elec} must exceed the work function E_0. An electron that has just absorbed a quantum of light energy has $E_{elec} = hf$. This electron can be ejected as a photoelectron if

$$[E_{elec} = hf] \geq E_0. \tag{33-5}$$

Thus a photoelectron current will flow if and only if the frequency of the light satisfies the inequality

$$f \geq \frac{E_0}{h} = f_0. \tag{33-6}$$

In other words, there is a *threshold frequency* $f_0 = E_0/h$ for the ejection of a photoelectron. If f is less than f_0, even by just a small amount, none of the electrons will have sufficient energy to escape no matter how intense the light. But even very weak light with $f \geq f_0$ will give a few electrons sufficient energy to escape *because each light quantum delivers all of its energy to one electron*. This threshold behavior is exactly what Lenard observed.

Note that the threshold frequency is directly proportional to the work function. Metals with large work functions, such as iron, copper, and gold, need high frequencies, and thus short-wavelength ultraviolet light, to exhibit the photoelectric effect. Photoemission occurs with lower-frequency visible light for metals with smaller values of E_0, such as sodium and potassium.

2. A more intense light delivers a larger number of light quanta to the surface. These quanta eject a larger number of photoelectrons and cause a larger current, exactly as observed.

3. We found in the previous section that the kinetic energy of a photoelectron is $K \leq E_{elec} - E_0$. There is a distribution of kinetic energies, because different photoelectrons require different amounts of energy to escape, but the *maximum* kinetic energy is

$$K_{max} = E_{elec} - E_0 = hf - E_0. \tag{33-7}$$

As we noted in Eq. 33-3, the stopping voltage V_{stop} measures K_{max}. If Einstein is right, we can predict that the stopping voltage is related to the light frequency by

$$V_{stop} = \frac{K_{max}}{e} = \frac{hf - E_0}{e}. \tag{33-8}$$

Notice that the stopping voltage does *not* depend on the intensity of the light. Both weak light and intense light will have the same stopping voltage, as Lenard had observed but which could not previously be explained.

4. According to classical physics, the electrons should increase their thermal energy slowly as light continues to fall on the surface. Only after some delay, which depends on the intensity, would any electron have sufficient energy to escape. But if each light quantum transfers its energy hf to just one electron, that electron *immediately* has enough energy to escape. The current should start flowing instantly, with no delay—again, exactly has Lenard had observed.

Not only do Einstein's hypotheses explain all of Lenard's observations, they make a new prediction that can be tested. According to Eq. 33-8, the stopping voltage should be a linearly increasing function of the light's frequency f. Using the threshold frequency $f_0 = E_0/h$, we can rewrite Eq. 33-8 as

$$V_{stop} = \frac{h}{e} \cdot (f - f_0). \tag{33-9}$$

A graph of V_{stop} versus f should start from zero, when $f = f_0$, then rise linearly with a slope of h/e. In fact, the slope of the graph provides a way to measure Planck's constant h.

Lenard had not measured the stopping voltage for different frequencies, so Einstein offered this as an untested prediction of his postulates. Confirming, or disproving, his prediction turned out to be a major challenge. Metals exposed to air have all kinds of contaminants and oxides on the surface, which can alter the work function. These difficulties stymied attempts to obtain accurate data on the stopping potential. Finally, in 1915, Robert Millikan—of the Millikan oil-drop experiment—devised a procedure by which he could cut soft alkali metals, such as sodium and potassium, by remote control *after* they were sealed in a vacuum tube. This exposed a clean metal surface with which he could experiment for a few hours before it began to oxidize.

Some of Millikan's data for a cesium cathode are shown in Fig. 33-6. As you can see, Einstein's prediction of a linear relationship between f and V_{stop} was fully confirmed. The threshold frequency for this sample is seen to be 4.39×10^{14} Hz, corresponding to a red-light wavelength of 683 nm. Light quanta, whether physicists liked the idea or not, were real. In Millikan's own words:

[Einstein's explanation of 1905] ignored and indeed seemed to contradict all the manifold facts of interference and thus to be a straight return to the corpuscular theory of light which had been completely abandoned since the time of Young and Fresnel. I spent 10 years of my life testing the 1905 equation of Einstein's, and, contrary to all my expectations, I was compelled in 1915 to assert its unambiguous experimental verification in spite of all its unreasonableness since it seemed to violate everything we knew about the interference of light.

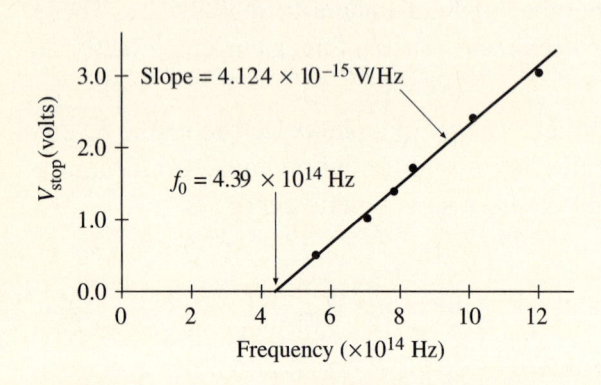

FIGURE 33-6 A graph of Millikan's data for the stopping potential as the light frequency is varied.

Millikan, as we see, had not really believed Einstein. The wave theory of light seemed far too well established to allow for Einstein's unorthodox ideas. But as an indication of what a remarkable scientist Millikan was, he was willing to spend ten years testing a theory that he found quite dubious! But the experimental evidence won out in the end. Einstein was right!

Millikan measured the slope of his graph ($\Delta V_{stop}/\Delta f$), which should be h/e, and multiplied it by the value of e—which he had measured a few years earlier in the oil-drop experiment—to find h. His value agreed with the value that Planck has determined in 1900 from an entirely different experiment.

We can now look at several examples of these ideas.

• **EXAMPLE 33-2** What is the energy of one quantum of light having a wavelength 500 nm?

SOLUTION Light of 500 nm wavelength has frequency

$$f = \frac{v}{\lambda} = \frac{c}{\lambda} = \frac{3.00 \times 10^8 \text{ m/s}}{500 \times 10^{-9} \text{ m}} = 6.00 \times 10^{14} \text{ Hz}.$$

One light quantum has energy

$$E = hf = 3.98 \times 10^{-19} \text{ J} = 2.49 \text{ eV}.$$

Because this is a typical wavelength (green-colored light), you can see how the electron volt is a unit of more appropriate size than the joule.
•

EXAMPLE 33-3 What are the threshold frequencies and wavelengths for photoemission from sodium and from aluminum?

SOLUTION From Table 33-1 we find that sodium has a work function $E_0 = 2.7$ eV and aluminum has $E_0 = 4.3$ eV. Using Eq. 33-6, along with h in units of eV s, we can calculate

$$f_0 = \frac{E_0}{h} = \begin{cases} 6.5 \times 10^{14} \text{ Hz} & \text{sodium} \\ 10.4 \times 10^{14} \text{ Hz} & \text{aluminum.} \end{cases}$$

These frequencies are converted to wavelengths with $\lambda = c/f$, giving

$$\lambda = \begin{cases} 460 \text{ nm} & \text{sodium} \\ 290 \text{ nm} & \text{aluminum.} \end{cases}$$

The photoelectric effect can be observed with sodium for $\lambda < 460$ nm ($f > 6.5 \times 10^{14}$ Hz), which includes blue and violet visible light but not red, orange, yellow, or green. Aluminum, with a larger work function, needs ultraviolet wavelengths $\lambda < 290$ nm.

EXAMPLE 33-4 What is the maximum photoelectron speed if sodium is illuminated with light of 300 nm?

SOLUTION The light frequency is $f = c/\lambda = 1.00 \times 10^{15}$ Hz, giving each quantum an energy $hf = 4.14$ eV. The maximum kinetic energy of a photoelectron is

$$K_{max} = hf - E_0 = 4.14 \text{ eV} - 2.7 \text{ eV} = 1.44 \text{ eV} = 2.30 \times 10^{-19} \text{ J}.$$

Because $K = \frac{1}{2}mv^2$, the maximum speed of a photoelectron leaving the cathode is

$$v_{max} = \sqrt{\frac{2K_{max}}{m}} = 7.11 \times 10^5 \text{ m/s,}$$

where m is the electron's mass, not the mass of the sodium atom. Note that we had to convert K_{max} to SI units of joules before calculating a speed in m/s.

33.4 Photons

Einstein was awarded the Nobel Prize in 1921 not for his theory of relativity, as many would suppose, but for his explanation of the photoelectric effect. The success of light quanta marked the end of classical physics. Energy was somehow quantized and light, despite the fact that it exhibited interference, came in some kind of particle-like packets of energy. Only many years later were these fundamental units of light given the name **photons**.

But just what are photons? Although particle-like, they clearly do not mesh with the classical idea of a particle. "Real" particles, when faced with Young's two-slit apparatus, would go through one hole or the other and make two light spots on the screen. What really happens, of course, is the appearance of interference fringes. So light seems to be *both* wave-like *and* particle-like at the same time. We observed, back in Chapter 18, that the usual interference pattern can be built up photon by photon if the light intensity is reduced to the point at which only one photon at a time is traversing the apparatus. This seems to indicate that a photon must, in some sense, go through *both* slits and interfere with itself! This is clearly not a classical particle.

A **wave packet**, such as shown in Fig. 33-7, is one way to think about a photon. It does have a wavelength and a frequency, yet it is also discrete and fairly localized. But this cannot be exactly what a photon is, because a wave packet would take a finite amount of time to be emitted or absorbed. This is contrary to much evidence that the entire photon is emitted or absorbed in a single instant; there is no point in time at which the photon is "half absorbed." The wave packet idea, although useful, is still too "classical" to represent a photon.

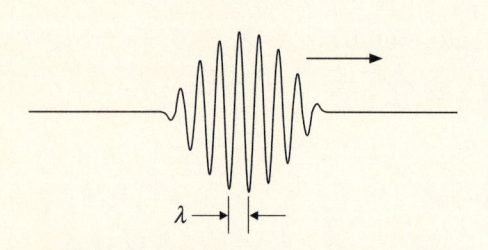

FIGURE 33-7 A wave packet has wave-like and particle-like properties.

The bottom line is that there simply is no "true" mental representation of a photon—hence Einstein's quote at the beginning of this chapter. Analogies such as raindrops or wave packets can be useful, but none are perfectly accurate. We can detect photons, measure the properties of photons, and put photons to practical use, but the ultimate nature of the photon remains a mystery. To paraphrase Gertrude Stein, "A photon is a photon is a photon."

Light, in the raindrop analogy, consists of a stream of photons. For monochromatic light of frequency f, N photons have a total energy $E_{light} = Nhf$. We are usually more interested in the *power* of the light, or the rate (in watts = joules per second) at which the light energy is being delivered. The power is

$$P = \frac{dE_{light}}{dt} = \frac{dN}{dt} \cdot hf = Rhf, \qquad (33\text{-}10)$$

where $R = dN/dt$ is the *rate* at which photons arrive or, equivalently, the number of photons per second.

• **EXAMPLE 33-5** The 1 mW light beam of a helium-neon laser ($\lambda = 633$ nm) shines on a screen. How many photons strike the screen each second?

SOLUTION The light-beam power, or energy delivered per second, is $P = 1$ mW $= 0.001$ W—very typical of real helium-neon lasers. The frequency of the light is $f = c/\lambda = 4.74 \times 10^{14}$ Hz. The number of photons striking the screen per second, which is the *rate* of arrival of photons, is

$$R = \frac{P}{hf} = 3.2 \times 10^{15} \text{ photons per second.}$$

That is a lot of photons per second, and a typical 60 W or 100 W light bulb emits far more. No wonder that we are not aware of individual photons!

•

Modern electronic photodetectors are descendants of the photoelectric effect. These range from simple "electric eyes" to the detector array in a video camera. Many of these detectors now use what is called a *photodiode*, in which the photoelectrons are emitted internally in a semiconductor, but they still have a threshold frequency, a stopping potential, and the other attributes of the photoelectric effect.

Very low light levels can be detected photon by photon with a device called a *photomultiplier tube*, or PMT. Figure 33-8a shows that a PMT consists of a cathode, an anode, and a number of intermediate electrodes sealed inside an evacuated glass tube. The cathode is coated with a low-work-function material, allowing it to respond to most or all visible wavelengths, and the tube is designed so that a "window" directs light onto the cathode. The cathode is at a fairly high negative voltage, whereas the anode, at the other end, is at essentially zero volts. The intermediate electrodes have steadily descending potentials, as shown in the figure.

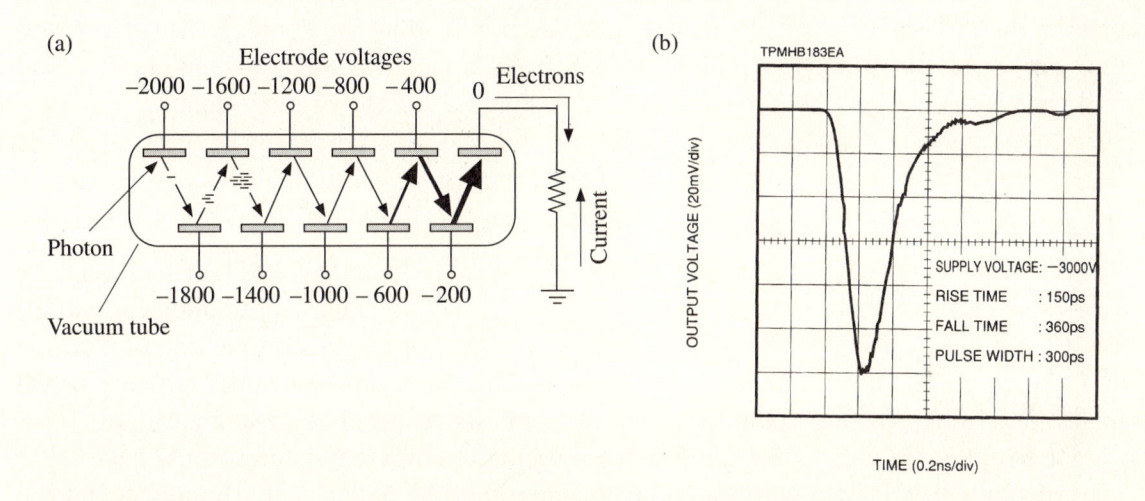

FIGURE 33-8 a) A photomultiplier tube. A photon ejects an electron from the cathode, via the photoelectric effect, and this electron is "multiplied" via secondary emission as it cascades from electrode to electrode. b) The voltage pulse due to a single photon as the output electrons flow through the resistor.

A photon of light ejects a photoelectron from the cathode. The electric field between the cathode and the first intermediate electrode accelerates that electron through a potential difference of about 200 V, and it then strikes this electrode at high speed. When a fast electron collides with a metal surface, it can kick out two or three other electrons called *secondary electrons*. The secondary electrons of the first electrode are accelerated to the second electrode, where they kick out more electrons that are accelerated to the third electrode, where they kick out yet more electrons that are accelerated to the fourth electrode, and so on. There is, in effect, a chain-reaction *multiplication* of electrons—1, 2, 4, 8, 16, etc.—as they move from the cathode toward the anode. The actual progression is not this smooth, because some electrons eject two secondary electrons, others one or three. But the number of electrons in the "bunch" does grow exponentially as they move down the tube. For a typical PMT, a single photon at the cathode leads to 10^6 or 10^7 electrons at the anode.

These electrons are collected by the anode, move through the metal wires, and pass through an external resistor. Because these are negative charge carriers, we would say that a current pulse *I* travels upward, the opposite direction, through the resistor. This creates a *negative* voltage across the resistor, given by $\Delta V = IR$, for the length of time that the current lasts. Figure 33-8b shows an actual measured pulse generated by a single photon. The horizontal scale is 0.2 ns/division, and the vertical scale is 20 millivolts (mV)/division. You can see that the "width" of the pulse is ≈ 0.3 ns and its "height" (measured downward from the baseline) is ≈ 120 mV = 0.12 V. This is not a large voltage, even after the multiplication, but it is a voltage easily detected with modern electronics.

The 0.3 ns that the pulse lasts is *not* an indication of how long the photon was. The photon absorption is instantaneous, but as the electron bunch grows in size the electron-electron repulsion causes the bunch to spread out some. The observed width is an artifact of the PMT, not a characteristic of the photon.

Photomultiplier tubes are interesting. You can "see" the detection of a single photon, making their existence convincing. They also nicely illustrate the photoelectric effect, and they give us an opportunity to practice thinking about electric potentials and circuits. You will have a chance, in a homework problem, to analyze a PMT in more detail.

33.5 De Broglie's Hypothesis: Matter Waves

Prince Louis-Victor de Broglie was a French graduate student in 1924. It had been 19 years since Einstein had shaken the world of physics with his ideas about photons and relativity. Einstein's introduction of light quanta blurred the distinction between a particle and a wave. As he thought about these issues, de Broglie began to wonder whether it still made sense to treat atomic particles, such as electrons, as purely classical particles.

De Broglie was merely speculating, but it seemed to him that nature should have some kind of symmetry. If light waves could have a particle-like nature, why shouldn't material particles have some kind of wave-like nature? In other words, could **matter waves** exist?

With no experimental evidence to go on, de Broglie reasoned by analogy with Einstein's equation $E = hf$ for the photon and with some of the ideas of his theory of relativity. The details need not concern us, but they led de Broglie to postulate that *if* a material particle of momentum $p = mv$ has a wave-like nature, its wavelength must be given by

$$\lambda = \frac{h}{p} = \frac{h}{mv}, \qquad (33\text{-}11)$$

where *h* is Planck's constant. This wavelength is called the **de Broglie wavelength**.

What would it mean for matter—an electron or a proton or a baseball—to have a wavelength? Would it obey the principle of superposition? Would it exhibit interference and diffraction? What is the medium of a matter wave? That is, what is doing the waving? (This, you will recall, was a serious question for light waves, finally resolved when it was discovered that they are waves in the electromagnetic field. But no other fields were known that could be the basis for matter waves.) And finally, if there was anything to de Broglie's suggestion, why had all the decades of research on cathode rays and atoms not revealed any evidence for wave-like aspects of matter? Although a clever idea, it seemed destined for quick oblivion.

EXAMPLE 33-6 What would be the wavelength of a 1 μm diameter oil droplet, such as Millikan observed in his experiment, moving with a speed of 100 μm/s? The density of oil is 900 kg/m^3.

SOLUTION De Broglie's hypothesis favors very light, very slow particles. If there was any hope of seeing wave-like aspects of macroscopic particles, Millikan's tiny oil drops would be likely candidates. The mass of a droplet is $m = \rho V$ where ρ is the density and $V = \frac{4}{3}\pi r^3$ is the volume. Calculation gives $m = 4.7 \times 10^{-16}$ kg—very tiny indeed. The speed of 100 μm/s converts to 10^{-4} m/s. The wavelength would thus be

$$\lambda = \frac{h}{mv} = 1.4 \times 10^{-14} \text{ m}.$$

This is roughly the size of an atomic nucleus and a factor of 10 million smaller than the wavelengths of visible light. There was not the slightest chance then, or even now, to observe wave-like properties for a wavelength this small.

EXAMPLE 33-7 What would be the wavelength of a 1 eV electron?

SOLUTION The energy 1 eV = 1.6×10^{-19} J is the kinetic energy $K = \frac{1}{2}mv^2$. Solving for the speed of such an electron gives

$$v = \sqrt{\frac{2K}{m}} = 5.9 \times 10^6 \text{ m/s}.$$

Although fast by macroscopic standards, this is a "slow electron" because it gains this speed by accelerating through a potential difference of a mere 1 volt. Its wavelength would be

$$\lambda = \frac{h}{mv} = 1.2 \times 10^{-9} \text{ m}.$$

The electron's predicted wavelength is small, but it is larger than the wavelengths of x rays and larger than the $\approx 10^{-10}$ m spacing of atoms in a crystal.

X-ray diffraction was known and understood by 1924, so observing the possible wave-like nature of electrons was not out of the question. De Broglie did not have to wait long. As we discussed in Chapter 18, the physicists Clinton Davisson and Lester Germer, at the Bell Telephone Laboratories in New York, had been studying the scattering of electrons by metal surfaces. They had begun this line of research in 1919, but nothing out of the ordinary happened until 1927, when they introduced a well-crystallized piece of nickel as their target. Metals exhibit a crystal lattice structure on the microscopic scale, but usually these crystals become fractured and disorganized when the metal condenses from the liquid. Special techniques are needed to preserve the crystalline order over macroscopic distances. Davisson and Germer had been interested in how electrons scatter from the surface, so the crystalline structure of the target had not been of interest. But something entirely new happened when they fired electrons at a good, well-formed crystal. Instead of seeing electrons scatter as if they were little particles, as they always had before, Davisson and Germer found multiple minima and maxima of scattering intensity. The electron behavior was very reminiscent of x-ray diffraction from crystals, and further analysis soon showed that the electrons were *diffracting* from the crystal with exactly the wavelength predicted by de Broglie.

The Davisson–Germer experiment was the first experimental evidence that matter can, indeed, act like a wave. De Broglie was correct—an electron, strange as it seems, could somehow be "divided" and later "recombined" so as to interfere with itself! Although we still refer to electrons and protons as subatomic *particles*, this no longer means a classical, well-defined particle but, instead, a fundamental unit of matter that exhibits both particle-like and wave-like behavior.

Modern Evidence of Matter Waves

The experimental evidence for matter waves has increased greatly since 1927, and it is interesting to look at some more modern examples. Figure 33-9a shows the pattern recorded when 50 keV electrons were passed through a double slit. The slits were 0.3 μm wide and spaced 1.0 μm apart in a 20 nm thick silver foil, formed with techniques similar to those used in the production of integrated circuits. The pattern is clearly a two-slit interference pattern, and the spacing of the fringes is exactly as predicted for a wavelength given by de Broglie's formula. Because the electron beam was weak, with a one electron at a time passing through the apparatus, it would appear that each electron somehow went through both slits, then recombined to interfere with itself!

Figure 33-9b shows the results from a similar two-slit experiment with neutrons. Although neutrons are much more massive than electrons, which would tend to decrease their wavelength, it is possible to generate "ultracold" neutrons that move very slowly. The neutrons used in this experiment had a wavelength of ≈ 2 nm, roughly a factor of 10 larger than atomic spacings and similar to the 1 eV electron we examined in Example 33-7. The results, once again, are exactly as predicted for a wavelength given by $\lambda = h/mv$. The wave-like properties of matter are not restricted just to electrons but extend to more massive particles as well.

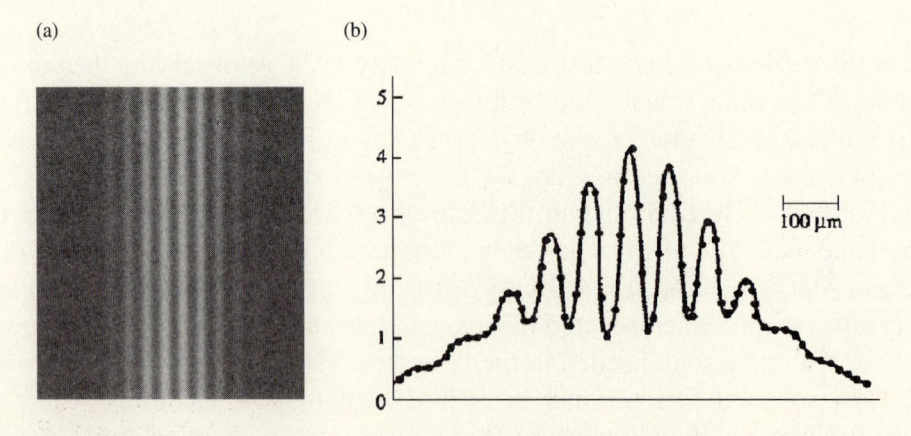

FIGURE 33-9 Measured two-slit interference patterns for a) electrons and b) neutrons.

It became possible in the 1970s to grow large crystals of silicon that are essentially perfect. That is, the spacing of the atoms remains constant throughout the entire sample with no dislocations, microscopic fractures, or any other imperfections. A few physicists soon realized that such a crystal, when used with ultracold neutrons having wavelengths >10^{-10} m, could be used as a *neutron interferometer.*

An interferometer, you will remember from Chapter 16, divides a wave into two halves, sends each half along a separate path, and then later recombines them to generate an interference pattern. Modifying one path of the interferometer changes the phase of the wave traveling along that half, thus altering or shifting the interference pattern. You should recall our analysis of the Michelson interferometer, which uses a beamsplitter to divide and recombine light waves. It can make very precise wavelength measurements by counting the changes between bright and dark fringes at the center of the interference pattern as one mirror is moved.

Figure 33-10 shows a neutron interferometer. It is made from a single crystal of silicon, about 10 cm long, which has been carefully machined to leave three thin, upright slabs of material with empty space between. A beam of ultracold neutrons is fired toward point A at an angle of about 20°. Each neutron is *diffracted* by the atomic planes in the crystal—the same process that causes the Bragg diffraction of x rays in a crystal! Thus a neutron is somehow "split" at point A. One portion continues along toward point B, but another portion emerges at a different angle and heads toward point C. Note that this is *not* a split into two material particles, because we cannot catch half of a neutron at points B or C, but the splitting of a *matter wave* into two smaller waves.

The same transmission/diffraction happens again at the second thin slab. A portion of each matter wave passes straight through, but another portion is diffracted by the crystal structure and heads toward point D. At point D, where the neutron matter waves meet at the third slab, the transmitted upper beam and the diffracted lower beam are recombined and move toward neutron detector 1. Detector 2 records the recombination

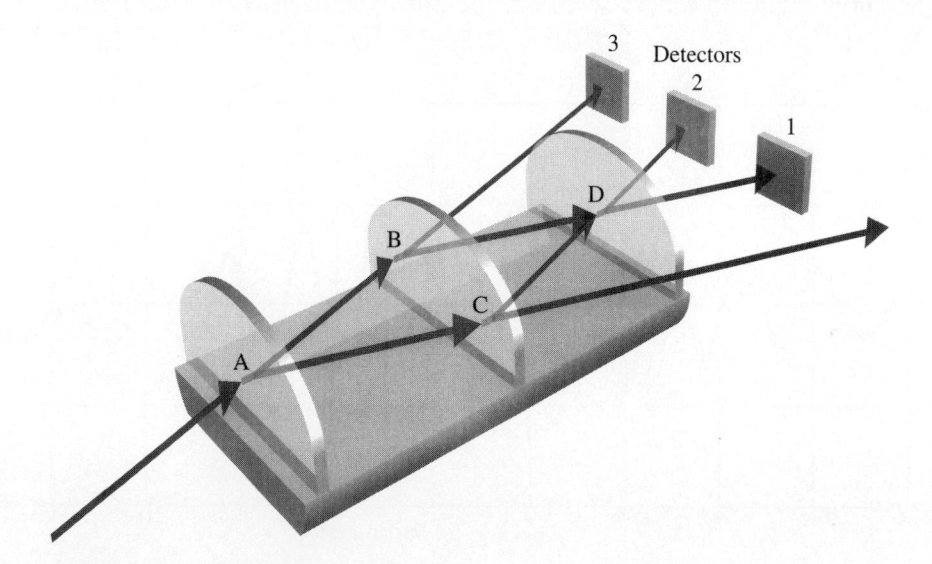

FIGURE 33-10 A neutron interferometer. A neutron is "split" at point A by diffraction, travels separate paths, and is then recombined at point D. Neutron detectors are numbered 1, 2, and 3.

of the diffracted upper beam with the transmitted lower beam. Detector 3 simply monitors the intensity of the original neutron beam.

The interferometer is designed such that detector 1 is placed at an interference maximum but detector 2 is at an interference minimum. So initially detector 1 detects neutrons and detector 2 does not. Think about what this implies. *One* neutron—a fundamental "particle" of matter—is divided into two parts at A. By the time they reach B and C, the two "parts" of this piece of matter are separated by more than 2 cm—hardly a mere microscopic separation. The pieces are recombined at D such that they interfere constructively along the path to detector 1, somehow "rebuilding" the neutron, but they interfere destructively and cancel the neutron along the path to detector 2. Detector 1 then records a single, particle-like neutron, but detector 2 sees nothing. If an extra piece of matter, or a magnetic field, or some other disturbance is placed in just *one* of the paths inside the interferometer, say near B, the neutron's phase is altered and the interference pattern shifts to where detector 2 records neutrons but detector 1 does not. A crazy way for matter to behave? Absolutely. But nature did not ask our opinion as to how matter "should" behave, and this is what it really does!

One particularly interesting experiment arranged the neutron interferometer such that the entire apparatus could be rotated about an axis passing through points A and D. The rotation causes point B to be at a slightly higher elevation than C. Consequently, the force of gravity on the neutrons prevents the ABD pathway from being a perfect mirror image of the ACD pathway. This changes the phase between the two beams and, when they are recombined, alters the interference pattern. Figure 33-11 shows the neutrons counted by the detector 1 as the interferometer is rotated from −40° to +40°, with 0° being level. We see a beautiful interference pattern. As the angle increases, so does the resulting phase shift between the two pathways. This causes *each* neutron as it is recombined on the path heading toward detector 1 to be alternately in phase (maximum intensity) and out of phase (minimum intensity). Not only do neutrons have wave-like properties, they can even be used to test how quantum theory works in the presence of gravitational fields.

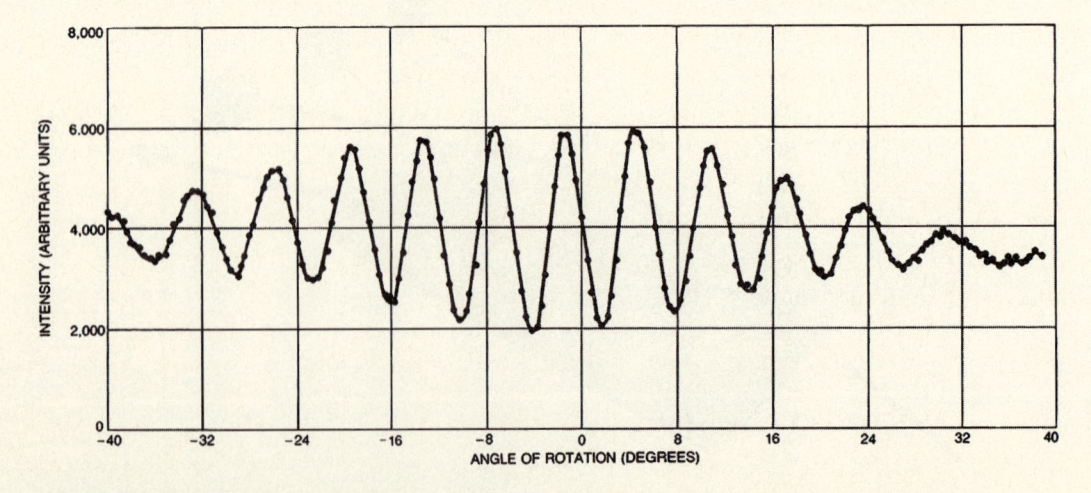

FIGURE 33-11 Interference fringes recorded by detector 1 when the neutron interferometer is rotated about a horizontal axis. The phases of the two neutron paths are altered by their different trajectories through the earth's gravitational field.

Electrons, even neutrons, are fundamental subatomic particles. Perhaps subatomic particles have wave-like aspects, but what about entire atoms—aggregates of many fundamental particles? Amazing as it seems, research during the 1980s demonstrated that whole atoms, and even molecules, can be "split," sent along separate paths, and recombined to give interference patterns. Figure 33-12 shows how an *atom interferometer* is built. First a laser is split into three beams (using beamsplitters), each of which is directed at a reflecting mirror. The light waves reflect to create three parallel *standing waves*, just like a wave on a string reflecting from a boundary. The nodes on the standing waves are spaced distance $\lambda/2$ apart.

Next, a beam of very slow sodium atoms is passed through the first standing light wave at point A. The light wave is set at a frequency such that it can exert a small force on each atom. Because the intensity along the standing waves alternates between high intensity at the antinodes and zero intensity at the nodes, the atom experiences a *periodic* force field. (This is analogous to a photon of light interacting with the periodic structure of a diffraction grating.) Part of each atom continues toward point B, but another part is *diffracted* by the laser beam standing wave and is directed along the path toward C. The atom has been divided!

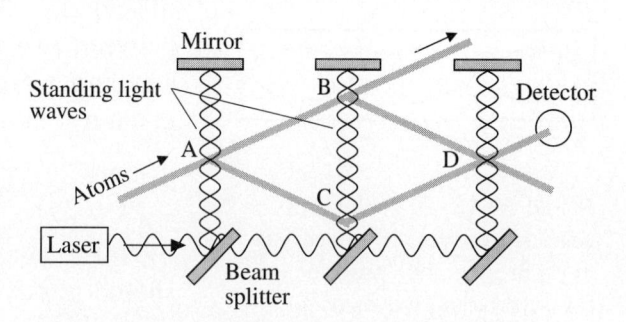

FIGURE 33-12 An atom interferometer. Sodium atoms are diffracted by the periodic standing light wave at A, travel different paths, and are recombined at D.

The atom-waves are diffracted again by the second standing wave at points B and C, directing them toward D where, with a third diffraction, they are recombined and move toward the detector. Depending on the phases of the waves as they recombine, the detector sometimes records atoms (constructive interference) but at other times does not (destructive interference). Altering the atom's environment along one of the paths, such as by applying an electric field in the region around B but not around C, shifts the phases of the waves and causes the detector to record interference fringes. This is *exactly* the same geometry and the same physics as the neutron interferometer of Fig. 33-10.

Notice that the atom interferometer completely inverts everything we previously learned about interference and diffraction. Thomas Young and others who studied the wave nature of light during the nineteenth century aimed light (a wave) at a diffraction grating (a periodic structure of matter) and found that it diffracted. Now we aim atoms (matter) at a standing wave (a periodic structure of light) and find that the atoms diffract. The roles of light and matter have been completely reversed!

33.6 Quantization of Energy

The previous examples considered the matter waves to be *traveling waves*. The experimental evidence is now overwhelming that electrons, protons, neutrons, and even entire atoms do exhibit wave-like properties under these circumstances. But suppose that a "particle" of matter is *confined* to a small region of space and cannot travel? How do the wave-like properties manifest themselves?

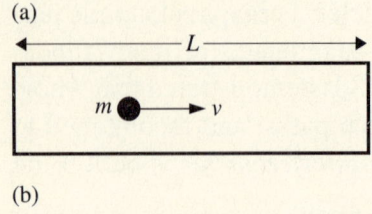

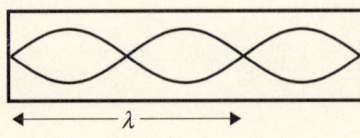

FIGURE 33-13 a) A classical particle in a box. b) According to de Broglie, the "particle" will set up standing waves as it reflects back and forth.

This is a problem we looked at in Chapter 18, where we considered a "particle in a box." Here we will briefly summarize that discussion. Figure 33-13 shows a particle of mass m moving in one dimension as it bounces back and forth with speed v between the ends of a box of length L. Now a wave, if it reflects back and forth between two fixed points, sets up a standing wave. A standing wave of length L has to have a wavelength given by

$$\lambda_n = \frac{2L}{n} \quad n = 1, 2, 3, 4, \ldots \quad (33\text{-}12)$$

If the confined particle has wave-like properties, it should satisfy both Eq. 33-12 *and* the de Broglie relationship $\lambda = h/mv$. That is, a confined "particle in a box" should obey the relationship

$$\frac{h}{mv} = \frac{2L}{n}$$

$$\Rightarrow v_n = \left(\frac{h}{2Lm}\right) \cdot n \quad n = 1, 2, 3, \ldots. \quad (33\text{-}13)$$

In other words, the particle cannot bounce back and forth with just *any* speed. Rather, it can have *only* those specific speeds v_n given by Eq. 13-13, for which the de Broglie wavelength creates a standing wave in the box.

The particle's energy, which is purely kinetic energy, is thus given by

$$E_n = \tfrac{1}{2} m v_n^2 = \frac{h^2}{8mL^2} \cdot n^2 \quad n = 1, 2, 3, \ldots. \quad (33\text{-}14)$$

De Broglie's hypothesis about the wave-like properties of matter has led us to the remarkable conclusion that the energy of a confined particle is **quantized**. The particle in a box can have an energy of $1 \times h^2/8mL^2$, or $4 \times h^2/8mL^2$, or $9 \times h^2/8mL^2$, etc. But it *cannot* have an energy between these values.

As you learned in Chapter 18, the possible values of the particle's energy are called **energy levels**, and the integer n that characterizes the energy levels is called the **quantum number**. Thus $n = 2$ is a quantum number describing the second energy level of the particle in a box. We can rewrite Eq. 33-14 in the useful form

$$E_n = n^2 E_1, \quad (33\text{-}15)$$

where

$$E_1 = \frac{h^2}{8mL^2} \quad (33\text{-}16)$$

is the **fundamental quantum of energy** for a particle in a box—analogous to the fundamental frequency f_1 of a standing wave on a string.

QUANTIZATION OF ENERGY 463

EXAMPLE 33-8 What is the fundamental quantum of energy for one of Millikan's 1 μm diameter oil droplets confined in a box of length 10 μm?

SOLUTION We found, in Example 33-6, that a droplet 1 μm in diameter has mass $m = 4.7 \times 10^{-16}$ kg. The confinement length is $L = 1.0 \times 10^{-5}$ m. It is a straightforward application of Eq. 33-16 to compute

$$E_1 = \frac{h^2}{8mL^2} = \frac{(6.63 \times 10^{-34} \text{ J s})^2}{8(4.7 \times 10^{-16} \text{ kg})(1.0 \times 10^{-5} \text{ m})} = 1.2 \times 10^{-42} \text{ J} = 7.3 \times 10^{-24} \text{ eV}.$$

This is such an incredibly small amount of energy that there is no hope of distinguishing among energies of E_1 or $4E_1$ or $9E_1$. For any macroscopic particle, even one this tiny, the allowed energies will *seem* to be perfectly continuous. We will not observe any quantization.

EXAMPLE 33-9 What are the first three allowed energies for an electron confined in a one-dimensional box of length 0.1 nm—about the size of an atom?

SOLUTION This is another basic application of Eqs. 33-15 and 33-16, with $m_{\text{elec}} = 9.11 \times 10^{-31}$ kg and $L = 1.0 \times 10^{-10}$ m. We find the fundamental quantum of energy to be $E_1 = 6.0 \times 10^{-18}$ J = 3.8 eV. Thus the first three allowed energies of an electron in a 0.1 nm box are

$$E_1 = 3.8 \text{ eV}$$
$$E_2 = 4E_1 = 15.2 \text{ eV}$$
$$E_3 = 9E_1 = 34.2 \text{ eV}.$$

An electron confined to a box of length 0.1 nm is a very simple "model" of an atom. It clearly is not an accurate model, so we do not expect the results of Example 33-9 to describe a "real" atom with much precision. Nonetheless, this simple model should give us *some* idea of how real atoms, which do confine electrons within a space of ≈0.1 nm, are likely to behave *if* the electrons have wave-like properties. The major result of this section is that confining a wave-like particle creates a standing wave, that a standing wave has only certain discrete wavelengths, and thus a confined particle can have only certain discrete energies. In other words, the confinement of a particle leads directly to the quantization of its energy. The particle in a box, although its physical significance is minimal, is a simple, calculable example to illustrate these ideas. An electron confined in a real atom will have to be a much more complex three-dimensional standing wave, with an equally complex formula for the allowed energies. But it will, just like the simple particle in a box, have quantized energies. Furthermore, we can expect the typical energy difference between adjacent energy levels to be a few electron volts.

Now this is an intriguing result. We found, in our analysis of the photoelectric effect, that visible and ultraviolet photons of light have energies of a few electron volts. We also know that atoms emit *discrete* wavelengths of visible and ultraviolet light, with photon energies of a few electron volts. Now we see that an electron confined in an atomic-size box has *discrete* energy levels spaced a few electron volts apart. Might there be a connection between these phenomena? We will explore this topic in the next chapter.

FIGURE 33-14 Scanning tunneling microscope picture of the electron standing waves produced in a "quantum corral" when 60 iron atoms are deposited in a circle.

Figure 33-14 is an interesting modern verification of the reality of electron standing waves. State-of-the-art techniques were used to deposit 60 iron atoms in a circle on a carbon plane. A scanning tunneling microscope was then used to measure the electron density in and near this "quantum corral." The circle of "towers" shows the confinement of electrons around each iron atom. But more interesting is the electron pattern in the *center* of the circle. There you can see a standing wave pattern of nodes and antinodes, very similar to the vibrations of a circular drum head. De Broglie was absolutely correct: Electrons have wave-like properties and, when confined, set up standing waves.

33.7 Wave–Particle Duality

Standing waves of matter and the quantization of energy are at complete odds with classical physics, but they appear to be primary characteristics of nature at the microscopic level of atoms. Planck, even if he did not quite believe it, introduced the idea that the atoms in a solid have quantized energies. Einstein carried this idea to light with his introduction of light quanta. Now we see that the quantization of energy goes hand-in-hand with de Broglie's suggestion about the wave-like nature of matter itself.

Scientists at the beginning of the twentieth century tried to extend their everyday classical ideas of particles and waves into the newly discovered realm of the atom. The dichotomy between particles and waves had been remarkably successful throughout the nineteenth century, as mechanics and electrodynamics became mature subjects. There was every reason to expect these concepts to apply equally well to atoms.

But as physicists delved further into the microscopic world of light and atoms, these familiar ideas failed to provide an explanation of the strange behaviors they found. Bit by bit, the distinction between particles and waves was eroded. "Particle" and "wave" are human concepts, and scientists eventually had to accept the fact that nature is not obligated to behave the way we think it should. The world of the atom turned out to be far more subtle and complex than could be described with the two-word language of particles and waves.

The experimental evidence shows that matter and light are neither particles nor waves but are something altogether new that can exhibit properties of *both* particles *and* waves. This unexpected aspect of matter and light is called *wave–particle duality*. Unfortunately for us, our minds seem limited to thinking of particles *or* waves—a wave–particle *dichotomy*. We cannot form a clear mental image of something that has properties of both, and this leads to many phenomena that seem quite paradoxical—such as the electron going through both slits in order to interfere with itself.

Einstein and de Broglie, like Balmer before them, found quantitative relationships—formulas—that successfully described certain phenomena. But their ideas were not theories. They did not propose general laws of nature, nor were their equations derived from any known laws of nature. They were simply hypotheses about specific situations—pulled, so to speak, out of a hat. Yet their successes indicated that Balmer, Rutherford, Einstein, and de Broglie were on the right track. They had pieces of a puzzle, but no one yet recognized how the completed puzzle would appear. Their challenge, which is now our challenge, was to find a general theory and a mathematical description for the behavior of matter and light at the microscopic scale of atoms. To do this, we still need one more crucial piece of the puzzle, which we will uncover in the next chapter.

Summary

Important Concepts and Terms

wave–particle duality	photon
photoelectric effect	wave packet
threshold frequency	matter wave
stopping potential	de Broglie wavelength
work function	quantized
thermal emission	energy level
Planck's constant	quantum number
light quantum	fundamental quantum of energy

Einstein's explanation of the photoelectric effect marked the first significant break with classical physics. It was not the photoelectric effect itself that called for an explanation. The ejection of electrons from metals under the proper circumstance was known and understood, but certain aspects of the photoelectric effect, particularly the existence of a threshold frequency f_0 and the essentially instantaneous appearance of the photoelectrons, defied the analysis of classical physics. Einstein postulated that light, rather than being a continuous wave, comes in discrete packets, called light quanta or photons, of energy

$$E = hf.$$

He also hypothesized that these photons are emitted and absorbed on an all-or-nothing basis and that one photon, when absorbed, transfers all its energy to *one* electron. This immediately gives the electron sufficient energy to escape the metal if $hf \geq E_0$, where E_0 is the work function of the metal. Thus there is a threshold frequency $f_0 = E_0/h$ the light must exceed for current to flow. If $f < f_0$, no current will flow, no matter how intense the light, because no electrons acquire sufficient energy to escape.

In reviewing the photoelectric effect you should make sure you understand what the photoelectric effect *is*; the characteristics of the effect as measured by Lenard; what classical physics can and cannot explain; and how Einstein's hypothesis of light quanta succeeded where classical physics had failed.

Einstein demonstrated that light can act like a particle as well as a wave. Two decades later, de Broglie proposed the complementary idea: Matter can act like a wave as well as a particle. The "wavelength" of a piece of matter is given by the de Broglie wavelength

$$\lambda = \frac{h}{p} = \frac{h}{mv}.$$

The idea that matter has a wavelength and wave-like properties, as strange as it seems, has amply been confirmed by experiment. Electrons, neutrons, and even entire atoms can interfere with themselves by somehow dividing, taking two different paths, then recombining in or out of phase. Be sure that you can explain how the experiments described in this chapter can be understood only as a wave-like phenomena, not by matter acting as a classical particle.

When de Broglie's hypothesis is applied to a confined "particle," we find that the particle's energy must be quantized. It can have only certain, discrete energies that correspond to standing waves of matter. For a particle in a one-dimensional box of length L, the allowed energies are given by

$$E_n = \frac{h^2}{8mL^2} \cdot n^2 = n^2 E_1 \quad n = 1, 2, 3, \ldots,$$

where n is the quantum number. Thus the quantization of energy, first introduced by Einstein for light quanta, extends to matter as well.

Exercises and Problems

Exercises

1. a. Determine the energy, in eV, of a 700 nm photon.
   b. Determine the wavelength of a 5 keV x-ray photon.
2. A photoelectric effect experiment finds a stopping potential of 2.0 V when light of 197 nm is used to illuminate the cathode.
   a. From what metal is the cathode made?
   b. What is the stopping potential if the intensity of the light is doubled?
3. What speed electron has a wavelength of 500 nm?
4. An electron in a box is observed, at different times, to have energies of 12 eV, 27 eV, and 48 eV. What is the width of the box?
5. a. Explain why the graphs of Fig. 33-3 are horizontal for $\Delta V > 0$.
   b. Explain why photoelectrons are ejected from the cathode with a range of velocities, rather than all electrons having the same velocity.
   c. Explain, in words, the reasoning by which we claim that the stopping potential V_{stop} measures the maximum kinetic energy of the electrons.

6. Use Millikan's photoelectric effect data in Fig. 33-6 to determine:
   a. The work function, in eV, of cesium.
   b. The value of Planck's constant.

Problems

7. a. What would the graph of Fig. 33-2 look like *if* classical physics were the correct description of the photoelectric effect? Draw the graph *and* explain your reasoning. Assume that the light intensity remains constant as its frequency and wavelength are varied.
   b. What would the graph of Fig. 33-3 look like *if* classical physics were the correct description of the photoelectric effect? Draw the graph, including curves for both weak light and intense light, *and* explain your reasoning.

8. An AM radio station broadcasts with a power of 10 MW at a frequency of 1000 kHz.
   a. How many photons does the antenna emit each second?
   b. Does this emission rate suggest that the broadcast should be treated as an electromagnetic wave or as discrete photons?

9. A ruby laser emits an intense pulse of light that lasts a mere 10 ns. The pulse has a wavelength of 690 nm and an energy of 500 mJ.
   a. How many photons are emitted in each pulse?
   b. What is the *rate* of photon emission, in photons per second, during the 10 ns that the laser is "on"?

10. Consider a photoelectric effect experiment in which both potassium and gold cathodes are used. For each type of cathode find:
    a. The threshold frequency.
    b. The threshold wavelength.
    c. The maximum photoelectron ejection speed if the light has a wavelength of 220 nm.
    d. The stopping potential if the wavelength is 220 nm.

11. The electron interference pattern of Fig. 33-9a was made by shooting 50 keV electrons through two slits spaced 1.0 μm apart. The fringes were recorded on a piece of film 1 m behind the slits. The whole experiment took place in a vacuum chamber.
    a. What was the speed of the electrons?
    b. The photograph of Fig. 33-9a is greatly magnified. What was the actual spacing on the film between adjacent bright fringes?

12. The neutron interference pattern of Fig. 33-9b was made by shooting neutrons with a speed of 200 m/s through two slits spaced 0.10 mm apart.
    a. What was the energy, in eV, of the neutrons?
    b. What was the de Broglie wavelength of the neutrons?
    c. The pattern was recorded by moving a neutron detector through the interference pattern, measuring the neutron intensity at different positions. Notice the 100 μm scale on the figure. By making appropriate measurements directly *on the figure*, determine the distance between the slits and the neutron detector.

13. Rutherford found the atomic nucleus to have a diameter of ≈10 fm. A simple model of the nucleus is that protons and neutrons are confined within a one-dimensional box of length 10 fm.
    a. What are the first three energy levels for a proton in such a box? Express your answer in MeV.
    b. What are the first three energy levels for a neutron in such a box?

14. Imagine that the box of Fig. 33-13a, confining a particle of mass m, is oriented vertically rather than horizontally. Also imagine the box to be on a neutron star where the gravitational field is so strong that a particle slows significantly, nearly stopping, before it impacts the top of the box and reflects back. Make a *qualitative* sketch, similar to Fig. 33-13b, of the $n = 3$ de Broglie standing wave of a particle in this box. (**Hint:** The nodes are *not* uniformly spaced.)

15. The data shown in the table are from a photoelectric effect experiment. The stopping potential was measured for several different wavelengths of incident light. Analyze this data to determine:
    a. The metal used for the cathode.
    b. An experimental value for Planck's constant.

λ (nm)	V_{stop} (volts)
500	0.19
450	0.48
400	0.83
350	1.28
300	1.89
250	2.74

16. The photomultiplier tube (PMT) of Fig. 33-8a consists of a cathode, which the photon strikes; an anode, where the electrons are collected; and a number of intermediate electrodes that are called *dynodes*. The tube shown in the figure has nine dynodes. Consider a PMT that has a cathode, N dynodes, and an anode. The cathode, when struck by a photon, ejects a single photoelectron. That electron is accelerated to the first dynode, where it causes (on average) the ejection of ε secondary electrons. The quantity ε is called the *secondary emission coefficient*. Each of these electrons ejects, on average, ε electrons from the second dynode, they each eject ε electrons from the third dynode, and so on until a large pulse of electrons is collected by the anode.
    a. Write an expression, in terms of ε and N, for the average number of electrons arriving at the anode due to a single photon striking the cathode. This is called the *gain* of the PMT.
    b. The graph of Fig. 33-8b shows the voltage pulse generated when the electron current passed through an external 50 Ω resistor. The *baseline* of the pulse is zero volts, and the voltage scale is 20 mV per division. What is the maximum *current* of this pulse?
    c. Because $I = dQ/dt$, the amount of charge delivered by a pulse of current can be found as $Q = \int I dt$, which can be interpreted geometrically as the area under the I-versus-t curve. The area of a bell-shaped curve is reasonably well approximated as its "height" multiplied by its "width" measured at half of its maximum height. Estimate the number of electrons in the current pulse shown in Fig. 33-8b.

d. The PMT that produced the pulse of Fig. 33-8b had 14 dynodes. By comparing your answers to parts a) and c), determine the secondary emission coefficient for this PMT.
   e. What is the potential difference ΔV between adjacent dynodes if the anode is at 0 V and the cathode is at –3000 V?
   f. Assuming that the speed of an ejected electron is "small," with what speed does an electron impact the next dynode?
17. During the 1980s, physicists learned how to use the light forces of a laser beam to "cool" a gas-phase sample of atoms to temperatures << 1 K. In the atom interferometer experiment of Fig. 33-12, laser-cooling techniques were used to cool a dilute vapor of sodium atoms to a temperature of 0.001 K = 1 mK. The cold atoms were allowed to pass through a series of collimating apertures, forming the *atomic beam* you see entering Fig. 33-12 from the left. They then intersected the standing light waves created from laser beams of wavelength 590 nm.
   a. What is the rms speed v_{rms} of a sodium atom ($A = 23$) in a gas at temperature 1 mK?
   b. Assume that the atomic beam is directed at right angles to the laser-beam standing wave, rather than obliquely as shown in the figure. The atoms intersect the standing wave at point A. By treating the laser beam as a diffraction grating, calculate the first-order diffraction angle of a sodium atom traveling with the rms speed of part a).
   c. How far apart are points B and C if the second laser-beam standing wave is 10 cm from the first?
   d. Each individual atom is apparently present at both point B *and* point C. Describe, in your own words, what this experiment tells you about the nature of matter.

[**Estimated 10 additional problems for the final edition.**]

Chapter 34

The Bohr Model of the Atom

LOOKING BACK Sections 32.6–32.8; 33.5, 33.6

34.1 The Atomic Structure Enigma

The existence of atoms was well established by the opening years of the twentieth century, and the research of Thompson, Millikan, Rutherford, and others was quickly moving into the realm of subatomic physics. Thompson's electron and Rutherford's nucleus made it clear that the atom has a *structure* of some sort. It is not, as Democritus had proposed so long ago, the indivisible, uncuttable ultimate constituent of matter.

The challenge at the beginning of the twentieth century was to deduce, from experimental evidence, the correct structure of the atom. The difficulty of this task cannot be exaggerated. Because atoms could not be seen, and certainly not directly manipulated, all of the evidence was very indirect—observing atomic spectra, observing the large-angle deviations of alpha particles shot through thin foils, and observing the behavior of cathode-ray tubes. These experiments were carried out with only the simplest of electronic measuring devices. Most observations were made by eye, and all calculations were carried out by hand. With these observations as a guide, physicists were attempting to construct a *model* of the atom. Their goal was to find a model that could successfully explain, via mathematical analysis, the various observations.

Rutherford's nuclear model, which we looked at in detail in Section 32.6, was the most successful of various proposals. His model explained both the scattering of alpha particles and the production of different ions as seen in mass spectroscopy. But as we have already seen, Rutherford's model failed to explain why atoms are stable or why their spectra are discrete. Although Rutherford seemed to be on the right track, his model could not be accepted as a completely correct description of atomic structure.

A missing piece of the puzzle, although not recognized as such for a few years, was Einstein's 1905 introduction of light quanta. If light comes in discrete packets of energy, which we now call photons, and if atoms emit and absorb light, what does that imply about the structure of atoms? That was the question posed by a young Dane named Niels Bohr.

34.2 Bohr's Model

Niels Bohr was born, educated, and spent most of his life in Denmark. Figure 34-1 shows Bohr as a young man. He later established an institute in Copenhagen that, for many decades, was the leading center for the development of quantum physics. Although few discoveries and theories bear Bohr's name, he was the intellectual driving force behind the development of quantum mechanics and the mentor of many of the young physicists who reshaped physics in the 1920s and 1930s. He worked tirelessly in his later years to promote international cooperation and the peaceful use of nuclear physics.

In 1905, when Einstein first introduced the idea of light quanta, Bohr was a 20-year-old student. After receiving his doctoral degree in physics in 1911, at the age of 26, he immediately went to England to work in Rutherford's laboratory. This brought Bohr into direct contact with the research started by Thompson, then continued by Rutherford, into the structure of the atom. Rutherford had just, within the previous year, completed his development of the nuclear model of the atom.

FIGURE 34-1 Niels Bohr.

Rutherford's model certainly contained a kernel of truth, but Bohr wanted to understand how a solar-system-like atom could be stable and not radiate away all its energy. He soon recognized that Einstein's light quanta had profound implications about the structure of atoms. In 1913, Bohr published a radically new model of the atom in which he added quantization to Rutherford's nuclear atom.

The basic assumptions of the **Bohr model of the atom** are as follows:

1. An atom consists of negative electrons orbiting a very small positive nucleus, as in the Rutherford model.

2. Atoms can exist only in certain **stationary states**. Each stationary state corresponds to a particular set of electron orbits around the nucleus. These states are discrete and can be numbered $n = 1, 2, 3, 4, \ldots$, where n is the *quantum number*. Each stationary state has a discrete, well-defined energy E_n, this being the energy (kinetic plus potential) of the orbiting electrons. That is, atomic energies are *quantized*.

3. The stationary states of an atom are numbered in order of increasing energy: $E_1 < E_2 < E_3 < E_4 < \ldots$. The lowest energy state of the atom, with energy E_1 is *stable* and can persist indefinitely. It is called the **ground state** of the atom. Other stationary states with energies $E_2, E_3, E_4, \ldots$ are called **excited states** of the atom.

4. An atom can "jump" from one stationary state to another, undergoing an energy change $\Delta E_{atom} = |E_f - E_i|$, by emitting or absorbing a photon of frequency

$$f_{photon} = \frac{\Delta E_{atom}}{h}, \qquad (34\text{-}1)$$

where h is Planck's constant. E_i and E_f are the energies of the initial and final states. Such a jump is called a **transition** or, sometimes, a **quantum jump**.

5. An atom can absorb energy in a collision with an electron or another atom, causing the electron to move from a lower energy state to a higher energy state. The energy absorbed has to be exactly $\Delta E = E_f - E_i$.

6. Atoms will seek the lowest energy state—the ground state. An atom in an excited state, if left alone, will jump to lower and lower energy states until it reaches the ground state.

Bohr's model builds upon the model of Rutherford, but it adds two crucial new ideas that are derived from Einstein's ideas of quanta. The first, expressed in assumption 2, is that only certain electron orbits are "allowed" or can exist. The second, in assumption 4, is that the atom can jump from one state to another, suddenly changing the electron orbits, by emitting or absorbing a photon of just the right frequency to conserve energy.

According to Einstein, a photon of frequency f has energy $E_{photon} = hf$. If an atom jumps from an initial state with energy E_i to a final state with *lower* energy E_f, losing energy $\Delta E_{atom} = |E_f - E_i|$, energy can still be conserved if the atom emits a photon with $E_{photon} = \Delta E_{atom}$. This photon must have exactly the frequency given by Eq. 34-1 if it is to carry away exactly the right amount of energy. Similarly, an atom can jump to a higher energy state, for which additional energy is needed, by absorbing a photon of frequency $f_{photon} = \Delta E_{atom}/h$. The total energy of the atom-plus-light system is conserved.

The implications of Bohr's model are profound. In particular:

1. Matter is stable. Once an atom is in its ground state, there are no states of any lower energy to which it can decay. It can remain in the ground state forever. This is in sharp contrast to Rutherford's model, in which the electrons continue losing energy as they spiral into the nucleus.

2. Atoms will emit and absorb a *discrete spectrum*—those frequencies and only those frequencies that match the energy *intervals* between the stationary states. Photons of other frequencies do not match the atom's possible energies, so they cannot be emitted or absorbed without violating energy conservation. A discrete spectrum is a consequence of the energy quantization of both the atom and the light.

3. Bohr's model explains the *origin* of photons. Einstein postulated that the light arriving at the cathode of a photoelectric effect experiment is quantized, but where do those photons originate? Bohr's model asserts that the photons are emitted naturally by atoms.

4. Bohr's model provides an explanation for how light is emitted from a discharge tube. Electrons moving through a low-pressure gas, as they carry the current through the tube, occasionally collide with the atoms. Most of the atoms are in their ground state, but energy transferred to the atom in the collision can kick it into a higher energy state n. The electron, to conserve energy, loses energy $\Delta E = E_n - E_1$. This process is called **collisional excitation** of the atom. Once in an excited state, the atom can emit photons of light—a discrete emission spectrum—as it jumps back down to lower energy states.

5. Similarly, Bohr's model explains why emission spectra differ from absorption spectra. Absorption wavelengths, you will recall, are a subset of the wavelengths in the emission spectrum. That is, all the lines seen in an absorption spectrum are also seen in emission, but many emission lines are *not* seen in absorption. The explanation, according to Bohr's model, is that most atoms, most of the time, are in their lowest energy

state—the ground state $n = 1$. The absorption spectrum thus consists of transitions in which the atom jumps from $n = 1$ to a higher value of n by absorbing a photon. Transitions $1 \rightarrow 2$, $1 \rightarrow 3$, $1 \rightarrow 4$, and so on, will be seen in absorption, but transition such as $2 \rightarrow 4$ will *not* be seen because there are only a very tiny fraction of the atoms in $n = 2$ at any instant of time.

On the other hand, atoms that have been excited to the $n = 4$ state by collisions can emit photons corresponding to transitions $4 \rightarrow 3$, $4 \rightarrow 2$, and $4 \rightarrow 1$. So the wavelength corresponding to $\Delta E = |E_4 - E_1|$ is seen in both emission and absorption, but transitions with $\Delta E = |E_4 - E_2|$ or $\Delta E = |E_4 - E_3|$ occur in emission only.

6. Lastly, Bohr's model explains why each element in the periodic table has a unique spectrum. The energies of the stationary states are really just the energies of the orbiting electrons. The atom has no other form of energy. Different elements, with different numbers of electrons, will have different stable orbits and thus different stationary states. States with different energies will, of course, emit and absorb photons of different wavelengths.

EXAMPLE 34-1 An atom has stationary states numbered j and k, with energies $E_j = 4$ eV and $E_k = 6$ eV. What is the wavelength of a photon emitted in a transition between these two states?

SOLUTION The atom can jump from the higher-energy state k to the lower-energy state j by emitting a photon of frequency

$$f = \frac{\Delta E}{h} = \frac{2 \text{ eV}}{4.14 \times 10^{-15} \text{ eV s}} = 4.83 \times 10^{14} \text{ Hz}.$$

The wavelength of this photon is

$$\lambda = \frac{c}{f} = 621 \text{ nm}.$$

EXAMPLE 34-2 An atom has states $E_1 = 0$ eV, $E_2 = 3$ eV, and $E_3 = 5$ eV. What wavelengths are in the absorption spectrum and in the emission spectrum of this atom?

SOLUTION This atom will absorb photons on the $1 \rightarrow 2$ and $1 \rightarrow 3$ transitions, having energies $\Delta E_{1 \rightarrow 2} = 3$ eV and $\Delta E_{1 \rightarrow 3} = 5$ eV. From $f = \Delta E/h$ and $\lambda = c/f$ we find the wavelengths in the absorption spectrum:

$1 \rightarrow 2$ $f = 3 \text{ eV}/h = 7.25 \times 10^{14}$ Hz $\lambda = 414$ nm (blue)

$1 \rightarrow 3$ $f = 5 \text{ eV}/h = 1.21 \times 10^{15}$ Hz $\lambda = 248$ nm (ultraviolet).

The emission spectrum will also have the 414 nm and 248 nm wavelengths due to the $2 \rightarrow 1$ and $3 \rightarrow 1$ quantum jumps from excited states 2 and 3 to the ground state. In addition, it will contain the $3 \rightarrow 2$ quantum jump of $\Delta E_{3 \rightarrow 2} = 2$ eV that is *not* seen in absorption because there are too few atoms in the $n = 2$ state to absorb. We learned in Example

34-1 that a 2 eV transition corresponds to a wavelength of 621 nm. Thus the emission wavelengths are:

$$2 \to 1 \quad \lambda = 414 \text{ nm (blue)}$$
$$3 \to 1 \quad \lambda = 248 \text{ nm (ultraviolet)}$$
$$3 \to 2 \quad \lambda = 621 \text{ nm (orange)}.$$

34.3 The Bohr Atom

Bohr's hypothesis was an intriguing and bold new idea. It certainly held out hope for understanding the behavior of atoms and the emission and absorption of light. Yet there was still one enormous stumbling block: What *are* the stationary states of an atom? Everything in Bohr's model hinges on the existence of these stationary states, of there being only certain electron orbits that are "allowed," but nothing in classical physics provides any basis for such orbits. And Bohr's model itself describes only the *consequences* of having stationary states, not how to find them. If such states really exist, we will have to go well beyond classical physics to find them.

In an effort to address this problem, Bohr did an explicit analysis of the hydrogen atom. The hydrogen atom, with only a single electron, was known to be the simplest atom because the singly ionized hydrogen ion H$^+$ was observed but doubly ionized H^{++} never was. Furthermore, as we discussed in Chapter 32, Balmer had discovered a fairly simple formula that characterized the wavelengths in the hydrogen emission spectrum. Anyone with a successful model of an atom would have to *derive* Balmer's formula for the hydrogen atom. Bohr's analysis of hydrogen is what we now call the **Bohr atom**. Note that the *Bohr atom*, which refers only to hydrogen and hydrogen-like ions, is distinct from the more general assumptions of the *Bohr model*, which apply to any atom.

The method that Bohr followed in his analysis is rather obscure. That is not surprising, because he had little to go on at the time. But our goal is a clear explanation of the ideas, not a historical study of Bohr's methods, so we are going to follow a different analysis using de Broglie's matter waves. De Broglie did not propose matter waves until 1924, eleven years after Bohr's paper, but with the clarity of hindsight we can see that treating the electron as a wave provides a more straightforward analysis of the Bohr atom. So although our route will be different from Bohr's, we will arrive at the same point and, in addition, will be in a much better position to understand the work that came after Bohr.

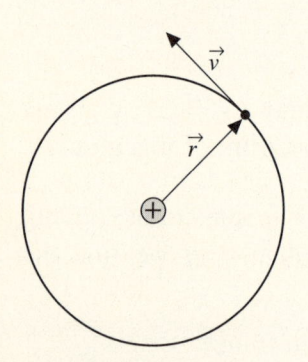

FIGURE 34-2 A Rutherford hydrogen atom. The size of the nucleus is greatly exaggerated. It would be a barely visible speck if drawn to scale.

The Stationary States of the Hydrogen Atom

Figure 34-2 shows a Rutherford hydrogen atom, with a single electron orbiting a nucleus that consists of a single proton. We will assume a circular orbit of radius r and speed v. We will also assume, to keep the analysis manageable, that the proton remains stationary while the electron revolves about it. This assumption is reasonable because the proton is roughly 1800 times as massive as the electron. With these assumptions, the atom's energy is exclusively the kinetic and potential energy of the electron, with the potential energy being that due to the electrical interaction between the electron and the proton. Thus the atom's energy is

$$E = K + U = \frac{1}{2}mv^2 + \frac{1}{4\pi\varepsilon_0} \cdot \frac{q_{elec}\, q_{proton}}{r}$$
$$= \frac{1}{2}mv^2 - \frac{e^2}{4\pi\varepsilon_0 r},$$
(34-2)

where we have used $q_{elec} = -e$ and $q_{proton} = +e$. Note that m is the mass of the electron, *not* the mass of the entire atom.

Now the electron, as we are coming the understand it, has both particle-like and wave-like properties. Each of these two aspects of its nature places a constraint on its behavior. First, let us treat the electron as a charged particle. It experiences a long-range force $\vec{F}_{elec}$ from the proton, as given by Coulomb's law:

$$\vec{F}_{elec} = \left(\frac{1}{4\pi\varepsilon_0} \cdot \frac{e^2}{r^2},\ \text{toward center}\right).$$
(34-3)

This force, according to Newton's second law, gives the electron an acceleration $\vec{a}_{elec} = \vec{F}_{elec}/m$ that also points to the center. Recall from our earlier study that this is a centripetal acceleration, causing the particle to move in its circular orbit. The centripetal acceleration of a particle moving in a circle of radius r at speed v *must* have magnitude v^2/r. Equating $|\vec{a}_{elec}|$ and v^2/r allows us to find a relationship between v and r for the orbiting electron:

$$|\vec{a}_{elec}| = \frac{|\vec{F}_{elec}|}{m} = \frac{e^2}{4\pi\varepsilon_0 m r^2} = \frac{v^2}{r}$$
$$\Rightarrow v^2 = \frac{e^2}{4\pi\varepsilon_0 m r}.$$
(34-4)

Equation 34-4 is, in effect, a *constraint* on the motion. It says that the electron cannot have just any values of v and r. They must, instead, obey the relationship of Eq. 34-4 if the electron is to move in a circular orbit. Note that this is not unique to electrons; we found a similar relationship between v and r in our Chapter 12 study of the orbits of satellites due to the gravitational force. Equation 34-4 is simply a consequence of treating the electron as a particle. We will return to this conclusion shortly.

Now let's treat the electron as a de Broglie wave. We considered a one-dimensional particle in a box in the previous chapter, where we found that the particle sets up a standing wave as it reflects back and forth. How can we apply this idea to a wave-like particle moving in a circle?

A standing wave, you will recall, consists of two traveling waves moving in opposite directions. When the round-trip distance in the box is equal to an integer number of wavelengths, the two oppositely traveling waves interfere constructively to set up the standing wave. Because the *round-trip* distance is $2L$, this condition for the existence of standing waves is

$$2L = n\lambda \quad \Rightarrow \quad \lambda = \frac{2L}{n} \quad n = 1, 2, 3, \ldots . \qquad (34\text{-}5)$$

This you should recognize as exactly our previous result for one-dimensional standing waves.

Suppose that instead of traveling back and forth along a line, our wave-like particle travels around the circumference of a circle of radius r. The particle will set up a standing wave, just like the particle in the box, if there are waves traveling in both directions and if the round-trip distance is equal to an integer number of wavelengths. This is the idea we want to carry over from the particle in a box: To confine the wave on a circle, we need to set up a standing wave, and the condition for a standing wave is that the round-trip distance be an integer number of wavelengths. Figure 34-3, for example, shows a standing wave around a circle with $n = 10$ wavelengths.

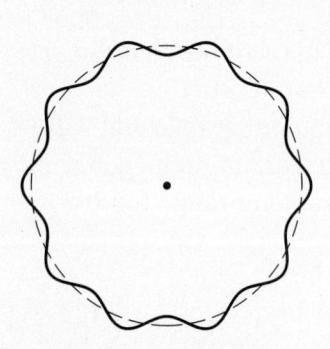

FIGURE 34-3 An $n = 10$ electron standing wave around the orbit's circumference.

The mathematical condition for a circular standing wave is analogous to Eq. 34-5, simply replacing the round-trip distance in a box of $2L$ with the round-trip on a circle of $2\pi r$. Thus

$$2\pi r = n\lambda \quad n = 1, 2, 3, \ldots . \qquad (34\text{-}6)$$

But the de Broglie wavelength for a particle *has* to be $\lambda = h/p = h/mv$. This gives

$$2\pi r = n \cdot \frac{h}{mv}$$

$$\Rightarrow v = \frac{nh}{2\pi mr} \quad n = 1, 2, 3, \ldots . \qquad (34\text{-}7)$$

The quantity $h/2\pi$ occurs so often in quantum physics that it is customary to give it a special name. We define the quantity $\hbar$, pronounced "h bar," as

$$\hbar = \frac{h}{2\pi} = 1.055 \times 10^{-34} \text{ J s} = 6.58 \times 10^{-16} \text{ eV s}.$$

With this definition, we can write Eq. 34-7 as

$$v = \frac{n\hbar}{mr} \quad n = 1, 2, 3, \ldots . \qquad (34\text{-}8)$$

This, like Eq. 34-4, is another relationship between v and r. This is the constraint that arises from treating the electron as a wave.

Now if the electron can act as both a particle *and* a wave, then it must be the case that both the Eq. 34-4 and the Eq. 34-8 constraints have to be obeyed. If they are both true, then v^2 as given by the Eq. 34-4 particle constraint has to be equal to v^2 as given by the Eq. 34-8 wave constraint. Equating these gives

$$\left[v^2 = \frac{e^2}{4\pi\varepsilon_0 mr} \right] = \left[v^2 = \frac{n^2\hbar^2}{m^2 r^2} \right]. \qquad (34\text{-}9)$$

Solving this for r yields

$$r_n = n^2 \cdot \frac{4\pi\varepsilon_0 \hbar^2}{me^2} \qquad n = 1, 2, 3, \ldots, \qquad (34\text{-}10)$$

where we have added a subscript n to the radius r to indicate that it depends on the integer n. The right-hand side of Eq. 34-10, except for the n^2, is just a collection of constants. Let's group them all together and define the **Bohr radius** a_B as

$$a_B = \text{Bohr radius} = \frac{4\pi\varepsilon_0 \hbar^2}{me^2} = 5.29 \times 10^{-11} \text{ m} = 0.0529 \text{ nm}. \qquad (34\text{-}11)$$

With this definition, Eq. 34-11 becomes

$$r_n = n^2 a_B \qquad n = 1, 2, 3, \ldots$$

$$= \begin{cases} 0.053 \text{ nm} & n = 1 \\ 0.212 \text{ nm} & n = 2 \\ 0.476 \text{ nm} & n = 3 \\ \vdots & \vdots \end{cases} \qquad (34\text{-}12)$$

We have discovered stationary states! That is, a hydrogen atom can exist *only* if the radius of the electron's orbit is one of the values given by Eq. 34-12. Intermediate values of the radius, such as $r = 0.100$ nm, cannot exist because the electron cannot set up a standing wave around the circumference. The possible orbits are *quantized*, with only certain orbits allowed.

The key step leading to Eq. 34-12 was the requirement that the electron have wave-like properties in addition to particle-like properties. You learned previously that de Broglie's idea applied to a particle in a box leads to quantized energy levels. The same idea, now applied to an electron moving in a circular orbit, leads to quantized orbits—what Bohr called stationary states. The integer n is thus the *quantum number* that numbers the various stationary states.

Now we can make progress quickly. Knowing the possible radii, we can return to Eq. 34-8 and find the possible electron speeds to be

$$v_n = \frac{n\hbar}{mr_n} = \frac{1}{n} \cdot \frac{\hbar}{ma_B} = \frac{v_1}{n} \qquad n = 1, 2, 3, \ldots, \qquad (34\text{-}13)$$

where

$$v_1 = \hbar/ma_B = 2.19 \times 10^6 \text{ m/s}$$

is the speed in the $n = 1$ orbit. The speed decreases as n increases.

Finally, we can determine the energies of the stationary states. From Eq. 34-2 for the energy, with Eqs. 34-12 and 34-13 for r and v, we have

$$E_n = \frac{1}{2}mv_n^2 - \frac{e^2}{4\pi\varepsilon_0 r_n} = \frac{1}{2}m\left(\frac{\hbar^2}{m^2 a_B^2 n^2}\right) - \frac{e^2}{4\pi\varepsilon_0 n^2 a_B}$$

$$= \frac{1}{n^2 a_B} \cdot \left(\frac{\hbar^2}{2ma_B} - \frac{e^2}{4\pi\varepsilon_0}\right). \qquad (34\text{-}14)$$

The term in parentheses can be simplified by using the Eq. 34-11 definition of a_B, giving

$$\begin{aligned} E_n &= \frac{1}{n^2 a_B} \cdot \left(\frac{1}{2} \cdot \frac{e^2}{4\pi\varepsilon_0} - \frac{e^2}{4\pi\varepsilon_0} \right) \\ &= -\frac{1}{n^2} \cdot \left(\frac{e^2}{4\pi\varepsilon_0 \cdot 2a_B} \right) \\ &= -\frac{E_1}{n^2}, \end{aligned} \qquad (34\text{-}15)$$

where

$$E_1 = \frac{e^2}{4\pi\varepsilon_0 \cdot 2a_B} = 13.60 \text{ eV}. \qquad (34\text{-}16)$$

This has been a lot of math, so we need to see where we are and what we have learned. To begin, Table 34-1 shows values of r_n, v_n, and E_n evaluated for quantum number values $n = 1$–5. We do indeed seem to have discovered stationary states of the hydrogen atom. Each state, characterized by its quantum number n, has a unique radius, speed, and energy. These values are displayed graphically in Fig. 34-4, in which the orbits are drawn to scale. Notice how the atom's diameter increases very rapidly as n increases. Its speed, at the same time, is decreasing.

TABLE 34-1 Radii, speeds, and energies for the first five states of the Bohr hydrogen atom.

n	r_n (nm)	v_n (m/s)	E_n (eV)
1	0.053	2.19×10^6	–13.60
2	0.212	1.09×10^6	–3.40
3	0.476	0.73×10^6	–1.51
4	0.846	0.55×10^6	–0.85
5	1.322	0.44×10^6	–0.54

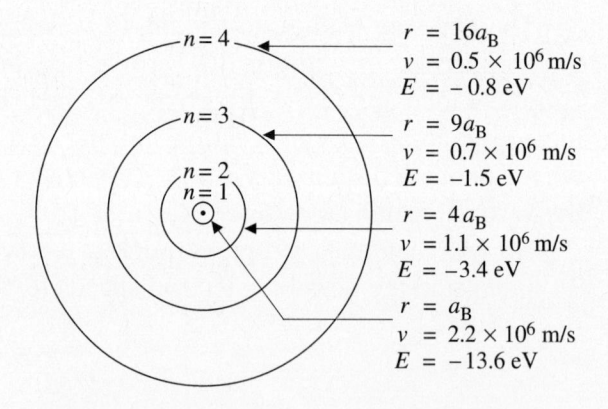

FIGURE 34-4 The first four stationary states, or allowed orbits, of the Bohr atom drawn to scale.

Energy of Stationary States

Note that the energies of the stationary states are negative. It is very important to understand what this means. Because the potential energy of two charged particles is $U = q_1 q_2 / 4\pi\varepsilon_0 r$, the zero of potential energy occurs at $r = \infty$, where the particles are infinitely far apart. The state of zero total energy ($E = K + U$) corresponds to having the electron at rest ($K = 0$) and infinitely far from the proton ($U = 0$). This situation—the case of two "free particles"—occurs in the limit $n \to \infty$, for which $r_n \to \infty$ and $v_n \to 0$. Now an electron and a proton bound into an atom have *less* energy than two free particles. We know this because we would have to do work—add energy—to pull the electron and proton apart. So if the bound atom has less energy than two free particles, and if the total energy of two free particles is zero, then it must be the case that the atom has a *negative* amount of energy.

The significance of this is not that the energy is negative. We could change that, if we wished, by redefining the zero of potential energy. In practice, we only use energy *changes* ΔE, and those are independent of where we place the zero. The significance is that E_n tells us the **binding energy** of the electron. In the ground state, where $E_1 = -13.60$ eV, we would have to add 13.60 eV to the electron to free it from the proton and reach the zero energy state of two free particles. So we can say that the electron in the ground state is "bound by 13.60 eV." An electron in an $n = 3$ orbit, where it is farther from the proton and moving more slowly, is bound by only 1.51 eV. That is the amount of energy you would have to supply to remove the electron from an $n = 3$ orbit.

Removing an electron leaves a positive ion behind—H$^+$ in the case of a hydrogen atom. (The fact that H$^+$ happens to be a proton does not alter the fact that it is also an atomic ion, just missing one electron from being a neutral atom.) The energy needed to ionize an atom is simply the amount of energy needed to remove a ground state electron— exactly $|E_1|$. As a result, the absolute value of the ground state energy is called the **ionization energy** of the atom. We now have a prediction, using the Bohr atom, that hydrogen has an ionization energy of 13.60 eV.

We can test this prediction by accelerating a beam of electrons through a potential difference ΔV and shooting them at hydrogen atoms. Electrons can transfer their kinetic energy to the atoms in collisions. A projectile electron can knock out an atomic electron if its kinetic energy K is greater than the atom's ionization energy, leaving an ion behind. But an electron will be unable to cause ionization if it has K less than the atom's ionization energy. This is a fairly straightforward experiment to carry out, and the evidence shows that the ionization energy of hydrogen is, in fact, 13.60 eV. This prediction is a first success of Bohr's model.

Energy-Level Diagrams

It is customary in atomic physics to represent the stationary state energies graphically in an **energy-level diagram**. Figure 34-5 shows an energy-level diagram for the hydrogen atom. This is not really a graph, rather more like a picture. The vertical axis does represents energy, but the horizontal axis is not a scale. Think of this diagram as a picture of a ladder, where the energies are the rungs of the ladder. The lowest rung is the ground state, with

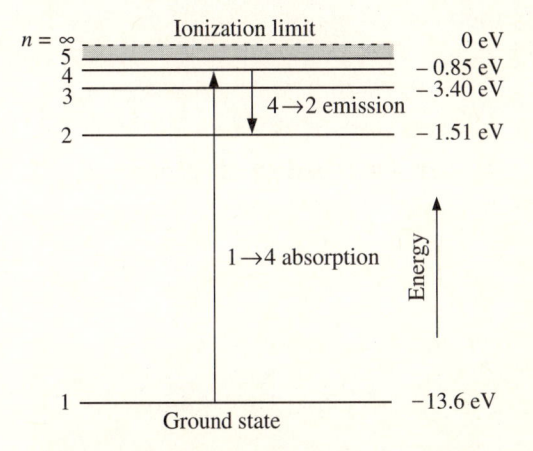

FIGURE 34-5 An energy-level diagram for the hydrogen atom. This is a ladder-like picture, not a graph. States $n = 6 \to \infty$ are shown as a shaded band. Vertical arrows show a quantum jump between states with the emission or absorption of a photon.

$E_1 = -13.6$ eV. The top rung, with $E = 0$, corresponds to a hydrogen ion in the limit $n \to \infty$. This top rung is called the **ionization limit**. There are, in principle, an infinite number of rungs, but only the lowest few are shown. Compare these to the energies shown in Table 34-1. The higher values of n are all crowded together just below the $n = \infty$ ionization limit.

It is not easy, in practice, to make n very large. A hydrogen atom in the $n = 10$ state is bound by only 0.136 eV. Such an atom is so fragile that it is easily ionized in collisions with other atoms. Physicists do, though, study states of very high n under high-vacuum conditions where collisions are negligible, and astronomers detect the emission of radio signals from very high n states of hydrogen in interstellar gas clouds. Both laboratory experiments and astronomers have observed hydrogen in states up to $n \approx 300$. These are macroscopic-size atoms, with diameters of approximately 10 μm = 0.01 mm! They are, however, bound by a mere 0.00015 eV.

EXAMPLE 34-3 Can an electron in a hydrogen atom have a speed of 3.60×10^5 m/s? If so, what is its energy and the radius of its orbit? What about a speed of 3.65×10^5 m/s?

SOLUTION To be in a stationary state of the hydrogen atom, the electron must have speed

$$v_n = \frac{v_1}{n} = \frac{2.19 \times 10^6 \text{ m/s}}{n},$$

where n is an integer. A speed of 3.60×10^5 m/s would require quantum number

$$n = \frac{2.19 \times 10^6 \text{ m/s}}{3.60 \times 10^5 \text{ m/s}} = 6.08.$$

This is not an integer, so the electron *cannot* have this speed. But if $v = 3.65 \times 10^5$ m/s, then

$$n = \frac{2.19 \times 10^6 \text{ m/s}}{3.65 \times 10^5 \text{ m/s}} = 6.$$

This is the speed of an electron in the $n = 6$ excited state. An electron in this state has energy

$$E_6 = -\frac{13.60 \text{ eV}}{6^2} = -0.38 \text{ eV},$$

and the radius of its orbit is

$$r_6 = 6^2 (5.29 \times 10^{-11} \text{ nm}) = 1.90 \times 10^{-9} \text{ m} = 1.90 \text{ nm}.$$

34.4 The Hydrogen Spectrum

Our analysis of the Bohr atom has revealed stationary states. But does this have any connection with reality? How do we know whether this model makes any sense? The most important experimental evidence that we have about the hydrogen atom is its spectrum, so the primary test of the Bohr atom is whether it correctly predicts the spectrum.

Energy-level diagrams are useful for showing transitions, or quantum jumps, in which a photon of light is emitted or absorbed. For example, Fig. 34-5 shows a $1 \to 4$ transition in which a photon is absorbed and a $4 \to 2$ transition in which a photon is emitted. Consider a transition between quantum state m and quantum state n, having energies E_m and E_n, where $n > m$. That is, E_n is the higher-energy state. An atom can *emit* a photon in an $n \to m$ transition or *absorb* a photon in an $m \to n$ transition. The frequency will be the same in both cases, but we will focus just on emission because it was the emission spectrum that Balmer was able to characterize with his formula.

According to the fourth assumption of Bohr's model, as given in Eq. 34-1, the frequency of the photon emitted in an $n \to m$ transition is

$$f = \frac{\Delta E_{\text{atom}}}{h} = \frac{E_n - E_m}{h}. \tag{34-17}$$

Our analysis of the Bohr atom has yielded an explicit result (Eq. 34-15) for the energies E_n and E_m in Eq. 34-17. Thus we predict that the emitted photon has frequency

$$\begin{aligned}f &= \frac{1}{h} \cdot \left\{ \left[-\frac{1}{n^2} \left(\frac{e^2}{4\pi\varepsilon_0 \cdot 2a_B} \right) \right] - \left[-\frac{1}{m^2} \left(\frac{e^2}{4\pi\varepsilon_0 \cdot 2a_B} \right) \right] \right\} \\ &= \frac{e^2}{4\pi\varepsilon_0 \cdot 2ha_B} \cdot \left(\frac{1}{m^2} - \frac{1}{n^2} \right).\end{aligned} \tag{34-18}$$

Because $m < n$, we have $1/m^2 > 1/n^2$ and the frequency, as expected, is a positive number.

We are more interested in wavelength than frequency, because wavelengths are the quantity measured by experiment. Using $\lambda = c/f$, the wavelength of this photon is

$$\lambda_{n \to m} = \frac{c}{f} = \frac{4\pi\varepsilon_0 \cdot 2hca_B / e^2}{\left(\dfrac{1}{m^2} - \dfrac{1}{n^2} \right)}. \tag{34-19}$$

This looks rather gruesome, but notice that the numerator is simply a collection of various constants. We can group these together, calling the numerator λ_0, and evaluate it. Doing so gives

$$\lambda_0 = \frac{4\pi\varepsilon_0 \cdot 2hca_B}{e^2} = 9.112 \times 10^{-8} \text{ m} = 91.12 \text{ nm}. \tag{34-20}$$

With this definition, we can write our prediction for the wavelengths in the hydrogen emission spectrum as

$$\lambda_{n \to m} = \frac{\lambda_0}{\left(\dfrac{1}{m^2} - \dfrac{1}{n^2} \right)}, \quad m = 1, 2, 3, \ldots \quad n = m+1, m+2, \ldots . \tag{34-21}$$

This result should look familiar. It is the Balmer formula from Chapter 32! There is, however, one *slight* difference: Bohr's model of the hydrogen atom has predicted $\lambda_0 = 91.12$ nm, whereas Balmer found, from experiment, that $\lambda_0 = 91.18$ nm. Could Bohr have come this close but then failed to predict the Balmer formula correctly? The problem, it turns out, is in our assumption that the proton remains at rest while the electron orbits it. In fact, as a more advanced analysis shows, *both* particles rotate about their common center of mass, rather like a dumbbell with a big end and a small end. The center of mass is very close to the proton, because it is far more massive than the electron, but the proton is not entirely motionless. The good news is that the more advanced analysis can account for the proton's motion. It changes the energies of the stationary states ever so slightly—about 1 part in 2000—but that is precisely what is needed to give a revised value,

$$\lambda_0 = 91.18 \text{ nm} \text{ when corrected for the nuclear motion.}$$

It works! Unlike all the previous atomic models, the Bohr atom predicts exactly and precisely the discrete spectrum of the hydrogen atom. Figure 34-6 shows the *Balmer series* and the *Lyman series* on an energy-level diagram. Only the Balmer series, which produces transitions ending on the $m = 2$ state, gives visible wavelengths, and this is the series that Balmer initially analyzed. The Lyman series, ending on the $m = 1$ ground state, is in the ultraviolet region of the spectrum and was not measured until a later date. These series, as well as others in the infrared that end on $m = 3$, $m = 4$, etc., are observed in a discharge tube where collisions with electrons excite the atoms upward from the ground state to state n. They then decay downward by emitting photons. Only the Lyman series is observed in the absorption spectrum because, as noted previously, essentially all the atoms in a quiescent gas are in the ground state.

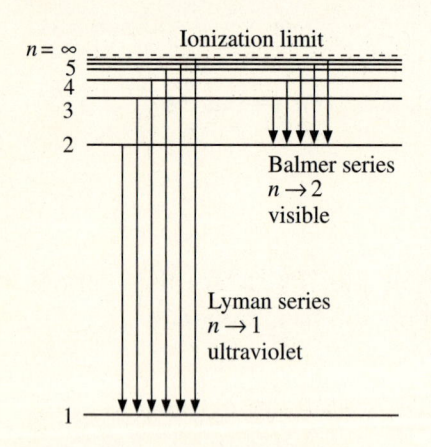

FIGURE 34-6 Transitions producing the Lyman series and the Balmer series of lines in the hydrogen spectrum.

EXAMPLE 34-4 Whenever astronomers look at distant galaxies, they always find that the light has been strongly absorbed at the wavelength of the $1 \to 2$ transition in the Lyman series of hydrogen. This absorption tells us that interstellar space is filled with vast clouds of hydrogen left over from the Big Bang. What is the wavelength of the $1 \to 2$ absorption in hydrogen?

SOLUTION Equation 34-21 predicts the *absorption* spectrum of hydrogen if we let $n = 1$. The absorption seen by astronomers is from the ground state of hydrogen ($n = 1$) to its first excited state ($m = 2$). From Eq. 34-21, the wavelength is

$$\lambda_{1 \to 2} = \frac{91.18 \text{ nm}}{\left(\frac{1}{1^2} - \frac{1}{2^2} \right)} = 121.6 \text{ nm}.$$

This wavelength is far into the ultraviolet. Ground-based astronomy cannot observe this region of the spectrum because the wavelengths are strongly absorbed by the atmosphere. But with space-based telescopes, first widely used in the 1970s, astronomers see 121.6 nm absorption in nearly every direction they look.

34.5 Hydrogen-like Ions

Any ion with a *single* electron orbiting Z protons in the nucleus is called a **hydrogen-like ion**. Z, you will recall, is the atomic number and describes the number of protons in the nucleus. The nuclear charge is thus $q_{\text{nuc}} = Ze$. So both He$^+$, with one electron circling a $Z = 2$ nucleus, and Li^{++}, with one electron and a $Z = 3$ nucleus, are hydrogen-like ions. So is U^{+91}, with one lonely electron orbiting a $Z = 92$ uranium nucleus.

Any hydrogen-like ion is simply a variation on the Bohr atom. The only difference between a hydrogen-like ion and neutral hydrogen is that the potential energy $-e^2/4\pi\varepsilon_0 r$ becomes, instead, $-Ze^2/4\pi\varepsilon_0 r$. Hydrogen itself is the $Z = 1$ case. If we repeat the analysis of the previous sections with this one change, we find

$$r_n = \frac{n^2 a_B}{Z}$$

$$v_n = Z \cdot \frac{v_1}{n}$$

$$E_n = -Z^2 \cdot \frac{13.60 \text{ eV}}{n^2} \qquad (34\text{-}22)$$

$$\lambda_0 = \frac{91.18 \text{ nm}}{Z^2}.$$

As the nuclear charge increases, the electron moves in to a smaller-diameter, higher-speed orbit. Its ionization energy $|E_1|$ increases significantly, and its spectrum shifts to smaller wavelengths. Table 34-2 compares the ground state atomic diameter $D = 2r_1$, the ionization energy $|E_1|$, and the first wavelength $3 \to 2$ in the Balmer series for hydrogen and the first two hydrogen-like ions.

TABLE 34-2 Comparison of hydrogen-like ions with $Z = 1$, 2, and 3.

Ion	Diameter $D = 2r_1$	Ionization Energy $|E_1|$	Wavelength of $3 \to 2$
H ($Z = 1$)	0.106 nm	13.6 eV	656 nm
He$^+$ ($Z = 2$)	0.053 nm	54.4 eV	164 nm
Li^{++} ($Z = 3$)	0.035 nm	125.1 eV	73 nm

34.6 The Quantization of Angular Momentum

Our study of orbital motion in this text has been limited to circular orbits. They have included the gravitational-force orbits of planets around the sun, or satellites around the earth, and now the analogous electrical-force orbits of an electron around a nucleus. In both cases, the force vector is always directed toward a fixed point—the sun in one case, the nucleus in another.

Circular orbits are actually rather rare. It turns out that elliptical orbits are also possible in these situations where the force is directed at a fixed point. As you may recall, from

Chapter 12, Kepler's first law about planetary motions is that planets move in *ellipses* with the sun at one focus of the ellipse. Similarly, electrons can follow an elliptical orbit about the nucleus. A circular orbit is simply a special case when the major and minor axes of the ellipse happen to be equal.

The mathematics of elliptical orbits gets rather complex, which is why we have stuck with the circular orbits, but there is one piece of information that we now need. Figure 34-7 shows an elliptical orbit with the force center at one focus. This orbit could be either a satellite moving in response to the gravitational force or an electron orbiting a nucleus. The important thing to notice is that, in general, the position vector $\vec{r}$ and the velocity vector $\vec{v}$ are *not* perpendicular to each other. The angle θ between them is 90° only for the special case of a circular orbit.

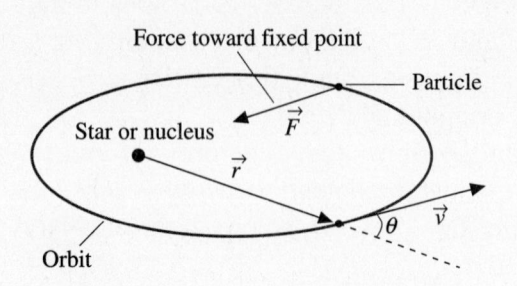

FIGURE 34-7 Elliptical orbit of a particle about a star or a nucleus. The force is always directed toward the fixed point.

Let's define a quantity called the **angular momentum** as

$$L = \text{angular momentum} = mvr\sin\theta. \qquad (34\text{-}23)$$

You should be able to convince yourself that angular momentum has SI units of J s. It can be shown, although we will not do so here, that the angular momentum is *conserved* for a particle moving under the influence of a force directed at a fixed point. As you know from our study of energy and momentum (strictly called *linear* momentum to distinguish it from angular momentum), conserved quantities take on special significance. Those of you who go on to take an advanced dynamics class will find that angular momentum plays a critical role in the analysis of rotational motion.

Orbital motion, whether of a planet or an electron, must conserve both energy *and* angular momentum. That is, the quantities r and v change as the particle moves around the ellipse to keep the product $vr\sin\theta$ constant. A circular orbit is a special case with $\theta = 90°$ and $\sin\theta = 1$ at all points, so $L = mvr$.

Now Bohr was fully aware of the significance of angular momentum to orbital motion. He used conservation of energy explicitly in his analysis of the hydrogen atom, but what role does conservation of angular momentum play? The condition that a de Broglie wave for the electron set up a standing wave around the circumference was given, in Eq. 34-7, as

$$2\pi r = n\lambda = n \cdot \frac{h}{mv}.$$

We can rewrite this as

$$mvr = n \cdot \frac{h}{2\pi} = n\hbar. \qquad (34\text{-}24)$$

But mvr is just the angular momentum L for a particle in a circular orbit, which is what Bohr was considering. The units of h and $\hbar$ are, in fact, J s—exactly those of angular momentum!

This led Bohr to the conclusion that the angular momentum of an orbiting electron cannot have just any value. Instead, it must satisfy

$$L = n\hbar \quad n = 1, 2, 3, \ldots . \tag{34-25}$$

That is, angular momentum is quantized! It can only have values that are integer multiples of $\hbar$. Bohr recognized that the quantization of angular momentum was likely to be as significant for the understanding of atoms as Einstein's quantization of energy had been for the understanding of light.

Many textbooks introduce the Bohr atom by *asserting* that angular momentum is quantized. This is a bit of a trick, because they provide no evidence for such an assertion. But we have seen good evidence that matter *does* have wave-like properties and is characterized by its de Broglie wavelength. The quantization of angular momentum, as expressed in Eq. 34-25, is a direct consequence of this wave-like nature of the electron. We will find that the quantization of angular momentum plays a big role in behavior of more complex atoms, leading to the idea of electron "shells" that you likely have studied in chemistry.

EXAMPLE 34-5 Show that the first three states of the Bohr atom have angular momentum $\hbar$, $2\hbar$, and $3\hbar$, respectively.

SOLUTION The radius and speed for the first three states were given in Table 34-1. The mass of the orbiting electron is $m = 9.11 \times 10^{-31}$ kg. So we can calculate

$n = 1 \quad r_1 = 5.29 \times 10^{-11}$ m $\quad v_1 = 2.19 \times 10^6$ m/s $\quad L = mv_1r_1 = 1.05 \times 10^{-34}$ J s $= \hbar$

$n = 2 \quad r_2 = 2.12 \times 10^{-10}$ m $\quad v_2 = 1.09 \times 10^6$ m/s $\quad L = mv_2r_2 = 2.10 \times 10^{-34}$ J s $= 2\hbar$

$n = 3 \quad r_3 = 4.76 \times 10^{-10}$ m $\quad v_3 = 7.30 \times 10^5$ m/s $\quad L = mv_3r_3 = 3.16 \times 10^{-34}$ J s $= 3\hbar$.

34.7 Success and Failure

Bohr's model of the hydrogen atom seemed to be a resounding success. By introducing the idea of stationary states, along with Einstein's ideas about light quanta, Bohr was able to provide the first solid understanding of discrete spectra and, in particular, the Balmer formula for the wavelengths in the hydrogen spectrum. The Bohr atom, because of the presence of a ground state, was also stable. Unlike the basic Rutherford model, the electron did not execute a death spiral into the nucleus. There was clearly some validity to the idea of stationary states.

But the Bohr atom was completely unsuccessful at explaining the spectra of any other atom. It did not work even for helium, the second element in the periodic table, with a mere two electrons. Something inherent in the assumptions of the Bohr atom seemed to work all right for a single electron but not in situations with two or more electrons. The difficulty was that the electrons in multielectron atoms repel one another with forces comparable in magnitude to their attraction to the nucleus. This greatly upsets their orbits, and the simple Bohr atom fails.

We made a distinction earlier between the Bohr *model*, as described in Section 34.2, and the Bohr *atom*, which is a specific description of the hydrogen atom. The Bohr model assumes that stationary states exist, but it does not say how to find them. The Bohr atom, on the other hand, finds the stationary states in a hydrogen atom by requiring that an integer number of de Broglie waves fit around the circumference of the orbit, setting up standing waves. The difficulty with more complex atoms is not the Bohr model, but with the method of finding the stationary states. The model remains valid, and we will continue to use it, but the procedure of fitting standing waves to a circle is just too simple to find the stationary states of complex atoms. We need to find a better procedure.

Einstein, Bohr, and de Broglie carried physics into uncharted waters. Their successes made it clear that the microscopic realm of light and atoms is governed by quantization, discreteness, and a blurring of the distinction between particles and waves. Although Bohr was clearly on the right track, his inability to extend the Bohr atom to more complex atoms made it equally clear that the complete and correct theory remained to be discovered. Bohr's theory was what we now call "semiclassical," a hybrid of classical Newtonian mechanics with the new ideas of quanta. Apparently what was still missing was a theory of motion and dynamics in a quantized universe—a *quantum* mechanics.

Summary

Important Concepts and Terms

Bohr model of the atom
stationary state
ground state
excited state
transition
quantum jump
collisional excitation
Bohr atom

Bohr radius
binding energy
ionization energy
energy-level diagram
ionization limit
hydrogen-like ion
angular momentum

Bohr introduced quanta into atoms, just as Einstein had introduced quanta to light. There are two crucial features of the *Bohr model* of atomic structure:

1. Atoms can exist only in certain stationary states, each having a specific energy E_1, E_2, E_3, ... The possible energies of an atom are quantized.

2. An atom can jump from one stationary state to another, undergoing an energy change ΔE, by the emission or absorption of a photon of frequency

$$f = \frac{\Delta E}{h}.$$

These aspects of the Bohr model explain the stability of matter and the existence of discrete spectra.

The Bohr model does not say anything about how to find the stationary states. Bohr went on to develop a specific model of the hydrogen atom, which we now call the *Bohr atom*. It is important to distinguish between the Bohr model, which applies to atoms in general, and the Bohr atom, which is specific to hydrogen and hydrogen-like ions (one-electron systems). By requiring the electron to be a standing de Broglie wave around the circumference of a circular orbit, we found the stationary states of hydrogen and hydrogen-like ions to be

$$r_n = \frac{n^2 a_B}{Z}$$
$$E_n = -\frac{13.60 Z^2 \text{ eV}}{n^2} \qquad n = 1, 2, 3, \ldots,$$

where Z is the number of protons in the nucleus ($Z = 1$ for neutral hydrogen) and a_B is the Bohr radius: $a_B = 0.0529$ nm. It is the radius of a neutral hydrogen atom in its ground state.

From these results, we were able to derive the Balmer formula for the wavelengths in the spectrum of hydrogen:

$$\lambda_{n \to m} = \frac{\lambda_0}{\left(\frac{1}{m^2} - \frac{1}{n^2}\right)} \qquad m = 1, 2, 3, \ldots \quad n = m+1, m+2, \ldots,$$

where $\lambda_0 = (91.18/Z^2)$ nm. We also introduced the idea of an energy-level diagram, a pictorial tool that we will use extensively in the next few chapters.

The Bohr atom for hydrogen was very successful, but it failed for atoms having more than one electron. The ideas of the Bohr model are valid, and we will continue to use them, but we need a better procedure for identifying the stationary states in more complex systems.

Exercises and Problems

Exercises

1. Show, by direct calculation, that the Bohr radius really is 0.0529 nm and that the ground state energy level of hydrogen really is −13.60 eV.

2. Determine the wavelengths of all the possible photons that can be emitted from the $n = 4$ state of a hydrogen atom.

3. a. What quantum number of the hydrogen atom comes closest to giving an electron orbit of diameter 500 nm?
   b. What is the electron's speed and energy in this state?

4. a. Calculate the de Broglie wavelength of the electron in the $n = 1, 2$, and 3 states of the Bohr atom. Use the information in Table 34-1.
   b. Show numerically that the circumference of the orbit for each of these stationary states is exactly equal to n de Broglie wavelengths.
   c. Sketch the de Broglie standing wave for the $n = 3$ orbit of the Bohr atom.

5. a. Find the radius of the electron's orbit, the electron's speed, and the energy of the atom for the first three stationary states of He$^+$.
   b. Show that the angular momentum in each state is equal to $n\hbar$.

Problems

6. The first three energy levels of the fictitious element X are shown in Fig. 34-8.
   a. What is the ionization energy of element X?
   b. What wavelengths are observed in the absorption spectrum of element X? Express your answers in nm.
   c. State whether each of your wavelengths in part b) corresponds to ultraviolet, visible, or infrared light.
   d. An electron with a speed of 1.4×10^6 m/s collides with an atom of element X. Shortly afterward, the atom emits a 1240 nm photon. What was the electron's speed after the collision? Assume, because the atom is so much more massive than the electron, that the recoil of the atom is negligible. (**Hint:** The energy of the photon is *not* the energy transferred to the atom in the collision. Think about this carefully.)

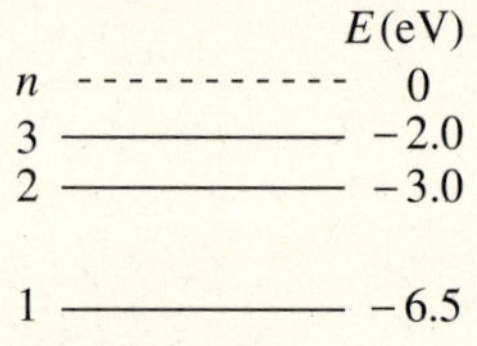

```
                              E(eV)
n  ----------  0
3  _____ -2.0
2  _____ -3.0

1  _____ -6.5
     FIGURE 34-8
```

7. Draw an energy-level diagram, similar to Fig. 34-5, for the He⁺ ion. Show the following on your diagram:
   a. The first five energy levels. Label each with the values of n and E_n.
   b. The ionization limit.
   c. All possible emission transitions from the $n = 4$ energy level.
   d. Calculate the wavelengths (in nm) for each of the transitions in part c) and show them alongside the appropriate arrow.

8. a. Calculate the orbital radius and the speed of an electron both in the $n = 99$ and in the $n = 100$ state of hydrogen.
   b. Determine the orbital frequency of the electron in each of these states.
   c. Calculate the frequency of a photon emitted in a $100 \rightarrow 99$ transition.
   d. Compare the photon frequency of part c) to the *average* of your two orbital frequencies from part b). By what percentage do they differ?

9. a. A beam of electrons is incident upon a gas of hydrogen atoms. What minimum speed must the electrons have to cause the emission of 656 nm light from the $3 \rightarrow 2$ transition of hydrogen?
   b. Through what potential difference must the electrons be accelerated to have this speed?

10. The *muon* is a subatomic particle with the same charge as an electron but a mass that is 207 times greater: $m_\mu = 207 m_e$. Physicists generally think of muons as "heavy electrons." The muon is not a stable particle. It decays, after an average lifetime of 1.5 μs, into an electron plus two neutrinos. Muons are created in the upper atmosphere by the impact of cosmic rays—typically protons moving at nearly the speed of light—on atmospheric molecules. Many of these muons reach the ground, where they can be "captured" by the nuclei of the atoms in a solid. A captured muon orbits this nucleus, like an electron, until it decays. Because the muon is often captured into an excited orbit ($n > 1$), its presence can be detected by observing the photons emitted in transitions such as $2 \rightarrow 1$ and $3 \rightarrow 1$.

Consider a muon captured by a carbon nucleus ($Z = 6$). Due to its mass, the muon orbits well *inside* the electron cloud. It "sees" the full nuclear charge Ze and, to a good approximation, is not affected by the distant electrons.

a. What are the orbital radius and speed of a muon in the $n = 1$ ground state? (**Hint:** The mass of the muon differs from the mass of the electron.)
b. What is the wavelength of the $2 \to 1$ muon transition?
c. What kind of photon is is emitted in the $2 \to 1$ transition—infrared, visible, ultraviolet, or x ray?
d. How many orbits will the muon complete during a 1.5 μs lifetime? Is this a sufficiently large number that the Bohr model "makes sense," even though the muon is not a stable particle?

[**Estimated 5 additional problems for the final edition.**]

Chapter 35

Wave Functions and Probabilities

> **LOOKING BACK** Sections 16.3, 16.4; 18.3, 18.4; 33.4, 33.5

35.1 First Steps Toward a Quantum Theory

As you have seen in the last three chapters, the classical mechanics of Newton and the classical electromagnetism of Faraday and Maxwell were unable to explain the new phenomena associated with light, electrons, and atoms. Scientific theories that triumphed during the eighteenth and nineteenth centuries were, at the beginning of the twentieth century, stumbling over the smallest specks of matter. Many scientists refused to accept these limitations of classical physics, thinking that it was only a matter of time until someone discovered how to apply classical ideas to atoms. Their hopes were to go unfulfilled.

At the same time, the new ideas put forward by Einstein, Bohr, and de Broglie began pointing the way toward a new theory of light and matter. This theory reconciled the wavelike and particle-like aspects of electrons and photons, and it predicted the quantization of energy and angular momentum. **Quantum mechanics**, as the theory came to be called, did not reach its completed form until the late 1920s, but it has since proven to be the most successful physical theory ever devised.

It may seem surprising how slowly we have been building up to quantum mechanics. Why not just write it down and start using it? The difficulty is twofold. First, quantum theory explains microscopic phenomena that we cannot directly sense or experience. It was important to look at the experimental evidence for how light and atoms behave so that we could understand why a new theory was needed. Second, the concepts used in quantum mechanics are quite abstract. Making the connection between theory and reality is a difficult issue in quantum theory, so, as a starting point, let's take a brief look at just what a theory is.

What does it mean to be a theory? Although this is not a class on scientific methodology, we can loosely say that a physical theory requires three basic ingredients:

1. A *descriptor*. This is a mathematical quantity used to describe our knowledge of a physical object.

2. A *causative agent*. This is what causes the descriptor to change. Without causative agents, the universe would be in unchanging equilibrium. For a physical theory, the essential properties of a causative agent need to be expressed mathematically.

3. One or more *laws* that govern the behavior of the descriptor.

For example, Newtonian mechanics is a theory of motion. The primary descriptor in Newtonian mechanics is a particle's *position* $x(t)$ as a function of time. This describes our knowledge of the particle at all times. (Mathematical manipulations of $x(t)$—taking derivatives—tell us the particle's velocity and acceleration, but these are not independent pieces of information.) The causative agents are *forces*, and the laws are *Newton's laws*. These laws, especially the second law, are mathematical statements of how the descriptor changes in response to various causative agents. If we predict $x(t)$ for a known set of forces, we can feel confident that an experiment carried out at time t will find the particle right where predicted.

This method for describing motion may seem obvious, but that has not always been the case. It would have made no sense at all prior to Descartes's invention of Cartesian geometry and Galileo's introduction of the concept of measuring $x(t)$ for moving objects. It seems obvious to us only because we have all grown up and been educated in a world in which mathematical laws and regularities are used to describe nature *and* because our everyday experience leads us to believe that the position of an object—a rock, a chair, ourselves—is a valid and well-defined description of the object.

Newton's theory of motion *assumes* that $x(t)$, which gives a well-defined position at every instant of time, is a valid description of motion. The difficulty facing physicists early in twentieth century was the astounding discovery that the position of an atomic-size particle is *not* well defined. For example, an electron in a double-slit experiment must, in some sense, go through *both* slits to produce an electron interference pattern. We cannot deduce which slit it goes through because it simply does not have a well-defined position $x(t)$ as it interacts with the slits. But if the position function $x(t)$ is not a valid descriptor for matter at the atomic level, what is?

The goal of this chapter is to learn about a new descriptor, called the *wave function*, that is valid in the microscopic realm of wave–particle duality. We will examine the properties of this descriptor and, more importantly, will consider how to *interpret* it. That is, we will look at how we can connect this mathematical descriptor to the reality of experimental measurement. Then, in the next chapter, we will introduce the "laws" that govern this descriptor. Our approach will be analogous to the beginning of Newtonian mechanics. There you spent several chapters learning how to *describe* motion, particularly how to use and interpret the difficult concept of acceleration, before we introduced the *laws* of motion.

It may seem to you, as we go along, that we are simply "making up" ideas. That is, indeed, at least partially true. The inventors and discoverers of entirely new theories use their existing knowledge as a guide, but ultimately they have to make an "inspired guess" as to what a new theory should look like. Newton and Einstein both made such leaps, and now the inventors of a successful quantum theory had to make such a leap. New theories are not derivable from the old—that is why they are new! We can attempt to make the new ideas *plausible*, and much of this chapter will attempt to do so, but ultimately a new theory is simply a bold new assertion that must be tested against experimental reality. Many other hypotheses can be, and were, put forward, but they did not meet the experimental tests and are now of interest mostly to historians. The descriptor of quantum mechanics that we will "invent" passed the only test that really matters in science—it worked!

35.2 Waves, Particles, and the Double-Slit Experiment

We will make extensive use in this chapter of the double-slit interference experiment, so you may find it helpful to review the portions of Chapter 16 indicated in the Looking Back section. The significance of the double-slit experiment arises from the fact that both light and matter exhibit the same interference pattern. Furthermore, the interference pattern of light is found whether we detect the light as a wave or as a collection of photons. The double-slit experiment, it seems, goes right to the heart of wave–particle duality. We are seeking a new descriptor to connect the particle-like and wave-like aspects of matter, so the double-slit experiment is a good place to focus our attention.

Regardless of whether we look at photons, electrons, or neutrons passing through the slits, their detection is a particle-like "event." They make a collection of discrete dots on a piece of film. Yet our understanding of how interference "works" is based on a wave analysis. Can we find a way of connecting these wave-like and particle-like behaviors? Let us begin by looking at the interference of light, then we will return to the interference of matter in the next section.

A Wave Analysis of Interference

We can analyze the interference of light from either a wave perspective or a photon perspective. Let's start with a wave analysis. Figure 35-1 shows a double-slit apparatus along with pictures and graphs to describe the situation. Notice that all the graphs and pictures are aligned vertically. You can see several wave fronts impinging on the screen, where they undergo constructive and destructive interference. Recall that the lines in a wave front diagram represent wave crests spaced one wavelength apart; the wave troughs are halfway between. Our goal is to find a mathematical description of the light intensity on the screen.

We analyzed the double-slit experiment in Chapter 16. The two waves traveling from

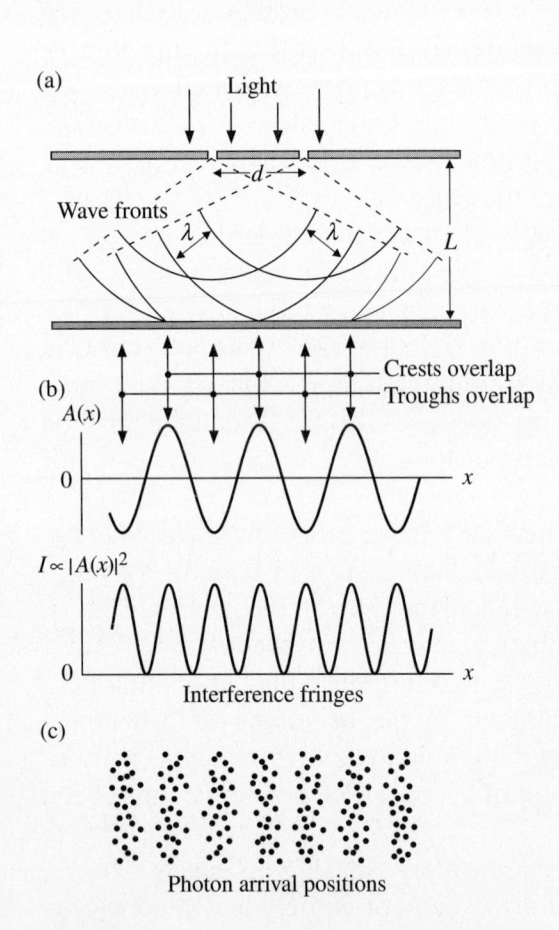

FIGURE 35-1 a) The double-slit experiment with light. b) The top graph shows the wave amplitude $A(x)$ on the screen, and the lower graph shows the light intensity, which is proportional to $A^2(x)$. c) A picture from the photon perspective.

the slits to the screen are traveling waves with displacements
$$D_1 = a\sin(kr_1 - \omega t)$$
$$D_2 = a\sin(kr_2 - \omega t),$$
where a is the amplitude, $k = 2\pi/\lambda$ is the wave number, and r_1 and r_2 are the distances from the two slits. For simplicity, we have assumed that the phase constants of the waves are both zero. Keep in mind that the "displacement" of a light wave is not a physical displacement, as in a water wave, but a change or displacement of the electromagnetic field.

According to the principle of superposition, these two waves will add together where they meet at a point on the screen. The net displacement at the screen (see Eqs. 16-16 and 16-17) is

$$D_{net} = D_1 + D_2 = a\sin(kr_1 - \omega t) + a\sin(kr_2 - \omega t)$$
$$= \left[2a\cos\left(\frac{\pi \Delta r}{\lambda}\right)\right]\sin(kr_{avg} - \omega t), \tag{35-1}$$

where $\Delta r = |r_1 - r_2|$ is the path difference of the two light waves. The sine term simply characterizes the rapid oscillations of the light wave traveling outward from the slits. We are interested in the term in brackets because this term describes the *amplitude* of the interference.

We found in Chapter 16 that Δr is related to the horizontal coordinate x on the screen by

$$\Delta r = \frac{dx}{L}, \tag{35-2}$$

where d is the spacing between the slits and L is the distance between the slits and the screen. Using Eq. 35-2, we can write the amplitude portion of Eq. 35-1 as

$$A(x) = 2a\cos\left(\frac{\pi d x}{\lambda L}\right). \tag{35-3}$$

The function $A(x)$ is called the *amplitude function* because it describes the wave's amplitude A as a function of the position x on the viewing screen. The top graph in Fig. 35-1b shows the amplitude function $A(x)$. Notice how it has crests where two crests from individual waves overlap, interfering constructively to make a larger crest. It also has troughs where two individual troughs overlap. $A(x)$ is zero at points where the two individual waves are out of phase and interfere destructively.

What we *observe* on the screen is not the amplitude but the light's *intensity*. A wave's intensity I (see Eq. 14-16) is proportional to the square of the wave's amplitude: $I \propto A^2$. Using Eq. 35-3 for the amplitude at each point, the intensity $I(x)$ as a function of position x on the screen is

$$I(x) = C\cos^2\left(\frac{\pi d x}{\lambda L}\right), \tag{35-4}$$

where C is a proportionality constant. The lower graph in Fig. 35-1b shows a graph of the intensity as a function of position along the screen. Here you can recognize the interference fringes, bright strips of maximum intensity alternating with dark strips of zero intensity. Note that there are bright fringes both where two wave crests overlap, making the amplitude $A(x)$ a maximum, *and* where two troughs overlap so that $A(x)$ is most negative.

A Photon Analysis of Interference

Now that we've done a wave analysis of the double-slit experiment, let's look at it from a photon perspective. We know, from experimental evidence, that the interference pattern is built up photon by photon if the overall light intensity is very weak. Figure 35-1c shows the pattern made on a film after the arrival of the first few dozen photons. It is clearly a double-slit interference pattern, but one made—as in newspaper photographs—by piling up dots in some places but not others.

The most important thing to recognize when looking at photons is that the arrival position of a photon is *unpredictable*. Nothing about how the experiment is set up or conducted allows us to predict exactly where the dot of an individual photon will appear on the detector. A second photon will not arrive at the same place as a previous photon, even if it is prepared in exactly the same way and sent through the apparatus in exactly the same way. Yet there is clearly an overall *pattern* to the dots on the screen. There are some positions at which a photon is *more likely* to be detected, other positions at which it is *less likely* to be found.

Suppose we conduct an experiment in which we fire individual photons, one at a time, toward a double slit. After recording the arrival positions of many thousands of these, we will be able to determine the *probability* of where a photon will be detected. For example, if 50 out of 50,000 photons land in one small area of the screen, we can deduce that each photon has a probability of 50/50,000 = 0.001 = 0.1% of being detected there. The probability will be zero at the points we call the *interference minima*, because no photons at all arrive at those points. Similarly, the probability will be a maximum at the *interference maxima*. The probability will have some in-between value on the "sides" of the interference fringes. Although we cannot predict the precise arrival point of a photon, we can determine the *probabilities* for various outcomes.

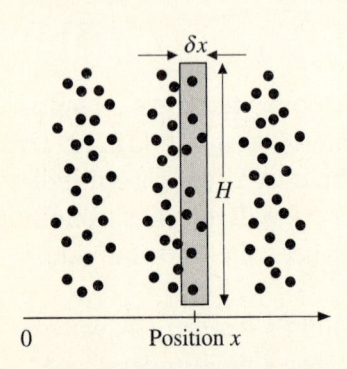

FIGURE 35-2 A strip of width δx at position x. Energy δE falls on this strip each second due to the arrival of $N = \delta E/hf$ photons.

Consider a small strip located at position x on the detection screen, as shown in Fig. 35-2. This strip has width δx and height H. We will assume that δx is very small in comparison with the fringe spacing, so the light's intensity over δx is very nearly constant. If N_{total} photons pass through the slits, the number that arrive in this small strip will be designated N(in δx at x). This number will vary from point to point, depending on the position x at which the strip is located.

On average, the number of photons arriving in this small strip is the total number of photons fired at the slits multiplied by the *probability* that any one photon ends up in the strip. That is,

$$N(\text{in } \delta x \text{ at } x) = N_{total} \cdot \text{Prob}(\text{in } \delta x \text{ at } x), \qquad (35\text{-}5)$$

where Prob(in δx at x) is the probability that *one* photon ends up making a dot within the small width δx at the position x. This is like predicting the number of sixes you will get if you throw N dice. Because the probability that any one die will be a six is 1/6, the expected number of sixes is $N_{total}/6$. Throwing 60 dice would, on average, yield 10 sixes. Although we cannot give a precise position to the photon, we can locate it within some *region* of space δx. Prob(in δx at x) tells us the probability of this happening.

Connecting the Wave and Photon Views

It is clear, from Fig. 35-1, that there is a correlation between the *intensity of the waves* and the *probabilities of the photons*. That is, photons are more likely to be detected at those points where the electromagnetic wave intensity is high and less likely to be detected at those points where the wave intensity is low. This suggests a link between the wave-like and the particle-like perspectives, so let us see if we can analyze this further.

The intensity of a wave was defined as $I = P/A$, the ratio of light power P (joules per second) to the area A on which the light falls. Intensity has units of watts per square meter, or joules per second per square meter. Consider the small strip shown in Fig. 35-2. This strip has area $A = H\delta x$. If the intensity at position x is $I(x)$, the amount of light energy falling onto this little strip during each second is

$$\delta E(\text{in } \delta x \text{ at } x) = I(x) \cdot A = I(x) \cdot H \delta x. \tag{35-6}$$

You might think of this narrow strip as being a detector to measure the energy. As you move the detector around, placing it at various locations on the screen, the energy it receives each second will be high if it is placed in a bright interference fringe. If you place it in a dark fringe of destructive interference, where $I(x)$ is low, it will receive very little energy each second. Thus the notation $\delta E(\text{in } \delta x \text{ at } x)$ refers to the energy received per second by this small detector, of width δx, if you place it at position x.

Now the light intensity is proportional to $|A(x)|^2$, the square of the amplitude function, so we can write

$$\delta E(\text{in } \delta x \text{ at } x) \propto |A(x)|^2 \delta x. \tag{35-7}$$

Because Eq. 35-7 contains an unspecified proportionality constant ($a \propto b$ *means* that $a = $ constant $\times b$), any multiplicative constants, such as the H in Eq. 35-6, can be incorporated into the proportionality constant.

From the photon perspective, energy δE is really a collection of N photons, each of energy hf, such that

$$\delta E = Nhf. \tag{35-8}$$

So instead of talking about how much energy falls each second on our little strip, we can talk about how many photons arrive each second on the strip. Because N and δE are directly proportional to each other, we have

$$N(\text{in } \delta x \text{ at } x) \propto |A(x)|^2 \delta x, \tag{35-9}$$

where the hf has been incorporated into the proportionality constant.

Combining Eqs. 35-5 and 35-9, with N_{total} pulled into the proportionality constant, gives

$$\text{Prob}(\text{in } \delta x \text{ at } x) \propto |A(x)|^2 \delta x. \tag{35-10}$$

In other words, the probability of a photon arriving at a particular point is *directly proportional* to the square of the amplitude function at that point or, equivalently, to the intensity. If the wave intensity at point A is twice that at point B, then a photon is twice as likely to land in a small strip at A as it is to land in an equal-width strip at B.

The ideas of the last few paragraphs, culminating in Eq. 35-10, are critically important. Equation 35-10 is the connection between the particle perspective and the wave perspective. It relates the probability of observing a particle-like event—the arrival of a photon—to the amplitude of a continuous, classical wave. This connection will become the basis of how we interpret the results of a quantum-mechanical calculation. Make sure that you understand what is being said, in particular what is meant by Prob(in δx at x).

We need one last definition. You will recall that we found the mass of a length L of a wire or string in terms of its linear mass density: $m = \mu L$. Similarly, we found the charge along a length L of a wire in terms of its linear charge density: $Q = \lambda L$. If the length had been very short, in which case we might have denoted it as δx, and if the density varied from point to point, we could have written

$$\text{mass(in length } \delta x \text{ at } x) = \mu(x) \cdot \delta x$$
$$\text{charge(in length } \delta x \text{ at } x) = \lambda(x) \cdot \delta x, \tag{35-11}$$

where $\mu(x)$ and $\lambda(x)$ are the densities at position x. Writing Eq. 35-11 this way separates the role of the density from the role of the small length δx.

With these as an analogy, let us define the **probability density** $P(x)$ such that

$$\text{Prob(in } \delta x \text{ at } x) = P(x) \cdot \delta x. \tag{35-12}$$

The probability density has SI units of m^{-1} so that when multiplied by a width, as in Eq. 35-12, it yields a dimensionless probability. Note that $P(x)$ is *not* a probability by itself, just as the linear mass density λ is not, by itself, a mass. You *must* multiply the probability density by a width, as shown in Eq. 35-12, to find an actual probability.

EXAMPLE 35-1 In an experiment, 1% of all the photons arrive at a 1-mm-wide strip located at position $x = 50$ cm. What is the probability density at $x = 50$ cm?

SOLUTION The probability that photon arrives at this particular strip is

$$\text{Prob(in 1 mm at } x = 50 \text{ cm)} = 0.010.$$

Using Eq. 35-12, the probability density at this position is

$$P(x = 50 \text{ cm}) = \frac{\text{Prob(in 0.001 m at } x = 50 \text{ cm)}}{0.001 \text{ m}} = \frac{0.01}{0.001 \text{ m}} = 1.0 \times 10^{-5} \text{ m}^{-1}.$$

Comparing Eq. 35-12 to Eq. 35-10, we find that the probability density for the arrival of a photon at position x is

$$P(x) \propto |A(x)|^2 \quad \text{or} \quad P(x) = C \cdot |A(x)|^2 \tag{35-13}$$

where C is a proportionality constant. The probability density, unlike the probability itself, is independent of δx and depends only on x.

We were led to this line of reasoning by noticing the correlation in Fig. 35-1 between the wave intensity and the photons' positions. But nothing in the analysis depends on the double-slit geometry, so our result, Eq. 35-13, is quite general. It says that for *any* experiment in which we detect photons, the probability density for detecting a photon is directly proportional to the square of the amplitude function of the corresponding electromagnetic wave. We now have an explicit connection between the wave-like and the particle-like properties of the light.

In the specific case of the double-slit experiment, for which we know $|A(x)|^2$ from Eq. 35-3:

$$P_{\text{two-slit}}(x) = C \cdot |A(x)|^2 = C \cdot \cos^2\left(\frac{\pi d x}{\lambda L}\right). \tag{35-14}$$

You will learn, in Section 35.4, how to determine the proportionality constant C. We will then be able to perform some numerical examples to make these ideas clear.

35.3 The Wave Function

Now let's look at the interference of matter. Electrons passing through a double-slit apparatus create the same interference patterns as photons. The pattern is built up electron by electron, but there is no way to predict where any particular electron will go. By measuring many individual electrons, we could establish the *probability* of an electron landing in a small strip of width δx. The probability, as you might guess, turns out to be exactly the same as the arrival probability of a photon that has the same wavelength.

For light, we determined the probability density $P(x)$ by connecting the photon probabilities to the intensity of an electromagnetic wave. But these is no wave for electrons that is analogous to electromagnetic waves for light. So how do we find the probability density for electrons?

We have reached the point where we must make an inspired leap beyond classical physics. Let us *assume* that there is some kind of continuous, wave-like function for matter that plays a role analogous to electromagnetic waves for light. We will call this function the **wave function** $\psi(x)$, where ψ is a lowercase Greek "psi."

To be precise, $\psi(x)$ is called the *time-independent wave function*. The complete wave function, called the *time-dependent wave function*, oscillates in time and is given by

$$\Psi(x,t) = \psi(x)\cos\omega t.$$

The full time-dependent wave function is very important in more advanced applications of quantum mechanics, but all the examples we will look at in this text can be analyzed with the simpler $\psi(x)$. The time-independent wave function $\psi(x)$ depends only on the position x and is analogous to the amplitude function $A(x)$ for light waves.

To connect the wave function with the real world, we will interpret $\psi(x)$ in terms of the *probability* of finding a particle at position x. If a particle, such as an electron, is described by the wave function $\psi(x)$, then the probability Prob(in δx at x) of locating the particle within a region of width δx at position x is

$$\begin{aligned}\text{Prob(in } \delta x \text{ at } x) &= |\psi(x)|^2 \, \delta x \\ &= P(x)\delta x.\end{aligned} \tag{35-15}$$

Equation 35-15 defines the probability density $P(x)$ for the particle as

$$P(x) = |\psi(x)|^2. \tag{35-16}$$

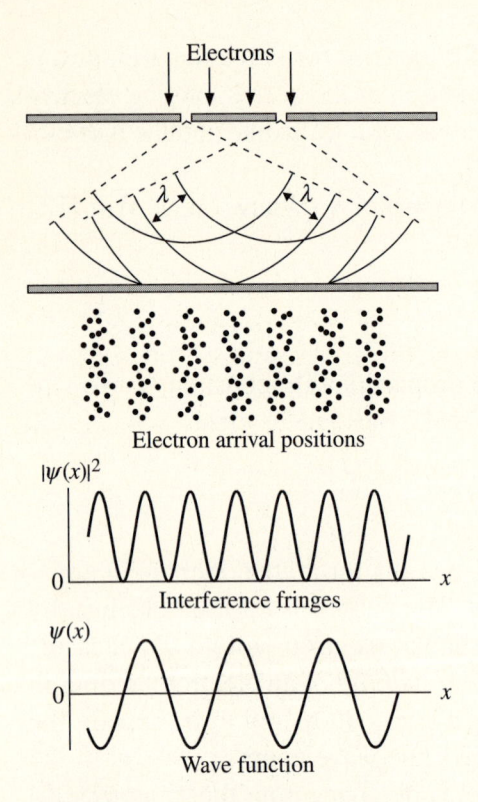

FIGURE 35-3 The double-slit experiment with electrons. We can experimentally determine the wave function from the arrival probabilities.

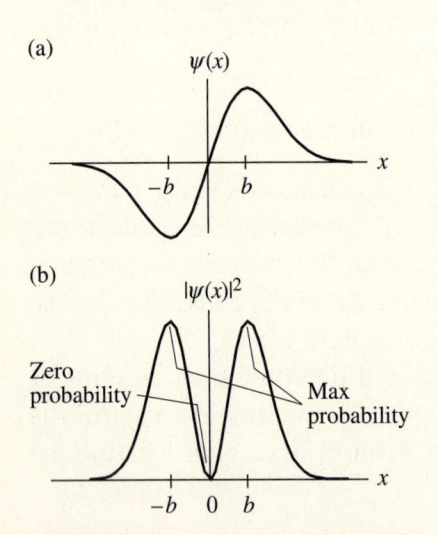

FIGURE 35-4 a) A typical electron wave function. b) The square of the wave function describes the probability of detecting the electron at various values of the position x.

With Eqs. 35-15 and 35-16 we are *defining* the wave function $\psi(x)$ to play the same role for material particles that the amplitude function $A(x)$ does for photons. The only difference is that we had $P_{\text{photon}}(x) \propto |A(x)|^2$, whereas Eq. 35-16 is $P_{\text{particle}}(x) = |\psi(x)|^2$. The difference arises due to the fact that the electromagnetic field amplitude $A(x)$ had previously been defined through the laws of electricity and magnetism. We found that $|A(x)|^2$ is *proportional* to the probability density for finding a photon, but it is not directly *the* probability density. But we don't have any preexisting definition for the wave function $\psi(x)$. Thus we are free to *define* $\psi(x)$ such that $|\psi(x)|^2$ is *exactly* the probability density. That is why we used = rather than $\propto$ in Eq. 35-16.

Figure 35-3 shows the double-slit experiment for electrons. This time, however, we work backward. From the observed distribution of electrons, which represents the probabilities of where they will land, we can deduce $|\psi(x)|^2$, the square of the wave function. Its square root is the oscillatory wave function $\psi(x)$ at the screen. Note the very close analogy with the amplitude function $A(x)$.

Figure 35-4a shows a different example of a wave function. After squaring it at each point, as shown in Fig. 35-4b, we find that this wave function represents a particle most likely to be detected at $x = -b$ or $x = +b$. These are the points where $|\psi(x)|^2$ is a maximum. There is zero likelihood of finding the particle right in the center. The particle is more likely to be detected at some positions than at others, but—unlike in classical physics—we cannot predict its exact location. The wave function, from which we can predict probabilities, is all we know about the particle.

Equation 35-16 defines the wave function $\psi(x)$ for a particle in terms of the probability of finding the particle at different positions x. But our interests go beyond merely characterizing experimental data. We would like to develop a new *theory* of matter. With that in mind, we assert that the wave function $\psi(x)$ is the *descriptor* of a particle in quantum mechanics, just as the position function $x(t)$ is the descriptor of a particle in Newtonian mechanics. That is, the wave function of a "particle," such as an electron, tells us everything we can know about the particle! In particular, it tells us the probability for finding the particle in a given region of space. Whether this hypothesis has any merit will not be known until we see if it leads to predictions that can be verified.

One of the difficulties in learning to use the concept of a wave function is coming to grips with the fact that there is no "thing" that is waving. There is no medium of any sort oscillating up and down. Some people like to call ψ a "probability wave," although that is really not accurate because probability is not something that can wave. It is simply a wavelike function that can be used to make predictions—probabilistic predictions—about atomic particles. Keep in mind that the purpose of this chapter is just to introduce the wave function as the descriptor of matter in quantum mechanics. Chapter 36 will then introduce the "laws" that tell us how to find the wave function in a particular situation.

35.4 Normalization

Suppose you blindly throw a ball 200 times toward buckets A, B, and C. It lands in bucket A 60 times, bucket B 120 times, and bucket C 20 times. All 200 throws are accounted for. What is the *probability* that your next throw will land in A, B, or C? Let $N_A = 60$, $N_B = 120$, and $N_C = 20$ be the number of balls in each bucket, with $N_{total} = 200$. The *percentage* of throws going into A was $N_A/N_{total} = 60/200 = 0.30 = 30\%$. Extending this reasoning, it seems clear that the probabilities are

$$P_A = \frac{N_A}{N_{total}} = 0.30 \qquad P_B = \frac{N_B}{N_{total}} = 0.60 \qquad P_C = \frac{N_C}{N_{total}} = 0.10.$$

But there's an assumption at work here. Because 100% of the balls ended up in one of the three buckets, we assumed that the sum of the three probabilities *must* add up to 1. That is, we required that $P_A + P_B + P_C = 1$, which is the mathematical way of saying that the ball *has* to go into one of the three buckets.

The counts N_A, N_B, and N_C are not themselves probabilities, but they are *proportional to* the probabilities. That is, $P_A = C \cdot N_A$, where C is a proportionality constant. It was not hard in this case to realize that the proper value of C, the value that makes $P_A + P_B + P_C = 1$, is simply $C = 1/N_{total}$. The process of finding the constant C, so that we have real probabilities rather than raw data, is called **normalization**. Normalization is a matter of scaling the data, with an appropriate multiplicative constant, to get probabilities that sum to 1.

We need to apply the normalization idea to wave functions. Consider a single electron passing through two slits and being detected at the screen. The probability of being detected in a small width δx at the specific position x_1 is

$$\text{Prob(in } \delta x \text{ at } x_1) = P(x_1)\delta x.$$

Similarly, $P(x_2)\delta x$ is the probability of arriving at a different position x_2 and $P(x_i)\delta x$ is the probability for arriving at an arbitrary position x_i. Now the electron must land *somewhere* on the screen, so the total probability of its being detected somewhere must be exactly 1— just as the probability was exactly 1 that the ball would land in one of the three buckets.

If we divide the entire screen between position x_1 and x_N into many adjacent strips of width δx, as shown in Fig. 35-5, then the probability of the electron landing *somewhere* in the range $x_1 - x_N$ is

$$\text{Prob(electron in range } x_1 - x_N) = \sum_{i=1}^{N} P(x_i)\delta x = \sum_{i=1}^{N} |\psi(x_i)|^2 \delta x. \qquad (35\text{-}17)$$

If we let the strips become narrower and narrower, then $\delta x \to dx$ and the sum becomes an integral. Then if we let the screen become infinitely wide, so that the probability of the electron arriving *somewhere* on the screen becomes 1, the integration limits become $\pm\infty$. The statement that the electron *has* to hit the screen *somewhere* is thus expressed mathematically as

$$\int_{-\infty}^{\infty} P(x)\,dx = \int_{-\infty}^{\infty} |\psi(x)|^2\,dx = 1. \qquad (35\text{-}18)$$

A proper wave function must satisfy Eq. 35-18, which is called the **normalization condition**. Some examples will clarify these ideas.

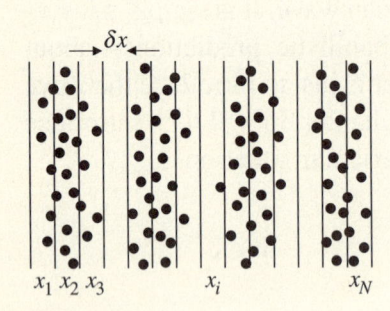

FIGURE 35-5 Dividing the entire screen into many small strips of width δx.

EXAMPLE 35-2 Figure 35-6 shows the wave function of a particle confined within the region between $x = 0$ and $x = L = 1$ nm. The wave function is zero outside this region.

a) Determine the value of the constant b, as defined in Fig. 35-6.

b) Draw a graph of the probability density $P(x)$.

c) Draw a "dot picture" showing where the first 40 or 50 particles might be found.

d) Calculate the probability of finding the particle in a region of width $\delta x = 0.01$ nm at positions $x_1 = 0$, $x_2 = 0.5$ nm, and $x_3 = 1$ nm.

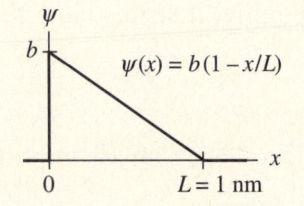

FIGURE 35-6 The wave function of Example 35-1.

SOLUTION a) The wave function is $\psi(x) = b(1 - x/L)$, decreasing linearly from $\psi = b$ at $x = 0$ to $\psi = 0$ at $x = L = 1$ nm. The constant b, which is the "height" of this wave function, is determined by requiring the wave function to meet the normalization condition of Eq. 35-18. That is, the particle *has* to be in the region $0 \le x \le L$, so the constant b has to be chosen to make this so. Because the wave function of Fig. 35-6 is zero outside the interval from 0 to L, the integration limits are 0 to L. Equation 35-18 is

$$1 = \int_0^L |\psi(x)|^2\,dx = b^2 \int_0^L \left(1 - \frac{x}{L}\right)^2 dx = b^2 \int_0^L \left(1 - \frac{2x}{L} + \frac{x^2}{L^2}\right) dx$$

$$= b^2 \cdot \left[x - \frac{x^2}{L} + \frac{x^3}{3L^2}\right]_0^L = \frac{1}{3}b^2 L.$$

Solving for b gives

$$b = \sqrt{\frac{3}{L}} = \sqrt{\frac{3}{1 \text{ nm}}} = 1.732 \text{ nm}^{-1/2}.$$

The units for b are unusual, but they are exactly what we need for the probability density $P(x)$ to have units of nm^{-1}. Although these are not SI units, we can correctly compute probabilities as long as δx has units of nm. Multiplicative constants such as b are often called *normalization constants* because their value is "adjusted" to normalize the wave function.

b) The wave function is

$$\psi(x) = 1.732 \text{ nm}^{-1/2} \cdot \left(1 - \frac{x}{1 \text{ nm}}\right).$$

Thus the probability density is

$$P(x) = |\psi(x)|^2 = (3 \text{ nm}^{-1})\left(1 - \frac{x}{1 \text{ m}}\right)^2.$$

This probability density is graphed in Fig. 35-7a.

c) Particles are most likely to be detected on the left edge of the interval, where the probability density $P(x)$ is maximum. The probability steadily decreases across the interval, becoming zero at $x = 1$ nm. Figure 35-7b shows how a group of particles described by this wave function might appear on a detection screen.

d) The probability of finding the particle in a region of width δx at the position x is

$$\text{Prob(in } \delta x \text{ at } x) = P(x)\delta x = |\psi(x)|^2 \delta x.$$

We are given $\delta x = 0.01$ nm, which is small in comparison with $L = 1$ nm, so we need to evaluate $|\psi(x)|^2$ at the three positions $x_1 = 0$, $x_2 = 0.5$ nm, and $x_3 = 1$ nm. Doing so gives

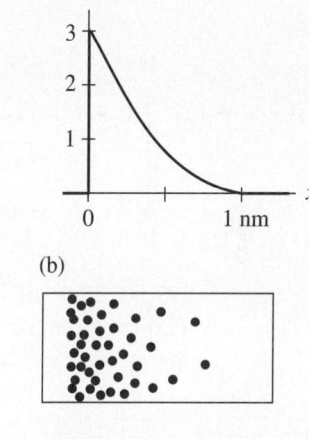

Prob(in 0.01 nm at $x_1 = 0.0$ nm) $= b^2(1 - x_1/L)^2 \cdot \delta x = 0.0300 = 3.00\%$

Prob(in 0.01 nm at $x_2 = 0.5$ nm) $= b^2(1 - x_2/L)^2 \cdot \delta x = 0.0075 = 0.75\%$

Prob(in 0.01 nm at $x_3 = 1.0$ nm) $= b^2(1 - x_3/L)^2 \cdot \delta x = 0$.

FIGURE 35-7 a) The probability density $P(x)$ for the wave function of Example 35-2. b) The detection of particles described by this wave function.

The calculation of probabilities in Example 35-2 made use of the fact that δx was "small." This made δx the equivalent of one of the strips in Fig. 35-5. However, we might need to know the probability for finding the particle within some not-so-small range $x_1 \leq x \leq x_2$. This is simply the sum over all the strips δx within this range. As $\delta x \to dx$, this becomes

$$\text{Prob(in range } x_1 \leq x \leq x_2) = \int_{x_1}^{x_2} P(x)dx = \int_{x_1}^{x_2} |\psi(x)|^2 \, dx. \quad (35\text{-}19)$$

We can interpret Prob(in range $x_1 \leq x \leq x_2$) as the *area* under the probability density curve between x_1 and x_2.

EXAMPLE 35-3 Figure 35-8a shows the wave function

$$\psi(x) = \begin{cases} 0 & x < 0 \\ ce^{-x/L} & x \geq 0, \end{cases}$$

where $L = 1$ nm.
a) Determine the value of the constant c.
b) Draw a graph of the probability density $P(x)$.
c) Calculate the probability of finding the particle in the region $x \geq 1$ nm.

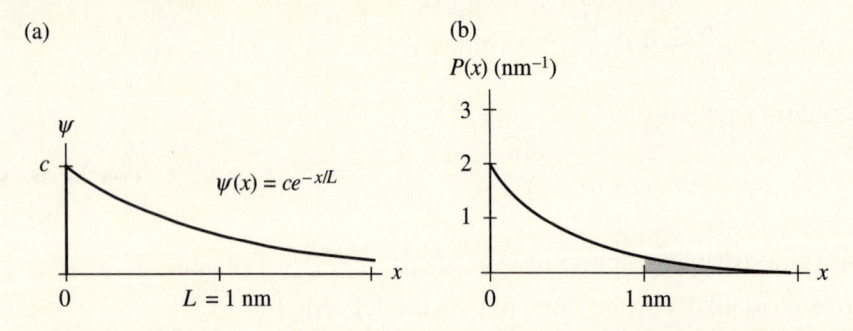

FIGURE 35-8 a) The wave function of Example 35-3. b) The probability density $P(x)$ for Example 35-3. The shaded area under the curve is the probability that the particle will be found at $x \geq 1$ nm.

SOLUTION a) The wave function is an exponential $\psi(x) = ce^{-x/L}$ that extends from $x = 0$ to $x = +\infty$. Equation 35-18 for the normalization is

$$1 = \int_0^\infty |\psi(x)|^2\, dx = c^2 \int_0^\infty e^{-2x/L}\, dx = -\frac{c^2 L}{2} e^{-2x/L}\Big|_0^\infty = \frac{c^2}{2L}.$$

We can solve this for the normalization constant c:

$$c = \sqrt{\frac{2}{L}} = \sqrt{\frac{2}{1\text{ nm}}} = 1.414 \text{ nm}^{-1/2}.$$

b) The probability density is

$$P(x) = |\psi(x)|^2 = (2\text{ nm}^{-1})e^{-2x/1\text{ nm}}.$$

The probability density is graphed in Fig. 35-8b.

c) The probability of finding the particle in the region $x \geq 1$ nm is Prob($1\text{ nm} \leq x \leq \infty$). This probability is the shaded area under the curve in Fig. 35-8b. Using Eq. 35-19, the probability is

$$\text{Prob}(x \geq 1\text{ nm}) = \int_{1\text{ nm}}^\infty |\psi(x)|^2\, dx = (2\text{ nm}^{-1})\int_1^\infty e^{-2x}\, dx$$

$$= -e^{-2x}\Big|_1^\infty = e^{-2} = 0.135 = 13.5\%.$$

There is a 13.5% chance of finding the particle beyond 1 nm, or a 86.5% chance of finding it within the interval $0 \leq x \leq 1$ nm, but—unlike classical physics—we cannot make an exact prediction of the particle's position.

EXAMPLE 35-4 Figure 35-9 shows a more realistic double-slit interference pattern for electrons. Unlike a purely cosine wave function, which extends to infinity, the interference fringes steadily decrease in intensity. The probability density shown in Fig. 35-9 is

$$P(x) = |\psi(x)|^2 = \begin{cases} C \cdot \left(1 - \dfrac{|x|}{5w}\right) \cos^2\left(\dfrac{\pi x}{w}\right) & |x| \leq 5w \\ 0 & |x| > 5w, \end{cases}$$

where w is the width of each fringe. Consider an experiment in which the fringe width is $w = 10$ nm. Determine the probability of an electron landing within the central maximum.

SOLUTION First we must determine the normalization constant C. The wave function is nonzero only within the range $-5w \leq x \leq 5w$. Integrating with absolute values can be tricky, as you know from calculus, but notice that $P(x)$ is symmetrical about $x = 0$. We can integrate just the positive side from 0 to $5w$, where the absolute value sign is not needed, then multiply by 2 to get the full value of the integral. The normalization condition of Eq. 35-18 is

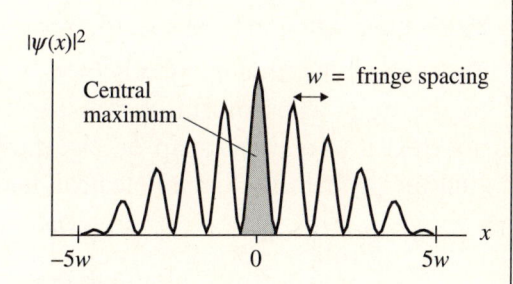

FIGURE 35-9 Probability density of a "real" double-slit electron interference. The shaded area is the probability that an electron will land within the central maximum of the interference pattern.

$$1 = \int_{-5w}^{5w} |\psi(x)|^2 \, dx = 2\int_{0}^{5w} |\psi(x)|^2 \, dx = 2C \int_{0}^{5w} \left(1 - \frac{x}{5w}\right) \cos^2\left(\frac{\pi x}{w}\right) dx$$

$$= 2C \left[\int_{0}^{5w} \cos^2\left(\frac{\pi x}{w}\right) dx - \frac{1}{5w} \int_{0}^{5w} x \cos^2\left(\frac{\pi x}{w}\right) dx \right].$$

These integrals are somewhat more complex than integrations we have needed up until now, but you have faced similar integrals in calculus. You can find these in integral tables or by using software that does symbolic integration. They are

$$\int_{0}^{5w} \cos^2\left(\frac{\pi x}{w}\right) dx = \frac{w}{2\pi} \left[\cos\left(\frac{\pi x}{w}\right) \sin\left(\frac{\pi x}{w}\right) + \frac{\pi x}{w} \right]_{0}^{5w}$$

$$= \frac{5w}{2}$$

and

$$\frac{1}{5w} \int_{0}^{5w} x \cos^2\left(\frac{\pi x}{w}\right) dx = \frac{w}{20\pi^2} \left[\frac{2\pi x}{w} \cos\left(\frac{\pi x}{w}\right) \sin\left(\frac{\pi x}{w}\right) + \cos^2\left(\frac{\pi x}{w}\right) + \left(\frac{\pi x}{w}\right)^2 - 1 \right]_{0}^{5w}$$

$$= \frac{5w}{4}.$$

Although gruesome looking, these both simplified nicely with the recognition that all the sine and cosine terms are either 0 or 1. Placing the integrals back into the normalization condition gives

$$1 = 2C\left[\frac{5w}{2} - \frac{5w}{4}\right] = \frac{5}{2}Cw$$

$$\Rightarrow C = \frac{2}{5w} = \frac{2}{5 \cdot 10 \text{ nm}} = 0.040 \text{ nm}^{-1}.$$

We first had to find the correct value for C, making the wave function the "right size" to give a total probability of 1 for finding the electron *somewhere* on the screen. With C now known, we can compute specific probabilities.

The central maximum extends from $x_L = -w/2 = -5$ nm on the left to $x_R = +w/2 = +5$ nm on the right. The probability density changes dramatically within this region, so we cannot treat it as a single strip δx. We must, instead, perform the integration of Eq. 35-19 to find the probability that an electron lands in the central maximum. This integration is

$$\text{Prob}\left(-\frac{w}{2} \leq x \leq \frac{w}{2}\right) = \int_{-w/2}^{w/2} |\psi(x)|^2 \, dx = 2\int_0^{w/2} |\psi(x)|^2 \, dx$$

$$= 2C \int_0^{w/2} \left(1 - \frac{x}{5w}\right) \cos^2\left(\frac{\pi x}{w}\right) dx$$

$$= \frac{4}{5w}\left[\int_0^{w/2} \cos^2\left(\frac{\pi x}{w}\right) dx - \frac{1}{5w}\int_0^{w/2} x\cos^2\left(\frac{\pi x}{w}\right) dx\right].$$

These are the same integrals as before, only with a different upper limit. Skipping over the algebraic details, evaluation gives

$$\text{Prob}\left(-\frac{w}{2} \leq x \leq \frac{w}{2}\right) = \text{Prob(central max)} = \frac{4}{5}\left[\frac{19}{80} - \frac{1}{20\pi^2}\right] = 0.186 = 18.6\%.$$

Thus we find that each electron we fire toward the double slit has a probability of just under 20% of landing in the central maximum of the diffraction pattern. This has been an extended example, but it shows how to obtain precise probability information from the wave function. We will do similar calculations in the upcoming chapters.

Summary

Important Concepts and Terms

quantum mechanics   normalization
probability density   normalization condition
wave function

Because atomic particles exhibit wave-like aspects as well as particle-like aspects, they do not have a well-defined position $x(t)$. Our knowledge of matter at the atomic level is limited to measuring the probabilities of various possible outcomes. Consequently, we have introduced a new *descriptor* of matter called the *wave function*. The wave function allows us to predict the probability of finding a particle in a specified region of space.

A particle of matter, such as an electron or a neutron, is characterized by a wave function $\psi(x)$. This is the spatial part of a continuous oscillatory function defined at all points along the x-axis. The probability of finding the particle in a small region of width δx at the position x is given by

$$\text{Prob}(\text{particle in width } \delta x \text{ at position } x) = |\psi(x)|^2 \, \delta x.$$

We defined the probability density $P(x)$ as

$$P(x) = |\psi(x)|^2,$$

so we can also write

$$\text{Prob}(\text{in } \delta x \text{ at position } x) = P(x)\delta x.$$

$P(x)$ has units of m^{-1}.

The wave function is normalized such that

$$\int_{-\infty}^{\infty} P(x)\,dx = \int_{-\infty}^{\infty} |\psi(x)|^2 \, dx = 1.$$

This is the mathematical statement that the particle must be found *somewhere* along the x-axis. The probability for finding the particle within the finite range $x_1 - x_2$ is

$$\text{Prob}(\text{in range } x_1 \leq x \leq x_2) = \int_{x_1}^{x_2} P(x)\,dx = \int_{x_1}^{x_2} |\psi(x)|^2 \, dx.$$

One of the major challenges to understanding quantum mechanics is learning to *interpret* the wave function. What does $\psi(x)$ mean? How is it connected to reality or to quantities that you might measure in an experiment? It is quite clear what the classical mechanics descriptor $x(t)$ means, and how to use it, but not at all clear what to do with the new quantum descriptor $\psi(x)$. That is why we have spent this entire chapter trying to trying to show what ψ means before we attempt, in the next chapter, to calculate wave functions.

The wave function tells us not "what is" but merely "what is probable." This seems a meager kind of knowledge in comparison with the precision of Newtonian mechanics, but the experimental evidence has been that you simply cannot know and cannot predict the precise outcome of an experiment with an atomic particle. Probabilities are all that we can know at the microscopic level.

As unsettling as this seems, you soon will find that we still can make very accurate calculations of the energies of electrons in atoms. And when large numbers of electrons flow through a device, its behavior is as predictable as the pressure in a gas with large numbers of molecules. The averages of large numbers of particles, be they electrons in a device or molecules in a gas, are very well defined even though we cannot say anything about one individual particle. So quantum mechanics will be a useful tool, one that allows us to understand the properties of the atomic world and to make practical use of that knowledge. But using quantum mechanics successfully requires a new way of thinking about nature, and that is why we have been taking such a cautious and slow approach.

Exercises and Problems

Exercises

1. What are the units of ψ? Explain.
2. What is the difference between the probability and the probability density?

3. For the electron wave function shown in Fig. 35-10, at what position or positions is the electron most likely to be found? Least likely to be found? Explain how you can tell.

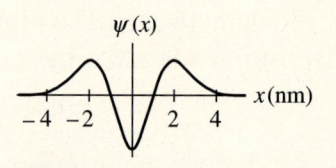

FIGURE 35-10

4. When performing an interference experiment with electrons, you find the most intense fringe at $x = 7$ cm, slightly weaker fringes at 6 and 8 cm, still weaker fringes at 4 and 10 cm, and two very weak fringes at 1 and 13 cm. No electrons at all are found for $x < 0$ cm or $x > 14$ cm.
   a. Sketch a graph of $|\psi(x)|^2$ for these electrons.
   b. Sketch a possible graph of $\psi(x)$.
   c. Are there other possible graphs for $\psi(x)$? If so, draw one.

5. An electron that is confined to $x \geq 0$ has the normalized wave function

$$\psi(x) = \begin{cases} 0 & x < 0 \\ (1.414 \text{ nm}^{-1/2})e^{-x/1 \text{ nm}} & x \geq 0. \end{cases}$$

   a. What is the probability of finding the electron in a region of width 0.01 nm at $x = 1$ nm?
   b. What is the probability of finding the electron in the interval $0.5 \text{ nm} \leq x \leq 1.5 \text{ nm}$?

Problems

6. Consider a single-slit diffraction experiment using electrons. (Single-slit diffraction was presented in Section 17.3, which you may wish to review.)
   a. Draw a picture, similar to Fig. 35-3, showing the arrival positions of the electrons.
   b. Draw a graph of $|\psi(x)|^2$ for the electrons on the detection screen.
   c. Draw a graph of $\psi(x)$ for the electrons. Keep in mind that ψ, as a wave, alternates between positive and negative.

7. Figure 35-11 shows the wave function of a particle confined within the region between $x = 0$ and $x = L = 1$ nm. The wave function is zero outside this region.
   a. Determine the value of the constant a, as defined in Fig. 35-11.
   b. Draw a graph of the probability density $P(x)$.
   c. Draw a "dot picture" showing where the first 40 or 50 particles might be found.
   d. Calculate the probability of finding the particle in the interval $0 \leq x \leq 0.3$ nm.

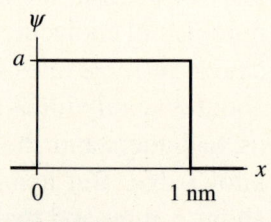

FIGURE 35-11

8. An experiment finds electrons to be uniformly distributed over the interval $0 \leq x \leq 2$ cm, with no electrons falling outside this interval.
   a. Draw a graph of $|\psi(x)|^2$ for these electrons.
   b. What is the probability that an electron will land within the interval 0.79–0.81 cm?
   c. If 10^6 electrons are detected, how many will be detected in the interval 0.79–0.81 cm?
   d. What is the probability density at $x = 0.80$ cm?

9. The graph in Fig. 35-12 shows $|\psi(x)|^2$ for the electrons in an experiment.
   a. Is the electron wave function normalized? Explain.
   b. Draw a graph of $\psi(x)$ over this same interval. Provide a numerical scale on both axes.
   c. What is the probability that an electron will be detected in a region of width 0.001 cm at $x = 0$ cm? At $x = 0.5$ cm? At $x = 1.0$ cm?
   d. If 10^4 electrons are detected, how many will be in the interval -0.3 cm $\leq x \leq 0.3$ cm?

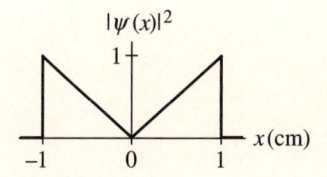

FIGURE 35-12

10. Consider the wave function

$$\psi(x) = \begin{cases} c\sqrt{1-x^2} & |x| \leq 1 \text{ cm} \\ 0 & |x| \geq 1 \text{ cm.} \end{cases}$$

   a. Determine the normalization constant c.
   b. Draw a graph of $\psi(x)$ over the interval -2 cm $\leq x \leq 2$ cm. Provide numerical scales on both axes.
   c. Draw a graph of $|\psi(x)|^2$ over the interval $-2 \leq x \leq 2$. Provide numerical scales.
   d. If 10^4 electrons are detected, how many will be in the interval $0 \leq x \leq 0.5$ cm?

[**Estimated 10 additional problems for the final edition.**]

Chapter 36

One-Dimensional Quantum Mechanics

I don't like it, and I'm sorry I ever had anything to do with it.

Erwin Schrödinger

LOOKING BACK Sections 10.6; 13.6; 33.5, 33.6; 34.2; 35.3, 35.4

36.1 Schrödinger and Quantum Mechanics

In the fall of 1925, just before Christmas, the Austrian physicist Erwin Schrödinger (Fig. 36-1) left his wife behind in Zurich, gathered together a few books and papers, picked up a former Viennese girlfriend, and headed off to a villa in the Swiss Alps. He had just recently learned of de Broglie's 1924 suggestion that matter has wave-like properties, and he wanted some time free from distractions—outside distractions at least!—to think about it. Before the trip was done, Schrödinger had discovered the primary law of quantum mechanics.

Schrödinger's goal was to learn how to calculate and predict the outcome of atomic experiments. The mathematical equation that he developed is now called the **Schrödinger equation**. For all intents and purposes, it is the "law" of quantum mechanics in much the same way that Newton's laws are the "laws" of classical mechanics. The Schrödinger equation tells us how to determine the wave function $\psi(x)$ of a particle in a specified situation. It would make perfect sense to call it "Schrödinger's law," but by tradition it is called simply "the Schrödinger equation."

The realm of quantum mechanics has steadily expanded since its introduction in the 1920s. Although initially developed to solve problems in physics, quantum mechanics is not just for physicists any more. The first applications of quantum mechanics were to chemistry, where the quantum mechanical model of the atom provided an understanding of

FIGURE 36-1 Erwin Schrödinger.

the electron shell structure of atoms, of the periodic table, and of the nature of the molecular bond. Starting in the 1930s, physicists began to apply quantum mechanics to the structure of solids. This developed into a whole field of study called *solid-state physics*. Quantum mechanics has provided an understanding of such diverse properties of solids as electrical conductivity, specific heat, magnetism, and superconductivity. Contemporary materials engineering relies heavily on these results from solid-state physics.

Over the past twenty years, quantum mechanics has become an essential tool in the design of semiconductor devices. As memory chips and microprocessors have became ever smaller and more powerful, the *size* of the features in integrated circuits has shrunk from a few microns in the 1970s to around 0.1 μm by the mid-1990s. The quantum nature of the atomic world has begun to manifest itself in structures this small. Whole new classes of devices, called *quantum-well devices*, have been designed and built to exploit the quantized energy levels. We will look at some examples in this chapter.

Another idea on the leading edge of engineering science is the design and manufacture of *nanostructures*—small machines or other devices only a few microns in size and shrinking as the technology improves. Quantum effects are important in devices this small. Some engineers envision a day in the near future when nanostructures will be constructed literally atom by atom to exacting specifications. Figure 36-2 shows a photograph of an early result of "atomic engineering" in which scientists at IBM's research laboratories built a curious symbolic structure by moving individual xenon atoms around on a metal surface.

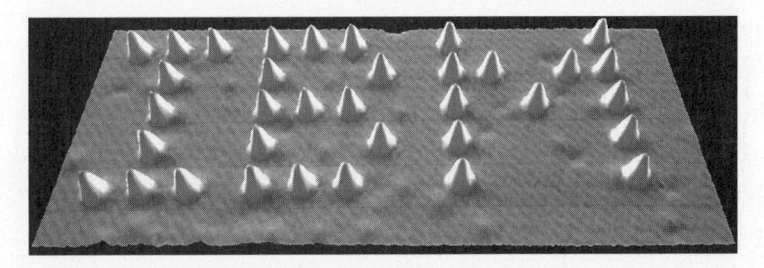

FIGURE 36-2 An example of atomic engineering. Thirty-five individual xenon atoms have been manipulated into position with the probe tip of a scanning tunneling microscope.

Our goal for this chapter is to introduce the Schrödinger equation and give examples of how it's used. We will discuss some of the mathematical aspects of solving the Schrödinger equation, and we will outline a problem-solving strategy for quantum mechanics. Then we will look at several applications that use the Schrödinger equation. Although the real world is three dimensional, we will limit our study of quantum mechanics to one dimension. This will allow us to focus on the essential ideas without becoming overwhelmed by mathematical complications.

36.2 Schrödinger's Equation: The Law of Psi

Let's consider an atomic particle of mass m whose interactions with the environment are described by a potential-energy function $U(x)$. In quantum mechanics, this particle is characterized by a wave function $\psi(x)$ that we can use to find the probability density for locating the particle. The Schrödinger equation for the particle's wave function is

$$\frac{d^2\psi}{dx^2} = -\frac{2m}{\hbar^2}[E - U(x)]\psi(x) \quad \text{(the Schrödinger equation)}, \tag{36-1}$$

where E is the particle's energy. This is a *differential equation* whose solution is the wave function $\psi(x)$ that we seek. Our goal in this section is to learn what this equation means and how it is used.

The Schrödinger equation cannot be derived or proved. It is not an outgrowth of any previous theory. Its success, as with any scientific theory, depended on its ability to explain the various phenomena that had mystified classical physics and to make new predictions that were subsequently verified.

Although the Schrödinger equation cannot be derived, the reasoning behind it can at least be made *plausible*. De Broglie had postulated a wave-like nature for matter in which a particle of mass m, velocity v, and momentum $p = mv$ has a wavelength

$$\lambda = \frac{h}{p} = \frac{h}{mv}. \tag{36-2}$$

Schrödinger's goal was to find a *wave equation* for which the solution would be a wave function having the de Broglie wavelength.

Recall from our earlier study of waves that the spatial part of a wave function having wavelength λ is

$$\psi(x) = \psi_0 \sin\left(\frac{2\pi x}{\lambda}\right), \tag{36-3}$$

where ψ_0 is the amplitude of the wave function. (Normally we would have to include some kind of $\cos\omega t$ term to describe the full wave, but here we are focusing on the spatial part only.) Suppose we take a second derivative of $\psi(x)$:

$$\frac{d\psi}{dx} = \frac{2\pi}{\lambda} \cdot \psi_0 \cos\left(\frac{2\pi x}{\lambda}\right)$$

$$\Rightarrow \frac{d^2\psi}{dx^2} = \frac{d}{dx}\frac{d\psi}{dx} = -\frac{(2\pi)^2}{\lambda^2} \cdot \psi_0 \sin\left(\frac{2\pi x}{\lambda}\right). \tag{36-4}$$

Using the Eq. 36-3 definition of $\psi(x)$, we can write the second derivative as

$$\frac{d^2\psi}{dx^2} = -\frac{(2\pi)^2}{\lambda^2} \cdot \psi(x). \tag{36-5}$$

A slight rearrangement gives

$$\frac{1}{\lambda^2} = -\frac{\dfrac{d^2\psi}{dx^2}}{(2\pi)^2 \psi(x)}. \tag{36-6}$$

Equation 36-6 relates the wavelength λ to a combination of the wave function $\psi(x)$ and its second derivative. Nothing about these manipulations has been specific to wave functions. Equation 36-6 was well known from the analysis of classical waves, and it applies equally well to sound waves or waves on a string.

Schrödinger's new insight was to identify λ with the de Broglie wavelength. From Eq. 36-2, we can write

$$\frac{1}{\lambda^2} = \frac{m^2 v^2}{h^2} = \frac{2m}{h^2}\left(\frac{1}{2}mv^2\right) = \frac{2mK}{h^2}, \qquad (36\text{-}7)$$

where K is the particle's kinetic energy. Using this expression for $1/\lambda^2$ in Eq. 36-6 gives

$$\frac{2mK}{h^2} = -\frac{\dfrac{d^2\psi}{dx^2}}{(2\pi)^2 \psi(x)},$$

which can be rearranged, using $\hbar = h/2\pi$, to read

$$\frac{d^2\psi}{dx^2} = -\frac{2m}{\hbar^2} K \psi(x). \qquad (36\text{-}8)$$

Equation 36-8 is a differential equation for the function $\psi(x)$. Its solution is the sinusoidal wave function of Eq. 36-3, where λ is the de Broglie wavelength for a particle having kinetic energy K. We know this because Eq. 36-8 has been specifically constructed so that Eq. 36-3 would be a solution.

Equation 30-8 includes an assumption that the kinetic energy K is constant. However, most particles are always changing speed as they convert kinetic energy to potential energy and vice versa, so their kinetic energy is *not* constant. The kinetic energy at position x is, in fact, given by

$$K = E - U(x), \qquad (36\text{-}9)$$

where E is the particle's total mechanical energy and $U(x)$ is the potential-energy function of the particle. That is, $U(x)$ is the particle's potential energy—gravitational or electrical or whatever kind it has—at the position x. If we use this expression for K in Eq. 36-8, we get

$$\frac{d^2\psi}{dx^2} = -\frac{2m}{\hbar^2}[E - U(x)]\psi(x),$$

which is Eq. 36-1—the Schrödinger equation for the wave function $\psi(x)$.

To be more precise, this equation is called the *time-independent Schrödinger equation in one dimension*. Many textbooks write it as

$$-\frac{\hbar^2}{2m}\frac{d^2\psi}{dx^2} + U(x)\psi = E\psi,$$

which is equivalent. This version appears a bit more "elegant" than Eq. 36-1, but we will find the Eq. 36-1 version to be more useful. This is analogous to our having written Newton's second law as $\vec{a} = \vec{F}/m$ rather than the more customary (and elegant) $\vec{F} = m\vec{a}$.

Note that this has not been a *derivation* of the Schrödinger equation. We have simply been "guessing" as to what seemed plausible, and only experience will show whether these guesses are correct.

Solving the Schrödinger Equation

The Schrödinger equation is a second-order differential equation, meaning that it is a differential equation for $\psi(x)$ involving its second derivative. This equation may look pretty complicated if you're not used to dealing with differential equations. But consider the following: The a in Newton's second law is really a second derivative: $a = d^2x/dt^2$. Furthermore, you may recall from Chapter 10 that the force on a particle is related to the slope of the potential energy curve by $F = -dU/dx$. So the "proper" way to write Newton's second law is

$$\frac{d^2x}{dt^2} = -\frac{1}{m}\frac{dU}{dx}.$$

This looks as bad as the Schrödinger equation! Fortunately, we were able to simplify this to the familiar $a = F/m$ by working with "special cases" in which F and a were constants.

The point is not to be overwhelmed with the mathematical appearance of the Schrödinger equation. The little bit of mathematical knowledge you need to work with the Schrödinger equation will be developed in this section. Just as we did with Newton's laws, we will restrict ourselves to situations where the necessary mathematical skills are those you have been developing in calculus.

The solution to an algebraic equation is simply a number. For example, $x = 3$ is the solution to the equation $2x = 6$. But the solution to a differential equation is a *function*. We can make this clear by "designing" a differential equation to which we know the solution. For example, consider the function

$$f(x) = x^3, \tag{36-10}$$

which has the second derivative

$$\frac{d^2f}{dx^2} = 6x. \tag{36-11}$$

Note that the second derivative is also a function of x, in this case the function $g(x) = 6x$.

We can rewrite $6x$ in terms of x^3 as

$$6x = \frac{6}{x^2} \cdot x^3. \tag{36-12}$$

Then using Eqs. 36-10 and 36-11, we can write Eq. 36-12 as

$$\frac{d^2f}{dx^2} = \left(\frac{6}{x^2}\right) f(x). \tag{36-13}$$

Equation 36-13 is a second-order differential equation. We know the solution to this differential equation—it is the function $f(x) = x^3$. This one was easy because we built the equation to have this solution. With the Schrödinger equation we do not know the solution in advance. The point, however, is that the solution to the Schrödinger equation is a *function* $\psi(x)$ that is defined at all values of x. We can evaluate $\psi(x)$ at any point on the x-axis or can display it as a graph of ψ-versus-x.

There are, however, restrictions on just which functions make *acceptable* solutions. There may be functions that satisfy the Schrödinger equation in a strictly mathematical sense but that are not physically meaningful. We have previously encountered restrictions in our solutions of algebraic equations in the sense that we insist, for physical reasons, that masses be positive rather than negative numbers, that positions be real rather than imaginary numbers, and so on. Because we want to interpret $|\psi(x)|^2$ as a probability density, we have to insist that the function $\psi(x)$ be one for which this interpretation is possible.

Because probabilities cannot be infinite, we have to insist that $\psi(x)$ be finite at all points. Because $\psi(x)$ has to be normalized, we have to insist that $\psi(x) \to 0$ as $x \to +\infty$ or $-\infty$. Otherwise the normalization integral would not converge. Because physically meaningful potential energies $U(x)$ are well-behaved functions, the second derivative of ψ must also be well behaved. This can only be true if $\psi(x)$ is a *continuous* function.

These conditions or restrictions on acceptable solutions are called the **boundary conditions**. You will see, in later examples, how the boundary conditions help us to choose the correct solution for $\psi(x)$. The primary conditions that the wave function must obey are:

1. $\psi(x) \to 0$ as $x \to +\infty$ or $-\infty$.
2. $\psi(x) = 0$ if x is in a region where it is physically impossible for the particle to be.
3. $\psi(x)$ is a continuous function.
4. $\psi(x)$ is a normalized function.

The last is not, strictly speaking, a boundary condition but is an auxiliary condition we require for the wave function to have a useful interpretation.

After we've established the boundary conditions, there are three general approaches that we can follow to solve the Schrödinger equation: 1) Use general techniques for solving second-order differential equations, 2) use a numerical technique to solve the equation on a computer, or 3) "guess" a solution.

More advanced courses make extensive use of the first and second approaches. Many of the wave functions we will present for discussion were found using one of these techniques. However, we are not assuming a knowledge of differential equations, so you will not be asked to use these methods. The last method, although it sounds almost like cheating, is widely used in simple situations where we can use physical arguments to infer the functional form of the wave function. You will see examples of this approach in the upcoming examples.

A differential equation often has several possible solutions. According to a theorem from differential equations, if $\psi_1(x)$ and $\psi_2(x)$ are two independent solutions of the Schrödinger equation, then the *general solution* of the equation is

$$\psi(x) = A\psi_1(x) + B\psi_2(x). \tag{36-14}$$

A and B are constants whose values are determined by the boundary conditions. By "independent solutions," we mean that $\psi_2(x)$ is not just a constant multiple of $\psi_1(x)$ (such as $3\psi_1(x)$) or an algebraic rearrangement of $\psi_1(x)$ but, instead, that $\psi_1(x)$ and $\psi_2(x)$ are totally different functions.

This is a powerful theorem, although one that will make more sense after you see it being applied in upcoming examples. The main point is that if we can find two solutions $\psi_1(x)$ and $\psi_2(x)$ by guessing, then Eq. 36-14 is *the* general solution to the Schrödinger equation.

Although the particle's total energy E appears in the Schrödinger equation, it is treated, as far as the equation is concerned, as an unspecified constant. As it turns out, however, there are *no* acceptable solutions for most values of E. That is, there are no functions $\psi(x)$ that satisfy both the Schrödinger equation *and* the boundary conditions. Acceptable solutions are found to exist only for a few *discrete* values of E. This implies, as we had hoped, that a particle's energy must be *quantized* in order for acceptable wave functions to exist! The solutions to the Schrödinger equation have quantization as a built-in feature, and the wave functions themselves thus describe the possible *stationary states* of the system.

36.3 Problem Solving in Quantum Mechanics

An important distinction between classical mechanics and quantum mechanics is the causative agent. In classical mechanics, it is *force* that causes acceleration and thus causes the velocity and position to change. Much of our problem-solving strategy in classical mechanics focused on identifying and using forces. In quantum mechanics, the causative agent is *energy*. It is the kinetic energy $K = E - U(x)$ in the Schrödinger equation that gives the wave function a nonzero second derivative. We need a new strategy focused on identifying and using the potential-energy function for a particle.

The critical step in a quantum-mechanical analysis is to determine the particle's potential-energy function $U(x)$. Finding an appropriate $U(x)$ involves analyzing the particle's interactions, making simplifying assumptions, then using the our knowledge about the potential energy of gravitational forces, spring forces, electrical forces, or whatever interactions are involved. This is the *physics* of the problem. Once the potential-energy function is known, it is "mere mathematics" to solve for the wave function.

We can summarize the problem-solving strategy for quantum mechanics as follows:

Strategy for Quantum Mechanics Problems

1. From the problem statement, identify a potential-energy function $U(x)$. Sketching a graph of $U(x)$ is almost always helpful. This step is the words-to-symbols transition, and identifying the potential-energy function is analogous to identifying forces in Newtonian mechanics. Force identification, you will recall, was often the most difficult part of solving a problem, and that is why we spent so much time on it. Identifying the potential-energy function is now the challenge, and this is not a step to be rushed through. You will obviously not find the correct wave functions or energies if you start out with an incorrect potential energy.

2. Establish the boundary conditions that the wave function must satisfy. They are different in each problem, so you must think carefully about what happens as $x \to \infty$ or at any positions where the potential energy changes abruptly.

3. Solve the Schrödinger equation, utilizing boundary conditions and normalization, to find the wave functions and their energies.

4. Interpret the solution! What information is obtained from the wave functions? In nearly all cases you will want to draw graphs of $\psi(x)$ and $|\psi(x)|^2$ for wave functions having low values of the quantum number. (Because the solutions to the Schrödinger equation are *functions*, not just numbers, we cannot emphasize too highly the importance of using graphs to understand the solutions.) Think about where the particle is likely to be found, what the energies are, and what wavelength photons would be emitted or absorbed. This, after all, was the whole point in solving the problem

The solutions to the Schrödinger equation are the stationary states of the system. Bohr had postulated the existence of stationary states, but he didn't know how to find them. It took de Broglie and Schrödinger to complete Bohr's ideas.

An important idea from the Bohr model remains valid in Schrödinger's quantum mechanics: the idea of transitions, or quantum jumps, between stationary states. The system can jump from one stationary state, characterized by wave function ψ_i and energy E_i, to another, characterized by ψ_f and E_f, by emitting or absorbing a photon of frequency

$$f = \frac{\Delta E}{h} = \frac{|E_f - E_i|}{h}.$$

Thus the solutions to the Schrödinger equation will allow us to predict the spectrum of a quantum system. Such predictions can be used to test the validity of Schrödinger's theory.

Now that we've talked about the mathematics and outlined the strategy, let's look at an extended example to see how these ideas are applied.

36.4 A Particle in a Rigid Box

In our earlier study of de Broglie waves, we considered the situation known as a "particle in a box." There we asserted that the particle would bounce back and forth, setting up a standing wave. Such an assertion was just a guess, with no real justification, because we had no theory as to how a wave-like particle ought to behave. We will now revisit this problem from the new perspective of the Schrödinger equation.

Consider a particle of mass m confined in a rigid, one-dimensional box of length L. What are the possible wave functions and energies of this particle? In which part of the box is the particle most like to be found? To answer such questions, we need to follow the steps of the problem-solving strategy.

Identifying a Potential-Energy Function

By a *rigid box*, we mean a box whose walls are so sturdy that they can confine a particle no matter how fast the particle moves. Furthermore, the walls are so stiff that they do not flex or "give" as the particle bounces. No real container has these attributes, so the rigid box is an ideal *model* of situations in which a particle is extremely well confined. Our first task is to characterize the rigid box in terms of a potential-energy function.

We introduced energy diagrams in Section 10.7, which you may wish to review. Recall that an energy diagram consists of a graph of the potential-energy function $U(x)$ and a total energy line E. At any position x, the distance between the potential-energy curve and the total energy line is the kinetic energy at that point: $K(x) = E - U(x)$. The turning points in the particle's motion occur where the total energy line crosses the potential-energy curve. The particle cannot move beyond these points, because it cannot have a negative kinetic energy. Regions where $U(x) > E$ are called **classically forbidden regions**. As the particle moves between the turning points, the force on it at any point is the negative of the *slope* of the potential-energy curve at that point: $F = -dU/dx$.

With these ideas in mind, let's establish a coordinate axis with the boundaries of the box at $x = 0$ and $x = L$. The potential energy of the rigid box is shown in Fig. 36-3. Three important characteristics identify this potential-energy curve as a rigid box:

1. The particle can move freely inside the box (slope = zero) with constant kinetic energy (constant speed).
2. No matter how much kinetic energy the particle has, the particle's turning points are at $x = 0$ and $x = L$ (vertical, infinitely high walls).
3. The regions $x < 0$ and $x > L$ are forbidden. The particle cannot leave the box.

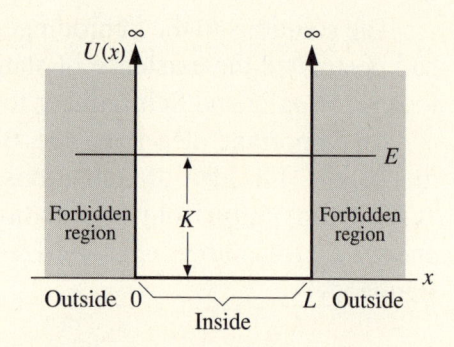

FIGURE 36-3 Potential energy of a particle in a rigid box of length L.

Keep in mind that Fig. 36-3 is *not* a "picture" of the box. It is a graphical representation of the particle's kinetic and potential energy in the box.

We express this potential-energy function mathematically as

$$U_{\text{rigid box}}(x) = \begin{cases} 0 & 0 \leq x \leq L \\ \infty & x < 0 \text{ or } x > L. \end{cases} \quad (36\text{-}15)$$

This is the potential energy $U(x)$ for which we want to solve the Schrödinger equation.

Establishing Boundary Conditions

We now need to establish the boundary conditions that the solution must satisfy. Because it is physically impossible for the particle to be outside the box, we require

$$\psi(x) = 0 \qquad x < 0 \text{ or } x > L. \quad (36\text{-}16)$$

That is, there is zero probability of finding the particle outside the box, so the wave function must be zero in these regions. Furthermore, the solution inside the box has to be *continuous* with the solution outside the box. Because the solution outside is zero everywhere, continuity requires the solution inside to obey

$$\psi(x = 0) = 0 \quad \text{and} \quad \psi(x = L) = 0. \quad (36\text{-}17)$$

In other words, the wave function must go to zero at the boundaries of the box if it is to be continuous with the wave function outside the box. Equations 36-17 are the boundary conditions for the wave function of a particle in a rigid box.

Finding the Wave Functions

Inside the box, where $U(x) = 0$ at all points, the Schrödinger equation is simply

$$\frac{d^2\psi}{dx^2} = -\frac{2mE}{\hbar^2} \cdot \psi(x). \quad (36\text{-}18)$$

The question to answer has two parts:

1. For what values of E does Eq. 36-18 have physically meaningful solutions?
2. What are the solutions $\psi(x)$ for those values of E?

To begin, let's simplify the notation by defining $\beta^2 = 2mE/\hbar^2$. The Schrödinger equation then reads

$$\frac{d^2\psi}{dx^2} = -\beta^2 \psi(x). \tag{36-19}$$

We're going to solve this differential equation by guessing! Can you think of a function whose second derivative is a *negative* constant times the function itself? Two such functions are

$$\psi_1(x) = \sin\beta x \quad \text{and} \quad \psi_2(x) = \cos\beta x. \tag{36-20}$$

These are both solutions to Eq. 36-19 because

$$\frac{d^2\psi_1}{dx^2} = \frac{d^2}{dx^2}(\sin\beta x) = -\beta^2 \sin\beta x = -\beta^2 \psi_1(x)$$

$$\frac{d^2\psi_2}{dx^2} = \frac{d^2}{dx^2}(\cos\beta x) = -\beta^2 \cos\beta x = -\beta^2 \psi_2(x).$$

These are two independent functions because ψ_2 is not a multiple or a rearrangement of ψ_1. Consequently, according to the theorem of Eq. 36-14, the general solution to the Schrödinger equation for the particle in a rigid box is

$$\psi(x) = A\sin\beta x + B\cos\beta x, \tag{36-21}$$

where

$$\beta = \frac{\sqrt{2mE}}{\hbar}. \tag{36-22}$$

The constants A and B must be determined by using the boundary conditions of Eq. 36-17. Starting at $x = 0$, we have:

$$\psi(x = 0) = A \cdot 0 + B \cdot 1 = 0$$
$$\Rightarrow B = 0 \tag{36-23}$$
$$\Rightarrow \psi(x) = A\sin\beta x.$$

In other words, the boundary condition at $x = 0$ can be satisfied only if $B = 0$. The $\cos\beta x$ term may satisfy the differential equation in a mathematical sense, but it is not a physically meaningful solution for this problem because it does not satisfy the boundary conditions. Then at $x = L$ we have:

$$\psi(x = L) = A\sin\beta L = 0. \tag{36-24}$$

This condition could be satisfied by $A = 0$, but then we wouldn't have a wave function. Fortunately that isn't necessary, because the boundary condition is also satisfied if

$$\beta L = n\pi \quad \Rightarrow \beta = \frac{n\pi}{L} \quad n = 1, 2, 3, \ldots. \tag{36-25}$$

Note that n starts with 1, not 0. The value $n = 0$ would give $\beta = 0$, which makes $\psi = 0$ at all points—a physically meaningless solution.

The solutions to the Schrödinger equation are thus

$$\psi_n(x) = A\sin\beta_n x = A\sin\left(\frac{n\pi x}{L}\right) \quad n = 1, 2, 3, \ldots. \tag{36-26}$$

There is, it appears, a whole *family* of solutions corresponding to different values of the integer n. The constant A remains to be determined.

We do need to note that there is an *assumption* in our solution that the energy E is greater than zero. That assumption enters because we have used the constant β as a positive real number. Now $E < 0$ is not physically meaningful because that would correspond to negative kinetic energy at all points, but what about $E = 0$? A classical particle could remain at rest in the box with $E = 0$, but can a quantum particle?

The Schrödinger equation for $E = 0$ is

$$\frac{d^2\psi}{dx^2} = 0. \tag{36-27}$$

Equation 36-21, which was valid for $E > 0$, is *not* a solution of this equation. But can we guess two independent solutions to this equation? Sure! Both $\psi_1(x) = 1$ and $\psi_2(x) = x$ are functions whose second derivative is zero, so they are both solutions of Eq. 36-27. Thus the general solution of the equation is

$$\psi(x) = C\psi_1(x) + D\psi_2(x) = C + Dx, \tag{36-28}$$

where C and D are constants. You can readily verify that the second derivative of $\psi(x)$ is zero, so it really is a solution of Eq. 36-27. Constants C and D are to be determined by the boundary conditions at $x = 0$ and $x = L$. At $x = 0$ we have

$$\psi(x = 0) = C = 0$$
$$\Rightarrow C = 0 \tag{36-29}$$
$$\Rightarrow \psi(x) = Dx.$$

Then at $x = L$ we have

$$\psi(x = L) = DL = 0$$
$$\Rightarrow D = 0. \tag{36-30}$$

Both constants are zero, so $\psi(x) = 0$ at all points if $E = 0$. But this does not represent a particle at all because there is zero probability of it being anywhere. We have to conclude that there is *not* a physically meaningful solution to the Schrödinger equation for $E = 0$. Although a classical particle could be at rest inside the box, with $K = E = 0$, a quantum particle cannot!

Finding the Energies

Now let's examine the energy of the particle corresponding to the wave functions of Eq. 36-26. Using the Eq. 36-22 definition of β in Eq. 36-25 gives

$$\beta_n = \frac{\sqrt{2mE_n}}{\hbar} = \frac{n\pi}{L}$$

$$\Rightarrow E_n = n^2 \cdot \frac{\pi^2 \hbar^2}{2mL^2} = n^2 \cdot \frac{h^2}{8mL^2} \quad n = 1, 2, 3, \ldots, \tag{36-31}$$

where, in the last step, we used the definition $\hbar = h/2\pi$. It is useful to write this as

$$E_n = n^2 E_1, \tag{36-32}$$

where E_n is the particle's energy in quantum state n and $E_1 = h^2/8mL^2$ is the energy of the $n = 1$ ground state.

The energy is quantized! Equation 36-31 for E_n, in fact, is exactly what we found before for the particle in the box by requiring the de Broglie waves to form standing waves. Only now we have a theory that both tells us the wave functions and that can be generalized to other situations as well. The integer n is the *quantum number* for each of the possible states of the particle in the box.

Normalizing the Wave Functions

We have not yet used the condition that the wave function must be normalized, but that is precisely the condition we need to determine the constant A in the wave functions of Eq. 36-26. The normalization condition was found in Chapter 35 to be $\int |\psi(x)|^2 dx = 1$. This is the mathematical statement that the particle *must*—with probability equal to 1—be *somewhere* on the x-axis. Because the wave function for a particle in a rigid box is zero outside the $0 \leq x \leq L$ range, the integration only needs to extend from 0 to L. Normalization thus requires

$$\int_0^L |\psi(x)|^2 \, dx = A_n^2 \int_0^L \sin^2\left(\frac{n\pi x}{L}\right) dx = 1$$

$$\Rightarrow A_n = \left[\int_0^L \sin^2\left(\frac{n\pi x}{L}\right) dx\right]^{-1/2}. \tag{36-33}$$

We have placed a subscript n on A_n because it is possible that the normalization constant is different for each wave function in the family. This is a standard integral. We will leave it as an exercise to show that its value, for any n, is simply $L/2$. Thus

$$A_n = \left(\frac{L}{2}\right)^{-1/2} = \sqrt{\frac{2}{L}} \quad n = 1, 2, 3, \ldots . \tag{36-34}$$

Interpreting the Solution

We have completed the quantum-mechanical solution to the problem of a particle in a rigid box. Our solution tells us that:

1. The particle can exist only if it has energy $E_n = n^2 E_1$, where $n = 1, 2, 3, \ldots$ is the quantum number and where $E_1 = h^2/8mL^2$ is the energy of the $n = 1$ ground state.

2. The wave function for a particle in the nth quantum state is

$$\psi_n(x) = \begin{cases} \sqrt{\dfrac{2}{L}} \sin\left(\dfrac{n\pi x}{L}\right) & 0 \leq x \leq L \\ 0 & x < 0 \text{ and } x > L. \end{cases} \tag{36-35}$$

These are the stationary states of the system

3. The probability density for finding the particle at position x inside the box is

$$P_n(x) = |\psi_n(x)|^2 = \frac{2}{L} \sin^2\left(\frac{n\pi x}{L}\right). \tag{36-36}$$

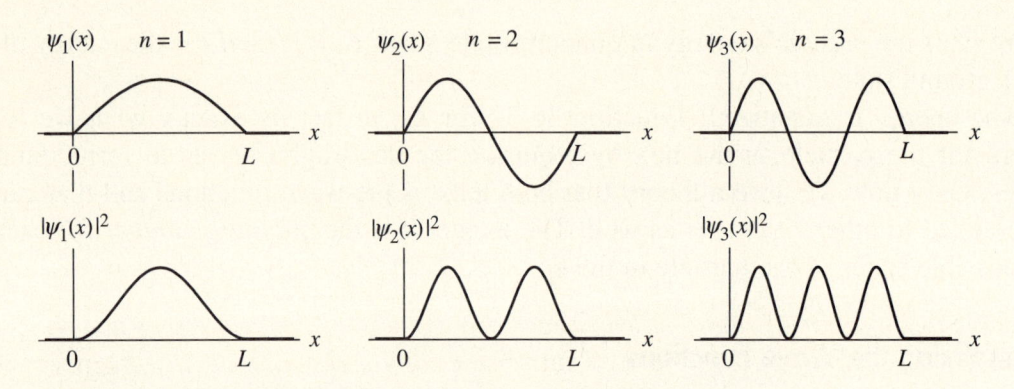

FIGURE 36-4 Wave functions and probability densities for the first three quantum states of a particle in a rigid box of length L. The wave function is zero for $x \leq 0$ and $x \geq L$.

A graphical presentation will help to make these results more meaningful. Figure 36-4 shows both the wave function $\psi(x)$ and the probability density $P(x) = |\psi(x)|^2$ for the first three quantum states $n = 1-3$. Notice how the wave functions go to zero at the boundaries—exactly as the boundary conditions specified for continuity—and are zero outside the box. You can see that the particle is more likely to be found at some positions than at others.

The wave functions $\psi(x)$ for a particle in a rigid box are analogous to standing waves on a string that is tied at both ends. That is not surprising if the wave function represents a standing matter wave. Notice that the number of *nodes* (zeros) in the wave function (including the ends) is $n + 1$ and that the number of antinodes (maxima and minima) is n. This will turn out to be a general result for any wave function, not just for a particle in a rigid box.

Figure 36-5 shows another way in which energies and wave functions are shown graphically in quantum mechanics. First, the graph shows the potential-energy function $U(x)$ of the particle. Second, the possible energies are shown as horizontal lines (total energy lines) across the potential-energy graph. These are labeled with the quantum number n and the energy E_n. Third—and this is a bit tricky—the wave function for each n is drawn *as if* the energy line were the x-axis. That is, the graph of $\psi_n(x)$ is drawn on top of the E_n energy line. This allows energies and wave functions to be displayed simultaneously, but it does *not* imply that ψ_2 is in any sense "above" ψ_1. Both are oscillating sinusoidally about zero, as Fig. 36-4 shows correctly.

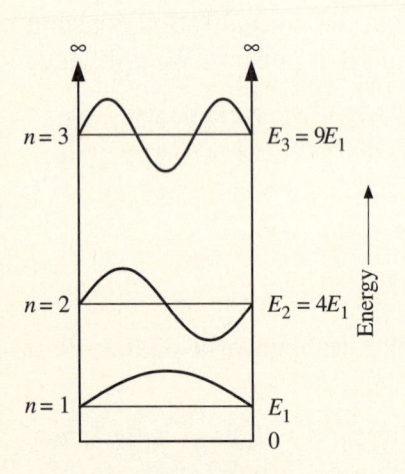

FIGURE 36-5 Alternative way to show the potential-energy diagram, the quantized energies, and the wave functions simultaneously. Each wave function really oscillates around zero.

• **EXAMPLE 36-1** A semiconductor device known as a *quantum-well device* is designed so as to "trap" electrons in a region 1 nm wide. Treating this as a one-dimensional problem: a) What are the energies of the first three quantum states of an electron? b) What wavelengths of light can this electron absorb?

SOLUTION a) If we model the confinement region as a rigid box with $L = 1$ nm and $m = m_{elec}$, we find

$$E_1 = \frac{h^2}{8mL^2} = 6.03 \times 10^{-20} \text{ J} = 0.377 \text{ eV}$$
$$E_2 = 4E_1 = 1.508 \text{ eV}$$
$$E_3 = 9E_1 = 3.393 \text{ eV}.$$

b) An electron will spend most of its time in the $n = 1$ ground state. According to Bohr's model of stationary states, the electron can absorb light and undergo a transition, or quantum jump, to $n = 2$ or $n = 3$ if the light has frequency $f = \Delta E/h$. The wavelengths are $\lambda = c/f = hc/\Delta E$. Calculation gives

$$\lambda_{1 \to 2} = \frac{hc}{E_2 - E_1} = 1098 \text{ nm}$$

$$\lambda_{1 \to 3} = \frac{hc}{E_3 - E_1} = 411 \text{ nm}.$$

Various complications usually make the $1 \to 3$ transition unobservable, but quantum-well devices do indeed exhibit strong absorption and emission at the $\lambda_{1 \to 2}$ wavelength. In this example, which is typical, the wavelength is in the near infrared. Devices such as these are used to construct the semiconductor lasers used in CD players and laser printers. •

The lowest energy state in Example 36-1—the $n = 1$ ground state— has $E_1 = 0.38$ eV, which corresponds to an electron speed of 3.6×10^5 m/s. There was no stationary state having $E = 0$. So unlike a classical particle, the quantum particle in a box *cannot be at rest*! No matter how much its energy is reduced, such as by cooling it toward absolute zero, it cannot have energy less than E_1. This resultant motion is called the **zero-point motion**, and it is a consequence of Heisenberg's uncertainty principle. If the particle were at rest, we would know its velocity and momentum to be exactly zero with *no* uncertainty. That is, the uncertainty in its momentum would be $\Delta p = 0$. Its position uncertainty, because it is somewhere in the box, is $\Delta x = L$. But then $\Delta x \Delta p = 0$, in violation of the Heisenberg uncertainty principle $\Delta x \Delta p \geq h/2$. Thus the particle cannot be at rest.

It is worth commenting on the fact that the *energies* of a quantum-mechanical system are exact. We keep mentioning how quantum particles do not have exact positions or velocities, hence the necessity of using probabilities. This is a consequence of wave–particle duality. But although the position and velocity are ill-defined, the particle's energy in each state can be calculated with a high degree of precision. This distinction between a precise energy and an imprecise position and velocity seems rather strange. It is, however, just our old friend the standing wave showing up again. To *have* a stationary state, the de Broglie waves have to appear as standing waves. Only for very precise frequencies, and thus precise energies, can the standing-wave pattern appear.

EXAMPLE 36-2 Protons and neutrons are tightly confined within the nucleus of an atom. If we use a one-dimensional model of a nucleus, what are the first three energy levels of a neutron in a rigid box 10 fm wide (1 fm = 10^{-15} m)?

SOLUTION Within the rigid box approximation with $L = 10$ fm and $m = m_{\text{neu}}$, we find

$$E_1 = \frac{h^2}{8mL^2} = 3.29 \times 10^{-13} \text{ J} = 2.06 \text{ MeV}$$
$$E_2 = 4E_1 \qquad\qquad\quad = 8.24 \text{ MeV}$$
$$E_3 = 9E_1 \qquad\qquad\quad = 18.54 \text{ MeV}.$$

Notice that an electron confined in an atomic-size space has energies of a few eV but a neutron confined in a nuclear-size space has energies of a few *million* eV.

EXAMPLE 36-3 Consider the ground state of a particle in a rigid box of length L. a) Where is the particle most likely to be found? b) Calculate the probability of finding the particle in an interval of width $0.01L$ at $x = 0, 0.25L, 0.50L, 0.75L$, and L. c) Calculate the probability of finding the particle in the center half of the box.

SOLUTION a) The ground state is $n = 1$. The particle is most likely to be found at the point where the probability density $P(x)$ is a maximum. You can see from Fig. 36-4 that the point of maximum probability for $n = 1$ is $x = L/2$.

b) For a *small* width δx, the probability of finding the particle in δx at position x is

$$\text{Prob(in } \delta x \text{ at } x) = P_1(x)\delta x = |\psi_1(x)|^2 \delta x = \frac{2}{L}\sin^2\left(\frac{\pi x}{L}\right)\delta x.$$

$\delta x = 0.01L$ is sufficiently small for this to be valid. Evaluating at the specified values of x gives

$$\text{Prob(in } 0.01L \text{ at } x = 0.00L) = 0.000 = 0.0\%$$
$$\text{Prob(in } 0.01L \text{ at } x = 0.25L) = 0.010 = 1.0\%$$
$$\text{Prob(in } 0.01L \text{ at } x = 0.50L) = 0.020 = 2.0\%$$
$$\text{Prob(in } 0.01L \text{ at } x = 0.75L) = 0.010 = 1.0\%$$
$$\text{Prob(in } 0.01L \text{ at } x = 1.00L) = 0.000 = 0.0\%.$$

c) The center half of the box stretches from $x = L/4$ to $x = 3L/4$. This is *not* a small interval, so we need to integrate to evaluate the probability:

$$\text{Prob(in interval } \tfrac{1}{4}L \text{ to } \tfrac{3}{4}L) = \int_{L/4}^{3L/4} P_1(x)dx = \frac{2}{L}\cdot\int_{L/4}^{3L/4} \sin^2\left(\frac{\pi x}{L}\right)dx$$

$$= \left[\frac{x}{L} - \frac{1}{\pi}\sin\left(\frac{\pi x}{L}\right)\cos\left(\frac{\pi x}{L}\right)\right]_{L/4}^{3L/4}$$

$$= \frac{1}{2} + \frac{1}{\pi}$$

$$= 0.818.$$

If a particle is in its $n = 1$ ground state, there is an 81.8% chance of finding it in the center half of the box. The probability is greater than 50% because, as you can see in Fig. 36-4, the probability density $P_1(x)$ is larger near the center of the box than near the boundaries.

This has been a lengthy presentation of the particle-in-a-box problem. However, it was important, as our first example of quantum mechanics, to explore the solution method completely. Future examples will now go more quickly because many of the issues discussed here will not have to be repeated. All this probably looks very complicated to you, as does any new topic the first time you see it. But rest assured that after seeing a few more examples, and practicing it yourself for homework, these procedures will soon seem no more complex than using Newton's laws to solve a problem in classical mechanics.

36.5 The Correspondence Principle

Quantum mechanics is the physics of microscopic, atomic-sized particles, and Newtonian mechanics is the physics of macroscopic objects. But suppose we confined an electron in a box that gets bigger and bigger and bigger. What started out as a purely quantum-mechanical problem, with wave functions and quantized energies, should eventually look like a classical physics situation. Or a Newtonian problem, such as two charged particles revolving about each other, should begin to exhibit quantum behavior as it is made smaller and smaller. There should be some in-between size, or energy, for which the quantum-mechanical solution corresponds to the Newtonian solution.

Niels Bohr formulated the principle that the *average* behavior of a quantum system should begin to look like the Newtonian solution in the limit that the quantum number becomes very large: $n \to \infty$. Because the radius of the Bohr atom is $r = n^2 a_B$, this corresponds to the atom becoming a macroscopic object. Bohr's idea, that the quantum world should blend smoothly into the classical world for high quantum numbers, is today known as the **correspondence principle**. It is worth seeing how this applies to a particle in a box.

Our quantum knowledge of a particle in the box is given by its probability density

$$P_{\text{quantum}}(x) = |\psi_n(x)|^2 = \frac{2}{L}\sin^2\left(\frac{n\pi x}{L}\right), \tag{36-37}$$

where we have, for the purposes of this section, added a subscript "quantum." We can also define a classical probability density $P_{\text{classical}}(x)$ that tells us the probability of finding a Newtonian particle at the position x.

For a classical particle, the probability of finding the particle within a small interval δx at the position x is directly proportional to the amount of time δt that it spends passing through δx. That is, the particle is more likely to be found in the intervals where it spends the most time. But the time δt to cross an interval of width δx is simply $\delta t = \delta x / v(x)$, where $v(x)$ is the particle's speed at position x. Thus the classical probability of finding the particle within the interval is:

$$\text{Prob}_{\text{classical}}(\text{in } \delta x \text{ at } x) \propto \delta t = \frac{\delta x}{v(x)}. \tag{36-38}$$

The probability is related to the probability density by

$$\text{Prob}_{\text{classical}}(\text{in } \delta x \text{ at } x) = P_{\text{classical}}(x)\delta x. \tag{36-39}$$

Thus we can see from Eq. 36-38 that

$$P_{\text{classical}}(x) \propto \frac{1}{v(x)} = \frac{C}{v(x)}. \tag{36-40}$$

Here C is a proportionality constant to be determined by the normalization condition

$$\int_{-\infty}^{\infty} P_{\text{classical}}(x)dx = C\int_{-\infty}^{\infty} \frac{dx}{v(x)} = 1. \tag{36-41}$$

The velocity *must* be expressed as a function of the position x in order to do the integration.

Now we can apply Eq. 36-41 to a classical particle in a box, where the particle's speed is a *constant* $v(x) = v_0$ as it bounces back and forth between the walls. The normalization condition, for which the integration limits extend only between 0 and L, becomes

$$\frac{C}{v_0}\int_0^L dx = \frac{CL}{v_0} = 1$$

$$\Rightarrow C = \frac{v_0}{L}. \tag{36-42}$$

The classical probability density is then

$$P_{\text{classical}}(x) = \frac{C}{v_0} = \frac{v_0/L}{v_0} = \frac{1}{L}. \tag{36-43}$$

A classical particle, it comes as no surprise, is equally likely to be found *anywhere* in the box because $P_{\text{classical}}(x)$ is independent of x.

Figure 36-6 shows both the quantum and the classical probability densities for the $n = 1$ and the $n = 10$ quantum states. You can see that the two densities are quite different for $n = 1$, indicating that the ground-state behavior for the quantum system will be very different from that of a classical system. But for $n = 10$ you can see that *on average* the quantum particle's behavior begins to look like the classical particle. As n gets even bigger, and the number of oscillations increases, the probability of finding the particle in some interval δx will be the same for both the quantum and the classical particle as long as δx is large enough to include several oscillations of the wave function. Only for the very tiniest of intervals, which get smaller as n increases, will there be any difference between the classical and the quantum descriptions. We can see, as Bohr predicted, that the quantum-mechanical solution "corresponds" to the classical solution in the limit $n \to \infty$.

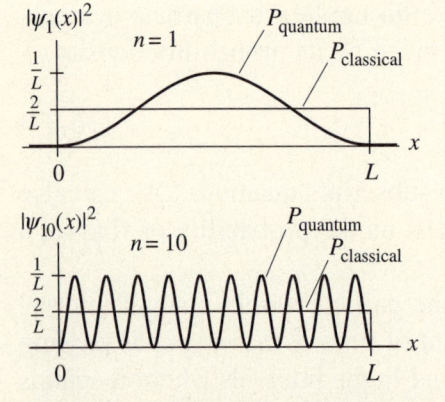

FIGURE 36-6 The quantum probability densities of a particle in a box for the $n = 1$ and $n = 10$ states, and the corresponding classical probability densities.

The correspondence principle has only limited applications. Nonetheless, our confidence in the Schrödinger equation is bolstered by seeing that quantum-mechanical solutions do, as they should, blend smoothly into classical solutions as the quantum-number n becomes large.

36.6 Quantum-Mechanical Models

The Schrödinger equation is a fairly abstract, mathematical statement about the wave function $\psi(x)$. Although we have given an *interpretation* of the wave function, it is hard to see how ψ and the Schrödinger equation are connected to reality. The purpose of the remaining sections of this chapter will be to apply quantum mechanics to real situations.

Long ago, in your initial study of Newtonian mechanics, you learned of the importance of *models*. To understand the motion of an object we made many simplifying assumptions: that the object could be represented by a particle, that friction could be described in a simple way, that air resistance could be neglected, that surfaces were perfect planes, and so on. In other words, we constructed a *model* of the situation that included the important features but ignored the details. This approach was essential if we were to end up with equations that we had any hope of solving. We recognized that some assumptions in the model—such as no air resistance—might not be truly valid. Thus our solutions gave a good, but perhaps not perfect, description of reality. In some cases we started with a basic "stripped-down model," then added extra features to improve the model's match with reality. Our goal throughout was to understand the primary features of an object's motion without getting lost in irrelevant details.

The same situation holds true in quantum mechanics. The exact description of an electron in a solid, for example, involves an astronomical number of interactions with all the ions as well as all the other electrons. Our only hope for using quantum mechanics effectively is to make a number of simplifying assumptions—that is, to make a **quantum-mechanical model** of the situation. Because the interaction terms in the Schrödinger equation are energies, rather than forces, quantum-mechanical modeling involves finding a potential-energy function that is an acceptable, although not perfect, description of a particle's interactions.

Much of the rest of this chapter will be about building and using quantum-mechanical models. The test of a model's success is its agreement with experimental measurement. Laboratory experiments cannot measure $\psi(x)$, and they rarely measure probabilities directly. Quantities that are measured include wavelengths, currents, times, temperatures, masses, and so on. Because quantum mechanics purports to be about the real world, it is important that we tie our models to measurable quantities.

The real world, of course, is three dimensional. Unfortunately, solving the Schrödinger equation in three dimensions raises the mathematical complexity beyond what is appropriate for this text. We will be forced to restrict our models to one dimension. This is an oversimplification, so we must expect that our predictions will not exactly match measured values. Nonetheless, there are many examples for which a one-dimensional model is reasonably good. Our goal in this text is to show how quantum-mechanical concepts are used, not to make highly accurate calculations. One-dimensional quantum mechanics will be sufficient for this purpose.

36.7 A Particle in a Capacitor

As our first practical example, let's consider the problem of "a particle in a capacitor." Many modern semiconductor devices are designed to confine electrons within a very thin layer only a few nanometers thick. If a potential difference is applied across the layer, the electrons act very much as if they are trapped within a microscopic capacitor. This is a variation of the particle-in-a-box problem of Section 36.4.

Figure 36-7a shows two capacitor plates separated by distance L and charged with a potential difference ΔV_0. The left plate is positive, so the electric field points to the right with strength $|\vec{E}| = \Delta V_0/L$. First consider electrons to be classical charged particles. An electron launched from the left plate experiences a *retarding* force, due to its negative charge, that slows it. The electron will make it across to the right plate if it starts with sufficient kinetic energy; otherwise it reaches a turning point and then is pushed back to the positive plate. We analyzed many similar situations in our earlier study of electricity.

This classical analysis is a valid model of a macroscopic capacitor. But if L becomes sufficiently small, comparable to the de Broglie wavelength of an electron, then the wave-like properties of the electron cannot be ignored. We need a quantum-mechanical model in which we find and interpret the wave function of an electron in the capacitor.

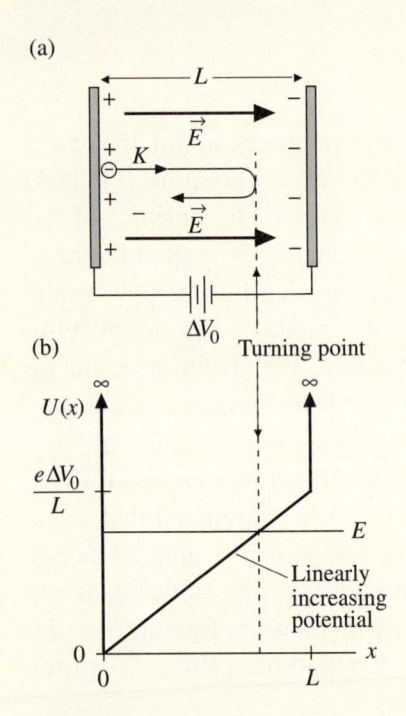

FIGURE 36-7 a) An electron in a capacitor. b) The corresponding potential-energy diagram. Note the turning point.

Let's establish a coordinate system with $x = 0$ at the left plate and $x = L$ at the right plate. Define the electric potential to be zero at the positive plate. The potential *decreases* in the direction of the field, so the potential inside the capacitor (see Section 27.6) is

$$V(x) = -|\vec{E}|x = -\frac{\Delta V_0}{L} \cdot x. \tag{36-44}$$

The electron, having charge $q = -e$, has a potential energy inside the capacitor

$$U(x) = qV(x) = +\frac{e\Delta V_0}{L} \cdot x \qquad 0 < x < L. \tag{36-45}$$

This is a linearly increasing potential energy in the interval $0 < x < L$. Because the electron cannot penetrate the capacitor plates, they act just like the walls of a rigid box. Thus $U(x) \to \infty$ at $x = 0$ and $x = L$.

The electron's potential-energy function is shown in Fig. 36-7b. It is similar to the particle-in-a-rigid-box potential, only now with a sloping "floor" due to the electric field. The figure also shows the total energy line E of an electron in the capacitor. The energy is purely kinetic at $x = 0$, where $K = E$, but it is converted to potential energy as the electron moves to the right. The right turning point occurs where the energy line E crosses the potential-energy curve $U(x)$. If the electron is a classical particle, it must reverse position at this point. The potential-energy diagram of Fig. 36-7b is a *model* for the potential energy of an electron confined within a small region in which there is an electric field due to a potential difference ΔV_0.

We need to solve the Schrödinger equation for an electron in the potential energy $U(x)$ of Fig. 36-7b. The wave function must be zero outside the capacitor, because it is physically impossible for it to be there, and the boundary conditions are the same as for a particle in a rigid box: $\psi(x = 0) = 0$ and $\psi(x = L) = 0$. It happens that the solutions to the Schrödinger equation for this situation are too complicated to find by guessing. Consequently, we have solved the Schrödinger equation numerically and presented the results graphically.

Figure 36-8 shows the wave functions and probability densities for the first five quantum states ($n = 1-5$) of an electron confined in a 5-nm-thick layer to which a 0.8 V potential difference has been applied. Each allowed energy is represented as a horizontal line, with the numerical values shown on the right. They range from $E_1 = 0.23$ eV up to $E_5 = 0.81$ eV. Note, as was the case for the particle-in-a-rigid-box, that the lowest energy state (the ground state) is *not* $E = 0$. Also keep in mind that each wave function in Fig. 36-8 has been graphed, as if the energy line were the x-axis, although each is really oscillating about zero.

It is worth reemphasizing that a particle can exist in the potential well *only* if it has one of the allowed energies shown in the figure. An electron simply cannot have $E = 0.3$ eV in this capacitor, because no de Broglie wave with that energy can match the necessary boundary conditions.

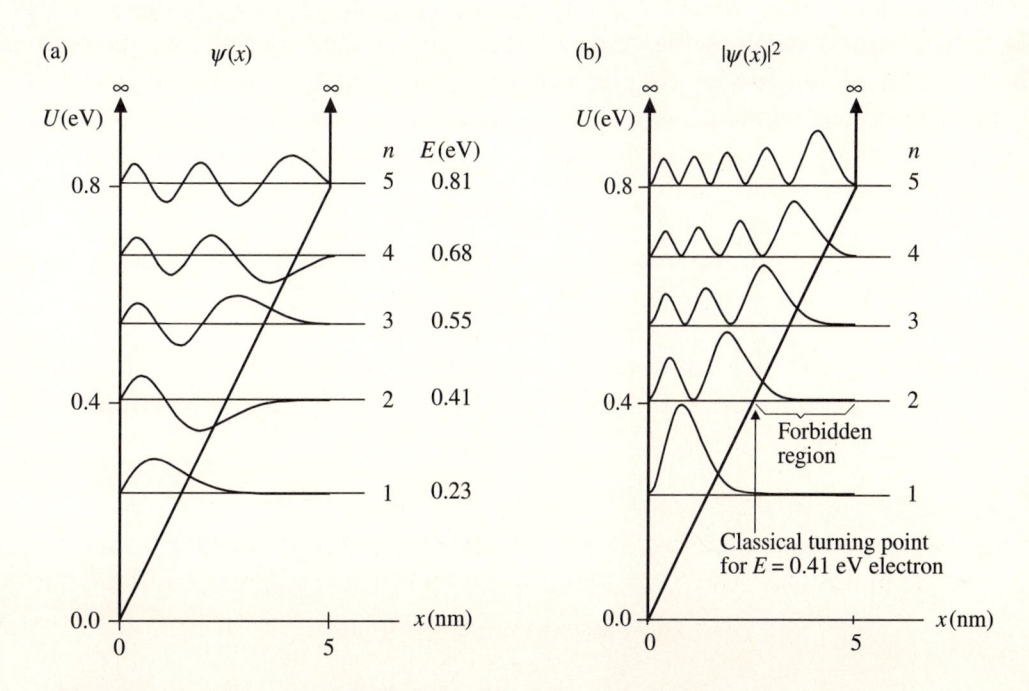

FIGURE 36-8 a) Energy levels and wave functions ($n = 1-5$) for an electron in a 5-nm-wide capacitor with potential difference 0.8 V. b) The probability densities $|\psi(x)|^2$. The classical turning point is indicated for $n = 2$.

We can make several observations about the Schrödinger equation solutions of Fig. 36-8:

1. The wave functions are *qualitatively* similar to those of a particle-in-a-rigid-box. The wave functions are oscillating standing waves, and the wave function with quantum number n has n antinodes and $n + 1$ nodes (including the ends).
2. The energies E_n become more closely spaced as n increases. This is in contrast to the particle in a box, for which E_n became more widely spaced (see Fig. 36-5).
3. The spacing between the nodes of a wave functions is not constant but increases toward the right. This is an important observation, but one that we can understand. An electron on the right side of the capacitor has less kinetic energy, and thus a slower speed, than an electron on the left. A slower speed implies a larger de Broglie wavelength $\lambda = h/mv$. Thus a standing wave inside the capacitor must have a wavelength that decreases toward the left and increases toward the right. That is what we observe.
4. The probability density $|\psi|^2$ increases toward the right. That is, we are more likely to find the electron on the right side of the capacitor than on the left. But this also makes sense if, classically, the electron is moving slower when on the right side and thus spending more time there than on the left side.
5. Last, but by far the most curious and unexpected of our observations, is that there is some probability for finding the electron *beyond* the classical turning point. For example, Fig. 36-8b shows that the classical turning point of an electron with $E = 0.41$ eV is 2.5 nm. But you can see from the figure that the probability density for an electron with the allowed energy $E_2 = 0.41$ eV extends to $x > 3$ nm. A quantum particle, it would appear, can penetrate into what would be a *classically forbidden* region! It is as though a tennis ball penetrated partly *through* the racket's strings before bouncing back, but without breaking the strings. Here is yet another example of a quantum phenomenon that totally defies our commonsense understanding of how objects "should" behave. However, you will soon see experimental evidence that quantum objects really are this strange.

EXAMPLE 36-4 What are the frequencies of photons emitted by electrons in the $n = 4$ state of Fig. 36-8?

SOLUTION Photon emission can occur as the electrons make $4 \rightarrow 3$, $4 \rightarrow 2$, and $4 \rightarrow 1$ quantum jumps. In each case, the photon frequency is $f = \Delta E/h$ and the wavelength is

$$\lambda = \frac{c}{f} = \frac{hc}{\Delta E}.$$

The energies can be read from Fig. 36-8a. We can either convert energies from eV to J or, what is easier in this case, use $h = 4.14 \times 10^{-15}$ eV s instead of its value in J s. Thus

$$\lambda_{4 \rightarrow 3} = 9500 \text{ nm} = 9.5 \text{ μm}$$
$$\lambda_{4 \rightarrow 2} = 4600 \text{ nm} = 4.6 \text{ μm}$$
$$\lambda_{4 \rightarrow 1} = 2800 \text{ nm} = 2.8 \text{ μm}.$$

The $n = 4$ electrons in this device emit three distinct infrared wavelengths.

36.8 Finite Potential Wells

The potential-energy diagram of Fig. 36-3 for a particle in a rigid box is an example of a **potential well**—so named because the graph of the potential-energy "hole" looks like a well from which you would draw water. The rigid box was an *infinite* potential well, with infinitely high walls, because there was no chance that a particle inside could escape.

A more realistic situation is that of a particle in a *finite* potential well such as the one shown in Fig. 36-9. This potential-energy function has a finite depth U_0. A particle can bounce back and forth inside the well—like a rock in the bottom of a pothole—only as long as its kinetic energy remains less than U_0. If $K > U_0$, the particle will fly out of the hole. Notice that the classical turning points for a particle with any energy $E < U_0$ are at $x = 0$ and $x = L$; the regions $x < 0$ and $x > L$ are classically forbidden. Quantum effects will become important when the width L is reduced to a microscopic distance You can, if you wish, call this the "quantum pothole problem."

We've made no mention of the *force* that is responsible for this potential well. For a true pothole, the gravitational potential energy of a rock is due to the gravitational force. But we are more interested in situations such as an electron confined within a semiconductor or a proton confined with the nucleus. These particles experience a potential energy due to electrical or nuclear forces, not gravitational forces. But the Schrödinger equation only cares about the *shape* of the potential energy function, not the cause of the potential energy. Thus *any* situation in which a force confines a particle to a well-defined region can be modeled as a finite potential well.

Although it is possible to solve the Schrödinger equation exactly for the finite potential well, the result is cumbersome and not especially illuminating. Let's look at a numerical solution, as we did for the particle in a capacitor, for the specific case of an electron in a potential well of width $L = 2$ nm and depth $U_0 = 1$ eV. These are reasonable parameters for an electron in a semiconductor device. Figure 36-10a shows the energies and wave functions. For comparison, Fig. 36-10b shows the first three energy levels and wave functions for a *rigid* box ($U_0 \to \infty$) of 2 nm width.

FIGURE 36-9 A finite potential well of width L and depth U_0.

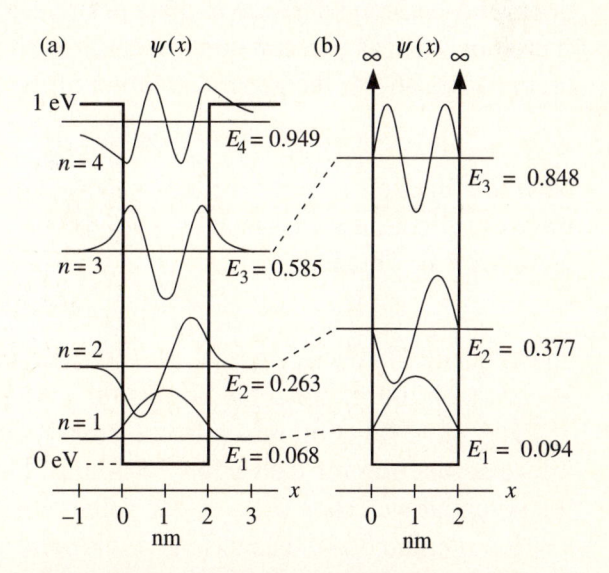

FIGURE 36-10 a) Energy levels and wave functions for a finite potential well of width 2 nm and depth 1 eV. All energies are given in eV. b) For comparison, the energies and wave functions of a rigid box 2 nm wide.

The first thing you should notice about the finite potential well is that there are only a finite number of **bound-state wave functions**—four in this example, although the number will be different for other examples. These wave functions represent electrons confined to, or bound in, the potential well. There are no stationary states with $E > U_0$ because such a particle would not remain in the well. Second, notice that the wave functions are similar to those of a particle in a rigid box. The energies are a bit lower, however, and the wave functions are more spread out.

Perhaps the most interesting thing we can see for the wave functions of Fig. 36-10a is that—like the particle-in-a-capacitor—they extend into the classically forbidden regions. For the perfectly rigid box of Fig. 36-10b the wave functions reach zero at $x = 0$ and $x = L$. That is, the particle cannot penetrate the walls of a perfectly rigid box. But in Fig. 36-10a the wave functions extend past the left wall, $x < 0$, as well as past the right wall, $x > L$. This is a highly nonclassical behavior, but one with interesting implications.

Let's consider the "forbidden region" $x > L$. The Schrödinger equation in this region is

$$\frac{d^2\psi}{dx^2} = -\frac{2m}{\hbar^2}(E - U_0)\psi(x) = \frac{2m}{\hbar^2}(U_0 - E)\psi(x) = \frac{1}{\eta^2}\psi(x), \qquad (36\text{-}46)$$

where

$$\eta^2 = \frac{\hbar^2}{2m(U_0 - E)} \qquad (36\text{-}47)$$

is a positive constant because $U_0 > E$ for any particle confined within the potential well. As a homework problem, you can confirm that η has units of m.

Equation 36-46 is one we can solve by guessing. We merely need to think of two functions whose second derivatives are a positive constant times the functions themselves. Two such functions, as you can quickly confirm, are $e^{x/\eta}$ and $e^{-x/\eta}$. Thus, according to the theorem of Eq. 36-14, the general solution of the Schrödinger equation in the region $x > L$ is

$$\psi(x) = Ae^{x/\eta} + Be^{-x/\eta}. \qquad (36\text{-}48)$$

Now the function $e^{x/\eta}$ diverges as $x \to \infty$. This function cannot meet the requirement that the wave function vanishes as $x \to \infty$. So the "boundary condition" at $x = \infty$ requires that we set $A = 0$. This leaves

$$\psi(x) = Be^{-x/\eta} \text{ for } x > L, \qquad (36\text{-}49)$$

an exponentially decaying function. You can, in fact, see that all the wave functions in Fig. 36-10a look like exponential decays for $x > L$.

The wave function must also be continuous. Thus at the $x = L$ boundary, the Eq. 36-49 wave function for the region $x > L$ must "match" the oscillating wave function within the potential well. Let ψ_{edge} be the value of the oscillating wave function when it reaches $x = L$. Equation 36-49 has to match this value at $x = L$, so

$$\psi(x = L) = Be^{-L/\eta} = \psi_{\text{edge}}. \qquad (36\text{-}50)$$

This boundary condition at $x = L$ implies that

$$B = \psi_{\text{edge}} e^{L/\eta}. \qquad (36\text{-}51)$$

If we use this value for the constant B in Eq. 36-49, we find that the wave function in the classically forbidden region is

$$\psi(x) = \psi_{\text{edge}} e^{-(x-L)/\eta} \qquad (x > L). \tag{36-52}$$

In other words, the wave function oscillates until it reaches the classical turning point at $x = L$, then it decays exponentially within the classically forbidden region.

Figure 36-11 shows that the wave function has decayed significantly by the time it reaches the position $x = L + \eta$. The wave function at that point has the value $e^{-1}\psi_{\text{edge}} = 0.37\psi_{\text{edge}}$. Although an exponential function does not have a sharp ending point, the parameter η measures "about how far" the wave function extends past the turning point before it has mostly decayed away. This distance is called the **penetration distance**:

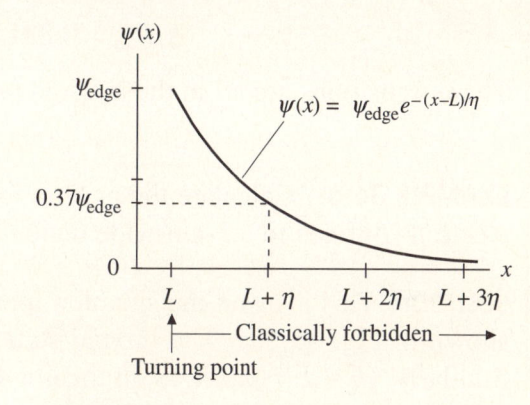

FIGURE 36-11 The exponentially decaying wave function has nearly vanished after penetrating a distance η into the classically forbidden region.

$$\text{penetration distance} = \eta = \frac{\hbar}{\sqrt{2m(U_0 - E)}}. \tag{36-53}$$

In making use of Eq. 36-53, you *must* use SI units of J s for $\hbar$ and J for the energies.

A classical particle reverses direction at the $x = L$ turning point. But atomic particles are not classical particles. Because of wave–particle duality, an atomic particle is "fuzzy," without a well-defined edge. Thus an atomic particle can spread a distance η into the classically forbidden region. This is the significance of the penetration distance.

The penetration distance is unimaginably small for any macroscopic mass, but it can be significant for atomic particles. Notice that wave functions with energies closer to the top of the well, and thus having smaller $U_0 - E$, have larger penetration distances. For example, you can see in Fig. 36-10a that the penetration distance for an electron in the $n = 1$ ground state is 0.20 nm, but it is a much larger 0.86 nm for the weakly bound $n = 4$ state.

EXAMPLE 36-5 What wavelengths of light would be absorbed by a semiconductor device in which electrons are confined in a 2-nm-wide region having a potential-energy depth of 1 eV?

SOLUTION Photons of light can be absorbed if a photon's energy hf exactly matches the energy difference ΔE between two energy levels of the system. The device in this example is exactly that for which the numerical solutions of the Schrödinger equation are shown in Fig. 36-10a, so we can look up the allowed energies there. Because most electrons will be in the lowest $n = 1$ energy level, the absorption transitions will be $1 \to 2$, $1 \to 3$, and $1 \to 4$. Calculating $f = \Delta E/h$ followed by $\lambda = c/f$ gives

$$\lambda_{n \to m} = \frac{hc}{\Delta E} = \frac{hc}{|E_n - E_m|}.$$

For this example we find

$$\Delta E_{1-2} = 0.195 \text{ eV} \quad \lambda_{1 \to 2} = 6.37 \text{ μm}$$

$$\Delta E_{1-3} = 0.517 \text{ eV} \quad \lambda_{1 \to 3} = 2.40 \text{ μm}$$

$$\Delta E_{1-4} = 0.881 \text{ eV} \quad \lambda_{1 \to 4} = 1.41 \text{ μm}.$$

These transitions are all in the infrared portion of the spectrum.

EXAMPLE 36-6 Calculate the penetration distances for the $n = 1$ and $n = 4$ states of an electron confined in a 2-nm-wide region having a potential-energy depth of 1 eV.

SOLUTION This is again the situation for which the energy levels and wave functions are shown in Fig. 36-10a. The ground state has $U_0 - E = 1.0 \text{ eV} - 0.068 \text{ eV} = 0.932 \text{ eV}$. Similarly, $U_0 - E = 0.051 \text{ eV}$ in the $n = 4$ state. Using these values in Eq. 36-53 gives

$$\eta = \frac{\hbar}{\sqrt{2m(U_0 - E)}} = \begin{cases} 0.20 \text{ nm} & n = 1 \\ 0.86 \text{ nm} & n = 4. \end{cases}$$

Now let's look at two applications of the finite potential well: quantum-well devices and nuclear physics.

Quantum-Well Devices

In our previous discussion of conductors, we noted that the outer electrons of the atoms form a loosely bound "sea of electrons" that are free to move about rather like the particles in a gas. You learned in thermodynamics that the typical speed of a particle is given by the rms velocity,

$$v_{\text{rms}} = \sqrt{\frac{3kT}{m}},$$

where k is Boltzmann's constant.

The rms velocity of an electron at room temperature is $v_{\text{rms}} \approx 1 \times 10^5$ m/s. Thus the de Broglie wavelength of a conduction electron is typically

$$\lambda \approx \frac{h}{mv_{\text{rms}}} = 6 \text{ nm}.$$

There will be a range of wavelengths, because the electrons have a range of velocities, but this is a typical value.

As you have learned, wave effects become significant only when structure sizes are comparable to or smaller than the wavelength. That is why the interference and diffraction of light are hard to observe and why the wave-like nature of matter becomes important only on microscopic scales. For electronic devices whose features are >100 nm in size, quantum effects are insignificant and the electrons can be treated as classical particles. That was how we analyzed current flow in Chapter 26. But when the features start becoming less than about 100 nm in size they begin to exhibit quantum effects. Integrated circuits made throughout the 1970s and 1980s were constructed of features typically 1 µm or larger in size, but by the mid-1990s the technology for producing high-density memory chips was approaching 100 nm features. Other kinds of devices, such as the semiconductor lasers used in fiber optic communications and to read CDs, incorporate features only a few nm in size. Quantum effects are very real and very important in these devices.

Figure 36-12a shows the construction of *semiconductor diode laser*. Although the operating principles of diodes goes beyond this text, we can note that a current flows from left to right through this device. In the center is a very thin layer of the semiconductor gallium arsenide (GaAs). It is surrounded on either side by layers of gallium aluminum arsenide (GaAlAs), and these in turn are embedded within the larger structure of the diode. When the current through the diode exceeds some *threshold current*, the electrons within the central GaAs layer begin to emit laser light out the ends of the device.

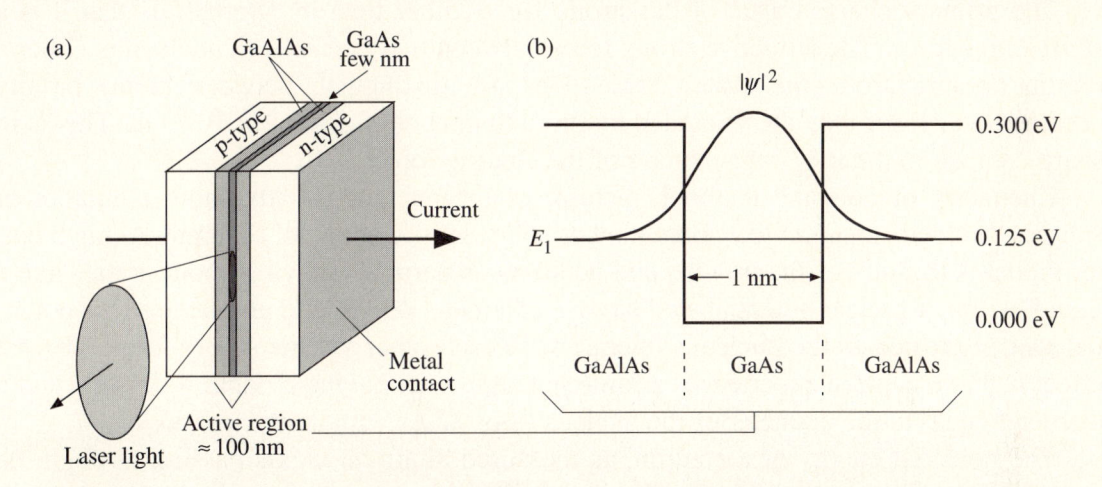

FIGURE 36-12 a) A semiconductor diode laser with a single quantum well. b) A 1-nm-wide, 0.3-eV-deep GaAs layer has only a single energy level. The probability density $|\psi(x)|^2$ is shown.

The potential energy of an electron is slightly lower in GaAs than in GaAlAs. We will assert this without proof because some understanding of solid-state physics would be required. Nonetheless, this makes the GaAs layer a potential well for electrons between the higher potential energies of the GaAlAs "walls" on either side. The electrons thus tend to be confined within the thin GaAs layer. If the GaAs layer is sufficiently thin, the electron energies in this potential well are quantized and the electron positions are governed by the wave function. Such a device is called a **quantum-well device**. These devices are becoming increasingly important for use as high-performance semiconductor diode lasers.

As an example, Fig. 36-12b shows a quantum-well device having a 1-nm-thick GaAs layer in which the electron's potential energy is 0.30 eV less than in the surrounding GaAlAs layers. A numerical solution of the Schrödinger equation finds that this potential well has only a *single* quantum state, $n = 1$ with $E_1 = 0.125$ eV. Every electron "trapped" in this quantum well has the same energy! The fact that the electron energies are so well defined—in contrast to bulk material in which the electrons have a wide range of energies—is what makes this a useful device. Also, as you can see from the probability density $|\psi|^2$ in Fig. 36-12b, the electrons are more likely to be found in the center of the layer than at the edges. This concentration of electrons makes it easier for the device to begin laser action.

Nuclear Physics

The nucleus of an atom consists of an incredibly dense assembly of protons and neutrons. The positively charged protons exert extremely strong electric repulsive forces against each other, so we might wonder how the nucleus keeps from exploding. During the 1930s physicists recognized that protons and neutrons exert an additional *attractive* force on each other. This is called the *strong force*, and it joins the gravitational force and the electromagnetic force as one of the fundamental forces of nature. It is the force that holds the nucleus together.

The primary characteristic of the strong force, other than its strength, is that it is a *short-range* force. The attractive strong force between two *nucleons* (a nucleon is either a proton or a neutron—the strong force does not distinguish between them) rapidly decreases to zero if they are separated by more than about 2 fm = 2×10^{-15} m. This is in sharp contrast to the long-range nature of the electric force.

Chemistry models and textbook pictures of the nucleus usually show a number of spheres "welded" together like some kind of solid. This is not at all accurate. A much better model is to think of the protons and neutrons as particles moving about, rather like a liquid, within a nuclear potential well having a diameter of a few femtometers. This potential well is created as the nucleons interact with each other via the strong force. The net effect is that a neutron experiences a finite potential well having a depth ≈50 MeV and a diameter equal to the diameter of the nucleus (this varies with atomic mass).

The potential energy of a neutron, as measured along an *x*-axis passing through the center of the nucleus, is shown in Fig. 36-13. The zero of energy has been chosen such that a "free" neutron, outside the nucleus, has $E = 0$. Then the potential energy inside the nucleus is −50 MeV. The 8 fm diameter shown is appropriate for a nucleus having atomic mass $A \approx 40$, such as argon or potassium. Lighter nuclei will be a little smaller and heavier nuclei will be somewhat larger. (The potential-energy diagram for a proton is very similar but is complicated a bit by the additional electric potential.)

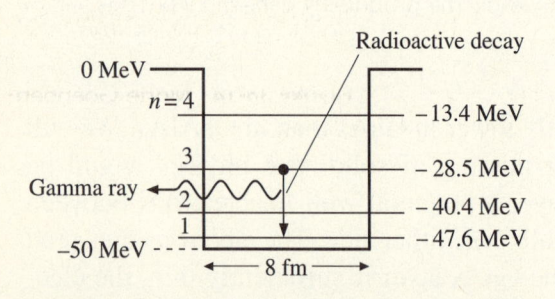

FIGURE 36-13 There are four allowed energy levels for a neutron in a potential well that is 8 fm wide and 50 MeV deep. Gamma rays are emitted in quantum jumps to lower levels.

If we treat this as a one-dimensional potential well, a numerical solution of the Schrödinger equation finds that there are four stationary states. These are shown on Fig. 36-13. The wave functions have been omitted, but they look essentially identical to the potential well wave functions of Fig. 36-10a. The major point to note is that the allowed energies differ by several *million* electron volts! These are enormous energies compared to those of an electron in an atom or a semiconductor. But you will recall that the energies of a particle in a rigid box were given by $E_n = n^2h^2/8mL^2$—proportional to $1/L^2$. We have been dealing with nanometer-size boxes, where the energies are in the eV range. By reducing the box size to femtometers, the relevant energies jump up into the MeV range.

It frequently happens that the nuclear decay of a radioactive atom leaves a neutron in an excited state of the nucleus. Figure 36-13, for example, shows a neutron left in the $n = 3$ state by a previous radioactive decay. This neutron can now undergo a quantum jump to the $n = 1$ ground state by emitting a photon having energy

$$E_{photon} = E_3 - E_1 = 19.1 \text{ MeV}$$

and wavelength

$$\lambda_{photon} = \frac{c}{f} = \frac{hc}{E_{photon}} = 6.50 \times 10^{-14} \text{ nm}.$$

This photon is $\approx 10^7$ times more energetic, and has a wavelength $\approx 10^7$ times smaller, than the photons of visible light! These extremely high-energy photons are called **gamma rays**. Gamma ray emission is, indeed, one of the primary processes in the decay of radioactive elements.

Real nuclei, of course, are three-dimensional. Our one-dimensional simplification cannot be expected to give accurate results for the energy levels or gamma ray energies of any specific nucleus. Nonetheless, our one-dimensional model does provide a good understanding of the energy level structure in nuclei and does—correctly—predict that nuclei can emit photons having energies of several million electron volts. This model, when extended to three dimensions, becomes the basis of the *shell model* of the nucleus. The nucleons can be grouped in various shells, analogous to the electron shells around an atom that you remember from chemistry and that we will study in the next chapter. The shell model of the nucleus, which has been very successful for predicting nuclear properties, was proposed in 1949 by Marie Goeppert-Mayer (Fig. 36-14), one of the first prominent women in physics. She was awarded the Nobel Prize in 1963.

FIGURE 36-14 Marie Goeppert-Mayer, who proposed the shell model of the nucleus.

36.9 The Quantum Harmonic Oscillator

Simple harmonic motion is an exceptionally important motion in classical physics. Not only does it serve as a "prototype" for more complex oscillations, it also forms the basis for the whole subject of wave physics. As you might expect, a microscopic oscillator—the **quantum harmonic oscillator**—is equally important as a model of oscillations at the atomic level.

The defining characteristic of simple harmonic motion is a linear restoring force: $F = -kx$, where k is the spring constant. The corresponding potential-energy function, as you learned in Chapter 13, is

$$U(x) = \tfrac{1}{2}kx^2. \tag{36-54}$$

A particle of mass m undergoes oscillations at an angular frequency $\omega = 2\pi f$ given by

$$\omega = \sqrt{\frac{k}{m}}. \qquad (36\text{-}55)$$

Classically, the particle oscillates back and forth between the two turning points where the energy line crosses the parabolic potential-energy curve. But, as you now know, this classical description will fail if m represents an atomic particle, such as an electron or an atom. For that case we need to solve the Schrödinger equation for a particle in the potential-energy function of Eq. 36-55. The Schrödinger equation for a quantum harmonic oscillator is

$$\frac{d^2\psi}{dx^2} = -\frac{2m}{\hbar^2}\left(E - \tfrac{1}{2}kx^2\right)\psi(x). \qquad (36\text{-}56)$$

The functions that solve this equation can be found, but the mathematics of doing so is beyond the level of this text. We will simply assert that the wave functions of the first three states are

$$\begin{aligned}
\psi_1(x) &= A_1 e^{-x^2/2b^2} \\
\psi_2(x) &= A_2 \frac{x}{b} e^{-x^2/2b^2} \\
\psi_3(x) &= A_3\left(1 - \frac{2x^2}{b^2}\right) e^{-x^2/2b^2},
\end{aligned} \qquad (36\text{-}57)$$

where b is

$$b = \sqrt{\frac{\hbar}{m\omega}}. \qquad (36\text{-}58)$$

The constant b has dimensions of length. We will leave it as a homework problem for you to show that b is the classical turning point of an oscillator in the $n = 1$ ground state. The constants A_1, A_2, and A_3 are normalization constants. For example, A_1 can be found by requiring

$$\int_{-\infty}^{\infty} |\psi(x)|^2 dx = A_1^2 \int_{-\infty}^{\infty} e^{-x^2/b^2} dx = 1. \qquad (36\text{-}59)$$

As expected, stationary states of quantum harmonic oscillator exist only for certain discrete energy levels—the quantum states of the oscillator. Solution of the Schrödinger equation shows that the energies are given by the very simple equation

$$E_n = \left(n - \tfrac{1}{2}\right)\hbar\omega \qquad n = 1, 2, 3, \ldots, \qquad (36\text{-}60)$$

where ω is the classical angular frequency of Eq. 36-55 and n is the quantum number for the state. Notice that the ground-state energy of the quantum harmonic oscillator is $E_1 = \tfrac{1}{2}\hbar\omega$. The ground-state energy is *not* zero! An atomic mass-on-a-spring *cannot* be brought to rest.

Figure 36-15 shows the first three energy levels and wave functions of a quantum harmonic oscillator. The parabolic shape of the potential-energy function is the "signature" of a harmonic oscillator. Notice that the energy levels are equally spaced, $\hbar\omega$ apart, in accordance with Eq. 36-60. The classical turning points are the intersections of the energy lines with the potential-energy curve. You can see that the wave functions, like those of the finite potential well, extend well into the classically forbidden region.

Figure 31-16 shows the probability density $|\psi(x)|^2$ for the $n = 11$ state of a quantum harmonic oscillator. There are two interesting features of this graph: The spacing between nodes increases as the distance $|x|$ increases, and the height of the antinode maxima increases as $|x|$ increases. We can understand these in terms of the correspondence principle, which says that the wave functions should behave more and more like a classical particle as n increases.

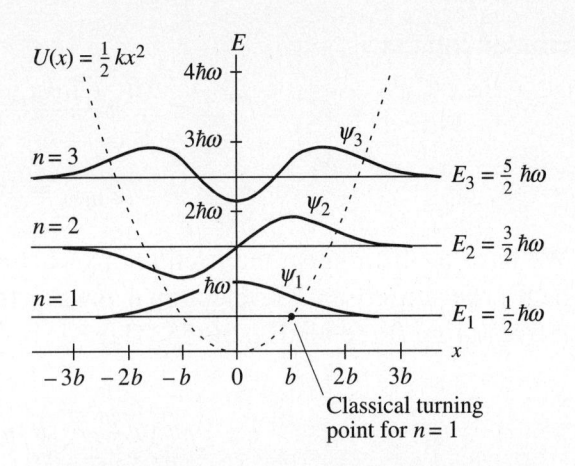

FIGURE 36-15 The first three energy levels and wave functions of a quantum harmonic oscillator.

A classical particle loses kinetic energy and moves more slowly as it approaches the turning points. Thus the particle's de Broglie wavelength increases as it moves away from the center, and this is why we see the node spacing increasing. In addition, a classical particle is more likely to be found in regions where it moves slowly than in regions where it moves quickly. In this situation, a classical particle is more likely to be found near the turning points than in the center.

We introduced the classical probability density $P_{\text{classical}}(x) = C/v(x)$ in Eq. 36-40 to describe where a classical particle is most likely to be found. Figure 36-16 shows $P_{\text{classical}}$ for a classical particle with the same total energy as the $n = 11$ quantum state. You can see that *on average* the quantum probability density $|\psi(x)|^2$ mimics the classical probability density. This is just what we expect, based on the correspondence principle.

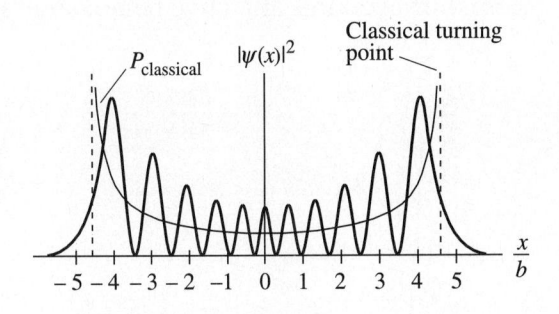

FIGURE 36-16 The quantum probability density $|\psi(x)|^2$ and the classical probability density $P_{\text{classical}}$ for the $n = 11$ state of a quantum harmonic oscillator. The horizontal axis is in units of x/b.

EXAMPLE 36-7 An electron in a harmonic potential well emits light of wavelength 600 nm as it jumps from one level to the next. What is the spring constant of the restoring force?

SOLUTION A transition from one level to the next is $n \to n - 1$. According to Eq. 36-60, *all* such transitions, regardless of the initial value of n, have $\Delta E = \hbar\omega$. This is because the energy levels of the quantum harmonic oscillator are equally spaced. The frequency of emitted light is

$$f_{light} = \frac{\Delta E}{h} = \frac{\hbar \omega_{elec}}{h} = \frac{\omega_{elec}}{2\pi}$$

$$\Rightarrow \omega_{elec} = 2\pi f_{light} = 2\pi \frac{c}{\lambda} = 3.14 \times 10^{15} \text{ s}^{-1}.$$

We have to distinguish the oscillations of the electron from the oscillations of the light, hence the subscripts. The electron's angular frequency is related to the spring constant of the restoring force via Eq. 36-55. Thus

$$\omega_{elec} = \sqrt{\frac{k}{m}}$$

$$\Rightarrow k = m\omega_{elec}^2 = 9.0 \text{ N/m}.$$

Now let's look at a real-world example of a quantum harmonic oscillator.

Molecular Bonds

The ball-and-stick models of molecules in chemistry classes suggest that the molecular bond between two atoms is rigid. However, this is not the case. Two nearby atoms are able to polarize each other and thereby exert *attractive* forces on each other in much the same way that a charged rod can polarize and pick up small pieces of paper. If the atoms get too close, however, this attractive polarization force is countered by a *repulsive* force between the negative electrons. This combination of attractive and repulsive forces creates a potential energy for two atoms as seen in Fig. 36-17. Here the potential energy $U(r)$ is graphed as a function of the separation r between the atoms. The situation is one of two atoms having a spring-like connection between them, as seen in the insert.

You can see that there is an equilibrium position at r_0, the position of minimum potential energy. Typically $r_0 \approx 0.15$ nm, and this is referred to as the *bond length*. A molecule will have a *rigid* bond of this length only if $E = 0$ and the molecule sits at rest in the bottom of the potential-energy well. But you have seen that quantum particles, even in their lowest energy state, have $E > 0$. Consequently, the molecule will *vibrate* as the two atoms oscillate back and forth along the bond. The oscillation amplitude is usually very small—much less than an atomic diameter—but it plays an important role in the properties of molecules and solids.

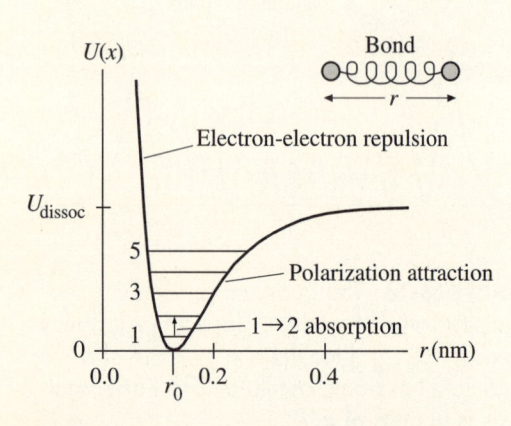

FIGURE 36-17 The potential energy of a molecule bond. The $n = 1-5$ energy levels are shown, along with a $1 \to 2$ absorption transition.

If the molecule's energy reaches U_{dissoc}, the energy of the flat portion of the potential-energy curve on the right, the molecule will *dissociate* and the two atoms will fly apart.

This can happen at very high temperatures or if the molecule has absorbed a high-energy (ultraviolet) photon, but under typical conditions the molecules have energy $E \ll U_{\text{dissoc}}$. Their energy is near the bottom of the potential well, where the molecular bond looks very much like the parabolic potential-energy curve of a harmonic oscillator.

Consequently, we can *model* a molecular bond as a quantum harmonic oscillator. As a result, the energy associated with the molecular vibration is quantized and can have *only* the values

$$E_{\text{vib}} \approx \left(n - \tfrac{1}{2}\right)\hbar\omega \qquad n = 1, 2, 3, \ldots \tag{36-61}$$

where ω is the angular frequency with which the "molecular spring" would vibrate if it were a classical spring. We've used an $\approx$ in Eq. 36-61 rather than an = because the molecular potential-energy curve is not exactly a harmonic oscillator. Nonetheless, the model is very good for low values of the quantum number n. The energy levels of Eq. 36-61 are called the **vibrational energy levels** of the molecule. The first few levels are shown on Fig. 36-17.

At room temperature, most molecules are in the $n = 1$ vibrational ground state. Their vibrational motion can be excited by absorbing photons of frequency $f = \Delta E/h$. This frequency is usually in the infrared region of the spectrum, and these *vibrational transitions* give each molecule a unique and distinctive infrared absorption spectrum.

Figure 36-18 shows the infrared absorption spectrum of acetone. The vertical axis in the graph is the percentage of the light intensity going all the way through the sample. The sample is essentially transparent at most wavelengths, but there are two prominent absorption features. The transmission drops to $\approx 75\%$ at $\lambda = 3.3$ μm and to a mere 7% at $\lambda = 5.8$ μm. The 3.3 μm absorption is due to a $1 \to 2$ transition in the vibration of a C–CH$_3$ carbon-methyl bond and the 5.8 μm absorption is the $1 \to 2$ transition of a vibrating C=O carbon-oxygen double bond.

Absorption spectra like this are known for thousands of molecules, and chemists routinely use absorption spectroscopy to identify unknown chemicals. Spectra can also be used to analyze the structure of molecules. A specific bond has the same absorption wavelength regardless of the larger molecule in which it is embedded, so the presence of that absorption wavelength is a "signature" that the bond is present within a molecule.

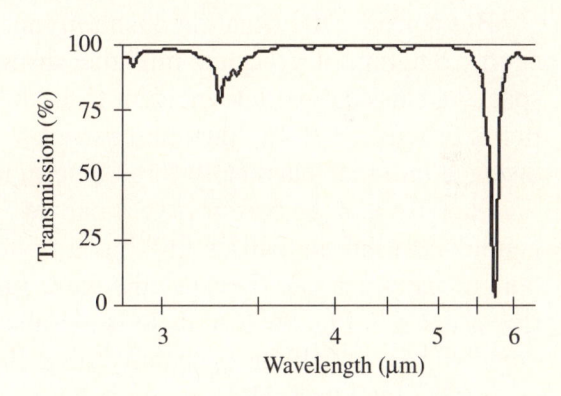

FIGURE 36-18 The absorption spectrum of acetone. There are prominent absorption features at 3.3 and 5.8 μm.

These quantum properties of molecular bonds are also very important for understanding how solids are bound together. The atoms in a solid vibrate back and forth about their equilibrium positions, and the energy of these vibrations is quantized. This quantization of the vibrational energies is necessary to understand various thermal properties of solids, such as their specific heats and their heat conductivity.

36.10 Quantum-Mechanical Tunneling

If a classical particle rolls up a hill, such as the bowling ball shown in Fig. 36-19a, it will roll only to the point where it has converted all of its initial kinetic energy into potential energy. At that point—the turning point—it has no choice but to reverse direction and roll back down from whence it came. It is impossible, according to the laws of classical physics, ever to find the ball on the right side of the hill.

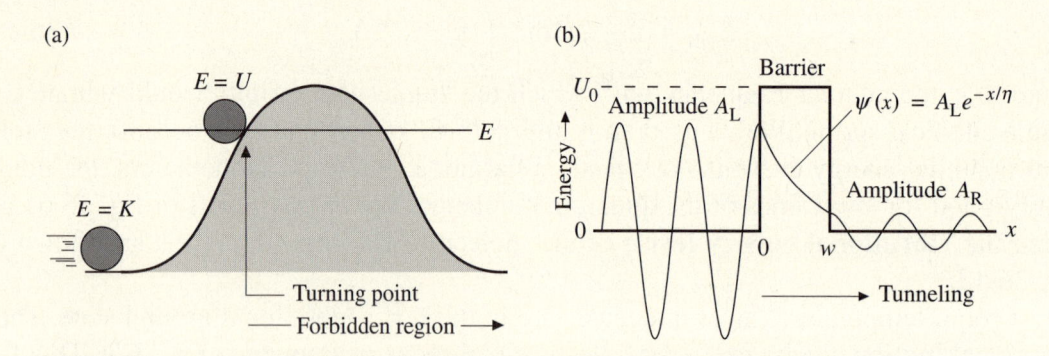

FIGURE 36-19 a) A bowling ball rolls up to the turning point, where its energy has been converted entirely to potential energy, then reverses direction. b) A quantum particle can penetrate *through* a potential-energy barrier because the wave function does not have a sharp edge.

But consider the situation quantum mechanically. Figure 36-19b shows the potential-energy diagram of a "square hill" that forms a barrier of width w between two regions of space. A classical particle incident from the left with $E < U_0$ would reflect from this barrier at exactly $x = 0$. But quantum particles, as we have already seen, can penetrate a short distance into the "classically forbidden region" of a barrier.

Suppose that the barrier is very narrow. The wave function, although decreasing exponentially within the barrier, still has a finite amplitude when it reaches $x = w$ and "pops out" of the other side. Because the wave function is nonzero on the right side of the barrier, as seen in Fig. 36-19b, there is a probability of finding the particle there. There is, in other words, a definite probability that the particle will pass *through* the barrier and emerge on the other side!

It is very much as if the bowling ball of Fig. 36-19a gets to the turning point and then, instead of rolling back down, *tunnels* its way through the hill and emerges at the same height on the other side. Although this is strictly forbidden in classical mechanics, it is apparently an acceptable behavior for quantum particles. The process is called **quantum-mechanical tunneling**.

The process of tunneling *through* a potential energy barrier, seemingly in defiance of the law of conservation of energy, is one of the strangest and most unexpected predictions of quantum mechanics. Yet you will soon see ample evidence that it does happen, and that it even has many practical applications.

The wave function to the left of the barrier in Fig. 36-19b is sinusoidal and has amplitude A_L. The wave function *within* the barrier is just the decaying exponential we found in Eq. 36-49,

$$\psi(0 \leq x \leq w) = A_L e^{-x/\eta}, \tag{36-62}$$

where the penetration distance η is given in Eq. 36-53 as

$$\eta = \frac{\hbar}{\sqrt{2m(U_0 - E)}}. \tag{36-63}$$

Note that you *must* use SI units with Eq. 36-63. Energies must be in J and $\hbar$ in J s, which gives the penetration distance η in units of m.

The wave function decreases exponentially through the barrier, but before it can decay to zero it emerges again on the right side ($x > w$) as an oscillation with amplitude

$$A_R = \psi(x = w) = A_L e^{-w/\eta}. \tag{36-64}$$

The probability that the particle is to left of the barrier is proportional to $|A_L|^2$ and the probability of finding it to the right of the barrier is proportional to $|A_R|^2$. The probability that a particle striking the barrier from left will emerge on the right is thus

$$P_{\text{tunnel}} = \frac{|A_R|^2}{|A_L|^2} = \left(e^{-w/\eta}\right)^2 = e^{-2w/\eta}. \tag{36-65}$$

This is the probability that a particle will tunnel through a potential-energy barrier of width w.

Now our analysis, we have to say, has not been terribly rigorous. We assumed, for example, that the oscillatory wave functions on the left and the right were exactly at a maximum where they connected to the barrier at $x = 0$ and $x = w$. There is no reason this has to be the case. We have taken other liberties, which experts will spot, but—fortunately—it really makes no difference. Our result, Eq. 36-65, turns out to be perfectly adequate for nearly all applications of tunneling.

Because the tunneling probability is an exponential function, it is *very* sensitive to the values of w and η. Even a small increase in the thickness of the barrier can substantially reduce the tunneling probability. The constant η, which measures how far the particle can penetrate into the barrier, depends both on the particle's mass and on $U_0 - E$, the "distance" of the particle's energy E below the top of the barrier at U_0. A particle with energy only very slightly below U_0 will have a larger value of η, and thus a larger tunneling probability, than an identical particle with less energy.

EXAMPLE 36-8 a) Find the probability that an electron will tunnel through a barrier of width 1 nm if the electron's energy is 0.1 eV below the top of the barrier. b) Find the tunneling probability if the barrier in part a) is widened to 3 nm. c) Find the tunneling probability if the electron in part a) is replaced by a proton with the same energy.

SOLUTION a) An electron with energy 0.1 eV below the top of the barrier has $U_0 - E = 0.1$ eV. Thus its penetration distance is given by Eq. 36-63:

$$\eta = \frac{\hbar}{\sqrt{2m(U_0 - E)}} = 0.618 \text{ nm}.$$

The probability of this electron tunneling through a barrier of width $w = 1$ nm is

$$P_{\text{tunnel}} = e^{-2w/\eta} = e^{-2(1 \text{ nm})/(0.618 \text{ nm})} = 0.039 = 3.9\%.$$

b) Changing the width to $w = 3$ nm has no effect on η. The new tunneling probability is

$$P_{\text{tunnel}} = e^{-2w/\eta} = e^{-2(3 \text{ nm})/(0.618 \text{ nm})} = 6.0 \times 10^{-5} = 0.006\%.$$

Increasing the width by a factor of 3 decreases the tunneling probability by a factor of 660!

c) A proton only 0.1 eV below the top of the barrier has $\eta = 0.014$ nm. Its probability of tunneling through a barrier 1 nm wide is

$$P_{\text{tunnel}} = e^{-2w/\eta} = e^{-2(1 \text{ nm})/(0.014 \text{ nm})} = 1 \times 10^{-64}.$$

For all practical purposes, the probability is zero that a proton will tunnel through this barrier. If the probability of a proton tunneling through a mere 1 nm is only 10^{-64}, you can easily recognize that you should not stay up nights waiting to see a macroscopic object tunnel through a macroscopic distance!

•

Quantum-mechanical tunneling seems so obscure that it is hard to imagine practical applications. Surprisingly, there are many. We will look at two: the scanning tunneling microscope and resonant tunneling diodes.

The Scanning Tunneling Microscope

The resolution of an optical microscope is limited, by diffraction, to viewing objects no smaller than about a wavelength of light—roughly 500 nm. This is more than 1000 times the size of atoms, so there is no hope of ever seeing atoms or molecules via optical microscopy. Electron microscopes are similarly limited by the de Broglie wavelength of the electrons. Their resolution is much better than an optical microscope, but still not quite to the level of "seeing" atoms.

This situation changed dramatically with the 1981 invention of the **scanning tunneling microscope**, or **STM** as it is usually called. The inventors were Gerd Binnig and

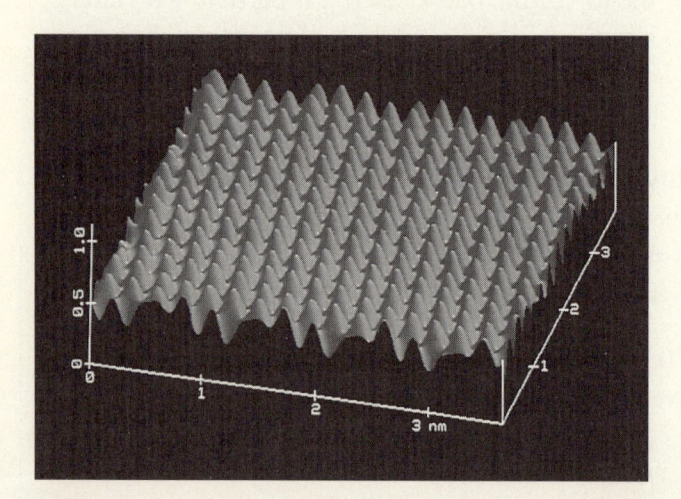

FIGURE 36-20 Two pictures made with a scanning tunneling microscope. On the left are individual atoms of carbon in graphite. On the right is a surface of silicon.

Heinrich Rohrer, physicists at IBM's research laboratory in Zurich, Switzerland. Their discovery allowed scientists, for the first time, to "see" surfaces literally atom by atom. The significance of this invention was so immense that Binnig and Rohrer were awarded the Nobel Prize a short five years later, in 1986.

Figure 36-20 shows two pictures taken with an STM. On the left you can see individual atoms of carbon on the surface of graphite. The right figure shows a somewhat less magnified surface of silicon. These pictures, and many others you have likely seen (but maybe did not know where they came from), are stupendous, but how are they made?

Figure 36-21a shows the basic idea behind the scanning tunneling microscope. A conducting probe with a *very* sharp tip—just a few atoms wide—is brought to within a few tenths of a nanometer from a surface. Preparing the tips and controlling the spacing are both difficult, but solvable, technical challenges. Once positioned, the probe can be mechanically scanned back and forth across the surface.

Recall from our analysis of the photoelectric effect that electrons are bound inside metals by an amount of energy called the *work function* E_0. A typical work function is 4 or 5 eV. This is the energy that must be supplied—by a photon or otherwise—to lift an electron out of the metal. In other words, the electron's energy in the metal is E_0 less than its energy outside.

This idea allows us to draw the potential-energy diagram of Fig. 36-21b. Here you can see that the small air gap between the sample and the probe tip represents a potential-energy barrier. Its height is $E_0 \approx 4$ eV above the energy level of an electron inside the sample. An absorbed photon of sufficiently high frequency could lift the electron *over* the barrier, from the sample to the probe. This is just the photoelectric effect. But Binnig and Rohrer recognized that electrons can also tunnel *through* the barrier if it is sufficiently narrow. This creates a *tunneling current* from the sample into the probe.

FIGURE 36-21 a) The probe tip of a scanning tunneling microscope is <1 nm from the surface. b) Potential-energy diagram of an electron.

In operation, the tunneling current is continuously recorded as the probe tip scans across the surface. The tunneling current, as we already saw, is extremely sensitive to the barrier thickness. As the tip scans over the position of an atom, the gap decreases by ≈ 0.1 nm and the current increases. Between atoms the gap is larger and the current drops. Today's STMs can sense changes in the gap of as little as 0.001 nm—about 1% of an atomic diameter! The final images you see are computer generated from the current measurements at each position.

The STM has revolutionized the science and engineering of surfaces. They are now used to study everything from how surfaces corrode and oxidize—a topic of great practical significance—to how biological molecules are structured. Another example of quantum mechanics working for you!

The Resonant Tunneling Diode

The semiconductor diode laser that we examined in Section 36.8 had a narrow GaAs layer surrounded by wider layers of GaAlAs. Because an electron's potential energy is ≈0.3 eV less in GaAs than in GaAlAs, this structure provides a quantum well in which electrons are confined with a single energy. You could see, in Fig. 36-12b, that the electron wave function drops to zero within a couple a nanometers after entering the potential-energy barrier. Because the GaAlAs layers in a laser are much thicker than the GaAs layer, the electrons are well confined within the quantum well.

Suppose, though, that the GaAlAs barrier is reduced in thickness to only a couple of nanometers. An electron's potential-energy diagram then looks like Fig. 36-22a. Now, because the GaAlAs layers are much thinner, an electron inside the potential well has the possibility of tunneling through to the outside. Conversely, an electron coming from the outside, and impinging on the GaAlAs barrier, might tunnel *into* the quantum well. The difficulty with tunneling in, though, is a serious energy mismatch. Tunneling may be a strange phenomenon, but energy does still have to be conserved.

An electron inside the quantum well *must* have one of the allowed energies. Typically there is a single allowed quantum state with $E \approx 0.15$ eV. Electrons on the outside have a thermal energy

$$E \approx \tfrac{3}{2}kT = 6.0 \times 10^{-21} \text{ J} = 0.04 \text{ eV}$$

at room temperature. An electron approaching the barrier with $E \approx 0.04$ eV cannot tunnel inside if there is no allowed state with this energy.

But suppose a potential difference ΔV is placed across the three layers, with the left side more positive. Because the electrons are negative, this will *lower* the potential energy on the left side. As ΔV is increased, as shown in Fig. 36-22b, it will reach a value $\Delta V_{\text{resonance}}$ at which the energy level inside the quantum well is exactly matched to the energy of an electron approaching from the right. We then have a *resonance*, much like when an external driving frequency matches the natural frequency of an oscillator, causing a large-amplitude response. In this case, an electron can easily tunnel into the quantum well from the right. Once in the well, it can tunnel through the opposite barrier and emerge on the left with kinetic energy $K \approx e\Delta V$.

In other words, matching the energy levels inside and outside of the quantum well allows electrons to flow from right to left by tunneling through the barriers. Thus a

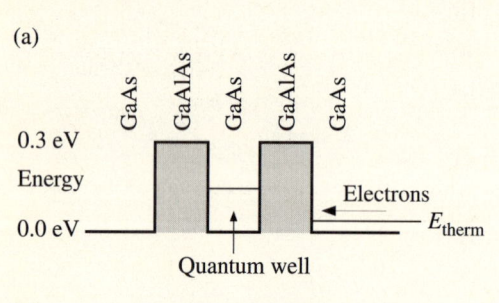

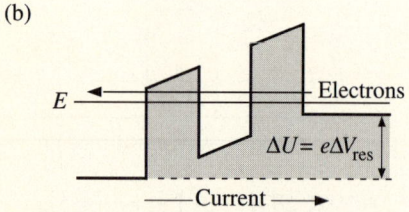

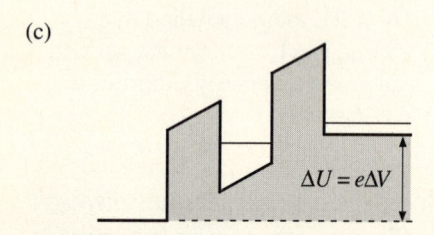

FIGURE 36-22 Potential energy of a resonant tunneling diode. a) Thermal energy electrons from the right cannot penetrate the barrier because of the energy mismatch with the state in the quantum well. b) The correct potential difference creates a resonance, causing a current to tunnel through. c) No current flows if the voltage is too high.

current flows through the device when the potential difference $\Delta V_{\text{resonance}}$ is applied. Consequently, this device is called a **resonant tunneling diode**.

Applying too much voltage, however, destroys the resonance. As Fig. 36-22c shows, a large ΔV drops the energy level in the quantum well too low, and again electrons from the right side have no matching energy level to tunnel into. Current flows through a resonant tunneling diode only for a small range of voltages near $\Delta V_{\text{resonance}}$.

Figure 36-23 shows an experimentally measured current-voltage graph for a device having a 4 nm GaAs quantum well surrounded by 10 nm wide GaAlAs barriers. There is a small range of voltages around 0.25 volts for which the current shoots up by a factor of 10, then it drops back to near zero by the time $\Delta V = 0.4$ V. So-called "normal" diode behavior begins for $\Delta V > 0.7$ V, but such a "high voltage" would not be used with this device.

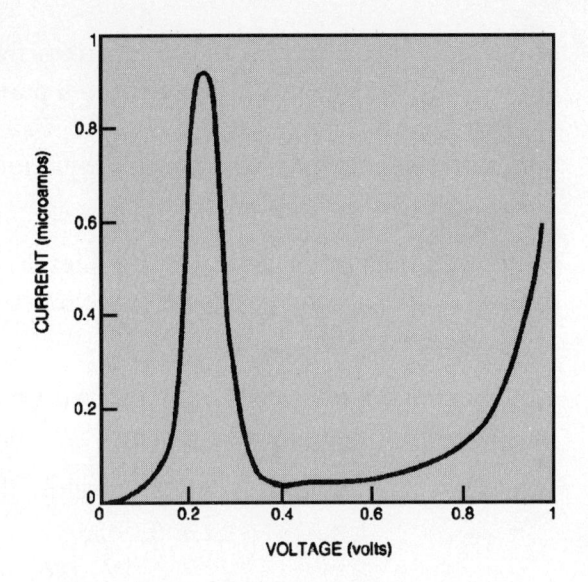

FIGURE 36-23 Experimental measurement of the current-voltage characteristics of a resonant tunneling diode.

The ability of a resonant tunneling diode to change its current drastically for just a small change in voltage makes it very useful in the digital electronic circuits of high-speed computers. They can also be used as very high-speed oscillators, creating oscillating voltages with frequencies as high as 500 GHz.

Summary

Important Concepts and Terms

Schrödinger equation
boundary conditions
classically forbidden region
zero-point motion
correspondence principle
quantum-mechanical model
potential well
bound-state wave function

penetration distance
quantum-well device
gamma rays
quantum harmonic oscillator
vibrational energy levels
quantum-mechanical tunneling
scanning tunneling microscope
resonant tunneling diode

The Schrödinger equation is

$$\frac{d^2\psi}{dx^2} = -\frac{2m}{\hbar^2}[E - U(x)]\psi(x).$$

The solutions to this equation are wave functions describing the stationary states and allowable energies of a particle of mass m in a region of space where its potential energy as a function of position is given by $U(x)$. The wave functions are the *descriptors* of quantum mechanics, with the square of the wave function $|\psi(x)|^2$ being interpreted as the probability density $P(x)$ for finding a particle at position x.

If $\psi_1(x)$ and $\psi_2(x)$ are two specific, independent solutions of the Schrödinger equation, then the *general solution* of the equation is

$$\psi(x) = A\psi_1(x) + B\psi_2(x),$$

where A and B are constants whose values are determined by the boundary conditions. The primary boundary conditions the wave function must obey are

1. $\psi(x) \to 0$ as $x \to +\infty$ or $-\infty$.
2. $\psi(x) = 0$ if x is in a region where it is physically impossible for the particle to be.
3. $\psi(x)$ is a continuous function.
4. $\psi(x)$ is a normalized function, meaning that

$$\int_{-\infty}^{\infty} |\psi(x)|^2 \, dx = 1.$$

These conditions can be stated as algebraic equations from which A and B can be determined.

The wave functions meet the boundary conditions only for discrete values of the energy E. These are the energies for which the wave function can form a standing wave. Hence quantization is an inherent feature of Schrödinger's theory. The wave function for the nth quantum state is an oscillatory function with n antinodes and $n + 1$ nodes (including the ends). The spacing between the nodes increases in regions where the kinetic energy of a classical particle would decrease.

The wave functions are the stationary states of the system. The system can undergo transitions from one stationary state to another (or from one energy level to another) by emitting or absorbing a photon of frequency

$$f = \frac{\Delta E}{h}.$$

The Bohr correspondence principle states that the *average* behavior of a quantum system should correspond to the classical physics solution when the quantum number becomes very large: $n \to \infty$.

The quantum-mechanical solution for a particle in a rigid box of length L showed:

1. The particle can exist only if it has energy $E_n = n^2 E_1$ where $n = 1, 2, 3, \ldots$ is the quantum number and where $E_1 = h^2/8mL^2$ is the energy of the $n = 1$ ground state.
2. The wave function for a particle in the nth quantum state is

$$\psi_n(x) = \begin{cases} \sqrt{\dfrac{2}{L}} \sin\left(\dfrac{n\pi x}{L}\right) & 0 \le x \le L \\ 0 & x < 0 \text{ and } x > L. \end{cases}$$

These are the stationary states of the system

The wave functions and energies of a particle in a finite potential well were similar to those of a particle in a rigid box. However, a finite potential well supports only a finite number of bound states. The finite potential well is a reasonable quantum-mechanical model of many real-world situations.

The quantum harmonic oscillator is another important and widely used model. The energy levels of a quantum harmonic oscillator are

$$E_n = (n - \tfrac{1}{2})\hbar\omega \qquad n = 1, 2, 3, \ldots,$$

where ω is the angular frequency with which a classical particle would oscillate. Molecular bonds in both molecules and solids are often modeled as quantum harmonic oscillators. This allows us to understand the infrared absorption spectrum of molecules.

Finally, a particularly important finding was that the wave function of a quantum particle can penetrate into regions of space that would be forbidden to a classical particle. The wave function decreases exponentially in such regions, obeying

$$\psi(x) = \psi_{\text{edge}} e^{-(x-L)/\eta} \qquad (x \geq L),$$

where the penetration distance is

$$\eta = \frac{\hbar}{\sqrt{2m(U_0 - E)}}.$$

The quantity $U_0 - E$ is the energy distance below the top of the potential-energy barrier. The penetration distance can be significant for atomic particles.

If a potential-energy barrier is sufficiently narrow, a quantum particle can penetrate all the way through and emerge on the other side. This is called quantum-mechanical tunneling. The probability for a particle of energy E to tunnel through a potential-energy barrier of height U_0 and width w is

$$P_{\text{tunnel}} = e^{-2w/\eta}.$$

Tunneling is one of the strangest phenomena of quantum physics, but it has become the basis for a variety of practical applications. The scanning tunneling microscope, in particular, has opened a new window into the microscopic realm of atoms and molecules.

There is irony in just how successful the Schrödinger equation is for predicting energy levels and the frequencies of photons. Schrödinger had taken his trip to the Alps for the purpose of trying to *save* classical physics. He did not like the quantum jumps that Bohr had introduced, and he thought that de Broglie's hypothesis offered a way out. If matter has a wavelength, Schrödinger reasoned, perhaps the reason why is that matter *is* a wave—not just with wave-like properties, as in wave–particle duality, but *really* a wave. Schrödinger envisioned his wave function as a classical wave describing the electron, just as classical electromagnetic waves described light. He tried to interpret $|\psi|^2$ not as a probability but as the actual mass density of the electron. But it did not work.

The interpretation of $|\psi|^2$ as a probability density was made the following year, in 1926, by Max Born. Schrödinger found it unacceptable. As he later told a colleague, "Had I known that we were not going to get rid of this damned quantum jumping, I never would have involved myself in this business." And look back at the quote that opened this chapter. Schrödinger never reconciled himself to the probabilistic aspects of quantum mechanics, and he spent his later years studying and writing about biology. Yet for all of Schrödinger's misgivings, the equation that bears his name stands at the center of one of the most successful scientific theories ever devised.

Exercises and Problems

Exercises

1. An electron in a rigid box absorbs light having a wavelength of 600 nm. How wide is the box?
2. The electrons in a rigid box emit photons of wavelength 1484 nm during the 3 → 2 transition.
   a. What kind of photon is this? Infrared, visible, or ultraviolet?
   b. How wide is the region in which the electrons are confined?
3. Sketch graphs of the $n = 8$ wave function in each of the two potential energy wells shown in Fig. 36-24. Assume that the particle's energy E_8 is greater than the height of the "step." (**Hint:** Keep in mind how the spacing between the nodes is related to the kinetic energy of a classical particle.)

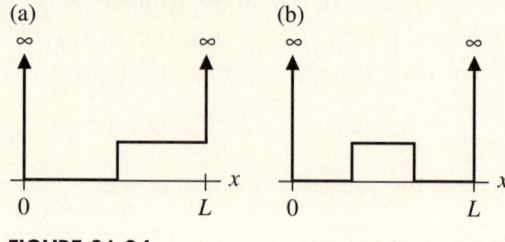

   FIGURE 36-24

4. The graph in Fig. 36-25 shows the potential-energy function $U(x)$ of a particle. Solution of the Schrödinger equation finds that the $n = 3$ level has $E_3 = 0.5$ eV and that the $n = 6$ level has $E_6 = 2.0$ cV.
   a. Redraw this figure and add to it the energy lines for the $n = 3$ and $n = 6$ states.
   b. Identify on your graph the turning points of the $n = 3$ level and the $n = 6$ level.
   c. Draw a graph of kinetic energy versus position for a classical particle with energy 2.0 eV.
   d. In what region will the de Broglie wavelength of a 2.0 eV particle be the smallest? The largest?
   e. Sketch the $n = 3$ and $n = 6$ wave functions. Show them as oscillating about the appropriate energy line.

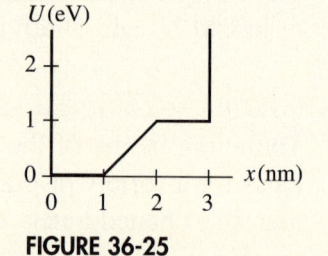

   FIGURE 36-25

5. a. Sketch graphs of the probability density $|\psi(x)|^2$ for the four states in the finite potential well of Fig. 36-10. Stack them vertically, similar to the Fig. 36-10 graphs of $\psi(x)$.
   b. What is the probability that a particle in the $n = 2$ state of the finite potential well will be found at the center of the well? Explain.
   c. Is your answer to b) consistent with what you know about waves? Explain.

6. Show that the penetration distance η has units of meters.

7. Consider a quantum harmonic oscillator.
   a. What happens to the spacing between the nodes of the wave function as $|x|$ increases? Why?
   b. What happens to the heights of the antinodes of the wave function as $|x|$ increases? Why?
   c. Sketch a reasonably accurate graph of the $n = 6$ wave function of a quantum harmonic oscillator.

8. An electron is confined in a harmonic potential well with a spring constant of 2.0 N/m.

a. What are the first three energy levels of the electron?
   b. What wavelength photon is emitted if the electron undergoes a 3 → 1 quantum jump?

Problems

9. Show that the normalization constant A_n for the wave functions of a particle in a rigid box has the value given in Eq. 36-34.

10. A particle confined in a rigid one-dimensional box of width 10 fm has an energy level $E_n = 32.9$ MeV and an adjacent energy level $E_{n+1} = 51.4$ MeV.
    a. Determine the values of n and $n + 1$.
    b. Draw an energy-level diagram showing all energy levels from 1 through $n + 1$. Label each level, and write the energy beside it.
    c. Sketch the $n + 1$ wave function on the $n + 1$ energy level.
    d. What is the wavelength of a photon emitted in the $n + 1 \rightarrow n$ transition? Compare this to a typical visible-light wavelength.
    e. What is the mass of the particle? Can you identify it?

11. A 2-μm-diameter water droplet is moving with a speed of 1 μm/s in a box of width 20 μm.
    a. Estimate the particle's quantum number.
    b. Use the correspondence principle to determine whether quantum mechanics is needed to understand the motion or whether it is "safe" to use classical physics.

12. Use the data from Fig. 36-18 to calculate the first three vibrational energy levels of a C=O carbon-oxygen bond.

13. Consider a particle in a rigid box of length L. For each of the states $n = 1$, $n = 2$, and $n = 3$:
    a. Sketch graphs of $|\psi(x)|^2$. Label the points $x = 0$ and $x = L$.
    b. Where, in terms of L, are the positions at which the particle is *most* likely to be found?
    c. Where, in terms of L, are the positions at which the particle is *least* likely to be found?
    d. Determine, by examining your $|\psi(x)|^2$ graphs, if the probability of finding the particle in the left one-third of the box is less than, equal to, or greater than 1/3. Explain your reasoning.
    e. *Calculate* the probability that the particle will be found in the left one-third of the box.

14. In a nuclear physics experiment, a proton is fired toward a $Z = 13$ nucleus that is modeled as in Fig. 36-13. A neutron, which was initially in its ground state, subsequently emits a gamma ray with wavelength 1.73×10^{-4} nm. What was the *minimum* initial speed of the proton? (**Hint:** Don't neglect the proton-nucleus collision.)

15. For the quantum well laser of Fig. 36-12, *estimate* the probability that an electron will be found within one of the GaAlAs layers rather than in the GaAs layer. Explain your reasoning.

16. Verify that the $n = 1$ wave function $\psi_1(x)$ of the quantum harmonic oscillator really is a solution of the Schrödinger equation. That is, show that the right and left sides of the equation are equal if you use the $\psi_1(x)$ wave function.

17. a. Show that the constant b used in the quantum harmonic oscillator wave functions has units of length.
    b. Show that the constant b is equal to the classical turning point of an oscillator in the $n = 1$ ground state.
    c. Determine the normalization constant A_1 for the $n = 1$ ground-state wave function of the quantum harmonic oscillator. Your answer will be in terms of b.
    d. Write an expression for the probability that a quantum harmonic oscillator in its $n = 1$ ground state will be found in the classically forbidden region.
    e. (Optional) Use numerical integration to evaluate your probability expression of part d). (**Hint:** Make a change of variables to $u = x/b$.)

18. Consider a particle in a rigid box having walls at $x = -L/2$ and $x = +L/2$.
    a. What is the wave function $\psi(x)$ for $x < -L/2$ and $x > L/2$? Explain.
    b. Write the Schrödinger equation in the region $-L/2 \leq x \leq L/2$ for a particle with energy E.
    c. Write down a general solution to the Schrödinger equation that is valid in the region $-L/2 \leq x \leq L/2$.
    d. What are the boundary conditions this wave function must satisfy?
    e. Apply the boundary condition to determine the allowed energy levels. Note that there are two different ways to satisfy the boundary conditions, each giving a different set of wave functions and energy levels.
    f. Compare your results to the box $0 \leq x \leq L$ that was analyzed in this chapter. In what ways are the results the same and in what ways are they different? Are any differences physically meaningful?

19. a. What is the ratio $P(d + L)/P(L)$ of the probability density for finding a particle at distance d *inside* a potential energy barrier to the probability density for finding it at the turning point?
    b. A typical electron in sodium has energy $-E_0$ compared to a free electron, where E_0 is the 2.7 eV work function of sodium. At what distance *beyond* the surface is the electron's probability density 10% of its value *at* the surface?
    c. Compare your answer to part b) to a typical atomic diameter.

20. a. What is the probability that an electron will tunnel through a 0.5 nm gap from a metal to a STM probe if the work function is 4 eV?
    b. The probe passes over a 0.05-nm-high atom. By what factor does the tunneling current increase?
    c. If a 10% current change is reliably detectable, what is the smallest height change the STM can detect?

21. Tennis balls traveling >100 mph routinely bounce off tennis rackets. At some sufficiently high speed, however, the ball will break through the strings and keep going. The racket is a potential energy barrier whose height is the energy of the slowest string-breaking ball. Suppose that a 100 g tennis ball traveling at 200 mph is just sufficient to break the 2-mm-thick strings. Estimate the probability that a 120 mph ball will tunnel through the racket without breaking the strings. Express your answer as a power of 10 rather than a power of e.

[**Estimated 10 additional problems for the final edition.**]

Chapter 37

The Structure of Atoms

LOOKING BACK Sections 34.2, 34.6; 35.3, 35.4; 36.2

37.1 The Atomic Structure Problem

The problem of discovering the structure of atoms is one that we have continued to revisit. Recall from Chapters 8 and 32 that the first model of an atom we looked at was purely classical. It consisted of a negatively charged electron orbiting a positive nucleus with the Coulomb electric force providing the centripetal acceleration. This model explained the incredibly high frequency with which the electron orbits, but otherwise the model had almost no agreement with experimental evidence about atoms! It could not explain their discrete spectra, nor could it even explain how atoms are stable.

The Bohr model of the hydrogen atom was a big step forward. The concept of stationary states, found by requiring de Broglie standing waves, provided a means of understanding both the stability of atoms and the quantum jumps that lead to discrete spectra. And Bohr's ability to derive the Balmer formula for the hydrogen spectrum indicated that he was on the right track. Yet, as we have noted, the Bohr model was not successful for any atom other than hydrogen.

In this chapter we will see whether the Schrödinger theory is better at explaining atomic structure than earlier models. To decide, let's first look at what we expect a successful atomic theory to do:

1. An atomic theory should correctly describe the spectra of all the elements.
2. An atomic theory should correctly describe the *ionization energies* of all the elements. Ionization energy, you will recall, is the energy required to remove an electron from an atom and leave a positive ion behind. As Fig. 37-1 shows, the ionization energy is not the same for all elements, but neither does it vary randomly. It increases quite steadily from ≈ 5 eV for the alkali metals, on the left edge of the periodic table, to ≥ 15 eV for the rare gases, then plunges back to ≈ 5 eV and starts again. Such a regular pattern cries out for an explanation.

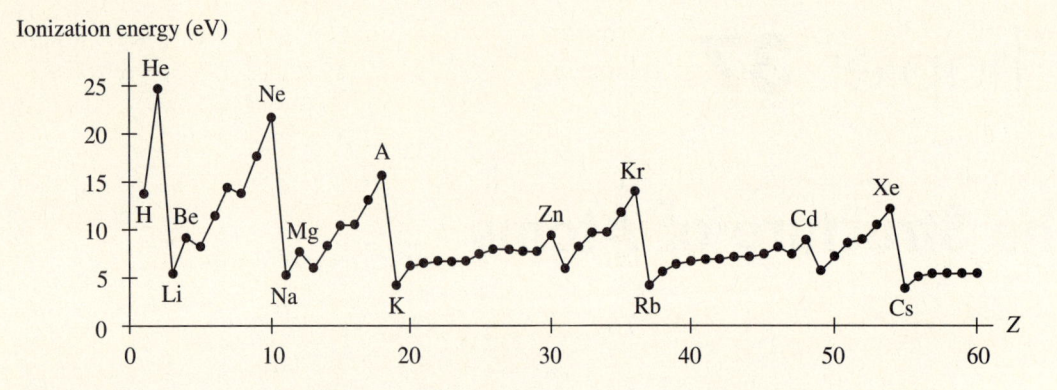

FIGURE 37-1 Ionization energies of the elements up to $Z = 60$. Notice the periodically recurring pattern. Closed shells occur at $Z = 2, 10, 18, 36,$ and 54.

3. The regularly recurring pattern seen in the ionization energies is also found in the chemical properties of the elements. Those elements with very low ionization energies are highly reactive metals, whereas the highest ionization energies belong to inert gases. These recurring patterns are the basis for the *periodic table of the elements*. A successful atomic theory should *explain* the periodic table.
4. Finally, a successful atomic theory should describe how molecular bonds form to join individual atoms together into molecules.

This seems a lot to ask of a theory. The Bohr atom did not succeed, but it was not really a *theory*—just an ad hoc collection of hypotheses. Schrödinger's quantum mechanics, however, really is a complete theory. It consists of a descriptor—the wave function—and an "equation of motion." It also has an *interpretation* of the wave function and a means of relating it to measured quantities. Quantum mechanics, from its inception, purported to be *the* theory of nature at the atomic level. If that claim was to be validated, quantum mechanics had to successfully "solve" the problem of atomic structure. And it did.

In this chapter we will focus attention on applying quantum mechanics to items 2 and 3 of our list. In the process, we will develop the *shell model* of atoms. You are likely familiar from chemistry with many ideas of the shell model, but we now want to place those ideas on a more rigorous footing. We are also keenly interested in the spectra of atoms, but we will defer that topic until the next chapter. We will leave it to more advanced courses to show that the wave function is also the proper way to understand molecular bonds.

37.2 A One-Dimensional Hydrogen Atom

Let's begin by applying Schrödinger's theory to an extremely simply model—a one-dimensional hydrogen atom. The potential energy of the electron in a hydrogen atom is

$$U(r) = -\frac{1}{4\pi\varepsilon_0}\frac{e^2}{r}. \tag{37-1}$$

A ONE-DIMENSIONAL HYDROGEN ATOM

The energy is negative because the charges of the proton and the electron are $+e$ and $-e$. This was the potential energy used in our analysis of the Bohr atom. It is, however, a three-dimensional potential energy because $r = (x^2 + y^2 + z^2)^{1/2}$ is the radial distance from the proton.

The nucleus is also three dimensional, but we modeled it in one dimension by considering the potential along a line passing through the center. Let us apply the same idea to a hydrogen atom, writing the one-dimensional potential energy along the x-axis as

$$U(x) = -\frac{1}{4\pi\varepsilon_0}\frac{e^2}{x}. \tag{37-2}$$

This is the potential-energy function for a one-dimensional hydrogen atom.

The Coulomb potential energy of Eq. 37-2, shown in Fig. 37-2a, is more complicated than those we have considered up until now. Nonetheless, the Schrödinger equation can be solved numerically, just as we did for several of the potentials in Chapter 36. The boundary conditions are $\psi(x \to \pm\infty) = 0$, because atoms are of finite size. Using these conditions, we can find the energies and wave functions for an electron in this potential-energy well. The first three energy levels are shown on Fig. 37-2a.

Note that the energy levels are *exactly* those we found in Chapter 34 for the Bohr atom! This seems—and is—very hard to believe because this is only a one-dimensional atom. Our one-dimensional model is far better than we had any reason to expect, although it takes a lot of very advanced mathematics to explain this curious result.

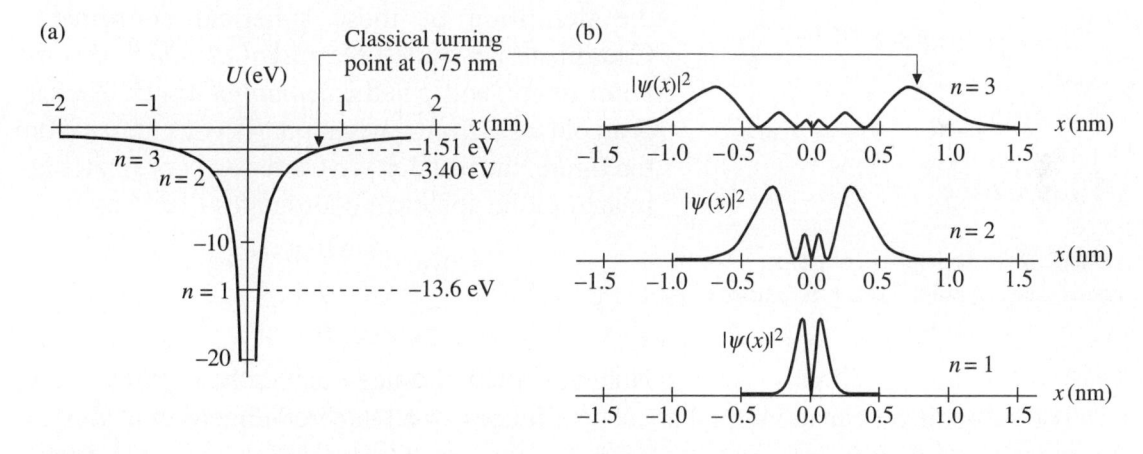

FIGURE 37-2 a) The Coulomb potential well of a one-dimensional hydrogen atom, with the first three energy levels shown. b) The probability density $|\psi(x)|^2$ of the first three stationary states.

The electron probability densities $|\psi(x)|^2$ for the first three stationary states are shown in Fig. 37-2b. As n increases, notice how rapidly the electron spreads out to larger values of x. This is due to the shape of the potential energy. Low-energy states are well confined, but the classical turning points of higher energy states are at much larger values of x. The wave functions of the higher energy states also penetrate well *beyond* the classical turning point, as you can see for $n = 3$.

Figure 37-2b is a particularly interesting result because it suggests that the electron exists as a "shell of probability" around the nucleus. That is, an electron in the $n = 1$ ground state is most likely to be found at a distance ≈ 0.05 nm from the nucleus. It is distinctly less likely to be found either closer to or farther from the nucleus than ≈ 0.05 nm. An $n = 3$ excited electron is most likely to be found at distance ≈ 0.7 nm, and an $n = 2$ electron is intermediate with a most likely distance ≈ 0.3 nm. Not that it is impossible for an $n = 2$ electron to be at other distances, but simply that we are far more *likely* to find it ≈ 0.3 nm from the nucleus than at any other distance. If this idea can be extended to three dimensions—and we will show in the next section that it can—then we can think of an $n = 2$ electron as forming a spherical "shell" of radius $r ≈ 0.3$ nm around the nucleus. This idea is the basis for the *shell model* of the atom, which we will introduce in this chapter.

Although our one-dimensional atom is very simple, it provides an initial explanation of the energy levels of atomic electrons and of their distribution within the atom. Thus, we are making progress toward understanding the *structure* of atoms!

37.3 The Three-Dimensional Hydrogen Atom

Atoms, molecules, and solids are, of course, three dimensional. These higher dimensions affect their quantum properties and change the quantum numbers needed for their description. The physical situation of particular interest to us is the hydrogen atom. In three dimensions, the electron's position is most easily described using **spherical polar coordinates** r, θ, and ϕ rather than Cartesian coordinates x, y, and z. Figure 37-3 shows the definition of these spherical coordinates. Coordinate r is called the *radial distance*, θ is the *polar angle*, and ϕ is the *azimuthal angle*. The last is an old term from navigation. As you can see from the figure, the usual x, y, z Cartesian coordinates are related to the spherical coordinates r, θ, ϕ by

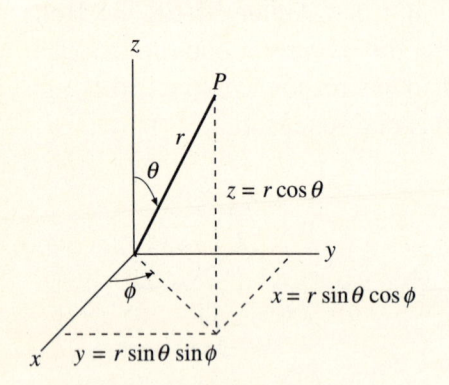

FIGURE 37-3 The spherical polar coordinates of point P are r, θ, and ϕ.

$$x = r \sin\theta \cos\phi$$
$$y = r \sin\theta \sin\phi \quad (37\text{-}3)$$
$$z = r \cos\theta.$$

Note the radial distance r cannot be negative: $r \geq 0$.

The Schrödinger equation in spherical coordinates, in all its three-dimensional glory, is

$$\frac{1}{r}\frac{\partial^2}{\partial r^2}(r\psi) + \frac{1}{r^2 \sin\theta}\frac{\partial}{\partial \theta}\left(\sin\theta \frac{\partial \psi}{\partial \theta}\right) + \frac{1}{r^2 \sin^2\theta}\frac{\partial^2 \psi}{\partial \phi^2} = \frac{2m}{\hbar^2}[U(r) - E]\psi, \quad (37\text{-}4)$$

where the wave function $\psi(r, \theta, \phi)$ is a function of all three coordinates. This equation is a *partial differential equation* that describes the behavior of the wave function in three dimensions. The mathematical task is to solve it, thereby finding the allowed energy levels E and the specific wave function associated with each energy.

Fortunately, we do not need to worry about the details of the solution. More advanced courses demonstrate that the wave function can be separated into a product of a **radial wave function** $R(r)$, dependent only on r, and two **angular wave functions** $\Theta(\theta)$ and $\Phi(\phi)$ that depend only on θ and ϕ. That is,

$$\psi(r, \theta, \phi) = R(r)\Theta(\theta)\Phi(\phi). \quad (37\text{-}5)$$

When this form of the wave function is used in Eq. 37-4, the partial differential equation becomes three separate equations, one each for the three functions R, Θ, and Φ. The equation for $R(r)$, which is called the **radial wave equation**, is the only equation in which the potential energy $U(r)$ appears, so the solutions to this equation are the only ones that we need to be concerned with in this text. We will look at the radial wave equation and its wave functions later in this section, but first we need to characterize the stationary states of the three-dimensional hydrogen atom.

Stationary States of Hydrogen

Solutions to the Schrödinger equation for the hydrogen atom exist only if three quantum conditions are satisfied:

1. The energy levels of the hydrogen atom are given by

$$E_n = -\frac{1}{n^2} \cdot \left(\frac{e^2}{4\pi\varepsilon_0 \cdot 2a_B} \right) = -\frac{13.60 \text{ eV}}{n^2} \qquad n = 1, 2, 3, \ldots, \qquad (37\text{-}6)$$

where $a_B = 4\pi\varepsilon_0 \hbar^2/me^2 = 0.0529$ nm is the Bohr radius. The integer n is called the **principal quantum number**. Note that these energies are the same as those in the Bohr atom.

2. The angular momentum L of the electron's orbit can only be

$$L = \sqrt{l(l+1)}\,\hbar \qquad l = 0, 1, 2, 3, \ldots, n-1. \qquad (37\text{-}7)$$

The integer l is called the **orbital quantum number**.

3. The z-component of the angular momentum, L_z, can only be

$$L_z = m\hbar \qquad m = -l, -l+1, \ldots, 0, \ldots, l-1, l. \qquad (37\text{-}8)$$

The integer m is called the **magnetic quantum number**.

Thus the full quantum physics solution for the hydrogen atom predicts *exactly* the same energy levels as the simpler Bohr model—levels that are confirmed by the hydrogen spectrum.

Note that we need *three* quantum numbers to describe, or characterize, each stationary state in a "real" three-dimensional atom. That is, each state is identified by a triplet of quantum numbers: (n, l, m). The possible values of the quantum numbers are:

        Principal quantum number $n = 1, 2, 3, 4, \ldots$
        Orbital quantum number $l = 0, 1, 2, \ldots, n-1$
        Magnetic quantum number $m = -l, -l+1, \ldots, l-1, l.$

Notice that the energy depends only on the principal quantum number n, not on l or m.

EXAMPLE 37-1 a) List all the possible states of a hydrogen atom that have principal quantum $n = 2$. b) What is the energy of each of these states?

SOLUTION a) An atom with principal quantum number $n = 2$ could have either $l = 0$ or $l = 1$, but $l \geq 2$ is ruled out. If $l = 0$, the only possible value for the magnetic quantum number m

is $m = 0$. If $l = 1$, then the atom could have $m = -1$, $m = 0$, or $m = +1$. Thus the possible quantum numbers are

$$n:\quad 2\quad\quad 2$$

$$l:\quad 0\quad\quad 1$$

$$m:\quad 0\quad -1\ 0\ 1.$$

There are four different states, or four combinations of quantum numbers, that have principal quantum number $n = 2$.

b) Because the energy depends only on the principal quantum number, each of these four states has the *same* energy:

$$E_2 = -\frac{13.60\text{ eV}}{2^2} = -3.40\text{ eV}.$$

•

The energy levels in the hydrogen atom depend only on the principal quantum number n. It will turn out that this is a unique situation. For all other elements, the energy levels depend on the *two* quantum numbers n and l (but not m). Thus it is useful to label the stationary states by their values of n and l. To avoid the possible confusion of having two integers, it is customary to label the various value of quantum number l with a lowercase letter. The labels for the first four values are:

$l = 0$ is called an s state

$l = 1$ is called a p state

$l = 2$ is called a d state

$l = 3$ is called an f state.

The labels themselves derived from spectroscopic notation in prequantum mechanics days, when some spectral lines were classified as *s*harp, others as *p*rincipal, and yet others as *d*iffuse. Using these labels, the ground state of the hydrogen atom, with $n = 1$ and $l = 0$, is called the $1s$ state. The $3d$ state has $n = 3$, $l = 2$. In Example 37-1 we found that there is one $2s$ state (with $m = 0$) but three $2p$ states (with $m = -1, 0$, and 1).

Angular Momentum

The quantum numbers l and m describe the angular motion of each stationary state of the atom. As you can see from Eqs. 37-7 and 37-8, l determines the angular momentum L of the electron's orbit and m determines the z-component L_z.

We introduced the angular momentum L of an orbiting particle in Chapter 34. Strictly speaking, the orbital angular momentum should be defined as a *vector* $\vec{L}$ that is perpendicular to the plane of the orbit. $L = |\vec{L}|$ is the magnitude of the angular momentum vector. This is shown in Fig. 37-4. The angular momentum vector has a z-component L_z along the z-axis. Here we have chosen to let the z-axis point upward even though the orbit is tilted out of the xy-plane.

The second two quantum conditions, Eqs. 37-7 and 37-8, have a curious implication. The magnitude $|\vec{L}|$ of the orbital angular momentum in quantum state l is

$$L = |\vec{L}| = \sqrt{l(l+1)}\,\hbar.$$

The z-component L_z can take on a range of values, from $-l\hbar$ up to a maximum value $+l\hbar$. We can interpret this as the $\vec{L}$ vector having different orientations in space, each with a different value of L_z.

Now *if* the angular momentum vector $\vec{L}$ were to point directly along the z-axis, then it would have *only* a z-component $L_z = |\vec{L}| = L$. But $l < \sqrt{l(l+1)}$, so $(L_z)_{\max} < |\vec{L}|$. That is, the maximum possible value of L_z is *less* than the magnitude of vector $\vec{L}$. So the angular momentum vector $\vec{L}$ *cannot* point straight along the z-axis but, instead, *must* have either an x- or a y-component (or both).

For example, let's consider an electron with orbital quantum number $l = 3$. The electron's angular momentum is

$$L = \sqrt{l(l+1)}\,\hbar = \sqrt{12}\,\hbar = 3.46\hbar.$$

The magnetic quantum number m has to be in the range $-3 \leq m \leq 3$, so the maximum possible value of L_z is $L_z = 3\hbar$. The vector $\vec{L}$ is restricted to those angles θ for which $L_z = m\hbar$. These are shown in Fig. 37-5. Even in its most vertical orientation, you can see that $\vec{L}$ has a component perpendicular to the z-axis. Thus we find not only that the magnitude of $\vec{L}$ is quantized, but also that its orientation in space is quantized!

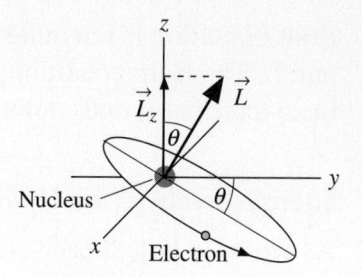

FIGURE 37-4 The orbital angular momentum $\vec{L}$. L_z is the z-component of the angular momentum vector. Note that the orbital plane is tilted below the xy-plane.

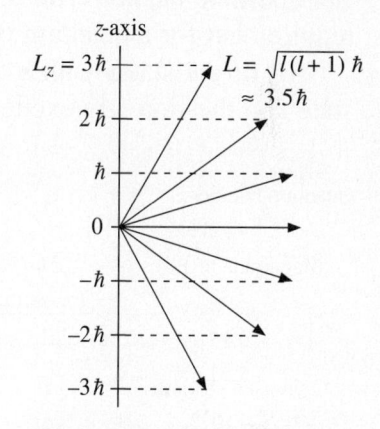

FIGURE 37-5 Seven possible orientations of the angular momentum vector for $l = 3$.

EXAMPLE 37-2 What is the angle between $\vec{L}$ and the z-axis for a hydrogen atom in the stationary state $(n, l, m) = (4, 3, 2)$?

SOLUTION: As you can see from Fig. 37-5, the angle θ between $\vec{L}$ and the z-axis is

$$\theta = \cos^{-1}\left(\frac{L_z}{|\vec{L}|}\right).$$

The state $(4, 3, 2)$ has $l = 3$ and $m = 2$, so $|\vec{L}| = L = \sqrt{12}\,\hbar$ and $L_z = 2\hbar$. Thus

$$\theta = \cos^{-1}\left(\frac{2}{\sqrt{12}}\right) = 54.7°.$$

One of the implications of the Bohr atom was the quantization of the angular momentum. We found, in Section 34.6, that $L = m\hbar$ where $m = 0, 1, 2, \ldots$. Now we see that the

Bohr condition is not quite right; it gives only $|L_z|$ and not the full orbital angular momentum L. The Bohr condition works if the orbit lies in the xy-plane, forcing $\vec{L}$ to point along the z-axis, but it does not allow for a more general situation such as that of Fig. 37-4.

Energy Levels of the Hydrogen atom

As we have noted, each state of the hydrogen atom is characterized by the values of the quantum numbers n, l, and m. The **ground state** of the hydrogen atom is the state $n = 1$, $l = 0$, $m = 0$. There is just a single state with $n = 1$, making the ground state unique. It has the lowest energy of any state, with

$$E_1 = -13.6 \text{ eV}.$$

The value $|E_1| = 13.6$ eV is called the **ionization energy** of the hydrogen atom. This is the *minimum* energy that would be needed to form a hydrogen ion H$^+$ by removing the electron from the ground state. Notice that the ground state has *no* angular momentum. A classical particle cannot orbit unless it has angular momentum, but apparently this is not a requirement for a quantum particle. We will examine this issue in the next section.

All of the states with $n > 1$ are called **excited states**. Figure 37-6 shows the ground state and the first few excited states in an **energy-level diagram** for the hydrogen atom. Each column of energy levels represents a single value of the quantum number l. The left column is all of the $l = 0$ s states, the next column is the $l = 1$ p states, and so on. Energy levels within each column are labeled by the principal quantum n, with the energy increasing as n increases. Because the quantum condition of Eq. 37-7 requires $n > l$, the s states begin with $n = 1$, the p states begin with $n = 2$, and the d states with $n = 3$. Stated another way, the lowest energy d state is $3d$ because states with $n = 1$ or $n = 2$ cannot have $l = 3$.

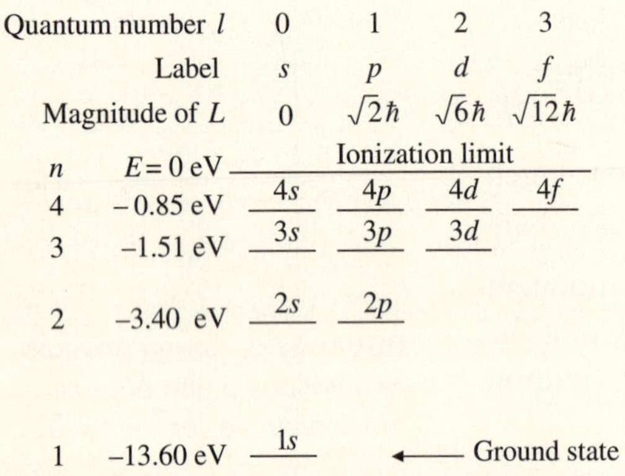

Quantum number l	0	1	2	3
Label	s	p	d	f
Magnitude of L	0	$\sqrt{2}\hbar$	$\sqrt{6}\hbar$	$\sqrt{12}\hbar$

n	E				
	$E = 0$ eV	Ionization limit			
4	-0.85 eV	$4s$	$4p$	$4d$	$4f$
3	-1.51 eV	$3s$	$3p$	$3d$	
2	-3.40 eV	$2s$	$2p$		
1	-13.60 eV	$1s$	← Ground state		

FIGURE 37-6 Energy-level diagram for the hydrogen atom. Energy is plotted upward, angular momentum to the right. The energy spacings are not to scale.

The line across the top is $E = 0$ and represents the energy of an H$^+$ ion in which the electron has been removed infinitely far away ($n \to \infty$). This line is called the *ionization limit*. Bound states of the electron have a negative total energy and are represented by the horizontal lines below the ionization limit. Only the first few energy levels are shown for each value of l, but it is important to keep in mind that there really are an infinite number of levels, as $n \to \infty$, crowding together beneath the ionization limit.

For the hydrogen atom, where the energy levels depend only on n and not on l, the energy-level diagram shows that the $3s$, $3p$, and $3d$ states all have equal energy. There seems little point in having the separate columns. As we have noted, though, the energy levels for atoms other than hydrogen will depend on both n and l. Then, as you will soon see, the columns for each l will be different.

Wave Functions of the Hydrogen Atom

The electron's radial wave function $R(r)$, which depends only on the radial coordinate r, is a solution of the radial wave equation

$$\frac{d^2R}{dr^2} + \frac{2}{r}\frac{dR}{dr} = -\frac{2m}{\hbar^2}\left(E - \frac{l(l+1)\hbar^2}{2mr^2} + \frac{e^2}{4\pi\varepsilon_0 r}\right)R(r), \tag{37-9}$$

where $U(r) = -e^2/4\pi\varepsilon_0 r$ has been used as the potential energy of the electron in a hydrogen atom. Notice that the radial equation depends explicitly on the orbital quantum number l, so the radial wave functions $R(r)$ will be labeled by both n and l. Keep in mind that $R(r)$ really *is* a wave function, similar to $\psi(x)$, even though we are using a different notation.

The solutions of Eq. 37-9 are a family of functions designated $R_{nl}(r)$. Each state of the atom—$1s$, $2s$, $2p$, and so on—is described by a different radial wave function. These are rather complicated functions of r, and we will leave a more complete description to advanced courses. For our purpose, we will simply list the first three:

$$\begin{aligned} R_{1s}(r) &= A_{1s}e^{-r/a_B} \\ R_{2s}(r) &= A_{2s}\left(1 - \frac{r}{2a_B}\right)e^{-r/2a_B} \\ R_{2p}(r) &= A_{2p}\left(\frac{r}{2a_B}\right)e^{-r/2a_B}, \end{aligned} \tag{37-10}$$

where a_B is the Bohr radius. The symbols A_{1s}, A_{2s}, and A_{2p} are normalization constants.

The radial wave functions for the $1s$ and $2s$ states shown in Fig. 37-7. The radial coordinate r runs from 0 to ∞, so the graph begins at $r = 0$, the position of the nucleus. A surprising observation is that the radial wave function is *nonzero* at the origin. According to our interpretation of the square of the wave function as a probability density, an electron in the $1s$ state is more likely to be found *at* the nucleus than at any other position! We can gain some understanding into this unexpected behavior by considering the expression for angular momentum: $L = mvr$ for a circular orbit of radius r. The $1s$ state has $l = 0$ and thus $L = 0$. The only way a classical particle can have $L = 0$ is to be *at the origin* with $r = 0$. This is impossible for a classical particle, which must maintain a nonzero angular momentum in order to orbit. A quantum particle, however, is not localized at a single point. A closer examination reveals that *all* s states, with $l = 0$, have wave functions that are nonzero at the origin. States with $l > 0$, however, have $R_{nl}(r = 0) = 0$.

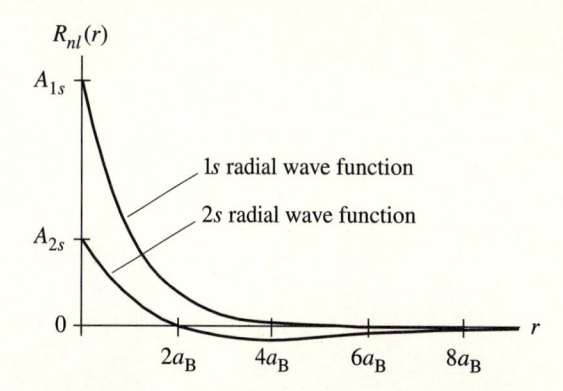

FIGURE 37-7 The $1s$ and $2s$ radial wave functions of hydrogen.

37.4 Visualizing Hydrogen

It is all fine and well to be able to write down the wave functions for hydrogen, but what do they *mean*? What have we learned about the hydrogen atom? And what is the best way to portray that knowledge?

You learned, in Chapter 35, that the probability of finding a particle in a small interval of width δx at the position x is given by

$$\text{Prob}(\text{in } \delta x \text{ at } x) = |\psi(x)|^2 \, \delta x = P(x) \cdot \delta x,$$

where

$$P(x) = |\psi(x)|^2$$

is the probability density. This interpretation of $|\psi(x)|^2$ as a probability density lies at the heart of quantum mechanics. However, $P(x)$ was for a one-dimensional wave function. Because we're now looking at a three-dimensional atom, we need to consider the probability of finding a particle in a small *volume* of space δV at the position described by the three coordinates (r, θ, ϕ). This probability is

$$\text{Prob}(\text{in } \delta V \text{ at } r, \theta, \phi) = |\psi(r, \theta, \phi)|^2 \, \delta V. \tag{37-11}$$

Equation 37-11 tells us the probability of finding the electron near a *point* in three-dimensional space. We can still interpret $|\psi(r, \theta, \phi)|^2$ as a probability density.

Figure 37-8 is an artist's rendering of the three-dimensional probability density $|\psi(r, \theta, \phi)|^2$ for the first three states of hydrogen: 1s, 2s, and 2p. Denser shading indicates regions of larger probability. Note that these pictures describe the full three-dimensional wave function, including the angular wave functions. Consequently, the probability density depends on the angles from the axes. As you can see, the probability densities in three dimensions create what is often called an **electron cloud** around the nucleus.

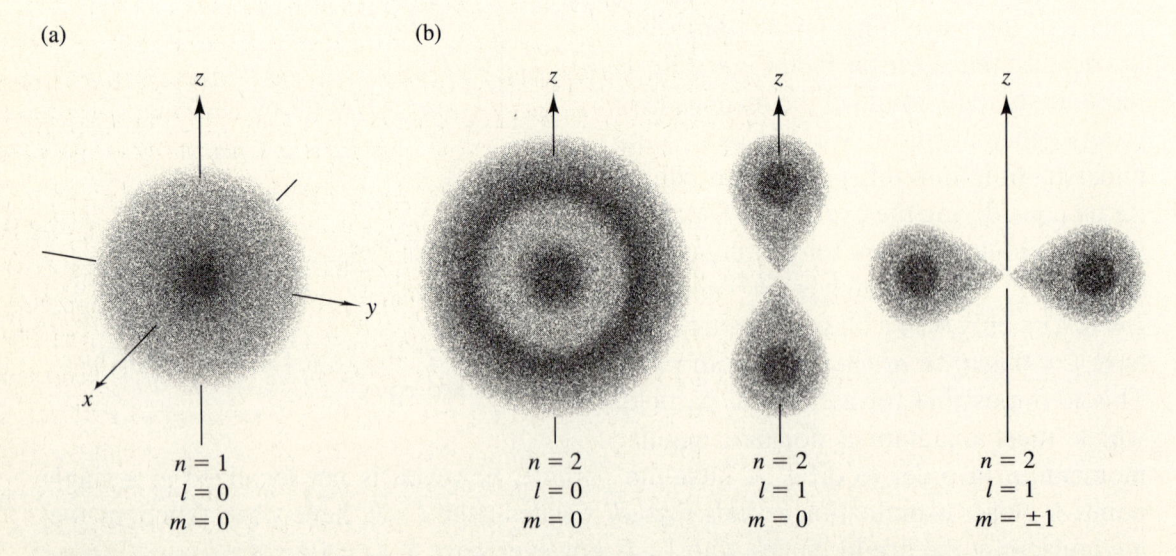

(a)   (b)

$n = 1$, $l = 0$, $m = 0$ $n = 2$, $l = 0$, $m = 0$ $n = 2$, $l = 1$, $m = 0$ $n = 2$, $l = 1$, $m = \pm 1$

FIGURE 37-8 a) The electron cloud of the 1s ground state of hydrogen. b) A cross-sectional view of the three $n = 2$ states, with 2s on the left and two 2p states, having $m = 0$ and $|m| = 1$, on the right.

(a) (b) (c)

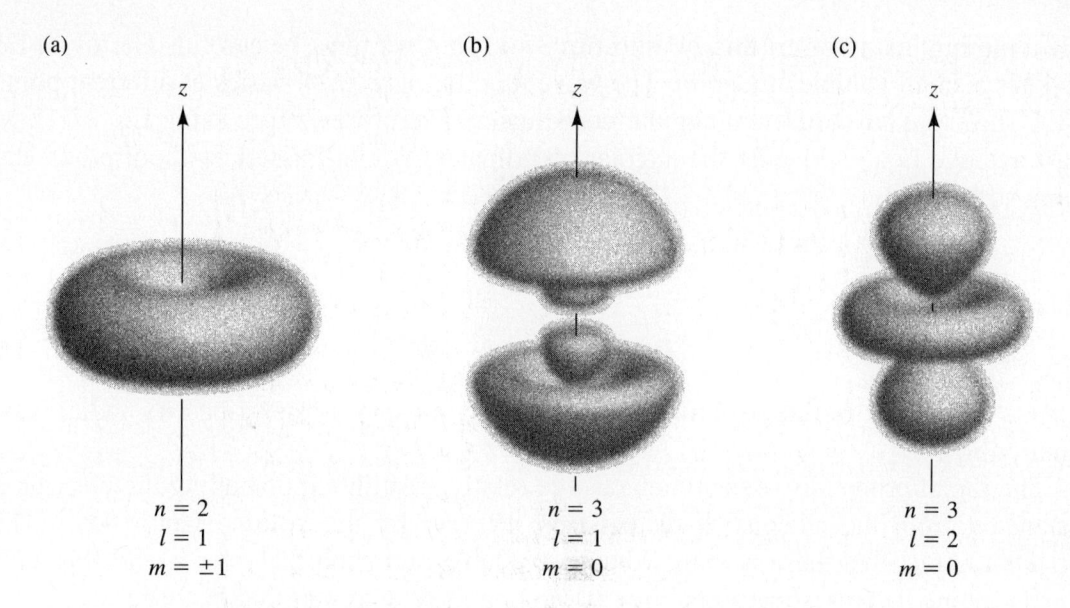

$n = 2$
$l = 1$
$m = \pm 1$

$n = 3$
$l = 1$
$m = 0$

$n = 3$
$l = 2$
$m = 0$

FIGURE 37-9 Hydrogen electron clouds for a) $2p$ $|m| = 1$, b) $3p$ $m = 0$, and c) $3d$ $m = 0$.

The $n = 2$ electron clouds of Fig. 37-8b are *axially symmetric*, meaning that you need to envision them as rotated about the z-axis. This makes the $2s$ state spherical, with a larger radius shell surrounding a smaller radius shell. The $2p$ state with $m = 0$ stretches along the z-axis in a cigar-like shape, whereas the $2p$ states with $|m| = 1$ are squeezed out in the xy-plane, forming a fat inner tube. (The two states with $m = \pm 1$ have the same probability density.)

Figure 37-9 shows another version of the $2p$ $|m| = 1$ probability density and also probability densities for two $n = 3$ states. Notice how these wave functions have *directional properties*, extending more along some directions than others. These directional properties allow atoms to reach out and touch each other, forming molecular bonds between them. The quantum mechanics of bonding goes beyond what we can study in this course, but the electron cloud pictures shown here give you a glimpse into how such bonds could form.

Drawings such as those of Figs. 37-8 and 37-9 are useful for "seeing" what the wave functions imply about the electron cloud, but these figures are hard to use. Often, we would simply like to know the probability of finding the electron a certain *distance* from the nucleus. That is, what is the probability that the electron is to be found within the small interval δr at the distance r?

Figure 37-10 shows a shell of radius r and thickness δr centered on the nucleus. Our question is equivalent to asking for the probability that the electron is found somewhere within this shell. The volume of a thin shell like this is its surface area multiplied by its thickness δr. The surface area of a sphere is $4\pi r^2$, so the volume of this thin shell is

$$\delta V = 4\pi r^2 \delta r, \tag{37-12}$$

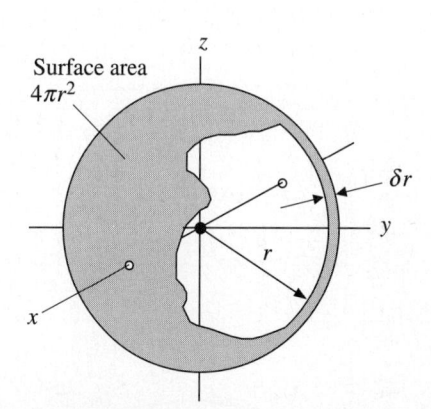

FIGURE 37-10 The radial probability density gives the probability of finding the electron in a spherical shell of thickness δr at radius r.

It is tempting just to insert this δV into Eq. 37-11, but we must be careful. The δV in Eq. 37-11 is a small volume at a *point*. The wave function $\psi(r, \theta, \phi)$ varies at different points in the shell that have different angular coordinates. Therefore, before using Eq. 37-12 we must average Eq. 37-11 over the angular coordinates. We will assert, without proof, that doing so leads to

$$\text{Prob(in } \delta r \text{ at } r) = 4\pi r^2 |R_{nl}(r)|^2 \, \delta r = P_r(r) \cdot \delta r, \tag{37-13}$$

where

$$P_r(r) = 4\pi r^2 |R_{nl}(r)|^2. \tag{37-14}$$

$P_r(r)$ is called the **radial probability density** for the state nl described by radial wave function $R_{nl}(r)$.

The radial probability density tells us the relative likelihood of finding the electron at distance r from the nucleus. It differs from $|R_{nl}(r)|^2$ by the volume factor $4\pi r^2$. This reflects the fact that there is more volume available in a shell of larger r, and this additional volume increases the probability of finding the electron at that distance.

Because we know the electron has to be found at *some* distance between 0 and ∞, the radial probability density must obey the normalization condition

$$\int_0^\infty P_r(r)\,dr = 4\pi \int_0^\infty r^2 |R_{nl}(r)|^2 \, dr = 1. \tag{37-15}$$

This is saying that the probability of finding the electron at distance r, when summed over all possible values of r, has to equal one. Equation 37-15 is the condition used to determine the values of the normalization constants A_{1s}, A_{2s}, and A_{2p} in the wave functions of Eq. 37-10. Doing so will be left as a homework problem.

Figure 37-11 shows the radial probability densities for the $n = 1$, 2, and 3 states of the hydrogen atom, all drawn to the same scale so that you can compare them to each other. The horizontal scale is in units of the Bohr radius a_B.

First, note that the 1s, 2s, and 3s states look remarkably similar to the Fig. 37-2 probability densities $|\psi(x)|^2$ of the one-dimensional atom. The one-dimensional atom gave the correct energy levels, and now we see that it also correctly describes the "real" wave

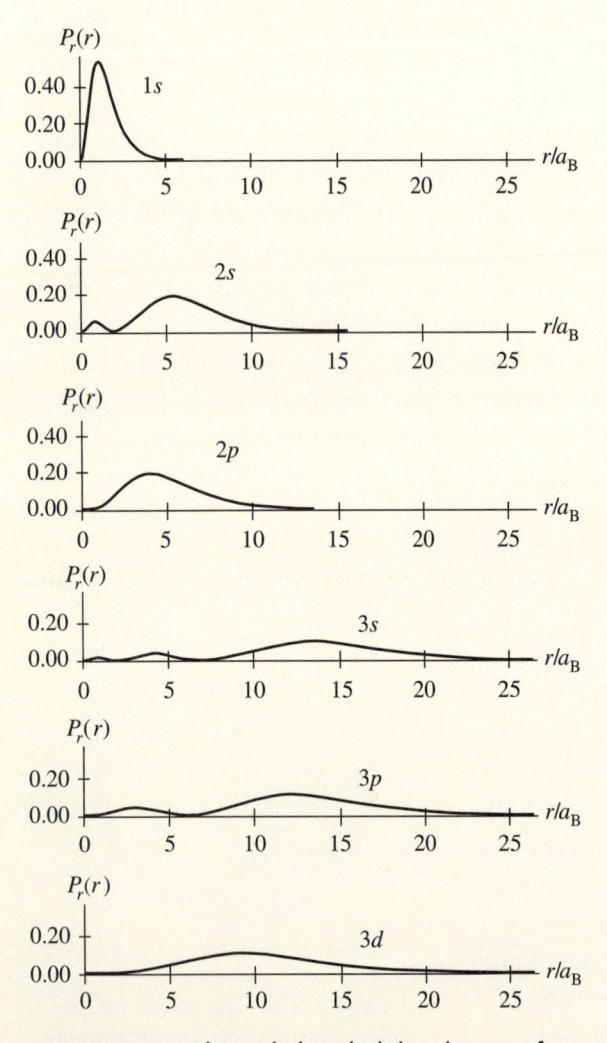

FIGURE 37-11 The radial probability densities for all states with $n = 1$, 2, and 3.

functions having $l = 0$. In three dimensions, though, we have the possibility for other angular momentum states with $l > 0$.

You can see that the $1s$, $2p$, and $3d$ states have maxima at a_B, $4a_B$ and $9a_B$. These maxima following the pattern

$$r_{\text{peak}} = n^2 a_B. \tag{37-16}$$

These are exactly the radii of the orbits in the Bohr hydrogen atom. There we simply had a one-dimensional de Broglie wave bent into a circle of that radius. Now we have a three-dimensional matter wave that is *most likely* to be found at the Eq. 37-16 distance from the nucleus, although it *could* be found at other values of r. The physical situation is quite different in quantum mechanics, but it is good to see that various aspects of the Bohr atom can be reproduced.

But why is it the $3d$ state that agrees with the Bohr atom rather than $3s$ or $3p$? A classical analysis of orbits shows that a circular orbit has a larger angular momentum than an elliptical orbit with the same energy. In the quantum mechanics of the hydrogen atom, all states with the same value of n form a collection of "orbits" having the same energy. The state with $l = n - 1$ has the largest angular momentum of the group, so it corresponds to a circular classical orbit. This matches the circular orbits in the Bohr atom. States with smaller l correspond to elliptical orbits that alternately swing in close to the nucleus, then move away again. You can see that the $3s$ and $3p$ states have a significant probability of being found in "close" to the nucleus, whereas the $3d$ electrons stays out around $9a_B$ at all times.

This distinction between circular and elliptical orbits will be important when we discuss the energy levels in multielectron atoms. But keep in mind that in quantum mechanics nothing is really orbiting. It is just that the probability densities of where the electron is likely to be, or not likely to be, mimic certain aspects of classical orbits.

Notice in Fig. 37-11 that an $n = 1$ electron is most likely to be found at a distance $\approx a_B$ from the nucleus. An $n = 2$ electron, whether it is $2s$ or $2p$, is most likely to be in the range $3a_B < 6a_B$, and an $n = 3$ electron is most likely to be found in the range $8a_B < r < 14a_B$. In other words, each value of n, regardless of the value of l, has a unique and nonoverlapping range of radii where the electron is most likely to be found. The radial probability density gives the impression of an "electron shell" surrounding the nucleus at a reasonably well-defined distance. Consequently, all the states with a given value of n taken together form what is called a **shell**.

The $1s$ ground state of Fig. 37-8a shows the electron cloud as a fuzzy sphere, densest at the center—where the electron is most likely to be found—and getting lighter at larger r. This picture, though presents a puzzle. A $1s$ electron cloud that is densest in the center agrees with the $1s$ radial wave function of Fig. 37-7, but it seems to be in sharp disagreement with the $1s$ graph of Fig. 37-11, where the radial probability density is *zero* at the nucleus and peaks at $r = a_B$. Resolving this puzzle requires distinguishing between the probability density $|\psi(r, \theta, \phi)|^2$ and the *radial* probability density $P_r(r)$. The wave function, and thus the probability density, really does peak at the nucleus. That is what Fig. 37-8a shows. But the *volume* of available space near the nucleus is very small. Although the probability density $|\psi(r, \theta, \phi)|^2$ is smaller at any *one* point having $r = a_B$, the volume of *all* points with $r \approx a_B$ is so large that the *radial* probability density P_r peaks at this distance.

To use a mass analogy, consider a fuzzy ball that is densest at the center. Even though the density decreases away from the center, a spherical shell of modest radius r can have *more total mass* than a small-radius spherical shell of the same thickness simply because it has so much more volume. Figures 37-7 and 37-8a display the density, but Fig. 37-11 displays the *radial* probability density. These are not the same, so the three figures are *all* correct.

EXAMPLE 37-3 Show that an electron in the 2p state is most likely to be found at $r = 4a_B$.

SOLUTION Using the 2p radial wave function from Eq. 37-10, the radial probability density is

$$P_r(r) = 4\pi r^2 |R_{2p}(r)|^2 = 4\pi r^2 \left[A_{2p} \left(\frac{r}{2a_B} \right) e^{-r/2a_B} \right]^2$$

$$= Cr^4 e^{-r/a_B},$$

where $C = \pi(A_{2p}/a_B)^2$. This is the function that was graphed in Fig. 37-11. To find the most probable value of r, we need to locate the maximum of this function. The maximum occurs at the point where the derivative is zero. Setting the derivative to zero gives

$$\frac{dP_r}{dr} = C\left(4r^3 e^{-r/a_B} + r^4 \cdot \frac{-1}{a_B} e^{-r/a_B} \right) = Cr^3 \left(4 - \frac{r}{a_B} \right) e^{-r/a_B} = 0.$$

This expression is zero only if

$$r = 4a_B.$$

This value of r locates the maximum of $P_r(r)$, so an electron in the 2p state is most likely to be found at $r = 4a_B$.

37.5 The Electron's Spin

Although the Schrödinger equation was able to predict many of the properties of atoms, there were still some experimental observations that defied explanation. One such set of observations was made by the German physicists Otto Stern and Walter Gerlach, who, just a few years prior to Schrödinger's theory, had developed a technique to measure the magnetic moments of atoms.

Recall from Chapter 30 that an electron orbiting a nucleus is a microscopic current loop. Figure 37-12, which is redrawn from Chapter 30, shows that this current loop has a north and south pole and that it generates a magnetic dipole field. These atomic magnets, as you learned previously, are called *magnetic moments* and symbolized by the vector $\vec{\mu}$.

Magnetic moments not only create magnetic fields, they also experience forces and torques in an external magnetic field—just like a current loop or a permanent bar magnet. Stern and Gerlach realized that they could use this property to separate atoms having different magnetic moments. Their apparatus, shown in Fig. 37-13, prepares an *atomic beam* by evaporating atoms out of a

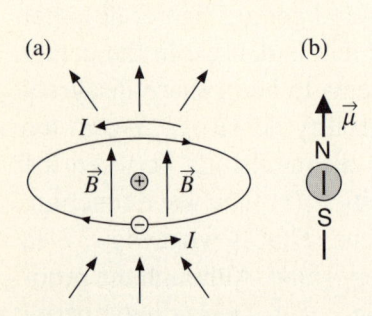

FIGURE 37-12 An orbiting electron generates a magnetic moment.

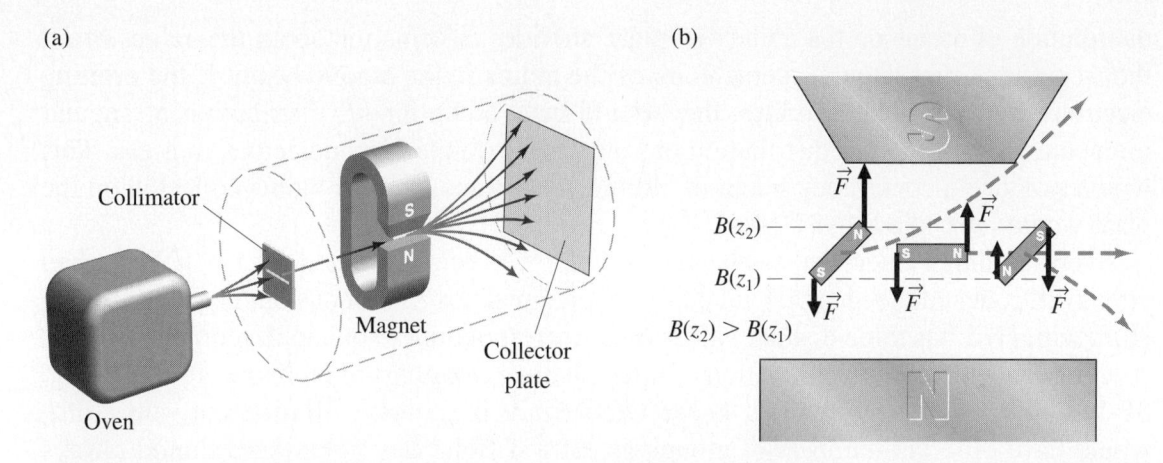

FIGURE 37-13 a) The Stern–Gerlach experiment. Atoms evaporated from an oven are collimated, then pass through a nonuniform magnetic field. b) The shape of the magnet pole tips makes the field stronger at z_2 than at z_1. The field exerts a net force on the magnetic moments, deflecting the atoms sideways.

hole in an "oven." These atoms pass through collimating apertures into a long tube from which all the air has been pumped. As they travel down the tube, they pass through a *nonuniform* magnetic field. The field is stronger toward the top of the magnet and weaker toward the bottom.

A nonuniform field exerts forces of *different* strength on the north and south poles of a magnetic moment, causing a *net force*. An atom whose magnetic moment vector $\vec{\mu}$ is tilted upward ($\mu_z > 0$, where μ_z is the z-component of $\vec{\mu}$) has an upward force on its north pole that is larger than the downward force on its south pole. Such an atom, as the figure shows, experiences a net upward force and is deflected in the upward direction as it passes through the magnet. A downward-tilted magnetic moment ($\mu_z < 0$) experiences a net downward force and is deflected downward. A magnetic moment perpendicular to the field ($\mu_z = 0$) feels no net force and passes through the magnet without deflection.

At the end of the apparatus is a collection plate on which the atoms stick. If the magnet is turned off, as in Fig. 37-14a, all the atoms land in a small spot at the center. With the magnet on, atoms are smeared vertically up and down. By examining the range of deflections found on the collection plate, and knowing the magnetic field characteristics, Stern and Gerlach could work backward to deduce the magnetic moments of the atoms.

It's not hard to show, although we will omit the proof, that an atom's magnetic moment is proportional to the orbital angular momentum of the electrons: $\vec{\mu} \propto \vec{L}$. The deflection of an atom depends on μ_z, the z-component of its magnetic moment. Thus the

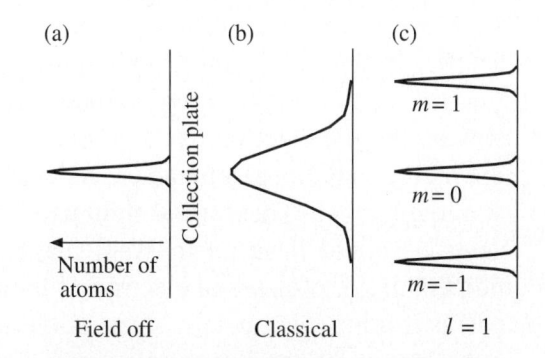

FIGURE 37-14 a) Distribution of the atoms on the collector plate if the field is off. b) Classically expected distribution when the field is turned on. c) Quantum mechanical prediction for $l = 1$ atoms.

distribution of atoms on the collection plate provides information about the values of L_z, the z-component of angular momentum, of the atoms in the atomic beam. If the orbiting electrons were classical particles, they would have a continuous distribution of angular momenta—similar to the distribution of velocities exhibited by molecules in a gas. This would produce a continuous smear of atomic deflections, and the atoms collected on the plate would look like Fig. 37-14b.

Now although quantum mechanics had not yet been invented in 1922, Bohr had already introduced the idea that angular momentum is not continuous but is quantized. An atom with $l = 1$ has three distinct values of L_z corresponding to quantum numbers $m = -1$, 0, and 1. This leads to the prediction of three distinct groupings of atoms, as shown in Fig. 37-14c, rather than a continuous smear. Other kinds of atoms, with different values of l, would have different numbers of groupings. But, if Bohr was right, there should always be an *odd* number of groups because there are $2l + 1$ values of L_z. Stern and Gerlach expected to find collection plates looking like Fig. 37-14c, rather than 37-14b, and this would have been a dramatic confirmation of Bohr's ideas about angular momentum quantization.

But unexpectedly, the Stern–Gerlach experiment produced the result shown in Fig. 37-15. They found not an *odd* number of groupings, as was expected, but an even two. Stern and Gerlach were using silver atoms for their experiment, so the issue was raised that maybe silver was just too complex an atom to understand correctly. Their experiment was repeated in 1927, by Phipps and Taylor, with a beam of hydrogen atoms. Because the ground state of hydrogen is 1s, with $l = 0$, the atoms should have *no* magnetic moment and there should be *no* deflection at all. However, Phipps and Taylor found exactly the same two-peaked distribution of Fig. 37-15.

Because the hydrogen atoms were deflected, they *must* have a magnetic moment. But where does it come from if $L = 0$? Even stranger was the deflection into two groupings, rather than an odd number. The deflection is proportional to L_z, and $L_z = m\hbar$, where m ranges in integer steps from $-l$ to $+l$. The experimental results would make sense only if $l = \frac{1}{2}$, allowing m to take on the two possible values $-\frac{1}{2}$ or $+\frac{1}{2}$. But according to the Schrödinger equation, the quantum numbers l and m must be integers.

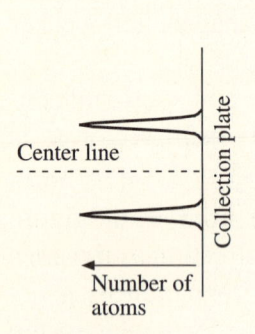

FIGURE 37-15 The outcome of the Stern–Gerlach experiment for hydrogen atoms.

To explain these anomalies, it was soon suggested, then confirmed, that the electron has an *inherent* magnetic moment. After all, the electron has an inherent gravitational character, its mass m_{elec}, and an inherent electric character, its charge $q_{elec} = -e$. These are simply part of what an electron is. Thus it is plausible that the electron should have an inherent magnetic character described by a built-in magnetic moment $\vec{\mu}_{elec}$. A classical electron, if thought of as a little ball of charge, could spin on its axis as it orbits the nucleus. This would be analogous to the earth spinning on its axis as it revolves about the sun. A spinning ball of charge is basically a current loop, so it would have a magnetic moment. It is this inherent magnetic moment of the electron that caused the unexpected deflection in the Stern–Gerlach experiment.

If the electron has an inherent magnetic moment, it must have an inherent angular momentum. This angular momentum is called the electron's **spin**, and it is designated by $\vec{S}$. The results of the Stern–Gerlach experiment tell us that the z-component of this spin angular momentum is

$$S_z = m_s \hbar, \quad \text{where } m_s = +\tfrac{1}{2} \text{ or } -\tfrac{1}{2}. \tag{37-17}$$

The term m_s is called the **spin quantum number**.

The z-component of the spin angular momentum vector is determined by its orientation. The $m_s = +\tfrac{1}{2}$ state, with $S_z = +\tfrac{1}{2}\hbar$, is called the **spin-up** state and the $m_s = -\tfrac{1}{2}$ state is called the **spin-down** state. It is convenient to picture a little angular momentum vector that can be drawn ↑ for a $m_s = +\tfrac{1}{2}$ state and ↓ for a $m_s = -\tfrac{1}{2}$ state. We will use this notation in the next section. Because the electron must be either spin-up or spin-down, a hydrogen atom in the Stern–Gerlach experiment will be deflected either up or down. This causes the two groups of atoms they observed. *No* atoms have $S_z = 0$, so there are no undeflected atoms in the center.

The equation for the spin angular momentum $S = |\vec{S}|$ is analogous to Eq. 37-7 for L:

$$S = \sqrt{s(s+1)}\hbar = \tfrac{\sqrt{3}}{2}\hbar, \tag{37-18}$$

where s is a quantum number with the single value $s = \tfrac{1}{2}$. S is the *inherent* angular momentum of the electron. Because of the single value of s, physicists usually say that "the electron has spin one-half."

Note the close analogy between Eqs. 37-17 and 37-18 for the spin angular momentum and Eqs. 37-7 and 37-8 for the orbital angular momentum. Both are angular momenta, but L is due to the orbital motion of the electron whereas S is due to the electron's own spin. A major distinction is that the orbital quantum number l can take on a range of values, corresponding to different orbits, but s *always* has the value $s = \tfrac{1}{2}$. That is because s characterizes an *inherent property* of the electron. Be sure to distinguish between m, which measures the z-component of $\vec{L}$, and m_s, which measures the z-component of $\vec{S}$.

Note that the term *spin* should be used with caution. Although a classical charged particle could generate a magnetic moment by spinning, the electron most assuredly is *not* a classical particle. It is not spinning in any literal sense. It simply has an inherent magnetic moment, just as it has an inherent mass and charge, and that magnetic moment makes it look *as if* the electron is spinning. But it is a convenient figure of speech, not a factual statement. The electron has a spin, but it is *not* a spinning electron!

The electron's spin has significant implications for atomic structure. The stationary states of the Schrödinger equation could be described by the three quantum numbers n, l, and m. But the Stern–Gerlach experiment implies that this is not a complete description. Knowing that a ground-state atom has quantum numbers $n = 1$, $l = 0$, and $m = 0$ is not sufficient to predict whether an atom will be deflected up or down. We need to add the spin quantum number m_s to make our description complete. (Strictly speaking, we also need to add the quantum number s, but because it never changes it provides no additional information.) So we really need *four* quantum numbers (n, l, m, m_s) to characterize an atomic electron. The spin orientation does not affect the atom's energy, so a ground-state electron in hydrogen could be in either the $(1, 0, 0, +\tfrac{1}{2})$ state or the $(1, 0, 0, -\tfrac{1}{2})$ state.

The fact that s has the single value $s = \tfrac{1}{2}$ also has other interesting implications. The correspondence principle tells us that a quantum particle begins to "act classical" in the limit of large quantum numbers. But s cannot become large! The conclusion—and this is important—is that the electron's spin is an intrinsically *quantum* property of the electron that has *no* classical counterpart.

37.6 Multielectron Atoms

The Schrödinger equation solution for the hydrogen atom correctly describes the experimental evidence, but so did the Bohr hydrogen atom. The real test of Schrödinger's theory is how well it works for multielectron atoms.

Recall that a neutral multielectron atom consists of Z electrons surrounding a nucleus that has Z protons and charge $+Ze$. The quantity Z is the *atomic number*, and it is the order in which elements are listed in the periodic table. Hydrogen is $Z = 1$, helium $Z = 2$, lithium $Z = 3$, neon $Z = 10$, and so on.

The potential-energy function of a multielectron atom consists of Z electrons interacting with the nucleus *and* of the Z electrons interacting *with each other*. It is this electron–electron interaction that makes the atomic structure problem more difficult than the solar system problem, and it proved to be the downfall of the simple Bohr model. Although the planets in the solar system do exert attractive gravitational forces on each other, they are so far apart and their masses are so much less than that of the sun that these planet–planet forces are insignificant for all but the most precise calculations. Not so in an atom. The electrons come very close to each other, so the electron–electron repulsion is just as important to atomic structure as the electron–nucleus attraction.

The potential energy for an atom of atomic number Z is

$$U = \sum_{i=1}^{Z} \frac{-Ze^2}{4\pi\varepsilon_0 r_i} + \frac{1}{2}\sum_{j\neq i}^{Z}\sum_{i=1}^{Z} \frac{e^2}{4\pi\varepsilon_0 r_{ij}}. \qquad (37\text{-}19)$$

The first term is the potential energy of each of the electrons, at radii r_i, with the nucleus of charge Ze. It is negative, of course, because each electron's charge is $-e$. The second term is the electron–electron interaction. Electrons i and j, which are distance r_{ij} apart, have positive potential energy $e^2/4\pi\varepsilon_0 r_{ij}$. This is summed over all pairs of electrons, requiring a double sum with the $j = i$ term excluded from the outer sum because an electron does not interact with itself. The double sum, however, counts each pair twice—once as r_{ij} and again as r_{ji}. This double counting is canceled by the factor of 1/2 in front.

Equation 37-19 is the potential energy that goes into the Schrödinger equation. The atom's wave function is a function of the coordinates of each of the Z electrons—a total of $3Z$ coordinates. For example, the wave function of the iron atom, which has $Z = 26$, is a function of 78 variables! Needless to say, the Schrödinger equation for a multielectron atom is not solved exactly. It is solved through a series of simplifying approximations and numerical calculations. Even so, it is still a very difficult problem.

We are obviously not going to solve the multielectron Schrödinger equation in this text. We can, though, note the major approximation used in its solution. The electron–electron interaction in Eq. 37-19 fluctuates rapidly in value as the electron–electron distances r_{ij} change. Rather than treat this in detail, it seems reasonable to consider each electron to be moving in an *average* potential due to all the other electrons. That is, electron i experiences a potential energy

$$U(r_i) = -\frac{Ze^2}{4\pi\varepsilon_0 r_i} + U_{\text{elec}}(r_i), \qquad (37\text{-}20)$$

where the first term is due to the electron–nucleus interaction and U_{elec} is the averaged potential due to all the other electrons. This approximation, it turns out, allows the

Schrödinger equation to be broken into Z separate equations, one for each electron. Because each of these separate equations treats its electron independently of the other Z – 1 electrons, this approach is called the **independent particle approximation**, or IPA.

A major consequence of the IPA is that each electron can be described with a wave function having the same four quantum numbers—n, l, m, and m_s—used to describe the single electron of hydrogen. Because m and m_s do not affect the energy, we will still refer to electrons by their n and l quantum numbers, using the same labeling scheme that we used for hydrogen.

A major difference, however, is that now an electron's energy depends on both n and l. Whereas the 2s and 2p states in hydrogen had the same energy, their energies are different in a multielectron atom. The difference arises from the electron–electron interactions, which the single electron in hydrogen did not have.

Figure 37-16 shows the energy-level structure for the electrons in a multielectron atom. The hydrogen atom energies are shown, for comparison, on the right-hand edge of the figure. The comparison is quite interesting. States in a multielectron atom that have small values of l are significantly lower in energy than the corresponding state in hydrogen. The energy increases as l increases, and the maximum-l state for each n has an energy very nearly that of the same n in hydrogen. For example, $E_{3s} < E_{3p} < E_{3d} \approx E$ (hydrogen $n = 3$). Can we understand this pattern?

Indeed we can. Recall, from Section 37.4, that states of lower l correspond to elliptical classical orbits and that the highest l states correspond to circular orbits. Except for the lowest values of n, an electron in a circular orbit would spend most of its time *outside* the electron cloud of the remaining $Z - 1$ electrons. This is illustrated in Fig. 37-17. When seen from the outside, a sphere of charge acts as if all the charge is located at the center. From the perspective of an outer electron in a circular orbit, the $Z - 1$ inner electrons "cancel" all but one of the Z protons, leaving a net charge of just $+e$—exactly the same as a hydrogen atom. The inner electrons, in effect, *shield* the outer electron, preventing it from seeing the full nuclear charge. So an electron in a maximum-l state is nearly indistinguishable from an electron in the hydrogen atom, and that is why its energy is very nearly that of a hydrogen energy level.

The low-l states, however, correspond to elliptical orbits. A low-l electron penetrates in very close to the nucleus, as seen in Fig. 37-17,

FIGURE 37-16 The energy-level structure of electrons in a multielectron atom. Unlike hydrogen, shown on the right for comparison, the different l states have different energies due to the electron–electron interaction.

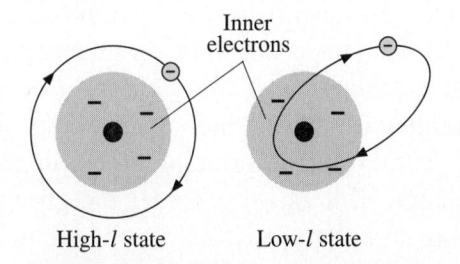

FIGURE 37-17 High-l states correspond to circular orbits. Their energies are nearly the same as hydrogen. Low-l states penetrate in near the core, lowering their energy.

where it is no longer shielded by the other electrons. Its interaction with the Z protons in the nucleus is much stronger than the interaction it would have with the single proton in a hydrogen nucleus. This *lowers* its energy in comparison to the same state in hydrogen.

Now a quantum electron does not really orbit in a classical sense. But the probability density of a 3s electron has in-close peaks that are missing in the probability density of a 3d electron, as you should confirm by looking back at Fig. 37-11. So a low-l electron really does have a likelihood of being at small r, where its interaction with the Z protons is strong, but a high-l electron remains farther from the nucleus. The probability interpretation of the wave function allows us to say that the electrons act as if they were in elliptical or circular orbits.

In a multielectron atom, just as in hydrogen, wave function solutions to the Schrödinger equation exist only for certain discrete values of the energy E. These are the quantized energy levels of the atom. The lowest possible energy level of an atom is its *ground state*. An atom in an excited state will eventually decay to its ground state, after emitting a photon or otherwise losing its extra energy, and there it will stay until either a collision or the absorption of another photon drives it back up to an excited state.

What is the ground state of an atom having Z electrons and Z protons? Because the 1s state is the lowest energy state in the independent particle approximation, it seems that the ground state should be one in which all Z electrons are in the 1s state. However, this hypothesis is not consistent with the experimental evidence. To understand this, we need one more piece of information—the *Pauli exclusion principle*.

The Pauli Exclusion Principle

In 1925, the young Austrian physicist Wolfgang Pauli proposed a radical idea—that no two electrons in a quantum system can be in the same quantum state. In other words, no two electrons can have exactly the same set of quantum numbers (n, l, m, m_s). If one electron is present in a state, it *excludes* all others. This statement is called the **Pauli exclusion principle**.

Wolfgang Pauli was a prodigy who burst onto the physics scene in 1921 when, at the age of 21, he wrote a masterful review article on Einstein's theory of relativity. He made many contributions to theoretical physics in the 1920s and the 1930s, including the hypothesis—not confirmed for many years—that certain modes of radioactive nuclear decay emit a *massless* particle called the *neutrino*, meaning "little neutral one." But he is most well known for his hypothesis that no two electrons can share the same quantum state. This turns out to be an extremely profound statement about the nature of matter. A full examination of this idea must be left to advanced courses. We can, however, make effective use of the Pauli exclusion principle without having to examine its origins.

The exclusion principle is not applicable to hydrogen, where these is a only a single electron. But as we consider the ground state of helium, with $Z = 2$ electrons, we must make sure that the two electrons are in different quantum states. This is not difficult. A 1s state, with $l = 0$, has the single possible value $m = 0$. But there are *two* possible values of m_s, namely $+\frac{1}{2}$ and $-\frac{1}{2}$. If a first electron is in the spin-up 1s state $(1, 0, 0, +\frac{1}{2})$, a second 1s electron can still be added as long as it is in the spin-down state $(1, 0, 0, -\frac{1}{2})$. This is shown schematically in Fig. 37-18a, where the circles represent electrons on the rungs of the "energy ladder" and the arrows represent spin-up or spin-down.

The Pauli exclusion principle does not prevent the two electrons of helium from both being in the 1s state as long as they have opposite values of m_s, so we predict this to be the ground state. This is called the **electron configuration** of the helium ground state, and it is written with the notation $1s^2$. The superscript 2 indicates two 1s electrons. An excited state of the helium atom might be the electron configuration $1s2s$. This state is shown in Fig. 37-18b. Here, because the two electron have different value of n, there is no restriction on their values of m_s.

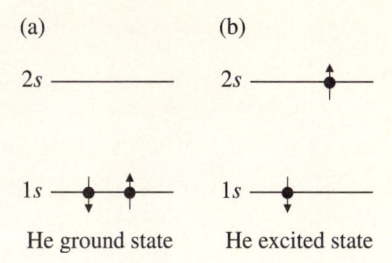

FIGURE 37-18 a) The ground state and b) the first excited state of helium.

The two states $(1, 0, 0, +\frac{1}{2})$ and $(1, 0, 0, -\frac{1}{2})$ are the only two states with $n = 1$. The ground state of helium has one electron in each of these states, so all the possible $n = 1$ states are filled. Consequently, the electron configuration $1s^2$ is called a **closed shell**.

The 1s state in helium is lower in energy than the 1s state in hydrogen because the electrons are attracted toward a nucleus with charge $2e$ instead of just e. This leads us to expect that helium will have a much larger ionization energy than hydrogen—that is, it will take a larger energy to remove an electron and leave an ion behind. The ionization energy of hydrogen was 13.6 eV. Measurements show that the ionization energy of helium is 24.6 eV, confirming our prediction. Because the two electron magnetic moments point in opposite directions, we can predict that helium has *no* net magnetic moment and will be undeflected in a Stern–Gerlach apparatus. This prediction is also confirmed by experiment.

The next element, with $Z = 3$ electrons, is lithium. The first two electrons can go into 1s states, with opposite values $m_s = \pm\frac{1}{2}$, but what about the third electron? The $1s^2$ shell is closed, and there are no additional quantum states having $n = 1$. The only option for the third electron is the next lowest energy state, $n = 2$. Now the 2s and 2p states had equal energies in the hydrogen atom, but this *not* the case in a multielectron atom. As Fig. 37-16 showed, a lower-l state has lower energy than a higher-l state with the same n. The 2s state of lithium is lower in energy than 2p, so lithium's third ground-state electron will be 2s. This requires $l = 0$ and $m = 0$ for the third electron, but the value of m_s is irrelevant because there is only a single electron in 2s. Figure 37-19a shows the electron configuration with the 2s electron being spin-up, but it could equally well be shown as spin-down. The electron configuration for the lithium ground state is written $1s^22s$. This indicates two 1s electrons and a single 2s electron.

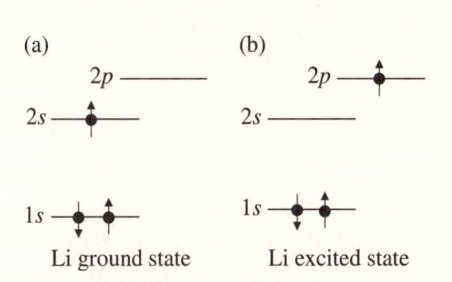

FIGURE 37-19 a) The ground state and b) the first excited state of lithium.

Figure 37-20a shows the probability density of a lithium atom in the $1s^22s$ ground state. You can see the 2s electron shell surrounding the inner $1s^2$ core. For comparison, Fig. 37-20b shows the *first excited state* of lithium, in which the 2s electron has been

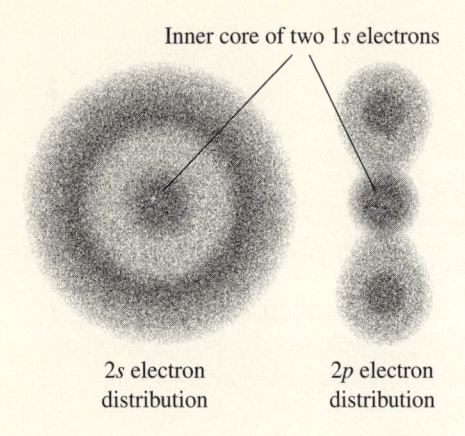

FIGURE 37-20 a) $2s$ and b) $2p$ electron clouds in the lithium atom.

excited upward to $2p$. This forms the $1s^22p$ configuration, also shown in Fig. 37-19b. We will defer discussing excited states of atoms until the next chapter.

The Schrödinger equation accurately predicts the energies of the $1s^22s$ and the $1s^22p$ configurations of lithium. But by itself, the Schrödinger equation could not predict which states the electrons would actually occupy. The electron spin and the Pauli exclusion principle were the final pieces of puzzle. Once these were added to Schrödinger's theory, the initial phase of quantum mechanics was complete. Physicists finally had a successful theory for understanding the structure of atoms. One of the earliest triumphs of the new theory was explaining the periodic table of the elements.

37.7 The Periodic Table of the Elements

The nineteenth century was a time when chemists were discovering new elements and studying their chemical properties. The century opened with the atomic model still not completely validated, with no clear distinction between atoms and molecules, and with no one having any idea how many different elements there might be. But chemistry developed quickly, and by midcentury it was clear that there were dozens of elements, but not hundreds.

Several chemists in the 1860s began to list the elements in order of increasing mass and to point out the regular recurrence of chemical properties. For example, there are obvious similarities between the alkali metals lithium, sodium, potassium, and cesium. But attempts at organization were hampered by the fact that many elements had yet to be discovered. As an example, none of the rare gases were discovered before 1870. So, although the recurrence of properties was recognized, the pattern did not appear as periodic as it does to us today.

The Russian chemist Dmitri Mendeleev was the first to propose, in 1867, a *periodic* arrangement of the elements. He did so by explicitly pointing out "gaps" where, according to his hypothesis, undiscovered elements should exist. He could then predict the expected properties of the missing elements. The subsequent discovery of these elements verified Mendeleev's organizational scheme, which came to be known as the *periodic table of the elements*.

Figure 37-21 shows a modern periodic table. A larger version is printed inside the front cover. The modern periodic table has evolved significantly from Mendeleev's first chart, but it still bears his overall organizational scheme. The significance of the periodic table to a physicist is the implication that there is a basic regularity or periodicity to the *structure* of atoms. Any successful theory of the atom needs to *explain* why the periodic table looks the way it does. This was one of the first challenges facing the new quantum mechanics in the late 1920s.

	Group	I	II												III	IV	V	VI	VII	VIII
Period	1	1 H																		2 He
	2	3 Li	4 Be											5 B	6 C	7 N	8 O	9 F	10 Ne
	3	11 Na	12 Mg			Transition elements								13 Al	14 Si	15 P	16 S	17 Cl	18 Ar
	4	19 K	20 Ca	21 Sc	22 Ti	23 V	24 Cr	25 Mn	26 Fe	27 Co	28 Ni	29 Cu	30 Zn	31 Ga	32 Ge	33 As	34 Se	35 Br	36 Kr
	5	37 Rb	38 Sr	39 Y	40 Zr	41 Nb	42 Mo	43 Tc	44 Ru	45 Rh	46 Pd	47 Ag	48 Cd	49 In	50 Sn	51 Sb	52 Te	53 I	54 Xe
	6	55 Cs	56 Ba	57–71 *	72 Hf	73 Ta	74 W	75 Re	76 Os	77 Ir	78 Pt	79 Au	80 Hg	81 Tl	82 Pb	83 Bi	84 Po	85 At	86 Rn
	7	87 Fr	88 Ra	89–103 †	104 Db	105 Jl	106 Rf	107 Bh	108 Hn	109 Mt	110	111	112						

Inner transition elements

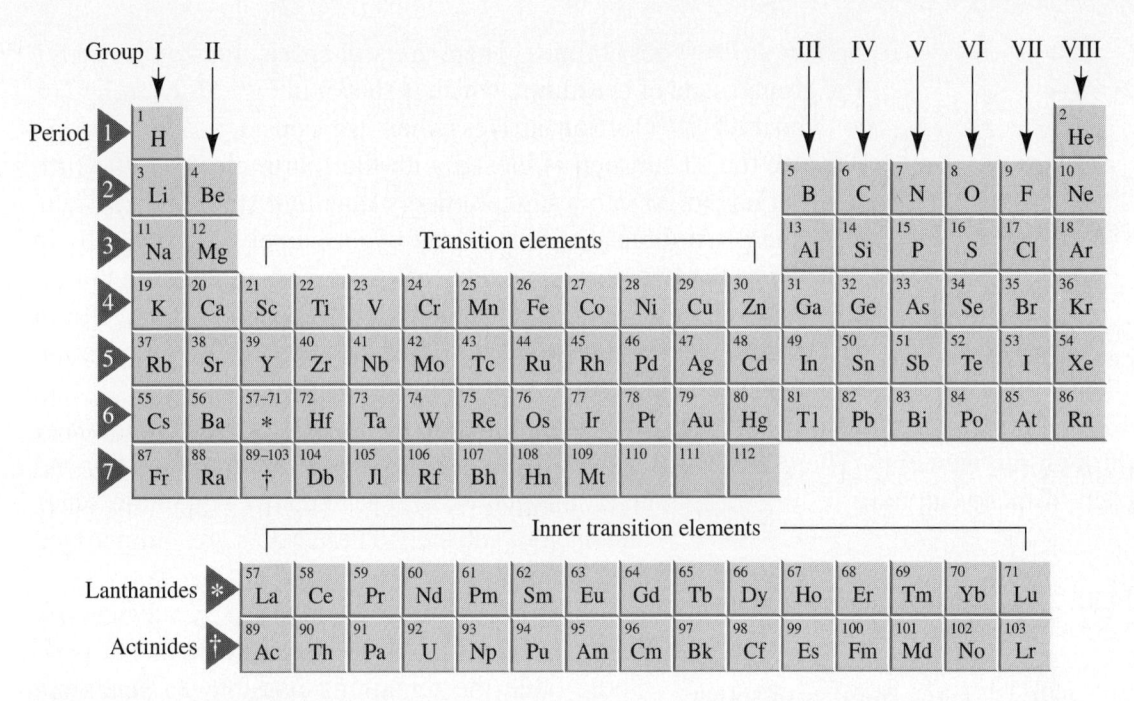

Lanthanides *	57 La	58 Ce	59 Pr	60 Nd	61 Pm	62 Sm	63 Eu	64 Gd	65 Tb	66 Dy	67 Ho	68 Er	69 Tm	70 Yb	71 Lu
Actinides †	89 Ac	90 Th	91 Pa	92 U	93 Np	94 Pu	95 Am	96 Cm	97 Bk	98 Cf	99 Es	100 Fm	101 Md	102 No	103 Lr

FIGURE 37-21 The modern periodic table of the elements, showing the atomic number Z of each. Only elements through uranium ($Z = 92$) occur naturally; those with $Z \geq 93$ are man-made. As is customary, the *inner transition elements* are detached from the main body of the table, at the boxes labeled * and †, and shown below it.

The First Two Rows

Quantum mechanics was able to explain the structure of the period table. We need three basic ideas to see how this works:

1. The energy levels of an atom are found by solving the Schrödinger equation. The results of this calculation were shown in the energy level diagram of Fig. 37-16. This is a very important figure for understanding the periodic table.

2. Each energy *level* in Fig. 37-16 is actually $2(2l + 1)$ different *states*, where l is the orbital quantum number. This is because there are $2l + 1$ possible values of the magnetic quantum number m and, for each of these, two possible values of the spin quantum number m_s. Each of these states has the same energy.

3. The electron configuration of the ground state is the lowest energy configuration that is consistent with the Pauli exclusion principle.

We used these ideas in the last section to look at the elements helium ($Z = 2$) and lithium ($Z = 3$). As we make our way through the elements, four-electron beryllium ($Z = 4$) comes next. The first two electrons go into $1s$ states, forming a closed shell, and the third goes into $2s$. There is room in the $2s$ level for a second electron as long as its spin is opposite that of the first $2s$ electron. Thus the third and fourth electrons occupy states $(2, 0, 0, +\frac{1}{2})$ and $(2, 0, 0, -\frac{1}{2})$. These are the only two possible $2s$ states. All of the states with the same values and n and l are called a **subshell**, so the fourth electron closes the $2s$ subshell. (The outer two electrons are called a subshell, rather than a shell, because they complete only

574 CHAPTER 37 THE STRUCTURE OF ATOMS

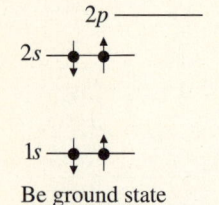

2p ———
2s ⤊⤋

1s ⤊⤋

Be ground state

FIGURE 37-22 The ground state of beryllium ($Z = 4$).

the $n = 2$, $l = 0$ possibilities. There are still spaces for $2p$ electrons.) The ground state of beryllium, which is shown in Fig. 37-22, is $1s^2 2s^2$.

The first $2p$ electron arrives when we consider $Z = 5$ boron. Because the $2s$ subshell is closed with the fourth electron, the fifth electron has to go into a higher energy quantum state, the $2p$ state. Now there are three possible values of m, namely -1, 0 and 1, in addition to the two possible values of m_s. Because neither m nor m_s affects the energy, we need not specify the values of m or m_s. Boron has a single $2p$ electron outside a closed subshell. This configuration is designated $1s^2 2s^2 2p$.

The principles of the preceding section can continue to be applied as we work our way through the elements. There are $2l + 1$ values of m associated with each value of l, and each of these can have $m_s = \pm\frac{1}{2}$. This gives, altogether, $2(2l + 1)$ distinct quantum states in each nl subshell. These states are summarized in Table 37-1.

TABLE 37-1 Number of states in each subshell of an atom.

Subshell	l	Number of states
s	0	2
p	1	6
d	2	10
f	3	14

Boron ($1s^2 2s^2 2p$) opened the $2p$ subshell. As we continue across the second row of the periodic table the remaining possible $2p$ states are filled. These are shown both pictorially and as electron configurations in Fig. 37-23. Upon reaching neon ($1s^2 2s^2 2p^6$), we have added six $2p$ electrons, the $n = 2$ shell is complete, and we have another closed shell. The second row of the period table is eight elements wide because of the two $2s$ electrons *plus* the six $2p$ electrons needed to fill the $n = 2$ shell.

The electrons in a closed shell spread out evenly around the nucleus, forming a spherically symmetric electron cloud. Each "sees" the full positive charge of the nucleus, so each electron is tightly bound within this shell. We can predict that a closed shell will have a high ionization potential and a low chemical reactivity. A closed-shell atom has no "handles" with which to interact with other atoms.

By contrast, an electron in an incomplete, or open, shell orbits outside the closed shells, which effectively shield it from the nucleus. The electrons in incomplete shells are called *valence electrons* in chemistry. Because the valence electrons are easily dislodged or moved about, we can predict that open-shell atoms will have low ionization energies and be chemically reactive.

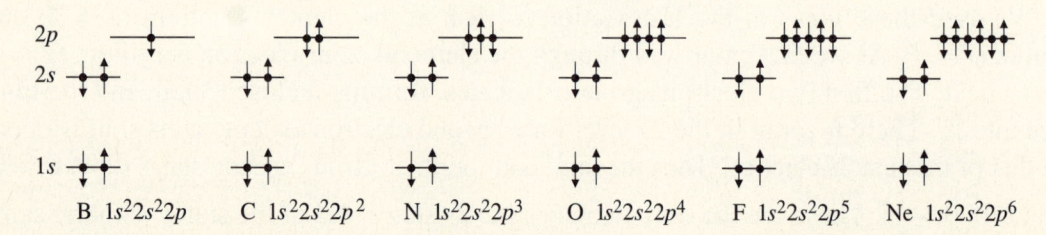

FIGURE 37-23 Filling the $2p$ subshell in the elements boron ($Z = 5$) through neon ($Z = 10$).

These predictions are accurate. The first two elements with closed shells are helium ($Z = 2$) and neon ($Z = 10$). These are both nonreactive monatomic gases. And if you look back at Fig. 37-1, showing experimental data that we wanted to explain, you can see that helium and neon have the two highest ionization energies of all the elements. But lithium ($Z = 3$) and sodium ($Z = 11$), which have a single electron outside a closed shell, are highly reactive and have among the lowest ionization energies.

Beryllium ($1s^2 2s^2$) has a closed $2s$ subshell, and you can see in Fig. 37-1 that this gives it an ionization energy much higher than that of lithium and slightly higher than that of boron ($1s^2 2s^2 2p$). The same is true for the closed $3s$ subshell of magnesium ($1s^2 2s^2 2p^6 3s^2$). However, a closed subshell is not nearly as tightly bound as a closed shell, so the ionization energies of beryllium and magnesium are much less than those of helium and neon. The electrons become more tightly bound as the p subshell fills, causing a steady rise in ionization energies. You can see that the basic idea of shells and subshells provides a good understanding of the recurring features in the ionization energies of Fig. 37-1.

Elements with $Z > 10$

The third row of the periodic table is similar to the second, filling first the two $3s$ states in sodium and magnesium. The arrangement of the periodic table, with two columns on the left (groups I and II), represents the two electrons that can go into an s subshell. Then the six $3p$ states are filled, one by one, in aluminum through argon. The six columns on the right (groups III–VIII) represent the six electrons of the p subshell. Argon ($Z = 18$, $1s^2 2s^2 2p^6 3s^2 3p^6$) is another inert gas, although perhaps this seems surprising given that the $3d$ subshell is still open.

The fourth row is where the periodic table begins to get complicated. We would expect the closure of the $3p$ subshell in argon to be followed, starting with potassium ($Z = 19$), by filling the $3d$ subshell. But if you look back at Fig. 37-16, where the energies of the different nl states are shown, you will see that the $3d$ state is slightly higher in energy than the $4s$ state. Because the ground state is the lowest energy state consistent with the Pauli exclusion principle, potassium finds it more favorable to fill a $4s$ state than to fill a $3d$ state. Thus the ground state configuration of potassium is $1s^2 2s^2 2p^6 3s^2 3p^6 4s$ rather than the expected $1s^2 2s^2 2p^6 3s^2 3p^6 3d$.

We begin, at this point, to see a competition between increasing n, which moves the electrons to larger radii and increases their energy, and decreasing l, which pulls the electron's elliptical orbit in closer to the nucleus and decreases its energy. The highly elliptical characteristic of the $4s$ state, such as was shown in Fig. 37-17, brings part of its orbit in so close to the nucleus that its energy is less than that of the more circular $3d$ state. The $4p$ state, though, reverts to the "expected" pattern. We find

$$E_{4s} < E_{3d} < E_{4p},$$

so the states are filled in the order $4s$, then $3d$, then finally $4p$.

That is what is happening across the fourth row. Potassium and calcium fill the two positions in the $4s$ subshell, but then we have ten slots devoted to filling the $3d$ subshell. Because there had been no previous d states, this subshell "splits open" the periodic table to form the ten-element-wide transition elements. Most commonly occurring metals are *transition elements*, and their metallic properties are determined by their partially filled d

subshell. The 3d subshell closes with zinc, at Z = 30, and you should notice in Fig. 37-1 that zinc has a larger ionization energy than its neighbors. The next six elements fill the 4p subshell up to Z = 36 krypton, another rare gas.

Because $E_{5s} < E_{4d}$, the same pattern occurs again in the fifth row. The 5s subshell is filled, followed by 4d, then 5p. As we make our way to Z = 54 xenon, at the end of the fifth row, we have seen a first row two elements wide (s subshell only), then two rows eight elements wide (2 + 6 for the s + p subshells), and now two rows eighteen elements wide (2 + 6 + 10 for the s + p + d subshells). Thus 2 + 8 + 8 + 18 + 18 = 54, the atomic number of xenon.

Things get even more complex starting in the sixth row, but the ideas are familiar. The l = 3f subshell becomes a possibility with n = 4, but we find from Fig. 37-16 that the 5s, 5p, and 6s states are all lower in energy than 4f. The $5s^2 5p^6$ subshell is filled with Z = 54 xenon, and the next two elements, cesium and barium, fill $6s^2$. Immediately after barium, the box in the table is marked * and you have to switch down to the *lanthanides* at the bottom of the table. (These are also called the rare earth elements.) The lanthanides fill in the fourteen 4f states.

Actually lanthanum itself has a ground-state configuration $6s^2 5d$ and, strictly speaking, might be considered one of the ten 5d transition elements. Indeed, it is often shown there on older periodic tables, but the present recommendation is to include it with the lanthanide group at the bottom. This makes the lanthanide group fifteen elements wide rather than the expected fourteen. The 4f subshell is complete with Z = 71 lutecium, which is actually $6s^2 5d 4f^{14}$ because the single 5d electron from lanthanum has continued. Then Z = 72 hafnium through Z = 80 mercury complete the transition element 5d subshell, followed by the 6p subshell in the six elements thallium through radon at the end of the sixth row. Radon, the last inert gas, has Z = 86 electrons and a ground state configuration

$$\text{radon } (Z = 86): 1s^2 2s^2 2p^6 3s^2 3p^6 4s^2 3d^{10} 4p^6 5s^2 4d^{10} 5p^6 6s^2 4f^{14} 5d^{10} 6p^6.$$

This is frightening to behold, but we can now understand exactly why it is!

The seventh row, which is incomplete, follows the same pattern as the sixth. Two 7s states are filled, followed by the 5f states in the actinide elements. Then the elements, all of which are now man-made rather than naturally occurring, return to the transition element category by filling 6d states. Elements through Z = 112 had been reported by 1996, completing the 6d subshell.

EXAMPLE 37-4 Predict the ground state electron configuration of arsenic.

SOLUTION The periodic table shows that arsenic (As) has Z = 33, so we must identify the states of 33 electrons. Arsenic is in the fourth row, following the first group of transition elements. Argon (Z = 18) filled the 3p subshell, then calcium (Z = 20) filled the 4s subshell. The next ten elements, through zinc (Z = 30) filled the 3d subshell. The 4p subshell starts filling with gallium (Z = 31), and arsenic is the third element in this group so it will have three 4p electrons. Thus the ground state configuration of arsenic is

$$1s^2 2s^2 2p^6 3s^2 3p^6 4s^2 3d^{10} 4p^3.$$

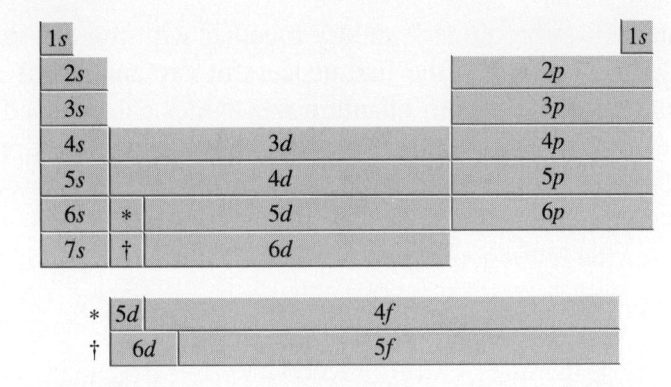

FIGURE 37-24 Summary of the order in which subshells are filled in the periodic table.

The entire periodic table is well explained by quantum mechanics. Figure 37-24 summarizes the results, showing the subshells as they are filled. This success gives us great confidence that our understanding of the structure of atoms is now correct. It is especially important to note the significance of the electron's spin. Although the introduction of the electron's spin and magnetic moment may have seemed obscure and unnecessary, we now find that the spin quantum number m_s is absolutely essential for understanding the periodic table.

The periodic table organizes information about the ground states of the elements. These are the states that are chemically most important because most atoms spend most of the time in their ground states. All the chemical ideas of valences, bonding, reactivity, and so on are consequences of these ground state atomic structures. But the periodic table does not tell us anything about the excited states of atoms. We cannot interpret the spectra of atoms without understanding their excited states, so that will be the topic to which we turn in the last chapter.

Summary
Important Concepts and Terms

spherical polar coordinates
radial wave function
angular wave function
radial wave equation
principal quantum number
orbital quantum number
magnetic quantum number
ground state
ionization energy
excited state
energy-level diagram
electron cloud

radial probability density
shell
spin
spin quantum number
spin-up
spin-down
independent particle approximation
Pauli exclusion principle
electron configuration
closed shell
subshell

Quantum mechanics—Schrödinger's theory together with the electron's spin and the Pauli exclusion principle—provided the first successful explanation of atomic structure. Although the calculations are difficult, quantum mechanics can be used to make precise calculations of the energy levels of all the elements. More generally, it provides an explanation for the structure of the periodic table, the ionization energies of atoms, and the chemical properties of the elements.

Wave functions for the hydrogen atom exist only if three conditions are met:

1. The energy levels of the atom are given by

$$E_n = -\frac{1}{n^2} \cdot \left(\frac{e^2}{4\pi\varepsilon_0 \cdot 2a_B}\right) = -\frac{13.60 \text{ eV}}{n^2} \qquad n = 1, 2, 3, \ldots.$$

The integer n is the principal quantum number.

2. The angular momentum L of the electron's orbit can only be

$$L = \sqrt{l(l+1)}\,\hbar \qquad l = 0, 1, 2, 3, \ldots, n-1.$$

The integer l is the orbital quantum number.

3. The z-component of the angular momentum, L_z, can only be

$$L_z = m\hbar \qquad m = -l, -l+1, \ldots, 0, \ldots, l-1, l.$$

The integer m is the magnetic quantum number.

The three quantum numbers n, l, and m, together with the spin quantum number m_s, identify each state in an atom. The energy levels for hydrogen depend only on n. In multielectron atoms, the energy depends on both n and l. States with lower values of l have lower energy than states of the same n that have higher values of l. This information was portrayed graphically in the energy-level diagram of Fig. 37-16.

The wave functions of hydrogen show that the electron clouds form shells around the nucleus. Each value of n is associated with a shell at a reasonably well-defined distance from the nucleus. A closed shell occurs when all the possible values of n have been filled.

It was discovered in the 1920s that the electron has an inherent magnetic moment that, along with its mass and charge, define what an electron is. We call this the electron spin. The spin quantum number m_s can have only the two possible values, $+\frac{1}{2}$ and $-\frac{1}{2}$, which we call the spin-up and spin-down states.

The Pauli exclusion principle states that no two electrons can occupy the same quantum state (n, l, m, m_s). This principle forms the basis for understanding the ground state configurations of the elements. The periodic table of the elements is structured around the maximum number of electrons that can go into each subshell and shell.

Exercises and Problems

Exercises

1. When all quantum numbers are considered, how many different quantum states are there for a hydrogen atom with $n = 1$? With $n = 2$? With $n = 3$? List the quantum numbers of each state.

2. A hydrogen atom has orbital angular momentum 3.65×10^{-34} J s.
   a. What letter (s, p, d, or f) describes the electron?
   b. What is the atom's minimum possible energy? Explain.
3. a. What is the difference between l and L?
   b.. What is the difference between s and S?
4. Why do electron energies depend on n and l but not on m or m_s?

Problems

5. a. Draw a diagram similar to Fig. 37-5 to show all the possible orientations of the angular momentum vector $\vec{L}$ for the case $l = 2$. Label each $\vec{L}$ with the appropriate value of m.
   b. What is the minimum angle between $\vec{L}$ and the z-axis?
6. There exist subatomic particles whose spin is characterized by $s = 1$, rather than the $s = \frac{1}{2}$ of electrons. These particles are said to have a spin of 1.
   a. What is the value of the spin angular momentum S for a particle with a spin of 1?
   b. What are the possible values of the spin quantum number?
   c. Draw a vector diagram similar to Fig. 37-5 to show the possible orientations of $\vec{S}$.
7. The hydrogen atom 1s wave function is a maximum at $r = 0$, as seen in Fig. 37-7 and the electron cloud of Fig. 37-8a. But the 1s radial probability density, shown in Fig. 37-11, peaks at $r = a_B$ and is zero at $r = 0$. Explain this paradox.
8. a. Determine the normalization constant A_{1s} of the 1s ground-state wave function of the hydrogen atom.
   b. What is the probability of finding a 1s electron at $r > a_B$?
9. Verify that $R_{1s}(r)$, as given in Eq. 37-10, really is a solution of the radial wave equation.
10. a. Graph the hydrogen radial wave function $R_{2p}(r)$ for $0 \leq r \leq 8a_B$.
    b. Determine the exact value of r (in terms of a_B) for which $R_{2p}(r)$ is a maximum.
    c. Example 37-3 and Fig. 37-11 showed that the radial probability density for the 2p state is a maximum at $r = 4a_B$. Explain why this differs from your answer to part b).
11. a. Prove that the radial probability density peaks at $r = a_B$ for the 1s state of hydrogen.
    b. Show that the *two* peaks in the radial probability density of the 2s state of hydrogen are at distances $r \approx 0.764$ nm and $r \approx 5.24$ nm..
12. In a multielectron atom, the lowest l state for each n (1s, 2s, 3s, etc.) is significantly lower in energy than the hydrogen state having the same n, as seen in Fig. 37-16. But the highest l-state for each n (2p, 3d, 4f, etc.) is very nearly equal in energy to the hydrogen state with the same n. Explain.
13. Draw a series of pictures, similar to Fig. 37-23, for the ground states of K, Sc, Fe, Ge, and Kr.

14. a. Predict the ground-state electron configurations of Mg, Sr, and Ba.
    b. Predict the ground-state electron configurations of Si, Ge, and Pb.

15. For each of the following electron configurations: First identify the element, then decide whether this configuration is the ground state or an excited state. Explain.
    a. $1s^2 2s^2 2p^5$.
    b. $1s^2 2s^2 2p^4 3d$.
    c. $1s^2 2s^2 2p^6 3s^2 3p^6 3d^8 4s^2$.
    d. $1s^2 2s^2 2p^6 3s^2 3p^6 3d^{10} 4s^2 4p$.

16. a. Which elements have the highest ionization potentials? Where are they located on the periodic table of the elements?
    b. Which elements have the lowest ionization potentials? Where are they located on the periodic table of the elements?
    c. The elements with the highest ionization potentials differ by just one atomic number from the elements with the lowest ionization potentials. How can such a small change in atomic number lead to such a large difference in atomic properties? Explain.

17. Figure 37-1 shows that cadmium ($Z = 48$) has a higher ionization potential than its neighbors. Why is this?

[**Estimated 10 additional problems for the final edition.**]

Chapter 38

Atomic Spectra

LOOKING BACK Sections 32.8; 34.2; 37.5–37.7

38.1 Light at the End of the Tunnel

In the late nineteenth century, scientists believed that atomic spectra—the patterns of light emitted or absorbed by atoms—were one of the keys to atomic structure. Recall, from Chapters 18 and 32, that every element in the periodic table has a unique, reproducible spectrum. A correct theory of how and why atoms emit light would allow scientists to unlock the atoms' structure and discern their inner workings. But the riddle of atomic spectra could not be explained by classical physics. Later, in 1913, Bohr provided an explanation of the hydrogen spectrum, but he was unable to extend his ideas to other elements. Thus one of the challenges facing quantum mechanics in the late 1920s was the need to explain atomic spectra.

In this last chapter, we want to bring all that you have learned to bear on the issue of atomic spectra. It took scientists over a century from the discovery of the discrete spectra of atoms to their full understanding in terms of the quantum physics of atoms. It was a remarkable scientific journey, leading to the recognition that quantum mechanics—including all the strange and unexpected phenomena of wave–particle duality—is the correct theory of nature at the microscopic level. As we put these final pieces into place we can, if you will excuse the pun, see the light at the end of the tunnel.

38.2 Excited States of Atoms

To understand atomic spectra, we must take a closer look at the quantum states of atoms. As we've noted before, the *ground state* of an atom is the stationary state with the lowest possible energy that is consistent with the Pauli exclusion principle. The ground state of hydrogen, for example, is $1s$. For lithium, the ground state is $1s^22s$. All states with an energy higher than the ground state but less than the ionization energy for that atom are called *excited states*. The periodic table of the elements organizes

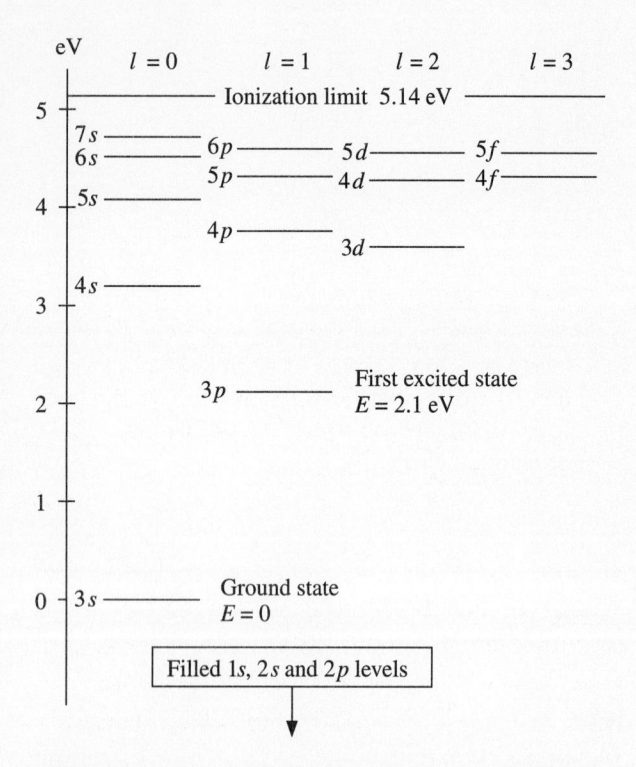

FIGURE 38-1 The [Ne]3s ground state of the sodium atom and some of the excited states.

information about the ground states of the elements, but it is their excited states that hold the key to spectra.

Consider the sodium atom ($Z = 11$). It is a multielectron atom that we will use as a prototypical atom. The ground-state electron configuration of sodium is $1s^2 2s^2 2p^6 3s$. The first ten electrons completely fill the $n = 1$ and $n = 2$ shells, creating a *neon core*. The 3s electron is a valence electron. It is customary to represent this configuration as [Ne]3s or, more simply, as just 3s. The sodium energy levels are shown in Fig. 38-1. Notice that the 1s, 2s, and 2p states of the neon core are not shown on the diagram. Because these states are filled and unchanging, energy-level diagrams always begin with the highest level of the ground-state configuration—3s in this case—and then show the *available* energy levels above it.

Notice something new about Fig. 38-1: The zero of energy has been shifted to the ground state. The zero of energy, as we have discovered many times, can be located where it is most convenient. Because we always work with energy *differences*, the location of the zero has no physical significance. When we solved the Schrödinger equation, it was most convenient to let zero energy represent the energy of an electron moved infinitely far away. But for analyzing spectra it is more convenient to let the ground state have $E = 0$. With this choice, all excited state energies will be positive and will tell us directly how far each state is above the ground state. The ionization limit now occurs at the value of the atom's ionization energy, which is 5.14 eV for sodium.

One possible way to create an excited state of the sodium atom would be to promote one of the ten core electrons to a higher energy level. For example, one of the two 2s electrons could be raised to a 4p level. The core electrons, however, are very tightly bound in the closed shell. The energy needed to excite an electron out of the core is far more than the 5.14 eV ionization energy.

The more meaningful excited states are those in which just the valence electron is raised to a higher energy level. These excited states, with the valence electron in quantum state nl, have the configuration [Ne]nl. For sodium, the first energy level above 3s is 3p, so the *first excited state* of sodium is $1s^2 2s^2 2p^6 3p$—written as [Ne]3p or, more simply, just 3p. This is followed, in order of increasing energy, by [Ne]4s, [Ne]3d, and [Ne]4p. Notice that the order of excited states is exactly the same order (3p–4s–3d–4p) that explained the fourth row of the periodic table.

This seems straightforward, but a careful look at the actual sodium spectrum (which we will do in Section 38.4) shows that things are somewhat more complicated. An orbiting electron, as we have already noted, is a microscopic current loop that generates a magnetic moment. The electron also has an inherent magnetic moment associated with its spin. These two magnetic moments, which are shown in Fig. 38-2, interact with each other just like two tiny bar magnets! This interaction of the two magnetic moments is called the **spin–orbit interaction**.

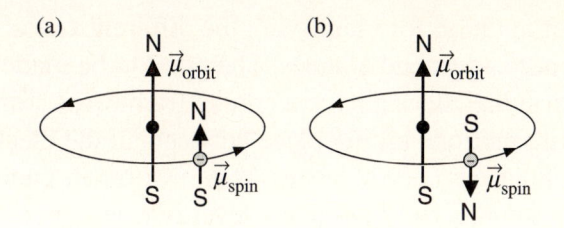

FIGURE 38-2 The magnetic moments of the orbital and the spin angular momenta interact with each other, creating the spin–orbit interaction. The antiparallel orientation of b) is energetically more favorable than the parallel orientation of a).

When the two magnetic moments are parallel to each other, as shown in Fig. 38-2a, the magnetic poles try to repel each other. But the antiparallel magnetic moments of Fig. 38-2b try to attract each other. These magnetic interactions within the atom cause the *p*-states in sodium to divide into two closely spaced energy levels. The antiparallel orientation, in which the magnetic moments attract each other, is slightly lower in energy than the parallel orientation.

Figure 38-3 shows a revised energy-level diagram for the excited states of sodium. As you can see, the spin–orbit interaction causes each *p*-state to divide into two levels. The "splitting" of each of the *p*-states is exaggerated for clarity—they differ by only 0.002 eV for 3*p* and less for 4*p*!—but the effects of this splitting are quite apparent in the spectrum. The *d* states also split into two levels, but the spin–orbit interaction is much less in *d*-states and is usually not important. The *s*-states, by contrast, do not have a spin–orbit interaction. They have no orbital magnetic moment, because $L = 0$, so the *s*-states remain a single energy level.

Other atoms with a single valence electron have energy-level diagrams similar to that of sodium. Things get more complicated when there are more than one valence electron. Not only are there spin–orbit interactions, there are also additional effects that enter depending on whether the spins of the valence electrons are parallel or antiparallel to each other. There is also a fairly complicated system

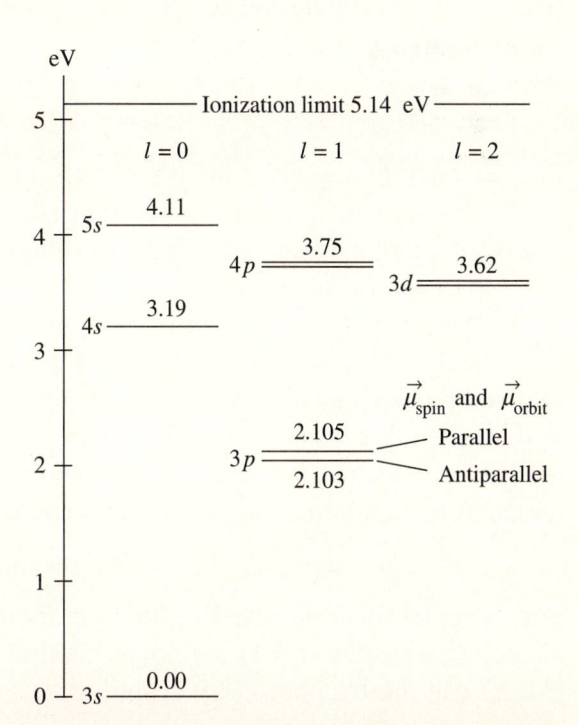

FIGURE 38-3 A revised energy-level diagram for sodium that includes the spin–orbit interaction.

of notation for identifying the different states. These are all details that will be deferred to more advanced courses. The point to be made here is that quantum mechanics really does provide the right conceptual framework for classifying and understanding the many interactions taking place between all the electrons in an atom. We can *utilize* information about the energy levels, as shown on an energy-level diagram, without having to understand precisely *why* each level is where it is.

38.3 Excitation

In the last section we looked at what it means for an atom to be in an excited state. But left to itself, an atom will be in its lowest-energy ground state. How does an atom get into an excited state? The process of doing so is called **excitation**, and there are two basic mechanisms: absorption and collision.

Excitation by Absorption

One of the postulates of the basic Bohr model states that an atom can jump from one stationary state, of energy E_1, to a higher-energy state E_2 by absorbing a photon of frequency

$$f = \frac{\Delta E}{h} = \frac{E_2 - E_1}{h}. \tag{38-1}$$

Because we are interested in spectra, it is more useful to write Eq. 38-1 in terms of the wavelength:

$$\lambda = \frac{c}{f} = \frac{hc}{\Delta E} = \frac{1240}{\Delta E(\text{in eV})} \text{ nm}. \tag{38-2}$$

The final expression gives the wavelength in nanometers *if* ΔE is given in electron volts.

Bohr's idea of quantum jumps remains an integral part of our interpretation of the results of quantum mechanics. By absorbing a photon, an atom is excited from its ground state up to one of its excited states.

A careful analysis of how the electrons in an atom interact with the electric and magnetic fields of a light wave shows that not every conceivable transition can occur. The **allowed transitions** must satisfy one or more **selection rules**. The only selection rule that will concern us says that a transition (either absorption or emission) from a state in which the valence electron has angular momentum quantum number l_1 to another with angular momentum quantum number l_2 is allowed only if

$$\Delta l = |l_2 - l_1| = 1 \quad \text{(selection rule for emission and absorption)}. \tag{38-3}$$

For example, this rule says that an atom in an *s*-state ($l = 0$) can absorb a photon and be excited to a *p*-state ($l = 1$) but *not* to another *s*-state or to a *d*-state. An atom in a *p*-state ($l = 1$) can emit a photon by dropping to a lower-energy *s*-state *or* to a lower-energy *d*-state, but not to another *p*-state.

EXAMPLE 38-1 What is the longest wavelength in the absorption spectrum of hydrogen?

SOLUTION The longest wavelength corresponds to the smallest energy change ΔE. Because the atom begins in the $n = 1s$ ground state, the smallest energy change occurs for absorption to the first $n = 2$ excited state. The energy change is

$$\Delta E = E_2 - E_1 = \frac{-13.6 \text{ eV}}{2^2} - \frac{-13.6 \text{ eV}}{1^2} = 10.2 \text{ eV}.$$

This gives a wavelength

$$\lambda = \frac{1240}{\Delta E} = \frac{1240}{10.2} = 122 \text{ nm}.$$

This is an ultraviolet wavelength. Because of the selection rule, the transition is to the $2p$ state, not to $2s$.

EXAMPLE 38-2
What is the longest wavelength in the absorption spectrum of sodium?

SOLUTION
The sodium ground state is [Ne]$3s$. The lowest excited state is the lower of the two $3p$ states. $3s \rightarrow 3p$ is an allowed transition ($\Delta l = 1$), so this will be the longest wavelength. You can see from the energy data in Fig. 38-3 that $\Delta E = 2.103$ eV for this transition. The wavelength for this transition is

$$\lambda = \frac{1240}{2.103} = 589.6 \text{ nm}.$$

This wavelength (yellow color) is a prominent feature in the absorption spectrum of sodium. Note that because the ground state has $l = 0$, absorption *must* be to a p state. The s-states and d-states of sodium cannot be excited by absorption.

Collisional Excitation

An electron traveling with a speed of 8.595×10^5 m/s has a kinetic energy of 2.103 eV. If an electron with this speed collides with a ground-state sodium atom, the electron could transfer its energy to the atom. The collision would excite the atom to the $3p$ state and, at the same time, bring the electron to rest. This process is called **collisional excitation** of the atom.

Collisional excitation differs from excitation by absorption in one very fundamental way. In absorption, the photon disappears. This means that *all* of the photon's energy must be transferred to the atom. But the electron is still present after collisional excitation. Consequently, the electron can carry away some kinetic energy; it does *not* have to transfer its entire energy to the atom. If the electron has an incident kinetic energy of 3.000 eV, which it would with an initial speed of 10.26×10^5 m/s, it could transfer 2.103 eV to the sodium atom, thereby exciting it to the $3p$ state, and still depart the collision with a speed of 5.61×10^5 m/s and an energy of 0.897 eV.

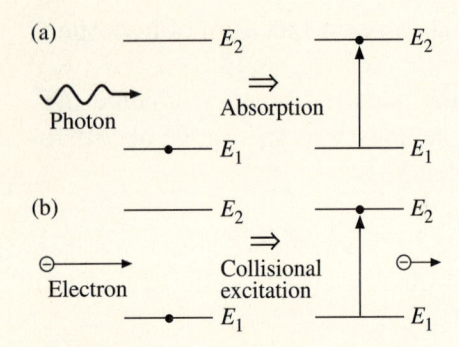

FIGURE 38-4 Excitation by a) photon absorption and b) electron collision.

The important idea is that the photon's frequency and energy have to match the atom's ΔE *exactly*, but the incident energy of the electron merely has to *exceed* ΔE. It is all a matter of energy conservation. Figure 38-4 shows the idea graphically. The electron can carry excess energy away after the collision, but there is no photon after the absorption to carry away energy.

Collisional excitation by electrons is the predominant method of excitation in electrical discharges. This is the excitation mechanism in all forms of fluorescent lights, street lights, and neon signs. A gas is placed in a tube at reduced pressure ($\approx$1 mm of Hg), then a fairly high voltage ($\approx$1000 V) is established between electrodes at the ends of the tube. This causes the gas to ionize, creating a current in which both ions and electrons are charge carriers. The mean free path of electrons between collisions is large enough for the electrons to gain several eV of kinetic energy as they accelerate in the electric field. This energy is then transferred to the gas atoms upon collision. The process does not work at atmospheric pressure, because the mean free path between collisions becomes too short for the electrons to gain enough kinetic energy to excite the atoms.

38.4 Emission

Although the absorption of light is an important process, it is the emission of light that really gets our attention. The overwhelming bulk of sensory information that we perceive comes to us in the form of light. The recognition and appreciation of light and colors has formed the basis of aesthetics and art since the days of prehistory. All of our knowledge about the cosmos comes to us, with the small exception of cosmic rays, in the form of light emitted in various processes. It was the discovery of discrete emission spectra that brought down classical physics, and it was the understanding of discrete emission spectra that provided the first major triumph of quantum mechanics.

The basic understanding of emission hinges upon two of Bohr's postulates: that quantum systems exist in stationary states having discrete, quantized energy levels, and that a system can "jump" from a higher-energy level to a lower-energy level by emitting a photon of frequency and wavelength given by Eqs. 38-1 and 38-2. Once we have determined the energy levels of an atom, we can immediately predict its emission spectrum. Conversely, we can use the measured emission spectrum to determine its energy levels.

The latter, in fact, is what spectroscopists do. Although it is possible in principle to calculate the energy levels of an atom or molecule directly from the Schrödinger equation, in practice the calculations are very difficult to perform. Instead, quantum mechanics is used to predict the *types* of energy levels to be expected, such as $3p$ or $4d$, and approximately what their energy should be. Spectroscopists can then use exceedingly accurate wavelength measurements to determine the energy levels experimentally. Energy levels for most elements are determined in this way to a typical accuracy of $\pm 1 \times 10^{-5}$ eV.

Emission spectra are more than just scientific curiosities. Most of today's artificial light sources—from fluorescent lights to street lights to lasers—are applications of emission spectra. We will look at several examples in this section.

Sodium

Figure 38-5 shows some of the transitions and wavelengths observed in the emission spectrum of sodium. This diagram makes the point that each wavelength represents a quantum jump between two well-defined energy levels. The fact that there are so many energy levels is why emission spectra have very large numbers of emission lines. Notice the selection rule $\Delta l = 1$ being obeyed in the sodium spectrum. The $5p$ levels, for example, can undergo quantum jumps to $3s$, $4s$, or $3d$ but *not* to $3p$ or $4p$.

Of particular interest are the two closely spaced wavelengths 589.0 nm and 589.6 nm, generated by quantum jumps of the two $3p$ levels back to the $3s$ ground state. As you learned in Section 38.2, these are the result of the spin–orbit interaction. Although these wavelengths differ by only one part in a thousand, they are easily resolved into two separate lines by a diffraction grating or a prism spectrometer. These two wavelengths are the most prominent feature in the emission spectrum of sodium—because the $3p$ states are the easiest to excite—and since the early nineteenth century they have been called the *sodium D-lines*. (The term originates from a period long before spectra were understood, when reproducible spectral lines were labeled simply A, B, C, etc.)

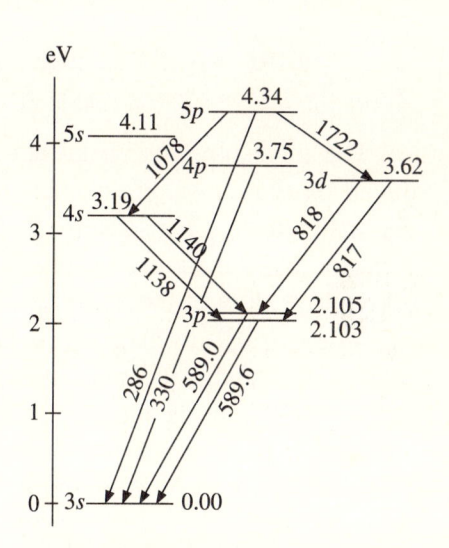

FIGURE 38-5 Some of the transitions in the emission spectrum of sodium. Wavelengths are given in nanometers, and energy levels are given in eV.

The $3d$, $4p$, and $5p$ states are also split into two different energy levels by the spin–orbit interaction. Their emission lines, such as the 330 nm emission from $4p$, are also pairs of two closely spaced wavelengths. However, the energy separations of each of these pairs of levels is much less than the separation in the $3p$ state. Although the 589.0/589.6 nm wavelengths are easy to resolve, high-resolution spectroscopy is necessary to resolve the two wavelengths in these other emission lines. Hence they have been shown as a single wavelength in Fig. 38-5.

Figure 38-6 shows the emission spectrum of sodium as it would be recorded in a spectrometer. By comparing the spectrum to Fig. 38-5, you can recognize that the spectral lines at 589 nm, 330 nm, 286 nm, 268 nm, etc. form a *series* of lines due to all the possible $np \rightarrow 3s$ transitions. They are the dominant features in the sodium spectrum. (Note that the two sodium D-lines at 589 nm are not resolved in this spectrum.) The other lines in the spectrum of Fig. 38-6 originate in higher excited states that are not seen in the rather limited energy-level diagram of Fig. 38-5.

588 CHAPTER 38 ATOMIC SPECTRA

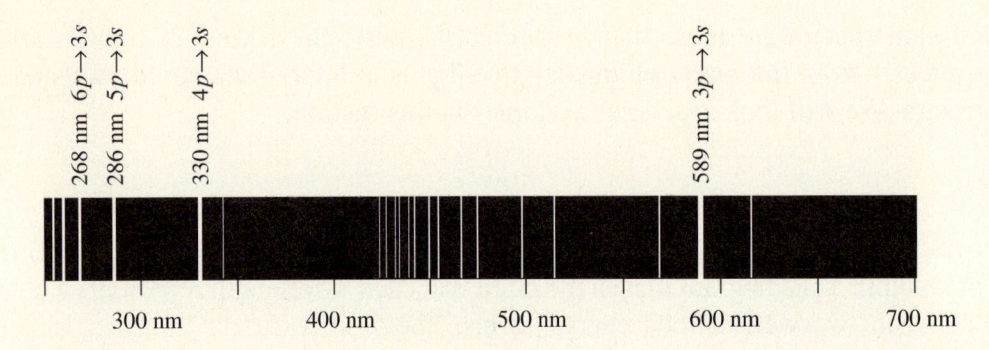

FIGURE 38-6 The emission spectrum of sodium.

The most obvious visual feature of sodium emission is its bright yellow color, produced by the emission wavelength of 589 nm. This is the basis of the *flame test* used in chemistry to test for sodium: A sample is held in a Bunsen burner, and a bright yellow glow indicates the presence of sodium. The 589 nm emission is also prominent in the pinkish-yellow glow of the common sodium-vapor street lights. These lights operate by creating an electrical discharge in sodium vapor. Most sodium-vapor lights use high-pressure lamps because this increases their light output. The high pressure, however, causes the formation of Na_2 molecules, and the molecules emit the pinkish portion of the light. There are a few cities that use low-pressure sodium lights, and these lamps emit the distinctively yellow 589 nm light of sodium. The low-pressure lights are especially common in cities close to astronomical observatories. The glow of city lights is a severe hazard to astronomers, but the very specific 589.0 nm and 589.6 nm emissions from sodium are easily removed with a *sodium filter*. The light from the telescope is passed through a container of sodium vapor, and the sodium atoms *absorb* just the unwanted 589.0 nm and 589.6 nm photons without disturbing any other wavelengths! This cute trick does not work, however, for all the other wavelengths emitted by high-pressure sodium lamps or light from other sources.

Mercury

Mercury-vapor lamps are a common alternative to sodium-vapor lamps for street lighting. They emit a bright, slightly greenish color. They are also widely used in laboratory settings as a source of intense light. And, as you will see, mercury atoms are the emitters in common fluorescent light bulbs, although the mercury emission in that case is altered by the fluorescent coating on the inside of the bulb. In all of these applications the mercury atoms are excited by electron collisions in an electrical discharge. They then emit photons of various wavelengths as the excited atoms jump down to lower energy levels.

Mercury ($Z = 80$) is a more complicated atom than sodium. The ground-state electronic configuration is $[Xe]4f^{14}5d^{10}6s^2$, where the [Xe] notation indicates a closed shell xenon-atom core. But the $4f$, $5d$, and $6s$ subshells are also closed. This makes mercury chemically unreactive. Mercury is the only metal that is a liquid at room temperature, indicating that the bonds between mercury atoms are too weak for it to coalesce into a solid. And at modest temperatures it becomes a fairly unreactive monatomic gas, not unlike the rare gases.

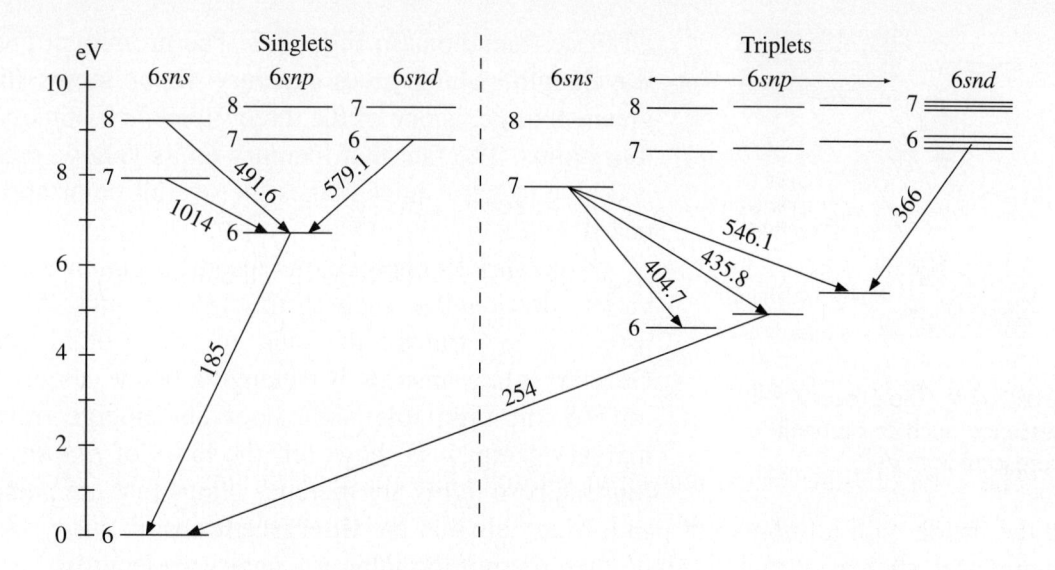

FIGURE 38-7 Energy-level diagram for mercury. Important transitions are indicated.

Because the 6s electrons are the least tightly bound, the excited states of mercury are formed by raising one of the 6s electrons to a higher level. The excited states are designated as 6snl, where the unchanging $[Xe]4f^{14}5d^{10}$ core is not mentioned. The first few excited states are illustrated in Fig. 38-7.

Having *two* valence electrons outside of this core creates a more complex energy-level structure than we found in sodium. The interactions within the atom now have to include not only the spin–orbit interactions but the interactions of the electrons with each other. We need not go into the rather complex details, but these interactions cause each 6snl level to divide into two sets of states that are fairly different in energy. One of these sets is further subdivided into three closely spaced energy levels. These two sets of states are labeled **singlets** (single energy level) and **triplets** (groups of three energy levels). For example, there is a singlet 6s6p energy level at 6.7 eV and a group of three triplet 6s6p energy levels at about 5 eV. The triplet 6sns levels are not split, just as the ns levels in sodium were not, because these states have no orbital angular momentum.

For our purposes we can simply use the measured energy levels *and* the $\Delta l = 1$ selection rule to predict the emission spectrum of mercury. A number of transitions are indicated in Fig. 38-7. You can see many of these in the mercury spectrum of Fig. 38-8. Observation of the mercury spectrum with a spectrometer reveals a very intense green wavelength at 546.1 nm, a bright yellow wavelength at 579.1 nm, and noticeable lines at 491.6 nm (green), 435.8 nm (blue), and 404.7 nm (violet). You should be able to identify

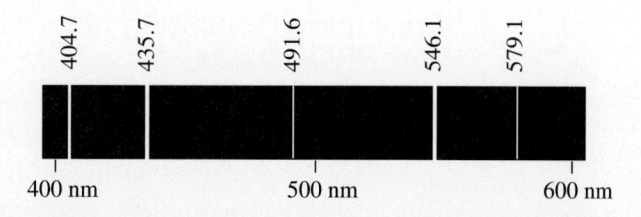

FIGURE 38-8 The emission spectrum of mercury.

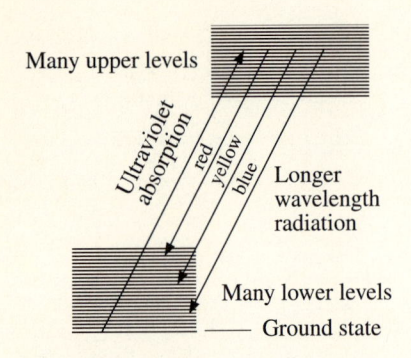

FIGURE 38-9 The process of fluorescence, such as occurs in a fluorescent light bulb.

all these transitions in Fig. 38-7. The prominent green wavelength, which gives mercury vapor lamps their greenish tinge, is one of the three triplet $7s \rightarrow 6p$ quantum jumps. The fact that mercury emits yellow, green, and blue spectral lines makes the overall color appear nearly white.

A mercury discharge lamp also emits intense ultraviolet wavelengths, such as the 185 nm and 254 nm $6s6p \rightarrow 6s^2$ transitions back to the ground state. Ordinary glass absorbs wavelengths below about 350 nm, so this ultraviolet light does not emerge from a mercury lamp. It is, however, the basis of *fluorescent lights*. These lights are mercury lamps that are painted on the inside with a fluorescent paint. Materials that are **fluorescent** absorb short wavelength light, such as ultraviolet light, then re-emit the light at longer wavelengths.

Figure 38-9 shows how fluorescence occurs. Fluorescent paints consist of large, complex molecules. Their energy-level structure contains many hundreds, or even thousands, of closely spaced energy levels grouped together. Although there are many lower levels, most molecules will be in the very lowest of the lower-energy levels—the ground state. A ground-state molecule easily absorbs an ultraviolet photon, exciting it into one of the upper energy levels. Although some excited molecules may emit an ultraviolet photon and jump directly back to the ground state, many will emit lower-energy, longer-wavelength visible photons as they jump to higher levels of the lower-energy group. There are so many possible transitions, each at a slightly different wavelength, that the overall effect is the emission of white light.

Neon

Another light source of practical importance is the electrical discharge in gases such as neon. Figure 38-10 shows a partial energy-level diagram for the neon atom. The neon ground state is the closed shell $1s^22s^22p^6$. It is not easy to excite an electron out of a closed shell, but this is the only way to create an excited state because neon has no outer valence electrons. The lowest available state is $3s$, so we expect the lowest excited state to be $1s^22s^22p^53s$. This is indeed the case, as Fig. 38-10 shows. Notice, however, that the energy is >16 eV. Most elements in the periodic table can be ionized for less energy than it takes to lift a neon atom to its first excited state! This demonstrates the difficulty of exciting an electron from within a closed shell.

The excited states of an atom such as neon are very complex. Not only is there an excited electron—$3s$ or $3p$—there is also now an inner shell with a *vacancy*. This leaves the incomplete $2s$ subshell with a net magnetic moment that interacts in a complicated fashion

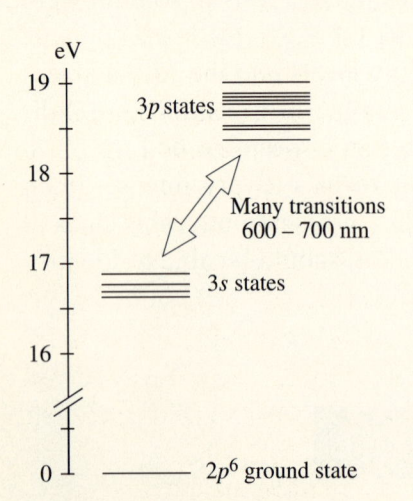

FIGURE 38-10 Energy-level diagram for neon.

with the orbital and spin magnetic moments of the excited electron. This results—analogous to the 3p doublet in sodium—in four distinct 3s excited states and ten different 3p states. There are a large number of possible transitions between these two groups of states, but they all have wavelengths near 600 nm. An emission spectrum of neon reveals many wavelengths in this region of the spectrum due to 3p–3s transitions, but the overall visual impression is the familiar orange-red color of a neon light.

Color in Solids

It is worth concluding this section with a few remarks about color in solids. Whether it be the intense multihued colors of a stained glass window, the bright colors of flowers or paint, or the deep luminescent red of a ruby, most of the colors we perceive in our lives come from solids rather than free atoms. The basic principles are the same, but the details are more complex for solids. The primary difference is that solids can undergo so-called **nonradiative transitions**.

An atom in an excited state has little choice but to give up that energy by emitting a photon. Its only other option—which is rare for free atoms—is to collide with another atom and transfer its energy into the kinetic energy of recoil. But the atoms in a solid are in intimate contact with one another at all times. Although an excited atom in a solid has the option of emitting a photon, it is often far more likely that the energy will be converted—via collisions with neighboring atoms—to the thermal energy of the solid. The atom is de-excited without radiating—a nonradiative transition—as the excitation energy is used to heat up the solid.

This is what happens in pigments, such as those in paints, plants, and dyes. Pigments are molecules that absorb certain wavelengths of light—raising the molecules to excited states—but not other wavelengths. The energy-level structure is complex, rather like that of fluorescent molecules, so the absorption consists of "bands" of wavelengths rather than discrete spectral lines. But instead of reradiating the energy by photon emission, as a free atom would, the pigment molecules undergo nonradiative transitions and convert the energy into increased thermal energy. That is why darker objects get hotter in the sun than lighter objects.

When light falls on an object, it can be either absorbed or reflected. If *all* wavelengths are reflected, the object is perceived as white. Wavelengths that are absorbed by the pigments are removed from the reflected light. A pigment with blue-absorbing properties converts the energy of blue-wavelength photons into thermal energy, but photons of other wavelengths are reflected without change. Thus a blue-absorbing pigment would reflect red and yellow wavelengths only. The object would be perceived as the color orange!

This creation of colors by the selective absorption of some wavelengths, followed by nonradiative transitions, is the mechanism at work in colored glass and plastic. Blue glass has molecules that absorb all wavelengths *except* blue. The absorbed energy is converted, by nonradiative transitions, to the thermal energy of the glass. Only blue wavelengths pass through and are seen. The glass acts as a *filter* that passes some wavelengths but filters out others.

Some solids, though, are a little different. The color of many minerals and crystals is due to so-called **impurity atoms** embedded in them. The gemstone ruby, for example, is a very simple and common crystal of aluminum oxide, called corundum, which happens

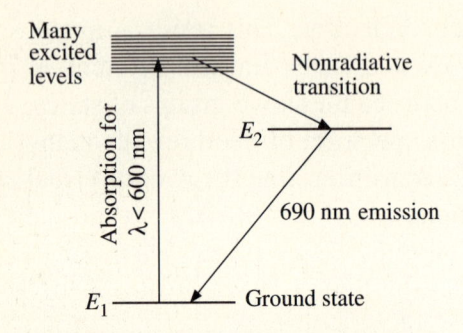

FIGURE 38-11 Shorter wavelengths are absorbed, then reradiated at 690 nm to give ruby its bright red color.

to have chromium atoms present at the concentration of about one part in a thousand. Pure corundum is transparent, so all of the ruby's color comes from these chromium impurity atoms. Figure 38-11 shows what happens. The chromium atoms have a group of excited states that absorb all wavelengths less than about 600 nm—that is, everything except orange and red. But unlike the pigments in red glass, which convert the absorbed energy into thermal energy, the chromium atoms dissipate only a small amount of heat as they undergo a small nonradiative transition to another excited state. From there, they emit a photon with $\lambda = hc/(E_2 - E_1) \approx$ 690 nm as they jump back to the ground state.

The net effect is that short-wavelength photons are *reradiated* as longer-wavelength photons rather than completely absorbed. This is why rubies sparkle and have such intense color, whereas red glass is a dull red color. The colors of other minerals and gems is due to different impurity atoms, but the principle is the same.

38.5 Lifetimes of Excited States

Excitation of an atom—either via absorption or collision—leaves it in an excited state. From there it jumps back to a lower energy level by emitting a photon. How long does this process take? There are actually two questions here. First, how long does an atom remain in an excited state before undergoing a quantum jump to a lower state? Second, how long does the transition last as the quantum jump is occurring?

Our best understanding of the quantum physics of atoms is that quantum jumps—both absorption and emission—are instantaneous. The absorption or emission of a photon is an all-or-nothing event; there is not a moment when a photon is "half-absorbed." The prediction that quantum jumps are instantaneous has troubled many physicists, but careful experimental tests have never revealed any evidence that the jump itself takes a finite amount of time.

The time spent in the excited state, waiting to make a quantum jump, is another story. Figure 38-12 shows experimental data for the length of time that doubly charged xenon ions Xe^{++} spend in a certain excited state. In this experiment a pulse of electrons was used to excite the atoms very rapidly to the excited state via collisional excitation. The number of excited state atoms was then monitored by detecting the photons emitted—one by one!—as the excited atoms jumped back to the ground state. The number of photons emitted at time t is directly proportional to the number of excited state atoms present at time t. As the figure shows, the number of atoms in the excited state decreased *exponentially* with time, and all had decayed within 25 ms of their creation.

Figure 38-12 has two important implications. First, atoms do spend a finite amount of time in the excited state before decaying back to a lower state. Second, the length of time spent in the excited state is not a constant value but, instead, varies from atom to atom. If every excited xenon ion lived for 5 ms in the excited state, then we would detect

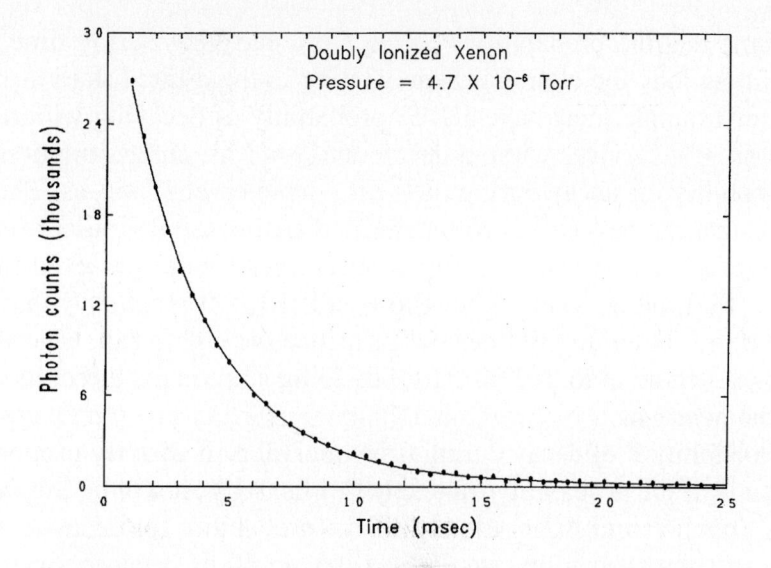

FIGURE 38-12 Experimental data for the photon emission rate from an excited state in Xe⁺⁺. The dots are data and the solid curve is an exponential function "fitted" to the data.

no photons for 5 ms, a big burst right at 5 ms as they all decay, then no photons after that. The data tell us that there is a *range* of times spent in the excited state. Some undergo a quantum jump and emit a photon after 1 ms, others after 5 ms or 10 ms, and a few wait as long as 20 or 25 ms.

Consider an experiment in which N_0 excited atoms are created at time $t = 0$. As the solid curve in Fig. 38-12 shows, the number of atoms remaining at time t is very well described by the exponential function

$$N(t) = N_0 e^{-t/\tau}, \qquad (38\text{-}4)$$

where τ is the point in time at which $e^{-1} = 0.368 = 36.8\%$ of the original atoms are left. Thus 63.2%—nearly two-thirds—of the atoms have decayed at time $t = \tau$. The specific length of time τ is called the **lifetime** of the excited state. From Fig. 38-12 we can deduce that the lifetime of this state in Xe⁺⁺ is ≈ 4 ms because that is the point in time at which the curve has decayed to 36.8% of its initial value.

This lifetime in Xe⁺⁺ is abnormally long, which is why the state was being studied. More typical excited state lifetimes are a few nanoseconds. Table 38-1 gives some measured values of excited state lifetimes. They differ significantly, but in all cases a group of excited state atoms exhibits the exponential decay of Eq. 38-4. Can we understand why this is?

Quantum mechanics is about probabilities. We cannot say exactly when an excited electron will decay any more than we can say exactly where the electron is located. However, we can use quantum mechanics to find the *probability P* that the electron will undergo a quantum jump during a time interval Δt.

TABLE 38-1 Some excited state lifetimes.

Atom	State	Lifetime
Hydrogen	2p	1.6 ns
Sodium	3p	17 ns
Neon	3p	20 ns

Let us assume that the probability P of an atom decaying during time interval Δt is *independent* of how long the atom has been waiting in the excited state. A newly created excited atom, for example, may have a 10% probability of decaying within the first 1 ns interval $0 \to 1$ ns. If it survives without decay until $t = 7$ ns, our assumption is that it still has a 10% probability for decay during the next 1 ns interval $7 \to 8$ ns. This is a reasonable assumption that can be justified with a detailed analysis. It is similar to flipping coins. The probability of a head on your first flip is 50%. Even if you flip seven heads in a row, the probability of a head on your eighth flip is still 50%. It is *unlikely* that you will flip seven heads in a row, but doing so does not influence the eighth flip. Likewise, it may be *unlikely* for an excited atom to live for 7 ns, but doing so does not affect its probability of decay during the next 1 ns.

Now the probability P of decay during time interval Δt is directly proportional to Δt *if* Δt is small. That is, if the probability for decay in 1 ns is 1%, it will be 2% in 2 ns or 0.5% in 0.5 ns. This fails if P and Δt get too big: If the probability for decay is 70% in 20 ns, we *cannot* say that the probability would be 140% in 40 ns because a probability >1 is meaningless. But because we are going to be interested in the limit $\Delta t \to dt$, the concept is valid and we can write

$$P = \text{probability of decay in } \Delta t = R\Delta t, \qquad (38\text{-}5)$$

where R is called the **decay rate**. It is a probability *per second*, with units of s^{-1}, and thus really is a *rate*. For example, if an atom has a 5% probability of decay during a 2 ns interval, its decay rate is

$$R = \frac{P}{\Delta t} = \frac{0.05}{2 \text{ ns}} = 0.025 \text{ ns}^{-1} = 2.5 \times 10^7 \text{ s}^{-1}.$$

If there are N_{exc} atoms in the excited state, the number that will decay and emit a photon during the interval Δt is just the N_{exc} multiplied by P, the probability of decay. So if $N_{\text{exc}}(t)$ excited atoms are present at time t, then the number decaying in the small interval of time $t \to t + \Delta t$ will be

$$\text{number of decays in } \Delta t \text{ at time } t = P(\text{in } \Delta t) \cdot N_{\text{exc}}(t) = RN_{\text{exc}}(t)\Delta t. \qquad (38\text{-}6)$$

Now the number of excited atoms *decreases* by the number of decays. That is, the *change* in N_{exc} is the *negative* of Eq. 38-6. Suppose, for example, there are 1000 excited atoms present at time t and that each has a 5% probability of decay in the next 1 ns. The number of photons emitted during the next 1 ns will be (on average!) $0.05 \times 1000 = 50$. Consequently, the number of excited atoms changes by $\Delta N_{\text{exc}} = -50$, with the minus sign indicating a decrease.

We can write the change in the number of atoms in the excited state as

$$\Delta N_{\text{exc}}(\text{in } \Delta t \text{ at } t) = -P(\text{in } \Delta t) \cdot N_{\text{exc}}(t) = -RN_{\text{exc}}(t)\Delta t$$
$$\Rightarrow \frac{\Delta N_{\text{exc}}}{\Delta t} = -RN_{\text{exc}}(t). \qquad (38\text{-}7)$$

Understanding the minus sign is very important. N_{exc} is not the number of decays or the number of photons emitted but, instead, the number of excited atoms present. That number is decreasing, so the *change* ΔN_{exc} is negative.

Now let $\Delta t \to dt$, in which case $\Delta N_{exc} \to dN_{exc}$. Equation 38-7 then becomes

$$\frac{dN_{exc}}{dt} = -RN_{exc}. \tag{38-8}$$

Equation 38-8 is called a **rate equation** because it describes the *rate* at which the excited state population changes. Rate equations are extremely common in science and engineering. Before solving it, let us think about what it says. First, the *rate* of decay depends on the rate constant R. If R is large, the population will decay at a rapid rate and will have a short lifetime. Conversely, a small value of R implies that the population decays slowly and will live a long time. Second, the rate of decay also depends on how many excited atoms are present. There will be many decays per second when N_{exc} is large, fewer decays per second at a later time when N_{exc} is small.

The rate equation is another differential equation, but this time one we can solve with elementary calculus. Rewrite Eq. 38-8 as

$$\frac{dN_{exc}}{N_{exc}} = -Rdt, \tag{38-9}$$

then integrate both sides:

$$\int \frac{dN_{exc}}{N_{exc}} = -R \int dt \tag{38-10}$$
$$\Rightarrow \ln N_{exc} + C = -Rt.$$

We can evaluate the integration constant C by using the initial condition: $N_{exc}(t=0) = N_0$. This is

$$\ln N_0 + C = 0$$
$$\Rightarrow C = -\ln N_0. \tag{38-11}$$

Inserting this result for C into Eq. 38-10 gives

$$\ln N_{exc} - \ln N_0 = \ln\left(\frac{N_{exc}}{N_0}\right) = -Rt. \tag{38-12}$$

We can finally solve for N_{exc} by taking the exponential of both sides:

$$\frac{N_{exc}}{N_0} = e^{-Rt} \tag{38-13}$$
$$\Rightarrow N_{exc}(t) = N_0 e^{-Rt}.$$

According to our model, the excited state population decays exponentially with time. This is exactly the experimental pattern seen in Fig. 38-12.

It will be more convenient to write Eq. 38-13 as

$$N_{exc}(t) = N_0 e^{-t/\tau}, \tag{38-14}$$

where

$$\tau = \frac{1}{R} = \text{the } \textit{lifetime} \text{ of the excited state.} \tag{38-15}$$

This is the value of t at which the population has decayed to $e^{-1} = 36.8\%$ of its initial value—exactly the definition of the lifetime we used in Eq. 38-4 to describe the experimental results. Now, however, we see explicitly that the lifetime is simply the inverse of the decay rate R.

EXAMPLE 38-3 The decay rate of the singlet $6s6p$ state in mercury is 7.7×10^8 s^{-1}. a) What is the lifetime of this state? b) If 10^{10} excited atoms are present at $t = 0$, how many photons will be emitted during the first 1 ns?

SOLUTION a) The lifetime is

$$\tau = \frac{1}{R} = \frac{1}{7.7 \times 10^8 \text{ s}^{-1}} = 1.3 \times 10^{-9} \text{ s} = 1.3 \text{ ns}.$$

b) If there are $N_0 = 10^{10}$ excited atoms at $t = 0$, the number still remaining at $t = 1$ ns is

$$N_{\text{exc}}(t = 1 \text{ ns}) = N_0 e^{-t/\tau} = 10^{10} \cdot e^{-1.0/1.3} = 4.63 \times 10^9.$$

This implies that 5.37×10^9 atoms have decayed during the first 1 ns. Each of these atoms emitted one photon, so the number of photons emitted during the first 1 ns is 5.37×10^9.

The decay rates for excited states can be calculated in quantum mechanics, although the calculation is difficult to do accurately. Calculated decay rates can be directly compared to experimentally measured lifetimes of excited states. The agreement is very good in those cases where the calculations are thorough, thus providing another validation of the quantum-mechanical description of atoms. Lifetimes for thousands of excited states have been measured and are tabulated in various handbooks. A knowledge of lifetimes, and thus of decay rates, is important for the detailed understanding of discharges, plasmas, flames, and astrophysical objects such as nebulae, stars, and supernovae.

38.6 The Stimulated Emission of Radiation

You have seen that an atom can jump from a lower-energy level E_1 to a higher-energy level E_2 by absorbing a photon of frequency $f = (E_2 - E_1)/h$. Figure 38-13a illustrates the basic absorption process, with a photon of frequency $f = \Delta E/h$ disappearing as the atom jumps from level 1 to level 2. Once in level 2, as shown in Fig. 38-13b, the atom can emit a photon of the

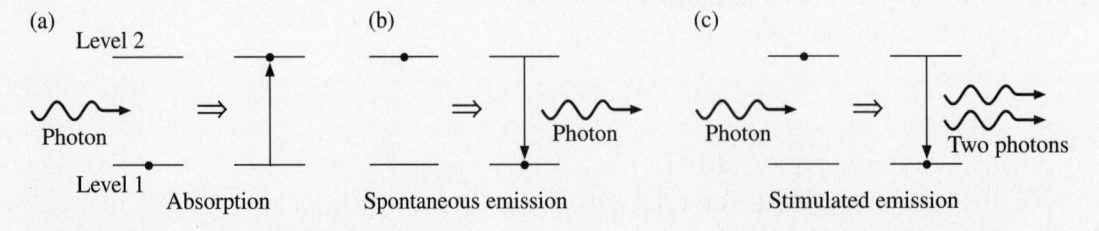

FIGURE 38-13 Three types of radiative transitions: a) absorption, b) spontaneous emission, and c) stimulated emission. The two photons in c) are identical.

same frequency as it jumps back to level 1. Because this transition occurs spontaneously, without the introduction of outside energy, it is called **spontaneous emission**.

In 1917, just two years after Bohr's proposal of stationary states in atoms but still prior to de Broglie and Schrödinger, a now-famous Einstein was puzzled by how quantized atoms reach thermodynamic equilibrium in the presence of electromagnetic radiation. Einstein found that the processes of absorption and spontaneous emission are not sufficient to allow a collection of atoms to reach thermodynamic equilibrium. To solve this problem, Einstein proposed a third mechanism for the interaction of atoms with light.

What would happen, Einstein wondered, if an *excited* atom were in a preexisting electromagnetic wave of exactly frequency $f = \Delta E/h$? This is the situation shown in the left half of Fig. 38-13c. If a photon could induce the $1 \rightarrow 2$ transition of absorption, Einstein reasoned, then it should also be able to induce a $2 \rightarrow 1$ transition. This is, in a sense, just a "reverse absorption." But to undergo a reverse absorption, the atom would have to *emit* a photon of frequency $f = \Delta E/h$. The net result, as seen in the right half of Fig. 38-13c, is an atom in level 1 plus *two* photons! Because the first photon induced the atom to emit the second photon, this process is called **stimulated emission**.

Stimulated emission can occur only if the first photon's frequency exactly matches the $E_2 - E_1$ energy difference of the atom. This is precisely the same condition that absorption has to satisfy. More interestingly, the emitted photon is absolutely *identical* to the incident photon. This means that as the two photons leave the atom they have exactly the same frequency and wavelength, are traveling in exactly the same direction, and are exactly in phase with each other.

Contrast this with the two photons emitted *spontaneously* by two different excited atoms. These two photons would have the same frequency, but they would be traveling in two different directions. Further, because their emissions are unrelated, the relation between the phases of their oscillations would be random: They might be in phase, out of phase, or anywhere in between. Whereas spontaneous emission produces two independent photons, stimulated emission produces a second photon that is an exact clone of the first.

Stimulated emission is of no importance in most practical situations. Atoms typically spend only a few nanoseconds in an excited state before undergoing spontaneous emission, so the atom would have to be in an extremely intense light wave for stimulated emission to occur prior to spontaneous emission. Ordinary light sources are not nearly intense enough for stimulated emission to be more than a minor effect. Thus it was many years before Einstein's prediction was confirmed. No one had doubted Einstein, because he had demonstrated clearly that stimulated emission was necessary to make the energy equations balance, but it seemed no more important than would pennies to a millionaire balancing her checkbook. At least, that is, until 1960, when a revolutionary invention appeared that made explicit use of stimulated emission: the laser.

38.7 Lasers

The word **laser** is an acronym for the phrase *l*ight *a*mplification by the *s*timulated *e*mission of *r*adiation. Lasers were an outgrowth of intense research with microwaves during the 1950s, which culminated in the invention of the *maser*, short for *m*icrowave *a*mplification by the *s*timulated *e*mission of *r*adiation. The first laser, a ruby laser, was demonstrated in

1960, and several other kinds of lasers appeared within a few months. The driving force behind much of the research was the American physicist Charles Townes. Townes was awarded the Nobel Prize in 1964 for the invention of the maser and his theoretical work leading to the laser.

By the mid-1960s, the laser was making dramatic changes in science and engineering. Today lasers do everything from being the light source in fiber optic communications to measuring the distance to the moon, from playing your CD to performing delicate eye surgery. The laser and the transistor are likely the two scientific inventions of the second half of the twentieth century that have most changed the technology of our everyday lives.

FIGURE 38-14 Charles Townes, inventor of the maser and contributor to the laser.

But what is a laser? Basically it is a device that produces a beam of highly *coherent* and essentially monochromatic (single color) light as a result of stimulated emission. **Coherent light** is light in which all the electromagnetic waves have the same phase, direction, and amplitude. It is the coherence of a laser beam that allows it to be very tightly focused or to be rapidly modulated for communications.

Let's take a look at how a laser works. Consider a system of atoms that have a lower energy level E_1 and a higher energy level E_2, as shown in Fig. 38-15. Suppose that there are N_1 atoms in level 1 and N_2 in level 2. Left to themselves, all the atoms would soon end up in level 1 because of the spontaneous emission $2 \rightarrow 1$. To prevent this, we can imagine some type of excitation mechanism—an electrical discharge, for example—is continuing to produce new excited atoms in level 2.

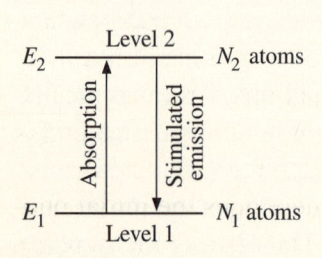

FIGURE 38-15 Energy levels 1 and 2, with populations N_1 and N_2. A population inversion occurs if $N_2 > N_1$.

Now suppose a photon of frequency $f = (E_2 - E_1)/h$ is incident on this group of atoms. Because it has the correct frequency, it could be absorbed by one of the atoms in level 1. But another possibility is that it could cause stimulated emission from one of the level 2 atoms. Ordinarily $N_2 \ll N_1$, so absorption events far outnumber stimulated emission events. Even if a few photons were generated by stimulated emission, they would soon be absorbed by the vastly larger group of atoms in level 1.

But what if we could somehow arrange to place *every* atom in level 2, leaving $N_1 = 0$. Then the incident photon, upon encountering its first atom, will cause stimulated emission. Where there was initially one photon of frequency f, now there are two. These will strike two additional atoms, again causing stimulated emission, and then there are four photons. As Fig. 38-16 shows, there would be a *chain reaction* of stimulated emission until all N_2 atoms had emitted a photon of frequency f.

But stimulated emission is more than just emitting a photon of frequency $f = \Delta E/h$. Each emitted photon is *identical* to the incident photon. The chain reaction of Fig. 38-16 will lead not just to N_2 photons of frequency f, but to N_2 *identical* photons, all traveling together in the same direction with the same phase. If N_2 is a large number, as would be the case in any practical device, the one initial photon will have been *amplified* into a gigantic, coherent pulse of light! Because the collection of excited state atoms amplifies the light from a small input to a large output, they are referred to collectively as an **optical amplifier**.

The stimulated emission is sustained by placing the *lasing medium*—the sample of atoms that emit the light—in an **optical cavity**. An optical cavity, shown in Fig. 38-17, consists of two facing mirrors spaced a reasonable distance apart. As the photons bounce back and forth between the mirrors, they interact repeatedly with the atoms in the medium. This repeated interaction is necessary for the light intensity to build up to a high level. One of the mirrors will be partially transmitting so that some of the light emerges as the *laser beam*.

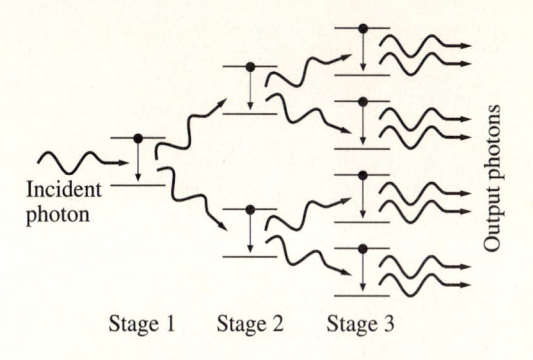

FIGURE 38-16 Stimulated emission creates a chain reaction of photon production in a population of excited atoms.

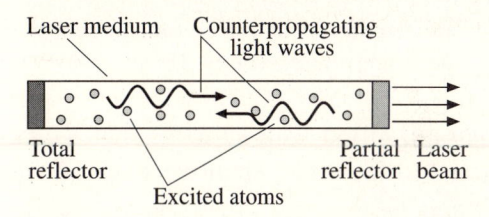

FIGURE 38-17 Lasing takes place in an optical cavity that reflects the light waves back and forth in order to sustain the stimulated emission.

Although the chain reaction of Fig. 38-16 illustrates the idea most clearly, it is not necessary for every atom to be in level 2 for amplification to occur. All that is needed is to have $N_2 > N_1$ so that stimulated emission exceeds absorption. Such a situation is called a **population inversion**. The process of obtaining a population inversion is called **pumping**, and we will look at two specific examples very shortly. Pumping is the technically difficult part of designing and building a laser because normal excitation mechanisms do not create population inversions. In fact, lasers would likely have been discovered accidentally long before 1960 if population inversions were easy to create.

One interesting issue is how the lasing action gets started. Where does the initial photon of frequency $f = \Delta E/h$ come from? This turns out not to be a problem. Once there are atoms in level 2, they are constantly emitting photons of frequency f by spontaneous emission. Most of these are emitted in the "wrong" direction and immediately leave the optical cavity. But it does not take long for a photon, just by chance, to be spontaneously emitted parallel to the axis of the optical cavity. That one photon is all it takes to start the chain reaction and initiate lasing.

Now let's look at two specific examples of lasers.

The Ruby Laser

The first laser to be developed was a ruby laser. Figure 38-18a shows the energy-level structure of the chromium atoms that gives ruby its optical properties. Normally the number of atoms in the ground-state level E_1 far exceeds the number of excited state atoms in level E_2. That is, $N_2 \ll N_1$. Under these circumstances 690 nm light is absorbed rather

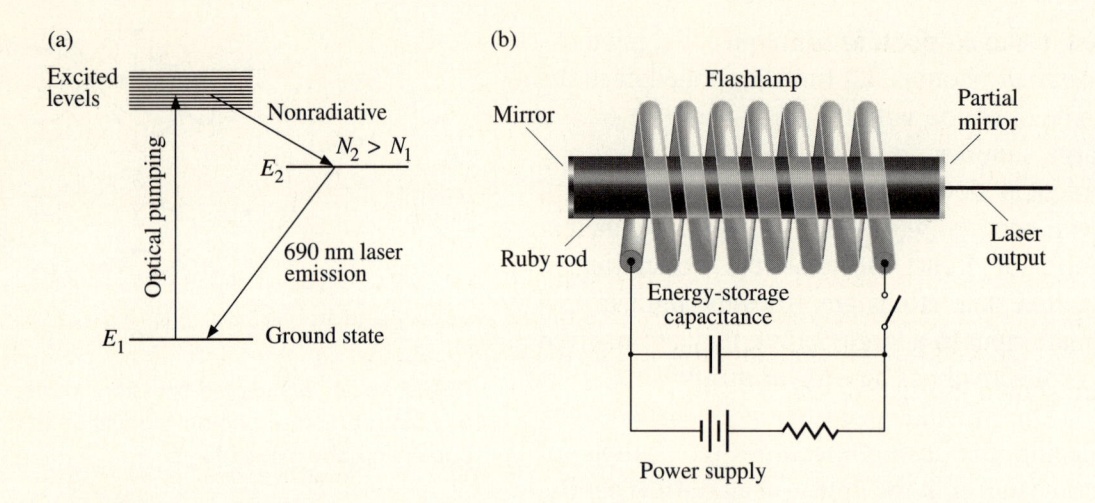

FIGURE 38-18 a) The chromium energy levels involved in a ruby laser. b) Construction of a flashlamp-pumped ruby laser.

than amplified. But suppose that we could *rapidly* excite more than half the chromium atoms to level E_2. Then we would have a population inversion ($N_2 > N_1$) between levels E_1 and E_2.

Population inversion can be accomplished by *optically pumping* the ruby with a very intense pulse of white light from a *flashlamp*. This is like a camera flash, only vastly more intense. In the basic arrangement, shown in Fig. 38-18b, a helical flashlamp is coiled around a ruby rod that has mirrors bonded to its end faces. The lamp is fired by discharging a high-voltage capacitor through it, creating a very intense light pulse that lasts just a few microseconds. Surrounding mirrors (not shown) help to focus this light onto the ruby. Because most of the light has wavelengths < 600 nm, it excites atoms from the ground state to the upper energy levels. From there they quickly ($\approx 10^{-8}$ s) decay nonradiatively to level 2. If the pumping light is sufficiently intense, it can drive more than half the chromium atoms to level 2. With $N_2 > N_1$ a population inversion has been created.

Once a photon initiates the laser pulse, the light intensity builds quickly into a giant pulse. It cannot be sustained because the stimulated emission soon destroys the population inversion, but large amounts of light energy can be extracted in a brief, incredibly intense burst of light. A typical output pulse has an energy of 1 J and lasts 10 ns = 10^{-8} s. This gives a *peak power* of

$$P = \frac{\Delta E}{\Delta t} = \frac{1 \text{ J}}{10^{-8} \text{ s}} = 10^8 \text{ W} = 100 \text{ MW}.$$

One hundred megawatts of light power! That is more than the electrical power consumed by a small city. The difference, of course, is that a city consumes that power continuously but the laser pulse lasts a mere 10 ns. The laser cannot fire again until the capacitor is recharged and the laser rod cooled. A few pulses per second is a typical firing rate, so the laser is "on" only a few billionths of a second out of each second.

Ruby lasers are no longer widely used. They have been replaced by other pulsed lasers that, for various practical reasons, are easier to operate. However, they all operate with the same basic idea of rapid optical pumping to upper states, rapid nonradiative decay to level 2 where the population inversion is formed, then rapid build-up of an intense optical pulse.

The Helium-Neon Laser

The familiar red laser used in lecture demonstrations, laboratories, and supermarket checkout scanners is the helium–neon laser, often called a HeNe laser. Its output is a *continuous*, rather than pulsed, wavelength of 632.8 nm. The medium of a HeNe laser is a mixture of ≈90% helium and ≈10% neon gases. They are sealed in a glass tube of ≈30 cm length and 1 mm diameter, then an electrical discharge is established down the bore of the tube. Other than the small diameter, this structure is no different from a conventional neon sign. Two mirrors are bonded to the ends of the discharge tube, one a total reflector and the other having ≈2% transmission so that the laser beam can be extracted. These features are shown in Fig. 38-19a.

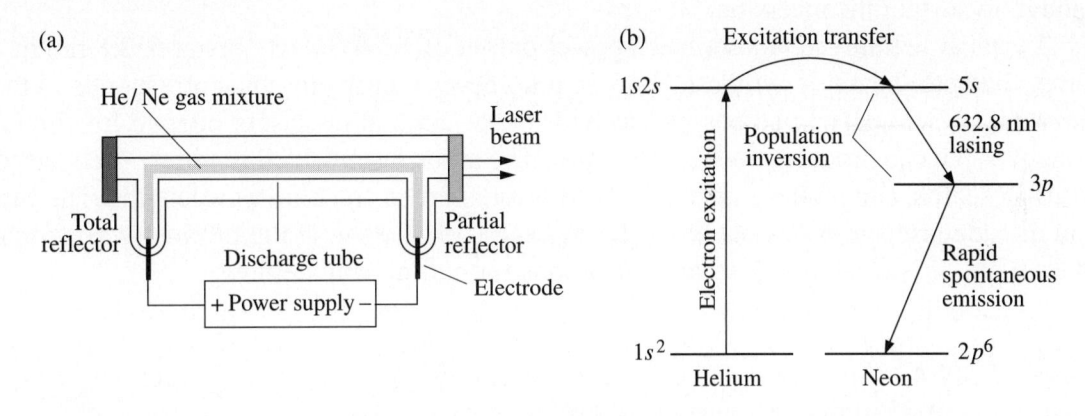

FIGURE 38-19 a) The construction of a HeNe laser. b) The physics of a HeNe laser.

The atoms that lase in a HeNe are the neon atoms, but the pumping method is a rather complex procedure involving the helium atoms. The electrons in the discharge easily excite the $1s2s$ excited state of helium. This state has a very small spontaneous decay rate (very long lifetime), so it is possible to build up a fairly large population (but not an inversion) of excited helium atoms in this state. The energy of the $1s2s$ state is 20.6 eV.

Interestingly, one of the four $5s$ excited states of neon also has an energy of 20.6 eV. If a $1s2s$ excited helium atom collides with a ground-state neon atom—as frequently happens—the excitation energy can be transferred from one atom to the other! Written as a chemical reaction, the process is

$$\text{He}^* + \text{Ne} \rightarrow \text{He} + \text{Ne}^*,$$

where the * superscript indicates the atom is in an excited state. This process, called **excitation transfer**, is very efficient for one particular $5s$ state of neon because the process is *resonant*—a perfect energy match. By the two-step process of collisional excitation of helium, followed by excitation transfer between helium and neon, the neon atoms are pumped into an excited $5s$ state. This is shown in Fig. 38-19b.

The $5s$ energy level in neon is ≈1.95 eV above the $3p$ excited states. The $3p$ states are very nearly empty of population, both because they are not efficiently populated in the discharge and because they undergo very rapid spontaneous emission to the $3s$ states. Thus the large number of atoms pumped into the $5s$ state creates a population inversion with respect to the lower $3p$ state! These are the necessary conditions for laser action.

In the ruby laser, the lower level of the laser transition normally has a large population. Over half of all the chromium atoms have to be pumped to the upper state to create a population inversion, and that inversion lasts only a very short time. Thus the ruby laser has a pulsed output, and it takes a large power supply and flashlamp to perform the intense pumping. In the HeNe laser, by contrast, the lower level of the laser transition is normally empty of population, or very nearly so. Placing only a tiny fraction of the neon atoms in the 5s state will create a population inversion. A very modest pumping action is sufficient to create the inversion and start the laser. So not only are HeNe lasers much smaller and more efficient, they can maintain a *continuous* inversion and thus sustain continuous lasing. The electrical discharge continuously creates 5s excited atoms in the upper level, and the rapid spontaneous decay of the 3p atoms from the lower level keeps its population low enough to sustain the inversion.

A typical helium–neon laser has a power output of 1 mW = 10^{-3} J/s at 632.8 nm in a 1 mm diameter beam. As you can show in a homework problem, this corresponds to the emission of 3.2×10^{15} photons per second. Other continuous lasers operate by similar principles but can produce much more power. The argon laser, which is widely used in scientific research, can produce up to 20 W of power at green and blue wavelengths. The carbon dioxide laser produces output power in excess of 1000 W at the infrared wavelength of 10.6 μm. It is used in industrial applications for cutting and welding.

EXAMPLE 38-4 An ultraviolet laser generates a 10 MW, 5-ns-long light pulse at a wavelength of 355 nm. How many photons are in each pulse?

SOLUTION The energy of each light pulse is the power multiplied by the duration:

$$E_{\text{pulse}} = P\Delta t = (10^7 \text{ W}) \cdot (5 \times 10^{-9} \text{ s}) = 0.050 \text{ J}.$$

Each photon in the pulse has energy

$$E_{\text{photon}} = hf = \frac{hc}{\lambda} = 3.50 \text{ eV} = 5.60 \times 10^{-19} \text{ J}.$$

Because $E_{\text{pulse}} = NE_{\text{photon}}$, where N is the number of photons,

$$N = \frac{E_{\text{pulse}}}{E_{\text{photon}}} = 8.9 \times 10^{16} \text{ photons}.$$

38.8 The Final Word

This final chapter has been less technical than the last few chapters. Even so, it has been very much an *application* of ideas you have learned throughout Part VI. Understanding atomic spectra requires knowledge of electrons and photons, of stationary states and energy levels, and of transitions and probabilities. And we've been able to end our tour of quantum physics with an important application of atomic spectra—lasers.

Quantum mechanics successfully explains *all* the many features of atomic spectra. Despite the uncertainties of wave–particle duality, the energies of a quantum system are very well defined because only very specific energies allow the creation of an electron standing wave. These energy levels, with all the information they contain about the atom's structure and interactions, are revealed to us in the unique discrete spectrum of each element.

Quantum mechanics also explains the spectra of molecules, the electrical properties of conductors and semiconductors, the strange behavior of liquid helium, the structure of the nucleus, and much more. Quantum mechanics has been, and continues to be, the most successful scientific theory every developed. But these stories will have to await another time and place.

Summary
Important Concepts and Terms

spin–orbit interaction	decay rate
excitation	rate equation
allowed transition	spontaneous emission
selection rule	stimulated emission
collisional excitation	laser
singlets	coherent light
triplets	optical amplifier
fluorescent	optical cavity
nonradiative transition	population inversion
impurity atoms	pumping
lifetime	excitation transfer

The unique discrete spectrum of each element arises because the energies of atoms are quantized and because the energy-level pattern of each element is unique. Physicists use spectra to determine the energy levels of an atom and thus to gain insight into the structure of the atom. For example, the magnetic interaction called the spin–orbit interaction is revealed when some spectral lines split into two closely spaced wavelengths.

The three types of transitions that can occur are absorption, spontaneous emission, and stimulated emission. The wavelength of a transition is

$$\lambda = \frac{hc}{\Delta E} = \frac{1240}{\Delta E(\text{in eV})} \text{ nm},$$

and each transition must obey the selection rule

$$\Delta l = |l_2 - l_1| = 1.$$

Emission must be preceded by excitation, which can be either by absorption of a photon or by collision with an electron or another atom. Many artificial light sources are based on collisional excitation of atoms by the electrons in an electrical discharge.

The average time that an atom spends in an excited state before decaying to a lower level is called the lifetime τ of the excited state. The time when a specific atom will decay cannot be predicted, but the total number of excited atoms in the excited state decays exponentially with time as

$$N_{\text{exc}}(t) = N_0 e^{-t/\tau}.$$

Exercises and Problems

Exercises

1. a. Is a $4p \to 4s$ transition allowed in sodium? If so, what is its wavelength? If not, why not?
   b. Is a $3d \to 4s$ transition allowed in sodium? If so, what is its wavelength? If not, why not?

2. 10^6 sodium atoms are excited to the $3p$ state at $t = 0$. How many of these atoms remain in the $3p$ state at
   a. $t = 10$ ns?
   b. $t = 20$ ns?
   c. $t = 50$ ns?
   d. $t = 100$ ns?

3. How many photons are emitted per second by a 1 mW helium–neon laser emitting a visible laser beam with a wavelength of 633 nm?

4. How many photons are emitted per second by a 1000 W carbon dioxide laser emitting an infrared laser beam with a wavelength of 10.6 μm?

5. Neon emits a bright reddish-orange spectrum. But a glass tube filled with neon is completely transparent. Why doesn't the neon in the tube absorb orange and red wavelengths?

Problems

6. a. What transitions are possible for a sodium atom in the $5s$ state? (See Figs. 38-3 and 38-5.)
   b. What are the wavelengths of each of these transitions?

7. a. Use the wavelengths listed in Fig. 38-7 to determine the energies, in eV, of the singlet $6s7s$ and the triplet $6s7s$ states of the mercury atom.
   b. What minimum speed must an electron have to excite the 546.1 nm green emission line in the Hg spectrum?

8. Figure 38-20 shows the first few energy levels of the lithium atom. Make a table showing:
   a. All the allowed transitions in the Li emission spectrum.
   b. The wavelength of each transition, in nm.
   c. An indication of whether each transition is in the infrared, the visible, or the ultraviolet spectral region.
   d. Whether each wavelength would be observed in the Li absorption spectrum.

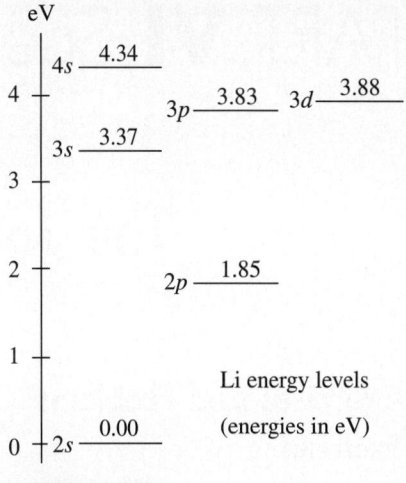

eV

4s —— 4.34

3p —— 3.83   3d —— 3.88

3s —— 3.37

2p —— 1.85

Li energy levels
(energies in eV)

2s —— 0.00

FIGURE 38-20

9. a. What is the decay rate constant R for the $2p$ state of hydrogen?
   b. During what interval of time will 10% of a sample of $2p$ hydrogen atoms decay?

10. A hydrogen atom is in the $2p$ state. How much time must elapse for there to be a 1% chance that this atom will decay to the ground state?

11. What is the *half life* of the $3p$ state of sodium?

12. An electrical discharge in a neon-filled tube maintains a *steady* population of 10^9 atoms in the excited $3p$ states. How many $3p \rightarrow 3s$ photons are emitted from the tube each second?

13. A ruby laser emits a 100 MW, 10 ns pulse of light with a wavelength of 690 nm. How many chromium atoms undergo stimulated emission to generate this pulse?

14. A laser emits 1.0×10^{19} photons per second from an excited state with energy $E_2 = 1.17$ eV.
    a. What is the wavelength of this laser?
    b. What is the power output of this laser?

[**Estimated 10 additional problems for the final edition.**]

PART VI Scenic Vista

The Journey Never Ends

Anyone who is not shocked by quantum theory has not understood it.

Niels Bohr

Bohr—who thought as deeply as anyone about quantum mechanics—was right on target with the remark quoted above. Quantum mechanics *is* shocking. Quantum mechanics replaces the absolute certainty and predictability of Newtonian physics with a mysterious world in which particles seem to suddenly materialize at a localized point on a film without previously having a well-defined position at all. Physical entities that by all rights should be waves sometimes act like particles. Electrons, neutrons, and even whole atoms—chunks of matter!—are somehow "divided" and later "recombined" to produce wave-like interference with themselves. These discoveries stood common sense on its head. How can we ever possibly understand matter—the solidity of a table or the predictability of a rocket trajectory—if the atoms are busily dividing, recombining, and waving about?

Shocking, indeed. Nonetheless, according to quantum mechanics, the wave function and its associated probabilities are *all we can know* about an atomic particle. This idea is so unsettling that many great scientists were reluctant to accept it. Einstein is well known for saying that "God does not play dice with the universe." Einstein thought that there *must* be more to reality than mere probabilities. He and others hoped that there were some underlying laws that, once discovered, would explain how quantum mechanics works. These deeper and more fundamental laws, it was thought, would restore our ability to make precise predictions about atomic particles.

But Einstein was wrong. Many very clever experiments to probe the foundations of quantum mechanics have all reached the same conclusion: There are *no* underlying laws that will restore the exactness and predictability of classical physics. All that we can predict about an experiment are the probabilities of various outcomes. As strange as it seems, this is the way that nature really is.

Whether you're shocked or not, quantum mechanics, on a practical level, is the most successful theory ever devised. It is a powerful tool for understanding and using semiconductors, lasers, superconductors, and a host of other modern technologies. Such emerging technologies as nanostructures, atomic force microscopes, and quantum computing rest on the concepts and principles of quantum theory.

Table SVI-1 summarizes what you have learned with a knowledge structure of quantum mechanics. The wave function is the descriptor of quantum mechanics, and our two major goals were to *determine* and then to *interpret* the wave function. The Schrödinger equation is the fundamental law of quantum mechanics, telling us how the wave function behaves as a particle interacts with its environment. Note that quantum mechanics views interactions in terms of energy, in contrast with the forces of Newtonian mechanics. The solutions of the Schrödinger equation describe the standing-wave stationary states of the particle.

TABLE SVVI-1 The knowledge structure of quantum mechanics.

Descriptor: The wave function $\psi(x)$ for a particle of mass m.

Governing law: The Schrödinger equation $\dfrac{d^2\psi}{dx^2} = -\dfrac{2m}{\hbar^2}[E - U(x)]\psi(x)$.

Interpretation
  Prob(in δx at x) = $|\psi(x)|^2 \delta x$
  Uncertainty principle: $\Delta x \Delta p \gtrsim h/2$

Quantization
  Energy levels: E_n
  Quantum jumps: $f = \Delta E/h$

  Specific situations
  Particle in a potential well
  Tunneling
  Atoms and atomic structure

The wave function is interpreted in terms of the probability density $P(x) = |\psi(x)|^2$ for finding the particle at position x. Although this is important for understanding what the wave function means, few experiments measure probabilities directly. More practical applications of quantum mechanics use the wave function to determine the energies of allowed states, the frequencies of photons emitted or absorbed in quantum jumps, or the probabilities of tunneling through various barriers. Quantum mechanics is unsurpassed at explaining the atomic structure of matter and at allowing us to put that knowledge to practical use.

In Conclusion

Long ago, in the Overview to Part I, this text was likened to a journey. The text may now have reached its final page, but your journey into physics never ends.

As with any long trip, it was sometimes difficult to know whether we were on the right road. I tried at the outset to provide a guide to the upcoming journey so that you would know where we were headed. Now that we've reached our destination, you can form your own judgments as to what were the most interesting, and least interesting, parts of the trip. As you review your journey, you'll probably be surprised—pleasantly so, I hope—as to the amount of physics you have learned. You have also become vastly more sophisticated in your ability to analyze situations and solve problems. You might find it interesting to reread the Part I Overview just to remind yourself how far you've come.

It is not easy to understand the atomic structure of matter. We had to study the basic concepts of motion and energy. We had to learn about waves and the properties of light. We had to introduce electric fields, magnetic moments, and other properties of charged particles. We had to discover how knowledge is obtained about the invisible realm of the atom. And lastly, we had to enter the strange world of wave functions, probabilities, and wave–particle duality. I say *we* because I have been your guide, but it is *you* who have learned all this!

Now our paths part. Some of you will go on to more advanced courses in physics. Others will make good use of your physics knowledge as you pursue careers in engineering, mathematics, or other sciences. And some of you, even if you move on to nontechnical fields, will carry with you a better appreciation for what science is and how scientific knowledge grows and changes. Whatever your chosen path and ultimate destination, you are now prepared and ready to continue the journey on your own.

As an old saying goes, I wish you calm seas and prosperous voyages.

Answers to Selected Exercises and Problems

Chapter 19
1. 639 N **2.** a) 22.6 m^3; b) 129.0 kPa
3. a) 7.0×10^{-21} J; b) 2050 m/s
4. a) 1.563 mol; b) 9.41×10^{23}; c) 7.49×10^{25} m^{-3}; d) 3.80 kg/m^3; e) 202 kPa
5. a) 3.05×10^{21}; b) 20.2 μg **6.** a) 9520 kPa
7. a) 97.8 cm^3 **11.** 3.7 mm **13.** c) 824°C
16. b) −43°C

Chapter 20
1. a) 245 J; b) 944 m/s **2.** a) 6110 N; b) 6110 N; c) 611 J; d) −611 J; e) 361 J added **3.** 883 cal **4.** a) 30°C; b) 95 cal
5. a) 3.08 atm; b) 975 cm^3 **7.** 25.3°C
9. a) 253°C **11.** a) 152 J

Chapter 21
1. 0.117, 0.246 **3.** a) 0.328; b) 3.2×10^{-4}; c) 0.410 **4.** a) 1000; b) 26; c) 2.6%
6. $10^{5,566,000}$ **9.** c) 0.161; g) 2.4; j) 5.8×10^9 years **10.** a) $P(5000) = 0.00798$

Chapter 22
1. a) 2.07×10^{-22} J; b) 249 m/s; c) 307 nm; d) 8.12×10^8 s^{-1} **2.** a) 2.28×10^5 J; b) 4.74×10^{10} s^{-1} **3.** a) He; b) 1364 m/s; c) 1.864 μm; d) 7.32×10^8 s^{-1} **4.** 30 g
6. b) $p_{He} = 3.11$ atm, $p_{Ar} = 3.45$ atm
8. c) 436 K **10.** b) 1.3×10^{10} s^{-1}

Chapter 23
1. a) 473°C; b) 1550 J; c) 3880 J; d) 2330 J
2. a) 0.5 atm; b) 1074 J; c) 1074 J; d) 0
3. a) 370°C; b) −3850 J; c) 0; d) 8.50
4. a) 5190°C; b) 0; c) 54,000 J; d) 20
5. a) 9.9; b) 460°C **6.** a) 40%; b) 2500 W; c) 1500 W **7.** a) 50.6 J; b) 25.3 J; c) −25.3 J **9.** c) 17.9% **13.** b) 9.0%
14. a) 464 K; c) 2.08 m

Chapter 24
4. a) $-9.8\,\hat{j}$ m/s^2; b) (0.00266 m/s^2, toward earth) **5.** b) 3.12×10^{10} **6.** a) 0.056 N; b) 2.9; c) 8.4×10^{10} **7.** $\vec{F}_{+10} = +0.0045\,\hat{j}$ N, $\vec{F}_{-20} = -0.0045\,\hat{j}$ N **8.** a) $\vec{E}(5\text{ cm}, 0\text{ cm}) = 36{,}000\,\hat{i}$ N/C, $\vec{E}(-5\text{ cm}, 5\text{ cm}) = (-12{,}700\,\hat{i} + 12{,}700\,\hat{j})$ N/C, $\vec{E}(-5\text{ cm}, -5\text{ cm}) = (-12{,}700\,\hat{i} - 12{,}700\,\hat{j})$ N/C **9.** a) $\vec{E}(5\text{ cm}, 0\text{ cm}) = -36{,}000\,\hat{i}$ N/C, $\vec{E}(-5\text{ cm}, 5\text{ cm}) = (12{,}700\,\hat{i} - 12{,}700\,\hat{j})$ N/C, $\vec{E}(-5\text{ cm}, -5\text{ cm}) = (12{,}700\,\hat{i} + 12{,}700\,\hat{j})$ N/C **11.** $(-0.00128\,\hat{i} - 0.00407\,\hat{j})$ N
15. d) 4.28×10^{10} m/s^2; e) 7.85×10^{13} m/s^2
16. b) (0 cm, 2 cm) **18.** c) 139 nC

Chapter 25
1. (0.00255 N, toward rod)
2. a) $2qx/4\pi\varepsilon_0(x^2 + s^2/4)^{3/2}\,\hat{i}$; b) $\vec{E}(10\text{ mm}) = 144{,}000\,\hat{i}$ N/C **3.** 50,000 N/C
4. a) 3.60×10^6 N/C; b) 8.30×10^5 m/s
6. a) (64,400 N/C, left); b) (6.44×10^{-5} N, right) **8.** a) $(-96{,}500\,\hat{i} - 92{,}400\,\hat{j})$ N/C
11. a) $-Q/4\pi\varepsilon_0 d(d+L)\,\hat{i}$
13. b) $(2\eta/4\pi\varepsilon_0)\ln[(2x+L)/(2x-L)]$
16. b) 37,500 N/C **18.** 1.40 cm

A-1

Chapter 26

1. a) 1.73×10^7 A/m^2; b) 5.31×10^{18} elec/s
2. 4.24×10^6 A/m^2 **3.** tungsten
4. a) 1.25×10^{17} elec/s; b) 1.64×10^{-3} N/C;
c) 1.10×10^{-5} m/s   **7.** b) 0.187 mm
9. d) 3.32×10^{-4} m/s top, 1.33×10^{-3} m/s middle

Chapter 27

3. a) 3140 V; b) 5.02×10^{-16} J   **4.** a) 2010 V; b) -3.22×10^{-16} J   **5.** c) 14.4 N; d) $v_2 = 21.9$ m/s, $v_4 = 10.95$ m/s   **7.** a) 5.58×10^7
9. a) 1000 V; c) 6.98×10^6 m/s

Chapter 28

1. a) 1.00; b) 0.50   **2.** a) 0.087 Ω; b) 3.5 Ω
3. tungsten   **4.** a) 7.1 pf; b) 0.71 nC   **5.** 27 V
7. d) 0.62 mm   **9.** a) $\vec{E}_3 = (66,000$ V/m, straight down)   **12.** $Q_1 = 2$ nC, $Q_2 = 4$ nC

Chapter 29

1. 34 Ω **2.** $I_{10} = I_{20} = 0.5$ A, $\Delta V_{10} = 5$ V, $\Delta V_{20} = 10$ V, $P_{10} = 2.5$ W, $P_{20} = 5.0$ W
3. a) 0.10 A, left to right   **4.** $P_5 = 45$ W, $P_{20} = 20$ W   **5.** 23.6 μm   **10.** 7.0 Ω
12. b) 20.25 W   **14.** $I_{6\,\Omega} = 1.333$ A

Chapter 30

2. a) 2.5 A, 250 A, 5000 A, 5×10^5 A; b) 4 cm, 0.4 mm, 20 μm, 200 nm   **3.** a) 20 A; b) 1.6 mm   **4.** a) 0; b) $(1.6 \times 10^{-15}$ T, into page); d) $(5.6 \times 10^{-16}$ T, out of page)
5. a) up; c) into page   **6.** a) out of page; b) up   **7.** a) 11 cm; b) 10.4 m
8. a) 1.6423 MHz; b) 1.4378 MHz; c) 1.6429 MHz   **11.** b) 8   **14.** b) $\sqrt{2}\mu_0 I / \pi R$
17. a) 110.07 V; c) 110.11 V   **19.** 238 A
23. 12.6 T

Chapter 31

1. a) 1.0×10^{-3} Wb; b) clockwise
2. a) 39 mA, clockwise; b) 39 mA, clockwise; c) 0   **3.** At 0 s: $B = 0, I = 1.6$ A; At 2 s: $B = 2.0$ T, $I = 0$ **4.** 3.14 V   **5.** 3.54×10^{-4} Wb
7. b) 0.04 N; c) 11°C   **9.** c) 35.4 A
12. 3.93 V   **13.** $v(0.02$ s$) = 1.35$ m/s

Chapter 32

3. a) 7.12 keV; c) 5.0 keV; e) 1.22×10^{19} eV

4. a) 5.93×10^6 m/s; b) 6.19×10^7 m/s
5. c) 365 nm   **6.** b) 2.3×10^{17} kg/m^3
7. a) 6.00×10^7 m/s; b) 8.4°   **9.** (0.0457 T, into page)   **11.** b) 2.25×10^7 V

Chapter 33

1. a) 1.76 eV; b) 0.248 nm   **2.** a) aluminum
3. 1450 m/s   **4.** 0.354 nm   **6.** a) 1.8 eV; b) 6.60×10^{-34} J s   **9.** a) 1.73×10^{18}
11. b) 5.5 μm   **13.** a) 2.1 MeV, 8.2 MeV, 18.5 MeV   **17.** c) 1.1 cm

Chapter 34

2. 1875 nm, 486 nm, 97.3 nm   **3.** a) 69; b) 3.2×10^4 m/s, -0.0029 eV   **4.** a) $\lambda_1 = 0.332$ nm, $\lambda_2 = 0.665$ nm, $\lambda_3 = 0.997$ nm
5. For $n = 1$: 0.026 nm, 4.38×10^6 m/s, -54.4 eV; For $n = 2$: 0.106 nm, 2.19×10^6 m/s, -13.6 eV   **8.** b) $f_{99} = 6.78 \times 10^9$ Hz, $f_{100} = 6.59 \times 10^9$ Hz; c) 6.68×10^9 Hz
10. a) 4.26×10^{-5} nm; 1.31×10^7 m/s; b) 0.0164 nm

Chapter 35

3. Most at 0 cm, least at ± 1 cm
5. a) 0.27%; b) 31.8%   **7.** d) 30%
9. d) 900   **10.** a) 0.866 cm$^{-1/2}$; d) 3440

Chapter 36

1. 0.74 nm   **2.** b) 1.5 nm   **4.** b) For $n = 3$: 0 and 1.5 nm; For $n = 6$: 0 and 3.0 nm; d) smallest in 0–1 nm, largest in 2–3 nm
5. b) 0   **8.** a) 0.49 eV, 1.46 eV, 2.44 eV; b) 637 nm   **10.** d) 6.72×10^{-5} nm
12. $E_1 = 0.107$ eV   **14.** 4.77×10^7 m/s
17. c) $(\pi b^2)^{-1/4}$ **19.** b) 0.136 nm
20. a) 3.4×10^{-5}

Chapter 37

1. 2 for $n = 1$, 8 for $n = 2$, 18 for $n = 3$
2. a) f; b) -0.85 eV   **6.** a) 1.48×10^{-34} J s
8. a) $(\pi a_B^3)^{-1/2}$; b) 67.7%   **15.** c) nickel, ground state

Chapter 38

1. a) 2.2 μm; b) no   **2.** a) 555,000; c) 53,000   **3.** 3.2×10^{15} s^{-1} **4.** 5.3×10^{22} s^{-1} **7.** b) 1.65×10^6 m/s
9. a) 6.25×10^8 s^{-1}; b) 0.17 ns
11. 12 ns   **14.** b) 1.87 W

Index

Absolute zero 15, 90
Adiabatic process 124–126
Alpha particles 426–427
Angular momentum 484, 556–557
   quantum number 555
   selection rule 584
   spin 566–567
Atmospheric pressure 18, 21
Atomic mass 12
Atomic number 432
Atoms
   Bohr atom 474–478
   Bohr model 471
   hydrogen atom 554–558
   ions 173, 429
   isotopes 433
   lifetime of excited states 593
   magnetic moment 358–359
   multielectron atoms 568–571
   polarization 179
   Rutherford model 172, 427–429
   spectra 435–436, 584–592
   transitions 471
Avogadro's number 12

Balmer formula 437, 481
Battery 287
   emf 287
   internal resistance 316
   power 307
Binomial approximation 207
Binomial distribution 72
Biot and Savart law 336–337
Bohr atom 474–478
Bohr model 471–473
Bohr radius 477

Calorimetry 53–55

Capacitor 221, 291–292
   capacitance 292
   electric field 222
   electric potential 266–268
   quantum mechanical 526
Carnot cycle 143–144, 147
Cathode rays 416–418
Charge 169
   charge density 210–211
   fundamental unit of charge 172, 425
   quantization of charge 172, 415
Charge carriers 175, 237, 239
Charging process 167–168
Circuits 301–305
   grounding 322–323
   Kirchhoff's laws 303
   power dissipated 307–309
   resistive 318–322
Conductivity 250–251
Conductor 170, 175, 241–242
   electrostatic equilibrium 283–284
Configuration
   electron 571, 574
   microstate 61
Conservation laws
   charge 174
   closed loops 283
   current 245
   energy (first law of thermodynamics) 48
Correspondence principle 523
Coulomb's law 181
Current 241–242, 247–248
   current density 249–250
   electron current 243–245
   induced current 372, 374
   magnetic field of 342–345
   magnetic force on 356–357
   relationship to potential 289–290

Current loop
  forces and torques on 356–357
  magnetic field of 342–345
Cyclotron motion 351–353

De Broglie wavelength 456, 510
Density
  charge 210–211
  conduction electrons 244
  current 249
  mass 10
  number 9
  probability 496–497
Dipoles
  electric 180, 206, 228
  magnetic 332, 361
  torque on 228, 334
Drift speed of electrons 242

Eddy current 378–379
Einstein, A. 448–449, 597
Electric charge *see* Charge
Electric circuit *see* Circuits
Electric current *see* Current
Electric dipole *see* Dipoles
Electric field 193–194
  continuous charge 210–211
  dipole 206–208
  force on charge 224
  induced 392
  inside conductor 283–284
  inside wire 246–247
  line of charge 212–214
  parallel-plate capacitor 221–222, 267
  plane of charge 217–219
  point charge 194
  relationship to potential 262, 276–278, 281–282
  sphere of charge 223
  vectors 195
Electric generator 388–389
Electric motor 358
Electric potential 261
  capacitor 266–267
  continuous charge 270
  emf 287
  equipotential surface 280–282
  point charge 264
  relationship to current 289–290
  relationship to field 262, 276–278, 281–282
  relationship to potential energy 261
  sources 286–288
Electric potential energy 257
  charge in capacitor 257
  two point charges 258–259
Electricity 166–180
  two-charge model 169–170
Electromagnetic induction (*see also* Induced current) 370, 394
Electromagnetic waves 394
Electron volt 430
Electrons 172
  charge 423–425
  conduction electrons in metals 239, 241, 244
  diffraction 458

  discovery 419–423
  sea of electrons 175
  spin 359, 564–567
  tunneling 542–545
Electrostatic constant 181
Emf 287
  battery 287
  induced 385
  motional 373–374
Energy
  conservation of 48
  electric potential energy 257–259
  internal 43
  ionization 479
  quantization of 462
  thermal 43, 96
  zero point 521
Energy level diagram 479, 558, 569
Energy levels 462, 478
  excited states 581
  hydrogen atom 555
  multielectron atoms 569
  vibrational 539
Entropy 106–107, 155–159
Equilibrium
  and randomness 81
  electrostatic 176
  thermal 101, 105–106
Equipartition theorem 97–98
Equipotential surfaces 280–282
Excitation
  by absorption 584
  by collision 472, 585
Excited states 558, 581–583
  lifetime of 593

Faraday, M. 187, 369–370, 415–416
Faraday's law 385
Field 186–189
  electric 193–194
  gravitational 191
  magnetic 335
First law of thermodynamics 48, 115
Fluctuations 63–64, 76–79
Fluids 8
Force
  electric field on charge 224
  magnetic field on charge 349
  magnetic field on current 354
  strong 534

Gases (*see also* Ideal gas) 22–27
  diatomic 25
  ideal gas law 28
  ideal gas model 25
  kinetic theory 86–95
  molar heat capacity 51–52, 121
  molecular speeds 26
  monatomic 25
Ground 178, 322–323

Heat 42, 45–47, 101, 109
Heat capacity
  heat capacity ratio 125

molar heat capacity   51–52
Heat engine   127, 140
   Brayton cycle   138
   Carnot cycle   143
   Diesel cycle   137
   Otto cycle   132
   refrigerator   138–139
Hydrogen atom
   Bohr model   474–478
   energy levels   478–479, 558
   hydrogen-like ions   483
   one-dimensional model   552
   quantum-mechanical model   554
   quantum numbers   555
   spectrum   480–482
   wave functions   559–562

Ideal gases
   ideal gas law   28
   ideal gas model   24–25
   ideal gas processes   31–35
Induced current   372, 374–375, 382, 385
Induced field
   electric   392
   magnetic   393
Insulator   170
Interaction parameters   42
Interference
   atoms   461
   electrons   458, 498
   light   492–496
   neutrons   458–460
Ionization energy   479, 551–552, 558
Ions   173, 429, 483
Isentropic process   124
Isobaric process   32, 121
Isochoric process   33, 119–120
Isothermal process   34, 122–123
Isotopes   433

Kinetic theory
   of electrical conduction   241–245
   of gases   86–95
Kirchhoff's laws
   junction law   246, 303
   loop law   303
Knowledge structures
   electricity and magnetism   400
   quantum mechanics   607
   thermal and statistical physics   154

Lasers   597–601
Lenz's law   381–382
Lifetime of excited states   592–595
Light
   photons   453–454, 494
   quantization   449
   waves   394, 492

Magnetic domains   361–362
Magnetic field   335–336
   Biot and Savart law   337
   charge   336–337
   current loop   342–345
   induced   393
   right-hand rule   334, 337
   solenoid   347
   uniform   346
   wire   336, 339–341
Magnetic flux   379
Magnetic force
   between current-carrying wires   349, 355
   on charge   349
   on current   354
   on current loop   356–357
Magnetic materials   331, 360–362
Magnetic moment   360–362
Magnetic poles   331–332
Magnetism   330–332, 360–362
   ferromagnetism   332
Magnets   330–332
   electromagnet   347
   permanent magnet   362
Mass spectrometer   367, 433
Matter waves   456–461
   standing waves   462, 520
Mean free path   93–94
Micro/macro connection   4, 65, 86, 153, 241–242
Microstate   60
Millikan oil-drop experiment   423–425
Models
   atomic   427, 470, 568
   conduction   241
   electricity   169–180
   field   186–189
   heat engine   140
   hydrogen atom, one dimension   552
   hydrogen atom, three dimensions   554
   quantum-mechanical   525
Molar heat capacity   51–52, 98–99, 121
Mole   12

Neutrons   433
Normalization of wave functions   499–500, 519
Nucleus   427, 429, 432–434, 534

Ohm's law   300
Order and disorder   80–81, 155–158

Particle in a box   462, 515–520
Pauli exclusion principle   570
Periodic table of the elements   572–577
Permeability constant   337
Permittivity constant   182
Photoelectric effect   443–452
   Einstein's explanation   449–451
Photons   453–454, 471–472, 494
Planck's constant   449
Point charge   182
   electric field   194
   electric potential   264
   force between   181–182
   magnetic field   336–337
Polarization   178–180
   polarization force   179
Potential   *see* Electric potential
Potential energy
   electric   257

harmonic oscillator 535
    particle in a box 516
Potential well 529
Power 307–310
Pressure 17
    gas 23, 86–90
    gauge pressure 19
    liquid 18
Probability 62, 66–67, 72
    interpretation of wave function 497–498
    of detecting photon 494–496
Probability density 496–497, 560
    normalization 500
    radial 562
Protons 432
pV diagram 31

Quantization
    angular momentum 485
    charge 172, 415
    energy 462
    light 449
Quantum harmonic oscillator 535–537
Quantum jump (*see also* Transition) 471, 592
Quantum numbers 462, 477, 519
    magnetic 555
    orbital 555
    principal 555
    spin 567
Quantum-well device 532–533

Radioactivity 426
    alpha rays 426
    beta rays 426
    gamma rays 535
Randomness 80–81, 155–158
Refrigerator 138–139
Resistance 241, 290, 298–300
    internal 316
Resistivity 250–251
Resistors 299–300, 311–313
    equivalent 311–313
    parallel combination 313
    power dissipation 308–310
    series combination 311
Resonant-tunneling diode 544–545
Right-hand rule
    magnetic field 334
    magnetic force 350
Rms velocity 91

Second law of thermodynamics 81, 107–109, 142, 156–157
Scanning tunneling microscope 542–543
Schrödinger's equation 509–513
Selection rule for transitions 584
Solenoid 347
Specific heat 49–50
Spectrum 434–436, 480–482
    discrete 435, 473
    emission 434, 473, 384–390, 586–590
    absorption 435, 473
Spin 359, 564–567
    spin-orbit interaction 583

spin quantum number 567
State variables 8, 41
Stationary states 471, 515
    Bohr hydrogen atom 474, 478
    quantum-mechanical hydrogen atom 555
Statistical mechanics 59
Stern-Gerlach experiment 564–566
Stimulated emission of radiation 597
Stirling approximation 74
STP (standard temperature and pressure) 30
Strategy for
    electric field of continuous charge 211
    electric field of multiple point charges 206
    quantum mechanics 514
Superconductivity 252

Temperature 14–15, 90
Thermal efficiency 130, 142
    maximum 146
Thermal energy 43, 96–98
Thermodynamic processes 114
    adiabatic 124–126
    cyclical 129
    isobaric 32, 121
    isochoric 33, 119–120
    isothermal 34, 122–123
Thermodynamics 114
    applications 124–148
    first law 48
    heat engines 124
    second law 107
Thompson, J. J. 419–421, 426, 443
Torque
    current loop 357
    dipole 228, 334
Transition 471–472, 515
    lifetime 593
    nonradiative 591
    selection rule 584
Tunneling 540–541

Wave functions 497–500
    bound states 530
    hydrogen atom 559
    interpretation 498
    law of (Schrödinger equation) 509
    normalization 500
    particle in a box 519
    probability density 497
    quantum harmonic oscillator 536
Wave-particle duality 442, 464
Waves
    electromagnetic 394
    matter 456
    wave functions 497
Work 42, 44, 116
    by electric field 308
    by system 115
Work function 445

X rays 419

Zero point energy 521

STUDENT EVALUATION FORM

Knight *Physics: A Contemporary Perspective*
Chapter 19   A Macroscopic Description of Matter

Name (Optional): _____   Date: _____

College: _____   Professor: _____

1. What did you like about this chapter? Why?

2. Were there any areas of particular difficulty? If yes, please specify those areas, and, if possible, explain why you found them to be difficult.

3. How do you rate the illustrations and photos in this chapter?
   (High)   5   4   3   2   1   (Low)
   What suggestions do you have for their improvement?

4. How do you rate the worked examples in this chapter?
   (High)   5   4   3   2   1   (Low)
   Which did you particularly like or dislike? Why?

   Any suggestions for improvement?

Continued on back

5. How do you rate the end-of-chapter problems for this chapter?

   (High)   5   4   3   2   1   (Low)

   Which did you particularly like or dislike? Why?

   Any suggestions for improvement?

6. How do you rate the Student Workbook exercises for this chapter?

   (High)   5   4   3   2   1   (Low)

   Which did you particularly like or dislike? Why?

   Any suggestions for improvement?

7. Please provide any additional comments you think would be helpful to the author for improving or strengthening the material in this chapter.

THANK YOU FOR YOUR FEEDBACK.
PLEASE GIVE THIS FORM TO YOUR PROFESSOR.
ADDISON-WESLEY PUBLISHING COMPANY

STUDENT EVALUATION FORM

Knight *Physics: A Contemporary Perspective*
Chapter 20   Heat, Work, and the First Law of Thermodynamics

Name (Optional): _____   Date: _____

College: _____   Professor: _____

1. What did you like about this chapter? Why?

2. Were there any areas of particular difficulty? If yes, please specify those areas, and, if possible, explain why you found them to be difficult.

3. How do you rate the illustrations and photos in this chapter?
   (High)   5   4   3   2   1   (Low)
   What suggestions do you have for their improvement?

4. How do you rate the worked examples in this chapter?
   (High)   5   4   3   2   1   (Low)
   Which did you particularly like or dislike? Why?

   Any suggestions for improvement?

Continued on back

5. How do you rate the end-of-chapter problems for this chapter?

   (High)   5   4   3   2   1   (Low)

   Which did you particularly like or dislike? Why?

   Any suggestions for improvement?

6. How do you rate the Student Workbook exercises for this chapter?

   (High)   5   4   3   2   1   (Low)

   Which did you particularly like or dislike? Why?

   Any suggestions for improvement?

7. Please provide any additional comments you think would be helpful to the author for improving or strengthening the material in this chapter.

THANK YOU FOR YOUR FEEDBACK.
PLEASE GIVE THIS FORM TO YOUR PROFESSOR.
ADDISON-WESLEY PUBLISHING COMPANY

STUDENT EVALUATION FORM

Knight *Physics: A Contemporary Perspective*
Chapter 21   A Statistical View of Gases

Name (Optional): _____   Date: _____

College: _____   Professor: _____

1. What did you like about this chapter? Why?

2. Were there any areas of particular difficulty? If yes, please specify those areas, and, if possible, explain why you found them to be difficult.

3. How do you rate the illustrations and photos in this chapter?
    (High)   5   4   3   2   1   (Low)
    What suggestions do you have for their improvement?

4. How do you rate the worked examples in this chapter?
    (High)   5   4   3   2   1   (Low)
    Which did you particularly like or dislike? Why?

    Any suggestions for improvement?

Continued on back

5. How do you rate the end-of-chapter problems for this chapter?

   (High)   5   4   3   2   1   (Low)

   Which did you particularly like or dislike? Why?

   Any suggestions for improvement?

6. How do you rate the Student Workbook exercises for this chapter?

   (High)   5   4   3   2   1   (Low)

   Which did you particularly like or dislike? Why?

   Any suggestions for improvement?

7. Please provide any additional comments you think would be helpful to the author for improving or strengthening the material in this chapter.

THANK YOU FOR YOUR FEEDBACK.
PLEASE GIVE THIS FORM TO YOUR PROFESSOR.
ADDISON-WESLEY PUBLISHING COMPANY

STUDENT EVALUATION FORM

Knight *Physics: A Contemporary Perspective*
Chapter 22   The Kinetic Theory of Gases

Name (Optional): _____   Date: _____

College: _____   Professor: _____

1. What did you like about this chapter? Why?

2. Were there any areas of particular difficulty? If yes, please specify those areas, and, if possible, explain why you found them to be difficult.

3. How do you rate the illustrations and photos in this chapter?
   (High)   5   4   3   2   1   (Low)
   What suggestions do you have for their improvement?

4. How do you rate the worked examples in this chapter?
   (High)   5   4   3   2   1   (Low)
   Which did you particularly like or dislike? Why?

   Any suggestions for improvement?

Continued on back

5. How do you rate the end-of-chapter problems for this chapter?

   (High)   5   4   3   2   1   (Low)

   Which did you particularly like or dislike? Why?

   Any suggestions for improvement?

6. How do you rate the Student Workbook exercises for this chapter?

   (High)   5   4   3   2   1   (Low)

   Which did you particularly like or dislike? Why?

   Any suggestions for improvement?

7. Please provide any additional comments you think would be helpful to the author for improving or strengthening the material in this chapter.

THANK YOU FOR YOUR FEEDBACK.
PLEASE GIVE THIS FORM TO YOUR PROFESSOR.
ADDISON-WESLEY PUBLISHING COMPANY

STUDENT EVALUATION FORM

**Knight *Physics: A Contemporary Perspective*
Chapter 23   Thermodynamics of Ideal Gases**

Name (Optional): _____   Date: _____

College: _____   Professor: _____

1. What did you like about this chapter? Why?

2. Were there any areas of particular difficulty? If yes, please specify those areas, and, if possible, explain why you found them to be difficult.

3. How do you rate the illustrations and photos in this chapter?
   (High)   5   4   3   2   1   (Low)
   What suggestions do you have for their improvement?

4. How do you rate the worked examples in this chapter?
   (High)   5   4   3   2   1   (Low)
   Which did you particularly like or dislike? Why?

   Any suggestions for improvement?

Continued on back

5. How do you rate the end-of-chapter problems for this chapter?

   (High)   5   4   3   2   1   (Low)

   Which did you particularly like or dislike? Why?

   Any suggestions for improvement?

6. How do you rate the Student Workbook exercises for this chapter?

   (High)   5   4   3   2   1   (Low)

   Which did you particularly like or dislike? Why?

   Any suggestions for improvement?

7. Please provide any additional comments you think would be helpful to the author for improving or strengthening the material in this chapter.

THANK YOU FOR YOUR FEEDBACK.
PLEASE GIVE THIS FORM TO YOUR PROFESSOR.
ADDISON-WESLEY PUBLISHING COMPANY

STUDENT EVALUATION FORM

Knight *Physics: A Contemporary Perspective*
Chapter 24   Electric Forces and Fields

Name (Optional): _____   Date: _____

College: _____   Professor: _____

1. What did you like about this chapter? Why?

2. Were there any areas of particular difficulty? If yes, please specify those areas, and, if possible, explain why you found them to be difficult.

3. How do you rate the illustrations and photos in this chapter?
   (High)   5   4   3   2   1   (Low)
   What suggestions do you have for their improvement?

4. How do you rate the worked examples in this chapter?
   (High)   5   4   3   2   1   (Low)
   Which did you particularly like or dislike? Why?

   Any suggestions for improvement?

Continued on back

5. How do you rate the end-of-chapter problems for this chapter?

   (High) 5   4   3   2   1 (Low)

   Which did you particularly like or dislike? Why?

   Any suggestions for improvement?

6. How do you rate the Student Workbook exercises for this chapter?

   (High) 5   4   3   2   1 (Low)

   Which did you particularly like or dislike? Why?

   Any suggestions for improvement?

7. Please provide any additional comments you think would be helpful to the author for improving or strengthening the material in this chapter.

THANK YOU FOR YOUR FEEDBACK.
PLEASE GIVE THIS FORM TO YOUR PROFESSOR.
ADDISON-WESLEY PUBLISHING COMPANY

STUDENT EVALUATION FORM

Knight *Physics: A Contemporary Perspective*
Chapter 25   Calculating the Electric Field

Name (Optional): _____   Date: _____

College: _____   Professor: _____

1. What did you like about this chapter? Why?

2. Were there any areas of particular difficulty? If yes, please specify those areas, and, if possible, explain why you found them to be difficult.

3. How do you rate the illustrations and photos in this chapter?
   (High)   5   4   3   2   1   (Low)
   What suggestions do you have for their improvement?

4. How do you rate the worked examples in this chapter?
   (High)   5   4   3   2   1   (Low)
   Which did you particularly like or dislike? Why?

   Any suggestions for improvement?

Continued on back

5. How do you rate the end-of-chapter problems for this chapter?

    (High)   5   4   3   2   1   (Low)

    Which did you particularly like or dislike? Why?

    Any suggestions for improvement?

6. How do you rate the Student Workbook exercises for this chapter?

    (High)   5   4   3   2   1   (Low)

    Which did you particularly like or dislike? Why?

    Any suggestions for improvement?

7. Please provide any additional comments you think would be helpful to the author for improving or strengthening the material in this chapter.

THANK YOU FOR YOUR FEEDBACK.
PLEASE GIVE THIS FORM TO YOUR PROFESSOR.
ADDISON-WESLEY PUBLISHING COMPANY

STUDENT EVALUATION FORM

Knight *Physics: A Contemporary Perspective*
Chapter 26   Current and Conductivity

Name (Optional): _____   Date: _____

College: _____   Professor: _____

1. What did you like about this chapter? Why?

2. Were there any areas of particular difficulty? If yes, please specify those areas, and, if possible, explain why you found them to be difficult.

3. How do you rate the illustrations and photos in this chapter?
   (High)   5   4   3   2   1   (Low)
   What suggestions do you have for their improvement?

4. How do you rate the worked examples in this chapter?
   (High)   5   4   3   2   1   (Low)
   Which did you particularly like or dislike? Why?

   Any suggestions for improvement?

Continued on back

5. How do you rate the end-of-chapter problems for this chapter?

   (High)   5   4   3   2   1   (Low)

   Which did you particularly like or dislike? Why?

   Any suggestions for improvement?

6. How do you rate the Student Workbook exercises for this chapter?

   (High)   5   4   3   2   1   (Low)

   Which did you particularly like or dislike? Why?

   Any suggestions for improvement?

7. Please provide any additional comments you think would be helpful to the author for improving or strengthening the material in this chapter.

THANK YOU FOR YOUR FEEDBACK.
PLEASE GIVE THIS FORM TO YOUR PROFESSOR.
ADDISON-WESLEY PUBLISHING COMPANY

STUDENT EVALUATION FORM

Knight *Physics: A Contemporary Perspective*
Chapter 27   The Electric Potential

Name (Optional): _____   Date: _____

College: _____   Professor: _____

1. What did you like about this chapter? Why?

2. Were there any areas of particular difficulty? If yes, please specify those areas, and, if possible, explain why you found them to be difficult.

3. How do you rate the illustrations and photos in this chapter?
   (High)   5   4   3   2   1   (Low)
   What suggestions do you have for their improvement?

4. How do you rate the worked examples in this chapter?
   (High)   5   4   3   2   1   (Low)
   Which did you particularly like or dislike? Why?

   Any suggestions for improvement?

Continued on back

5. How do you rate the end-of-chapter problems for this chapter?

   (High)   5   4   3   2   1   (Low)

   Which did you particularly like or dislike? Why?

   Any suggestions for improvement?

6. How do you rate the Student Workbook exercises for this chapter?

   (High)   5   4   3   2   1   (Low)

   Which did you particularly like or dislike? Why?

   Any suggestions for improvement?

7. Please provide any additional comments you think would be helpful to the author for improving or strengthening the material in this chapter.

THANK YOU FOR YOUR FEEDBACK.
PLEASE GIVE THIS FORM TO YOUR PROFESSOR.
ADDISON-WESLEY PUBLISHING COMPANY

STUDENT EVALUATION FORM

Knight *Physics: A Contemporary Perspective*
Chapter 28   Potential and Field

Name (Optional): _____   Date: _____

College: _____   Professor: _____

1. What did you like about this chapter? Why?

2. Were there any areas of particular difficulty? If yes, please specify those areas, and, if possible, explain why you found them to be difficult.

3. How do you rate the illustrations and photos in this chapter?
   (High)   5   4   3   2   1   (Low)
   What suggestions do you have for their improvement?

4. How do you rate the worked examples in this chapter?
   (High)   5   4   3   2   1   (Low)
   Which did you particularly like or dislike? Why?

   Any suggestions for improvement?

Continued on back

5. How do you rate the end-of-chapter problems for this chapter?

   (High)   5   4   3   2   1   (Low)

   Which did you particularly like or dislike? Why?

   Any suggestions for improvement?

6. How do you rate the Student Workbook exercises for this chapter?

   (High)   5   4   3   2   1   (Low)

   Which did you particularly like or dislike? Why?

   Any suggestions for improvement?

7. Please provide any additional comments you think would be helpful to the author for improving or strengthening the material in this chapter.

THANK YOU FOR YOUR FEEDBACK.
PLEASE GIVE THIS FORM TO YOUR PROFESSOR.
ADDISON-WESLEY PUBLISHING COMPANY

STUDENT EVALUATION FORM

Knight *Physics: A Contemporary Perspective*
Chapter 29   Fundamentals of Circuits

Name (Optional): _____   Date: _____

College: _____   Professor: _____

1. What did you like about this chapter? Why?

2. Were there any areas of particular difficulty? If yes, please specify those areas, and, if possible, explain why you found them to be difficult.

3. How do you rate the illustrations and photos in this chapter?
   (High)   5   4   3   2   1   (Low)
   What suggestions do you have for their improvement?

4. How do you rate the worked examples in this chapter?
   (High)   5   4   3   2   1   (Low)
   Which did you particularly like or dislike? Why?

   Any suggestions for improvement?

Continued on back

5. How do you rate the end-of-chapter problems for this chapter?

   (High)   5   4   3   2   1   (Low)

   Which did you particularly like or dislike? Why?

   Any suggestions for improvement?

6. How do you rate the Student Workbook exercises for this chapter?

   (High)   5   4   3   2   1   (Low)

   Which did you particularly like or dislike? Why?

   Any suggestions for improvement?

7. Please provide any additional comments you think would be helpful to the author for improving or strengthening the material in this chapter.

THANK YOU FOR YOUR FEEDBACK.
PLEASE GIVE THIS FORM TO YOUR PROFESSOR.
ADDISON-WESLEY PUBLISHING COMPANY

STUDENT EVALUATION FORM

Knight *Physics: A Contemporary Perspective*
Chapter 30   The Magnetic Field

Name (Optional): _____   Date: _____

College: _____   Professor: _____

1. What did you like about this chapter? Why?

2. Were there any areas of particular difficulty? If yes, please specify those areas, and, if possible, explain why you found them to be difficult.

3. How do you rate the illustrations and photos in this chapter?
   (High)   5   4   3   2   1   (Low)
   What suggestions do you have for their improvement?

4. How do you rate the worked examples in this chapter?
   (High)   5   4   3   2   1   (Low)
   Which did you particularly like or dislike? Why?

   Any suggestions for improvement?

Continued on back

5. How do you rate the end-of-chapter problems for this chapter?

   (High) 5   4   3   2   1 (Low)

   Which did you particularly like or dislike? Why?

   Any suggestions for improvement?

6. How do you rate the Student Workbook exercises for this chapter?

   (High) 5   4   3   2   1 (Low)

   Which did you particularly like or dislike? Why?

   Any suggestions for improvement?

7. Please provide any additional comments you think would be helpful to the author for improving or strengthening the material in this chapter.

THANK YOU FOR YOUR FEEDBACK.
PLEASE GIVE THIS FORM TO YOUR PROFESSOR.
ADDISON-WESLEY PUBLISHING COMPANY

STUDENT EVALUATION FORM

Knight *Physics: A Contemporary Perspective*
Chapter 31   Electromagnetic Induction

Name (Optional): _____   Date: _____

College: _____   Professor: _____

1. What did you like about this chapter? Why?

2. Were there any areas of particular difficulty? If yes, please specify those areas, and, if possible, explain why you found them to be difficult.

3. How do you rate the illustrations and photos in this chapter?
   (High)   5   4   3   2   1   (Low)
   What suggestions do you have for their improvement?

4. How do you rate the worked examples in this chapter?
   (High)   5   4   3   2   1   (Low)
   Which did you particularly like or dislike? Why?

   Any suggestions for improvement?

Continued on back

5. How do you rate the end-of-chapter problems for this chapter?

    (High)   5   4   3   2   1   (Low)

    Which did you particularly like or dislike? Why?

    Any suggestions for improvement?

6. How do you rate the Student Workbook exercises for this chapter?

    (High)   5   4   3   2   1   (Low)

    Which did you particularly like or dislike? Why?

    Any suggestions for improvement?

7. Please provide any additional comments you think would be helpful to the author for improving or strengthening the material in this chapter.

THANK YOU FOR YOUR FEEDBACK.
PLEASE GIVE THIS FORM TO YOUR PROFESSOR.
ADDISON-WESLEY PUBLISHING COMPANY

STUDENT EVALUATION FORM

Knight *Physics: A Contemporary Perspective*
Chapter 32   The End of Classical Physics

Name (Optional): _____   Date: _____

College: _____   Professor: _____

1. What did you like about this chapter? Why?

2. Were there any areas of particular difficulty? If yes, please specify those areas, and, if possible, explain why you found them to be difficult.

3. How do you rate the illustrations and photos in this chapter?
   (High)   5   4   3   2   1   (Low)
   What suggestions do you have for their improvement?

4. How do you rate the worked examples in this chapter?
   (High)   5   4   3   2   1   (Low)
   Which did you particularly like or dislike? Why?

   Any suggestions for improvement?

Continued on back

5. How do you rate the end-of-chapter problems for this chapter?

   (High) 5   4   3   2   1 (Low)

   Which did you particularly like or dislike? Why?

   Any suggestions for improvement?

6. How do you rate the Student Workbook exercises for this chapter?

   (High) 5   4   3   2   1 (Low)

   Which did you particularly like or dislike? Why?

   Any suggestions for improvement?

7. Please provide any additional comments you think would be helpful to the author for improving or strengthening the material in this chapter.

THANK YOU FOR YOUR FEEDBACK.
PLEASE GIVE THIS FORM TO YOUR PROFESSOR.
ADDISON-WESLEY PUBLISHING COMPANY

STUDENT EVALUATION FORM

Knight *Physics: A Contemporary Perspective*
Chapter 33   Waves, Particles, and Quanta

Name (Optional): _____   Date: _____

College: _____   Professor: _____

1. What did you like about this chapter? Why?

2. Were there any areas of particular difficulty? If yes, please specify those areas, and, if possible, explain why you found them to be difficult.

3. How do you rate the illustrations and photos in this chapter?
   (High)   5   4   3   2   1   (Low)
   What suggestions do you have for their improvement?

4. How do you rate the worked examples in this chapter?
   (High)   5   4   3   2   1   (Low)
   Which did you particularly like or dislike? Why?

   Any suggestions for improvement?

Continued on back

5. How do you rate the end-of-chapter problems for this chapter?

   (High)   5   4   3   2   1   (Low)

   Which did you particularly like or dislike? Why?

   Any suggestions for improvement?

6. How do you rate the Student Workbook exercises for this chapter?

   (High)   5   4   3   2   1   (Low)

   Which did you particularly like or dislike? Why?

   Any suggestions for improvement?

7. Please provide any additional comments you think would be helpful to the author for improving or strengthening the material in this chapter.

THANK YOU FOR YOUR FEEDBACK.
PLEASE GIVE THIS FORM TO YOUR PROFESSOR.
ADDISON-WESLEY PUBLISHING COMPANY

STUDENT EVALUATION FORM

Knight *Physics: A Contemporary Perspective*
Chapter 34   The Bohr Model of the Atom

Name (Optional): _____   Date: _____

College: _____   Professor: _____

1. What did you like about this chapter? Why?

2. Were there any areas of particular difficulty? If yes, please specify those areas, and, if possible, explain why you found them to be difficult.

3. How do you rate the illustrations and photos in this chapter?
   (High)   5   4   3   2   1   (Low)
   What suggestions do you have for their improvement?

4. How do you rate the worked examples in this chapter?
   (High)   5   4   3   2   1   (Low)
   Which did you particularly like or dislike? Why?

   Any suggestions for improvement?

Continued on back

5. How do you rate the end-of-chapter problems for this chapter?

   (High)   5   4   3   2   1   (Low)

   Which did you particularly like or dislike? Why?

   Any suggestions for improvement?

6. How do you rate the Student Workbook exercises for this chapter?

   (High)   5   4   3   2   1   (Low)

   Which did you particularly like or dislike? Why?

   Any suggestions for improvement?

7. Please provide any additional comments you think would be helpful to the author for improving or strengthening the material in this chapter.

THANK YOU FOR YOUR FEEDBACK.
PLEASE GIVE THIS FORM TO YOUR PROFESSOR.
ADDISON-WESLEY PUBLISHING COMPANY

STUDENT EVALUATION FORM

Knight *Physics: A Contemporary Perspective*
Chapter 35   Wave Functions and Probabilities

Name (Optional): _____   Date: _____

College: _____   Professor: _____

1. What did you like about this chapter? Why?

2. Were there any areas of particular difficulty? If yes, please specify those areas, and, if possible, explain why you found them to be difficult.

3. How do you rate the illustrations and photos in this chapter?
   (High)   5   4   3   2   1   (Low)
   What suggestions do you have for their improvement?

4. How do you rate the worked examples in this chapter?
   (High)   5   4   3   2   1   (Low)
   Which did you particularly like or dislike? Why?

   Any suggestions for improvement?

Continued on back

5. How do you rate the end-of-chapter problems for this chapter?

   (High)   5   4   3   2   1   (Low)

   Which did you particularly like or dislike? Why?

   Any suggestions for improvement?

6. How do you rate the Student Workbook exercises for this chapter?

   (High)   5   4   3   2   1   (Low)

   Which did you particularly like or dislike? Why?

   Any suggestions for improvement?

7. Please provide any additional comments you think would be helpful to the author for improving or strengthening the material in this chapter.

THANK YOU FOR YOUR FEEDBACK.
PLEASE GIVE THIS FORM TO YOUR PROFESSOR.
ADDISON-WESLEY PUBLISHING COMPANY

STUDENT EVALUATION FORM

Knight *Physics: A Contemporary Perspective*
Chapter 36   One-Dimensional Quantum Mechanics

Name (Optional): _____   Date: _____

College: _____   Professor: _____

1. What did you like about this chapter? Why?

2. Were there any areas of particular difficulty? If yes, please specify those areas, and, if possible, explain why you found them to be difficult.

3. How do you rate the illustrations and photos in this chapter?
   (High)   5   4   3   2   1   (Low)
   What suggestions do you have for their improvement?

4. How do you rate the worked examples in this chapter?
   (High)   5   4   3   2   1   (Low)
   Which did you particularly like or dislike? Why?

   Any suggestions for improvement?

Continued on back

5. How do you rate the end-of-chapter problems for this chapter?

   (High)   5   4   3   2   1   (Low)

   Which did you particularly like or dislike? Why?

   Any suggestions for improvement?

6. How do you rate the Student Workbook exercises for this chapter?

   (High)   5   4   3   2   1   (Low)

   Which did you particularly like or dislike? Why?

   Any suggestions for improvement?

7. Please provide any additional comments you think would be helpful to the author for improving or strengthening the material in this chapter.

THANK YOU FOR YOUR FEEDBACK.
PLEASE GIVE THIS FORM TO YOUR PROFESSOR.
ADDISON-WESLEY PUBLISHING COMPANY

STUDENT EVALUATION FORM

Knight *Physics: A Contemporary Perspective*
Chapter 37   The Structure of Atoms

Name (Optional): _____   Date: _____

College: _____   Professor: _____

1. What did you like about this chapter? Why?

2. Were there any areas of particular difficulty? If yes, please specify those areas, and, if possible, explain why you found them to be difficult.

3. How do you rate the illustrations and photos in this chapter?
   (High)   5   4   3   2   1   (Low)
   What suggestions do you have for their improvement?

4. How do you rate the worked examples in this chapter?
   (High)   5   4   3   2   1   (Low)
   Which did you particularly like or dislike? Why?

   Any suggestions for improvement?

Continued on back

5. How do you rate the end-of-chapter problems for this chapter?

   (High)   5   4   3   2   1   (Low)

   Which did you particularly like or dislike? Why?

   Any suggestions for improvement?

6. How do you rate the Student Workbook exercises for this chapter?

   (High)   5   4   3   2   1   (Low)

   Which did you particularly like or dislike? Why?

   Any suggestions for improvement?

7. Please provide any additional comments you think would be helpful to the author for improving or strengthening the material in this chapter.

THANK YOU FOR YOUR FEEDBACK.
PLEASE GIVE THIS FORM TO YOUR PROFESSOR.
ADDISON-WESLEY PUBLISHING COMPANY

STUDENT EVALUATION FORM

Knight *Physics: A Contemporary Perspective*
Chapter 38   Atomic Spectra

Name (Optional): _____   Date: _____

College: _____   Professor: _____

1. What did you like about this chapter? Why?

2. Were there any areas of particular difficulty? If yes, please specify those areas, and, if possible, explain why you found them to be difficult.

3. How do you rate the illustrations and photos in this chapter?
   (High)   5   4   3   2   1   (Low)
   What suggestions do you have for their improvement?

4. How do you rate the worked examples in this chapter?
   (High)   5   4   3   2   1   (Low)
   Which did you particularly like or dislike? Why?

   Any suggestions for improvement?

Continued on back

5. How do you rate the end-of-chapter problems for this chapter?

    (High) 5   4   3   2   1   (Low)

    Which did you particularly like or dislike? Why?

    Any suggestions for improvement?

6. How do you rate the Student Workbook exercises for this chapter?

    (High) 5   4   3   2   1   (Low)

    Which did you particularly like or dislike? Why?

    Any suggestions for improvement?

7. Please provide any additional comments you think would be helpful to the author for improving or strengthening the material in this chapter.

THANK YOU FOR YOUR FEEDBACK.
PLEASE GIVE THIS FORM TO YOUR PROFESSOR.
ADDISON-WESLEY PUBLISHING COMPANY

Useful Data

m_p	Mass of the proton (and the neutron)	1.67×10^{-27} kg
m_e	Mass of the electron	9.11×10^{-31} kg
e	Fundamental unit of charge	1.60×10^{-19} C
c	Speed of light in vacuum	3.00×10^8 m/s
h	Planck's constant	6.63×10^{-34} J s
		4.14×10^{-15} eV s
$\hbar$	Planck's constant	1.05×10^{-34} J s
		6.58×10^{-16} eV s
a_B	Bohr radius	5.29×10^{-11} m
k	Boltzmann's constant	1.38×10^{-23} J/K
R	Gas constant	8.31 J/mol K
N_A	Avogadro's number	6.02×10^{23} particles/mol
T_0	Absolute zero	$-273.15°C$
K	Coulomb's law constant ($1/4\pi\varepsilon_0$)	8.99×10^9 N m²/C²
ε_0	Permittivity constant	8.85×10^{-12} C²/N m²
μ_0	Permeability constant	1.26×10^{-6} T m/A
G	Gravitational constant	6.67×10^{-11} N m²/kg²
g	Acceleration due to gravity	9.80 m/s²
M_e	Mass of the earth	5.89×10^{24} kg
R_e	Radius of the earth	6.37×10^6 m

1 u = 1.661×10^{-27} kg
1 cal = 4.186 J
1 standard atmosphere = 101.3 kPa = 760 mm of Hg = 14.7 lb/in²
1 eV = 1.60×10^{-19} J

SOME ATOMIC DATA

Atom	Z	A (in u)
proton	–	1.007
^{1}H	1	1.008
^{2}H	1	2.014
^{4}He	2	4.003
^{12}C	6	12.0000 exactly
^{14}C	6	14.003
^{14}N	7	14.003
^{16}O	8	15.995
^{238}U	92	238.051

COMMON PREFIXES

Prefix	Meaning
femto-	10^{-15}
pico-	10^{-12}
nano-	10^{-9}
micro-	10^{-6}
milli-	10^{-3}
centi-	10^{-2}
kilo-	10^3
mega-	10^6
giga-	10^9

Binomial approximation: $(1+x)^n \approx 1 + nx$ if $x \ll 1$

Small angle approximation: $\sin\theta \approx \tan\theta \approx \theta$ and $\cos\theta \approx 1$ if $\theta \ll 1$ radian